Durability of Building Materials and Components

Other books on Durability and Building Materials

Building Failures: Diagnosis and Avoidance
W.H. Ransom

Corrosion of Steel in Concrete
RILEM Report
Edited by P. Schiessl

Defects and Deterioration in Buildings
B.A. Richardson

Durability of Geotextiles
RILEM Report

The Maintenance of Brick and Stone Masonry Structures
Edited by A.M. Sowden

Microbiology in Civil Engineering
Edited by P. Howsam

Protection of Concrete
Edited by R.K. Dhir and J.W. Green

Vegetable Plants and their Fibres as Building Materials
Edited by H.S. Sobral

Publisher's Note
This book has been compiled from camera ready copy provided by the individual contributors. This method of production has allowed us to supply finished copies to the delegates at the Conference.

Durability of Building Materials and Components

Proceedings of the Fifth International Conference held in Brighton, UK, 7–9 November 1990

Sponsored by
American Society for Testing and Materials (ASTM)
Building Research Establishment, UK (BRE)
International Council for Building Research Studies and Documentation (CIB)
National Institute of Standards and Technology, USA (NIST)
National Research Council of Canada (NRCC)
International Union of Testing and Research Laboratories for Materials and Structures (RILEM)

Edited by

J.M. BAKER, P.J. NIXON,
A.J. MAJUMDAR and H. DAVIES
Building Research Establishment

E. & F.N. SPON
An imprint of Chapman and Hall
LONDON · NEW YORK · TOKYO · MELBOURNE · MADRAS

UK	Chapman and Hall, 2–6 Boundary Row, London SE1 8HN
USA	Van Nostrand Reinhold, 115 5th Avenue, New York NY10003
JAPAN	Chapman and Hall Japan, Thomson Publishing Japan, Hirakawacho Nemoto Building, 7F, 1-7-11 Hirakawa-cho, Chiyoda-ku, Tokyo 102
AUSTRALIA	Chapman and Hall Australia, Thomas Nelson Australia, 480 La Trobe Street, PO Box 4725, Melbourne 3000
INDIA	Chapman and Hall India, R. Seshadri, 32 Second Main Road, CIT East, Madras 600 035

First edition 1991

Printed in Great Britain at
the University Press, Cambridge

ISBN 0 419 15480 9   0 442 31260 1 (USA)

British Library Cataloguing in Publication Data
Available

Library of Congress Cataloging-in-Publication Data
Available

Contents

Poster Presentations

Poster Presentations

Poster Presentations

Steering Committee

Professor S.L. Lee, (Chairman) National University of Singapore
Dr R.D. Browne, RILEM
Dr G.G. Litvan, National Research Council of Canada
Dr C. Sjöström, CIB
Dr L. Masters, ASTM and NIST
Professor C.T. Tam, (Secretary) National University of Singapore

Organizing Committee

Mr J.M. Baker, (Chairman) BRE
Dr P.J. Nixon, BRE
Professor P.L. Pratt, Imperial College, London
Dr C.D. Pomeroy, British Cement Association
Dr A.J. Majumdar, BRE
Dr J.F.A. Moore, BRE
Mrs P.M. Rowley, BRE
Mr H. Davies, (Secretary) BRE

Preface

Interest in the durability of the built environment has been growing over the last decade, and is likely to remain a major concern for the foreseeable future. The concept of durability is complex, and its understanding requires an appreciation of many physical and chemical properties of the materials in question. The assessment of such specific materials properties is an advanced science. Yet we hear regularly of the premature deterioration and even failure of buildings and structures worldwide. Our understanding of the basic properties of building materials is clearly more advanced than our understanding of their long-term behaviour. Our ability accurately to predict the service life of both materials and the structural components which they comprise is limited: enhancement of this ability depends largely on understanding the mechanisms of deterioration.

The development and application of new materials in construction continually adds to the choices and decisions facing clients, designers and all responsible for building and construction. They continually seek greater and more reliable information about serviceability in order that they may meet more stringent design, safety and economic criteria.

The series of International Conferences on the Durability of Building Materials and Components, which started in 1978, seeks to bring together all those scientists engaged in research into the durability of construction. The wide-ranging interest in the subject is shown by the range of organizations giving scientific sponsorship to the conference, and by the degree of support given to the conference by organizations in the UK.

The major themes of this, the fifth conference in the series, seek to advance our understanding of deterioration mechanisms, of *in-situ* behaviour and assessment and the role of manufacturing practice in achieving greater durability. The diversity of the contributions and contributors is a measure of the wide ranging importance of durability. It is our wish that this record of the conference will be a useful reference to all those engaged in the study and provision of greater durability.

The organization of any international conference is a major undertaking, and the task would be impossible without the support of the sponsors, the enthusiasm of the authors, and the labours of the staff of the BRE Seminar Office, to all of whom we would like to express our sincere thanks.

J.M. Baker
P.J. Nixon
A.J. Majumdar
H. Davies

Garston, UK
July 1990

Acknowledgements

The Organizing Committee are grateful for financial support received from the UK construction industry in the arrangement of the conference. The generous assistance of the following is particularly acknowledged.

AACPA – Aircrete Bureau
Appleby Slag Co. Ltd
Blue Circle Industries PLC
British Cement Association
British Steel PLC
Elkem Materials Ltd
Fosroc Technology
Instron Ltd
Lafarge Special Cements
Mott MacDonald Group, Special Services Division
National Council of Building Material Producers
Rugby Cement
Harry Stanger Ltd
Taywood Engineering Ltd

SESSION A

DESTRUCTIVE MECHANISMS IN BUILDING MATERIALS AND MODELLING

The keynote speech to start the 5th International Conference on the Durability of Building Materials and Components was given by Professor P.L. Pratt of Imperial College, London. This paper examined the themes of the conference and attempted a synthesis of some of the major topics reported by the contributors. For this reason it was inappropriate to prepare a detailed text in advance and the following synopsis simply outlines the main framework of Professor Pratt's keynote speech.

J.M. Baker

Conference Chairman

Building Research Establishment, Garston, UK

1 SYNOPSIS OF KEYNOTE ADDRESS: THE DURABILITY OF BUILDING MATERIALS

P.L. PRATT
Department of Materials, Imperial College, London, UK

The building materials of interest to this Fifth International Conference include the traditional bricks and mortar, timber and stone, together with concrete, metals, polymers and a range of fire-reinforced composite materials. Surprisingly in 1990, new examples of limited durability and of failure of some of these materials are reported, requiring remedial measures and repair sometimes even before the buildings have gone into service. Part of the problem lies in poor workmanship and inadequate quality control, part in a lack of scientific understanding of the effects upon the durability of the environment in which the building exists, and part in a lack of communication between the materials scientitists and engineers, who develop and produce the building materials, and the designers, engineers and contractors who are responsible for selecting and for using the materials properly.

This paper looks at some examples of ancient building materials which have proved durable enough to attract the attention of those involved in the disposal of nuclear waste, and at some examples of modern building materials which have not proved durable. Deterioration may occur for a variety of reasons, some external to the materials or structure involving the chemical or physical environment in which the materials must operate. Aggressive chemical environments may lead to attack of the material leading to progressive deterioration and ultimate failure; aggressive physical environments, including cyclic changes of moisture content, temperature or pressure, may cause mechanical failure or may change the microstructure of the material in such a way as lead to failure. Deterioration may also occur for reasons that are internal to the material where metastable microstructures become unstable causing expansion or shrinkage.

In practice numerous causes of deterioration can act simulataneously and it can be difficult to decide which is the most important. In order to identify the cause and to suggest appropriate methods for repair and for control of the deterioration, mathematical modelling techniques are being developed with growing success. Making use of models of each mechanism the rate-controlling steps can be determined and the rates of degradation compared with those found in practice. In this way a better understanding of the processes of deterioration can be obtained and the models can be used to predict the remaining service life of the building with increasing confidence.

2 PREDICTION OF SERVICE LIVES OF REINFORCED CONCRETE BUILDINGS BASED ON THE CORROSION RATE OF REINFORCING STEEL

S. MORINAGA
Research Institute of Shimizu Corporation, Tokyo, Japan

Abstract
The corrosion rate of reinforcing steel due to carbonation, due to chloride included in concrete, and the allowable limit of corrosion of reinforcing steel were investigated. By combining the allowable limit and the rate of corrosion, the method to predict the life of reinforced concrete structures under various conditions were established.
Keywords: Reinforced Concrete, Service Life, Life Prediction, Corrosion Rate of Reinforcing Steel, Allowable Limit of Corrosion, Carbonation, Chloride Content.

1 Introduction

The life of reinforced concrete structure is mainly determined either by the degradation of concrete itself or by the corrosion of reinforcing steel. The subject of this paper was set up to investigate the latter case, and the method to predict the service life in connection with corrosion due to carbonation and corrosion due to chloride was investigated.

2 Concept of life prediction

Figure 1 is the conceptual diagram showing the process of degradation of reinforced concrete structures due to corrosion of reinforcing steel. When concrete contains chloride, the reinforcing steel will corrode at a certain rate as a curve from point O to A according to the various conditions such as chloride content, mix proportion, cover thickness, environmental conditions and others. As the amount of corrosion of reinforcing steel increases, expansive pressure will be generated around the reinforcing steel due to the volume increase as a result of rust formation. Being accompanied by this process, the structure begins to be adversely affected by various phenomena such as exudation of rust juice, occurrence of concrete cracking along the reinforcing steel, the spalling of cover concrete, deflection of members, and in an extreme situation, it will lead to a collapse of the structure.

Accordingly, a certain limit must exist as for the amount of corrosion which is allowable to the structure from structural,

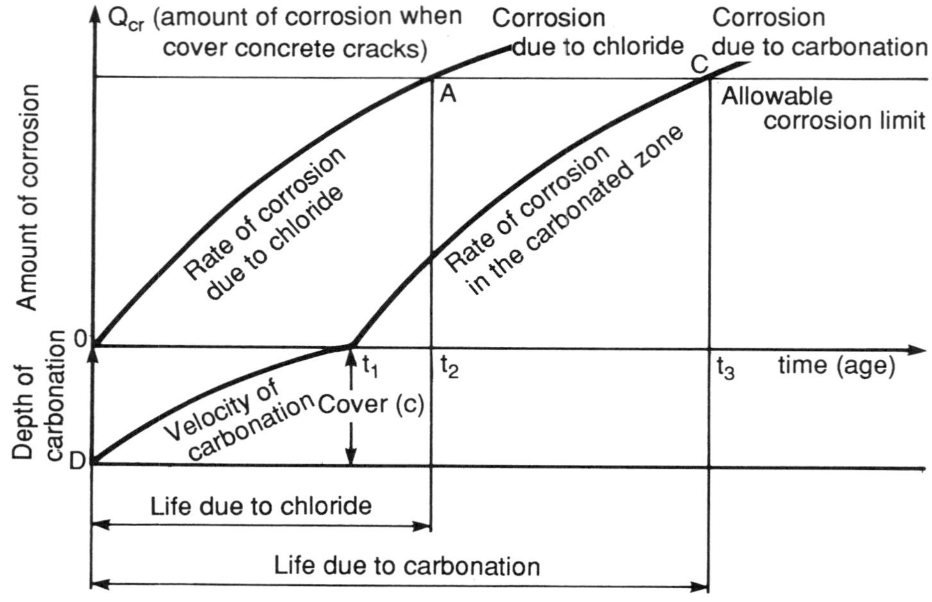

Fig.1 Conceptual diagram of life (degradation model)

functional and other necessary points of view. If this allowable limit of corrosion is determined and also the life is defined as the time at which amount of corrosion comes to the allowable limit, life could be estimated by knowing the rate of corrosion due to chloride.

In the case where the life is determined by carbonation, a similar concept could be applied. In Fig. 1, the ordinate starting from point D upwards shows the depth of carbonation of concrete. It is assumed that the carbonation front reaches the reinforcing steel at time t_1 (point B), and that the corrosion due to carbonation is initiated after this time. The curve from point B to C corresponds to the rate of corrosion in the carbonated zone. The amount of corrosion will reach the allowable limit with the lapse of time similarly as in the case of corrosion due to chloride.

To predict the life in accordance with the above-mentioned concept, the following items are necessary to be investigated under various conditions.

a) The rate of corrosion due to chloride.
b) The carbonation velocity of concrete.
c) The rate of corrosion in the carbonated zone of concrete.
d) The allowable limit of corrosion.

In the following sections these items will be investigated in sequence. However only the substance will be mentioned due to limitation of space. As for details, refer to the reference.

3 Corrosion rate due to chloride

In order to investigate this item, two series of experiments were carried out. One was the experiment to investigate the influence of design conditions, such as mix proportion, cover thickness and others on the corrosion rate of reinforcing steel, and the other was to investigate that of environmental conditions such as ambient temperature, relative humidity and others.

3.1 Influence of design conditions

3.1.1 Experiment

The specimens of concrete containing chloride and with reinforcing steel embedded were exposed outdoors for approximately 10 years, and the rates of corrosion were determined through measurements of time-dependent weight losses due to corrosion. The factors and levels are shown in Table 1.

Table 1. Factors and levels used for the test of design conditions

Factors	Levels
Kind of concrete	Ordinary, Light-weight
Water-cement ratio (%)	40, 55, 70
Chloride content (%) (NaCl by weight of mixing water)	0, 0.1, 0.5, 1.0, 1.5, 3.0
Direction of reinforcing steel	Horizontal, Vertical
Diameter of reinforcing steel (mm)	9, 25
Cover thickness (mm), for 9 mm steel	4, 6, 9, 12, 16, 22, 29, 37
for 25 mm steel	5, 8, 13, 20, 27, 37

3.1.2 Results

By this experiment, the equation (1) was derived. The effects of factors, kind of concrete and direction of reinforcing steel, were not significant. The coefficient of determination of the equation (1) was 0.953 and the ratio of variance was quite significant with the significance level of 0.5 %. The equation (1) can be applied to estimate the rate of corrosion of the reinforcing steel embedded in concrete and exposed to the outdoor conditions of normal climate.

$$q_2 = \{-0.51 - 7.60 \times N + 44.97 \times (W)^2 + 67.95 \times N \times (W)^2\}\ d/c^2 \qquad (1)$$

where q_2: rate of corrosion ($\times 10^{-4}$ g/cm²/year)
N : NaCl by weight of mixing water (%)
W : water-cement ratio (%/100)
d : diameter of reinforcing steel (mm)
c : cover thickness (mm)

3.2 Influence of environment

3.2.1 Experiment

The specimens of reinforcing steel coated by cement paste containing chloride were exposed to various environmental conditions with different temperatures, relative humidities, oxygen concentrations and moisture conditions. The rates of corrosion were determined through measurements of time-dependent weight losses for about 8 years. The factors and levels are shown in Table 2. The levels of chloride content were chosen as the same values shown in Table 1.

Table 2. Factors and levels used for the test of envoronmental

Factors		Levels
Temperature	(℃)	20, 40
Relative humidity	(%)	0, 51, 62, 100, and underwater
Oxygen concentration	(%)	0, 10, 20

3.2.2 Results

By this experiment, the equation (2) was derived. The coefficient of determination of the equation (2) was 0.835, and the equation (2) was regarded as to be quite significant with the significance level of 0.5 %. It was found by this test that the corrosion was completely prevented irrespective of chloride content, temperature and oxygen concentration, when relative humidity was kept less than 45%. This is the reason why the variable X_2 for relative humidity is transformed to be $X_2=(RH-45)/100$ in the equation (2), where RH: relative humidity(%).

$$q_3=2.59 -0.05X_1 -6.89X_2 -22.87X_3 -0.99X_4 +0.14X_5 +0.51X_6 +0.01X_7 +60.81X_8 +3.36X_9 +7.32X_{10} \quad (2)$$

where
q_3 : rate of corrosion ($\times 10^{-4} g/cm^2/year$)
X_1 : temperature (℃)
X_2 : relative humidity ((R-45) %/100)
X_3 : oxygen concentration (%/100)
X_4 : chloride content(% NaCl by weight of mixing water)
X_5 : $X_1 \times X_2$ (interaction between X_1 and X_2)
X_6 : $X_1 \times X_3$ (interaction between X_1 and X_3)
X_7 : $X_1 \times X_4$ (interaction between X_1 and X_4)
X_8 : $X_2 \times X_3$ (interaction between X_2 and X_3)
X_9 : $X_2 \times X_4$ (interaction between X_2 and X_4)
X_{10}: $X_3 \times X_4$ (interaction between X_3 and X_4)

4 Corrosion rate due to carbonation

In order to investigate this item, two series of experiments were carried out. One was the experiment to investigate the velocity of carbonation and the influence of finishing materials on it. The other was the experiment to investigate the rate of corrosion of reinforcing steel which was in the carbonated zone of concrete.

4.1 Velocity of carbonation

4.1.1 Experiment

Concrete prisms were stored in a carbon dioxide gas chamber and the depth of carbonation was measured at several ages. To investigate the influence of finishing materials on the velocity of carbonation, nine kinds of finishing meterials were tested.

4.1.2 Results

By this experiment, the equation (3) and (4) were derived. The value of R in the equation (3) and (4) means the carbonation velocity ratio of specimen with finishing material to the one without finishing material. Therefore the value of R is taken to be 1 when finishing material is not used.

In the case when $w/c \leqq 60$ %

$$x = \sqrt{C/5} \cdot 2.44 \cdot R\ (1.391 - 0.174\ RH + 0.0217\ T)(4.6\ W - 1.76)\ \sqrt{t} \qquad (3)$$

In the case when $w/c \geqq 60$ %

$$x = \sqrt{C/5} \cdot 2.44 \cdot R\ (1.391 - 0.174\ RH + 0.0217\ T) \times [4.9\ (W - 0.25) / \sqrt{1.15 + 3\ W}\ ]\ \sqrt{t} \qquad (4)$$

where
- x : depth of carbonation (mm)
- C : concentration of carbon dioxide gas (%)
- R : ratio of carbonation velocity of finishing material
- T : temperature (℃)
- RH : relative humidity (%)
- W : water-cement ratio (% / 100)
- t : time (day)

4.2 Corrosion rate due to carbonation

4.2.1 Experiment

A piece of reinforcing steel was embedded in a mortar prism. The specimen was carbonated to the depth of reinforcing steel and thereafter was exposed to various environmental conditions. The weight loss of the reinforcing steel was measured periodically, and the rate of corrosion in the carbonated zone of concrete and the influence of environmental conditions on it were investigated.

4.2.2 Results

By this experiment, the equation (5) was derived. It was found that the weight loss increased steadily with time even under the conditions where the oxygen concentration was zero, and the corrosion product which was created in the carbonated zone of concrete was thought not to be the so-called iron hydroxide but a sort of iron carbonate. The coefficient of determination of the equation (5) was 0.963 and the ratio of variance was regarded to be quite significant with the significance level of 0.5 %.

$$q_1 = 21.84 - 1.35X_1 - 35.43X_2 - 234.76X_3 + 2.33X_4 + 4.42X_5 + 250.55X_6 \qquad (5)$$

where
- q_1: rate of corrosion due to carbonation ($\times 10^{-4}$ g/cm²/year)
- X_1: temperature (℃)
- X_2: relative humidity (%/100)
- X_3: oxygen concentration (%/100)
- X_4: $X_1 \times X_2$ (interaction between X_1 and X_2)
- X_5: $X_1 \times X_3$ (interaction between X_1 and X_3)
- X_6: $X_2 \times X_3$ (interaction between X_2 and X_3)

5 Influence of corrosion on properties of reinforcing steel

The reinforced concrete structures are designed structurally under the assumptions that reinforcing steel has specified tensile properties such as yield point, tensile strength, elongation capacity and bond strength between concrete and reinforcing steel. Therefore these structural properties of the reinforcing steel must not be damaged due to the corrosion of the steel. Not only the structural properties but also the other functional properties such as safety, comfortability and appearance are required for the reinforced concrete structures. Accordingly these functional properties must not be spoiled by phenomena such as the occurrence of cracking, spalling of cover concrete and the exudation of rust juice as a result of the corrosion of the reinforcing steel.

From a reinforced concrete building degraded due to corrosion of reinforcing steel, concrete specimens including reinforcing steels were cut out by a diamond saw. The influence of corrosion on the above-mentioned structural and functional properties was investigated on about nine hundred pieces of reinforcing steel. The main results obtained by this investigation are summarized as follows.

1) The most important structural properties such as the yield point, the tensile strength and the bond strength remain unchanged unless the cover concrete cracking due to the corrosion is found.

2) If the degree of corrosion increases to the level which causes cracking of cover concrete, various detrimental effects begin to arise.

3) The occurrence of cover concrete cracking accelerates the corrosion rate of reinforcing steel to a great extent, and this becomes a direct cause to injure the structural and functional properties of the reinforced concrete structures.

4) Therefore the amount of corrosion which causes cracking to cover concrete is the very important and distinct measure to judge the life and soundness of the reinforced conrete structures.

6 Conditions to cause cover concrete cracking

In order to investigate the conditions when cover concrete cracks due to corrosion of reinforcing steel, two series of experiments were carried out. One was the experiment to determine the tensile stress when cover concrete cracks. A hollow concrete cylinder was used as a specimen, simulating the internal diameter of the hollow cylinder as the diameter of the reinforcing steel and the wall thickness of the hollow cylinder as the thickness of cover concrete. Oil pressure was applied to the inner surface of the hollow, and the influence of bar diameter, cover thickness and tensile strength of the concrete on the maximum oil pressure at failure of the specimen was investigated.

The other was the one to determine the tensile strain when cover concrete cracks. A piece of reinforcing steel was embedded at the center of the concrete cylinder. The specimen was immersed in a salt solution and direct current was applied to the reinforcing steel and was forced to corrode electrolytically. After the concrete cylinder cracked, the reinforcing steel was weighed, and the amount of corrosion or the volume of corrosion products was determined.

Combining the results of these two experiments, the the equation (6) to estimate the amount of corrosion when cover concrete cracks due to corrosion was obtained.

$$Qcr = 0.602 \left(1 + \frac{2c}{d}\right)^{0.85} d \tag{6}$$

where

Qcr : amount of corrosion when concrete cracks ($\times 10^{-4}$g/cm^2)

c : cover thickness of concrete (mm)

d : diameter of reinforcing steel (mm)

7 Method of life prediction

7. 1 Definition of life

In section 5, it was shown that the amount of corrosion which causes cracking to cover concrete is a very important and distinct measure to judge the life and the soundness of reinforced concrete structures, and this amount can be regarded as the allowable limit of corrosion. Therefore in this report, the life is defined as the point of time at which the amount of corrosion has reached the level to cause cover concrete cracking.

7. 2 Method of life prediction

The life of reinforced concrete structures can be predicted as follows using the results obtained until now, and the degradation model shown in Fig. 1

a) Flow of life prediction

The basic idea is shown in the main flow in Fig. 2.

Firstly, the amount of corrosion which causes cover concrete cracking Qcr is detremined. Qcr is a function of the bar diameter and the cover thickness, and is determined irrespective of the carbonation velocity or the chloride content.

Secondly, the life due to carbonation t_3 in Fig. 1 is determined. In the case where the concrete does not contain chloride and there is no risk of corrosion due to chloride, t_3 is chosen as the life.

Thirdly, in the case where the concrete contains a questionable amount of chloride, the life due to chloride t_2 in Fig. 1 is obtained. And the smaller one between t_2 and t_3 is chosen as the life.

In this idea of the life prediction, it is assumed that the carbonation velocity and the corrosion rate in the carbonated zone of concrete are not affected by the chloride content. In the experiment described in section 3, it is confirmed that the carbonation velocity is not affected by the chloride content.

On the other hand, the rate of corrosion in the carbonated zone of concrete will be strongly affected by the chloride content. However as is shown later, the time t_1 and the time t_3 are usually very close because the rate of corrosion in the carbonated zone of concrete is quite high, and the life due to carbonation is practically determined mainly by the carbonation velocity and not by the rate of corrosion in the carbonated zone of concrete. Therefore the above assumption does not cause any problems.

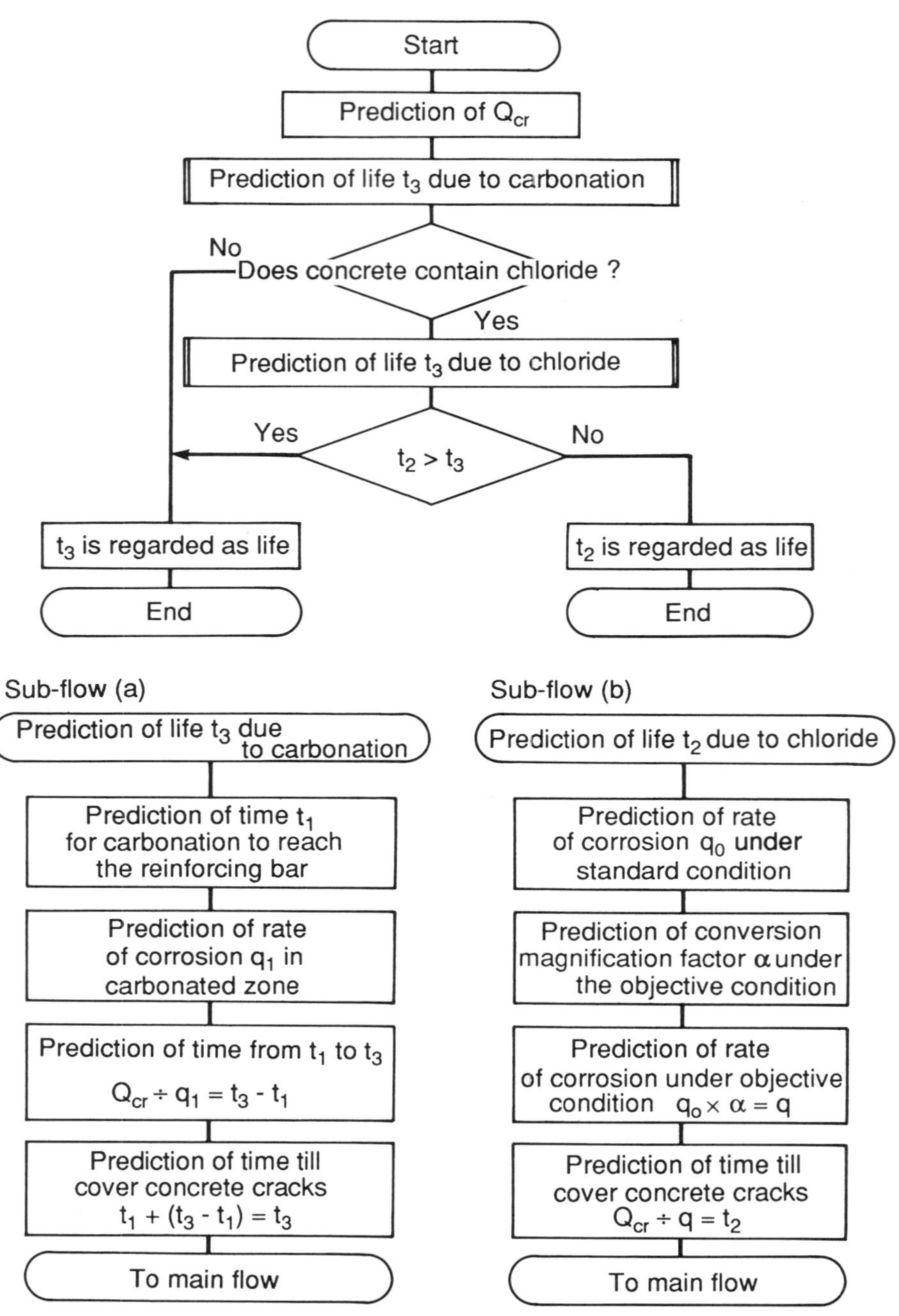

Fig.2 Flow of life prediction

b) Prediction of life due to carbonation

The basic idea to predict the life due to carbonation is shown in the sub-flow (a) in Fig. 2.

1) Time till carbonation reaches reinforcing steel

The time from the placement of concrete of the structure to the time when the carbonation reaches the reinforcing steel, t_1 in Fig. 1, is determined by the equation (3) or (4) according to the w/c ratio.

2) Time from commencement of corrosion to cover concrete cracking

The corrosion of reinforcing steel begins to take place after the carbonation reaches the reinforcing steel. If the rate of corrosion in the carbonated zone of concrete is taken to be q_1 and the amount of corrosion when cover concrete cracks is taken to be Qcr then the time t from the commencement of corrosion to the time of the cover concrete cracking is obtained by the equation (7).

$$t = t_3 - t_1 = Qcr \div q_1 \qquad (7)$$

Qcr is determined by the equation (6) and q_1 is determined by the equation (5).

In the case where the concrete is covered by a finishing material having the ratio of carbonation velocity of R, the oxygen concentration X_3 in the equation (5) is substituted for $X_3 = 0.2 \times R^2$. See reference.

3) Life due to carbonation

The life due to carbonation t_3 is determined by adding the two periods of time obtained above.

$$t_1 + t = t_1 + (t_3 - t_1) = t_3 \qquad (8)$$

c) Prediction of life due to chloride

The basic idea to predict the life due to chloride is shown in the sub-flow (b) in Fig. 2.

1) Rate of corrosion under standard environmental condition

The rate of corrosion q_2 shown in the equation (1) was derived from the experiments conducted under the environmental conditions with the temperature of 15 ℃, relative humidity of 69% and the oxygen gas concentration of 20 % (exposed to the open air without finishing material). This environmental condition is called "the standard condition" hereafter. The rate of corrosion q_2 is determined by the equation (1), and let q_2 be replaced temporarily by q_0.

2) Conversion of corrosion rate from standard to other environmental condition

The equation (2) was derived mainly for the purpose of investigating the influence of the environmental conditions on the rate of corrosion. The rate of corrosion q_3 is determined by substituting the temperature, relative humidity, oxygen concentration and the chloride content of the objective structure into the X_1, X_2, X_3 and X_4 of the equation (2) respectively. Let q_3 be replaced temporarily by q_3'.

Another q_3 is determined by substituting the temperature, relative humidity and oxygen concentration of "the standard condition", and the chloride content of the objective structure into the X_1, X_2, X_3 and X_4 of the equation (2) respectively. Let the another q_3 be replaced temporarily by q_3''.

If the ratio of q_3' to q_3'' is taken to be $\alpha = q_3' / q_3''$, then α implies the magnification of the corrosion rate when the environment changes from the standard condition to the another one. The α is called the conversion magnification factor hereafter.

The rate of corrosion of the objective structure q is determined by multiplying the q_0 by the conversion magnififation factor α.

$$q = q_0 \times \alpha \tag{9}$$

When the equation (2) is used, the oxygen concentration X_3 is substituted for $X_3 = 0.2 \times R^2$. See reference.

3) Life due to chloride

The life due to chloride t_2 is determined by dividing the Qcr by the value of q in the equation (9)

7.3 Example of life prediction

Fig. 3 shows some results of life which were predicted according to the method described above. The following parameters were used in this example.

Temperature	20℃
Relative humidity	70%
Oxygen concentration	20% (no finish, R=1.00)
Carbon dioxide gas concentration	0.03% , 0.10 %
Water-cement ratio	65%
Cover thickness	20mm, 30mm, 40mm
Chloride content of concrete (as chlorine ion kg/ m^3 of concrete)	0.3 ~ 5.0 kg/ m^3
Diameter of reinforcing steel	9 mm

The abscissa of the Fig. 3 shows the time from the placement of concrete, and the ordinate shows the weight loss of the reinforcing steel and Qcr. It is obvious that the rate of corrosion after the carbonation reached the reinforcing steel is quite high, and the rising of the corrosion curve is almost vertical. In other words, the rate of corrosion in the carbonated zone of concrete is far greater than the rate of corrosion when the concrete contains a very large amount of chloride.

Therefore it is concluded that the life due to carbonation is practically determined by the carbonation velocity and not by the rate of corrosion in the carbonated zone of concrete, and that the time when carbonation reached the reinforcing steel can be regarded as the life under this condition.

7.4 Factors affecting the life

As shown in Fig. 3, the factors such as the chloride content, carbon dioxide gas concentration and cover thickness have great influence on the life of the reinforced concrete structures. Though the influence of the factors such as water-cement ratio, finishing material, diameter of reinforcing steel, temperature, relative humidity and oxygen concentration is not shown in Fig. 3, it is confirmed by the above-mentioned method that these factors also have strong influence on the life.

Above all the cover thickness is the fundamental and extremely

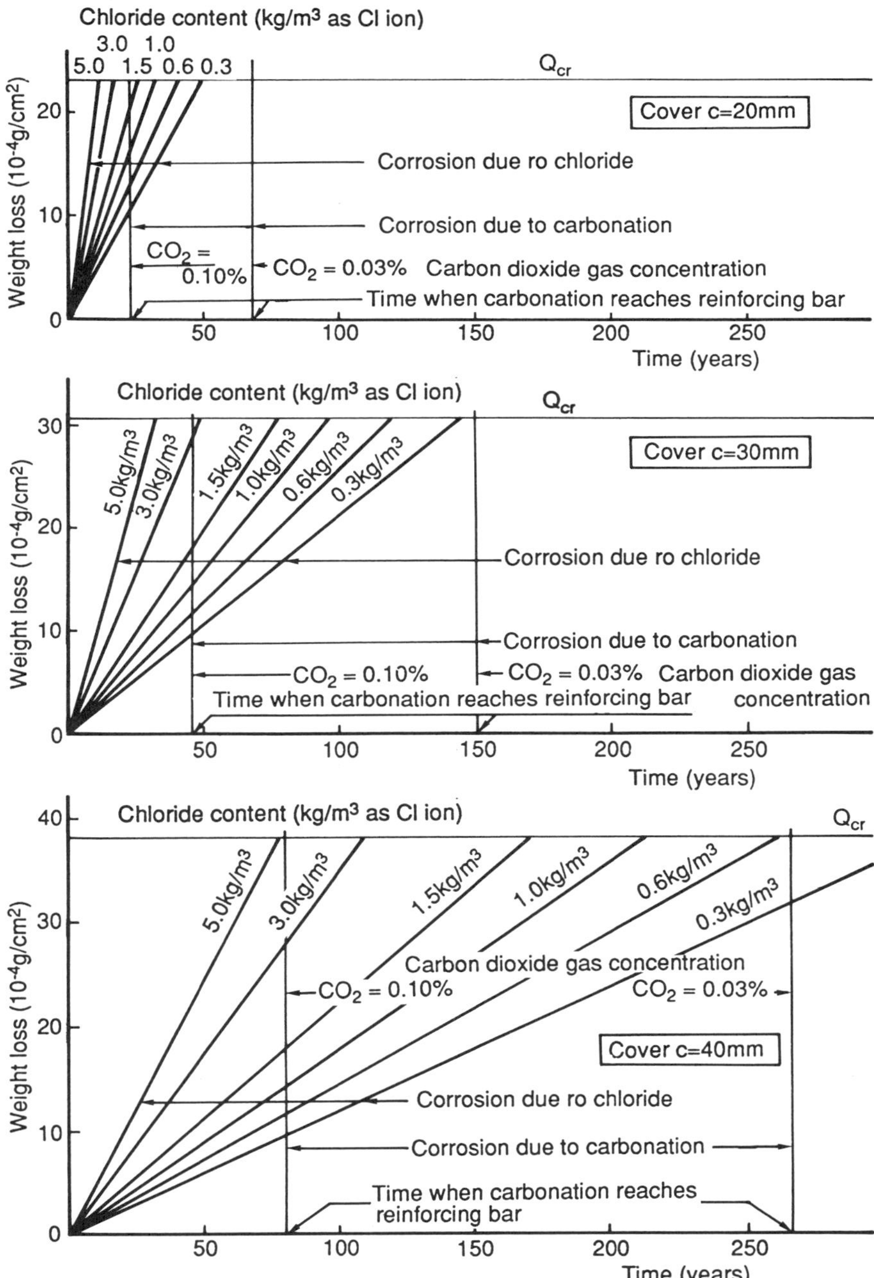

Fig.3 Examples of life prediction

important factor which determines the life. The insufficiency of the cover thickness is the main and decisive cause of the early and heavy degradation of the reinforced concrete structures.

The measures to improve the durability, such as keeping the low water-cement ratio, lessening the chloride content, the use of finishing materials and others are effective only when the cover thickness is kept enough and sufficient. In order to make the reinforced concrete structures durable, it is quite important to avoid making the portion where the cover thickness is insufficient.

8 CONCLUSION

There are many causes which degrade the reinforced concrete structures. One of them is the corrosion of reinforcing steel. In this paper, the corrosion of reinforcing steel due to carbonation of concrete and the corrosion due to chloride included in concrete were investigated. And the method to predict the life of the reinforced concrete structures determined by the corrosion of reinforcing steel was established. The main conclusions obtained by this work are as follows.

1) The reinforced concrete structures reveal many levels of degradation according to the extent of corrosion of reinforcing steel. The influence of corrosion on the properties which are required for reinforcing steel was investigated. And it was elucidated that the amount of corrosion Qcr at which cover concrete cracks was regarded as the allowable limit of corrosion from the structural and functional points of view.

2) Regarding the corrosion due to carbonation, the velocity of carbonation of concrete and the rate of corrosion of reinforcing steel in the carbonated zone of concrete were investigated. And a process of corrosion was formulated as a function of water-cement ratio, carbon dioxide gas concentration, oxygen gas concentration, temperature and relative humidity of the environment.

3) Regarding the corrosion due to chloride, the influence of the quality of concrete, the environmental conditions and others on the corrosion process of reinforcing steel embedded in concrete containing chloride was investigated. And the rate of corrosion was formulated as a function of chloride content, bar diameter, cover thickness, water-cement ratio, temperature, relative humidity and oxygen concentration of the environment.

4) The life was defined as the point of time at which the amount of corrosion reached the allowable limit Qcr. By combining the above-mentioned results together, it was made possible to predict the life of reinforced concrete structures under various conditions and to evaluate quantitatively the factors which have influence on the life.

9 Reference

Morinaga, S. " Prediction of service lives of reinforced concrete buildings based on rate of corrosion of reinforcing steel", Special report of institute of technology, Shimizu Corporation, No. 23, June, 1988, Tokyo, Japan.

3 SOLAR UV, WETNESS AND THERMAL DEGRADATION MAPS IN JAPAN

T. TOMIITA
Advanced Construction Technology Center, Tokyo, Japan

Abstract
Japan is located in sub-arctic, temperate and sub-tropical zones surrounded by seas. In the center of the Japanese Islands, there are high mountains which divide Japanese climate into several zones. Durability of building materials is greatly influenced by the climates which can be quantified by the deterioration factors. In this report, estimated deterioration factors: solar UV energy, wetness time / wet-dry cycle, thermal degradation and daily temperature difference are shown on the maps.
Keywords: Degradation Factor, Solar UV, Wetness Time, Wet-dry Cycle, Thermal Degradation, Daily Temperature Difference, Map

1 Introduction

The durability of building materials is greatly influenced by the deterioration factors which are deeply related to the climatic conditions. The author has estimated degradation factors from the meteorological data and plotted them on the maps. By refering to these deterioration factor maps, local differences of the service life of various building materials can be discussed.

The Japanese Archipelago is located mainly in the temperate oceanic zone, but the northern area is in the sub-arctic zone and the southern area is in the sub-tropical area due to its north-south length of 3,000 km. The summer is hot and humid with south-easterly winds. In winter, it is dry and clear on the Pacific side, while it snows heavily on the Japan Sea side. This difference comes from the existance of ranges of high mountains running through in the center of the islands. In addition, warm ocean currents and cold ocean currents running along the islands also influence the climate. The geography of Japan and yearly average of ambient temperature are shown in Fig.1.

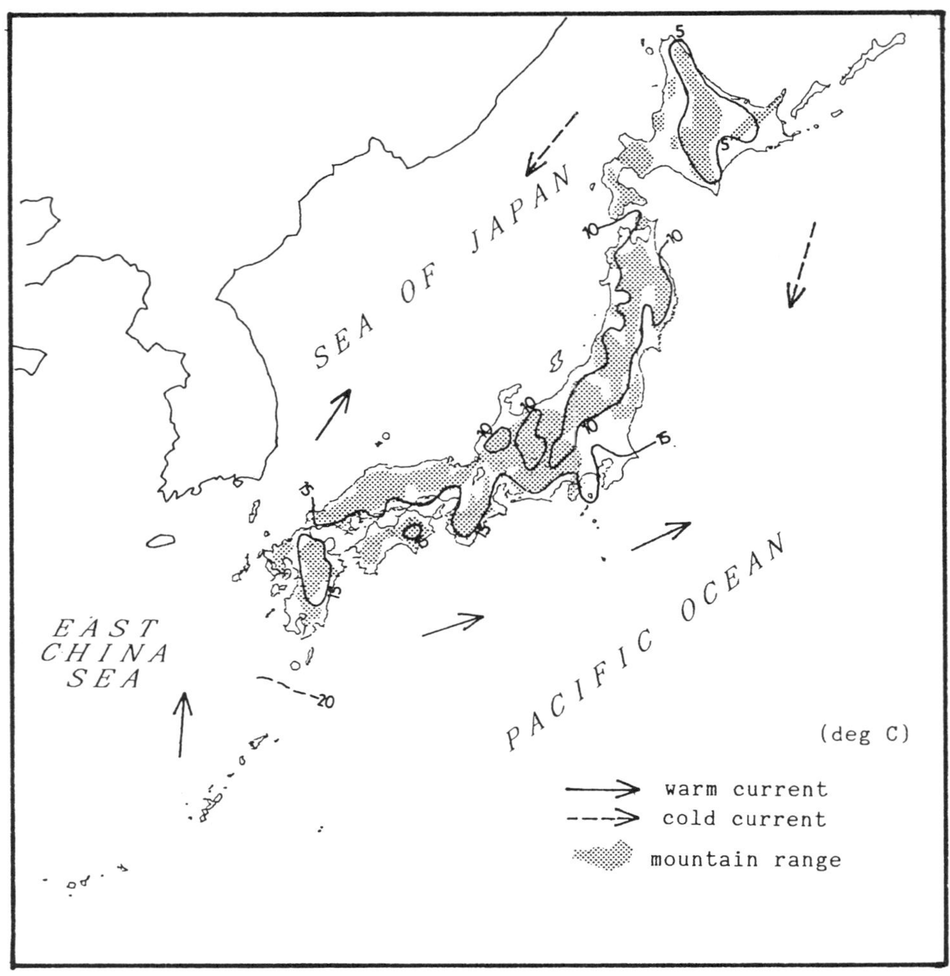

Fig.1. Geography of Japanese Archipelago and yearly average temperature

2 Solar Ultraviolet Energy

Polymeric building materials such as paint, sealants, roofing and adhesives are deteriorated by ultraviolet rays in solar radiation. Some researchers are using chemical dosemeters to quantify UV energy around the world. However, the author used a photo-diode sensor, EKO's MS-140 (wave range: 305-395 nm) to measure hourly solar UV energy on the horizontal plane in the outdoor exposure field of Building Research Institute in Tsukuba, Japan for a year and expressed it as a function of the entire range of solar radiation observed by a pyranometer, EKO's MS-801-305 (305-2800 nm) and solar altitude as follows:

$$U_{hour} = -0.894\ Z^{-0.229}\ S_{hour} + 0.796\ Z^{-0.165}\ S_{hour}^{0.968} \quad (1)$$

where, U_{hour}: hourly solar UV energy ($MJ/m^2/hour$),
Z: solar altitude (degree) and
S_{hour}: hourly entire range solar energy ($MJ/m^2/hour$)

By applying eq.(1) to the hourly solar energy calculated from the percentage of sunshine at 141 points in Japan, hourly solar UV energy was estimated. The yearly amount of solar UV energy was calculated by integration and is shown in Fig.2.

3 Wetness Time / Wet-dry Cycle

Water is a factor involved in the corrosion of metals and the hydrolysis of organic materials. Substances such as stabilizers or plasticizers contained in materials dissolve in water. ISO/TC156/WG4 have used the criterion that vapor in air condenses into dew when relative humidity is more than 80 (%) and ambient temperature is higher than 0 (deg C).

Relative humidity is observed at 3, 9, 15, 21 o'clock, and ambient temperature is measured at 3, 6, 9, 12, 15, 18, 21, 24 o'clock every day by the Meteorological Agency in Japan. Hourly data were estimated by internal division of the adjacent data. When relative humidity and ambient temperature satisfied the dew criterion, one hour was added to wetness time. The averages of yearly total of wetness time at 154 points during 1961-1985 were obtained and plotted in Fig.3.

When vapor condenses into dew or dew dries, a thin water layer exists on the surface of material. Oxygen in air and water combine together to deteriorate the material in this condition. Therefore, the number of wet-dry cycle is an index of these complex deteriorations. The author, Tomiita (1989a) counted the number of such cycles when dew condensed on the dry surface. Fig.4. is the map of the yearly total of wet-dry cycles.

4 Thermal Degradation

Heat accelerates degradation reaction of materials. The property change of a building material is expressed as a function of temperature based on Arrhenius' Equation.

$$\ln(P/P_0) = C_h \exp(-E_h/RT)\ t \quad (2)$$

where P: material property,

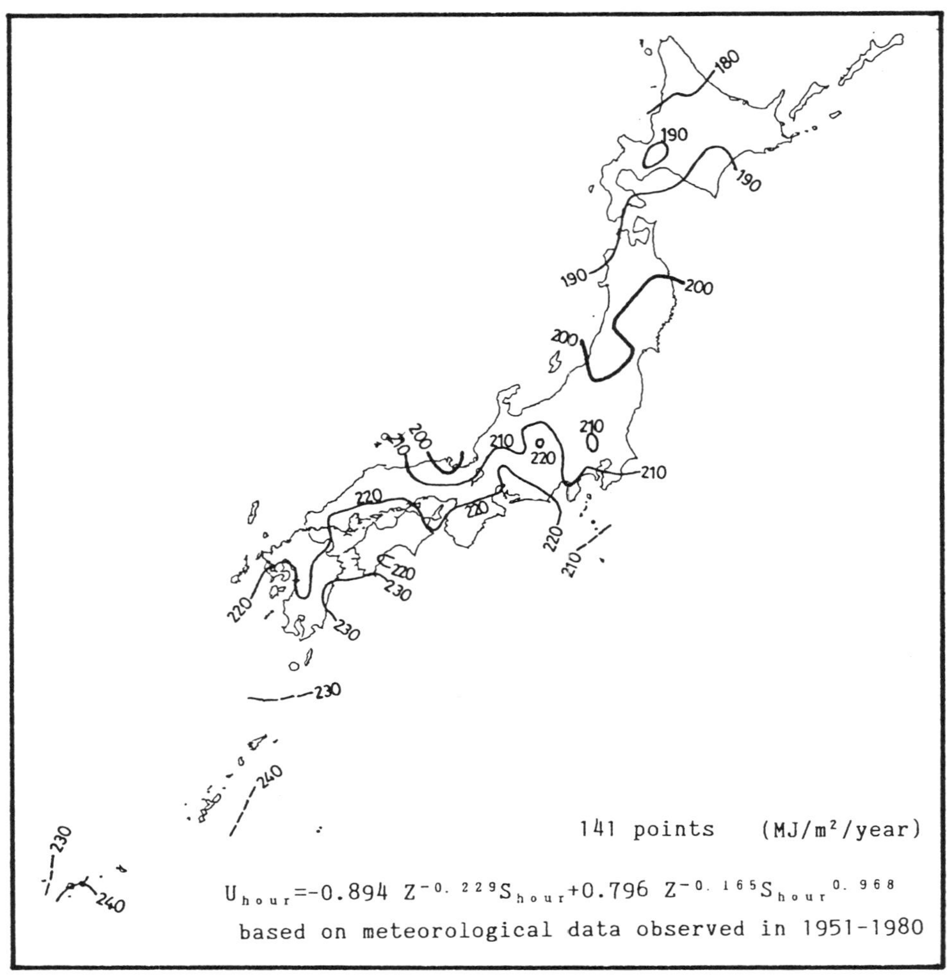

Fig.2. Solar ultraviolet energy

P_0: initial material property
C_h: thermal deterioration material constant
E_h: activation energy of thermal deterioration (KJ/mol),
R: gas constant, 8.314×10^{-3} (KJ/mol/K),
T: absolute temperature of material (K)
t: elapsed time

Assuming the following performance over time,

1) Elevation of temperature from 20 (deg C) to 30 (deg C) doubles deterioration rate.
2) Initial property decreases to 80 (%) at 20 (deg C) in

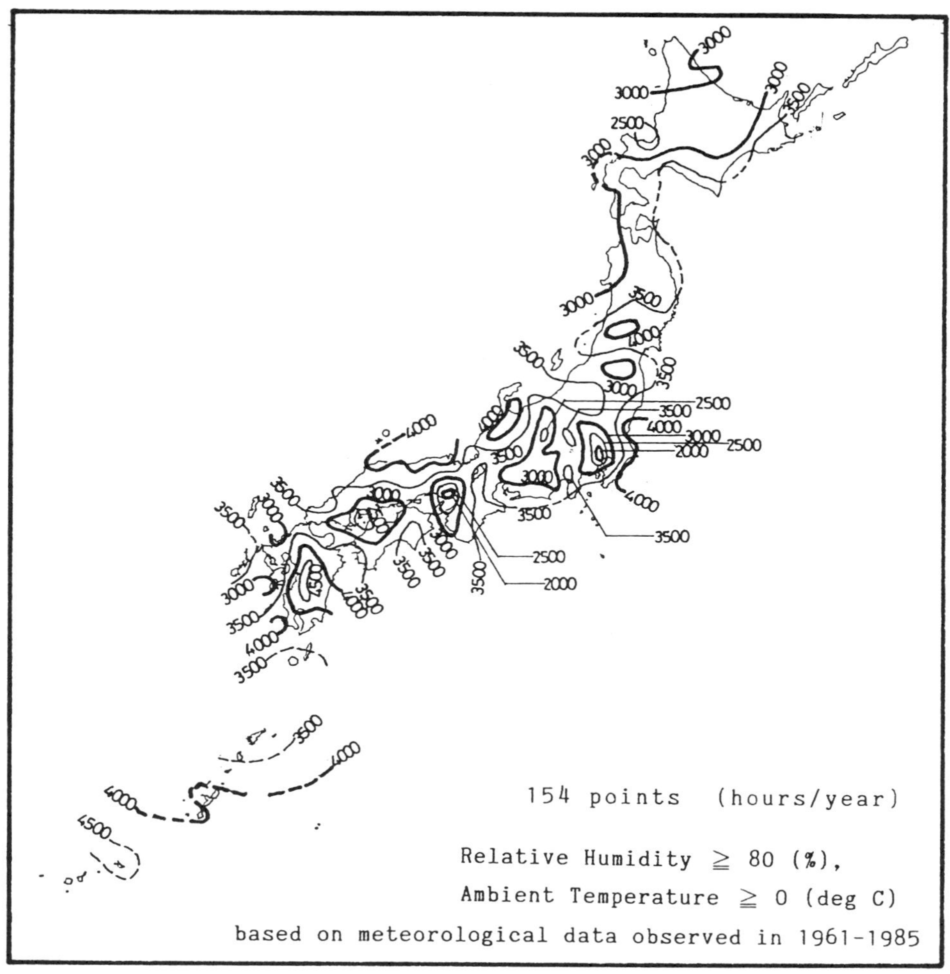

Fig.3. Wetness time

3653 (days).

the material deterioration constants were determined as follows:

$$C_h = -8.16 \times 10^4 \ (1/\text{day}); \quad E_h = 51.2 \ (\text{KJ/mol})$$

The author, Tomiita(1989b) measured the black panel temperature (BPT) every 10 minuites outdoors. The BPT is an index of the severest thermal degradation conditions. This was a 1.0 mm thick stainless steel panel painted matt black with a thermo-couple set on the surface. It was exposed

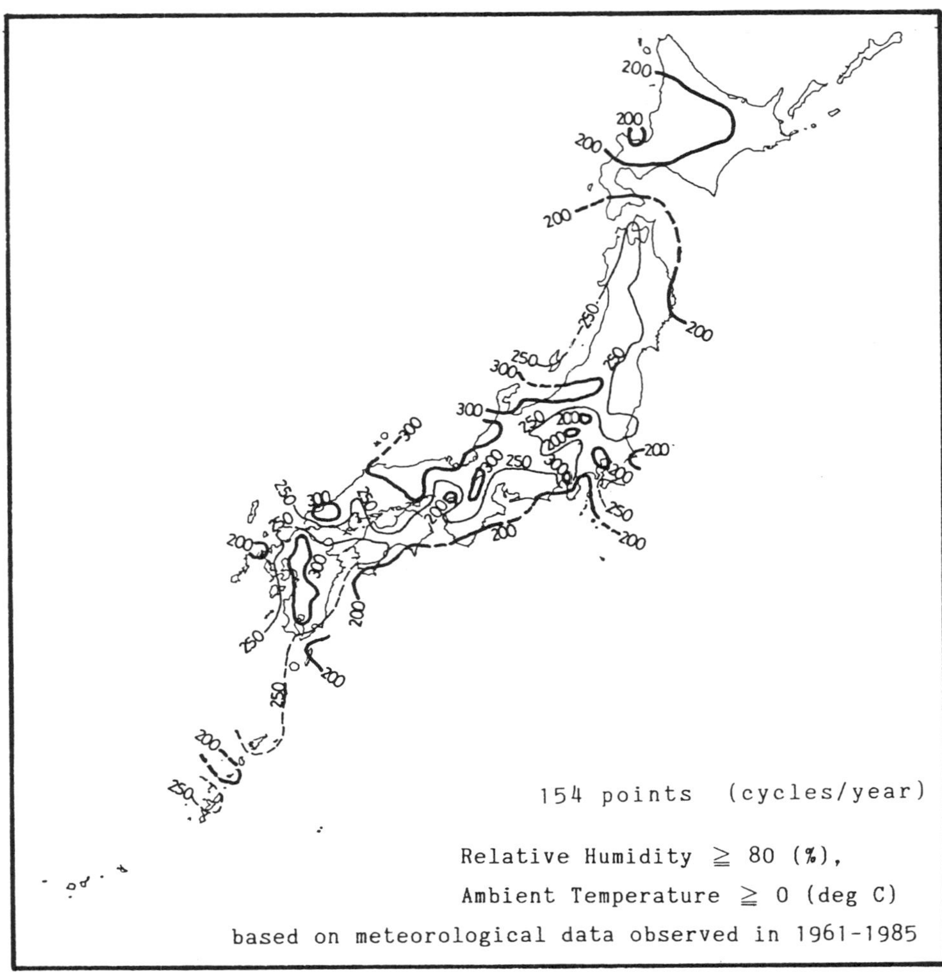

Fig.4. Wet-dry cycle

horizontally. When the property degrades from P_1 to P_2 in a day, the ratio of P_2 to P_1 can be expressed as the sum of 144 tiny thermal degradation a day and written as eq.(3).

$$\ln(P_2/P_1) = \sum^{144} C_h \exp\{-E_h/R(T_P+273)\} \times (1/144) \tag{3}$$

In order to estimate the property change at various points in Japan, it is neccesary to use meteorological data including daily total of solar radiation energy: S_D (MJ/m^2/day) provided by the Meteorological Agency. Here the author introduced a concept, daily equivalent BPT, T_E

(deg C), which was determined as a virtual constant temperature that causes the same amount of thermal degradation as by the actual fluctuating temperature. Based on this determination,

$$\ln(P_2/P_1) = C_h \exp \{ -E_h/R(T_E+273) \} \qquad (4)$$

Eq.(3) and eq.(4) have the same left hands, then

$$\sum^{144} C_h \exp \{ -E_h/R(T_P+273) \} \times (1/144) = C_h \exp \{ -E_h/R(T_E+273) \} \qquad (5)$$

This means that the daily equivalent BPT can be calculated by the actual BPT measured at every 10 minutes. The daily equivalent BPT was expressed as a function of climatic elements.

$$T_E = C_1 T_D + C_2 T_N + S_D (C_3 + C_4 W_D) + C_5 \qquad (6)$$

where C_1, C_2, C_3, C_4 and C_5: coefficients,

T_D: daily average of ambient temperature in daytime (deg C),

$T_D = (T_6/2 + T_9 + T_{12} + T_{15} + T_{18}/2) /4$

sub number means observed o'clock

T_N: daily average of ambient temperature in night (deg C) and

$T_N = (T_3 + T_6/2 + T_{18}/2 + T_{21} + T_{24}) /4$

W_D: daily average of wind speed in daytime (m/s)

$W_D = (T_6 + T_9 + T_{12} + T_{15} + T_{18}) /5$

The coefficient of each term in eq.(6) was obtained by applying the least square error method for 676 days data observed in Tsukuba for 2 years.

$$T_E = 0.646 T_D + 0.357 T_N + S_D (0.736 -0.133 W_D) -2.449 \qquad (6')$$

The daily equivalent BPT at 66 points from 1976 to 1985 were estimated by eq.(6'). The change of property in 10 years is the sum of daily change of property calculated by inputting the daily equivalent BPT into eq.(4). Obtained results are shown in Fig.5.

5 Daily Temperature Difference

Building materials are heated by sunshine in the daytime and are cooled by emission of solar radiation and by greater extraterrestrial radiation on a cloudless night. The fluctuations of temperature induce joint movements between building elements or cracks in concrete. Sealants and exterior finishing materials suffer from repeating tension, compression and shear stresses and are fatigued sufficiently to deteriorate. The daily temperature differ-

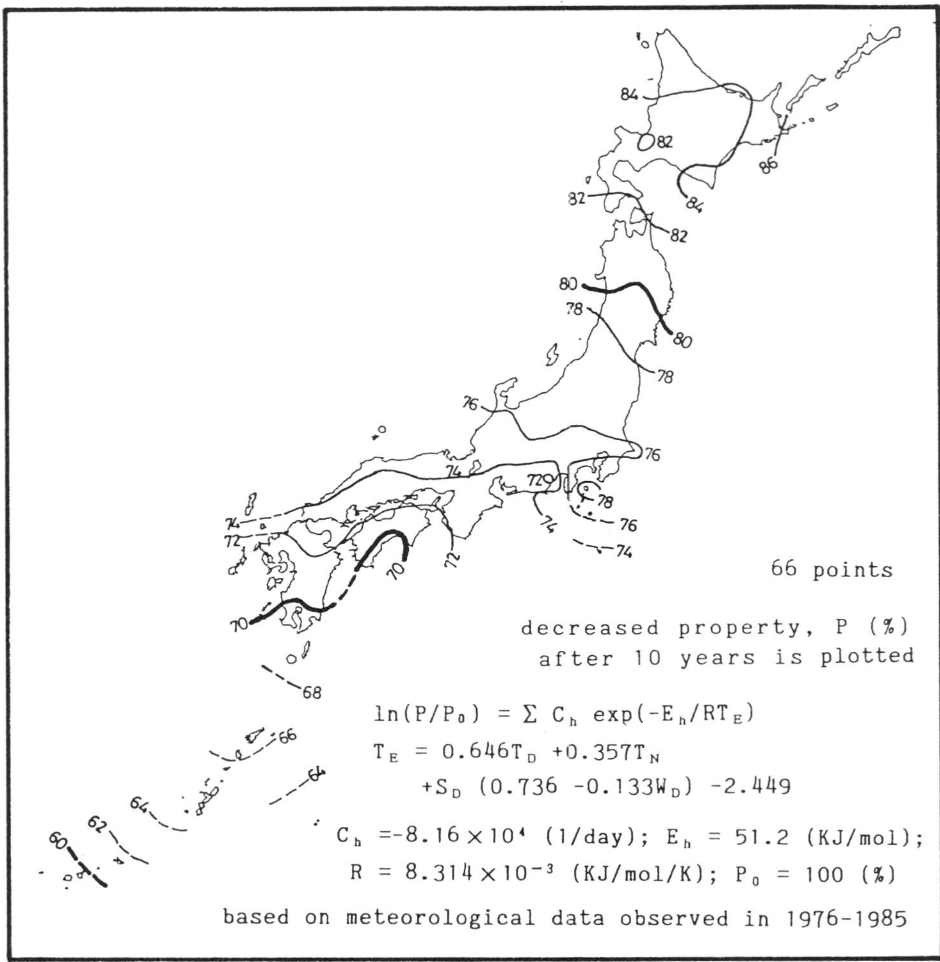

Fig.5. Thermal degradation introduced black panel temperature in 10 years

ence of building elements can be treated as a mechanical deterioration index. The author, Tomiita(1989c) collected daily maximum / minimum BPT, $T_{P\ max}$ (deg C) / $T_{P\ min}$ (deg C) and expressed them as functions of climatic data.

$$T_{P\ max} = 1.143\ T_{max} - 2.244\ W_{12} + 0.928\ S_D + 5.214 \tag{7}$$

$$T_{P\ min} = 1.049\ T_{min} + 0.022\ W_N - 2.389\ C_I - 3.056 \tag{8}$$

where T_{max}: daily maximun ambient temperature (deg C),
W_{12}: wind velocity at 12 o'clock (m/s),

T_{min}: daily minimun ambient temperature (deg C),
W_N: daily average of wind speed in night (m/s),
$W_N = (W_3 + W_6/2 + W_{18}/2 + W_{21} + W_{24}) /4$
C_I: clearness index (-), the ratio of daily total of solar radiation to the extra-terrestrial solar radiation.

By introducing meteorological data observed at 66 points during 1976-1985 into eq.(7) and eq.(8), daily maximum / minimum BPT was estimated. The yearly averages of daily differences of BPT are plotted in Fig.6. By multiplying the length of a building element by its thermal expansion rate, the daily movement in the joint or crack can be calculated.

6 Conclusion

Solar ultraviolet ray, water and heat are common chemical and mechanical deterioration factors for many building materials used in outdoor environments. The author has estimated the intensity or index of these factors from meteorological data and expressed them in the following maps.

Solar UV energy,
Wetness time,
Wet-dry cycle,
Thermal degradation based on BPT and
Daily temperature difference of BPT

The local differences in the ways or speed of the deterioration of building materials can be generally explained by citing these maps in this paper.

In addition to black panel temperature the temperature of a black panel which was insulated on the reverse side was measured. Based on insulated BPT, thermal degradation and daily temperature difference maps were reported by Tomiita(1989a).

There are synergistic effects of multi degradation factors. The author and his colleague, Tomiita and Kashino (1989c) have estimated temperature modified wetness time / wet-dry cycle and plotted on maps. In the near future, the UV degradation accelerated by heat will be quantified.

7 References

ISO/TC156/WG4 Corrosion of metals and alloys **Classification of atmospheres deducation methods of determination**

Tomiita, T. (1989a) Thermal degradation map based on black panel temperature. **Transactions of A.I.J.**, No.395, 13-20.

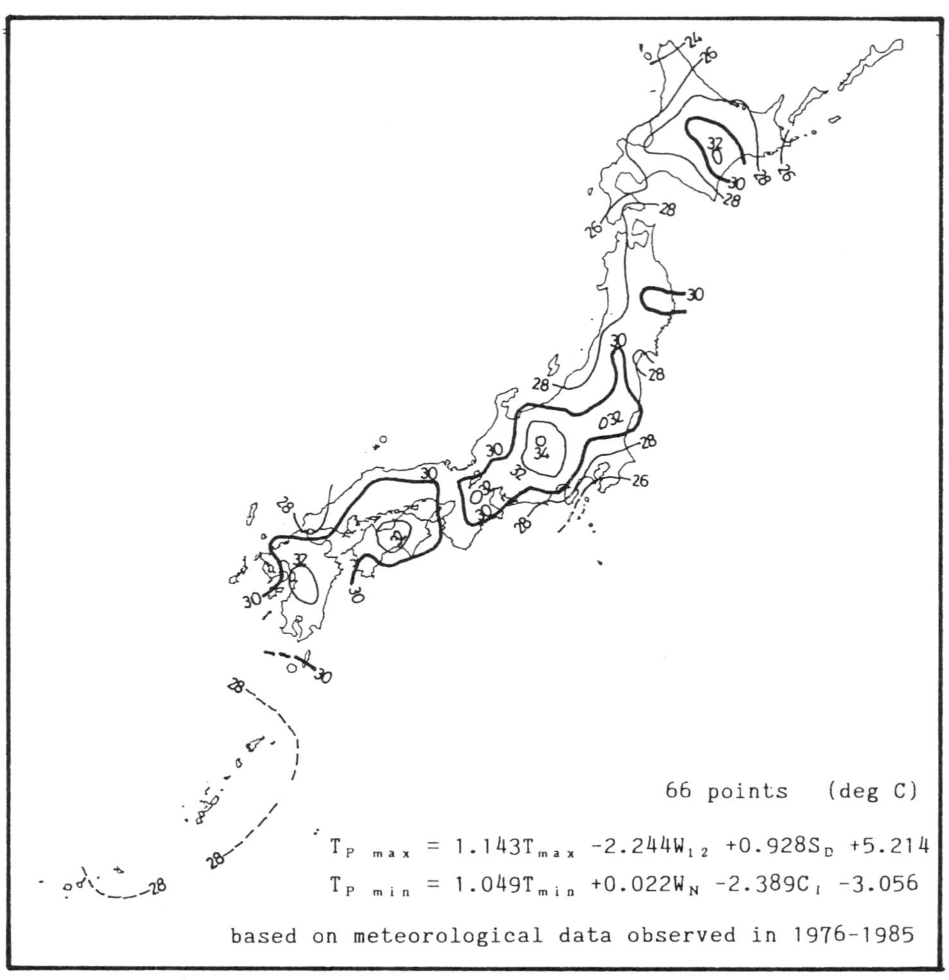

Fig.6. Daily temperature difference

Tomiita, T. and Kashino, N. (1989b) Wetness time and wet-dry cycle map in Japan. **Transactions of A.I.J.**, No.395, 21-27.

Tomiita, T. and Kashino, N. (1989c) Temperature modified wetness time and wet-dry cycle map in Japan. **Transactions of A.I.J.**, No.405, 1-7.

4 FUNDAMENTAL CONCRETE CARBONATION MODEL AND APPLICATION TO DURABILITY OF REINFORCED CONCRETE

V.G PAPADAKIS
Institute of Chemical Engineering and High Temperature Chemical Processes and Department of Chemical Engineering, University of Patras, Greece
M.N. FARDIS
Department of Civil Engineering, University of Patras, Greece
C.G. VAYENAS
Institute of Chemical Engineering and High Temperature Chemical Processes and Department of Chemical Engineering, University of Patras, Greece

Abstract
In reinforced concrete carbonation of concrete leads to depassivation of the reinforcing bars, and hence to initiation of corrosion. In this paper, the physicochemical processes of concrete carbonation are modeled mathematically. For the usual range of parameter values, the resulting complex mathematical model can be simplified, yielding a carbonation front and the rate at which this front advances within the concrete volume. This rate is expressed in terms of the effective diffusivity of CO_2 in concrete, the molar concentrations of $Ca(OH)_2$ and CSH in concrete, and the ambient concentration of CO_2. The model is verified experimentally, by comparison with the results of normal and accelerated carbonation tests, and then used to perform parametric studies on the effect of various composition parameters (water-cement and aggregate-cement ratios, chemical composition and type of cement) and of the ambient relative humidity and CO_2 concentration on the corrosion initiation period for given concrete cover.
Keywords : Carbonation, Cement hydration, Concrete durability, Diffusivity, Porosity, Pozzolans, Reinforcement corrosion, Service life.

1 Introduction

Reinforcement corrosion is the most important durability problem of reinforced concrete structures today. Reinforcing bars are protected from corrosion by a thin ferrous oxide layer which forms on their surface due to the high alkalinity (pH≈12.5) of their environment. If the pH-value of this environment drops below 9, or if chloride ions penetrate up to the surface of the reinforcement, the oxide layer is destroyed, and steel corrosion starts. The high alkalinity of the concrete mass is due to the $(OH)^-$ ions in the pore water, provided by the dissolution of $Ca(OH)_2$ from the solid phase of cement gel into the pore water. A very small concentration of $Ca(OH)_2$ in this solid phase is enough for its dissolution in the aqueous phase of the pores at the equilibrium concentration of $(OH)^-$, which in turn guarantees a pH-value equal to 12.5. Carbonation of concrete is the reaction of the $Ca(OH)_2$ which is dissolved in the pore water with the atmospheric CO_2, and its conversion into $CaCO_3$, which gives to the pore-water an equilibrium pH-value around 8.3, and therefore signals the onset of steel corrosion. Due to the importance of the emerging problem of concrete durability, a lot of research has been devoted to carbonation, mainly during the last decade, but also earlier (Hamada, 1969; Schiessl, 1976; Tuutti, 1982; Nagataki et al., 1988; Richardson, 1988). Some of this work has led to empirical models of the evolution of carbonation. Being empirical, these models cannot cover adequately the entire spectrum of conditions and all combinations of parameters that affect carbonation, and hence cannot serve as the basis of a comprehensive quantitative design process against carbonation - initiated corrosion. This work aims at filling exactly this gap, by developing a fundamental yet simple model of the carbonation process and applying it for the selection of design parameters, such as concrete cover of the reinforcement, concrete composition, etc.

2 Physicochemical processes

2.1 General

With the exception of the diffusion of CO_2 from the atmosphere into the concrete volume, which takes place in the gaseous phase of the concrete pores, all other physicochemical processes involved in carbonation take place in the pore water. So, water in the pores plays a key role in carbonation: On one hand it provides the medium within which all chemical reactions take place, but on the other it hinders diffusion of CO_2 by blocking the pores, partly or fully. For the estimation of the amount of water in the pores, it is assumed herein that hygrothermal equilibrium has been established between the environment and the pore system of concrete. This assumption allows determination of the amount and distribution of water in this pore system, from known values of the ambient relative humidity and temperature. Another assumption made herein (but not in the general version of the carbonation model presented elsewhere: Papadakis et al., 1989, 1990 a, c) is that cement hydration and any pozzolanic activity is complete. This assumption essentially limits the validity of the proposed model to the steady state achieved after about 3 months following concrete casting. Finally, the present work covers not only ordinary Portland cement (OPC), but also pozzolanic cements, with fly ash blended with the clinker, as an additive.

2.2 Cement hydration and pozzolanic activity

The constituents of hardened paste of OPC or pozzolanic cement which are subject to carbonation are $Ca(OH)_2$ and calcium silicate hydrate (CSH), both of which are products of hydration and of pozzolanic action, and the yet-unhydrated amount of calcium silicates, C_3S and C_2S. For the quantitative description of the carbonation process, knowledge of the concentration of these carbonatable constituents is necessary. These concentrations can be obtained from the initial composition of cement, and the chemical reactions of hydration and pozzolanic action, on the basis of stoichiometric considerations.

The main hydration reactions of OPC have been reviewed by many authors, e.g. Brunauer and Copeland (1964), Lea (1970), Bensted (1983), Frigione (1983), and Taylor (1986). Using the notation of cement technology (C: CaO, S: SiO_2, A: Al_2O_3, F: Fe_2O_3, H: H_2O, $\bar{S}$: SO_3, C_3S: $3CaO.SiO_2$, CH: $Ca(OH)_2$, $C_3S_2H_3$ or CSH: $3CaO.2SiO_2.3H_2O$, $C\bar{S}H_2$: $CaSO_4.2H_2O$, etc.) the hydration reactions of OPC in the presence of gypsum ($C\bar{S}H_2$) are the following:

$$2C_3S + 6H \rightarrow C_3S_2H_3 + 3CH \quad (1)$$

$$2C_2S + 4H \rightarrow C_3S_2H_3 + CH \quad (2)$$

$$C_4AF + 2CH + 2C\bar{S}H_2 + 18H \rightarrow C_8AF\bar{S}_2H_{24} \quad (3)$$

$$C_3A + C\bar{S}H_2 + 10H \rightarrow C_4A\bar{S}H_{12} \quad (4)$$

After all the gypsum has been consumed, reactions (3) and (4) are replaced by the following:

$$C_4AF + 4CH + 22H \rightarrow C_8AFH_{26} \quad (5)$$

$$C_3A + CH + 12H \rightarrow C_4AH_{13} \quad (6)$$

Pozzolanic activity, i.e. the reaction of the oxides of the pozzolan with $Ca(OH)_2$, has not been studied as much as the hydration of clinker. Recent progress in the study of pozzolanic activity has been reviewed by Mehta (1983) and Sersale (1983). On the basis of this recent work a simple model is proposed herein for the pozzolanic activity: On the basis of the observation that pozzolans consist of the same oxides as clinker (i.e., C, S, A and F), but in different proportions and with different mineralogical compositions, it is assumed that the products of pozzolanic activity are the same as those of the hydration of OPC. The main difference is assumed to be that not the entire amount of these oxides reacts with $Ca(OH)_2$, but only the fraction γ_i of oxide i which is in noncrystalline (amorphous) form. The reactions describing the pozzolanic activity are then proposed to be:

$$C + H \rightarrow CH \quad (7)$$

$$2S + 3CH \rightarrow C_3S_2H_3 \quad (8)$$

$$A + 4CH + 9H \rightarrow C_4AH_{13} \quad (9)$$

$$A + F + 8CH + 18H \rightarrow C_8AFH_{26} \quad (10)$$

$$A + C\bar{S}H_2 + 3CH + 7H \rightarrow C_4A\bar{S}H_{12} \quad (11)$$

The $Ca(OH)_2$ required by Eqs. (8)-(11) originates from the hydration reactions, Eqs. (1) and (2), and from the CaO of the pozzolan itself (see Eq. (7)).

Let us denote the weight fraction of clinker, pozzolan and gypsum in cement by p_{cl}, p_{po} and p_{gy} respectively. The weight fraction of oxide i=C, S, A, F and R (with R denoting all other oxides and impurities) in the clinker (j=c) and in the pozzolan (j=p) is denoted by $p_{i,j}$, and the active fraction of oxide i by γ_i. The weight fraction of the unhydrated constituents C_3S, C_2S, C_4AF and C_3A of clinker are given, according to Lea (1970), by :

$$p_{C_3A,c} = 2.65p_{A,c} - 1.69p_{F,c} \quad (12)$$

$$p_{C_4AF,c} = 3.04\ p_{F,c} \quad (13)$$

$$p_{C_3S,c} = 4.07\ p_{C,c} - (7.6\ p_{S,c} + 6.72\ p_{A,c} + 1.43\ p_{F,c}) \quad (14)$$

$$p_{C_2S,c} = 2.867\ p_{S,c} - 0.754\ p_{C_3S,c} \quad (15)$$

Eqs. (1)-(6) and stoichiometric considerations give the total amount of $Ca(OH)_2$ and CSH produced by complete hydration, expressed as a weight fraction of clinker:

$$p_{CH,c} = 0.487\ p_{C_3S,c} + 0.215\ p_{C_2S,c} - 0.610\ p_{C_4AF,c} - 0.274\ (p_{C_3A,c} - 1.569\ p_{gy}), \text{ if } p_{C_3A,c} > 1.569\ p_{gy} \quad (16a)$$

$$p_{CH,c} = 0.487\ p_{C_3S,c} + 0.215\ p_{C_2S,c} - 0.610\ p_{C_4AF,c}, \text{ if } p_{C_3A,c} \leq 1.569\ p_{gy} \quad (16b)$$

$$p_{CSH,c} = 0.750\ p_{C_3S,c} + 0.994\ p_{C_2S,c} \tag{17}$$

The total amount of $Ca(OH)_2$ consumed by 100% pozzolanic reaction of the active oxides in the pozzolan is:

If $p_{C_3A,c} > 1.569\ p_{gy}$:

$$p_{CH,p} = 1.850\ \gamma_S p_{S,p} + 3.712\ \gamma_F p_{F,p} + 2.907(\gamma_A p_{A,p} - 0.638\ \gamma_F p_{F,p}) \tag{18a}$$

whereas if $p_{C_3A,c} \leq 1.569\ p_{gy}$

$$p_{CH,p} = 1.850\ \gamma_S p_{S,p} + 3.712\ \gamma_F p_{F,p} + 2.907\ [\gamma_A p_{A,p} - 0.638\gamma_F p_{F,p} -$$

$$- 0.592(p_{gy} - 0.637\ p_{C_3A,c}] + 1.291(p_{gy} - 0.637\ p_{C_3A,c}),$$

$$\text{if } \gamma_A p_{A,p} > 0.638\ \gamma_F p_{F,p} + 0.592(p_{gy} - 0.637\ p_{C_3A,c}) \tag{18b}$$

or $p_{CH,p} = 1.850\ \gamma_S p_{S,p} + 3.712\ \gamma_F p_{F,p} + 2.180\ (\gamma_A p_{A,p} - 0.638\ \gamma_F p_{F,p})$

$$\text{if } \gamma_A p_{A,p} \leq 0.638\ \gamma_F p_{F,p} + 0.592(p_{gy} - 0.637\ p_{C_3A,c}) \tag{18c}$$

The amount of CSH produced from pozzolanic reaction (8) is:

$$p_{CSH,p} = 2.850\ \gamma_S p_{S,p} \tag{19}$$

Finally, after complete pozzolanic reaction and clinker hydration, a pozzolanic cement contains the following weight fractions of $Ca(OH)_2$ and CSH (in kg/kg cement):

$$p_{CH} = p_{CH,c} + 1.321\ \gamma_C p_{C,p} - p_{CH,p} \tag{20}$$

$$p_{CSH} = p_{CSH,c} + p_{CSH,p} \tag{21}$$

To convert the above weight fraction of constituent k (k:CH, CSH), into molar concentration [k] in concrete (mol of k/m^3 concrete), one can use the expression:

$$[k] = \frac{\rho_c (1 - \varepsilon_{air})}{MW_k \left(1 + \frac{w}{c}\frac{\rho_c}{\rho_w} + \frac{a}{c}\frac{\rho_c}{\rho_a} \right)} p_k \tag{22}$$

in which, w/c and a/c: water to cement and aggregate to cement ratio, respectively, ρ_c, ρ_w, ρ_a : mass denstity of cement, water and aggregates, respectively (kg/m^3), MW_k: molar weight of constituent k (kg/mol), and ε_{air}: volume fraction of entrapped or entrained air in concrete which depends on the maximum aggregate size.

2.3 Carbonation of $Ca(OH)_2$ and CSH

The carbonation reactions can be written as:

$$Ca(OH)_2 + CO_2 \rightarrow CaCO_3 + H_2O \tag{23}$$

$$(3CaO.2SiO_2.3H_2O) + 3CO_2 \rightarrow (3CaCO_3.2SiO_2.3H_2O) \tag{24}$$

These reactions consist of the following elementary steps:

1. The diffusion of atmospheric CO_2 in the gaseous phase of the concrete pores. The CO_2 occurs at a molar concentration $[CO_2]$ (in mol/m^3) in the atmosphere and diffuses through the concrete pores with an effective diffusivity D_{e,CO_2} (in m^2/s, see Sect. 2.5). The concentration can be estimated from the volume fraction, y_{CO_2} of CO_2 in the air, using the perfect gas law:

$$[CO_2] = y_{CO_2}\, p / R\, T \tag{25}$$

where p: atmospheric pressure (in atm), T: temperature (in K), and R: gas constant (=82.06x10^{-6} m^{3}atm/Kmol). The CO_2 is then dissolved in the pore water.

2. The dissolution of solid $Ca(OH)_2$ from cement gel into the pore water and the diffusion of dissolved $Ca(OH)_2$ from regions of high alkalinity to those of low.

3. The reaction of dissolved CO_2 with dissolved $Ca(OH)_2$ in the pore water (Eq. (23)).

4. The reaction of dissolved CO_2 with CSH (Eq. (24)).

A side-effect of reactions (23) and (24) is the reduction of pore volume (see Sect. 2.4).

The reaction rates of Eqs. (23) and (24) are given in detail elsewhere (Papadakis et al.,1989, 1990 a,c).

2.4 Concrete porosity

After the completion of hydration and of pozzolanic activity, the porosity ε of concrete equals:

$$\varepsilon = \varepsilon_o - \Delta\varepsilon_H - \Delta\varepsilon_p - \Delta\varepsilon_c \tag{26}$$

in which ε_o is the initial porosity of fresh concrete, given by:

$$\varepsilon_o = \frac{\frac{w}{c}\frac{\rho_c}{\rho_w}(1 - \varepsilon_{air})}{(1 + \frac{w}{c}\frac{\rho_c}{\rho_w} + \frac{a}{c}\frac{\rho_c}{\rho_a})} + \varepsilon_{air} \tag{27}$$

and $\Delta\varepsilon_H$, $\Delta\varepsilon_p$ and $\Delta\varepsilon_c$ is the reduction in porosity due to hydration, pozzolanic activity and (complete) carbonation, respectively. In non-carbonated concrete this latter term is zero, whereas $\Delta\varepsilon_p$ is zero for OPC. These reduction terms are due to the fact that the molar volume of the solid products of hydration, pozzolanic action, and carbonation exceed that of the solid reactants. These terms are equal to:

if $p_{C_3A} > 1.569\ P_{gy}$:

$$\Delta\varepsilon_H = [C_3S]\Delta\overline{V}_{C_3S} + [C_2S]\Delta\overline{V}_{C_2S} + [C_4AF]\,\Delta\overline{V}_{C_4AF} + [C\overline{S}H_2]\,\Delta\overline{V}_{C\overline{S}H_2} +$$

$$([C_3A] - [C\overline{S}H_2])\,\Delta\overline{V}_{C_3A} \tag{28a}$$

and

$$\Delta\varepsilon_p = \gamma_S\,[S]\,\Delta\overline{V}_S + \gamma_F\,[F]\,\Delta\overline{V}_F + (\gamma_A[A] - \gamma_F\,[F])\Delta\overline{V}_A + \gamma_C[C]\Delta\overline{V}_C \tag{28b}$$

whereas if $p_{C_3A} \leq 1.569\ p_{gy}$:

$$\Delta\varepsilon_H = [C_3S]\,\Delta\overline{V}_{C_3S} + [C_2S]\,\Delta\overline{V}_{C_2S} + [C_4AF]\,\Delta\overline{V}_{C_4AF} + [C_3A]\,\Delta\overline{V}_{C\overline{S}H_2} \tag{29a}$$

$$\Delta\varepsilon_p = \gamma_S[S]\,\Delta\bar{V}_S + \gamma_F\,[F]\,\Delta\bar{V}_F + ([C\bar{S}H_2] - [C_3A])\,\Delta\bar{V}_{A,\bar{S}} + [\gamma_A[A] -$$

$$- \gamma_F\,[F] - ([C\bar{S}H_2] - [C_3A])]\,\Delta\bar{V}_A + \gamma_C\,[C]\,\Delta\bar{V}_C,$$

$$\text{if} \quad \gamma_A\,p_{A,p} > 0.638\,\gamma_F p_{F,p} + 0.592\,(p_{gy} - 0.637\,p_{C_3AC}) \tag{29b}$$

or

$$\Delta\varepsilon_p = \gamma_S[S]\,\Delta\bar{V}_S + \gamma_F\,[F]\,\Delta\bar{V}_F + (\gamma_A[A] - \gamma_F\,[F])\,\Delta\bar{V}_{A,\bar{S}} + \gamma_C\,[C]\,\Delta\bar{V}_C,$$

$$\text{if} \quad \gamma_A p_{A,p} \leq 0.638\,\gamma_F\,p_{F,p} + 0.592\,(p_{gy} - 0.637\,p_{C_3A,\,c}) \tag{29c}$$

In addition, for fully carbonated concrete:

$$\Delta\varepsilon_c = [CH]\,\Delta\bar{V}_{CH} + [CSH]\,\Delta\bar{V}_{CSH} \tag{30}$$

where, [k]: concentration of constituent k (mol/m^3) given by Eq. (22), and $\Delta\bar{V}_k$: difference in molar volumes between solid reaction products and solid reactants (m^3/mol), given in Table 1.

Table 1. Molar volumes differences $\Delta\bar{V}_k$ x 10^6 (m^3/mol)

	Hydration reaction of					Pozzolanic reaction of					Carbon. of	
k	C_3S	C_2S	C_4AF	$C\bar{S}H_2$	C_3A	S	F	A	$A,\bar{S}$	C	CH	CSH
	53.28	39.35	~230	155.86	149.82	-1.90	179.39	114.26	120.30	16.19	3.85	15.39

2.5 Effective diffusivity of CO_2 in concrete

The effective diffusivity of CO_2 in carbonated concrete, D_{e,CO_2} (m^2/s), is determined by the pore-size distribution of concrete, its total porosity, ε, and the degree of saturation of the various pore sizes by water. It was found that for the purposes of the determination of D_{e,CO_2}, these characteristics of concrete can be effectively summarized in the porosity of the hardened cement paste, ε_p, and in the ambient relative humidity, RH. The effective diffusivity of concrete, was measured using a Wicke-Kallenbach type of apparatus (Papadakis et al., 1990 b).

Experimental values were fitted by the following empirical expression:

$$D_{e,CO_2} = 2.1\;10^{-6}\,[\varepsilon_p\,(1 - RH/100)]^{2.2} \tag{31}$$

in which the porosity of hardened cement paste, ε_p, is given by the expression:

$$\varepsilon_p = \varepsilon\left(1 + \frac{\dfrac{a}{c}\dfrac{\rho_c}{\rho_a}}{1 + \dfrac{w}{c}\dfrac{\rho_c}{\rho_w}}\right) \tag{32}$$

3 Mathematical model

In a previous work (Papadakis et al., 1989), a general and comprehensive mathematical model has been proposed for the physicochemical processes involved in concrete carbonation. This model, which can be solved only numerically, is based on the mass conservation of CO_2, $Ca(OH)_2$ and CSH in any control volume of concrete mass. It has been shown that, within the usual range of relative humidities, i.e., for $50\% \leq RH \leq 90\%$ certain simplifying assumptions can be made, which lead to the formation of a carbonation front seperating the concrete volume in two regions: a completely carbonated and a non-carbonated region. The evolution with time, t, of the distance of this front from the concrete surface, called carbonation depth and denoted by x_c, is given by a simple analytical expression, in terms of the compositional parameters of cement and concrete and of the environmental conditions:

$$x_c = \left(\frac{2[CO_2]\, D_{e,CO_2}}{[CH] + 3[CSH]}\, t \right)^{1/2} \tag{33}$$

The information required for the application of Eq. (33) for predictive purposes comprises: a) the composition of cement, i.e., the weight fraction of clinker, pozzolan and gypsum, the oxide analysis of clinker and pozzolan and the mineralogical analysis of pozzolan; b) the composition of concrete, i.e., its water-cement ratio, w/c, and aggregate-cement ratio, a/c; and c) the CO_2 concentration and the relative humidity of the atmosphere. If all these parameters are known, one can find the carbonation depth from Eq. (33), using also Eqs. (12) - (22) and (25) - (32).

4 Experimental results and comparison with theory

Because of the very long time needed to perform carbonation experiments under normal exposure conditions (CO_2 mol fraction between 0.03% and 0.05%), an accelerated test apparatus has been constructed, in which fully hydrated concrete speciments are exposed to 50% CO_2 at constant and controlled temperature and relative humidity. Carbonation depths are measured using a phenolphthalein pH-indicator. The apparatus, test procedure, and materials are described in detail by Papadakis et al. (1990 a, b). Some specimens have also been subjected to thermogravimetric analysis, to determine the amount of $Ca(OH)_2$ and $CaCO_3$.

Fig. 1 shows the evolution of carbonation depth with time, in concrete made with OPC or with fly-ash cement (denoted by AHPC and having a fly-ash content of 20%) under accelerated carbonation conditions. (Both cements have a gypsum content of 5%). The solid lines on the figure correspond to Eq. (33), and are in excellent agreement with the experimental results. The same Figure shows the effect of water-cement-ratio, of aggregate-cement ratio and of cement type. Fig. 2 shows carbonation depth results obtained in this work and by previous works (Nagataki et al., 1986, 1988), under normal CO_2 exposure and under accelerated carbonation conditions. When plotted vs. the square root of an effective time scale $t' = y_{CO_2} t$ all results fall on the same line. The agreement with the predictions of Eq. (33) verifies the validity of the accelerated tests. Fig. 3 shows the experimentally measured effect of relative humidity both for Portland and for a pozzolanic cement. The agreement with Eq. (33) is good only as long as the rate-controlling step is the diffusion of CO_2, so that a sharp carbonation front is formed. This is the case for $RH \geq 50\%$. As RH decreases, the water content of the pores decreases and reactions (23) and (24) become slow and control the overall carbonation rate. Under such conditions one has to solve numerically the general mathematical model. Such a solution, presented in detail by Papadakis et al. (1990c), gives no sharp front for values of RH below about 50%. For such low values of relative humidity, the predictions of the general model for the depth at which the value of pH falls below 9 are in good agreement with experimental results (Fig. 3).

5 Parametric studies

The time required by the carbonation front (or, equivalently, a pH-value below 9) to reach the reinforcement can be interpreted as the corrosion initiation period, as when a pH-value lower than 9 in the vicinity of the steel bars causes dissolution of the thin oxide layer protecting the reinforcement, and signals the beginning of the deterioration period. This time, denoted by t_c, can be obtained from Eq. (33) by setting x_c equal to the concrete cover of the reinforcement, c, and solving for t. In this way a sensitivity analysis has been performed of the effect of various parameters on the length of the corrosion initiation period, for various values of the concrete cover, ranging from 5 to 30 mm. This sensitivity analysis (see Table 2) refers to fully hydrated cement, no entrained or entrapped air, and a constant temperature of 20°C. The baseline case for the sensitivity analysis is an OPC (65% CaO in the clinker) concrete with an aggregate-cement ratio of 5, a water-cement ratio of 0.65, at a constant RH relative humidity of 65% and in an environment with 0.05% CO_2 in the air (corresponding to urban conditions). The very significant effect of varying the water-cement ratio, shown in Fig. 4, is due to the influence of this factor on the porosity of cement paste, and through it on the effective diffusivity of concrete. The effect of varying the aggregate-cement ratio, shown in Fig. 5, is minor and is due to the reduction of the molar concentration of $Ca(OH)_2$ and CSH in the concrete, with the increase of this ratio. Changing the CaO-content of the clinker within its usual range of variation has also a negligible effect (Fig. 6). On the contrary replacing part of the OPC with a fly ash with a low CaO-content decreases drastically the initiation period, due to the combined effect of the increased porosity and of the reduced concentration of carbonatable materials in the concrete (Fig. 7). In the opposite direction but to a much lesser degree, works the replacement of part of the aggregates by the same pozzolan (Fig. 8). The very profound effect of relative humidity for RH>50%, shown in Fig. 9, is due to its effect on D_{e,CO_2}. Finally, Fig. 10 shows the moderate effect of changing the ambient CO_2-content within its usual range of variation.

Table 2. Parametric studies scheme

	w/c	a/c	CaO(%)	repl. of cem. by pozz. (%)	repl. of aggr. by pozz. (%)	RH(%)	y_{CO_2}(%)
Baseline	0.65	5	OPC	0	0	65	0.05
Fig. 4	.45-.85	5	OPC	0	0	65	0.05
Fig. 5	0.65	3-7	OPC	0	0	65	0.05
Fig. 6	0.65	5	60-70	0	0	65	0.05
Fig. 7	0.65	5	OPC	0-40	0	65	0.05
Fig. 8	0.65	5	OPC	0	0-8	65	0.05
Fig. 9	0.65	5	OPC	0	0	50-90	0.05
Fig.10	0.65	5	OPC	0	0	65	0.03-0.1

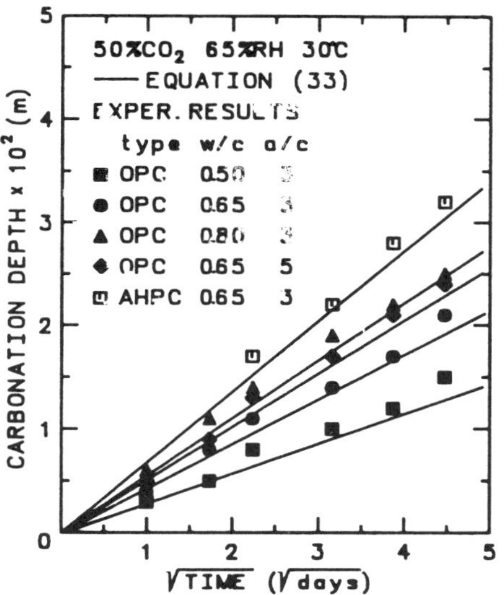

Fig. 1. Effect of exposure time on carbonation depth and comparison with model predictions.

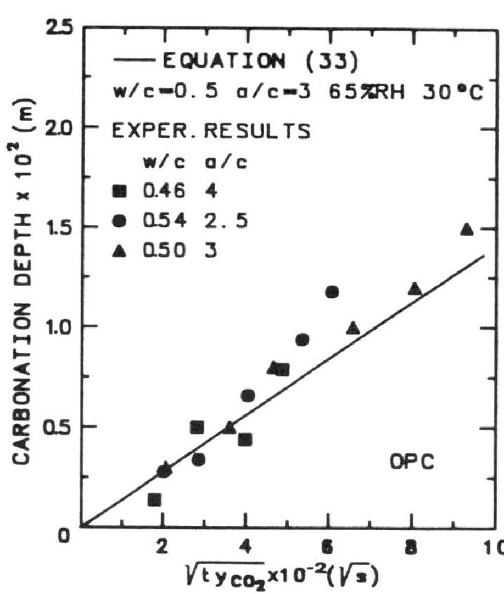

Fig. 2. Test results for various CO_2-contents in air, and comparison with model predictions, (■) indoor conditions, Nagataki et al. (1986), (●)7% CO_2, 60%RH, Nagataki et al. (1988), (▲) 50% CO_2, 65% RH, this work.

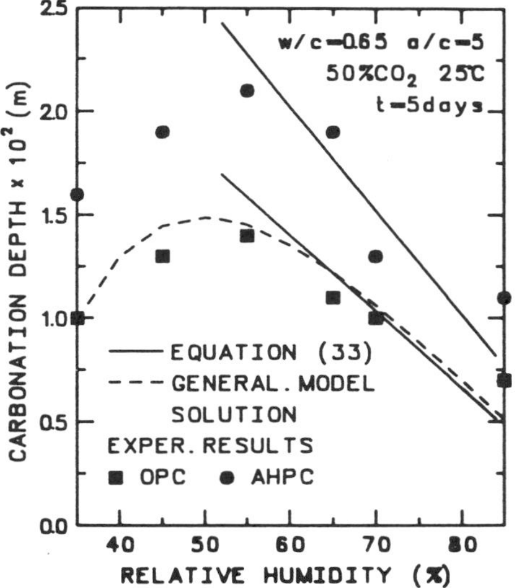

Fig. 3. Effect of relative humidity on carbonation depth and comparison with model predictions.

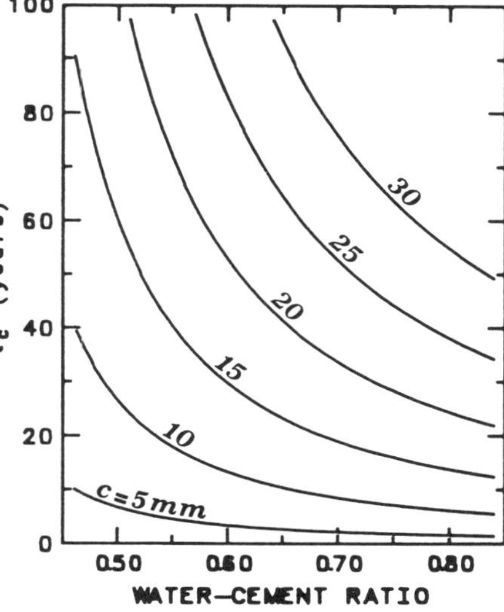

Fig.4. Effect of water-cement ratio on corrosion initiation time (OPC).

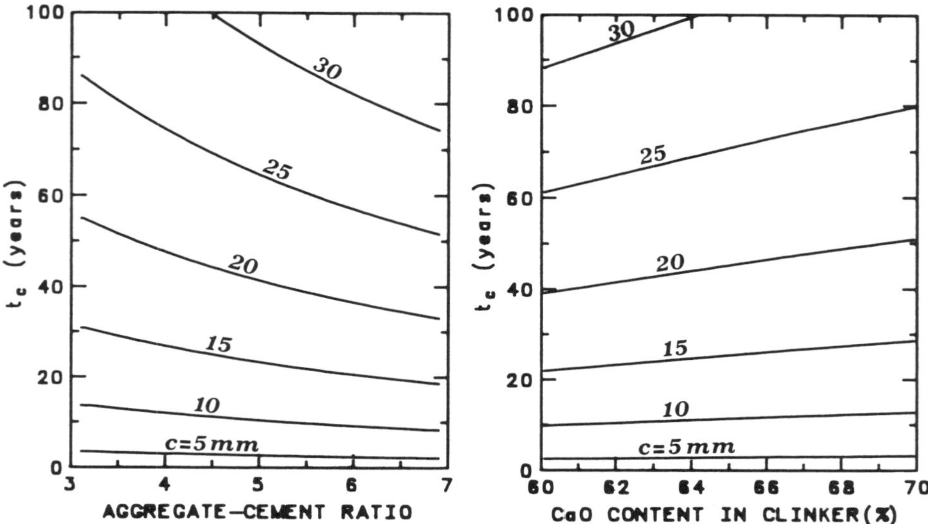

Fig.5. Effect of aggregate-cement ratio on corrosion initiation time (OPC).

Fig.6. Effect of CaO content in the clinker of Portland cement on corrosion initiation time (OPC).

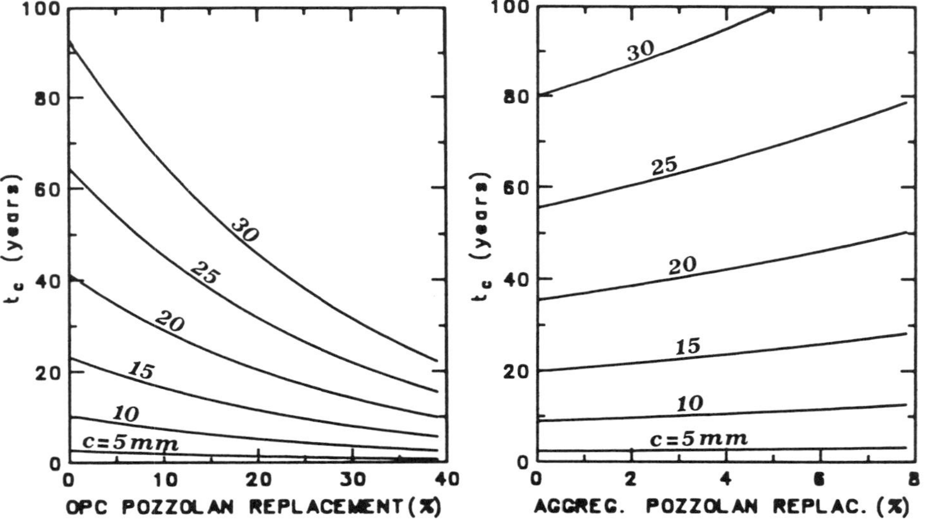

Fig.7. Effect of the % replacement of OPC by fly ash on corrosion initiation time (OPC).

Fig.8. Effect of the % replacement of aggregates by fly ash on corrosion initiation time (OPC).

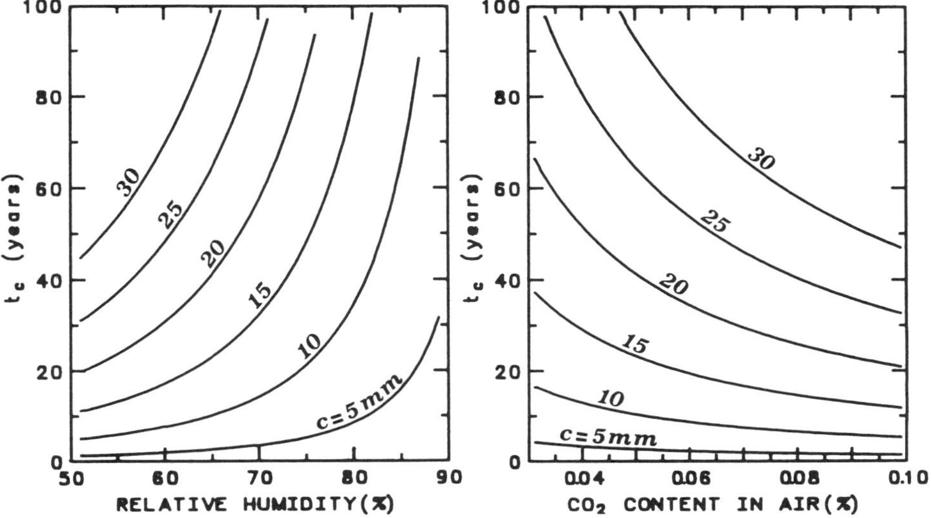

Fig.9. Effect of ambient relative humidity (>50%) on corrosion initiation time (OPC).

Fig.10. Effect of ambient CO_2-content on corrosion initiation time (OPC).

6 Conclusions

The physicochemical processes of concrete carbonation are described and modeled mathematically using fundamental principles of chemical reaction engineering. For the usual relative humidity conditions the phenomenon can be described by a simple expression which is in good agreement with test results obtained under normal or accelerated carbonation conditons. So, it can be used to reliably predict the initiation period of corrosion of steel bars embedded in concrete, as a function of composition parameters of cement and concrete and of the environmental conditions.

7 Acknowledgement

The Hellenic General Secretariat for Research and Technology has provided financial support to this work.

8 References

Bensted, J. (1983) Hydration of Portland cement, in **Advances in Cement Technology** (ed S.N. Ghosh), Pergamon Press, New York, pp. 307-347.

Brunauer, S. and Copeland, L.E. (1964) The chemistry of concrete. **Sci. Amer.**, 210, pp. 80-92.

Frigione, G. (1983) Gypsum in cement, in **Advances in Cement Technology** (ed S.N. Ghosh), Pergamon Press, New York, pp. 485-535.

Hamada, M. (1969) Neutralization (carbonation) of concrete and corrosion of reinforcing steel, in **Proceedings of 5th Inter. Symp. on the Chem. of Cem.,** Tokyo, pp. 343-369.

Lea, F.M. (1970) **The Chemistry of Cement and Concrete**. Edward Arnold (Publishers) Ltd., London.

Mehta, K. (1983) Pozzolanic and cementitious by-products as mineral admixtures for concrete - A critical review, in **Proceedings of 1st Inter. Conf. on the Use of Fly Ash, Silica Fume, Slag and Other Mineral By-Products in Concrete,** ACI SP-79, Detroit, pp. 1-46.

Nagataki, S. Ohga, H. and Kim, E.M. (1986) Effect of curing conditions on the carbonation of concrete with fly ash and the corrosion of reinforcement in long-term tests, in **Proceedings of 2nd Inter. Conf. on the Use of Fly Ash, Silica Fume, Slag and Natural Pozzolans in Concrete**, ACI SP-91, Detroit, pp. 521-539.

Nagataki, S. Mansur, M.A. and Ohga, H. (1988) Carbonation of mortar in relation to ferrocement contruction. **ACI Mat. J.**, 85, pp. 17-25.

Papadakis, V.G. Vayenas, C.G. and Fardis, M.N. (1989) A reaction engineering approach to the problem of concrete carbonation. **AIChE J.,** 35, pp. 1639-1650.

Papadakis, V.G. Vayenas, C.G. and Fardis, M.N. (1990 a) Fundamental modeling and experimental investigation of concrete carbonation. submitted to **ACI Mat. J.**

Papadakis, V.G. Vayenas, C.G. and Fardis, M.N. (1990 b) Physical and chemical characteristics affecting the durability of concrete. submitted to **ACI Mat. J.**

Papadakis, V.G. Fardis, M.N. and Vayenas, C.G. (1990 c) A reaction engineering analysis of the concrete carbonation problem: experiment and theory, in **Proceedings of 11th Inter. Symp. on Chem. React. Eng.**, Toronto.

Richardson, M.G. (1988) **Carbonation of Reinforced Concrete**. Citis Ltd., Dublin.

Schiessl, P. (1976) Zur Frage der zulässigen Rissbreite und der erfoderlichen Betondeckung im Stahlbetonbau unter besonderer Berücksichtigung der Karbonatisierung des Betons. **Deutscher Ausschuss für Stahlbeton**, 255.

Sersale, R. (1983) Aspects of the chemistry of additions, in **Advances in Cement Technology** (ed S.N. Ghosh), Pergamon Press, New York, pp. 537-567.

Taylor, H.F.W. (1986) Chemistry of cement hydration, in **Proceedings of 8th Inter. Cong. on the Chem. of Cement**, Rio de Janeiro, 1, pp. 82-110.

Tuutti, K. (1982) **Corrosion of Steel in Concrete**. CBI forskning research, Swedish Cement and Concrete Research Institute, Stockholm.

5 FUNDAMENTAL STUDIES OF FROST DAMAGE IN CLAY BRICK

W. PROUT, W.D. HOFF
Department of Building Engineering, UMIST, Manchester, UK

Abstract
This paper describes some results from an investigation of the fundamental physics of frost damage particularly in clay brick ceramic. The experimental work involved the production of frost damage in a unidirectional freezing procedure. Direct strain measurements proved to be a more sensitive method of damage detection than simple visual observation. The damage began, or was most severe, on the rear face of the specimens.

Measurements of amounts of water/ice extruded from saturated specimens during freezing tend to suggest that significant amounts of water remain unfrozen in brick ceramic even at -15°C, and the authors suggest that this water is entrapped by surrounding ice. The pressures built up in this way can account for the observed damage and suggest a new theory of frost action.
Keywords: Frost Damage, Brick Ceramic, Porous Media

1 Introduction

Frost damage in brick masonry is one of its most serious and unsightly forms of deterioration and is a risk wherever bricks are exposed to a wet and cold climate. The manufacturers of clay bricks produce a range of brick types which are designated as suitable for all conditions of exposure and which are 'frost resistant'. Typically, highly vitrified, high strength and thus low porosity bricks (as, for example, those of engineering standard) are regarded as not being liable to frost damage in practice. At the other extreme it is recognised that weaker, more porous bricks generally suffer severe damage if they are frozen and thawed repeatedly in wet conditions. The fascinating observation in practice, however, is that some of the weak, porous bricks are very resistant to damage even under the most severe of conditions. This is recognised by, for example, some handmade bricks (which are relatively porous) being designated frost resistant in the relevant British Standard.

Scientific work having the objective of explaining the processes and consequences of freezing porous materials which are partially or totally saturated with water has attracted major attention in the literature over the last 70 years. However there has been no orderly progression towards a theory of frost damage. Perhaps the most emphatic immediate conclusion to be drawn from the literature is that frost damage does not occur simply as a result of the 9% expansion in molar volume when water turns to ice.

This paper reports the results of some aspects of a research study of the fundamentals of frost damage which are particularly relevant to an understanding of the effects of freeze thaw cycling on clay brick.

2 Theoretical Background

A detailed review of the literature suggests two distinct theories of frost action in porous media.

The first of these is the thermodynamic or capillary theory and in a detailed exposition of this theory Everett (1961) derives an excess pressure due to freezing in a capillary. This excess pressure is needed in the ice phase because the ice water interface is curved. This theory fully explains the phenomenon of frost heave since the thermodynamic analysis predicts the formation of ice lenses in unconsolidated media such as soils. The formation of such lenses has been observed in practice and has been noted in the literature as long ago as 1765 in the work of Runeberg quoted by Everett. The thermodynamic theory is modified by Litvan (1980) who argues that rather than freeze in situ under an excess pressure capillary water will desorb through the vapour phase to the surface where it then freezes. Any damage is caused by resistance to desorption by such properties as low permeability. A highly important point to note about the thermodynamic theory is that the 9% volume expansion on freezing of water is not required to generate the damaging pressure.

The second of the theories of frost action is a hydraulic theory and is based on the fact that the freezing of water is accompanied by 9% increase in volume. Essentially this theory states that water must flow away from the zone within the sample while the water is freezing to accommodate the volume increase. This generates a hydraulic pressure within the fluid which is responsible for any damage to the matrix. Powers (1945) was among the first to develop this theory quantitatively, but his analyses are based on concrete rather than stone or brick ceramic. Concrete generally has a much lower permeability than brick ceramic.

3 Experimental Work

In our programme of laboratory work a freezing test procedure was developed in which single brick specimens were frozen unidirectionally from one exposed stretcher face. Most of the specimens tested were fully vacuum saturated with water prior to testing; in some cases where lower water contents were required the water contents were adjusted by partial drying followed by equilibration in sealed bags.

The apparatus used consisted of an open topped plywood box lined with a 5cm thickness of expanded polystyrene. The brick specimen to be tested was sealed in a polythene bag and placed in the box with a 2cm thickness of loose, dry sand packed around each of the insulated sides. The box was placed in a freezing cabinet so that cold air was free to circulate over the exposed stretcher face of the brick.

A series of experiments in which the temperature distributions through the brick specimens were measured under two different freezing air temperatures (-5°C and -15°C) established that although there was some curvature of the advancing 0°C isotherm at the lower freezing temperature the test is nevertheless a valid unidirectional freezing test.

A detailed analysis indicated that a freezing time duration of 48 hours was fully adequate to ensure complete freezing through the brick specimens. At the end of each freeze cycle the specimens were thawed under water at room temperature for 24 hours which was shown to be a fully sufficient time for a saturated frozen sample to thaw completely. This freeze thaw procedure afforded the chance to inspect specimens between freeze and thaw parts of the cycle and advantage was taken of this to record the amount of water which had extruded from the sample and then frozen during the freezing cycle. To do this the specimens were weighed at the end of the freeze cycle and then dipped into warm (40°C) water to melt any surface ice. The excess surface water was then removed with a damp cloth and the specimen re-weighed.

In our research programme we used a range of techniques in addition to straightforward visual inspection to assess damage on freeze thaw cycling. These included measurements of dimensional changes, of tensile strength changes and of microstructural changes using mercury intrusion porosimetry. In addition fissure analysis was used in an attempt to follow crack growth and sonic methods were also used to non-destructively detect damage.

In this paper we report some of the results from measurements of dimensional changes on freeze thaw cycling together with visual observations of frost damage and book-keeping measurements of water migration on freezing. To measure the dimensional changes the technique used involved fixing reference studs onto the brick specimens so that the length of each face of the bricks could be measured. The reference points were arranged so that the strain was measured diagonally across each face thus incorporating as much as possible of each face in the measurement. A Demec dial gauge was used for the strain measurements which gave a high degree of accuracy. The procedure followed was to measure each bed face and each stretcher face twice in each freeze thaw cycle; once at the end of freeze and once at the end of the thaw. The strains on the header faces were occasionally measured as well.

4 Experimental Results

Some 13 types of brick were tested in our research programme. Some representative results are described in this section.

4.1 Strain Measurements

The data from the strain measurements on three representative types of brick are shown graphically in figures 1 and 2. These bricks have been designated types I, II, III on these figures and in Table 1. The strains on the bed and stretcher faces are shown on these graphs. The uppermost stretcher face through which heat is removed in the freezing cycle is designated face A and the lower stretcher face is designated C. Faces B and D are the bed faces, and faces E and F are the header faces.

The results in figure 1 and those in figure 2(a) were obtained using bricks which were initially fully vacuum saturated with water (100% saturation). Brick type I was designated M (moderately frost resistant) in the British Standard and the strain results shown in figure 1(a) are typical in general form of those we obtained for bricks which are susceptible to serious frost damage when saturated. Brick type II was an extruded brick of class B engineering standard which was designated F (frost resistant) and the strain results for this brick are shown in figure 1(b). Brick type III was a handmade brick also designated F and figure 2(a) shows the strain results for this brick.

In figure 2(b) strain results are shown for brick type I, but in this case the initial water content of the brick was 92% saturation.

4.2 Visual assessment of damage

The most notable result for all of the brick types tested was that damage tended to begin and indeed to be concentrated on the lower, unexposed, stretcher face. (One type of brick of a strong laminar structure showed

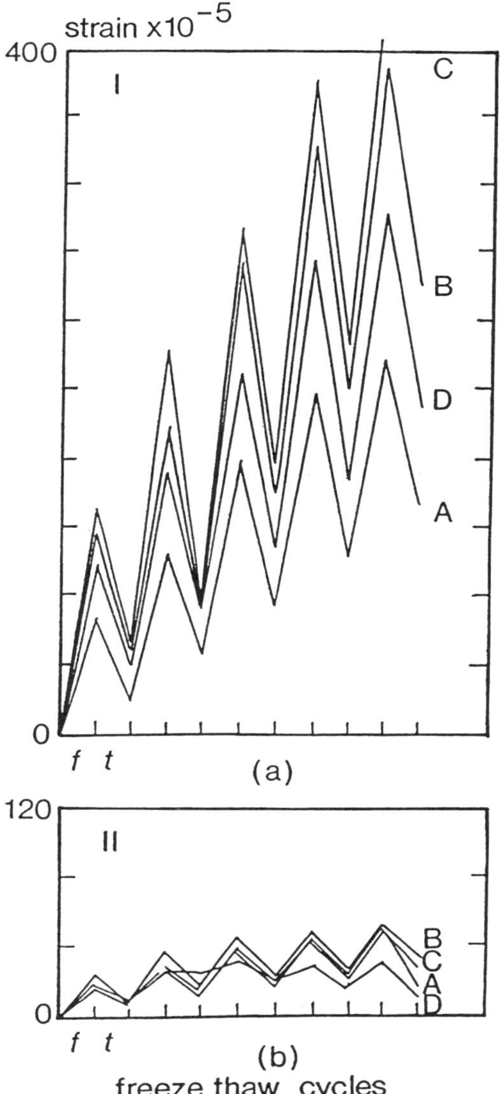

Fig.1. Strain measurements on freeze thaw cycling for (a) type I and (b) type II brick at 100% initial saturation.

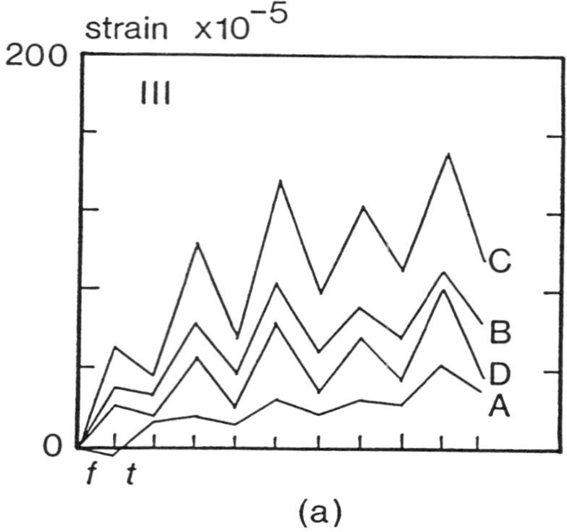

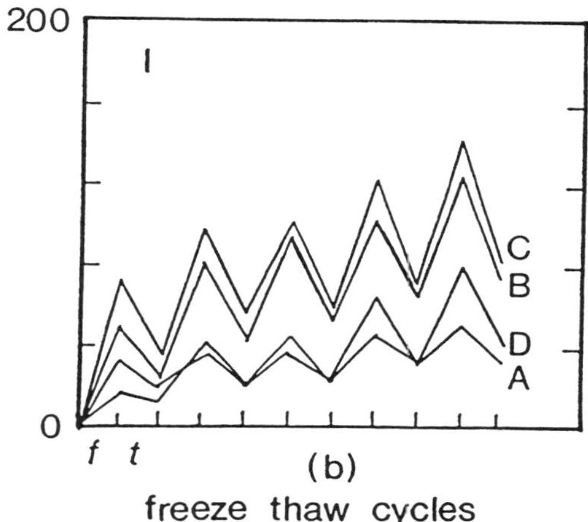

Fig.2. Strain measurements on freeze thaw cycling for (a) type III and (b) type I brick. Brick type III was initially 100% saturated. Brick type I was initially 92% saturated.

spalling of the front face after approximately 11 freeze thaw cycles, but even this brick showed initial spalling from the lower face after only 8 cycles.) The results of visual observation for the three brick types shown in figures 1 and 2 are shown in table 1.

Table 1. Summary of visual assessments

Brick reference & water content	Description of damage
Brick I 100% saturation	Slight cracking in C after 3 cycles, damage increasing in C on further cycling. Damage appearing in B after 7 cycles, in E & F after 8 cycles. Detachment of pieces from the middle of C after 9 cycles. C badly cracked and distorted after 10 cycles.
Brick II 100% saturation	No damage observed up to 20 cycles. Experiment stopped at this point.
Brick III 100% saturation	Very small pieces detached from middle of C after 3 & 5 cycles. Slight cracking in C after 6 cycles and very slight crumbling in the middle of C after 7 & 8 cycles. Very slight further damage in C at 9 & 11 cycles. No sign of further damage thereafter up to 50 cycles.
Brick I 92% saturation	Very fine crack in middle of C after 2 cycles. No sign of further damage up to 16 cycles. Slight cracking in A around frog edge at 18 cycles and cracking in A,C & D around arrises at 20 cycles. No sign of further damage until 26 cycles when cracking becomes slightly extended. Cracking in B around edges of frog at 28 cycles. At 40 cycles damage is limited to the arrises.

4.3 **Book-keeping measurements of water migration on freezing**

The results of measurements of water extruded from the brick specimens during the freeze part of the cycle are presented in table 2. It includes a measure of the volume increase of the specimens during freezing based on the average strain measurements. Knowledge of the mass of water extruded made it possible to calculate the expansion of each specimen that would have occurred if all the water in the specimen had frozen. The measured expansions were less than these calculated figures and these differences are expressed as shortfalls in volume expansion. From these figures it is possible to calculate the amount of water that is apparently unfrozen in the specimens. In the calculations if the water content of the specimens increased with cycle number then it was assumed to be due to a permanent expansion of the specimens. If the water content reduced on cycling then the figures in parenthesis in table 2 assume that the difference between actual water content volume and maximum water content volume is available to help accommodate expansion of water on freezing.

Table 2. Water extrusion and specimen expansion on freezing for brick specimens at 100% saturation

Number of cycles	Mass of water in specimen (g)	Mass of water extruded (g)	Volume expansion of specimen (%)	Shortfall in volume expansion (%)	Amount of water unfrozen (%)
Brick I					
1	363	14	0.35	1.28	44.9
5	367	10	0.41	1.58	55.3
Brick II					
1	151	3	0.05	0.08	73.5
5	153	5	0.10	0.60	54.4
10	146	2	0.10	0.79	75.4
				(0.23)	(22.0)
Brick III					
1	553	37	0.09	0.93	23.3
5	540	30	0.11	1.38	35.5
				(0.34)	(8.8)

5 Discussion

The underlying objective of the experiments described in this paper was to gain further understanding of the fundamental causes of frost damage in brick masonry. To do this extremely severe freeze thaw testing procedures were adopted generally using vacuum saturated specimens. The authors must emphasise that damage in clay bricks in these extremely severe testing procedures does not imply, and must not be taken to imply, failure to meet the frost resistant classification in the appropriate British Standard.

In general the results reported in this paper are typical of those we obtained over a wide range of brick types and broadly agree with the practical observations of frost durability. Thus, in terms of visual assessment, the high strength, highly vitrified bricks tended to be resistant to damage, some of the more porous weaker bricks were moderately or seriously damaged, while some relatively porous weak bricks were only very slightly damaged in our tests. However, even in terms of simple visual assessment some slight cracking was produced in some of the strong bricks of the frost resistant classification in the more severe tests adopted.

Comparison of figures 1 and 2 with Table 1 shows that strain measurements were a more sensitive method of detecting damage than visual assessment. It was found that the majority of brick types expanded on freezing and contracted on thawing. This is in agreement with the observations on the freezing of porous media by other workers, for example, Litvan (1978) and Watson (1964). The contraction on thawing was insufficient to return the specimens to their original volumes and this residual expansion is indicative of damage having occurred. Particularly notable in our work is the fact that the expansion and contraction of the specimens was not uniform and tended to be most pronounced on the face which froze last (face C) and least on the face which froze first (face A). This observation was reinforced by the visual assessments which showed that in general the rear stretcher face damaged first and was sometimes the only face to damage. Interestingly, we have corroborated this laboratory observation in a very limited field study of one damaged wall. The practical implications of this result are potentially quite serious since it might be taken to imply that in most building applications the only visible face of the masonry is the one least likely to be damaged.

The data collected from the freeze thaw testing of bricks having water contents below 100% (typified by the results of figure 2(b)) show similar expansions and contractions but smaller in size than those of saturated bricks. This correlates with a reduced amount of visible damage.

The reason for the expansions observed in the specimens in freezing is stress caused by the adsorbed water freezing. Any water which turns to phase I ice will result in a 9% volume expansion. To accommodate this expansion either the specimen has to expand or water has to leave the specimen. Table 2 shows typical results from this book-keeping exercise and these results suggest a significant amount of water remaining unfrozen within the specimen. Capillary theory predicts that at -15°C water held in pores of radius 3×10^{-9} m or larger will be frozen. Mercury intrusion porosimetry predicts that virtually none of the adsorbed water is held in pores of this size. Thus the non-freezing of water cannot be explained by capillary theory. The alternative then is to assume that this water does not freeze because it is entrapped by surrounding ice. If the entrapped water is assumed to behave as bulk water then the pressure required for it to remain unfrozen at -15°C can be calculated to be 1.6×10^8 N/m^2. This is considerably larger than the tensile strengths of the brick specimens which we measured in our work. However it is reasonable to explain the ability of the brick materials to withstand these pressures due to the presence of ice in some of the pores having a reinforcing effect. It is clear, however, that entrapment of water by ice can lead to high stresses more than sufficient to cause damage to most brick ceramic.

Table 3. Typical stresses on freezing saturated brick type I

Measured tensile strength (N/m^2)	3.0×10^6
Stress required to explain observed strain (N/m^2)	3.0×10^7
Maximum hydraulic stress (N/m^2)	5.0×10^5
Thermodynamic excess pressure (N/m^2)	1.6×10^5
Entrapment pressure (N/m^2)	1.6×10^8

Typical stresses predicted by the various theories of frost damage are shown in Table 3. The stress predicted by the thermodynamic theory has been calculated using a representative pore radius deduced from mercury intrusion porosimetry. The hydraulic stress has been calculated using measurements of saturated hydraulic conductivity made on brick specimens in our laboratories. The tensile strength figure is taken from measurements made in our laboratories on brick cores. The stress required to explain observed strains was calculated using measurements of dynamic Young's modulus. The main point to note from this table is that neither the thermodynamic pressure nor the hydraulic pressure are sufficiently large to exceed the tensile strength of the brick or to produce the stress required to explain the observed strain.

Thus the analysis of the experimental results obtained in this work suggests that neither the thermodynamic theory nor the hydraulic theory can fully explain frost damage to brick masonry. This does not mean that either of these theories is wrongly conceived because there is evidence in the experimentation to support both of them. The hydraulic theory is supported by the appearance of ice lenses on the lower faces of the specimens. Experiments allowing free drainage from the lower face indicate that the water forming these lenses leaves the specimens in the liquid phase. The thermodynamic theory is supported by the fact that in this work condensed droplets of ice are found on the upper surfaces of the specimens after the freezing cycle. This indicates that water in the vapour phase has moved towards the colder zones and condensed as ice on the outer faces of the specimens.

There is, however, further evidence to support the conclusion that these two theories cannot account for the damage that occurs to the specimens. Firstly, the visual and strain assessment of the damage shows that the majority of the damage occurs towards the lower face of the specimen. Hydraulic theory predicts that the larger pressure will be towards the upper part of the specimen. Considering the thermodynamic theory, ideally, the pressure in the ice phase will be equal throughout the pore structure of the solid provided it is continuous but this seems unlikely to happen in reality. The higher pressures are going to be found in the smaller pores. However our experimental data from mercury intrusion porosimetry and fissure analysis tend to suggest that damage occurs in the larger cracks rather than in the smaller ones.

The book-keeping type of calculations provide some evidence for the argument that there is entrapment of water by ice. The pressures required, however, to keep this water unfrozen at temperatures as low as -15°C are so large they considerably exceed the tensile strength of the brick ceramic and so some ice reinforcement must be postulated.

Further consideration of an entrapment theory requires an explanation of how a unidirectional freeze can produce entrapment. It is clear that the capillary bundle model most often used to describe masonry porosity is not a satisfactory model in respect of explaining entrapment. Entrapment can, however, be shown to occur if a two-dimensional pore network model is assumed. In reality, of course, the pore network will be three-dimensional but it is convenient to limit the discussion to a two-dimensional model at this stage. As such a saturated pore network cools thermodynamic theory predicts that water in certain pores will freeze before that in others. The freezing will progress into the specimens. It will nucleate initially in the uppermost pores and subsequently in those which have at least one end at a node at which there is ice present. Generally the larger pores freeze first. In figure 3 there is a schematic representation of water entrapment in a two-dimensional pore model. In this idealised system as it cools, so a temperature is reached at which the outer medium-sized pores freeze. The water in the large pore cannot, however, freeze because its freezing temperature is controlled by the narrow pores between it and the medium sized pores. Once the outer,

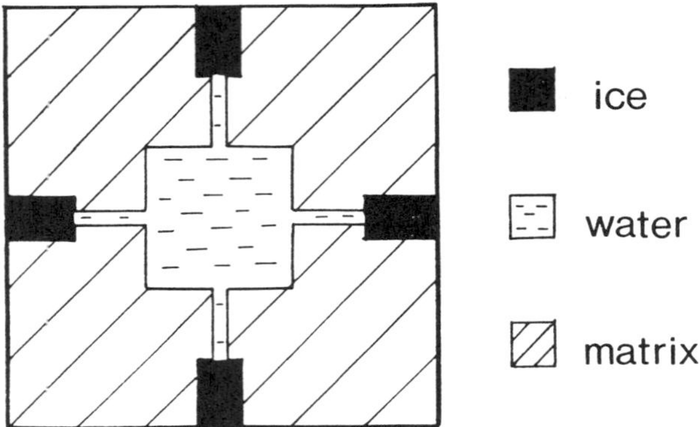

Fig.3. Schematic representation of entrapment

medium sized, pores have frozen then the water in both the small and large pores is trapped.

It is possible to carry out a freezing analysis on 2D pore networks as, for example, the network published by Mann et al. (1981) for a limestone. Whilst such a network model would predict that water can be entrapped in the way shown in figure 3 nevertheless the total volume of water entrapped is some 20 to 30 times smaller than the amounts suggested by our book-keeping analysis. It may be that in three-dimensions there is more possibility of entrapment. It may be in real solid there are a significant number of blind pores which would also lead to entrapment. The interesting fact is that theory and experiment both tend to suggest that some entrapment of water is occurring even on unidirectional freezing and clearly such entrapment will lead to very significant stresses.

6 Conclusions

In order to test specimens of masonry materials for frost damage susceptibility it is essential that all specimens to be compared are treated identically under defined conditions and particularly that the water content of the specimens is precisely known. It is also essential that a unidirectional freezing test procedure is used.

The results of this programme of research show that strain measurements are the most sensitive method of assessing frost damage. Visual assessment was found to be unreliable as a method of following damage.

Strain measurements and the visual assessment of damage in unidirectional freezing tests indicated that damage tends to begin and to occur more severely at the rear of the brick specimens. This has been corroborated by a very limited amount of field study and more field observations would be of interest.

The results of book-keeping measurements of water flows on freezing and the nature of the damage observed in these experiments lead the authors to suggest tentatively that entrapment of water by ice is occurring on unidirectional freezing. This entrapment will produce stresses fully sufficient to cause the observed damage. The thermodynamic and hydraulic theories of frost damage, whilst explaining some phenomena, do not predict sufficient stresses to explain the damage that occurs, and this might be taken to support the view that entrapment is occurring. If certain assumptions were made about the pore size distributions the entrapment theory could explain all the observed damage phenomena in rigid porous materials.

Acknowledgement

The authors thank the Science and Engineering Research Council for financial support.

7 References

Everett, D.H. (1961) The thermodynamics of frost damage to porous solids. Trans. Faraday Soc., 57, 1541-1551.

Litvan, G.G. (1980) Freeze thaw durability of porous building materials. ASTM STP 691, 455-463.

Litvan, G.G. (1978) Adsorption systems at temperatures below the freezing point of the adsorbate. Advances in Colloid and Interface Science, 9, 253-302.

Mann, R. et al. (1981) Application of stochastic network pore model to oil bearing rock with observations relevant to oil recovery. Chem. Eng. Sci., 36, 337-347.

Powers, T.C. (1945) A working hypothesis for further studies of frost resistance. Proc. Amer. Conc. Inst., 41(3), 245-272.

Watson, A. (1964) Laboratory tests and the durability of bricks. VI The mechanism of frost action on bricks. Trans. Brit. Ceram. Soc., 63, 655-680.

6 DURABILITY OF CONCRETES MIXED WITH SEA WATER

A. YEGINOBALI
Civil Engineering Department, Middle East Technical University, Ankara, Turkey
F.S. MUJAHED
Al-Amar Elkem Materials, Jeddah, Saudi Arabia

Abstract
Series of mortar and concrete mixes were prepared to have combinations of three types of cement: ordinary, pozzolanic and sulfate-resisting; three types of mixing water: tap, Red sea and its 50 % diluted version. From each mix cubic specimens were cast, each containing pairs of steel bars having different covers. The specimens were divided into groups and were subjected to different exposure conditions. At the age of one year they were tested for compressive strength and weight losses in steel bars. According to the results mixing with sea water increased the strengths of the specimens stored in dry air while causing slight reductions in most of the other specimens. In general ordinary portland cement performed better. Smaller cover, mixing with sea water, leaner mix and wetting-drying in sea water were corrosion increasing factors. Diluting the sea water for mixing could reduce corrosion to half or even less. The use of pozzolanic portland or pozzolan replacement of cement generally were not helpful in reducing steel corrosion.
Keywords: Concrete, Durability, Sea Water, Cements, Reinforcement Corrosion.

1 Introduction

Mixing concrete with sea water is discouraged by building codes such as American Concrete Institute 318 (1983) and British Standard 3148 (1980) due to the increased risk of reinforcing steel corrosion and possible long term strength losses. However, severe water shortages in arid regions along sea shores can make sea water a serious alternative for mixing concrete. It is also well known that the risks can be minimized by using the proper cement and admixture as well as by following good concrete making practices.

In a recent expert group meeting of United Nations Economic and Social Commission for Western Asia region (ESCWA 1987) one of the recommendations was 'to investigate the prospects of using sea and underground water in the preparation of concrete ...'. One of the main seas in the region is the Red sea which is surrounded by Jordan, Saudi Arabia, Yemen, Sudan, Egypt and Israil. Its shores are generally extensions of arid planes and deserts. Therefore, a research project was initiated in Jordan to study the properties of concretes mixed with Red sea water as affected by the type of local cements, admixture and exposure conditions. Part of the study dealt with the durability aspects involving long term strengths and reinforcing steel corrosion. The findings are summarized in this paper.

2 Materials

Three types of local cement; pozzolanic, ordinary and sulfate resisting portlands were used in the study. Their chemical properties are shown in Table 1.

Local natural sand and crushed limestone aggregate in three particle size groups were used in preparing the mortar and concrete mixes. Their gradations were adjusted to be within proper limits and were kept constant throughout the experiments.

Three types of mixing water; tap, Red sea and 50 % diluted sea water were used. The composition of the Red sea water is given in Table 2.

Table 1. Compositions of the cements

Chemical analysis (%)	Type of cement		
	Pozzolanic	Ordinary	Sulfate resisting
CaO	56.0	62.0	63.4
SiO_2	21.4	19.7	21.8
Al_2O_3	6.2	5.4	4.2
Fe_2O_3	3.8	2.9	4.5
SO_3	2.9	3.0	1.6
MgO	4.6	3.5	2.5
Na_2O	0.4	0.1	0.1
K_2O	0.8	0.8	0.5
$3CaO.SiO_2$	-	53.6	53.5
$2CaO.SiO_2$	-	16.1	22.2
$3CaO.Al_2O_3$	-	9.4	3.4
$4CaO.Al_2O_3.Fe_2O_3$	-	8.8	13.7

Table 2. Composition of Red sea

Principal ions (g/L)								
Na	K	Mg	Ca	Cl	Br	SO_3	HCO_3	pH
12.2	0.53	2.11	0.51	22.7	-	3.00	0.15	8.2

In some of the mortar mixes the effects of pozzolanic admixture were investigated by separately using a ground natural pozzolan. It was the same material used in the production of the commercial pozzolanic portland cement and consisted of a blend of local basalt tuffs. Its properties are given in Table 3.

3 Procedure and results of experiments

3.1 Experiments with mortars

The effects of mixing with sea water on compressive strengths of mortars were investigated by following the general procedures of the relevant American and British standard test methods. The first method involved the preparation of mortar mixes having cement: sand ratios of 1: 2.5 and water-cement ratios around 0.62, adjusted to provide a standard consistency with the natural sand used. Cubic specimens having 50 mm sides were cast and wet cured until the age of testing. According to the British method the mortar mixes had a constant water-cement ratio and were cast by vibration into cubic molds having 70.7 mm sides. The specimens were again cured in water. The compressive strength values obtained at various ages are presented in Tables 4 and 5.

The effect of mixing with sea water on the corrosion of steel reinforcement embedded in mortar was investigated using 150 mm cubic specimens each containing 8 mm diameter plain steel bars. The bars were initially cleaned with diluted HCl acid and their initial weights were determined. The mortar mixes had cement: sand: water proportions of 1: 1.75: 0.55. While casting each specimen two steel bars were inserted; one to have 20 mm mortar cover and the other to have 40 mm cover. From each mix three specimens were cast. After 28 days of moist curing two of the specimens were immersed to mid-height in sea water while the third specimen remained in the moist storage. The ones in sea water were inverted on a weekly basis to produce wetting-drying cycles. They were also capped with sulfur at their top and bottom faces to prevent water penetration towards the ends of the embedded bars. After one year of storage, the specimens were crushed under compression. The bars were removed, visually inspected and pickled with the diluted HCl acid to remove all the rust stains. Afterwards,

Table 3. Properties of the ground natural pozzolan

Chemical analysis	(%)	Chemical analysis	(%)
SiO_2	40.3	CaO	9.10
Al_2O_3	13.7	Na_2O	1.32
Fe_2O_3	11.1	K_2O	1.10
MgO	10.3	SO_3	-
Ignition loss	10.0	Insoluble	-
Specific surface (m^2/kg): 504			
Specific gravity : 2.60			

Table 4. Compressive strength of mortars with 50 mm cubes (MPa)

Cement	Mixing water	Age (days)			
		7	28	90	360
Pozzolanic portland	Tap	23.0	30.7	32.3	42.7
	Diluted sea	24.3	29.8	33.1	40.4
	Sea	23.2	27.3	31.4	39.2
Ordinary portland	Tap	20.3	26.1	26.4	35.1
	Diluted sea	21.8	27.7	29.9	36.3
	Sea	23.5	26.9	29.2	37.1
Sulfate resisting	Tap	22.0	34.2	37.9	49.9
	Diluted sea	26.2	33.6	35.2	42.0
	Sea	26.8	31.3	33.1	42.5

Table 5. Compressive strength of mortars at 28 days with 70.7 mm cubes (MPa)

Cement	Mixing water			
	Distilled	Tap	Diluted sea	Sea
Pozzolanic portland	42.0	42.0	42.5	41.5
Ordinary portland	38.0	38.5	40.0	39.0
Sulfate resisting	45.5	46.5	42.5	44.5

the dry weight of each bar was recorded. The weight losses due to corrosion were calculated after making the necessary corrections for the acid effect on clean bars. The results are shown in Table 6.

3.2 Experiments with concretes

The effects of mixing concretes with sea water on their long term strengths and on the reinforcing steel corrosion were investigated using 15 cm cubic specimens. Tests were repeated for two types of concretes. One group was relatively rich, having cement contents of 364 kg/m³ and water-cement ratios of 0.55. The corresponding values for the other group, which was leaner, were 245 kg/m³ and 0.73, respectively. From each individual mix six cubic specimens were cast, each containing four 8 mm diameter steel bars. The bars were arranged so that two of them had 4 cm concrete cover while the other two had only 2 cm cover. After 28 days of moist curing the specimens, in groups of two, were stored to be subjected to three different exposure conditions: storage in curing room at high humidity, storage in air at

Table 6. Weight losses of steel bars in mortar (%)

Mixing water	Storage condition	Cover (cm)	Type of cement*				
			P	O	S	OP	SP
Tap	Sea water	2	0.25	1.52	0.22	0.57	0.44
		4	0.00	0.00	0.00	0.00	0.00
	Moist room	2	0.00	0.00	0.00	0.00	0.00
		4	0.00	0.00	0.00	0.00	0.00
Diluted sea	Sea water	2	0.70	0.74	0.34	1.14	2.08
		4	0.09	0.39	0.13	0.24	0.46
	Moist room	2	0.07	0.06	0.08	0.10	0.26
		4	0.08	0.17	0.18	0.10	0.10
Sea	Sea water	2	3.15	1.54	0.68	1.82	1.20
		4	0.56	0.84	0.77	1.35	0.65
	Moist room	2	0.28	0.16	0.65	0.24	0.44
		4	0.40	0.12	0.61	0.55	0.24

* P: pozzolanic, O: ordinary, S: sulfate resisting, OP: ordinary mixed with 15 % pozzolan, SP: sulfate resisting mixed with 15 % pozzolan

low humidity and immersed to mid height in sea water and inverted weekly to have wetting-drying cycles. The last group was again capped with sulfur on faces perpendicular to steel bars. After one year of storage the specimens were tested under compression applied perpendicular to the axes of the bars. The compressive strength values obtained for different mixes are listed in Table 7. The weight losses in steel bars were determined by the same procedure used for the bars embedded in mortar. The results are used to prepare Tables 8 to 11.

The effects of mixing concretes with sea water and the type of cement used on the reinforcing steel corrosion were also investigated following the (ASTM C 876-1987) method. From each mix 15x15x55 cm beam specimens were cast. Four 16 mm steel bars were inserted into each beam to have 20 mm and 40 mm covers in pairs and 20 mm protruding sections at one end. Half cell potential readings were taken on the specimens which were stored in the moist room on a weekly basis. For this purpose a copper-copper sulfate half cell consisting of a tube with a porous end containing the sulfate solution and a copper electrode was used. To study the corrosion activity at a certain bar a circuit would be formed by connecting the protruding end of the bar to a sensitive voltmeter while placing the tube on the concrete surface over the steel bar and connecting the electrode to the positive terminal of the voltmeter. Fig.1 shows the results obtained on specimens made with ordinary portland cement.

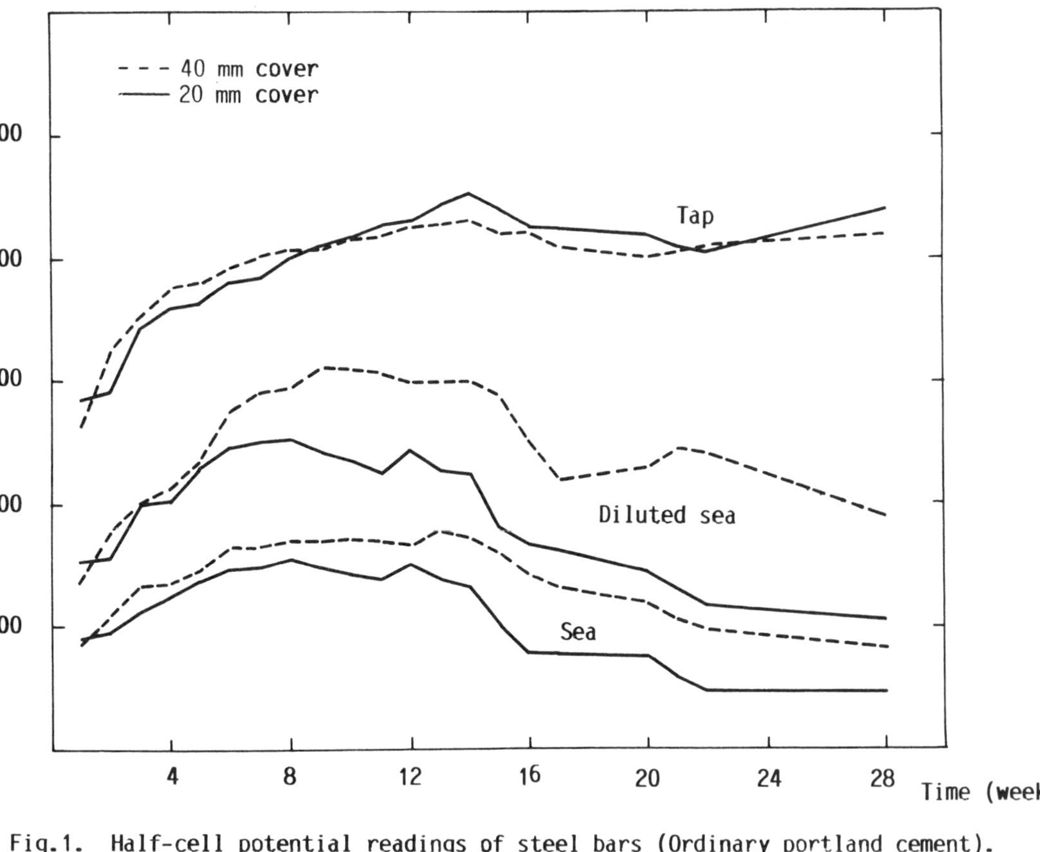

Fig.1. Half-cell potential readings of steel bars (Ordinary portland cement).

Table 7. Compressive strength of reinforced concrete cubes after one year

Cement	Mixing water	Storage condition	Compressive strength (MPa)	
			Rich mix	Lean mix
Pozzolanic	TAP	Sea water	45.6	33.2
		Moist room	52.2	36.3
		Dry air	34.7	28.4
	Diluted sea	Sea water	44.3	32.0
		Moist room	48.4	35.0
		Dry air	34.5	31.9
	Sea	Sea water	44.5	31.3
		Moist room	50.2	34.1
		Dry air	35.0	32.3
Ordinary	Tap	Sea water	45.9	32.5
		Moist room	50.5	34.4
		Dry air	37.1	28.6
	Diluted sea	Sea water	43.2	33.5
		Moist room	46.4	33.1
		Dry air	40.6	31.1
	Sea	Sea water	44.0	31.0
		Moist room	46.8	32.7
		Dry air	41.5	33.3
Sulfate res.	Tap	Sea water	48.3	39.0
		Moist room	53.8	39.3
		Dry air	29.3	26.2
	Diluted sea	Sea water	44.5	34.8
		Moist room	50.0	36.2
		Dry air	35.5	32.4
	Sea	Sea water	44.9	34.0
		Moist room	46.3	35.4
		Dry air	36.8	33.8

4 Discussion and conclusion

4.1 Discussion

The effects of mixing with sea water on the compressive strengths of mortars varied according to the type of cement used. According to Table 4 one year strengths increased in ordinary portland mortars, slightly decreased in pozzolanic portland mortars and significantly decreased in mortars containing sulfate resisting cement with sea water

Table 8. Effects of storage condition and cement type on steel corrosion in rich concretes (% weight loss)

Concrete cover (cm)	Storage condition	Cement type			All combined
		Pozzolanic	Ordinary	Sulfate res.	
2	Sea water	0.42	0.36	0.35	1.13
	Moist room	0.16	0.03	0.08	0.27
	Dry air	0.44	0.16	0.37	0.97
4	Sea water	0.22	0.15	0.21	0.58
	Moist room	0.12	0.02	0.06	0.20
	Dry air	0.48	0.17	0.37	1.02
Total weight loss (%)		1.84	0.89	1.44	4.17

Table 9. Effects of mixing water and cement type on steel corrosion in rich concretes (% weight loss)

Type of cement	Mixing water	Concrete cover		Cumulative
		2 cm	4 cm	
Pozzolanic	Tap	0.15	0.02	0.17
	Diluted sea	0.34	0.29	0.63
	Sea	0.53	0.51	1.04
Ordinary	Tap	0.11	0.05	0.16
	Diluted sea	0.19	0.12	0.31
	Sea	0.25	0.17	0.42
Sulfate res.	Tap	0.07	0.06	0.13
	Diluted sea	0.35	0.29	0.64
	Sea	0.38	0.29	0.67

mixing. A similar trend could be observed among the 28 day strength values listed in the same table and in Table 5. The effect of diluting the sea water on mortar strength was not clear.

As for the corrosion of steel bars embedded in mortar, the following observations could be made from Table 6: As expected, wetting-drying cycles in sea water and smaller cover were factors increasing the weight losses in steel bars. The use of ordinary portland cement seemed to reduce the corrosion in case of moist room storage, especially in sea water mixed mortars. This could be explained by the high $3CaO.Al_2O_3$ content of this cement resulting in a relatively good chloride binding capacity. On the other hand, under more severe conditions of sea water storage sulfate resisting and pozzolanic portland cements seemed to provide better protection against corrosion with 20 mm and 40 mm covers, respectively. In this case probably the quality and density of the mortars were more dominant factors than the cement

Table 10. Effects of storage condition and cement type on steel corrosion in lean concretes (% weight loss)

Concrete cover (cm)	Storage condition	Cement type			All combined
		Pozzolanic	Ordinary	Sulfate res.	
2	Sea water	1.73	0.95	0.60	3.28
	Moist room	1.80	1.51	0.24	3.55
	Dry air	0.73	0.39	0.30	1.42
4	Sea water	0.92	0.45	0.43	1.80
	Moist room	0.47	0.10	0.10	0.67
	Dry air	0.94	0.21	0.39	1.54
Total weight loss (%)		6.59	3.61	2.06	12.26

Table 11. Effects of mixing water and cement type on steel corrosion in lean concretes (% weight loss)

Type of cement	Mixing water	Concrete cover		Cumulative
		20 mm	40 mm	
Pozzolanic	Tap	0.26	0.14	0.40
	Diluted sea	2.13	1.00	3.13
	Sea	1.87	1.19	3.06
Ordinary	Tap	0.54	0.26	0.80
	Diluted sea	0.20	0.19	0.39
	Sea	2.11	0.31	2.42
Sulfate res.	Tap	0.27	0.16	0.43
	Diluted sea	0.49	0.39	0.88
	Sea	0.38	0.37	0.75

composition. The effect of separate additions of pozzolan to these cements was not very clear. Diluting the sea water with tap water before mixing the mortars reduced the rate of corrosion in general.

Table 7 gives detailed information on the strength development of reinforced concrete specimens made with different cements and mixing waters after one year of storage under various conditions. Considering the rich mixes, moist room storage was the most and dry air was the least beneficial for strength development. Mixing with sea water or its diluted version resulted in strength increases in the specimens stored in dry air, especially with ordinary and sulfate resisting portlands. This could be explained by the hygroscopic nature of the sea salts. Strengths of the concretes mixed with sea water were reduced under the other storage conditions. However, the reductions were generally less than by 10 %. Concretes made with sulfate resisting cement exhibited higher strength changes. Similar comments could be made for the lean concrete specimens. Again, in specimens stored in dry air the compressive strength increased as salinity of the mixing water increased. The increases were higher than those obtained with the rich concretes.

Test results on steel reinforcement embedded in concrete are arranged in Tables 8 to 11. Tables 8 and 10 are prepared by adding up all weight losses regardless of the type of mixing water to show the direct effects of storage conditions and type of cement on corrosion. According to these tables corrosion was less with rich mixes and larger covers. Moist room storage was the least detrimental to steel bars especially with rich mixes and larger concrete covers. This could be explained by the improved hydration conditions causing relatively dense covers and limiting the oxygen penetration. Storage in dry air could cause relatively more steel corrosion than one would expect in the same group of specimens. This may be due to the presence of abundant oxygen and the hygroscopic salts providing the necessary moisture in cases of sea water mixing. Similar observations were made by Hime and Erlin (1985). In general, ordinary portland cement with its high chloride binding capacity provided the best protective environment for the steel bars. In lean mixes sulfate resisting cement also performed well. Pozzolanic portland cement was the most detrimental to steel possibly due to the lower lime content and lower alkalinity of its paste.

Test results were also combined to show the direct effects of mixing water and cement types used on the corrosion of the embedded steel. As seen from Tables 9 and 11 the weight losses of steel bars generally increased with increasing salinity of mixing water. Diluting the sea water did not seem beneficial in lean mixes, except the ones containing ordinary portland cement. In rich mixes diluting the sea water was more effective against corrosion and could reduce weight losses by up to 40 % in concretes made with ordinary portland.

According to (ASTM C 876-1987) half cell potentials of steel bars in concrete only indicate the probability of the occurence of steel corrosion. If potential readings are numerically less than -200 mV, there is a greater than 90 % probability that no corrosion activity of the reinforcing steel is taking place. On the other hand, readings more negative than -350 mV indicate corrosion activity with the same level of probability. However, according to Hime and Erlin (1987) such interpretations can be misleading in the presence of chloride ions and more negative readings do not necessarily indicate that more rust is present over the steel. In any case, as seen from Fig.1 mixing with sea water resulted in more negative half cell potential readings in specimens made with ordinary portland cement. The slopes of these curves start to decrease after 14 weeks of storage possibly indicating the initiation stage for the corrosion activity in saline water mixed concretes. Diluting the sea water had a positive effect as seen from the figure. Increasing the concrete cover was more effective in the case of saline water mixing. The results obtained with the other cements followed a similar pattern, but with more negative potential readings, in general.

4.2 Summary and conclusion

Durability of mortars and concretes mixed with Red sea water has been studied considering their strengths at the age of one year and corrosion of steel bars embedded in them. Possible effects of cement type, exposure conditions, cover and composition have also been investigated.

Mixing with sea water could increase the mortar or concrete compressive strengths at the age of one year in cases of using ordinary portland cement and under dry air exposure conditions. The strength reductions would not be by more than 10 % if sulfate resisting or pozzolanic portlant cements were used and/or other exposure conditions were applied.

As determined by weight losses, using sea water for mixing mortars and concretes would increase the corrosion of embedded steel bars. Using larger cover for the bars, richer mixes and diluting the sea water could reduce corrosion. Steel in concretes mixed with sea water and exposed to dry air could develop more corrosion than expected. In general, the use of ordinary portland cement provided the best protection for the embedded steel while the use of pozzolanic portland or pozzolanic admixtures were not helpful in this respect. Half cell potential measurements also confirmed the findings.

Acknowledgements

The authors would like to thank the Deanship of Scientific Research and Graduate Studies at Jordan University of Science and Technology and Jordan Cement Factories Company for their financial support and encouragement during this study.

References

American Concrete Institute, (1983) **Building Code Requirements for Reinforced Concrete.** ACI 318-83, Detroit, Michigan.

American Society for Testing and Materials, (1987) Half-Cell potentials of uncoated reinforcing steel in concrete C-876 in **Annual Book of ASTM Standards**, Section 4, Phila., PA, pp.563-570.

British Standards Institution, (1980) **Tests for Water for Making Concrete (including notes on the suitability of the water)**, BS 3148, London.

Erlin, B. and Hime, W. (1985) Chloride-induced corrosion, **Concrete International**, 7, No.9, pp.23-25.

Erlin, B. and Hime, W. (1987) Some chemical and physical aspects of phenomena associated with chloride-induced corrosion, **Corrosion, Concrete and Chlorides**, ACI SP-102, Detroit, pp.1-12.

United Nations Economic and Social Commission for Western Asia-ESCWA, (1987) **Proc. of the expert group meeting on energy efficient building materials for low cost housing**, Amman, pp.5-7.

7 STUDY OF MECHANISM AND DYNAMICS OF CONCRETE CORROSION IN SUGAR SOLUTIONS

B. SKENDEROVIC
Faculty of Civil Engineering, Subotica, Yugoslavia
L. OPOCZKY
Institute for Technology of Silicate Materials,
Budapest, Hungary
L. FRANC
Institute of Chemical Industry 'Zorka', Subotica, Yugoslavia

Abstract
Chemism and mechanism of concrete corrosion in sugar solutions of different concentrations were investigated. The experiments were carried out in laboratory conditions on samples of hardened cement paste which were exposed to the effect of sugar solution of different concentrations (0,5,10,15 and 20%). It is established that the corrosion speed in solutions with sugar to 10% is proportional to the sugar concentration in the solution, nevertheless, by further increase of the concentration of sugar the corrosion process does not change or even slows down. In sugar solutions, besides free $Ca(OH)_2$, some ferrites, aluminates and silicates from hardened cement paste are dissolved. In more concentrated sugar solutions, under determined conditions, some poorly soluble compounds are formed between Ca^{2+} (dissolved from cement paste) and some sugar components. The formation of these compounds is, probably, the cause of slowing down the corrosion process in more concentrated sugar solutions.
Keywords: Concrete, Cement Paste, Corrosion, Sugar Solution, Dynamics of Corrosion, Chemism and Mechanism of Corrosion.

1 Introduction

Sugar solutions are present in many products and in waste waters of food industry. They are aggressive to concrete constructions and can make serious problems in food factories. Therefore, it is peculiar that we know so little about the chemism and mechanism of concrete corrosion in sugar solutions. We do not know enough, either, about how different factors influence the dynamics of this kind of corrosion.

The process of concrete corrosion in sulphate solutions has been studied thoroughly. Special cements are commercially produced which are resistant to such corrosion, and so we are able to make concrete resistant to sulphate corrosion. In order to be able to make concrete resistant to sugar corrosion we must know much more about the chemism of this process and also, how different factors influence the dynamics of this kind of corrosion. This was the reason we started our research and we tried to find out what really happens in concrete when it is attacked by aggressive solutions of different concentrations of sugar.

2 Materials and methods

The experiments were carried out in laboratory conditions at room temperature on samples made of hardened cement paste. Thin prisms (0,5.4.16 cm) and crushed cement paste (particle size of 0,5 - 1 mm) were used as test samples which were cured in water for 28 days.

In the first experiment thin prisms were kept in water (control samples), and in sugar solutions (10,20%). Mass relation of test prisms and solutions was 1:10. The treatment period lasted for 9 months, during this period the water and sugar solutions were replaced every month.

The change of mass was measured in the prisms, and also the change of the concentration of Ca^{++} in test solutions during the treatment period of nine months. The porosity (measured by mercury intrusion porosimeter) and the changes in cement pastes compositions (measured by X-ray powder diffraction, DTA, DTG and TG methods) were analysed after the treatment period of nine months.

In the second experiment crushed cement paste (0,5-1 mm) was used, it was kept for 30 days in water (control sample) and in solutions with different (5, 10, 15, 20%) sugar concentrations. Solutions were replaced after: 2, 5, 10, 20 days, and pH and the concentrations of: Ca^{2+}, Mg^{2+}, Na^{+}, K^{+}, Si^{4+}, Al^{3+} and Fe^{3+} were measured in them. After 10 days of treatment the change of mass and the chemical composition of samples kept in water and in sugar solutions were measured. Their composition was also analysed by X-ray powder diffraction and by TG, DTG and DTA methods.

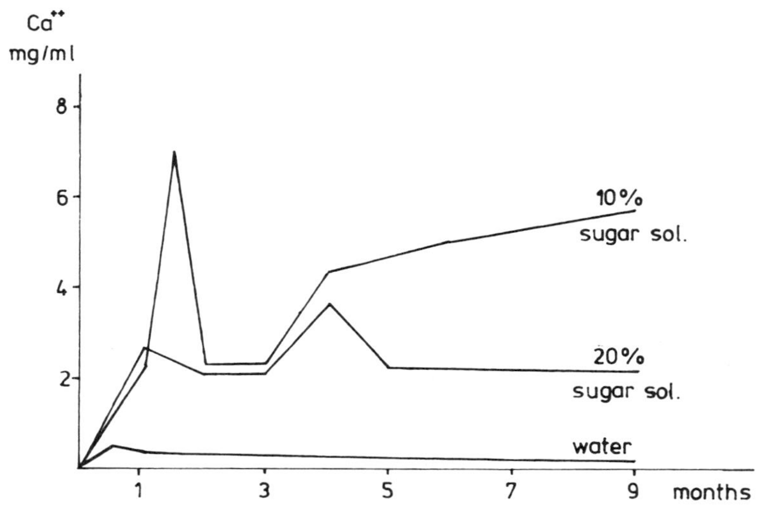

Figure 1. Dynamics of Ca^{++} dissolving in sugar solutions.

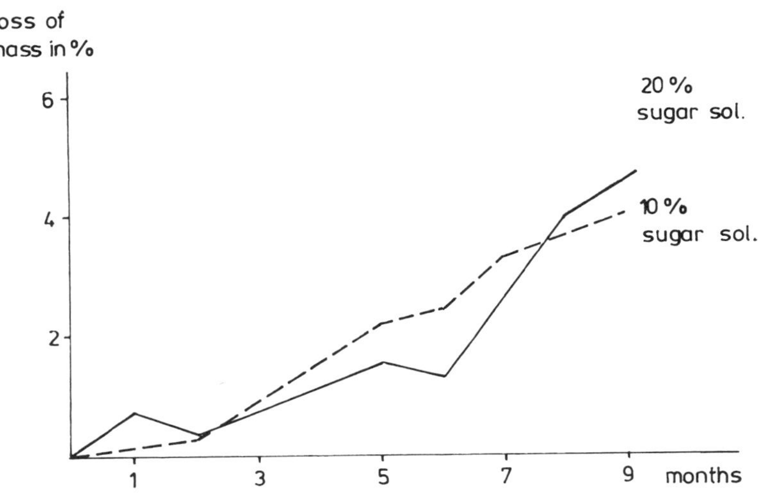

Figure 2. Loss of mass of test prisms in sugar solutions.

3 Results and discussion

3.1 Experiments with thin test prisms

Test prisms were kept in water (control) and in sugar solutions (10, 20%). In order to follow the dynamics of cement paste corrosion the loss of mass of test prisms was measured, and also the change of the concentration of Ca^{++} in test solutions during the whole treatment period of nine months. The results of these measurements are presented in Figures 1 and 2.

From Figure 1 it is evident that we got expected results during the first month of treatment when extraction of Ca^{++} in 20% was faster than in 10% sugar solution, what we considered as logical. Later, faster and more intensive extraction of Ca^{++} from cement paste specimens in 10% than in 20% sugar solutions occurred. At the same time it can be seen from Figure 2 that there is no big difference in weight loss of test prisms in 10% and 20% sugar solutions. It is not possible to give satisfactory explanation of these unexpected results. One of the possible explanation could be that some insoluble compounds were formed at high concentration of Ca^{++} and sugar in solution. In favour of such assumption speaks the fact that the formation of some white precipitate was noticed after 4 month of treatment of prisms in sugar solutions, especially in 20% solution. But it was impossible to find any sign of such new compounds by X-ray diffraction and DTA, DTG and TG analyses of test prisms kept nine months in 20% sugar solutions (Figures 3 and 4).

As it is seen from these figures there is less $Ca(OH)_2$ in samples from sugar solution than in control samples kept in water.

The results of porosity measurements of prisms kept nine months in 20% sugar solution and in water are presented in Table 1, and pore distribution curves in Figure 5.

Table 1. Porosity of cement pastes kept in water and in 20% sugar solution

% of sugar in solution	Total volume of open pores		Spec surface of all pores	
	cm3/g	Difference in %	m2/g	Diff in %
0	0,0334	-	11,75	-
20	0,0546	63,5	17,01	44,8

Total volume of open pores is, of course, much larger in samples of cement paste kept nine months in 20% sugar solution. From Table 1 and Figure 5 it is also evident that there are many more large pores (>500mm) in corroded samples from sugar solution. The quantity of very small pores (<10 mm) is also larger in corroded samples.

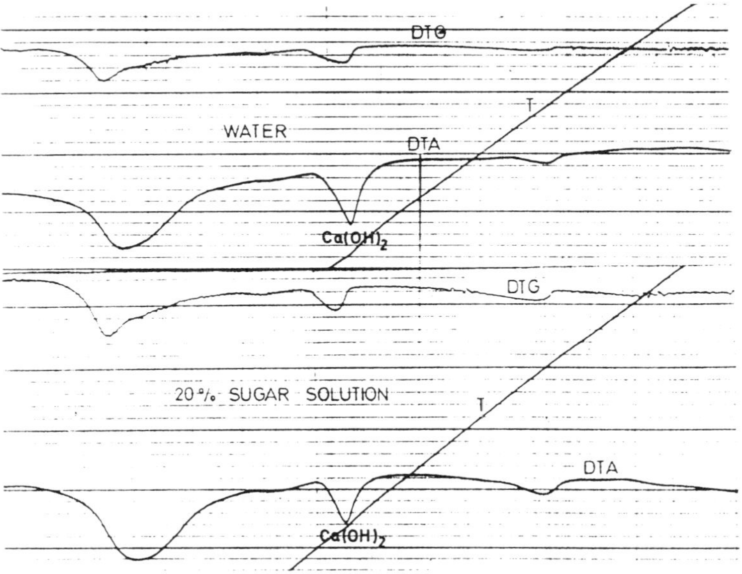

Figure 3. DTA and DTG curves of samples from water and 20% sugar solutions.

3.2 Experiments with crushed cement paste samples

There were too many questions without answers in connection with the results of experiments with thin prisms. This was the reason we decided to undertake new experiments with samples of crushed (particle size 0,5 - 1 mm) hardened cement paste. In the first experiment hardened cement paste contained about 7% of pores, and in the second one the total volume of pores was about 18%.

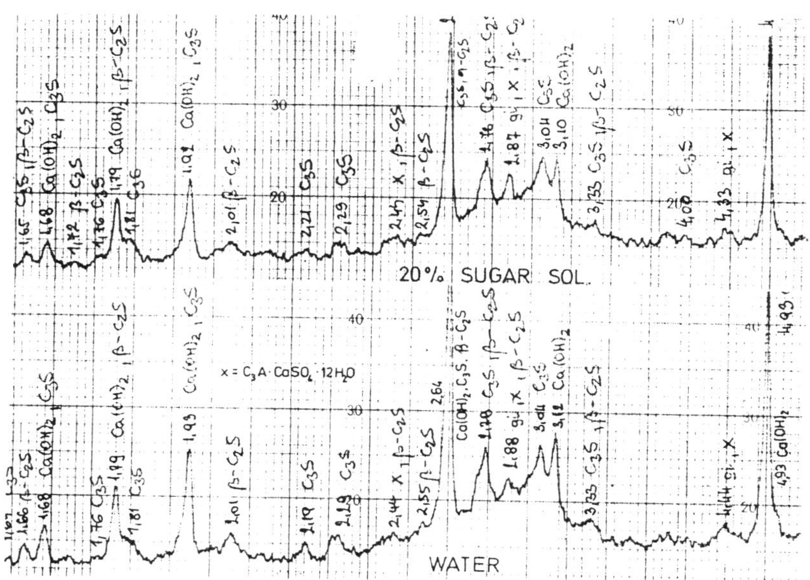

Figure 4. X-ray diffraction curves of samples from water and 20% sugar solution.

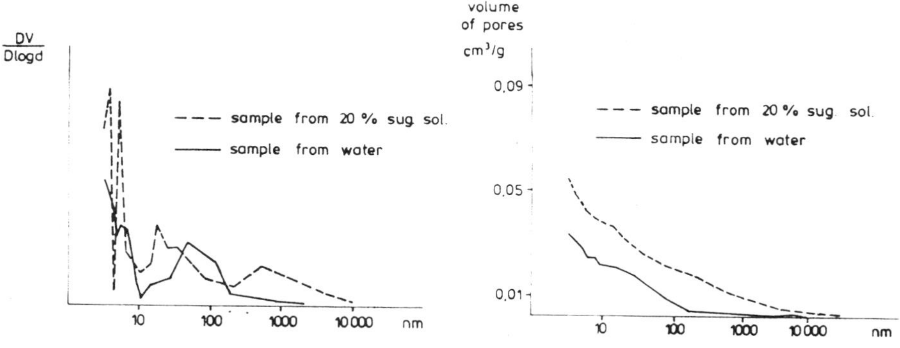

Figure 5. Pores distribution curves of samples from water and 20% sugar solution.

Table 2. Quantities of dissolved ions in the period of 10 days in sugar solutions of varions concentrations.

Sorts of diss.ions	Quantities (mg) of dissolved ions in sugar sol. 0	5	10	20 %
Na^+	8,3	8	8,3	8,3
K^+	27	23,5	28	29
Ca^{2+}	1.245	4.550	6.270	2.332
Mg^{2+}	0	17,3	36,3	30
Si^{4+}	0	68	179	156
Al^{3+}	0	66	238	244
Fe^{3+}	0	101	253	239

In Figure 6 the dynamic of extractions of different ions (Ca^{2+}, Si^{4+}, Al^{3+}, Fe^{3+}) in solutions with different sugar concentrations (0, 5, 10, 15, 20%) are presented. It is obvious that the dynamics of extraction of Ca^{2+} is proportional to sugar concentration of 10% in the solution, but in more concentrated sugar solutions (15, 20%) there was less extracted Ca^{2+} than in 10% sugar solution. During the period of the experiment (from the second till the fifth day) a high concentration of Ca^{2+} appeared, and white precipitate was observed in all sugar solutions (more in 20% than in 5% solution).

There were found only Na^+, K^+ and Ca^{2+} in water, but in all sugar solutions Mg^{2+}, Si^{4+}, Al^{3+} and Fe^{3+} ions were extracted as well. The extraction of these ions was faster and stronger in solutions with higher sugar concentration. In Table 2 there are presented the quantities of extracted ions during the period of 10 days in solutions with various (0, 5, 10, 15, 50%) sugar concentrations. From the results presented in Figure 6 and Table 2 it is clear that concrete corrosion in sugar solutions is not a simple extraction of free $Ca(OH)_2$ from hardened cement paste. Obviously, some other components of cement paste, which contain Mg, Si, Al and Fe, are also dissolved in sugar solutions. We do not know from which compounds are extracted these ions, but it is possible that some silicates (from CSH-gel), aluminates (from CAH) and ferrites (from CFH), which were produced during cement hydration, are partly soluble in sugar solutions. It is also obvious that ferrites are dissolved best in all sugar solutions, aluminates are the next and silicates seem to be the least soluble in sugar solutions.

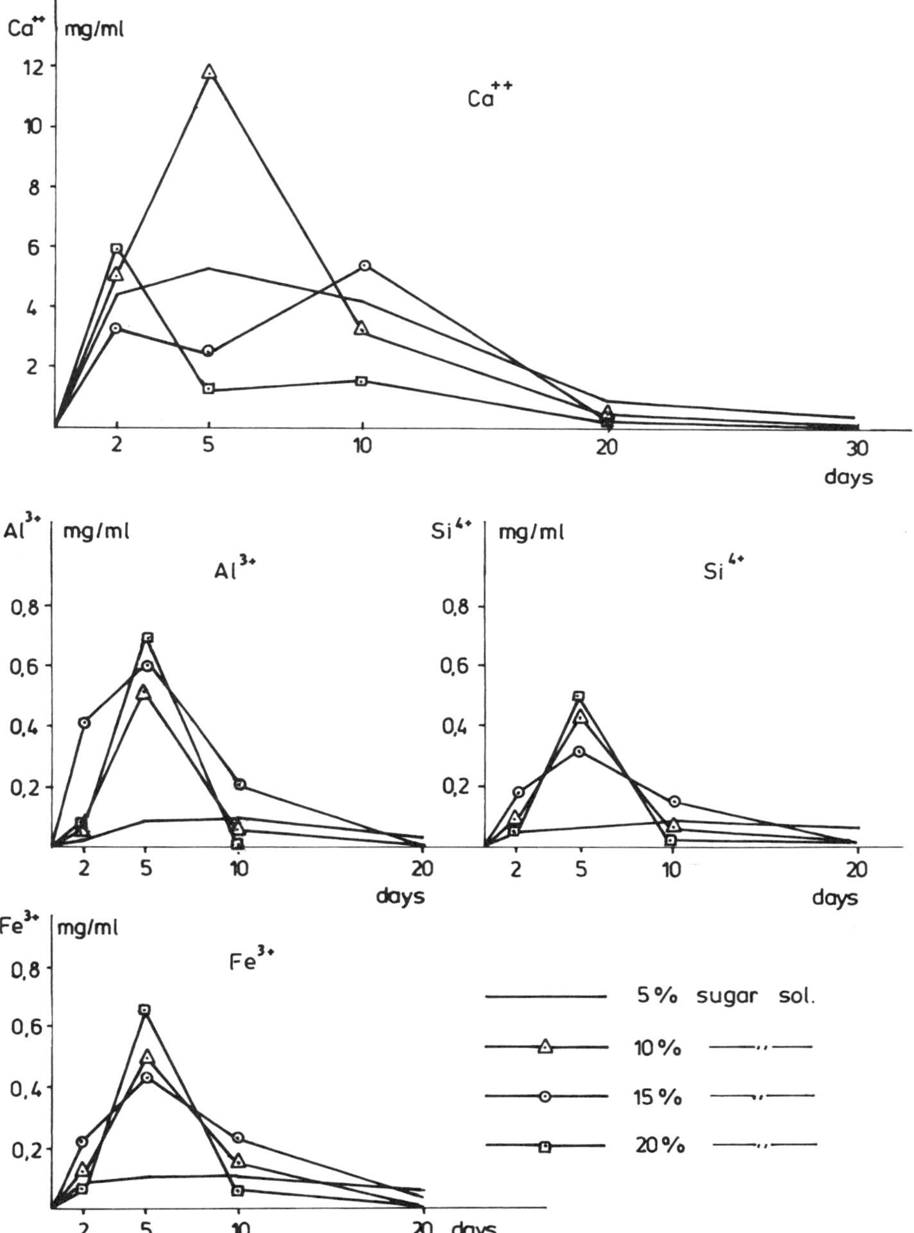

Figure 6. The dynamics of extractions of different ions in solutions.

This conclusion is more pronounced if we take into account the fact (Table 3) that there are the largest quantities of silicates (SiO_2 about 19%), much less quantities of aluminates (Al_2O_3 about 7%), and very small quantities of ferrites (Fe_2O_3 about 2%) in cement paste.

The composition of cement paste samples was analysed after 10 days of treatment. The results of chemical analyses are given in Table 3. The loss of mass of samples by combustion at 550°C after 10 days treatment in sugar solutions is much greater than in control samples from water. These differences are proportional to sugar concentration in test solutions up to 10% sugar, and it is small between samples which were kept in 10% and 20% sugar solutions. This loss of mass at temperature of 550°C is the result of the release of chemically bound water from different cement paste compounds ($Ca(OH)_2$, CSH-gel) and it can also be the result of the combustion of some organic compounds.

We belive that the part of this loss of mass is the consequence of an organic compound which was formed as the result of the interaction between Ca^{2+} and sugar.

From Table 3 the loss of mass of elements (CaO, MgO, Fe_2O_3, Al_2O_3) which are extracted by sugar solution is evident.

Table 3. Results of chemical analyses of cement paste samples after 10 days of treatment

Compounds	Quantities (in %) of samp from sol with sugar con			
	0	5	10	20%
CaO	45,2	42,3	41,6	41,8
MgO	1,70	1,63	1,62	1,55
SiO_2	18,1	18,9	19,7	19,2
Al_2O_3	6,91	7,01	6,85	7,00
Fe_2O_3	2,10	2,03	1,74	1,68
Loss of mass as 550°C	8,05	11,57	15,66	16,04
Loss of mass of samples in solutions	2,97	12,33	18,69	18,55
pH	12,25	12,19	11,77	11,75

Figure 7 shows DTA and DTG curves. They show that free $Ca(OH)_2$ after 10 days of treatment is present only in cement paste samples from water, and in very small quantities in samples from 5% sugar solution. $Ca(OH)_2$ is completely extracted from other samples by sugar solutions. The increase of SiO_2% in samples from sugar solutions is only relative, and is a consequence of the decrease of other compounds ($Ca(OH)_2$, Fe_2O_3) and does not mean that silicates are not dissolved in sugar solutions.

The loss of mass of samples treated in water and sugar solutions of 10 days is increased by the growth of sugar concentration up to 10% in the solution, by further increase of sugar concentration it practically does not change.

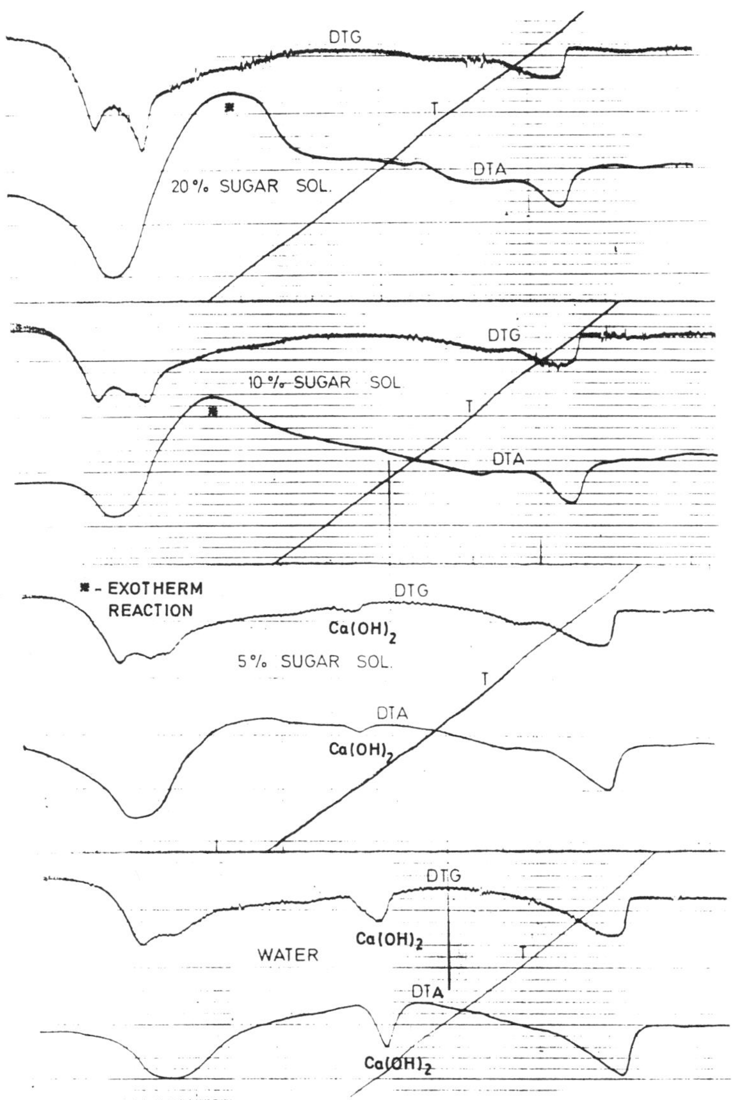

Figure 7.

Results of DTG and DTA analyses of cement paste samples after 10 days of treatment in sugar solutions of different concentrations.

From the shapes of DTA curves (Figure 7) it is evident that some exothermic reaction takes place in samples kept for 10 days in 10 and 20% sugar solutions. These cahnges are not present in control samples from water, and in the sample which was kept in 5% sugar solution the change is hardly noticeable. This change of DTA curve takes place on temperatures below 500°C (lower than the temperature for releasing water from $Ca(OH)_2$), and it is, probably, caused by combustion of organic compound formed by the reaction of Ca^{++} and sugar components. From DTA and DTG curves (Figure 7) it is also evident that there is no free $Ca(OH)_2$ any more in the samples of cement paste which were kept 10 days in 10 and 20% sugar solutions as it is completely washed out. There is a small quantity of free $Ca(OH)_2$ in samples kept 10 days in 5% sugar solution, but there is a considerable amount only in control sample from water.

In Figure 8 the curves obtained from the analysis of test samples by X-ray diffraction method after 10 days being exposed to water (control samples) or to sugar solutions (5, 10 and 20%) are shown. According to these results it can be positively stated that there is no free $Ca(OH)_2$ after 10 days in samples which were exposed to 10 and 20% of sugar solutions, but very little in samples from 5% sugar solution and in control sample from water there is a remarkable quantity, what has already been concluded on the basis of DTA and DTG analyses. Unfortunately, there is neither evidence on forming of new calcium compounds nor on decrease of CSH, CAH or CFH compounds in samples exposed to the effect of sugar solution. Only the presence of C_3AH_6 and C_4AH_{13} is proved, but without the essential differences in their quantities between control and test samples. However, some evidence seems to prove that in samples from 20% sugar solution there is slightly increased quantity of ettringites ($C_3A.3\ CaSO_4$) or some similar compounds.

4 Conclusions

On the basis of the investigation results it may be concluded:

Corrosion of cement paste (concrete) in sugar solutions is taking place by dissolving of large quantities of Ca^{++}, but in sugar solutions there are also considerable quantities of Mg^{2+}, Fe^{3+}, Al^{3+} and Si^{4+}, it means that besides free $Ca(OH)_2$ in sugar solutions some silicates, aluminates and ferrites of calcium and magnesium are dissolved as well. The most soluble compounds from cement paste in sugar solutions are ferrite compounds and the least the least silicate ones.

The intensity and dynamics of concrete corrosion is increased by the increase of sugar concentration in solution up to about 10%. By further increase of sugar concentration in the solution the process of corrosion stagnates or even may slow down.

The cause of slowing down the process of concrete corrosion in more concentrated sugar solutions is, probably, the partly filling of pores by poorly soluble compounds which are formed at determined conditions by mutual reaction of Ca^{++} and sugar (or products of hydrolysis of sugar).

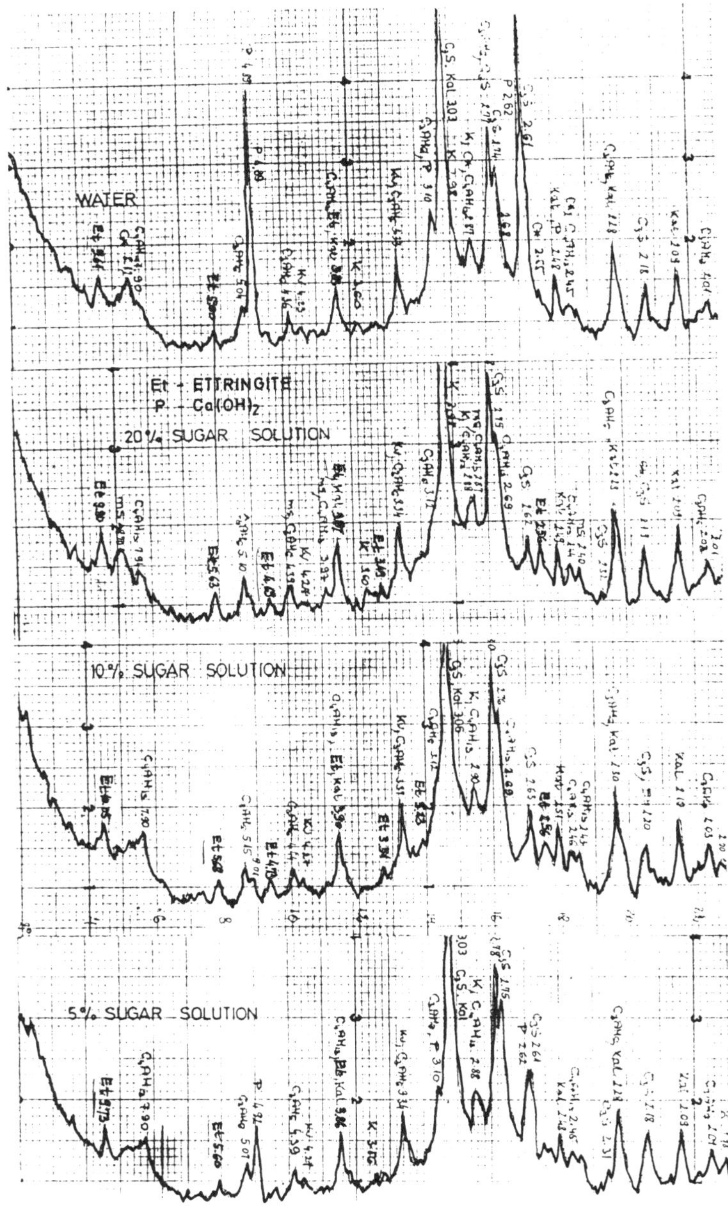

Figure 8.
Results of the analysis of cement paste samples by X-ray diffraction method.

5 References

Biczok, I. (1964) Concrete Corrosion Protection. House of the Hungarian Academy of Science, Budapest.

Godulla, P. and Plath, M. (1978) Oberflachenschutz der Bechalterinenwand von Weisszuckersilos. Lebensmittelind, 6, pp 267-268.

Skenderovic, B, Fisang, Lj and Takac, E. (1982) Izucavanje hemizma i dinamike korozije cementnog kamena u rastvorima secera. XVII. Kongress JUDIMK-a, Sarajevo, pp 43-59.

Steineger von, H. (1973) Schutz von Beton gegen sehr starke chemische Angriffe. Beton, 1, pp 18-20.

8 COMMON FACTORS AFFECTING ALKALI REACTIVITY AND FROST DURABILITY OF AGGREGATES

P.P. HUDEC
Geology Department, University of Windsor, Windsor, Ontario, Canada

Abstract
Petrographically similar rock aggregates are involved both in alkali reactivity and frost durability of aggregates. Examples are some cherts, argillaceous dolomites, and phyllites. Although alkali reactivity is essentially a chemical, and freeze-thaw durability a physical process, it appears that some common factors affect both.

The two common factors that have been identified are: a. the very fine grain size and pore size, and the attendant high surface area, and b. highly active mineral surfaces. The active, commonly negatively charged surfaces attract alkali cations and polar water (and other) molecules. The concentration of alkalies in the small pores results in osmotic differential and pressure - the cause of physical deterioration. The same concentration also initiates a chemical reaction if suitable chemistry exists (i.e., alkali - silica or alkali - carbonate reaction).

Results of alkali reactivity and freeze-thaw durability experiments confirm the above relationship. Some aggregates degrade during unconfined freezing and thawing in the presence of de-icing salts, and also expand as a result of accelerated alkali reactivity tests. Both expand isothermally on saturation. High internal surface area, and sorption of both water and ions are the common properties to non-durable and alkali reactive rocks. It should, however, be stressed that not all frost sensitive aggregates are alkali reactive, and not all alkali reactivity aggregates have frost durability problems.
Keywords: Aggregates, Frost durability, Alkali reactivity, Grain size, Pore size, Surface area, De-icing salts, Expansion, Adsorption, Internal surface area.

1 Introduction

Alkali reactivity of aggregates in concrete is basically a chemical reaction. In alkali silica reaction (ASR), the alkalies of primarily Na and K react with the siliceous aggregate to produce a silica gel (Hobbs, 1988). The gel fills available pores. Two main theories have been proposed to explain the expansion of the concrete: the pressure of the gel as it fills the pores (Vivian, 1950), and pressure due to osmotic difference between the water in the gel and outside of the gel (Hobbs, 1979). In the latter, the gel imbibes water, swells, and creates expansive stress on the pore walls. The concrete expands, and eventually fails in tension by microcracking. The larger the surface area of the reactive silica or silicate and greater its surface activity, the greater the production of the gel and expansion (assuming sufficient alkalies).

In the case of alkali-carbonate reaction (ACR), the reaction process is quite different. The most reactive carbonate rocks are partially dolomitized argillaceous limestones. In these rocks, the clay particles tend to be concentrated in the vicinity of the small dolomite rhombs. Process of de-dolomitization in the presence of strong alkaline solution is thought to be the cause, but not the reason for expansion (Swenson and Gillot, 1960). The expansion in argillaceous dolomitic limestone is attributed to water uptake by dry clay minerals enclosed originally in dolomite, but now exposed by the de-dolomitization reaction (Gillot and Swenson, 1969). The reactive carbonate rocks are characterized by very fine grain size, low absorption, and low porosity (Dolar-Mantuani, 1983). The adsorption of water into the very small pores causes expansive stresses within the rock, and eventual failure of concrete in tension.

In both ASR and ACR reactive rocks, the high internal surface area of the reactive minerals is primarily responsible for the initiation of the expansion. The alkali reactivity can also be initiated in presence of any alkali salts, not just NaOH. Road salt (NaCl) has been shown to initiate both types of reaction (Larbi and Hudec, 1990). NaCl is also well known as an accelerator of physical damage during wetting, freezing, thawing, and drying cycles. It can also be shown that the rocks that tend to deteriorate rapidly in the presence of salt under the above climatic cycles have large internal surface area and small pore and grain size (Hudec, 1989, 1987). This paper will show that indeed a relationship exists between freeze-thaw durability and alkali reactivity for some, but not for all rock types.

2 Common Sorption Properties of AR and Freeze-Thaw Sensitive Rocks

Figure 1 gives the most common rock (and mineral) types that are alkali reactive, and those that are frost and weathering sensitive. To be alkali-silica

Alkali Reactive Rocks	Frost Sensitive Rocks
Chert, flint, chalcedony	Chert (weathered)
Opal in variety of rocks	Volcanics (weathered)
Strained quartz & Iridimite in variety of rocks	Fine grained argillaceous dolomitic limestones
Volcanic rocks glassy & tuffaceous	Argillaceous rocks
Greywacke containing volcanic fragments	Aphanitic rocks
Fine grained argillaceous dolomitic limestone	
Aphanitic siliceous rocks	

Figure 1 Rock Types

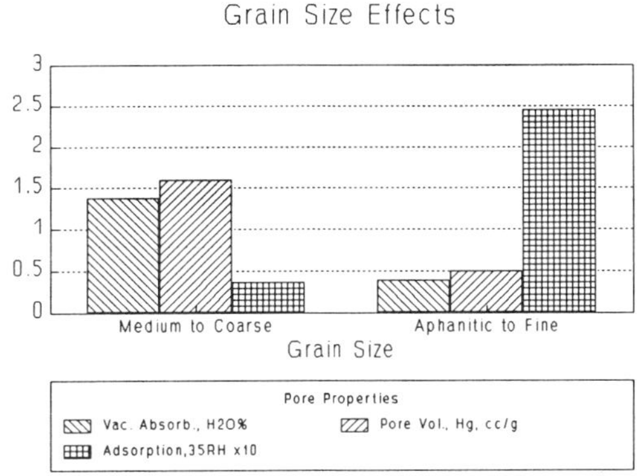

Figure 2 Effect of grain size on sorption properties of rock.

reactive, the rock must, of course, contain silica or silicate. That is not a requirement for frost sensitive rocks. For ACR rock, the requirement is that it be fine grained, argillaceous, and dolomitic. All but the last appear to be the requirements for frost sensitivity among carbonates.

Fine grained to aphanitic texture is a common property for both AR and frost sensitive (FS) rocks. The fine grain of the minerals in the rock provides both larger surface internal area and small pore diameters. Figure 2 shows the effect of grain size, on vacuum absorption of water, injection volume of mercury, and the adsorption of water at 35% relative humidity (RH) for medium to coarse grained and aphanitic to fine grained group of rocks respectivelly. The rocks are mostly carbonates. Vacuum absorption and mercury intrusion give approximately the same results, and represent the total pore volume of the rock. Adsorption of water at 35% RH results in one to two layer coverage of the internal surface by the water molecule, and therefore gives a measure of the internal surface area. At 35%RH, water meniscus cannot exist, and here is no capillary filling; the water is adsorbed directly to the mineral surface. The results clearly show that the fine grained rocks have over five times the internal surface area of the coarser grained rocks. In AR, the increased surface area promotes reaction. In FS, the adsorbed water has lower vapour pressure, and acts as an osmotic pump for ingress of higher vapour pressure of bulk water. Expansion occurs under both conditions of AR and FS sensitivity.

The relationship between sorption properties and FS is further explained in Figure 3. The amount of water absorbed depends on the salinity of the water. In this case, more of the 3% NaCl solution was absorbed compared to fresh water in the same high and low FS rocks. NaCl is thought to concentrate by ionic sorption on the mineral surface, and attracts more water into the pore by osmosis. The concentration of alkalis at the mineral surface by adsorption will initiate the AR process under proper circumstances.

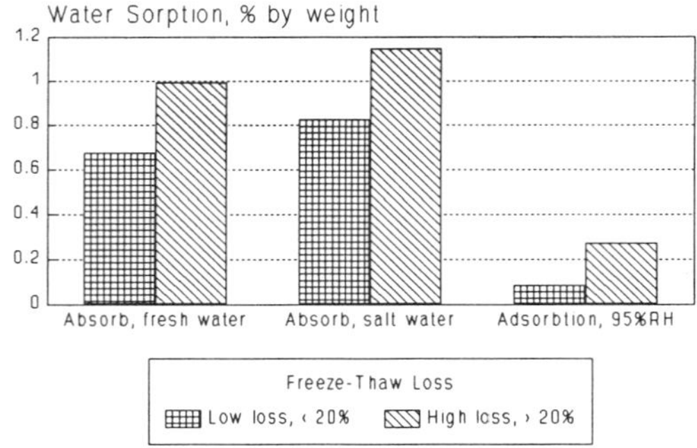

Figure 3 Absorption and Adsorption of water by low- and high freeze-thaw loss rocks.

In Figure 4, some potentially AR reactive rocks have been selected for comparison of the physical properties that influence durability. The line drawn through the figure gives approximate limit above which the rock may be potentially alkali silica reactive, mainly because of its higher internal surface area as measured by adsorption. It is seen that the rocks above this limit also show significantly higher freeze-thaw losses. The above rocks have not been specifically tested for AR, but belong to the rock types known for their AR potential.

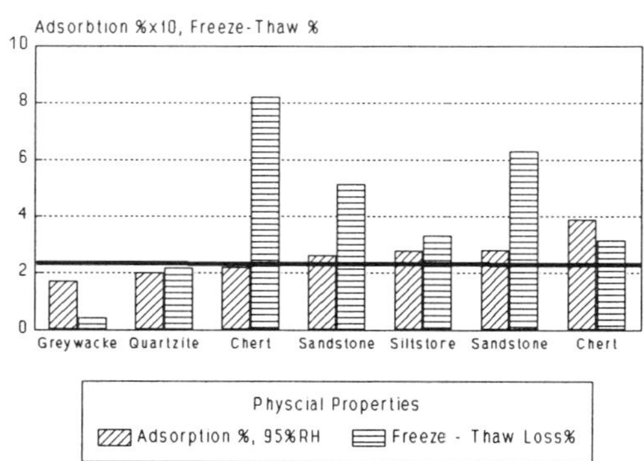

Figure 4 Relationship of ADsorption and Freeze-Thaw loss of some potential AR rocks.

3 Expansive Properties of AR and FS Rocks

The failure of concrete and rock by either AR or FS ultimately involves internal pressures and increase of volume. The AR is measured by monitoring the length change of the specimen as a function of time. Expansion also takes place during freezing-thawing cycles. The type of cracking observed in mass concrete is virtually indistinguishable whether AR or FS is involved. Both produce 'map' cracking pattern. It is therefore instructive to consider the expansive potential of rocks suspected of AR and FS.

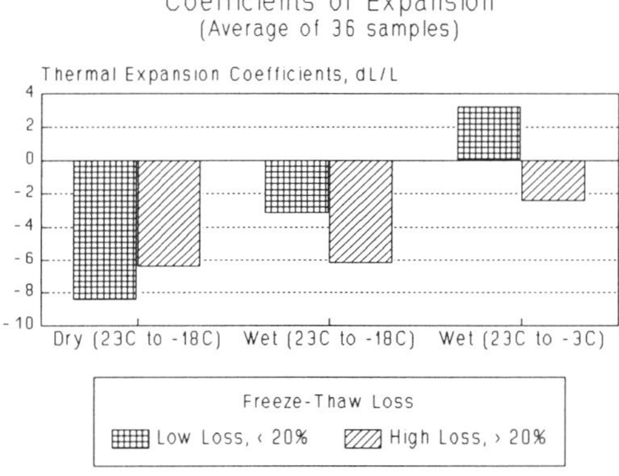

Figure 5 Thermal Expansion of Freeze-Thaw sensitive and non-sensitive aggregates in wet and dry state.

Water sorption properties were shown to influence FS and possibly AR in the previous section. NaCl, which is both an FS and an AR agent, is shown to influence the sorption properties. Figures 5 and 6 illustrate the length change observed upon immersing the dry cores of rocks in water as compared to immersion in 3% NaCl solution. Some rocks contract upon wetting, others expand. Generally, those rocks that expand tend to be the frost sensitive ones, and some also AR sensitive. Both isothermal contractions and expansions upon wetting are equivalent to thermal expansions of several tens of degrees Centigrade - i.e., equivalent of rapidly heating or cooling the rock over this temperature range. Wetting and drying can significantly stress the rock and the surrounding concrete paste. The stress can results in the microcracking of both aggregate and paste, which allows greater access of solutions, and accelerates the breakdown process.

The difference in isothermal expansion on immersion in salt solution is also illustrated in Figures 5 and 6. The same rocks, after immersion in water were immersed (after bench drying) in a 3% NaCl solution, bench dried, then immersed in water, and finally, without drying, placed back in the salt solution. latter step resulted in net shrinking of the specimen. The results illustrate that the

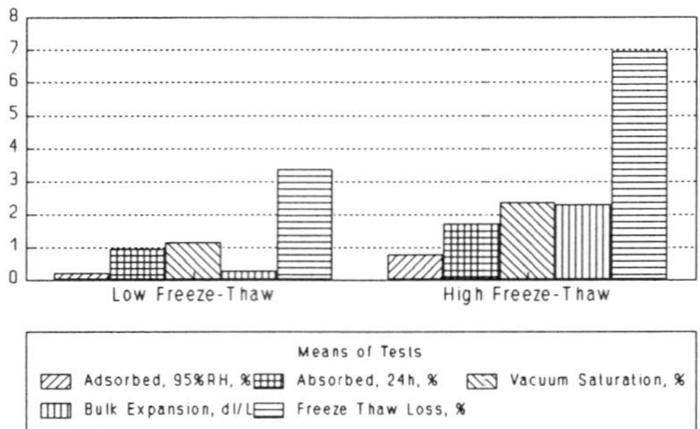

Figure 6 Physical durability properties: comparison of low and high freeze-thaw sensitive rocks.

nature of the ionic solution in the pores affects the isothermal expansion or contraction of the rock. This is to be expected, since it is basically the osmotic difference between the pore water and the surrounding environment that drives the expansion or contraction of the system upon wetting. The presence of de-icing salt solution changes not only the pore water chemistry, but the physical behaviour of the pore-water-rock-paste system. Both the physical changes (expansion and microcracking), and chemical changes (concentration of alkalies at the pore and microcrak walls) combine to produce the AR reaction.

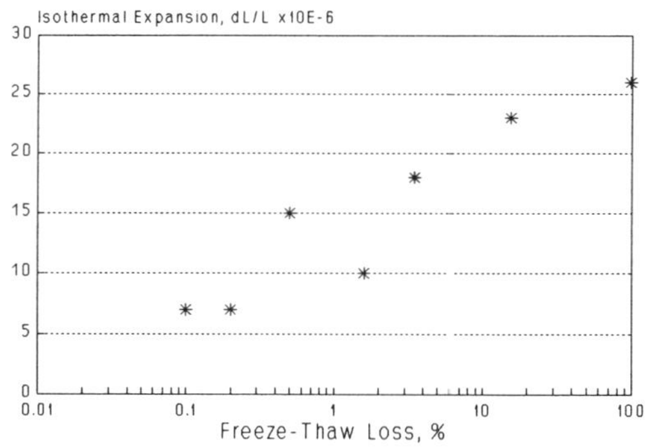

Figure 7 Log-Normal relationship between isothermal expansion and freeze-thaw loss aggregate.

Figure 7 gives the average thermal coefficients of expansion for frost sensitive and non-sensitive rocks. It is seen that the thermal coefficients differ, depending on the saturation state of the rock, and whether the rock is allowed to freeze. The saturated coefficients are lower, perhaps because the rock has already undergone volume change on immersion. Lowest coefficients are observed when saturated rock is cooled to -3°C. This is likely due to the volume change experienced on wetting. The bars in the diagram represent the average length changes. Thus, in any one group there are both contractions and expansions experienced, and the value of the bars in the figure is the numerical average. As a result, the diagram has to be interpreted with some caution. However, the trends observed are valid.

The summary of the physical durability properties are shown in Figure 8. Adsorption, absorption, vacuum saturation, bulk expansion and freeze thaw loss

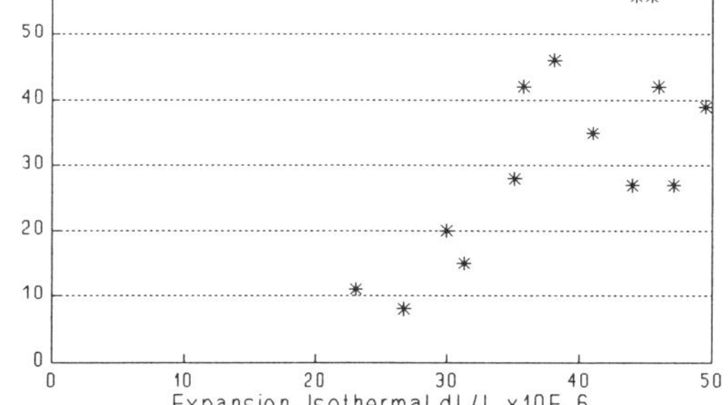

Figure 8 Comparison of expansion due to alkali reactivity and isothermal expansion on immersion.

values are given for frost sensitive and non-frost sensitive rocks. The bulk expansion is isothermal expansion determined on a crushed, bulk materials by immersion in water. It can be seen that all values are higher for the FS rocks compared to non-frost sensitive rock. The rocks in this group represent a variety of rock types used as aggregates in SW Ontario. Sorption and isothermal expansion correlate well with freeze-thaw loss as measured by direct freezing and thawing test on unconfined aggregate. The test uses 3% NaCl solution as an accelerator for deterioration.

That isothermal expansion is directly related to freeze-thaw resistance is shown in Figure 9. A strong log -normal relationship exists between the two parameters. A similar relationship can be expected between isothermal expansion and wetting-drying deterioration. The relationship suggests that even small isothermal expansions can result in significant physical deterioration, if

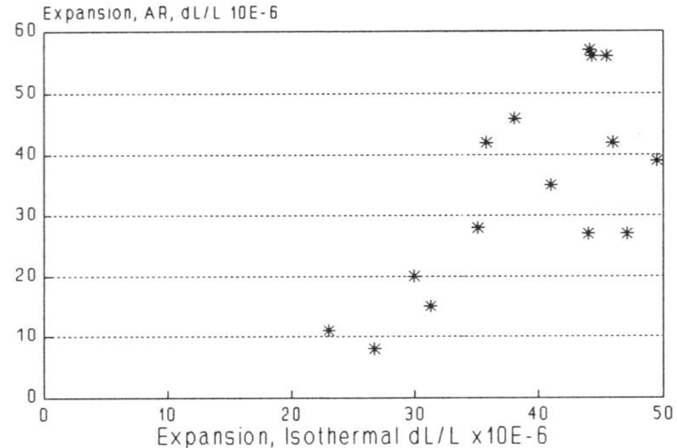

Figure 9 Comparison of expansion due to alkali reactivity and isothermal expansion on immersion.

sufficient number of either freezing and thawing or wetting and drying cycles are involved. Freezing can be considered as a drying cycle, since water is effectively removed from the system to form ice. The expansion due to ice formation may accelerate the process, but does not seem to be the dominant process.

The last figure in this paper, Figure 10, strongly supports the connection between the expansion due to alkali reactivity and physical expansion due to

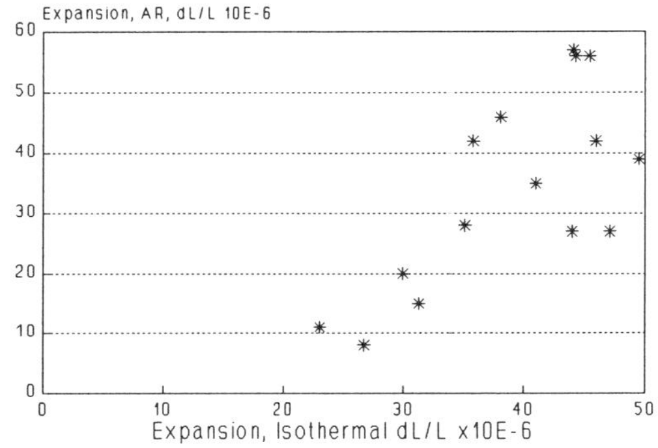

Figure 10 Comparison of expansion due to alkali reactivity and isothermal expansion on immersion.

wetting. The data was obtained on cores of concrete containing various proportions and sizes of reactive chert aggregate. Before exposing the cores to the rapid alkali reactivity test, the isothermal expansion upon wetting was determined. Then, the concrete was subjected to 21 days of exposure to hot (80°C) 1N NaOH solution. It is interesting to note that the overall magnitude of expansions noted were similar for both expansive processes. It may be possible that isothermal expansion provides the space into which silica gel can migrate and continue the expansion as alkali reactivity proceeds. Isothermal expansion is reversible (with some hysteresis) upon drying; however, AR products may prevent the contraction on drying, and the concrete stays expanded.

4 Discussion and Conclusions

The aggregates involved in alkali-silica and alkali-carbonate reaction are usually very fine grained, and therefore have very small pores. The fine grain results in high internal surface area. The high internal surface area is required for the AR reaction to proceed. High internal surface area was also found in aggregates that fail during freezing and thawing cycles, especially in presence of de-icing salts.

Expansion of the aggregates and concrete during AR and freezing and thawing process results in microcracking and eventual deterioration of the concrete. Adsorption of water and alkali ions on the internal surfaces of pores promote osmotic pressures that result in expansion and microcracking. Evidence was presented that aggregates that are frost sensitive have higher internal surface area, higher water adsorption and absorption, and expand more on wetting than do the more durable aggregates. Finally, it was shown that expansion of concrete containing alkali reactive aggregate when immersed in water or NaCl solution is similar in magnitude to expansion of the same concrete when subjected to accelerated alkaline reactive environment.

The results of this work indicate that isothermal expansion of aggregate, or concrete or mortar can be used as a quick screening method for detecting potential alkali reactivity.

The study was done on a limited number of alkali reactive aggregates, and must be expanded before the above observations can be universally applied to all AR materials.

6 Acknowledgement

The work described in this paper has been supported by grants from the National Research and Engineering Council of Canada (NSERC). Some of the experimental data has been gathered over the last several years by both undergraduate and graduate students working under author's direction. The financial support by NSERC and the diligent work by the students is gratefully acknowledged.

7 References

Dolar-Mantuani, L. (1983) **Handbook of Concrete Aggregates.** Noyes Publications, Park Ridge, N.J. USA.

Gillot, J.E. and Swenson, E.G. (1969) Mechanism of Alkali-Carbonate Rock Reaction. **Qart. J. Eng. Geology,** 2, pp.7-23.

Hobbs., D.W. (1978) Expansion of concrete due to the alkali-silica reaction: an explanation. **Mag. Concr. Res.**, 30, pp. 215-220.

Hobbs, D.W. (1988) **Alkali-silica reaction in concrete.** Thomas Telford Ltd., London, UK.

Hudec, P.P. (1987) Deterioration of aggregates - the underlying causes. **Concrete durability.** Katharine and Bryant Mather Int. Conference, Amer. Conc. Inst. SP-100, 2, pp. 1325-1342.

Hudec, P.P. (1989) Durability of rock as function of grains size, pore size, and rate of capillary absorption of water. **Jour. Materials in Civil Engg.** 1/1, pp. 3-9.

Larbi, J.A. and Hudec, P.P. (1990) A study of alkali-aggregate reaction in concrete: Measurement and Prevention. **Cement and Concrete Research,** vol. 20 (in press).

Swenson, E.G., and Gillot, J.E. (1960) Characteristics of Kingston Carbonate Rock Reaction. **Highway Res. Board, Bull. pp.** 275, 18-31.

Vivian, H.E. (1950) **Studies in cement-aggregate reaction. XV: The reaction product of alkalis and opal.** CSIRO, Melbourne, Australia, Bull. 259, pp.60-82.

9 DETERIORATION MECHANISMS IN SANDSTONE

C.L. SEARLS, S.E. THOMASEN
Wiss, Janney, Elstner Associates, Inc., Emeryville, California, USA

Abstract
Many sandstone buildings were constructed in the United States around the turn of the century. Sandstone was readily available and easily carved into ornamental shapes. It is also notorious for its tendency to decay. Sandstone buildings today suffer deterioration in the form of surface weathering, cracking, exfoliation of surface layers and spalling of large pieces of stone.

A local sandstone called "Colusa Sandstone" was used to construct many prominent San Francisco buildings. Deterioration on Colusa Sandstone buildings is most severe in areas experiencing frequent wet-dry cycles. On-site inspections of several buildings and a laboratory testing program revealed two ongoing deterioration mechanisms. First, the outer surface of the stone is hardened by the redeposition of binding material in the exterior crust as water evaporates, and a disaggregated zone, lacking in binding material, is formed behind this crust. Secondly, materials in the surface crust expand when in contact with water, resulting in separation of the crust from the underlying stone.

Deterioration of the stone results from natural weathering processes and intervention can attempt to slow down these processes and extend the service life of the building.
Keywords: Sandstone, Deterioration.

1 Introduction

Sandstone has been a popular building material in the United States, where it was widely used in the east for 19th century brownstone houses. On the west coast, many buildings were clad with local sandstone, especially following the 1906 San Francisco earthquake and fire. A great belt of sandstone and shale is found in California north of San Francisco. One particular type of local sandstone, "Colusa Sandstone" was used to clad such prominent San Francisco structures as the Westin St. Francis Hotel, the James Flood Building, the Ferry Building, and the Kohl Building. Today, approximately eighty years later, the sandstone cladding on these buildings is showing signs of severe deterioration.

2 Composition of Sandstones

Sandstones are sedimentary rocks composed of grains of sand and gravel held together by a cementing matrix. These rocks are formed by the accumulation and solidification of layers of sediments consolidated by the pressure of overlying beds. The properties of sandstones vary according to the mineral composition, texture and structure of grains, and the character of the cementing matrix. Sandstones are classified by the type of cementing matrix Weiss (1982)

(a) a calcareous sandstone has a cementing matrix of calcium carbonate
(b) a kaolintic sandstone has kaolinite cementing matrix
(c) an argillaceous sandstone has a clayey matrix
(d) a ferruginous sandstone has a high iron content in the matrix
(e) a siliceous sandstone has a cementing matrix of silica.

3 Evidence of Sandstone Decay

3.1 General Durability

Sandstone is easily worked, and it is notorious for its tendency to decay. Signs of deterioration sometimes occur within 20 years from the date of construction, Architectural Record (1986). A building may contain one or more of the following types of deterioration.

3.2 Surface Weathering

Weathering consists of disintegration of the surface of the stone. It is most noticeable as a loss of detail on intricately carved surfaces.

3.3 Exfoliation

A surface layer of stone, usually from 3 mm to 12 mm thick, separates in a sheet from the underlying substrate and falls off the building. (Fig. 1). Exfoliation can occur both on vertical wall surfaces and from the top of horizontal ledges. In stones having defined bedding planes, exfoliation often occurs along these planes.

Fig. 1. Exfoliation of stone at building parapet

3.4 Blind Exfoliation

A surface crust is formed but it remains loosely attached. Blind exfoliation can be detected by tapping the stone with a rubber mallet and listening for a hollow sound.

3.5 Cracking

Fractures in the stone may be found along edges of blocks or in a random pattern at corners. Cracks are sometimes seen around areas of blind exfoliation and may be a precursor of exfoliation or spalling.

3.6 Spalling

Loss of large pieces of stone material, not necessarily along bedding, planes. Spalling often occurs at edges and corners of projecting watertables, cornices and window sills. (Fig. 2).

4 Causes of Deterioration

4.1 Original Craftsmanship

The most common flaw in sandstone construction is "face bedding", the practice of setting stones with the bedding planes parallel to the face of the building. Layers of the stone can easily peel off the building along these planes of inherent weakness resulting in large scale exfoliation. "Edge-bedding", where planes are perpendicular to the wall can also result in deterioration. The proper method is to set the stones with the bedding planes horizontal, the same orientation as the stone has in the quarry.

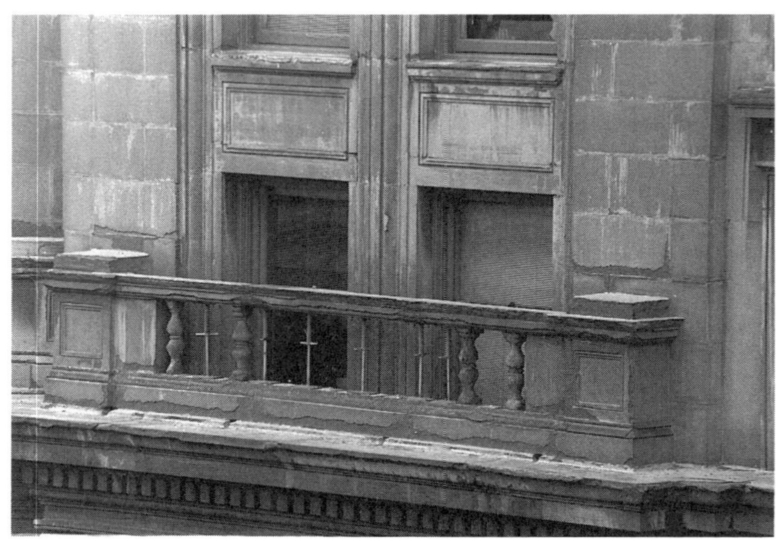

Fig. 2. Spalling of balusters was so severe that many were removed

Deterioration can result from selection of stones with inherent defects or inferior quality or from stones damaged by blasting during quarrying operations and by improper finishing techniques. Such defects may not become apparent until after the stone is installed.

4.2 Environmental Agents

Water is the most damaging environmental agent. It can enter the stone through the porous surface, or at open mortar joints, cracks and spalls. As deterioration progresses, increasing amounts of water can enter the stone, accelerating the process.

Expansion of freezing water can split the stone. Polluted water can dissolve the binder within the stone resulting in general disintegration. Water borne pollutants can react chemically with the binder, forming a hard surface that later spalls. Water can cause swelling of clays in the stone. It can transport salts which will crystallize and expand, causing cracking and spalling.

5 Case Study: Colusa Sandstone

5.1 Characteristics of Colusa Sandstone

Colusa sandstone was quarried from a rural area near the town of Sites in Colusa County, California, by the McGilvray Stone Company and the Colusa Sandstone Company. (Fig. 3). The stone was used along the California Coast and in the Hawaiian Islands. Physical and analytical tests conducted in 1890 described the stone as "a very superior building stone", with compressive strengths of 594 to 629 Kgf/cm^2. Aubury (1906).

Fig. 3. Location of Colusa Sandstone Quarry

Petrographic analyses of stone from examined buildings found that it is a graywacke, or low grade sandstone, with a siliceous - argillaceous matrix. The matrix contains the clays kaolinite, illite and montmorillinite and small amount of carbonate. The stone has fine grains of primarily quartz and feldspars, Erlin (1986), and it is moderately soft, quite porous and readily absorbs water. Exact composition of the stone varies from building to building, but all Colusa Sandstone examined shows similar modes of deterioration.

5.2 Observed Deterioration

The extent of deterioration depends on the location of the stone on the buildings, with the most severe decay occurring in areas experiencing frequent wet-dry cycles. These locations include the parapet and the windward elevations at upper stories. The deterioration is characterized by surface crust formation, surface cracking, shallow subflorescence, blind exfoliations, and shallow spalling.

5.3 Mechanism of the Deterioration

As water enters the stone it dissolves the elements in the binding matrix, and the acidic pollutants in the water alter the calcite portion of the binder to gypsum. (Fig. 4). As the water evaporates the gypsum is deposited at

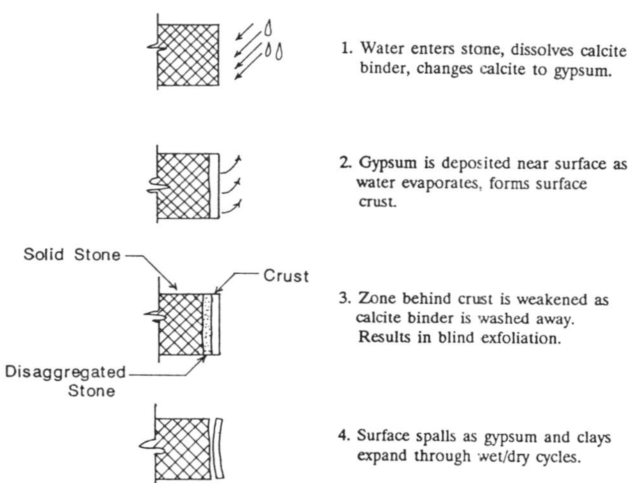

Fig. 4. Sandstone deterioration mechanism

the surface of the stone where a roughly 2 mm thick crust is formed during repeated wet-dry cycling. Inside the crust a layer of disaggregated stone is formed as the calcite binder is lost. While the crust is not as hard as the original stone, it is much harder than the layer of disaggregated stone.

After it is formed the surface crust will continue to expand in all directions due to two mechanisms: (1) the gypsum expands as it incorporates water in its crystal structure, and (2) the montmorillinite clay expands in the presence of water. At the flat ashlar blocks, where the surface expansion is restrained by adjacent blocks, cracks form around the perimeter parallel to the edges of the stone. (Figure 5).

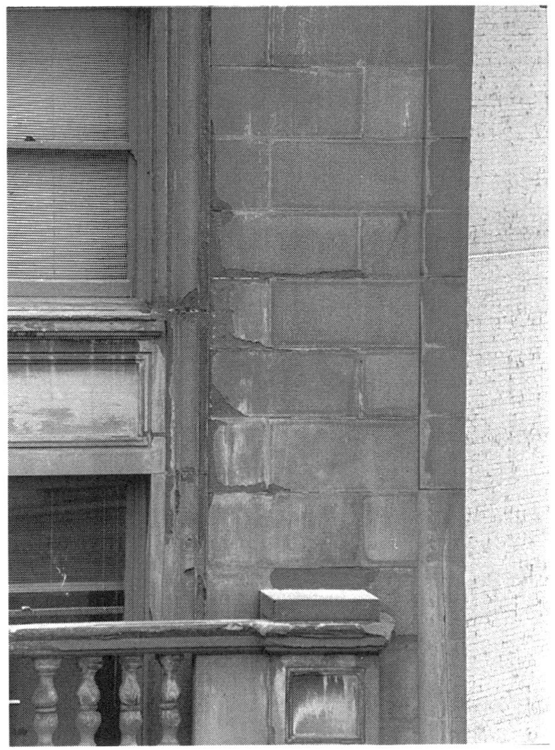

Fig. 5. Cracking and spalling at stone edges from expansion of surface crust

At masonry with projecting courses, longitudinal cracks develop, the projecting edges push against adjacent blocks and spall off. At stones with several sides exposed, such as column capitals or medallions, reticular (network) cracks occur.

Cracks allow more water to enter the stone and this accelerates the process of surface induration. The disaggregated zone becomes an area of blind exfoliation which increases in size until the crust falls away. (Fig. 6).

5.4 Preservation Treatments

The deterioration of sandstone is a natural weathering process. The deterioration mechanism of surface crust formation and exfoliation is accelerated by wet-dry cycles. Since the stone on a building is exposed to the elements, the deterioration process can never be completely stopped. The process can, however, be slowed and the service life of the stone extended by: 1) repairing already damaged areas, 2) reducing amount of moisture entering the stone by chemical surface treatments, flashings, and sloping horizontal surfaces, and 3) implementing a periodic maintenance program.

Fig. 6. Spalling of surface crust at ballustrade railing and watertable edge

6 Acknowledgements

The authors acknowledge the work of Richard Piper and his report, "Stone Deterioration of the Westin St. Francis Hotel, San Francisco, California," June 12, 1988 and Daniel Eilbeck for his work on the Westin St. Francis Hotel and James Flood Building.

7 References

1. Weiss, N., Teutonico, J., Matero, F., and Pepi, R, (1982) Sandstone Restoration Study. New York Landmarks Conservancy, New York, New York.
2. (May 1986) Saving Sandstone. Architectural Record.
3. Aubury, L., (1906) The Structural and Industrial Materials of California. California State Mining Bureau, San Francisco, California.
4. Erlin, Hime Associates, (1986) Petrographic Studies of Five "Sandstone" cores for Zephyr, Inc.

10 DURABILITY TESTS FOR NATURAL BUILDING STONE

K.D. ROSS, D. HART, R.N. BUTLIN
Building Research Establishment, Garston, UK

Abstract
This paper describes the main causes of stone decay, and the factors which influence stone durability. Various durability test methods are reviewed, with further discussion on the tests which are currently used to assess building stones in the U.K.

1 Introduction

BRE has been involved in testing the durability of building stone for a number of years. A series of BRE reports have been published concerning the availability and durability of different stone types used for building (Honeyborne(1982), Leary(1983,1986), Hart(1988)). These reports are intended primarily to enable architects and other stone specifiers to select the stone which is most suitable for the end use in mind. This suitability will depend on a number of factors including the durability of the stone, the degree of exposure in the building, the pollution levels in the environment, and the likelihood of frost. The reports also contain colour photographs of the stones, enabling the architect to select a stone which meets both his/her technical and aesthetic requirements. Known behaviour of different stone types in actual buildings is also taken into account.

2 Causes of stone decay

The main causes of stone decay are:

2.1 Frost

The mechanism of frost damage is complicated and not yet fully understood. There are three main theories of frost damage described in the literature; (a) the volumetric expansion of water on freezing; (b) the crystallisation theory, and (c) the hydraulic pressure theory.

The volumetric expansion theory attributes frost damage to the fact that when water freezes to form ice, the change is accompanied by an expansion of approximately 10% by volume. If this expansion is restricted by the pore walls, then a pressure will develop eventually leading to damage. The expansion theory was fairly widely accepted until the work of Taber (1929,1930). Taber worked on frost heave in soils and demonstrated in the laboratory that heave was due to the formation of an ice lens. He also showed that the ice lens could be made to grow under pressure, and that water could be drawn through the unfrozen soil from a reservoir to allow continued growth of the lens. In his 1929 paper he says:

> 'Surface heaving appreciably in excess of that which would result from the expansion in volume of the water contained in the soil means that additional material must have been introduced as a result of the freezing.'

This has important implications as far as frost damage to stone is concerned. Provided that the water in a stone is in a continuous phase, then it is able to flow to a single ice crystal growing elsewhere in the material. Also, because fluid must flow through quite small channels in the stone to get to the growing crystal, this mechanism would require a relatively slow growth rate and so a relatively slow freezing rate.

The idea that frost damage is due to a crystallisation process, rather than simple expansion on freezing, is strongly supported by the fact that benzene and nitro-benzene can both produce damage on freezing, despite the fact that they both contract on freezing (Taber 1930, Honeyborne and Harris 1958).

The third mechanism of frost damage (ie the hydraulic pressure theory) does rely on the volume expansion associated with the freezing of water, but in this case the ice is not the damaging agent. Two mechanisms are described in the literature. As ice grows in a pore, because of the expansion due to freezing, unfrozen water must be expelled from the pore. Powers (1945) describes a mechanism which says that if the water is expelled into fine pores at a high rate, then there will be considerable resistance to flow leading to a hydraulic pressure in the fluid which could cause damage to porous materials. This mechanism is also described by Chatterji and Christensen (1979) who suggest a second mechanism for materials totally saturated with water. In this second case if the water cannot escape (e.g. if the entire outer surface of the stone is sealed with ice) then a hydrostatic pressure will develop causing eventual damage. The difference between the two mechanisms is that the first one requires a high rate of freezing, whereas the second does not.

It is unlikely that any single mechanism of frost damage will completely explain observations of damage in buildings. There appears to be two modes of failure in practice; the first is characterised by thin sheets of stone blistering off the surface, and the second by fractures which go right through a large block. These are illustrated in plates 1 and 2 respectively. In practice, damage can probably occur by any of the mechanisms described above, but one mechanism being dominant depending on the prevailing freezing conditions. Thus under slow cooling conditions the crystallisation model is probably the dominant one leading to the decay shown in plate 1, and under more rapid cooling conditions the hydraulic mechanisms will dominate, leading to the decay pattern shown in plate 2.

Plate 1. Frost damage attributed to the formation of ice lenses.

Plate 2. Frost damage attributed to the entrapment mechanism.

2.2 Salt crystallisation

This is the main cause of stone decay in the U.K. and occurs when stone becomes contaminated with salts. Reasons for salt contamination include exposure of stone in coastal or polluted environments, and inadequate care in the use of cleaning preparations which contain, or can react with the stone to produce soluble salts; this may cause subsequent salt crystallisation if the solution or salts are not properly removed.

For limestones the most common occurrence of salt crystallisation is in polluted environments. Sulphur oxides and water in the atmosphere react with carbonate to form sulphates. For example, the principal mineral in limestone is calcite, which leads to gypsum formation:

$$CaCO_3 + SO_3 + 2H_2O + CO_2 = CaSO_4.2H_2O + CO_2$$

The gypsum can crystallise in the near surface pores to form a crust at the surface, or react and dissolve in rain-water. Thus black gypsum crusts, often incorporating soot and particulate matter, tend to form in sheltered areas which are not regularly washed by rain-water. Blistering and flaking occurs in these areas when the growing crystals cannot be accommodated by the available pore space, and the restraining force of the pore walls is not large enough to prevent further crystal growth. Areas washed by rain-water tend to suffer gradual erosion.

There are a number of possible mechanisms whereby salt can cause damage to stones. Crystallisation of salts from a saturated solution (which is normally accompanied by an increase in volume (Schaffer 1932)) can cause damage by crystals forcing grains of stone apart; this theory is supported by the fact that crystals can grow under pressure (Correns 1945, Winkler and Singer 1972). There are two ways whereby salt can be induced to crystallise from solution. Because the solubility of most salts increases with temperature, a drop in temperature can cause a solution to become saturated and so lead to nucleation of salt crystals. Evaporation of water can give the same result.

Many salts can exist in a number of hydration states. The change from one hydration state to another is usually accompanied by a volume change leading to the possibility of damage. Bonnell and Nottage (1939) have demonstrated that considerable forces can be generated in porous materials when salts hydrate.

It is possible for the same salt to cause damage by either a crystallisation mechanism or by a hydration mechanism. Sperling and Cooke (1980,1980a) carried out theoretical and experimental work on salt weathering in a climatic chamber using sodium sulphate as the weathering agent. Sodium sulphate has a stable deca-hydrate, a stable anhydrate and a meta-stable hepta-hydrate. By altering the conditions in the cabinet they were able to study the effects of hydration and crystallisation separately and showed that hydration played a somewhat minor role in weathering compared to crystallisation.

Sodium sulphate is also used in the crystallisation tests carried out at the BRE (see later). In this test, samples are alternately soaked in sodium sulphate solution and dried at 105°C. The transition temperature for the reaction

$$NaSO_4.10H_2O = Na_2SO_4 + 10H_2O$$

is 32.4°C. Thus, because the drying temperature is 105°C, the anhydrate will be formed in the drying part of the cycle, which is the time when crystallisation must occur. The Authors have never observed damage to the samples during drying period - damage always occurs during the soaking part of the cycle. This indicates that in the crystallisation test it is the hydration of the anhydrate to the deca-hydrate that causes damage. This is substantiated by Price (1978), who reported the results of some work carried out in the 1930s by Webster and Warlow. They carried out the crystallisation test in the standard way, but varied the temperature of the soaking part of the cycle. They found that as the soaking temperature increased the amount of damage decreased, and at temperatures near or above 32.4°C, no damage was observed (see figure 1).

2.3 Acid deposition

As described above, sulphur oxides in the atmosphere in conjunction with rain-water can lead to salt crystallisation in limestones. However, gypsum is more soluble than calcite and tends to dissolve at the surface of rainwashed areas, rather than being deposited in the pore structure, leading to a gradual erosion of the limestone surface. Although this effect is undesirable, it is not always considered a severe problem. The rates of erosion are relatively small - Trudgill et al (1989) measured the rate of erosion at St Paul's Cathedral to be 0.078 mm/annum. This mechanism is however more serious with calcareous sandstones. Because the sandstone grains are cemented together with calcite, many grains can be dislodged by the dissolution of a small amount of calcite, leading to rapid decay.

Sulphur oxides can also have a deleterious effect on some slates. If they contain calcite, gypsum can form as in the case of limestones and calcareous sandstones, but some contain iron rich chlorites and sulphides which can react to form hydrated sulphates of iron and aluminium (Shayan and Lancucki 1987), the increase in volume leading to disruption of the slate layers.

3 Stone durability

A durable building stone is one which resists the weathering elements in the atmosphere, without suffering decay in some way. Some aspects of durability are within our control; for example new stone in a building is given the best chance of long life if it is correctly bedded (ie with the sedimentary layers of stone lying perpendicular to the face of the building), and if the detailing and work standards are of a high quality. However, some building stones are intrinsically more susceptible to weathering than others. Their resistance to weathering depends mainly on the following factors:

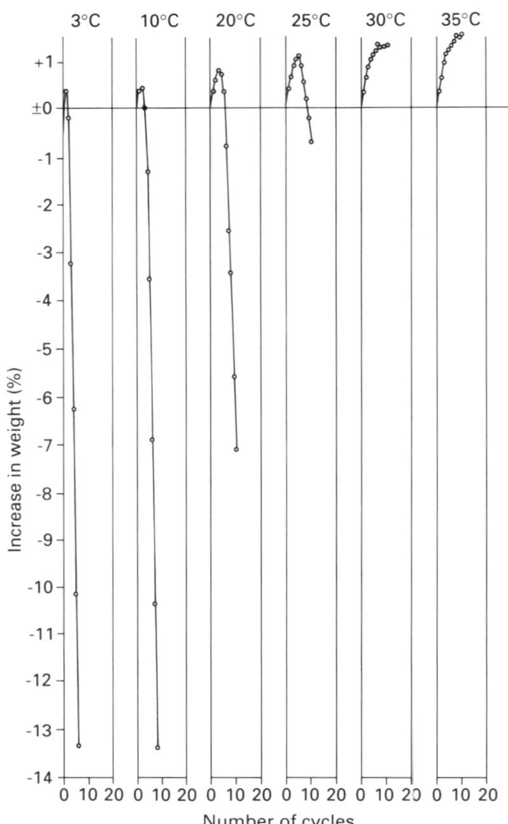

Figure 1 Influence of variation of soaking temperature on crystallisation loss

a) Mineralogy

Building stones which are composed principally of stable silicate minerals are generally highly resistant to weathering, for example sandstones (quartz), slates (mica, quartz), granites (quartz, feldspar). However, stones which contain carbonate minerals are more susceptible to weathering as described above for calcite bearing rocks. Dolomite ($CaMg(CO_3)_2$) is also a carbonate mineral, and the primary mineral in the British magnesian limestones; it can also be a cementing material in sandstone.

In polluted environments, rocks containing dolomite tend to weather more rapidly than those which only contain calcite. This is due to the difference in the decay products. Calcite bearing rocks produce gypsum which is only moderately soluble, and only has one stable hydrate at ambient conditions. Dolomite on the other hand produces not only gypsum but also magnesium sulphate which is a much more damaging salt due to its higher solubility and different hydrates. Magnesium sulphate can exist as the anhydrate, mono-hydrate or hepta-hydrate under ambient conditions.

Pyrite and certain clay minerals can also affect the durability of slates and shales in polluted environments or in zones of high exposure, as already discussed.

b) Pore structure

Durability is not only dependent on the mineralogy of rocks; durability can vary markedly within a single type of rock of almost identical chemical and mineralogical composition. This is particularly true in the case of limestones. Previous work at BRE (Schaffer 1932) has shown that the likely cause of this variability is the pore structure. Limestones that have a very open pore structure, ie the pores are relatively large, tend to be more durable than limestones which have the same volume of pore space made up of finer pores. This is because finer pores have a greater suction and tend to achieve a higher saturation than stones with a more coarse pore structure. This phenomenon is used to good effect in the measurement of saturation coefficient as an indicator of durability (see below).

4 Durability testing

Methods for testing the durability of stone are based on two principles. They either:

a) measure the stone's properties upon which durability is dependent. Tests in this category include the Microporosity test (Honeyborne and Harris 1958, Croney, Coleman and Bridge 1952), d10 - a measure of the proportion of small pores calculated from mercury intrusion data - (Centre Scientifique et technique de la Construction 1970), saturation coefficient and capillarity coefficients. The latter two tests are the ones most often used and they are described in more detail below.

or

b) monitor the stone's performance in conditions which simulate the causes of decay. Acid resistance, freeze/thaw, and crystallisation tests fall into this category.

4.1 Saturation coefficient

This is defined as the amount of pore space that will be filled by soaking in water for a given period of time, expressed as a percentage of the total pore space. In the U.K., 24 hours is specified as the soaking period, but 48 hours is sometimes used on the Continent. Despite the large difference in soaking time, the value of the coefficients obtained do not differ greatly because most of the water is taken up in the first 24 hours.

The saturation coefficient is also known as the Hirschwald coefficient after the man who developed it (Hirschwald 1912). It was first developed as a rapid frost resistance test on the theory that because water expands by approximately 10% on freezing, a stone must have at least 10% of its pore space empty to be able to accommodate the expansion. This means that a saturation coefficient greater than 0.9 would indicate that a stone is susceptible to frost. In practice an upper value of 0.8 indicates susceptibility to frost, and a value below 0.6 indicates resistance, but one cannot say with absolute certainty in either case. Between these values it i[s] not possible to say with any confidence what the resistance is likely to be, and since the saturation coefficients of many stones lie between these two values, the test is of limited value.

4.2 Capillarity tests

These are basically water absorption tests which measure the rate of uptake o[f] water through one face of the stone, and are used mainly in France and Belgium. The test is specified in national standards for both countries (NF 10 502, NBN B 05 201) and involves measuring the rate of uptake of water through the base of a prism. A graph is plotted of moisture content against the square root of time (in minutes). In the French test the moisture conten[t] is expressed as 100M/a where M is the weight increase in grams and 'a' is the area of the base of the sample (in square centimetres), and their capillarity coefficient is the slope of the graph. The value of the capillarity coefficient is not used by itself as a 'durability factor', but is used in conjunction with figure 2 and other information (see table 1) to decide where in a building the stone may be used.

Table 1. Use of French capillarity coefficient and other properties.

Characteristics required by stone	Positions where stone can be used
Zone 1 of figure 2; capillarity <2; lose<32 units in the standard wear test	Exterior paving, dock linings, bridge piers
Zone 1 or 2 of figure 2; capillarity < 5	Base courses, balconies
Zone 1, 2 or 3 of figure 2; capillarity < 15	Cornices, string courses, sills splash courses
Porosity < 47%	Elevations under projections

In the Belgian test the moisture content is expressed as the percentage saturation calculated from the expression

$$\%\ \text{saturation}\ (S) = 100(Wt - Wo)/(Ws - Wo)$$

where Wt is the weight after a time t, Wo is the dry weight, and Ws is the weight after vacuum saturation with water. A typical plot of S vs $\sqrt{t}$ is giv[en] in figure 3. St is the percentage saturation of the stone at the point of inflection, and $\alpha 1$ and $\alpha 2$ are the gradients of the two parts of the graph. Sometimes no point of inflection occurs during the time of the test and in this case St is taken to be the percentage saturation at the point where $\sqrt{t}$ = 100.

Two capillarity coefficients are calculated from this graph, namely G and GC. G is defined as St + 178$\alpha 2$ - 81. If there is no point of inflection G i[s] taken as St + 178$\alpha 1$ - 81. Stones are considered frost resistant if G < 0.

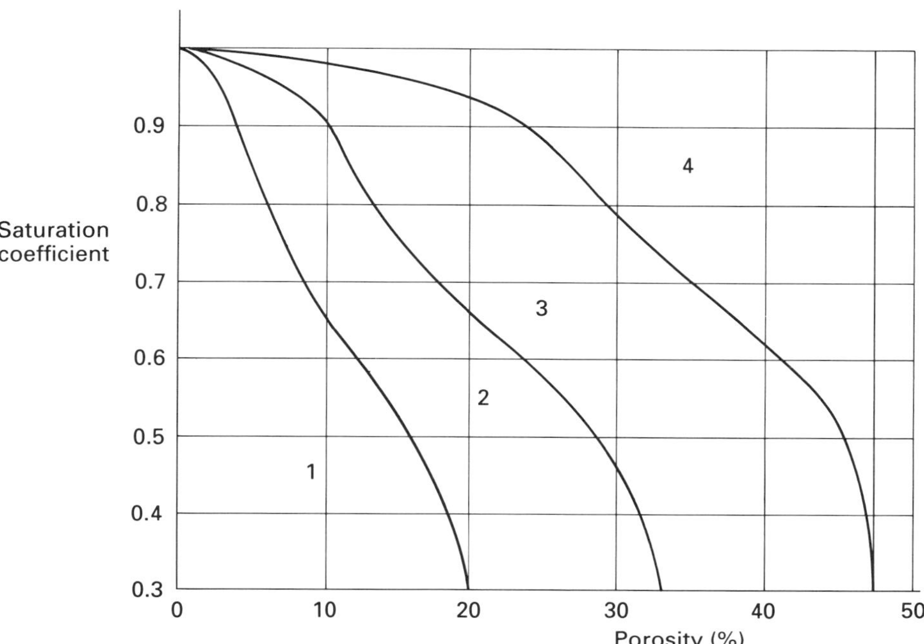

Figure 2 Definition of building zones in table 1

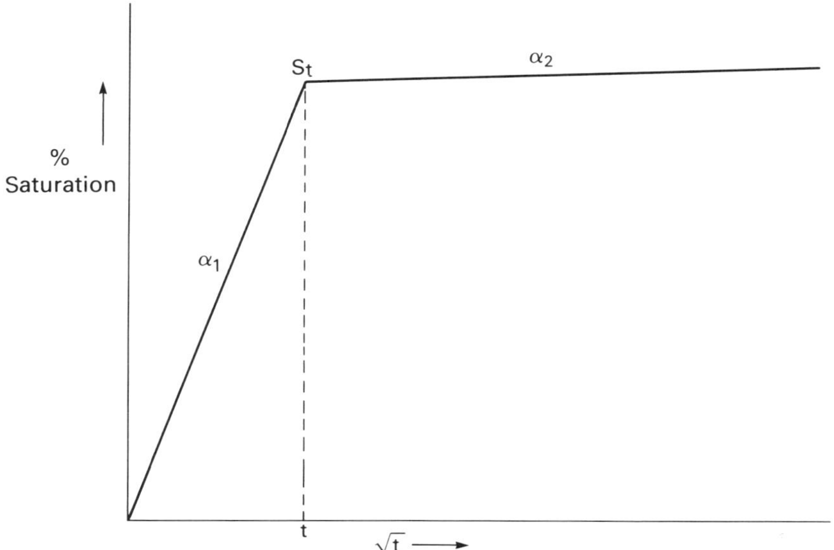

Figure 3 Typical form of capillarity curve

This obviously implies that the more negative the value of G the more durable the stone will be.

GC is defined as $0.203St - 14.53 - 0.31\alpha l$, or $21.47\alpha l - 6.35$ for stones with no point of inflection. Stones are graded into five classes according t the value of GC (see table 2). Again, the more negative the value of GC the

Table 2. Classification of stones according to Belgian Coefficient GC.

GC < -2.5	Class D - stone survives four winters in a tray of sand or at least two winters at Bertrix
GC > -2.5 to -0.95	Class C - stone survives four winters as a paving stone
GC > 0.95 to 0	Class B - stone survives free standing for four winters
GC > 0 to 4.5	Class A - stone survives in a vertical wall for four winters
GC > 4.5	Class 0 - stone does not survive in a vertical wall

more durable the stone.

Leary (1981) carried out capillarity tests on a number of limestones and compared the results with those from the BRE crystallisation test. She found that although a correlation exists between crystallisation loss and capillarity coefficients, the capillarity coefficients are far less sensitive to durability than crystallisation loss.

4.3 Acid resistance tests

These fall into two types - those in which samples are immersed in a solution of sulphuric acid, and those in which samples are exposed to an atmosphere of sulphur dioxide. Immersion tests are used in the U.K. for slates (BS680, BS5642, BS743) and sandstones (Ross and Butlin 1989), whereas vapour tests ar used in Belgium (STS 34 03.06) and Germany (DIN 52 206) for slate. The tests are intended to identify those materials which would suffer decay in a polluted environment. In practice this means those with a substantial amount of Calcite in the matrix. The French do not appear to have an acid resistanc test, but measure the calcium carbonate content directly by reacting a powdered sample with hydrochloric acid and calculating the carbonate content from the amount of gas given off. This approach has two disadvantages compared to a direct test - there are other minerals which react with hydrochloric acid to generate a gas (e.g. pyrite), and there are some cases where the calcite may not be in a particularly damaging form (this is the cas in some slates).

4.4 Freeze/thaw tests

Freeze/thaw tests are probably the most widely used durability tests in the world - and probably the most diverse. On the face of it, freeze/thaw testi would seem to be a very logical thing to do, namely to subject samples to th very action which causes stone to decay in nature.

Frost tests usually involve taking small samples of the material, soaking in water for a given length of time and subjecting the sample to cycles of freezing in air and thawing in air or water. Things get rather complicated when one tries to develop a realistic test however. Thomas (1938) found that the severity of a freeze/thaw test was very susceptible to the test conditions and the initial moisture content of the stone. This means that a test which is used in a particular country or region has probably been developed to give reasonable results in that area, and so cannot be used elsewhere where different conditions prevail.

Also in the 1930s and 1940s much work on the frost resistance of bricks was carried out at the BRE. From this work - which is fully described in the Transactions of the British Ceramic Society (Butterworth 1934, Butterworth 1964, Llewellyn and Butterworth 1964, Butterworth 1964a, Butterworth and Baldwin 1964, Watson 1964a) - it was concluded that, in the U.K. at least, the only reliable frost tests were natural exposure tests.

A different approach to frost testing is being developed by the British Ceramic Research Association (West, Ford and Peake 1984). In this new work, tests are carried out on panels of brickwork which are frozen from one face only. This is obviously more realistic than a small sample as far as plain walling is concerned, but does not reflect the more severe exposure of a coping.

4.5 Crystallisation tests

It was recognised by Brard (see de Thury et al 1828) that salt and frost damage were very similar in nature, and he suggested that a crystallisation test would be a good, rapid method for assessing the frost resistance of stone. Crystallisation tests are still in use today to assess the durability of porous building materials, particularly building stone and aggregates for concrete. They do not differ fundamentally from the test used by Brard in that they all involve subjecting samples to cycles of soaking in the salt solution and drying. The BRE have developed a crystallisation test, which is in general use in the U.K., to determine the durability of limestones for building. Relative durability is based on the weight loss suffered during the test, which is determined partly by the conditions of the test. Price(1978) carried out some work on the effect of altering the conditions of the test and concluded that the severity of the test was increased by the following factors:

a) an increase in the solution concentration;
b) a decrease in the soaking temperature;
c) an increase in the drying rate.

Although these parameters affected the severity of the test, they did not affect the ranking order observed when a number of different stones were tested together. The ranking order was only changed when the samples were not completely dried between cycles. For these reasons the BRE test includes reference samples (ie stones of known relative performance) in the test as an internal check to ensure sensible results.

5 Stone testing in the U.K.

Current tests in use for building stone are fully described in the BRE report 'Durability tests for building stone' (Ross and Butlin 1989). The tests whic are normally carried out are summarised in table 3 according to stone type an end use.

5.1 Testing limestones

Normally only the BRE crystallisation test is needed for limestones; if however, one needs to match the physical properties of one stone to those of another, then saturation coefficient and porosity can also be measured. As indicated above, saturation coefficient can be used to give a rapid indicatic of durability in some cases, but the crystallisation test should always be preferred.

5.2 Testing sandstones

For general building work there is no need to test sandstones unless they are to be used in a polluted environment. If this is the case, an acid immersion test can be carried out. Saturation coefficient and porosity can be carried out to match the properties of sandstones, but unlike the case for limestones the saturation coefficient should not be used as a rapid durability test for sandstones.

If the sandstone is to be used in a very severe environment (e.g. some coastal applications such as sea defence walls) or particularly long life is required, then a crystallisation test can be used. It should be noted, however, that the crystallisation test for sandstones is a much more severe specification than that used for limestones. For this reason results from crystallisation tests on limestones and sandstones should never be compared : one is attempting to choose between the two types of stone.

Table 3 Summary of tests needed for different building stones

Type of stone	End use	Crystallisation test	Saturation coefficient	Acid immersion	Porosity	Wet/ dry	Water absorption
Limestone	General	●	★	—	★	—	—
Sandstone	General	★	★	★	★	—	—
Sandstone	Severe exposure	●	—	●	—	—	—
Slate	Roofing	—	—	★	—	●	●
Slate	Copings	—	—	●	—	●	—
Slate	Damp course	—	—	●	—	●	—

● These tests should always be carried out for the stone in question

★ These tests may be required for certain applications of the stone

Note The test conditions may vary for different stones

5.3 Testing slates

Slate is used in a number of applications in building; those which are covered by standard tests are coping units, slate for damp-proof courses and roofing slates. For all these slates a wetting and drying test is usually carried out. This test identifies those slates which contain certain damaging minerals such as pyrite. An acid immersion test should be carried out on all three types of slate (although the strength of the acid varies according to the type of slate the only exception being roofing slate which is to be used in an unpolluted environment. Slate for roofing should also be subjected to a water absorption test.

5.4 Testing granites and marbles

There are no standard durability tests for granites and marbles.

6 Discussion

The tests described in this paper are mainly those which are used at present in the U.K., as well as some continental tests. The bulk of testing in the U.K. is carried out using 'direct' accelerated weathering tests such as the microporosity test. Testing on the continent relies less heavily on direct tests and more on indirect tests such as capillarity and saturation coefficient.

With the harmonisation of European standards in 1992 the trend will probably be towards more indirect testing and an increased number of tests being required to satisfy the needs of different construction techniques. For example, capillarity tests, which can be used to calculate the thickness a wall needs to be resist rain penetration, are likely to be carried out more often because cavity construction is less common on the Continent.

7 References

Bonnell, D.G.R. and Nottage, M.E. (1939) Studies in porous materials with special reference to building materials. I. The crystallisation of salts in porous materials. **J. Soc. of the Chem. Industry**, 58, 16-21.

BS 680: Part 2: 1971. **Specification for roofing slates.** British Standards Institution, London.

BS 5642. **Sills and Copings.** British Standards Institution, London. Part 1: 1978 **Specification for window sills of precast concrete, cast stone, clayware, slate and natural stone.** Part 2: 1983 **Specification for copings of precast concrete, cast stone, clayware, slate and natural stone.**

BS 743: 1970 **Specification for materials for damp-proof courses.** British Standards Institution, London.

Butterworth, B. (1934). **Trans. Brit. Ceram. Soc.**, 33, 495 et seq.

Butterworth, B. (1964) Laboratory tests and the durability of bricks, II. The recording, comparison and use of outdoor exposure tests. **Trans. Brit. Ceram. Soc.**, 63,(11), 615-628.

Butterworth, B. (1964a) Laboratory tests and the durability of bricks, IV. Th indirect appraisal of durability. **Trans. Brit. Ceram. Soc.** 63, (11), 639-646.

Butterworth, B. and Baldwin, L.W. (1964) Laboratory tests and the durability of bricks, V. The indirect appraisal of durability (continued). **Trans. Brit. Ceram. Soc.** 63, (11), 647-661.

Centre Scientifique et Technique de la Construction. (1970) **Pierres blanches naturelles.** Note d'Information Technique 80. CSTC, rue de la Violette 5, 1000 Brussels, Belgium.

Correns, C.W. (1949) Growth and dissolution of crystals under linear pressure Discussions of the Faraday Society, 5, **Crystal Growth.** Butterworths, 267-271.

Chatterji, S., and Christensen, P. (1979). A mechanism of breakdown of limestone nodules in a freeze/thaw environment. **Cement and Concrete Research,** 9, 741-746.

Croney, D., Coleman, J.D. and Bridge, P.M. (1952) The suction of moisture hel in soil and other porous materials. **Road Research Technical Paper No. 24,** HMSO, London.

de Thury, H., et al. (1828) On the method proposed by Mr. Brard for the immediate detection of stones unable to resist the action of frost. **Annal de Chemie et de Physique,** 38, 160-192.

DIN 52 206, March 1975. **Testing of roofing slates.** Acid resistance test.

Hart, D. (1988) **The building magnesian limestones of the British Isles.** HMSO, London.

Hirschwald, J. (1912) **Handbuch der bautechnischen gesteinprufung.** Borntraege Berlin.

Honeyborne, D.B. (1982) **The building limestones of France.** HMSO, London.

Honeyborne, D.B., and Harris, P.B. (1958) The structure of porous building stone and its relation to weathering behaviour. **Proc. 10th Symposium of t Colston Research Society, Bristol 24-27 March 1958,** 343-359.

Leary, E. (1981) A preliminary assessment of capillarity tests as indicators of the durability of British Limestones. **Proc. Int. Symp. on Stone Conservation, Bologna.**

Leary, E. (1983) **The building limestones of the British Isles.** HMSO, London.

Leary, E. (1986) **The building sandstones of the British Isles.** HMSO, London.

Llewellyn, H.M., and Butterworth, B. (1964) Laboratory tests and the durability of bricks, III. Some conventional laboratory freezing tests. **Trans. Brit. Ceram. Soc.,** 63, (11), 629-637.

Norme Belge (1976) **Resistance of materials to freezing - Water absorption by capillarity.** NBN B 05 201 IBN.

Norme Francaise (1973) **Quarry products - Limestones - Measurement of water absorption by capillarity.** NF B 10 502 AFNOR.

Powers, T.C. (1945) A working hypothesis for further studies of frost resistance of concrete. **J. of the American Concrete Institute**, 16, no. 4, 245-272.

Price, C.A. (1978) The use of the sodium sulphate crystallisation test for determining the weathering resistance of untreated stone. Presented at the Rilem/UNESCO Symposium, Paris.

Ross, K.D. and Butlin, R.N. (1989) **Durability tests for building stone.** BRE, Garston.

Schaffer, R J. (1932) **The weathering of natural building stones.** Building Research Establishment Special Report No 18, Garston, BRE, (Facsimile reprint 1985).

Shayan, A and Lancucki, C J. (1987) Deterioration of slate tiles containing iron sulphides. **Fourth International Conference of Building Materials and Components, Singapore.**

Sperling, C.B.H. and Cooke, R.U. (1980) Salt weathering in arid environments I. Theoretical considerations. **Papers in Geography** No.8, University of London, Bedford College.

Sperling, C.B.H. and Cooke, R.U. (1980a) Salt weathering in arid environments II. Laboratory studies. **Papers in Geography** No.9, University of London, Bedford College.

STS 34.(1987) Couvertures de batiments. 03.06 Ardoises naturelles. Brussels.

Taber, S. (1929) Frost heaving. **Journal of Geology**, 37, 428-461.

Taber, S. (1930) The mechanics of frost heaving. **Journal of Geology**, 38, 303-317.

Thomas, K N. (1938) **Experiments on the freezing of certain building materials.** Building Research Technical Paper No 17, HMSO, London.

Trudgill, S.T., Viles, H.A., Inkpen, R.J., and Cooke, R.U. (1989) Remeasurement of weathering rates, St. Paul's Cathedral, London. Earth **Surface Processes and Landforms**, 14, 175-196.

Watson, A. (1964) Laboratory tests and the durability of bricks, VI. The mechanism of frost action on bricks. **Trans. Brit. Ceram. Soc.**, 63, (11), 663-680.

Watson, A. (1964a) Laboratory tests and the durability of bricks, VII. Frost dilatometry as a routine test. **Trans. Brit. Ceram. Soc.**, 63, (11), 681-695.

Winkler, E.M. and Singer, P.C. (1972) Crystallisation pressure of salts in stone and concrete. **Bul. Geol Soc. Am.**, 83, 3509-3513.

11 SUGGESTIONS FOR A LOGICALLY-CONSISTENT STRUCTURE FOR SERVICE LIFE PREDICTION STANDARDS

G. FROHNSDORFF, L. MASTERS
National Insitute of Standards and Technology, Gaithersburg, Maryland, USA

Abstract
Ability to predict the service life of building materials, components, and systems is needed to improve the selection process. Evaluation of durability using existing standards does not give adequate service life information. Because service life prediction is more complex than current durability evaluations, its standardization will require a new body of standards to be put in place. The standards must define a general methodology, and essential components of the methodology. These are environmental characterization, characterization of the item whose service life is to be predicted, identification of the mechanisms and kinetics of the degradation processes, development of mathematical models of degradation, application of the models in service life prediction, and reporting of the results. It is proposed that the needed standards must comprise a hierarchy with the highest level being the general methodology, the second level defining the essential components of the methodology, and the third and lower levels describing the application of the generic standards to specific materials, components, or systems. The development of the proposed hierarchy will require a well-coordinated activity which cuts across the interests of many different standards committees.

1 Introduction

We owe much to Sereda [Sereda and Litvan (1980)] for leadership in establishing the International Conferences on Durability of Building Materials and Components. His hope was that the conferences would promote the development of a literature covering the durability of non-metallic building materials analogous to that which the corrosion literature provides for metals. The success of the conferences and the usefulness of their published proceedings suggests that Sereda's hopes will become reality. But even more must be done if durability knowledge is to be used widely and effectively in selecting building materials and components.

In view of the important role of standards in construction decisions, and the rudimentary nature of durability standards, it is time to decide on the preferred structure of durability standards, particularly standards for prediction of service life. This should not be thought of as a dull or rather routine exercise. In fact, the necessity of coupling a wide range of scientific and engineering disciplines to the standards process will challenge the system and its participants. It will require knowledge of materials science and engineering, environmental characterization, engineering statistics, and techniques of mathematical modeling. Success in establishing a sound, standardized basis for service life predictions will be important in aiding conservation of the earth's resources and improving the competitiveness of organizations which use service life prediction most effectively. It will also provide a needed tool for use by product approval systems in evaluating new products.

As pointed out earlier [Frohnsdorff and Masters (1980)], because durability is a vague concept, it is more fruitful to think in terms of service life of a material or component under specific conditions. Prediction of service life [Masters (1985), Sjostrom (1985)] is knowledge intensive [Fagerlund (1985)] and will undoubtedly improve with growth in the knowledge base and ability to handle knowledge with computers. Possibilities for using computers in developing an integrated knowledge base for concrete technology, including service life prediction, have been discussed by one of us [Frohnsdorff, Clifton, Jennings, Brown, Struble, and Pommersheim (1988)]. The integrated knowledge system envisioned would consist of interfaced databases, image bases, mathematical models, and expert systems, with access to the system being provided by computer networks [Frohnsdorff (1989)]. The possibility of developing such systems for all building materials gives hope for significant improvements in service life prediction.

The new ways of handling knowledge brought about by computers will make possible great changes in the nature of standards. It is not just that standards can be distributed in electronic form, but that knowledge stored in computers, and ways of using it, will be able to be standardized [Frohnsdorff (1989)]. Whereas, in the past, standard specifications had to be simple because of the difficulty of passing on the available knowledge in a practical way, it is now possible to think of an intermediate product specification in terms of, for example, a complex mathematical model which calculates specification limits for use of the product in different end product applications. This possibility must be borne in mind where matters of service life are concerned. It may, in fact, be the essential element in making service life prediction viable. In a related matter, the growing recognition of the need for national and international product acceptance systems [Gross (1989)] will increase the need for durability criteria based on predicted service lifes under expected conditions of use.

In selecting a material, component, or system for almost any application, whether or not in buildings, durability (along with performance and cost) is a major consideration. Thus, it must always be asked, "Is there a high enough probability that the item selected will perform satisfactorily for its design life?" Unfortunately, it is often difficult to answer this question adequately. It will always be challenging, but it would be easier to deal with if a coherent body of standards for service life prediction were in place.

The purpose of this paper is to draw attention to the needed body of standards and to recommend actions which could help build it.

2 Elements of a General Methodology for Service Life Prediction

A general methodology for service life prediction should be applicable to any item, whether a material, component, or system. It should state, in general terms, what steps should be taken in any logical prediction of service life. The steps are likely to be roughly as follows:

1. Define the failure criteria which will be used to establish the end of the item's service life.
2. Define the environmental stresses to which the item is likely to be exposed in service.
3. Define the composition and microstructure of the item, and its parts, if any, in terms relevant to its degradation.
4. Determine the mechanisms and kinetics of the degradation of the item and its parts, if any, in sufficient detail to allow prediction of rates of degradation under likely exposure conditions.
5. Develop and validate models for predicting the service life of the item, and its parts, if any.
6. Using the knowledge of the environment, the composition and microstructure of the item and its parts, and the failure criteria, apply the models to predict the service life.
7. Report the predictions for the range of environmental stresses likely to be encountered, stating how the predictions were made, with explicit comments about the assumptions on which they were based.

The essential steps in prediction of service life can be described concisely in a single document. This was done in ASTM E-632, Standard Practice for Development of Accelerated Short-Term Tests for Prediction of Service Life of Building Materials and Components [ASTM E-632 (1982)] and, more recently, in the RILEM Recommendation No. 64, Systematic Methodology for Service Life Prediction of Building Materials and Components [Masters and Brandt (1989)]. However, because of the complexity of applying service life prediction principles to actual problems, no single, concise document, such as these, can give more than broad guidance on the approach to be taken. More detailed guidance is needed to assure that the approach is applied in a relatively uniform way to different items that might be in competition for a given application.

To standardize service life prediction, it appears that the most practical approach is to develop a hierarchical body of standards which, together, can provide logically-consistent, detailed guidance for the predictions. Referring to Table 1, the highest level (Level 1) in the hierarchy would only contain one standard -- a standard such as ASTM E-632 (1982)] or RILEM Recommendation No. 64 -- providing the most general description of a service life prediction methodology; standards in Level 2 would give generic guidance on how to carry out the various steps in the general methodology, without specifying any particular material, component, or system; standards in Level 3 and lower levels would give guidance on application of the various steps in the general methodology to specific materials, components, and systems. A standard in any lower level must, of course, be consistent with the standards in the levels above.

TABLE 1. The Levels in the Proposed Hierarchy of Standards for Service Life Prediction

Level	Content
1	General methodology
2	Item-independent standards amplifying the parts of the general methodology
3	Standards for applying the general methodology to to specific classes of item (materials, components and systems)
4	Standards for applying the general methodology and below to subsets of the next higher level

Our present body of durability standards is not coherent. For the most part, each standard has been developed independently to meet a need to compare the "durabilities" of items of a given type [Masters and Wolf (1974)]. The tests usually use somewhat arbitrarily selected exposure conditions. Their purpose is usually limited to ranking rates of degradation, as indicated by changes in some easily-determined property under the arbitrary exposure conditions; it is not to predict service life under expected in-service exposure conditions. The present standards are, of course, useful, and it is possible that many of them could be modified to fit in to a scheme such as that proposed in this paper. Examples to illustrate the scheme will be given in the following sections.

3 Service Life Prediction for a Specific Item (Material, Component, or System)

The methodology for service life prediction, ASTM E-632 (which was developed in ASTM Committee E06 on Performance of Building Constructions, and is now under the jurisdiction of ASTM G03 on Durability of Non-Metallic Materials), and RILEM Recommendation No. 64, are too general to give detailed guidance on their application to individual items (materials, components, or systems). Their importance is in providing a framework for a body of more detailed standards for use in predicting the service life of any item. For example, for concrete, the relevant, logically-consistent body of standards might include generic, material-independent, standards in Levels 1 and 2 of the hierarchy, with standards specific to concrete and concrete products being in Level 3 and lower levels. Similarly, the logically-consistent standards relevant to prediction of the service life of an organic coating would include the same Levels 1 and 2 standards as for concrete, or any other item, but would have coatings-specific standards in Levels 3 and lower. Thus, the standards relevant to concrete and organic coatings could be listed as in Appendix 1.

The examples of groups of standards to be included in Levels 3 and lower illustrate how a self-consistent body of standards to guide the prediction of the service life of any item could be developed. Whereas the lower level standards are the ones which normally get attention, a logically-consistent body of standards for use in service life prediction will not be developed unless higher-level standards are in place and consistency with the higher-level standards is sought. Thus, much depends on the willingness of standards-writing committees to recognize the importance of building a hierarchical structure and to produce standards which will become part of it.

With this as background, we shall now discuss what we think should be included in the standards of each type in the scheme illustrated by the examples in Appendix 1. Standards for environmental characterization, standards for characterization of microstructures, standards for determination of kinetics and mechanisms of degradation, and standards for mathematical modeling of degradation processes will be discussed in turn.

4 Standards for Environmental Characterization

The single Level 1 standard, which should describe the general methodology, should present the principles to be followed in identifying important environmental factors and list those which should generally be considered. The Level 2 standard on environmental characterization should provide more detailed guidance on what information on environmental characterization should be included in the Level 3 standards concerned with prediction of service life of specific items. The environmental factors to be covered at Level 2 should be all those that can, in any real way, affect the service life of a material, component, or system. The standard should provide guidance on how quantitative descriptions of environmental factors may be produced in terms relevant to service life prediction. These will almost certainly have to recognize the variability of the environments to which most items are exposed and suggest how such variability should be dealt with statistically. It should recommend use of meteorological data wherever this is practical; this is because it is available to all and because it provides a common starting point for all materials exposed outdoors. It seems obvious that those who draft or use the Level 2 standard should work with national weather services to make sure these services know what data will be needed by the building community.

In general, the Level 2 standard should indicate that information is needed on the temperatures and movements of all the fluids and solids with which an item under consideration will come into contact, and all the radiation that will fall on the item. It should also give guidance on ways of expressing the data in appropriate forms. For example, if frost action is likely to be an important degradation process, the data might be used to determine the number of air temperature excursions through the freezing point of water or, if thermal degradation is of concern, a more complete description of the likely history of surface temperatures of exposed surfaces of the item might be needed. Saunders, Jensen, and Martin [Saunders, Jensen, and Martin (1990)] have recently shown how the short- and long-term variations in the temperatures of exposed painted panels can be represented by a Fourier series.

A Level 3 standard for service life prediction should only require characterization of those environmental factors in the Level 2 standard that are relevant to the specific type of material, component or system it addresses. It should be specific in describing how suitable information can be obtained, whether from published data, including maps, or from measurements made specifically for the purpose at hand, with comments on the preferred approaches and the potential errors associated with each.

5 Standards for Characterization of Materials, Components, and Systems

The Level 1 standard, being the most general, should only indicate that the composition and structure of the item to be considered must be known if reliable predictions of service life are to be made. The Level 2 standard for characterization of materials, components, and systems should provide more detailed, but still general, guidance on what characteristics should be included in a Level 3 standard for prediction of the service life of a specific type of item. The characteristics to be included will differ depending on the item. For a material, the main factors to be considered will usually be the overall chemical composition, and the compositions and distributions of the bulk and interfacial phases present (i.e. the microstructure of the material, taking into account cracks and voids). For a component or system, the characterization will usually be more complex because of the need to consider the interactions between different materials and the greater number of geometric factors which must be taken into account. These include the shapes and dimensions of the item and its parts, and their orientations (and movements, if any).

The appropriate description of a material may be in relatively simple terms, assuming uniformity of composition as for homogeneous materials, or it may be in such detail as to require information on spatial variations in composition, as for a composite such as concrete or a fiber-reinforced plastic. The Level 2 standard should comment on this and provide guidance on different approaches which might be used. For a porous material likely to be exposed to frost action or chemical attack, information on the pore system would probably be needed. The Level 2 standard should indicate types of microstructural features which may be important for service life prediction. It should also indicate ways of obtaining quantitative information about the microstructure for various types of material. Level 3 and lower level standards for service life prediction should give detailed guidance on preferred ways of characterizing the type of item to which each standard applies, with comments on the precautions to be taken to get valid results and minimize the possible errors.

6 Standards for Determination of Mechanisms and Kinetics of Degradation

The Level 1 standard should only mention the need for information on the mechanisms and kinetics of degradative processes in the items under consideration as being essential for service life prediction. The Level 2 standard for determination of mechanisms and kinetics of degradation will provide guidance as to what information on mechanisms and kinetics should be included in a Level 3 or lower-level standard for prediction of the service life of a specific type of item The mechanisms will be different for different materials, including the different materials in a composite. The Level 2 standard must, for example, include corrosion of metals, photochemical and thermal degradation of organic materials, fatigue effects, and cracking due to localized stresses resulting from differential volume changes in the item.

7 Standards for Mathematical Models of Degradation

The Level 1 standard should indicate the need for mathematical models of degradation in service life prediction without stating what the models should be like. Standardization of the modeling of degradative processes will be particularly challenging. At this stage in the evolution of degradation modeling as a component of service life prediction, only broad generalisations can be made. Much will depend on developments in computer hardware and software, and in telecommunications. The models needed will have to provide scientifically and technically sound representations of the relevant chemical, physical and mechanical processes leading to degradation. This implies complexity but, as for any standard, the standards for developing mathematical models should be neither more, nor less, complex than necessary. A standard practice for the development and use of models of degradation should probably provide recommendations on:

- o formats of statements of objectives of individual models and submodels
- o symbols to be used in flow charts and program code
- o programming language(s) to be used
- o program structure
- o qualifications of developers and users of models

- o selection of data for use in model calculations
- o methods for testing the models
- o documentation of models and assessment of their limitations
- o format and content of reports on model outputs.

Insight into the needs is being obtained through operation of the Cementitious Materials Modeling Laboratory in the NIST Center for Building Technology. The Laboratory, which is a step towards a more broadly-based Building Materials Performance Modeling Laboratory, is linked to other participants in the Center for Advanced Cement-Based Materials (ACBM) headquartered at Northwestern University to provide for the sharing of models among the ACBM members. It seems likely that, because of the complexity of models of degradation, some will be so complex that sharing through a central computer may be the most practical way of providing the support needed for service life prediction.

8 Standards for Service Life Prediction

The standards outlined in Sections 4 through 7 should provide the basis for service life predictions. The predictions themselves should be carried out and reported in standard ways. The Level 1 standard should give general guidance, and the Level 2 standard should give more detail on generic aspects of how predictions should be made. As usual, Level 3 and lower-level standards for service life prediction should apply the guidance of the higher-level standards to various types of items. Because of the many steps in making predictions, it is important that the reporting of the results should be standardized to aid interpretation. Other obvious requirements are that the assumptions should be clearly stated, and that the errors should be estimated.

9 Recommendations for Standards Development

In view of the importance of improving the reliability of service life predictions, and the large demands the required activities would place on research resources, several actions are recommended to the organizers and sponsors of these conferences:

a) the scope of the Conferences should be broadened to include "Developments in Standards for Service Life Prediction" as an explicitly-stated topic;

b) standards and prestandards organizations, such as ASTM, CIB, and RILEM, should establish, or strengthen, committees dealing with generic aspects of service life prediction, including characterization of environments, and encourage their interaction, at least on an advisory basis, with committees dealing with specific materials and components;

c) national building research organizations, should cooperate in planning and implementing the development of models of degradation of building materials and components suitable for standardization, and the development of the needed databases.

10 Summary

It has been pointed out that, if it is to be widely accepted as an aid to selection of building materials, service life prediction must be standardized. Because of the complexity of service life prediction, a large and logically-consistent body of interrelated standards is needed. Present standards do not have such a structure.

A possible structure for a body of service life prediction standards is outlined to promote discussion about the direction of future developments in durability standards. It is recognized that development of the proposed standards structure would require an unusual degree of coordination of the scientific and technical efforts of interested parties, preferably on an international scale. This would include improvements in characterization of environments and in mathematical modeling of degradation processes. However, benefits to be obtained from more reliable performance of building materials and components appear to warrant the effort being made.

11 Acknowledgements

The authors gratefully acknowledge the many contributors to our thoughts regarding standards for service life prediction, including the members on ASTM E06.22 on Durability Performance of Building Constructions, members of CIB W80/RILEM 71-PSL and 100-TSL on Prediction of Service Life, and our colleagues at NIST.

12 References

ASTM E-632, Standard Practice for Developing Accelerated Tests to Aid Prediction of Service Life of Building Components and Materials, ASTM, Philadelphia, PA (1982).

Fagerlund, G., Essential Data for Service Life Prediction, in Masters, L.W. (ed.), Problems in Service Life Prediction of Building and Construction Materials, pp. 113-138.

Frohnsdorff, G., Integrated Knowledge Systems for Concrete Science and Technology, in Skalny, J.P. (ed.), Materials Science of Concrete, American Ceramic Society, Columbus, OH (1989).

Frohnsdorff, G., Clifton, J.R., Jennings, H.M., Brown, P.W., Struble, L.J., and Pommersheim, J.M., Implications of Computer-Based Simulation Models, Expert Systems, Data Bases, and Networks for Advancing Cement Research, Bull. Amer. Ceram. Soc., V. 67, pp. 1368-1371, 1988.

Frohnsdorff, G., and Masters, L.W., The Meaning of Durability and Durability Prediction, in Sereda, P.J. and Litvan, G.G. (eds.), Durability of Building Materials and Components, STP 691, ASTM, Philadelphia (1980), pp. 17-30.

Gross, J.G., International Harmonization of Standards: Done With or Without Us, The Building Official and Code Administrator, Sept./Oct. 1989, pp. 46-47.

Masters, L.W. (ed.), Problems in Service Life Prediction of Building and Construction Materials, NATO ASI Series E: Applied Sciences No. 95, Martinus Nijhoff, Dordrecht, 1985.

Masters, L.W., and Brandt, E., Systematic Methodology for Service Life Prediction of Building Materials and Components, Materials and Structures, v. 22, pp. 385., 1989.

Masters, L.W., and Wolfe, W.C., The Use of Weather and Climatological Data in Evaluating the Durability of Building Components and Materials, NBS Technical Note 838, NIST, Gaithersburg, MD 20899 (1974).

Saunders S., Jensen, and Martin J.W., A Study of Meteorological Processes Important in the Degradation of Materials through Surface Temperature, NIST TN (in press), NIST, Gaithersburg, MD 20899 (1990).

Sereda, P.J., and Litvan, G.G., (eds.), Durability of Building Materials and Components, STP 691, ASTM, Philadelphia, PA (1980).

Sjostrom, C., Overview of Methodologies for Prediction of Service Life, in Masters, L.W. (ed.), Problems in Service Life Prediction of Building and Construction Materials, pp. 3-20.

Appendix 1. PROPOSED HIERARCHY OF STANDARDS FOR SERVICE LIFE PREDICTION

Level 1 (Generic to all items)

1. Standard practice for prediction of the service life of any item (material, component, or system).

Level 2 (Generic to all items)

2.1 Standard practice for characterizing the environment to which any item (material, component, or system) will be exposed in service.

2.2 Standard practice for characterizing any item (material, component, or system).

2.3 Standard practice for determining the dominant degradation processes of any item (material, component, or system), and their mechanisms and kinetics.

2.4 Standard practice for developing mathematical models for predicting rates of degradation of any item (material, component, or system).

2.5 Standard practice for using knowledge of the service environment, the characteristics of the item (material, component, or system) and its parts, degradation models, and the failure criteria, for predicting the service life of the item and reporting the results.

Level 3 (for concrete)

* 3.1.1 Standard practice for characterizing the environment to which a concrete item (material, component, or system) will be exposed in service.

3.1.2 Standard practice for characterizing a concrete item (material, component, or system).

3.1.3 Standard practice for determining the dominant degradation processes of a concrete item (material, component, or system), and their mechanisms and kinetics.

[Footnote: * The numbering system used in this appendix is for convenience in showing relationships. The first number designates the level in the proposed hierarchy and the last the relationship to the parts of Levels 2. The middle number (or letter) is used to distinguish between different sets of standards at a given level.]

3.1.4 Standard practice for development of mathematical models of degradation processes of a concrete item (material, component, or system).

3.1.5 Standard practice for using knowledge of the service environment, the characteristics of a concrete item (material, component, or system) and its parts, degradation models, and the failure criteria, for predicting the service life of the item and reporting the results.

<u>Level 3</u> (for an organic coating)

3.2.1 Standard practice for characterizing the environments to which an organic coating will be exposed in service.

3.2.2 Standard practice for characterizing an organic coating.

3.2.3 Standard practice for determining the dominant degradation processes of an organic coating, and their mechanisms and kinetics.

3.2.4 Standard practice for development of mathematical models for prediction of service life of an organic coating.

3.2.5 Standard practice for using knowledge of the service environment, the characteristics of an organic coating, degradation models, and the failure criteria, for predicting the service life of the coating and reporting the results.

It has already been implied that the item-specific standards will not all be in Level 3. For example, waterproofing membranes for roofing must be viewed as part of the larger category of organic-matrix sheet materials. It is logical to put the standards for prediction of the service life of organic-matrix sheet materials in Level 3, with any additional standards specific to roofing membranes going in Level 4 or lower levels. Thus, the relevant Level 3 standard, and some Level 4, 5, and 6 standards, for roofing membranes might be:

<u>Level 3</u> (for organic-matrix sheet materials)

3.3.1 Standard practice for characterizing the environment to which an organic-matrix sheet material will be exposed in service.

3.3.2 Standard practice for characterizing an organic-matrix sheet material.

3.3.3 Standard practice for determining the dominant degradation processes in an organic-matrix sheet material, and their mechanisms and kinetics.

3.3.4 Standard practice for development of mathematical models for degradation of an organic-matrix sheet material.

3.3.5 Standard practice for using knowledge of the service environment, the characteristics of an organic-matrix sheet material, degradation models, and the failure criteria, for predicting the service life of the material and reporting the results.

Level 4 (for roofing membranes)

4.x.1 Standard practice for characterizing the environment to which a roofing membrane will be exposed in service for use in service life prediction.

4.x.2 Standard practice for characterizing a roofing membrane for service life prediction.

4.x.3 Standard practice for determining the dominant degradation processes in a roofing membrane, and their mechanisms and kinetics.

4.x.4 Standard practice for development of mathematical models for prediction of service life of a roofing membrane.

4.x.5 Standard practice for using knowledge of the service environment, the characteristics of a roofing membrane, degradation models, and the failure criteria, for predicting the service life of the membrane and reporting the results.

Level 5 (for built-up roofing membranes)

5.y.1 Standard practice for characterizing the environment to which a built-up roofing membrane will be exposed in service for use in service life prediction.

5.y.2 Standard practice for characterizing a built-up roofing membrane for service life prediction.

5.y.3 Standard practice for determining the dominant degradation processes in a built-up roofing membrane, and their mechanisms and kinetics.

5.y.4 Standard practice for development of mathematical models for prediction of service life of a built-up roofing membrane.

5.y.5 Standard practice for using knowledge of the service environment, the characteristics of a built-up roofing membrane, degradation models, and the failure criteria, for predicting the service life of the membrane and reporting the results.

Level 5 (for single-ply roofing membranes)

5.z.1 Standard practice for characterizing the environment to which a single-ply roofing membrane will be exposed in service for use in service life prediction.

5.z.2 Standard practice for characterizing a single-ply roofing membrane for service life prediction.

5.z.3 Standard practice for determining the dominant degradation processes in a single-ply roofing membrane, and their mechanisms and kinetics.

5.z.4 Standard practice for development of mathematical models for prediction of service life of a single-ply roofing membrane.

5.z.5 Standard practice for using knowledge of the service environment, the characteristics of a single-ply roofing membrane, degradation models, and the failure criteria, for predicting the service life of the item and reporting the results.

Level 6 (for elastomeric roofing membranes)

6.w.1 Standard practice for characterizing the environment to which an elastomeric roofing membrane will be exposed in service for use in service life prediction.

6.w.2 Standard practice for characterizing an elastomeric roofing membrane for service life prediction.

6.w.3 Standard practice for determining the dominant degradation processes in an elastomeric roofing membrane, and their mechanisms and kinetics.

6.w.4 Standard practice for development of mathematical models for prediction of service life of an elastomeric roofing membrane.

6.w.5 Standard practice for using knowledge of the service environment, the characteristics of an elastomeric roofing membrane, degradation models, and the failure criteria, for predicting the service life of the item and reporting the results.

12 TRIAL MANUFACTURE OF AN APPARATUS FOR TESTING THE DETERIORATION OF BUILDING MATERIALS DUE TO RAINFALL

Y. KITSUTAKA, K. KAMIMURA
Utsunomiya University, Utsunomiya, Tochigi, Japan
T. SHIIRE
Kanagawa University, Yokohama, Kanagawa, Japan

Abstract
This paper deales with the manufacture of the rainfall apparatus for evaluating the deterioration of materials due to rainfall. The first step was to develop an apparatus which produced vertically falling waterdrops. It was designed to produce four different types of waterdrops similar to actual rainfall in regards to the falling properties; falling velocity, drop diameter,etc. The second step was to develop an apparatus which produced driving rain. This was achieved by adding an air fan to the apparatus producing vertically falling waterdrops. The similarity of the condition between actual driving rain and waterdrops produced by the apparatus was discussed.
Keywords: Deterioration, Rainfall testing apparatus, Acidic rain, Soiling, Driving rain.

1 Introduction

External building materials are subjected to the deteriorating action of various environmental factors such as UV radiation or carbonation etc. And the action of rainfall can be raised as one of the leading factors among them. For instance, degradation, soiling of the materials are generated mainly caused by the collision and flow of rain and the chemical reaction of acidic rain.

This study was carried out for the purpose of trial manufacture of an apparatus for testing the deterioration due to rainfall so that the action of rainfall causing the deterioration of materials can be ivestigated experimentally.

2 Basic concept of the apparatus

In this study, the authors divided rainfalls into 4 types by their intensity (Table 1) and undertook to reproduce them. Also, as the action of driving rainfall works on the external wall surfaces, a mechanism for generation of such raindrops was also provided. In the case of a driving rain, wind pressure under complicated conditions works simultaneous by on the external wall, but in this apparatus, the reproduction is confined only to the action of raindrops.

Table 1. Types of rainfall

Symbol	Types	Intnsity(mm/hr)	Maximum diameter(mm)
R1	Weak rain	0 - 3	2
R2	Medium rain	3 - 15	3
R3	Heavy rain	15 - 60	4
R4	Severe rain	60 -	6

3 Reproduction of vertically falling raindrops

3.1 Outline of the apparatus

The outline of the rainfall testing apparatus is shown in Fig.1. The following two systems are taken as the method for reproduction of the vertically falling raindrops :

- Natural falling system -

This system reproduces the conditions of rainfall with small falling velocities. As distilled water is supplied to the waterdrop producer from a pump, waterdrops are generated from 14 burettes(inner diameter 3mm). The falling waterdrops are dispersed by the wire netting at the water drop disperser. The properties of the waterdrops are adjusted by the size of meshes and the number of layers of the wire netting and the quantitiy of waterdrops by the burettes.

- Spraying system -

To get the conditions of rainfall with a higher speed, pressurized water is ejected intermittently from a nozzle and dispersed by way of wire nettings. The quantity of the falling waterdrops is adjusted by the solenoid valve attached to the nozzle by operating it itermittently. The on/off interval of the solenoid valve is controlled through the action of the relay by detecting the on/off light of an arbitrary cycle which is transmitted from the luminous metronome. The properties of the waterdrops are adjusted by the water pressure, nozzle diameter, size of the wire netting etc.

3.2 Properties of the falling waterdrops obtained by this apparatus

The waterdrops falling within the measuring range are illuminated intermittently by the stroboscope and their picture is taken with a camera, thus the velocity of a waterdrop is measured.

The falling waterdrops are received by a sheet of filter paper on which Water Blue is applied before hand, to observe the status of dispersion of the waterdrops as well as to make comparison with the measurement results of the traces of raindrops recorded in the post.

From measurements, the following trends are observed.

In the case of the natural falling systems: The larger the size of meshes of the wire netting, waterdrops of larger diameter can be obtained, but the conditions of dispersion get worse. The smaller the size of meshes of the wire netting, the falling velocity of the water drops gets closer to the natural rainfall, and the condition of dispersion is improved, but the waterdrops of large diameter becomes difficult to gain.

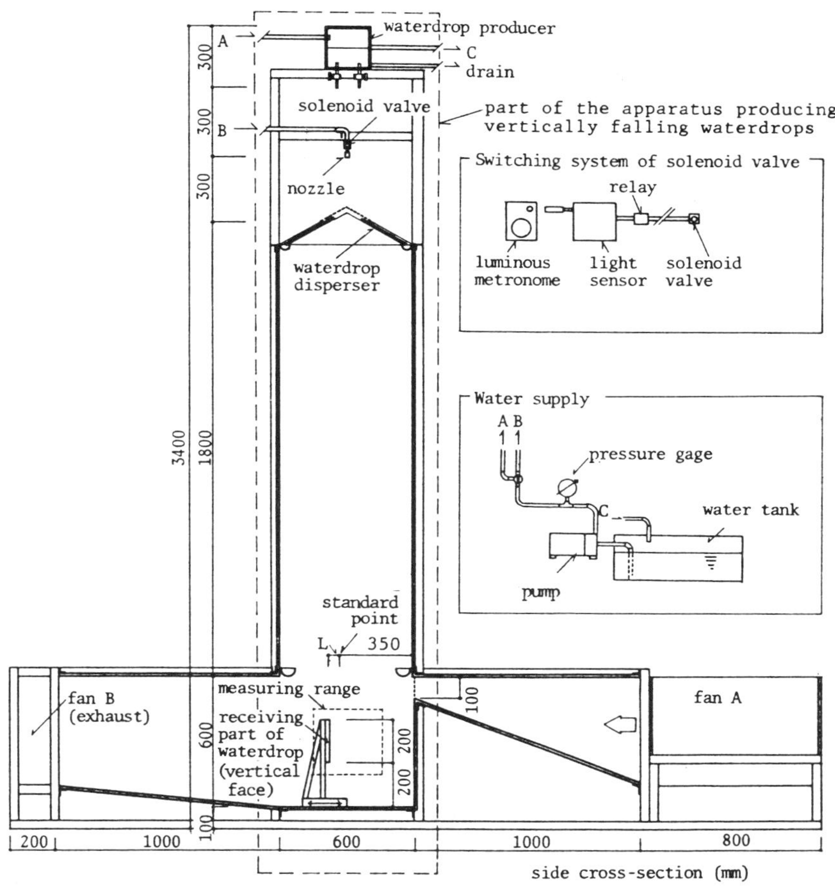

Fig.1. The rainfall testing apparatus

Table 2. Specification of the rainfall testing apparatus

	R1	R2	R3	R4
Method	Natural fall		Spray	
Mesh size of netting (mm)	upper=0.6 inter=1.2 lower=0.6	upper=1.2 lower=0.6	lower=1.2	lower=4.6
Water pressure (kgf/cm^2)	-	-	1.0	1.0
Diameter of nozzle (mm)	2.0	2.0	2.0	3.0
Falling interval of waterdrops (sec)	2.0	1.5	1.5	1.5
Rainfall intensity (mm/hr)	2.9	10.8	26.0	64.2
Coefficient of variation (%) *	11.9	12.4	14.8	16.9

* Variation of the amount of water by area(15x15cm)

In the case of spraying system: The larger the diameter of the nozzle is made, falling waterdrops of the larger diameter are gained, but dispersion in the precipitation gets larger. When the water pressure is raised, the falling velocity increases and a larger diameter is obtained which is closer to that of actual rainfall.

From these investigations, the testing conditions that resemble the conditions of actual raindrops most closely were determined. The result is shown in Table 2.

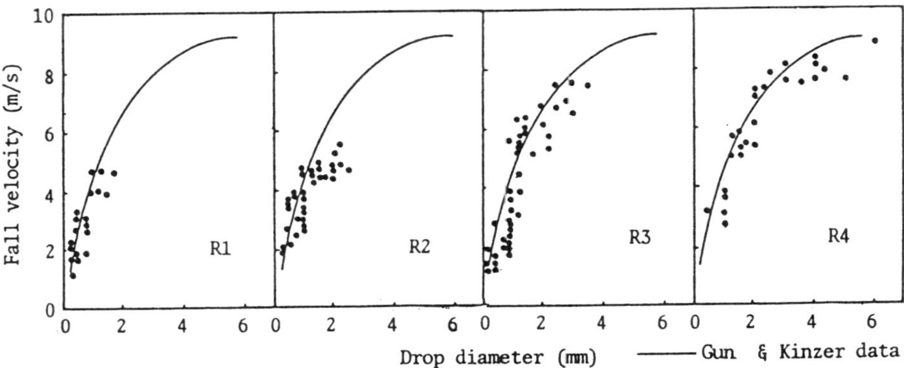

Fig.2. Relation between drop diameter and fall velosity

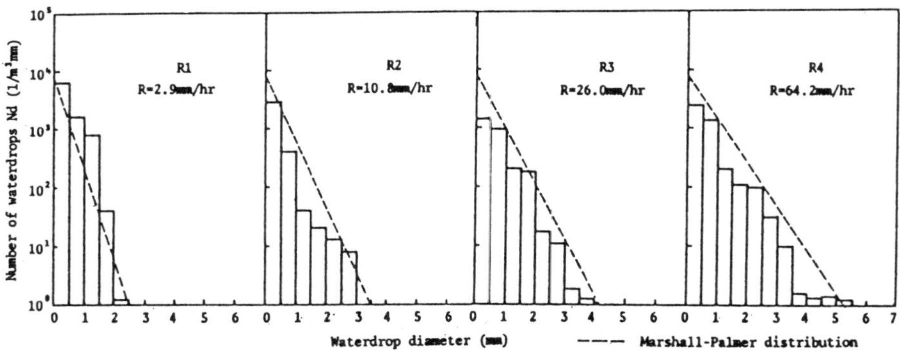

Fig.3. Distribution of waterdrop size

The relation between the drop diameter and the falling velocity under these conditions is shown in Fig. 2. On every condition, the relation between the diameter of raindrops and the falling velocity resembles the result of the experiment by Gun and Knzer (1949) and the maximum diameter almost satisfies the assumed conditions.

Also, the measured result of the distribution of the diameter of the waterdrops is shown in Fig. 3 together with the M-P distribution (1948). As for the distribution of the diameter of the waterdrops, it is seen to be almost approximated by the M-P distribution.

4 Reproduction of the driving rain

4.1 Outline of the apparatus

By applying the horizontal wind force to the vertically falling waterdrops, it was intended to get the slanting waterdrops that fall forming a group. Since the location that the falling waterdrops accelerated by the fan finally reach is lower than the position of the air inlet (10cmx20cm), the air inlet is provided at the level of the top of the receiving surface of waterdrops.

The quantity of arrival of the waterdrops varies depending upon the position where they are received. In Fig.4, the relations between the position of the receiving surface and the water quantity accepted by the vertical surface for 4 conditions of the intensity of rainfall are presented. From these diagrams, the optimum position of the vertical receiving face for waterdrops was determined (Table 3).

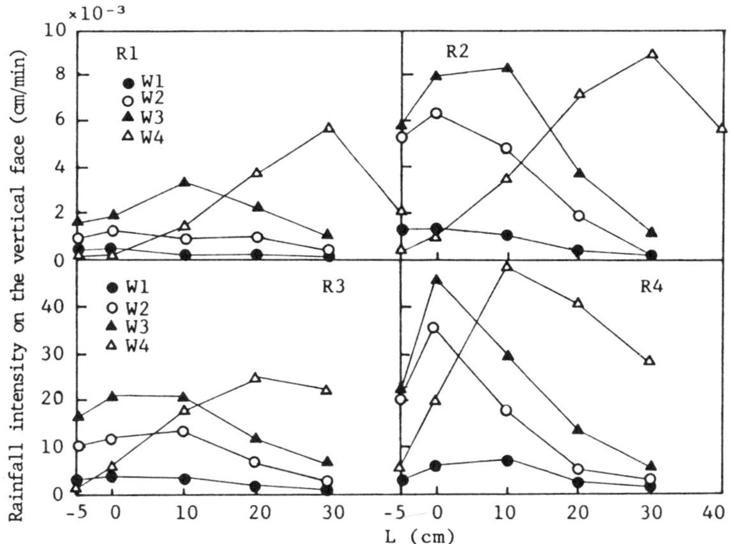

Fig.4. Relation between the distance from the standard point to the vertical face (L) and rainfall intensity on the face

Table 3. The distance from the standard point to the vertical face, L (cm)

	W1	W2	W3	W4
R1	0	0	10	30
R2	0	0	10	30
R3	0	10	10	20
R4	0	0	0	10

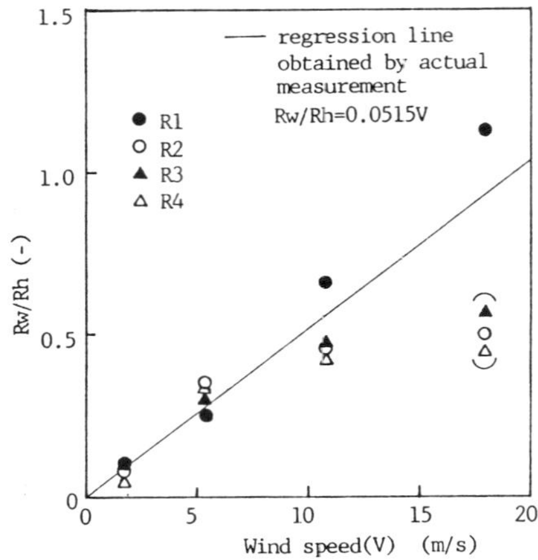

Fig.5 Relation of wind speed to Rw/Rh

4.2 Relationship with actual driving rain

With regard to the driving raindrops, a lct of information about measurement is available. In any of the results of measurement, the ratio of the precipitation on the wall surface Rw(mm) to that on the horizontal surface Rh(mm), namely, Rv/Rh, is proportional to the component of the average wind velocity along the normal direction to the wall surface.

In Fig. 5, the relation between Rh/Rw and the average wind velocity at the standard point measured in this rainfall apparatus is presented. In this diagram, an example of the result of mersurement on the actual driving rain by Itho (1984) is shown. Except for the conditions of intensity of rainfall R2-R4 (parenthesis in the diagram) in the wind condition W4, the relations between Rh/Rw and the wind condition are very close to the examples actually measured.

In order to reproduce the conditions of stronger wind and heavier rain, the authors are studying enlarging the receiving area for falling waterdrops and the area of the wind port and increasing the wind velocity.

5 Summary

In this study, in order to investigate the action of raindrops that causes deterioration of the external wall materials experimentally, a rainfall testing apparatus was manufactured for trial. And the conditions which resemble four kings of actual vertical rainfalls could be obtained. Besides, under several conditions, waterdrops that reached the vertical surface with close approximation to the examples of measurement of driving rainfalls could be obtained.

6 References

Gunn, R. and Kinzer, G.D. (1949) The terminal velocity of fall for water droplets in stangnant air. **J. Meteorology**, 6, 243-248

Marshall, J.S and Palmer, W.M. (1948) The distribution of raindrops with size. **J. Meteorology**, 5, 165-166

Itho, H. and Nishida, K. (1984) Distribution Map of driving rain acting external wall. **A. M. Arch. Inst. JAPAN**, 275-276

13 STATISTICAL FIELD SURVEY OF EXTERIOR BUILDING MATERIALS DEGRADATION

N. TOLSTOY, G. ANDERSSON, C.SJÖSTRÖM
Materials and Structures, The National Swedish Institute for Building Research, Gävle, Sweden
V. KUCERA
The Swedish Corrosion Institute, Stockholm, Sweden

Abstract
The paper presents an investigation of external materials on buildings in the greater Stockholm area. The investigation had two main goals; to make an inventory of the amount of different exterior building materials and, to inspect and account for the observed deterioration of the materials. This article mainly presents the results on the materials degradation.

The investigation was performed as inspections of stastistically sampled buildings. The random sampling of the buildings was for example performed in strata with different SO_2 concentrations. The investigation method, i.e. both the inspection routines and the sampling procedure, is reviewed briefly.

Degradation to roofings, facades and windows is estimated for different materials and surface finishes. The degradation for certain materials, distributed in age classes and also in some cases by areas with different environmental pollution level, is accounted for. The results showed increased corrosion in areas with high concentration of pollution. This could be observed for renderings, painted galvanized steel and painted wooden windows.

Maintenance periods are statistically estimated in two different ways from the results of the field survey. One way by using the classification of the condition of surface finishes and materials distributed in different age classes. The other way by making an estimate using accumulated age curves.
Keywords: Building Materials, Degradation, Air Pollutants, Field Survey, Statistical Sampling

1 Introduction

Corrosion damage to several important building materials presents clear connections with concentrations of air pollutants. The corrosive impact of, above all, sulphur dioxide (SO_2) has been studied in both laboratory and field experiments. (e.g. Kucera, 1986)

The cost of corrosion damage inflicted by sulphurous emissions on building structures has been a subject of numerous studies in many different countries. An improved methodology for assessing the quantity and distribution of different materials in building structures in an urban area will make a big difference to the dependability of input data for calculating the cost entailed by emissions of acid pollution.

This article is based on a project (Tolstoy et al., 1989) forming part of a joint Nordic research effort initiated under the auspices of the Nordic Council of Ministers as part of the programme "Calculation of reduced corrosion damage resulting from a reduction of SO_2 emissions". A similar study was conducted simultaneously in Sarpsborg by the Norwegian Air Research Institute (NILU). The project, conducted between 1986 and 1988, was funded by the National Swedish Institute for Building Research (SIB), the Swedish National Council for Building Research (BFR) and the Nordic Council of Ministers (NMR).

2 Purpose

The purpose of the project was to develop a methodology for the represen- tative inventory of material quantities and corrosion damage in building structures.In this article the presentation of results will focus mainly on the assessment of actual degradation and not so much on the inventory of material quantities.

3 Method

3.1 Sampling methodology

In a sample survey, one is usually interested in presenting results for certain categories. In order for the results not to be subject to unacceptably large random errors, it is important to have sufficient observations for every category. The sampling groups employed in SIB's surveys are different types of building, some of which are subdivided into building-year classes.

In the present study, where the focus of attention is on Greater Stockholm, a geographical division was made into three areas with different atmospheric concentrations of sulphur dioxide.

Once the building population has been determined, a number of real estates are calculated in strata and sampling groups. The highest total number of real estates which can be accommodated is determined with the aid of a costing and time frame for the work of inspection. The random error can be estimated by estimating the standard deviation for a required variable in the different sampling groups. Using the estimated random errors in the different sampling groups, the sample is then divided into sampling groups in such a way as to minimise the random error of the total sample (optimal allocation).

3.2 Status inspection

The condition or technical status of a building can be described in various ways. Sampling of test samples and laboratory evaluation studies or measurements on the spot supply exact data concerning the degradation of a material, given a knowledge of the corresponding values for that material as new. The Danish Technology Council, for example, distinguishes between four levels of status examination:

Level 1 = inspection.
Level 2 = inspection using simple aids.
Level 3 = inspection using non-destructive measurement.
Level 4 = inspection using destructive measurement.

Costs rise by a factor of ten between levels 1 and 4 according to the Technology Council. For example, if level-1 inspection of a concrete structure costs DKR 4 000, a level-4 inspection will cost 40 000. Statistical sample surveys involve the investigation of a large number of buildings. If the state of all roofs, facades and window materials in a large number of buildings is to be described, then unfortunately the cost of sampling and measurement is prohibitive. It then only remains to carry out ocular inspections and to make visual assessments. Aids such as checklists and graphic standards exist to facilitate assessments of this kind. Materials determination plays an important part during inspection. Simple aids such as a magnet, knife, certain chemicals and matches can be used here to simplify the classification of materials and surface finish.

4 Strategy and implementation

4.1 Variables investigated

The primary variables in the survey are material quantities on outer surfaces in the groups of accounting (types of building and land areas). Secondary survey variables include, for example, the state of a particular roofing material in the different land areas or the ageing of paint on sheet metal.Amount of materials and surface finish were recorded for every part of the building.

For roof, windows and walls, an assessment was made of the status of surface finish and underlay. Both were evaluated on a three-point scale: 0 = intact, 1 = minor damage (no repairs needed), 2 = repairs advisable. Record was also made of cause of status and age of material as well as surface treatment. In addition, certain environmental factors were described for each item. The elevation of the surfaces, proximity to traffic, local pollution source, sulphur dioxide area, degree of fouling, proximity to water, etc. were recorded.

The area investigated is Greater Stockholm, i.e. the County of Stockholm not including Södertälje, Norrtälje and Nynäshamn.

4.2 Groups of accounting

The primary survey groups comprise three different area divisions and ten building types and building-year classes.The area divisions were based on Stockholm sulphur dioxide graphs for winter 1979-1980 (Orrskog, 1982). The numbering complies with the corrosive environment classification proposed by ISO with reference to atmospheric sulphur dioxide (ISO DIS 9223).

Table 1 Area divisions

Area no.	SO_2 content, winter average, $\mu g/m^3$	Land area, km^2	Inhabitants/km^2
2	>60	28	7 437
1	20-60	444	1 658
0	<20	2 985	165

The types of building selected were single-family dwellings, multi-family dwellings, non-housing, industrial buildings and agricultural buildings.
Dwelling houses were divided into three age classes: built before 1920, 1921-1960 and 1961 or later. This division referred to original year of building.

4.3 Generalisations to the entire housing stock

Generalisation of the results from the sample to the entire housing stock was effected using a weighting procedure. Each building was allotted an upward adjustment factor (weight) showing exactly how many buildings in Greater Stockholm this particular building represents. The method has been used in previous surveys at SIB and is described by Waller, 1977. The size of the weight depends on the area to which the real estate belonged when the sample was drawn, the sampling group, the number of buildings on the real estate and the size of the nonresponse.

4.4 Inspection procedure

The inspection normally began with the personal contact being interviewed. After being interviewed and unlocking doors the personal contact would generally leave the inspectors, who then began measuring, calculating and assessing the building. Data obtained from drawings were checked with reality. All materials were quantified and recorded. Assessments were made of the state of roofs, windows and facades. Instructions were supplied for material determination. Time input per inspection, not including preparations and travelling time, varied between about one hour and two days. The average was probably about three hours. Photography was another important aspect of documentation.

4.5 Environment characterisation

As previously, buildings were classified with respect to SO_2 content, proximity to traffic, proximity to local pollution source and proximity to salt water. For every façade, a fouling assessment was made. All status- assessed surfaces were plotted by points of the compass. Subsequently an attempt was made to classify NO_2 loads on inner-city buildings with reference to a vehicular exhaust map. NO_x load can also be characterised by proximity to street traffic.

Table 2 Environmental parameters used in characterising real estates and for assessing the status of surfaces.

Environmental parameters	Classes	Basis
SO_2	0, 1, 2, (ISO DIS 9223)	Orrskog,1982
Elevation	S, E, W, N	Inspection
Proximity to traffic	1-5 (distance classes)	Inspection
Local pollution source	1-4 (distance classes)	Inspection
Degree of fouling	1-3	Inspection
Proximity to salt water	1-5	Inspection
NO_x	1-4	Exhaust map

5 Results

5.1 Inventory of materials

This survey is the first survey in the Nordic countries in which an inventory of the total quantity of external materials on buildings in a large built-up area has been made by inspecting the actual buildings.

Material quantities are stated by different types of building and with reference to geographical areas having different pollution situations. Single-family dwellings account for nearly half of all external area (48%) on residential buildings. Wood (23.5%), metal (17.8%) and rendering (14.6%) are the three materials having the largest total surface area in Greater Stockholm. The total external area of buildings in Greater Stockholm is estimated at 189.3 mill. m^2, including 61.9 mill. m^2 cladding materials and 57.2 mill. m^2 roofing materials. Material density ranges from 0.64 m^2 material area per m^2 land area in the centre of Stockholm to 0.033 m^2/m^2 on the outer fringes of Greater Stockholm. Material area per resident is least in the centre, at 86 m^2 per resident, and highest on the outer fringes of Greater Stockholm, at 198 m^2 per resident.

The method of inspecting houses in a statistical sample so as to describe material quantities has proved viable. The data obtained concerning quantities of material exposed in buildings are unique of their kind.

5.2 Status of materials in different environments

The surface-finish status of three different materials can be shown to be inferior in areas with heavier sulphur dioxide concentrations. Greater Stockholm is divided into three sulphur dioxide areas, in keeping with ISO classification. Connections between SO_2 area and status are shown for coarse stuff rendering, site-painted, zinc-coated sheet steel and painted wooden windows. In the case of untreated metallic materials and natural stone, there were too few observations available for analyses of connections to be possible.A slight connection can be discerned between the status of coarse stuff and NO_x load assessed from a vehicle exhaust map. Here, however, there are probably other factors involved as well, e.g. SO_2 and soot. For painted wooden windows, a connection was obtained between window orientation and status. The emergence of this well-known relation in the survey indicates that the method of status assessment employed functions successfully. Proximity to salt water revealed no connections with the status of materials. The low saline content of the Baltic Sea, the protective influence of the archipelago and the small number of buildings inspected in exposed positions probably helped to account for this. The degree of fouling was separately assessed. Strong connections were found here with SO_2 content, with distance from road traffic and with distance from a local pollution source. In some cases, then, the method of assessing material status by ocular inspection, using checklists and assessment standards, has proved to furnish a basis on which to establish descriptive relations between environmental impact and material degradation. This technique can be improved by using still more objective methods of inspection, through better environmental characterisation and through the inspection of larger samples of buildings, but this is not least a matter of resources.

5.3 Maintenance periods

The status of certain roof, façade and window materials is presented with reference to different age classes. On the basis of these status descriptions, we can calculate an average maintenance interval. It has been assumed that the material is ripe for maintenance when repairs are advisable to about 50% of the area of an age class. This method entails an over-estimate of the length of the maintenance intervals. Even so, short maintenance periods are obtained for two materials. The interval for painting zinc-coated steel roofing is estimated at seven years and that for painting wooden windows at seven years. Roof tiles, like coarse stuff rendering, should be replaced after forty years. The surface finish of bitumen felt roofing lasts for ten years and a bitumen felt roof should be replaced after 20 years. A wooden facade ought, on average, to be repainted after twelve years. Age distributions have been estimated for roof, façade and window materials. Leaving aside areas which ought to be repaired and areas which have recently been treated, one is left with a distribution over maintained surface areas. From these curves we have calculated more exact maintenance intervals. The average for painting a wooden facade, for example, is estimated at 9.4 +- 0.7 years, and that for painting wooden windows at 7.6 +- 0.3 years. In the inner city of Stockholm, with its heavy atmospheric concentrations of sulphur dioxide, the average for repainting zinc-coated sheet steel is estimated at 6.3 years, while in the cleaner outer area the repainting period is estimated at 8.0 years.

6 References

ISO DIS 9223 Corrosion of metals and alloys - Classification of corrosivity of atmospheres.

Kucera, V.(1986) Influence of Acid Deposition on Atmospheric Corrosion of Metals: A review, page 104-118, Materials Degradation Caused by Acid Rain (Ed R Baboian), ACS Symposium series 318, American Chemical Society, Washington DC.

Orrskog, G, 1982, Luften i stor-Stockholm, Svaveldioxid år 1990 och 2000 - Ett underlag för regional planering. SMHI rapport 1982:3, uppdrag från stor-Stockholms Energi AB (STOSEB), Norrköping.

Tolstoy, N., Andersson G., Kucera V. and Sjöström C. (1989) External Building Materials - Quantities and Degradation. (Publ in Swedish 1989 The National Swedish Institute for Building Research, Gävle) (submitted for publication in English 1990.)

14 SERVICE LIFE PREDICTION SYSTEM OF POLYMERIC BUILDING MATERIALS USING A PERSONAL COMPUTER

T. TOMIITA
Building Research Institute, Ministry of Construction, Tsukuba, Japan

Abstract
It has been eagerly desired to establish service life prediction system of building materials. Mathematical model based on chemical reaction theories can express deterioration phenomena as a function of degradation factors, solar UV and heat. Author has developed a personal computer software system which determines deterioration material constants in the model and predicts property change over time.
Keywords: Service Life Prediction, Computer Software, Mathematical Deterioration Model, Polymer, UV Deterioration, Thermal Deterioration

1 Introduction

Solar ultraviolet ray is known as a severe degradation factor for polymeric building materials. The results of weather-o-meter test are compared with the results of natural outdoor exposure test, and are discussed reffering to UV energy. However, there is another degradation mechanism. Neglecting the effect of heat, which dominates oxidation of polymer and accelerates photo-chemical reactions, causes less preciseness. In this paper author proposes a service life prediction system of polymeric building materials based on a mathematical deterioration model expressing the property change as a function of UV, heat and elapsed time. By introducing this model, a personal computer software system has been developed. Computer calculates the deterioration material constants in the model and estimates the property decrease in service environment.

2 Mathematical Deterioration Model

Generally, the decrease of property, P of some material is expressed as an exponential function of elapsed time, t (hour):

$$P = P_0 \exp(-At) \qquad (1)$$

where P: material property,
P_0: initial material property and
A: deterioration material constant (1/hour)

In this equation, decreasing material properties are mechanical strength and elongation, gross of paint, concentration of anti-degradation agent and so on. Converting eq.(1), we get:

$$\ln(P/P_0) = -At \qquad (2)$$

This can be applied for the deterioration in constant circumstance. Replacing the right side of eq.(2) with sum of two terms expressing heat and UV deterioration, then:

$$\ln(P/P_0) = C_h \exp(-E_h/RT) t + C_u U^{\alpha} \exp(-E_u/RT) t \qquad (3)$$

where C_h: thermal deterioration material constant (1/hour),
E_h: active energy of thermal deterioration (KJ/mol),
R: gas constant, 8.314×10^{-3} (KJ/mol/K),
T: absolute temperature of material (K),
C_u: UV deterioration material constant (1/hour),
U: UV irradiation energey (MJ/m^2),
α: UV deterioration material constant (-) and
E_u: active energy of UV deterioration (KJ/mol)

This model was proposed by Koike and Tanaka, originally. The first term of the right side is same as Arrhenius' Equation, and the power number: α is introduced into the second term of UV deterioration accelerated by heat. The deterioration material constants are obtained from artificial thermal and UV deterioration tests, respectively.

Author presumes DSET's lifetime prediction models in COSTER I Program is based on another UV degradation model:

$$\ln(P/P_0) = C\ U \exp(-E_h/RT) t \qquad (4)$$

There is no power number or heat deterioration term. Active energy may be obtained from Arrhenius' Plot of thermal degradation test results.

3 Artificial Accelerated Deterioration Test

The thermal deterioration material constants, C_h and E_h are obtained from the thermal deterioration tests. Input the values of property after the aging test at several constant temperatures and testing periods into a personal computer, then material constants are outputted. Three or more temperature levels are recommended. The deterioration

$$+ C_u \sum_{m=1}^{n} U_m{}^{\alpha} \exp(-E_u/RT_m) \triangle t \qquad (5\text{-}4)$$

Eq.(5-4) means that material property at time, t_n can be estimated by integration of every fluctuating deterioration factors, UV energy, U_1, U_2, ..., U_n and temperature, T_1, T_2, ..., T_n. Let time interval, $\triangle t$ to be one hour, the term of time interval is not expressed.

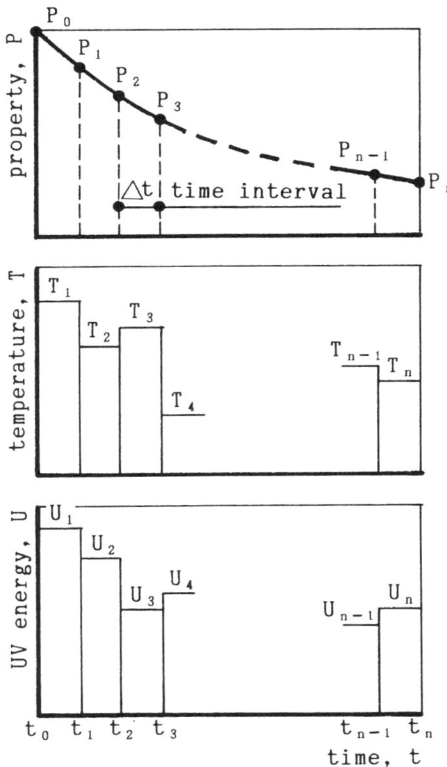

Fig.1. Concept of deterioration simulation

UV energy was hourly observed outdooor and stored in database. Also hourly average temperatures of the other specimen similar to the testing material in specification were collected and stored previously. Input the deterioration material constants obtained from artificial accelerated deterioration tests into computer, the machine calculates the deterioration and outputs the chart showing the relation between the property and elapsed time. Fig.2. is an example of output. The service life is correspond

behaviours of polymeric materials above the glass transition temperature differ from those below the transition point. Therefore, the testing temperature should be checked from this viewpoints.

At the second stage, the effect of UV is analysed using a weather-o-meter. Varing the test conditions of the dose of UV, temperature of specimen and testing period, the data of property change is collected. Here, the temperature of specimen is not represented by the black panel temperature in the test chamber. A Japanese weather-o-meter maker is developing a new model installed temperature monitoring facilities of the surface temperatures of samples and multi-level UV irradiation system using metal mesh screens. Computer outputs the UV deterioration material constants, C_u, α and E_u.

4 Deterioration Simulation Method

The environment where polymeric building materials are used is not constant. In this section, the concept of service life prediction system in fluctuating deterioration condition such as outdoor exposure test is explained.

Elapsed time is devided into many tiny time intervals, $\triangle t$ shown in Fig.1. Assuming the temperature of material, T_1 and the UV energy, U_1 are constant in the first time interval, t_0 to t_1, initial property, P_0 decreases to be P_1:

$$\ln(P_1/P_0) = \ln(P_1) - \ln(P_0) = C_h \exp(-E_h/RT_1)\triangle t + C_u U_1^{\alpha} \exp(-E_u/RT_1)\triangle t \qquad (5\text{-}1)$$

In the next time interval, t_1 to t_2:

$$\ln(P_2/P_1) = \ln(P_2) - \ln(P_1) = C_h \exp(-E_h/RT_2)\triangle t + C_u U_2^{\alpha} \exp(-E_u/RT_2)\triangle t \qquad (5\text{-}2)$$

....

As like as above, in the last n-th time interval, t_{n-1} to t_n:

$$\ln(P_n/P_{n-1}) = \ln(P_n) - \ln(P_{n-1}) = C_h \exp(-E_h/RT_n)\triangle t + C_u U_n^{\alpha} \exp(-E_u/RT_n)\triangle t \qquad (5\text{-}3)$$

Summing up from Eq.(5-1) to eq.(5-3):

$$\ln(P_n/P_0) = \ln(P_n) - \ln(P_0) = C_h \sum_{m=1}^{n} \exp(-E_h/RT_m)\triangle t$$

to the time when the property decreasing curve crosses the allowable level.

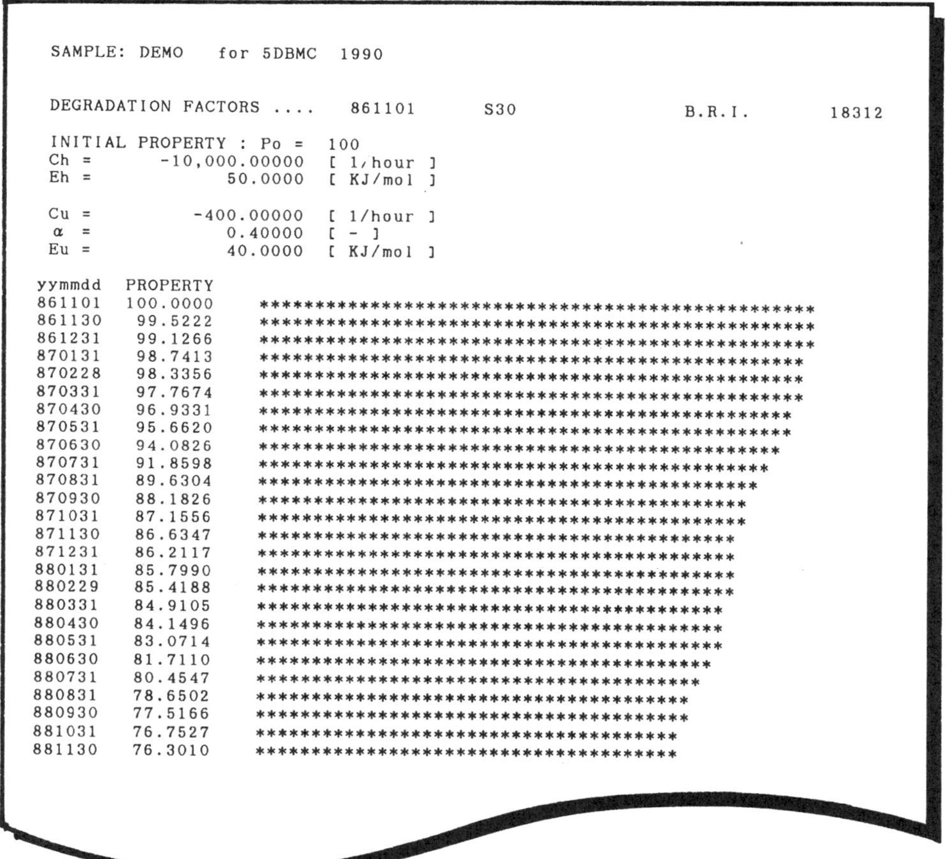

```
SAMPLE: DEMO    for 5DBMC  1990

DEGRADATION FACTORS ....    861101       S30                B.R.I.       18312

INITIAL PROPERTY : Po =   100
Ch =        -10,000.00000   [ 1/hour ]
Eh =               50.0000  [ KJ/mol ]

Cu =            -400.00000  [ 1/hour ]
α  =              0.40000   [ - ]
Eu =             40.0000    [ KJ/mol ]

yymmdd   PROPERTY
861101   100.0000    **************************************************
861130    99.5222    **************************************************
861231    99.1266    **************************************************
870131    98.7413    *************************************************
870228    98.3356    *************************************************
870331    97.7674    *************************************************
870430    96.9331    ************************************************
870531    95.6620    ************************************************
870630    94.0826    ***********************************************
870731    91.8598    **********************************************
870831    89.6304    *********************************************
870930    88.1826    ********************************************
871031    87.1556    ********************************************
871130    86.6347    *******************************************
871231    86.2117    *******************************************
880131    85.7990    *******************************************
880229    85.4188    *******************************************
880331    84.9105    ******************************************
880430    84.1496    ******************************************
880531    83.0714    ******************************************
880630    81.7110    *****************************************
880731    80.4547    ****************************************
880831    78.6502    ***************************************
880930    77.5166    ***************************************
881031    76.7527    **************************************
881130    76.3010    **************************************
```

Fig.2. An example of output of deterioration simulation

5 Personal Computer Software

Author has developed a personal computer software system loaded on NEC's PC-98 serious. This prediction system is supported by N88BASIC/MS-DOS. Operators can input the data guided by the message shown on the display. The following is the function of each software unit.

DGM-1 Calculation of deterioration material constants from artificial accelerated deterioration test results.

Step I: Available for heat deterioration tests. Substitutional for Arrhenius' Plot.

Step II: For UV deterioration tests. It is required that the thermal deterioration material constants are obtained previously.

DGM-2 Deterioration simulation. Output the property decreasing chart.

6 Conclusion

Author has developed a personal computer software system for **Service Life Prediction** of polymeric building materials. Using mathematical deterioration model, property decrease can be expressed as sum of the heat deterioration and the synergistic deterioration by UV and heat in tiny time intervals. Computer calculates the deterioration material constants suitable for the artificial weathering test results and simulates the property decrease of material. Author believes this proposed system can combine the accelerated aging tests and the outdoor exposure tests.

UV and thermal deterioration of polymeric materials predicted by this system may be lower than the outdoor exposure test result. This difference is due to the third degradation factor, water, which is not discussed in this paper. Author is now monitering the dew condensation outdoor to quantify the wetness time.

Refferences

Koike, M. and Tanaka, K. (1978) Light and heat effects on rubber and plastic sheets and films. **Report of the Research Laboratory of Engineering Materials, Tokyo Institute of Technology** No.3, 179-188

DSET Laboratories **COSTER I Multi client research program**

15 EFFECTS OF PULVERIZED FUEL ASH ON CORROSION OF STEEL IN CONCRETE

Y.H. LOO, S.K. TING
Department of Civil Engineering, National University of Singapore, Singapore

Abstract
This paper reports the findings of a preliminary investigation into the effect of pulverized fuel ash (pfa) on the corrosion of steel in mortar specimens using comparative accelerated corrosion tests. For indication of the protection afforded by the covering material against corrosion, three parameters were chosen: the normalised weight loss of steel due to accelerated corrosion achieved by applying a constant voltage across the specimens until cracks were detected, the amount of electrical charge imposed within the test duration, or the 'cracking charge', and the resistivity of the mortar. Other quality indices such as the pH values, and the compressive strength of the mortar were also monitored. It was observed that replacement of cement by pulverized fuel ash of up to 30% generally reduced the electrical energy required to crack the test specimens. The estimated pH values of mortar decreased with the increase of pfa replacement at a water/cement ratio of 0.6, but was relatively unaffected by pfa at water/cement ratio of 0.4. The replacement of cement by pfa in the mortar initially reduced its resistivity (measured at 14 days), but with prolonged curing, was shown to have increased its resistivity at 91 days of age. This phenomenon was observed for both precracked and uncracked specimens alike. Consequently, the measured weight loss of steel due to accelerated corrosion was found to be increased by pfa replacement at 14 days but decreased at 91 days. The importance of curing for durability of concrete with pfa replacement was thus clearly illustrated.
Keywords: Pulverized Fuel Ash, Corrosion, Embedded Steel, Mortar, Cracking Charge, Normalised Weight Loss, Resistivity, Durability.

1 Introduction

Corrosion of reinforcing steel in concrete is often the limiting factor in the service life of concrete structures (Cady [1977]). This ought not to be so as good quality concrete covers were known to have protected steel bars against corrosion for long periods of time even under severe exposure conditions. In addition to providing an effective physical barrier against the invasion of deleterious substances, a good quality concrete cover provides a highly alkaline environment conducive for the passivation of embedded steel. There are two general mechanisms by which this highly alkaline environment and the accompanying passivating effects may be destroyed:

(a) Reduction of alkalinity by leaching of alkaline substances or partial neutralization by reaction with carbon dioxide.

(b) Electrochemical reactions involving chloride ions in the presence of oxygen.

Thus, the loss of protection is often caused by excessive ingress of chlorides and other aggressive ions in the case of coastal structures, and extensive carbonation in general. To prevent these happenings, specifications for concrete covers often stipulate maximum water/cement ratios and minimum cement contents, which effectively limit the permeability of concrete, for various cover thicknesses to ensure durability. In addition, supplementary cementing materials have often been utilized to promote durability of concrete structures.

Pulverized fuel ash (pfa), also known as fly ash, is an artificial pozzolan which has gained a growing field of usefulness. It is a by-product of the combustion of pulverized coal in thermal power stations. By itself, it possesses little or no cementing value, but in its finely divided form, at normal temperatures in the presence of sufficient moisture, it reacts chemically with the calcium hydroxide released in the hydration of Portland cement, forming additional calcium silicate hydrates. As this reaction takes place after the formation of the cement paste structural system, the additional hydrates reduce the pore sizes and the porosity, and hence the permeability of the cementitious system. As a logical consequence, the protection afforded by concrete with pfa against corrosion would be higher than for ordinary concrete.

In order to verify this assumption, it is necessary to compare the corrosion characteristics of test specimens made with plain concrete and concrete with pfa replacement. Due to the slow rate of normal corrosion process, however, accelerated corrosion tests were performed in which an electric current was passed through the test specimens under a constant voltage until the specimens were cracked. This investigation is designed to study the effects of pfa replacement at different percentages on relevant parameters including the weight loss of embedded steel due to corrosion and the resistivity of the cover material.

2 Experimental program

2.1 Test specimens

The test specimens used for this investigation were mortar cylinders of 66 mm diameter and 150 mm length with a mild steel bar of 13 mm diameter embedded in the centre of each cylinder (Fig. 1). The cement used in this investigation was ordinary Portland cement conforming to SS 26 [1984] or BS 12 [1978]. The pulverized fuel ash was imported from the coal power station of Hong Kong Electric Co. Ltd. and was expected to comply with BS 3892 [1982]. The fine aggregate used was natural sand passing BS 2.36 mm sieves.

A total of six mortar mixes comprising two water/(cement+pfa) ratios and three replacement percentages of pfa was used for the preparation of these cylindrical specimens and their companion mortar cubes. The two water/-(cement+pfa) ratios used were 0.4 and 0.6 while the three replacement percentages of pfa used were 0%, 15% and 30% by weight. The sand/cement ratios of the six mixes were kept within the practical range of 1 to 3 with flow measurements of 80% to 90% for mixes with water/(cement+pfa) ratio of 0.4, and 95% to 105% for those of 0.6.

For each mix, twelve cylindrical specimens and twelve 70.7mm cubes were cast, making a total of 72 cylindrical specimens and 72 cubes. Of the twelve cylindrical specimens, six were 'precracked' by means of 0.2 mm thick plastic sheets while the remaining specimens were uncracked. The specimens were moist cured after demoulding at 1 day till they were subjected to accelerated corrosion. For each group of six specimens, three were tested at the age of 14 days while the remaining three were tested at the age of 91 days. The mortar cubes were also tested at these two ages but they were divided into moist cured and air cured groups of three cubes instead.

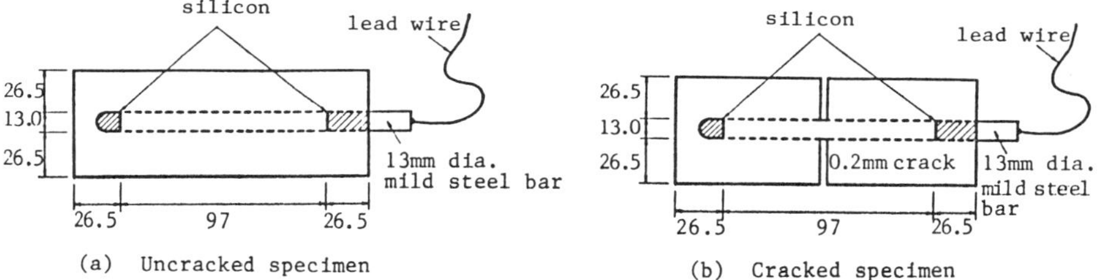

Fig. 1. Cylindrical Test Specimens.

2.2 Test Procedures

At the ages of 14 days and 91 days, the cylindrical specimens were subjected to accelerated corrosion by imposing a constant voltage across parallel circuits formed by the specimens through a 5% sodium chloride solution in which the specimens were immersed, and submerged cathodes in the form of steel cages. To monitor the current passing through each circuit, the voltage across a 1 Ohm resistor incorporated into each circuit was recorded by a TML TDS-301 portable data logger. A schematic diagram of the test set up is shown in Fig. 2.

Generally, the current intensity decreased with time as the buildup of oxidation products tended to obstruct the flow of ions. The passing of current was stopped when a sudden increase in current intensity was observed, indicating the cracking of test specimen. After the cracking was confirmed by visual inspection, the specimen was crushed to release the corroded steel bar which was then cleaned with Clarke's solution and weighed to obtain the weight loss. To keep the duration of testing within two days, the voltages applied were 20V, 30V and 60V for different specimens.

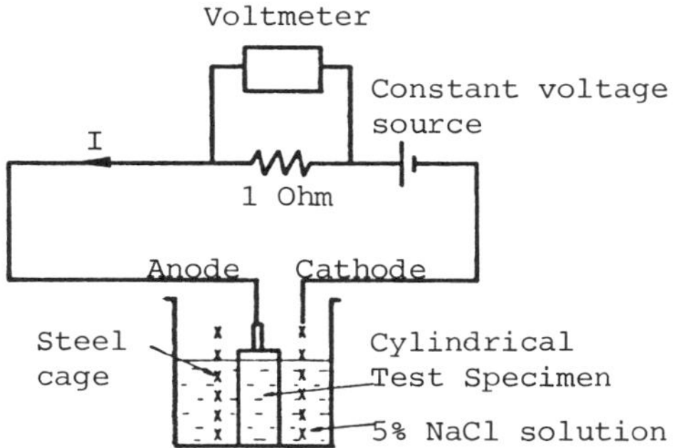

Fig. 2. Set up of the Accelerated Corrosion Test.

As companion specimens, the mortar cubes were tested for compressive strengths, after which, pH values were obtained for the moist cured specimens. Lacking standard testing method for pH measurement of concrete pore solution, the method reported by Szilard and Wallevik [1975] was adopted. Essentially, the pH value of a slurry consisting of 100 gm of crushed mortar, which passes a BS 1.40mm sieve, in 50 ml of de-ionized water was measured 15 minutes after its preparation.

3 Results and Discussions

3.1 Compressive strengths and pH values

As expected, compressive strength decreased with an increase in the percentage of cement replaced by pfa. This decrease was more apparent for the 14 day specimens and became less significant for the 91 day specimens (Fig. 3). This was due to the slower rate of pozzolanic reaction in comparison to cement hydration.

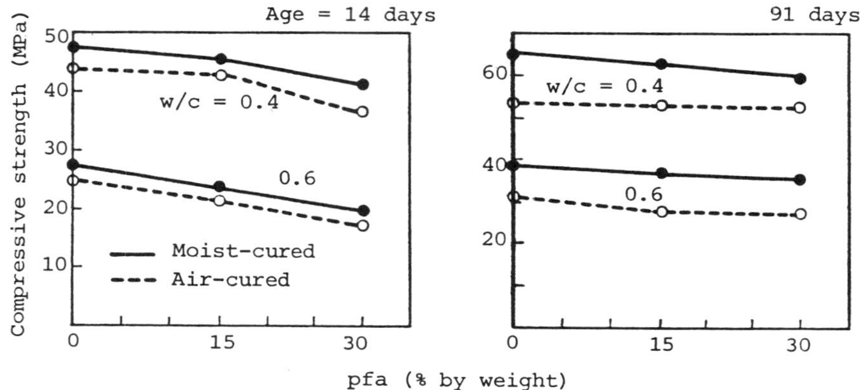

Fig. 3. Variation of Compressive Strength.

On the other hand, the effect of pfa replacement on the pH value apparently depended on the water/(cement+pfa) ratio (Fig. 4). The depletion of calcium hydroxide due to pozzolanic reaction might have caused the reduction in pH value in the case of the higher ratio of 0.6, but did not have much effect in the case of the lower ratio of 0.4, in which the amount of calcium hydroxide crystals was likely to be more than sufficient for replenishment.

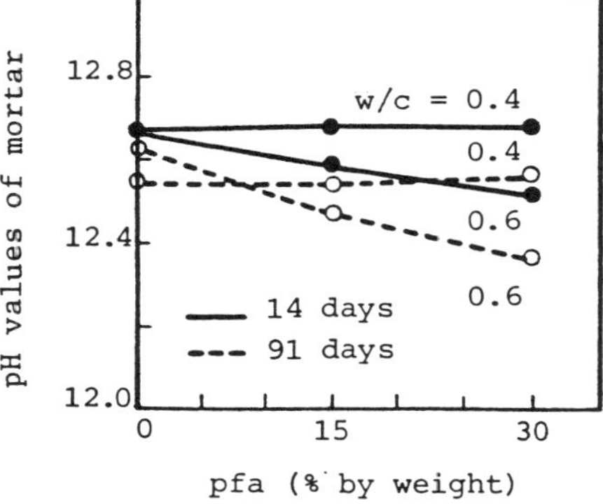

Fig. 4. Variation of pH Values.

3.2 Results of the accelerated corrosion tests

From the data collected in the accelerated corrosion tests on the cylindrical specimens, three parameters were determined as indicators of the degree of protection afforded by the cover mortar.

The total amount of electricity passed before the cracking of a test specimen, determined as the area under the current-time curve up to the point of cracking, was an indication of its mortar's resistance against corrosion current and cracking. It was abbreviated as the "cracking charge" with units of Ampere-hour (A-Hr). Results of the cracking charges are summarised in Fig. 5 from which it can be seen that the replacement of cement by pfa of the same weight generally resulted in a reduction of the cracking charge. The decreasing trend was, however, arrested in the case of the 91 day well cured specimens, thus underlining the importance of curing especially for pfa concrete.

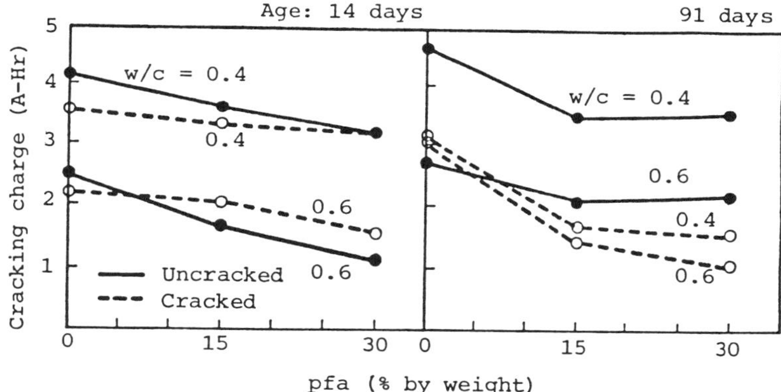

Fig. 5. Variation of Cracking Charge.

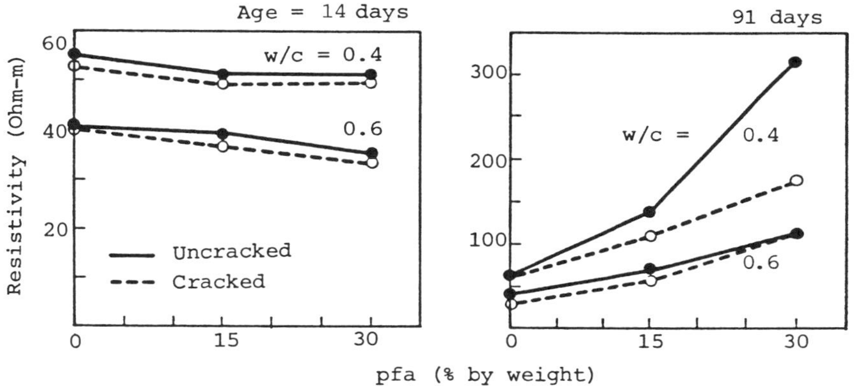

Fig. 6. Variation of Resistivity.

As corrosion of reinforcement is basically an electrochemical process, the resistivity of the cover concrete is a useful indicator of its protection against corrosion. In this investigation, resistivity of the cylindrical specimen was calculated as the product of average circuit resistance and the mid-depth cylindrical surface area divided by the cover thickness. Results of the mortar resistivity are presented in Fig. 6 from which the importance of curing can again be safely inferred. While the pfa replacement resulted in slight decrease in resistivity for the 14 day specimens, the increase in resistivity due to pfa replacement at 91 days was significant.

The third parameter is the normalised weight loss of steel with units of gram per Ampere-hour. An earlier study has shown that its value for a 13mm bare steel bar in 5% NaCl solution was 1.002 gm/A.h. When encased in mortar, it was less than 0.5 gm/A-Hr (Fig. 7). The replacement of cement by pfa showed beneficial effects for the 91 day specimens but not at the age of 14 days.

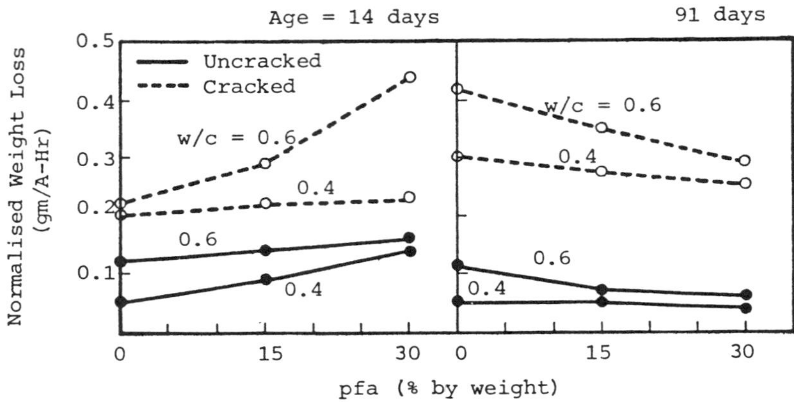

Fig. 7. Variation of Normalised Weight Loss.

4 Concluding remarks

The beneficial effects of pfa replacement on the corrosion resistance of concrete are probably the results of reduced porosity due to pozzolanic reaction. In order to reap the benefits, proper curing seemed inevitably important. As shown in this study, pfa replacement up to 30% by weight increased the resistivity and reduced the normalised weight loss of steel in accelerated corrosion tests on 91 day specimens. As the reduction in pH value was apparent for specimens with a water/(cement+pfa) ratio of 0.6 but insignificant for specimens with ratio of 0.4, the importance to control the water/(cement+pfa) ratio carefully was also illustrated.

5 References

BS 3892 : Part 1 [1982] Specification for pulverized-fuel ash for use as a cementitious component in structural concrete. British Standards Institution, London, U.K.

BS 12 [1978] Portland cement (ordinary and rapid-hardening). British Standards Institution, London, U.K.

Cady, P.D. [1977] Corrosion of reinforcing steel in concrete - a general view of the problem, in **Chloride Corrosion of Steel in Concrete,** ASTM STP 629 (eds D.E. Tonini and S.W. Dean Jr.), American Society for Testing and Materials, pp. 3-11.

SS 26 [1984] Portland cement (ordinary and rapid-hardening). Singapore Institute of Standards and Industrial Research, Singapore.

Szilard, R. and Wallevik, O. [1975] Effectiveness of concrete cover on corrosion protection of prestressing steel, in **Corrosion of Metals in Concrete,** ACI SP-49 (eds L. Pepper, R.G. Pike and J.A. Willett), American Concrete Institute, pp. 47-67.

16 REBAR CORROSION ON BLAST FURNACE SLAG CONCRETE EXPOSED TO CHLORIDES

A. PORRO, A. SANTAMARIA, E. MZ. De ITURRATE,
I. RECALDE, J. BILBAO, J.M. BARCENA
Labein, Bilbao, Spain
T. FERNANDEZ
UPV, F. Ciencias, San Sebastián, Spain

Abstract
Blast furnace slag concrete behaviour against chlorides was examined with regard to rebar corrosion. A series of microconcrete cubic specimens were fabricated with five different blast furnace slag contents, and three depths of cover to the rebars. After a short exposure in an industrial environment with a low chloride content, specimens were located in a salt spray cabinet, according to ASTM-B117. Instantaneous corrosion rates were measured on each rebar by linear polarization. Different behaviours are described based on blast furnace slag content and cover thickness.
Keywords: Rebar Corrosion, Blast Furnace Slag, Chlorides, Concrete, Cover.

1 Introduction

Corrosion of rebars embedded in concrete depends on permanence of passive surface state. Chloride ion diffusion is one of the more important causes of corrosion of reinforcing steel in concrete. There are oppinions about a lower chloride ion diffusion, when a blast furnace slag concrete (BFSC) is used. This could lead to the use of BFSC in situations where rebars are vulnerable to corrosion failure. Some of the concrete made in North Spain is relatively poor in cement content and is located next to the sea shore. This situation could to lead rebars corrosion if BFSC is used.

2 Experimental procedure

The study was conducted by monitoring rebar corrosion in specimens with different amounts of blast furnace slag by weight of cement, exposed to chloride. Cubic microconcrete specimens were made from different cement mixes (as Table 1 shows) with sand and gravel whose biggest sizes were avoided. The mix was as follows:

Table 1. Composition of cements and porosity of specimens.

Clinker (%)	B.F.S. (%)	gypsum (%)	porosity (%)
94	0	6	21,5
80	15	5	21,8
75	20	5	19,5
60	35	5	16,0
30	65	5	23,0

density: 2250 kg/m^3, cement content 250 kg/m^3, W/C=0.75, C/S= 1 : 7.8. Each specimen had four rebars and two graphite counter electrodes. They were cured for 28 days (100% RH, 20±2°C), and then exposed to an industrial atmosphere with a low chloride content (12 mg NaCl/m^2 day, 75% RH, 15.2°C) for 6 months.

Exposure to chloride ion was achieved by placing the specimens in a salt spray cabinet under controlled operation (ASTM B-117). Exposure was divided in two parts, 1 month each, and 1 month exposure to laboratory conditions between them. Every seven days a linear polarization experiment was conducted to estimate icorr, using a PAR M-273 potentiostat, with current interruption option to compensate ohmic drop. Values of 26 and 52 mV were used for the constant, B.

Natural weathering was carried out simultaneously for one year to asses gravimetric and electrochemical average thickness losses.

3 Results

Gravimetric and electrochemical data from natural weathering show good agreement, as other authors have noted. In accelerated tests the icorr measured at regular periods (weekly cycles) is shown in tables 2 to 4 for each depth of cover, and BFS content. When a high BFS content on cement is associated with a high icorr, the electrochemical method of testing used is not suitable.

There is an increase in icorr as the BFS content increases, at 1 cm depth of cover, as Fig. 1 shows. Higher depths (2 and 3 cm) do not show such a clear icorr increase, otherwise there is a slight minimum for 35% BFS content, but this is related to a lower concrete porosity, together with a higher surface to rebar distance.

Table 2. icorr values ($\mu A/cm^2$) at 1 cm depth.

BFS (%)	0	15	20	35	65
Cycle 0	0,37	0,43	0,53	0,46	--
1	2,11	4,50	2,42	2,65	--
2	2,66	3,32	2,07	3,02	--
3	2,62	4,69	3,69	3,76	--
4	2,20	2,26	5,32	4,76	--
dryness	2,31	3,73	5,36	4,61	--
cycle 5	4,33	9,05	11,44	11,84	--
6	4,74	8,47	13,62	12,07	--
7	4,85	8,47	12,04	17,02	--
8	4,95	9,44	11,30	12,27	--

Table 3. icorr values ($\mu A/cm^2$) at 2 cm depth.

BFS (%)	0	15	20	35	65
Cycle 0	0,22	0,26	0,49	0,27	0,32
1	1,63	1,23	1,79	1,05	2,21
2	1,78	1,09	2,14	1,26	2,52
3	1,64	1,50	2,83	0,90	3,18
4	1,76	2,67	3,53	0,99	4,89
dryness	0,89	4,09	3,48	0,85	5,16
cycle 5	2,22	6,05	6,42	2,64	16,25
6	1,94	6,85	7,26	2,62	13,43
7	2,58	6,00	6,77	3,39	14,28
8	2,76	5,81	7,21	2,83	13,29

Table 4. icorr values ($\mu A/cm^2$) at 3 cm depth.

BFS (%)	0	15	20	35	65
Cycle 0	0,32	0,23	0,79	0,35	0,81
1	1,77	2,07	4,50	0,94	1,39
2	2,68	1,67	4,20	1,79	2,79
3	1,76	2,53	3,58	1,37	3,13
4	1,44	2,20	2,95	1,60	3,29
dryness	1,49	2,77	1,25	0,79	2,19
cycle 5	2,33	4,48	5,79	3,22	10,96
6	2,07	4,23	5,93	2,81	40,22
7	3,47	3,87	5,96	3,79	25,98
8	2,76	4,06	8,26	2,52	28,94

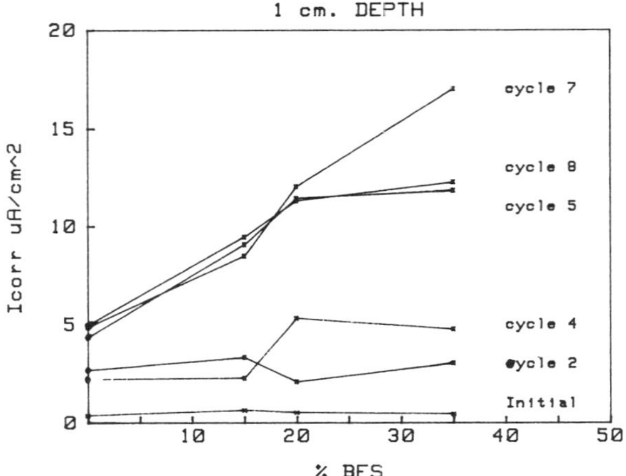

Fig. 1. Values of icorr versus BFS content.

Surface to rebar influence can be observed at Fig. 2. There are not noticeable differences between icorr values from 2 cm or 3 cm depth rebars in circumstances we were working.

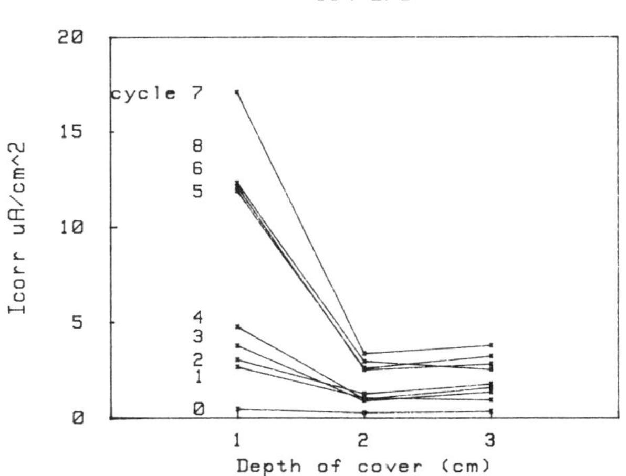

Fig. 2. Values of icorr versus depth of cover.

In the first part of the experiment (cycles 0-4) icorr shows an increase and then a trend to stabilize or slight increase. After one month of dryness this behaviour is seen again (cycles 5 to 8) but the first value of icorr had grown in regard to fourth cycle icorr (fig. 3).

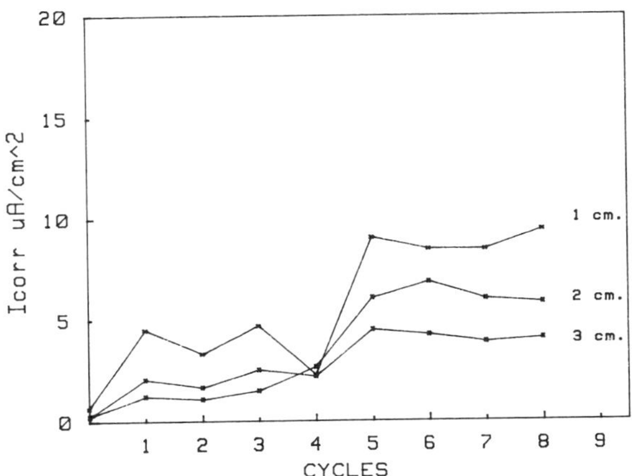

Fig. 3. Values of icorr versus number of cycles.

This situation is associated with the water saturated state of specimens in the cabinet and consequently with a limited oxygen diffusion. During dryness period the oxygen flow increases strongly. This alternance of wetness-dryness condition causes a larger increase in icorr.

4 Conclusions

From experimental data and according to test conditions (poor concrete quality, high porosity, exposure to salt spray), we can conclude:

The BFS content slightly affects corrosion rates.

Corrosion rate increases as the BFS content increases for high porosity concrete, specially at low depths of cover.

Porosity has a clear influence, remarkable at 2-3 cm depth.

There is a cathodic control of corrosion process at medium to high depths, as a consequence of low oxygen diffusion.

As expected, alternating wet/dry conditions are more aggresive than moisture saturation condition.

5 Acknowledgments

We wish to thank Comisión Interministerial para la Ciencia y la Tecnología and Gobierno Vasco for the financial support provided. We are also grateful to Cementos Lemona and Cementos Rezola for the cement samples supplied.

6 References

Andrade, C. Castelo, V. Alonso, C. and González, J.A. (1986) The Determination of the Corrosion Rate of Steel embedded in Concrete by Polarization Resistance and AC Impedance Methods, in **Corrosion Effect of Stray Currents and the Techniques for Evaluating Corrosion of Rebars in Concrete**, ASTM STP 906 (ed. V. Chaker). Philadelphia, pp. 43-63.

Gjorv, O.E. Vennesland, O. and El-Busaidy. (1986) Diffusion of dissolved oxygen through concrete. **Materials Performance**, December, 39-44.

Muravljov, M. and Jevtic, D. (1987) Resistance to Corrosion of Portland Cement Concrete and Concrete with Slag. **Structural Faults and Repair**, 1, 245-248.

Page, C.L. Short, N.R. and El Tarras A. (1981) Chloride ions in hardened cement pastes. **Cement and Concrete Research**, 11, 395-406.

Rose, J. (1987) The Effect of Cementitions Blast-Furnace Slag on Chloride Permeability of Concrete, in **Corrosion, Concrete, and Chlorides. Steel Corrosion in Concrete: Causes and Restraints.** SP 102-7 (ed. F.W. Gibson). American Concrete Institute, Detroit, pp. 107-125.

17 DESIGN OF MODELS AND FINITE-ELEMENT-CALCULATION FOR LOAD DISTRIBUTION IN SEDIMENTARY ROCKS INCLUDING THE INFLUENCE OF SILICIC-ACID-STRENGTHENERS

K. HÄBERL, A. RADEMACHER
IWB–Universität Stuttgart, Germany
G. GRASSEGGER
Lanesden Kmalaut B.–W., FMPA, Stuttgart, Germany

1 Abstract

Nowadays the conservation of deteriorated sandstones is mainly carried out using acid esters (SAE). The SAE are applied in a liquid form which, after a hydrolizing reaction, transform to an amorphous gel ($SiO_2 \cdot nH_2O$). The gel production is only 35 % by weight and is supposed to strengthen the stone fabric. There are hardly any experimental and no theoretical investigations into the structure of the gel in the pores of the stones, or what form the mechanical connections to the grains should have to increase the strength. For this reason each stone and case of stone deterioration has to be tested separately to determine the strengthening effect.

It is the aim of this study to present simplified models of natural stone fabrics and load distribution models of the stone within the micro-range. Models of different fabric types and their changes due to weathering were designed. Based upon these, mechanical effects of the strengthening were calculated. The models have delivered boundary conditions and quantitative results about the stresses on single grains submitted to compressive loads.
Keywords: Models, Stone fabrics, Mechanical effect of stone strengthening, silic acid esters, Finite-element-calculations.

2 Simplified fabric types of sedimentary rocks

For common types of sandstones and limestones the fabrics may be differentiated between:

1. Sandstones, there exist mainly 3 types of "cement" (fig. 1): contact cement, pore-filling cement and a floating grain structure.
2. For the limestones there exist, besides the influence of impurities, only direct calcite-calcite-grain contacts. Since the basic material is monomineralic, as a first approximation an isotropic fabric was assumed which is mainly influenced by pores.

Fabrics of sandstones

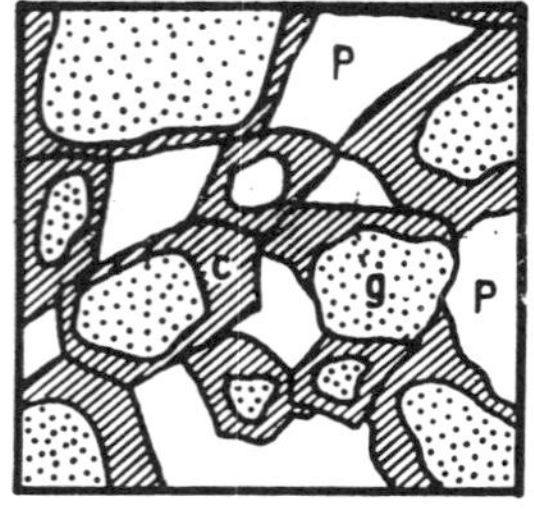

g = grains
c = cement
p = pore system

contact cement

g = grains
c = floating grain structure
p = pore system

floating grain structure

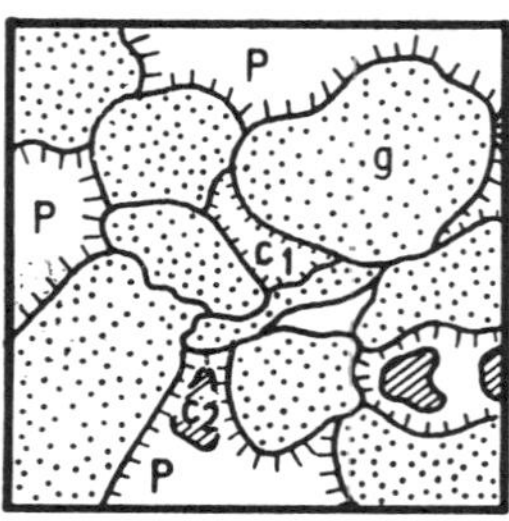

g = grains
c 1 = cement 1
c 2 = cement 2
p = pore system

pore cement

Fig. 1 Schematic representation of different types of cement fabrics. Contact cement means that the grains are directly connected. Pore cement fills the pores. In the floating grain structure the grains are not directly in contact with one another.

3 Development of load distribution models from different structure types

3.1 Sandstones with a floating grain structure (wackes)

The model developed for such sandstones (Fig. 2) has bearing behaviour largely corresponding to the model for concrete (Rehm, Diem, Zimbelmann, 1977). It contains stiff grains in a soft matrix, as for normal concrete the load distribution takes place along the shortest distance between the grains; the resulting tensile stresses between the grains as a result of load distribution are carried by the matrix. The more equal the elasticity modulii of the grains and the matrix are, the lower is this influence. The properties of the matrix are decisive.

3.2 Models for sandstones with pore cements and contact cements

For contact cements a model with contact areas consisting of a sphere packing may be applied. The properties of the model vary with the bonding areas (fig. 3).

The load distribution takes place from sphere to sphere or grain to grain. The tensile stresses occuring under compressive load must be absorbed by the contact cement

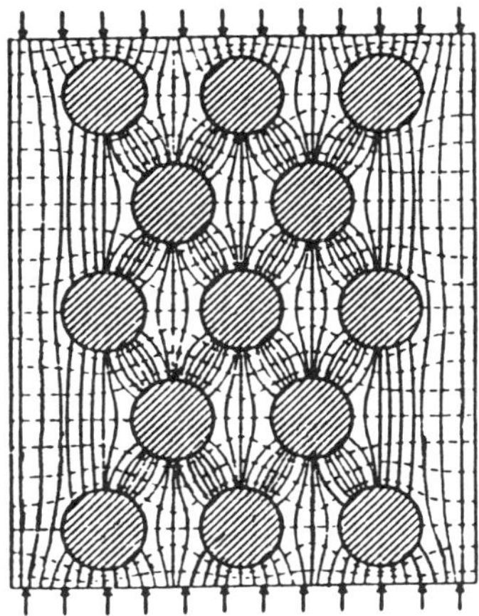

Fig. 2 Matrix model with stiff grains and relatively soft matrix which simulates a floating-grain structured sandstone. Representation of the stress distribution with compression and tension stress lines according to Rehm, Diem, Zimbelmann (1977). (Normally used for study of the bearing capacity of "normal" concrete)

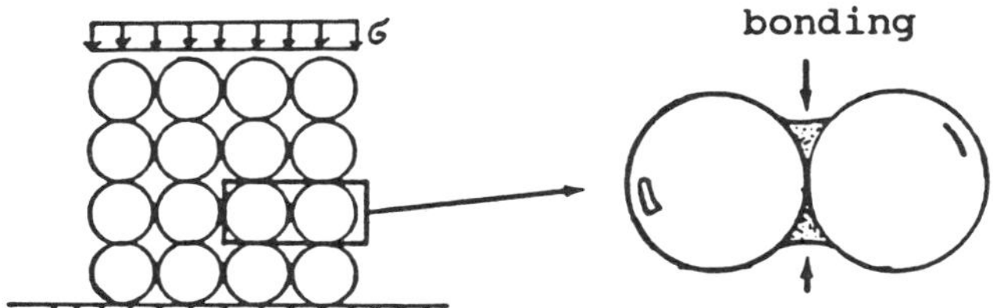

Fig. 3 Development of the sphere model for a sandstone fabric with contact cement (point contact).

and the direct contact between the grains.

Pore cements are normally too soft to contribute to load distributions therefore their influence may be neglected (Pore cements are those types of diagnetic cements which appear as coatings, fibres or granular cements in pore spaces).

3.3 Limestones and highly porous sandstones

The model for limestones and highly porous sandstones with a floating grain structure is based on a change in the

Representation of the derivation of the porous matrix model

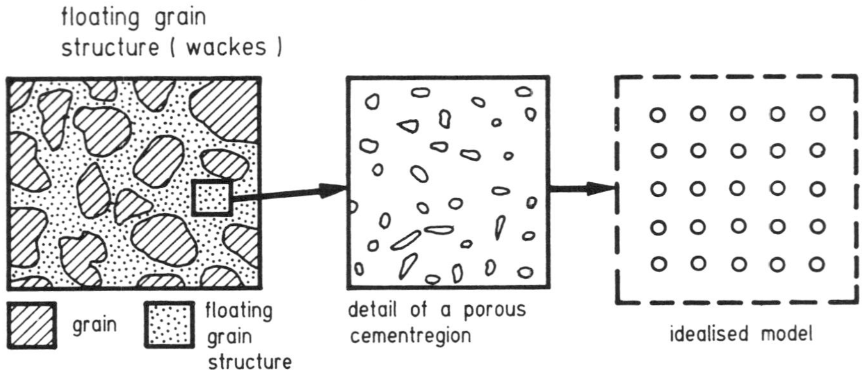

Fig. 4 Model of a porous matrix used to simulate the load distribution in a floating grain structure sandstone with a high pore content, which has a large influence on the bearing capacity.

strength of a matrix with porosity.

A model with a porous matrix (fig. 4) was developed. Under external compressive stress, forces are distributed around the pores.

This model can be applied to limestones with an isotropic matrix which contains a variable proportion of pores, besides the dependancy of the grain size. Other heterogenities such as texture effects were not investigated because the stiffness of the carbonate particles should be very similar even though they might have a different genesis (Dreyer, 1966)

4 Models for the deterioration due to weathering and strengthening with silic acid esters

Based on 3 models of grain contact (fig. 5) the load distribution of grain contacts was calculated by means of finite-element-analysis before and after a strengthening treatment. The Ansys-program, a finite-element-program under linear-elastic conditions was used.

4.1 Mechanical boundary conditions

As boundary conditions for the calculation it was assumed that the system follows the mechanical relations described below. That is to say that during tensile stress = 1 strain differences and, thus, different trajectory vectors appear as a function of the difference of E-moduls (fig. 6 and 7). Differences in the direction of the trajectories

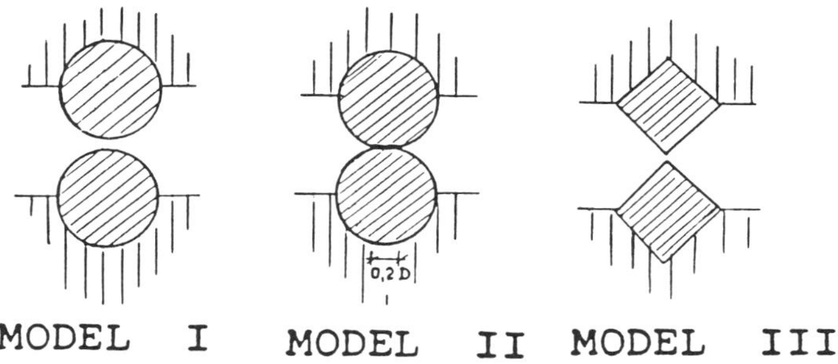

Fig. 5 Grain contact model for calculation of stress distribution before and after a SAE strengthening
I : Spherical elements which have to be bridged
II : Spherical elements with about 20 % contact, the contact area is enlarged.
III : Square elements with contact at the corners which have to be bridged.

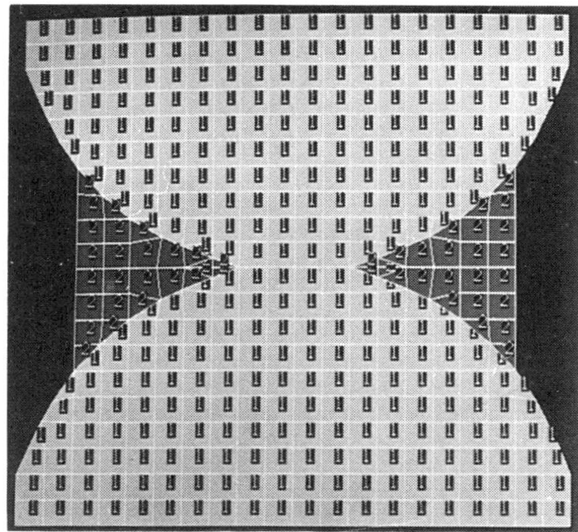

Fig. 6 Finite element simulation of the load distribution between strengthening gel (dark grey) and quartz (fair grey). The grain contact model II is used and the gussets are filled with increasing numbers of gel layers. The major stress trajectories are shown, they indicate that stresses in the gel decrease with the rising number of layers.

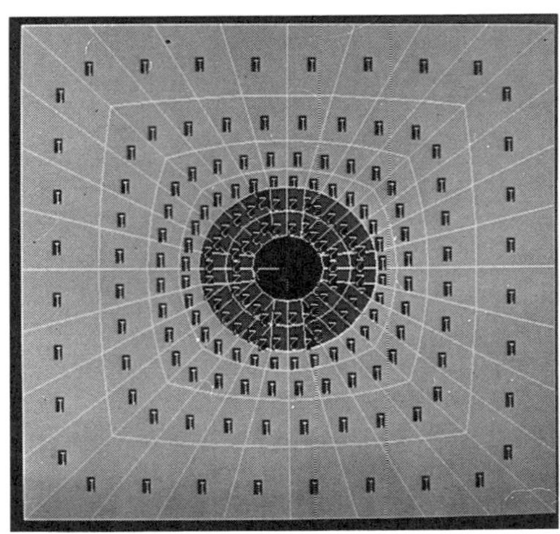

Fig. 7 Simulation of a porous matrix model with SAE gel filling using the ANSYS-program. The arrows show the direction of the major stress trajectories during loading. The stone fabric is shown in light gray and gel is dark gray. The length of the trajectories are proportional to the stresses and show that the stresses in the gel are very small.

of main stresses and the size of the stresses produce shear stresses in the contact area between gel and grain (fig. 6).

The E-modulii of the separate materials were taken into consideration because they can be regarded as being important: quartz 96, feldspars 63-75, silic-acid esters 1,5-15kN/mm^2. A comparative calculation was carried out for (theoretical) SAE E-moduli of 7,5 and 100. The deformation characteristics was estimated from SAE bound mortars to be 5 to 15 kN/mm^2 (see Ettl, 1987 and various measurements on glass). The contact gel-grain is assumed to be solidly connected. In these areas the strengthener may be assumed to be homogenous because the distances beween the cracks are normally larger than the crack widths, this is also indicated by the tensile strength values of Ettl, 1987. According to the calculations the Poissons's ratio (0,3) has only a minor influence.

4.2 Bridgings and strengthening of grain contacts

Fig. 8 shows a comparison of the three models (fig. 5): the open contact is closed, the existing contacts area is reinforced and acute-shaped grain contacts were bridged by gel layers.

The results show that for model I the largest additional forces may be carried by SAE. Above a gusset filling of 50 % by volume the tension area and the peak tensions change only inconsiderably so that no appreciable increase can be attained. Also for model II no further increases may be attained when more than 40 to 50 % by volume of the gussets are filled . Modell III shows that up to a filling of 36 - 64 % by volume the main effect already occurs.

Fig. 8 represents a summary of all results including a variation of the E-moduls of the strenghtener. The difference of the strengthening "quality" between the different types of grain contacts (models I to III) may be seen.

The relative transferable force in the stone fabrics by the fomation of gels (silicic - acid esters) upon grain contacts according to model I - III

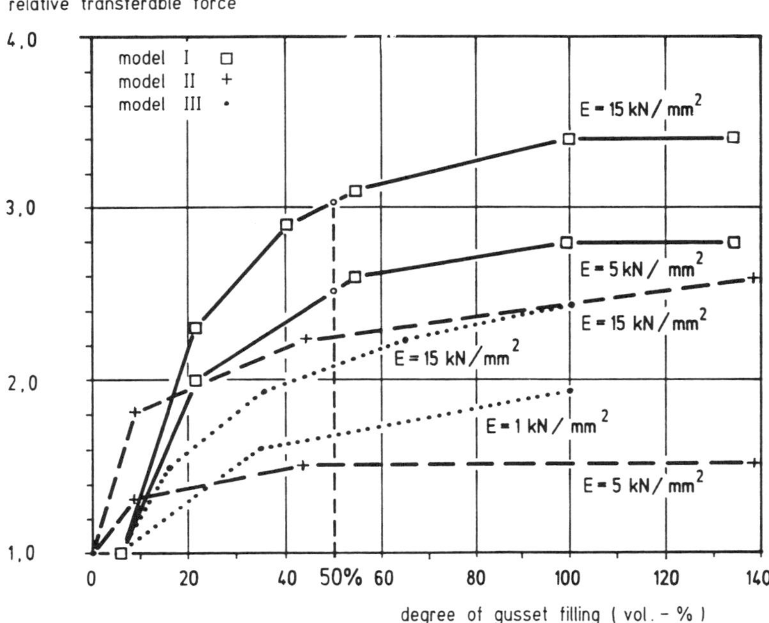

Fig. 8 Results of the finite element calculations of bridging in grain contact models (fig.5). The increase in load bearing in the gel was calculated relative to the model without gel. During the calculations the gels were assumed to be layered fillings in gussets.

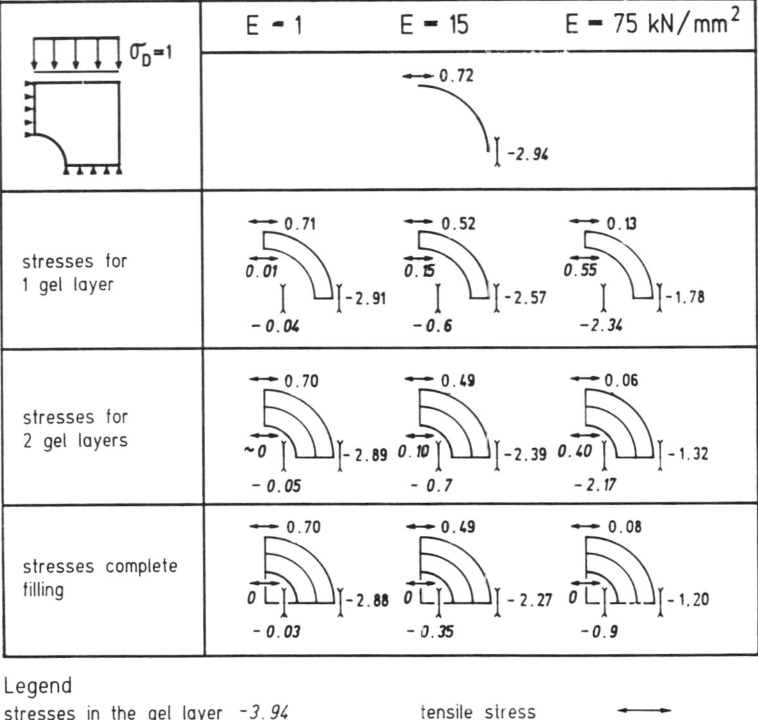

Fig. 9 Summary of the results of the stresses in the porous matrix models under compressive loading (compressive stresses negative). The neglectable effect of increasing the pore filling in spherical pores with SAE gel is shown. The calculations were carried out using the ANSYS finite element program.

4.3 The effects of pore filling

For limestones or sandstones with a high proportion of matrix the effect of gradually filling the pores (fig. 7) was studied using a finite-element-calculation. The results are shown in fig. 9. It can be seen that with a low E-modulus of the gels in relation to the stone the pore filling has hardly any stress-reducing effect. Above a theoretical E-modulus of 75 kN/mm^2 the gels assume a bearing function.

5 Conclusions for practical applications

1. On the basis of test results from stone strengthenings using several applications it was found that above a certain amount of strengthener, no further increase in strength may be attained (Ettl and Schuh, 1987, Schuh, 1987). The explanation for this effect as determined by the calculations was that the first gel layers in the gusset-position (up to 50 %) are decisive for the increase of strength. Further quantities of strengtheners have no additional effect but only fill the pores or form coatings, which from the mechanical point of view are unimportant.
2. The maximum increase of strength for a stone can be clarified theoretically by the number of grain-grain-contacts which are bridged by intermediate fillings. The type of grain contacts (see models I to III) which have to be bridged plays an additional role.
3. For deteriorated stone areas whose open grain contacts and crack widths are determined in advance by microscopic investigations, a prediction as to whether a strengthening can be achieved is possible.
4. From the mechanical point of view stiff stone strengtheners should be prefered to restore grain-contacts, if the intention is to have load bearing systems again.
5. An "overstrengthening" creating a stiff outside layer in comparison to the stone, cannot occur according to the calculations if SAE is used, because the gels have a lower E-modulus than the original cement.

6 References

Dreyer, W. (1966) Quantitative Untersuchungen über die Festigkeit einfach strukturierter Gesteinsarten in Korrelation zu den Gefügeparametern und dem Mineralgehalt der Akzessoren. Proc. 1 congess of the soc. of rock mechanics, Lisbon 133-142.

Ettl, H. (1987) Kieselsäureestergebundene Steinersatzmassen. Münchener Geowiss. Abh., Reihe B, 5, Verlag Friedrich Pfeil, München.

Ettl, H. and Schuh, H. (1987) Konservierende Festigung von Sandsteinen mit Kieselsäurereethylester, B+B, 12, 35-38.

Rehm, G. Diem P. and Zimbelmann, R. (1977) Technische Möglichkeiten zur Erhöhung der Zugfestigkeit von Beton , Deutscher Ausschuß f. Stahlbeton Bau, Heft 283.

18 DURABILITY OF GLASS

N.A. NILSSON
Lund Institute of Technology, Sweden

Abstract
This paper illustrates the effect of the size, the duration of the load and the environment on the rupture strength for window glass and other structural glazing. A glass failure prediction model will be presented here which takes the size dependence, the time depence and the effect of the environment into consideration.
<u>Keywords:</u> Glass, Structural glazing, Window glass, Durability, Size dependence, Time dependence.

1 Introduction

The concept durability has become a concern of time. The durability of glass is its ability to maintain its original characteristics during a certain period of time. Because glass can be affected by both loads and the environment in which it is operating its characteristics may change in time. The durability of a glass structure should not be seen as something absolute. Instead it is more appopriate to consider the time period during which the main characteristics will not fall below certain prescribed limits.

Today glass window design is based only on the window size and the value of the load. The designer works with manufacturer charts and recommendations which often are based on emperical information. It is a fact that the duration of the load and the environment in which glass windows operate is of importance in determining the rupture strength and service life. The effect of the environment consist of many parts. The temperature and the humidity will alter the creep characteristics and environmental agents may give rise to deterioration processes. Both aspects are of importance by themselves but also the combination is sometimes even more important. A glass failure prediction model will be presented here which takes the size dependence, the time depence and the effect of the environment into consideration.

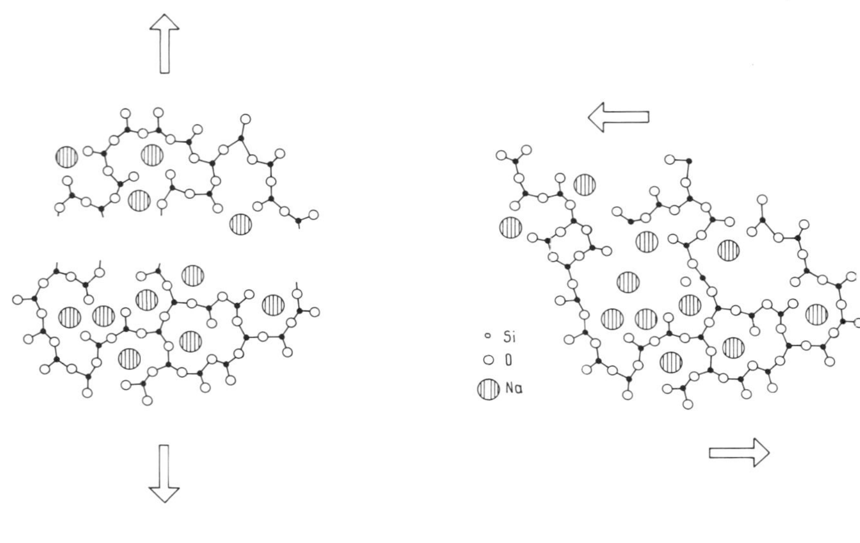

Fig.1. Volumetric and distortion failure in glass.

2 The nature of glass

Glass is a supercooled liqued. The basic unit of the glass structure are tetrahedral groups of SiO_4 in which the central silicon atom is surrounded by four oxygen atoms. These tetrahedra are joined together at their corners only, not along their edges or faces. The angles between the bonds and the distances between the atoms vary.

One feature of the glassy and crystalline structures is that since they are built up from similar polyhedra joined only at their corners, the structures are relatively open and contain relatively large voids. The sodium ions in a Na_2O-SiO_2 glass are situated in the voids in the network formed by the SiO_4 tetrahedra. The oxygens ions introduced together with the sodium ions in the form of Na_2O are incorporated by the rupture of some of the Si-O-Si linkages which are present in the SiO_2 glass.

When a body is subjected to stress, these stresses will deform the body. The deformation which takes place can be characterized as volumetric strain or distortion strain. The first type of strain reflects an increased or decreased distance between the constituents dependent on the stress field while the second type of strain is a shear like movement which is accompanied with internal continious deformation. Volumetric failure and distortion failure in glass are shown in figure 1.

The theoretical tensile strength of glass is about ten thousends times larger than the practical value. It has been proved that microscopic or submicroscopic surface scratches or flaws reduce the strength of glass considerably. The drawing of small diameter glass fibers in a controlled atmosphere is one way to avoid these flaws. The resulting fibers can demonstrate tensile strengths approaching the theoretical atomic bond strength of the material. This helps to make them excellent reinforcing fibers for composite systems.

When flat glass breaks, the fracture nearly always originate from a single point situated somewhere on the surface or at the edge. It is unusual to find a fracture starting from an internal point, unless the glass is of so poor quality that it contains foreign inclusions. It is important to differ between the causes to the failure, i.e. if the flaws are internal or on the surface or the edge.

Glass is usually regarded as resistant to chemical corrosion and it is used in applications such as enamel coatings for metals and as linings for water heaters and vessels to contain certain chemical reactions. It is therefore surprising that small amounts of water vapour, normally found in the atmosphere, react with glass under stress to cause a time dependent reduction in strength known as static failure or creep to failure. One example of this is the fact that window glass compositions are not completely resistant to atmospheric attack. In a stationary atmosphere containing water vapour, the film of water which is formed on the glass surface becomes highly alkaline and very corrosive as sodium ions are leached out from the glass.

Glass has also a healing effect. One bigger flaw will under good conditions change to several small flaws.

3 Glass strength characterization

The rupture strength of a glass window or some other glassware can be characterized with a stress or strain method. The stress approach, which today is the most common in standard engineering methods, does not contain any information about the strain behaviour.

There are some standard methods available with which the strain behaviour can be analysed. A new alternative method developed by Sentler [5] based on stochastic strain solution will be used here.

The rupture behaviour is analysed according to the principle of minimum strain. Failure is assumed to occur when the volumetric strain or the distortion strain, or a combination of both reaches some minimum value. The volumetric strain correspond to normal stresses and the distortion strain correspond to shear stresses. The minimum value will reflect the flaw characteristics of the material and can not be defined in exact terms. Therefore the rupture surface in time and space will be of stochastic nature.

For a homogeneous material, e g glass, subjected to a tensile stress it is assumed that the ultimate strain, ϵ_u, will depend on the the surface area under stress, A, the volume under stress, V, the duration of this stress, D, the temperature, T, and the humidity in the atmosphere, H. In general therms this can be expressed as

$$\epsilon_u = g(\sigma;A,V,D,T,H) \tag{1}$$

there g() is a strain function. The strain in a material can take place in two different ways, as volumetric strain or as distortion strain.

Because of the existence of flaws, the ultimate creep behaviour is not determined by the average material behaviour. Instead the ultimate strain will reflect the weakest part of the material. This behaviour is considered in the most rational way within the statistical theory of extreme values.

Sentler's original formulation is based on a limitation of the strain behaviour. The corresponding stress is obtained by taking the mean value with respect to the stress introduced in a body. Because σ can be expressed in terms volumetric stress or distortion stress there will be several solutions. Here two solutions are presented.

$$\sigma = \sigma_0 \cdot \left[a + b \cdot \left[\frac{A_0}{A} \right]^{1/kA} \cdot \left[\frac{D_0}{D} \right]^{1/h} \right] \tag{2}$$

$$\sigma = \sigma_0 \cdot \left[a + b \cdot \left[\frac{T_0}{T} \right]^{1/p} \left[\frac{D_0}{D} \right]^{1/h} \right] \tag{3}$$

4 Parameter estimation

Information concerning the parameters in the stochastic strength theory is not allways available in a form which makes it directly applicable, but as information which can be used for parameter estimates.

The surface and time dependence is determined by the two constants kA and h. They can be estimated with traditional statistical methods with information from table 1. The manufacture procedure gives a residual stress on about ten MPa which must be taken into consideration. The values will be estimated to $kA \approx 4$ and $h \approx 16$. Figure 2 gives information to estimate the temperature dependence

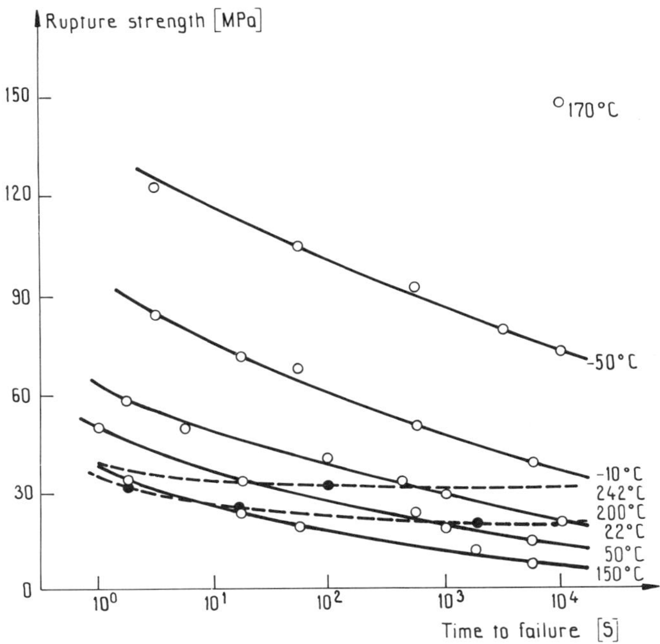

Fig.2. Creep to failure for glass [1].

constant, which is estimated graphically to a approximately value about $p \approx 0.32$. The mean failure surfaces by the equations (2) and (3) are shown in figure 3.

Table 1. Test data for glass panels of the size 1525 mm x 2440 mm x 6 mm [3].

Serie	Rate of loading (kPa/s)	Rupture strength Mean (MPa)	Rupture strength SD (MPa)	Time to failure Mean (s)	Time to failure SD (s)
ONT-A	0.15	66.2	18.9	36.2	6.8
ONT-B	1.5	76.8	18.9	4.4	1.0
ONT-C	15.	80.5	25.4	0.44	0.09
ONT-D	0.0025	43.7	12.1	1364.	491.
ONT-E	0.025	54.3	13.4	168.	43.
ONT-F	0.25	62.4	19.6	22.7	4.1
ONT-G	2.5	67.5	23.2	2.4	0.7
ONT-H	25.	72.1	19.4	0.21	0.07

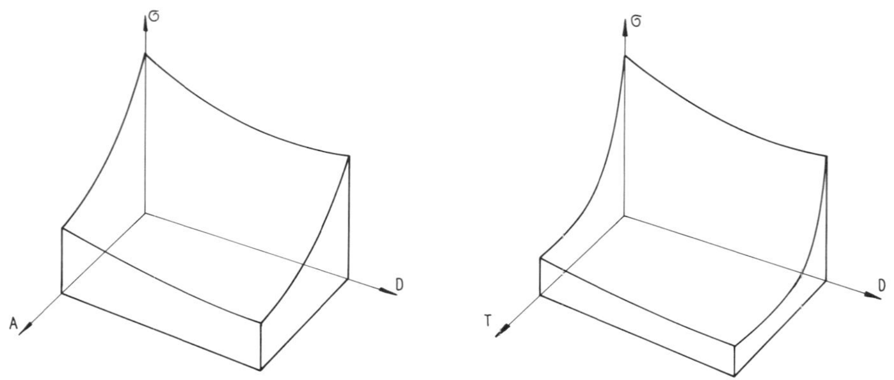

Fig.3. The mean failure surfaces.

5 Conclusions

The design process for window glasses in buildings is needed to undergo a major change. Today several changes in the design environment has made it increasingly more difficult for the designer to rely entirely upon the original glass design charts and manu-facturer recommendations. Glass is being used in larger sizes in higher buildings which are sited in more severe wind environments. The atmosphere has become more aggresive which creates problems with the durability.

It is possible to consider the durability of glass by the combination of both loads and environment with the statistical theory of extreme values. Solutions can be derived which take the effect by the size, the duration of the load, the temperature and the humdity in the atmosphere into consideration.

6 References

[1] Charles, R.J. (1958) Static Fatigue of Glass II. J.Appl.Physics., Vol 29, No 12, pp 1554-1560.
[2] Minor, J.E. (1981) Window Glass Design Practices: A Review. J.Struct.Div., Vol 107, pp 1-12.
[3] Ontario Research Foundation (1982) Dynamic Fatigue of Flat Glass. Report 67039 and 67049.
[4] Rawson, H. (1980) Properties and Applications of Glass. Elsevier Science Publishers B.V., Amsterdam.
[5] Sentler, L. (1987) A Strength Theory for Viscoelastic Materials. Swed.Coun. for Building Research, Stockholm.

19 STUDY ON THE PROTECTION OF REINFORCED CONCRETE BY PENETRATIVE CORROSION INHIBITER

S. USHIJIMA, A. KOBAYASHI, I. KAMURO, M. KOSHIKAWA
Shibuya-ku, Tokyo, Japan

Abstract
Concrete structures have been damaged mainly by the corrosion of the reinforcing bars due to chloride included in the concrete. Deterioration appears in the form of cracks and spalling of the concrete cover. This paper describes a coating method applied to the finished concrete surface with a special penetrative corrosion inhibiter which can penetrate and diffuse into the concrete by density-diffusion. The method proposed for preventing deterioration is focused on the ratio between NO_2^- given by the coat and Cl^- contained in concrete and NO_2^- density diffusion achieved by the volume of coating allowing for the existing cracks.
Keywords: Salt Damages, Repair, Penetrative Corrosion Inhibiter, Corrosion, Monitoring.

1 Introduction

Chloride-induced damage to concrete is classified into the following.
(1) Damage caused by chloride included in the fresh concrete and reinforcing bars before/during construction.
(2) Damage caused by chloride infiltrating into the finished concrete from the environment after constraction.

Preventives measures against salt damage have been recently developed in special mixes by adding a large amount of high performance rust. The cost of the special mix is, in general, 20 to 25 % higher than the ordinary mix and it often experiences a problem of slump loss caused by calcium nitrite which is the main component of high performance rust preventive. The proposed method is to ápply a penetrative coat onto the finished concrete surface where the penetrative corrosion inhibiter and diffuses into the concrete around the reinforcing bars through density-diffusion, consequently corrosion is prevented.

2 Anti-corrosion effects on rust-free re-bars

2.1 Outline of tests

Details of prepared specimens are as shown in Fig. 1 using the design mix with W/C ratio of 0.70 (Cement; 247kg/m³, Water; 173 kg /m³), sand coarse aggregate ratio of 0.50 and NaCl mixed according to the required Cl^- content levels such as 0, 600, 1200 and 1800 g/m^3 All specimens are reinforced with rust-free reinforcing bars with 2 cm cover and the proposed corrosion inhibiter is coated on the nearest face to the reinforcement. An accelerated corrosion test is then conducted in four cycles using the alternate immersion and autoclave technique.

2.2 Test results

Required NO_2^-/Cl^- mole ratio for anti-corrosion effects is approximately 0.6 to 0.7 as shown in Fig. 2 irrespective of Cl^- content and is smaller than the figure of 0.82 which is theoretically estimated on the basis that corrosion inhibiter is mixed in. This could be explained by the fact that NO_2^- is of a high solubility and submersion under the high pressure during autoclave process may cause NO_2^- to accumulate near the reinforcing bars.

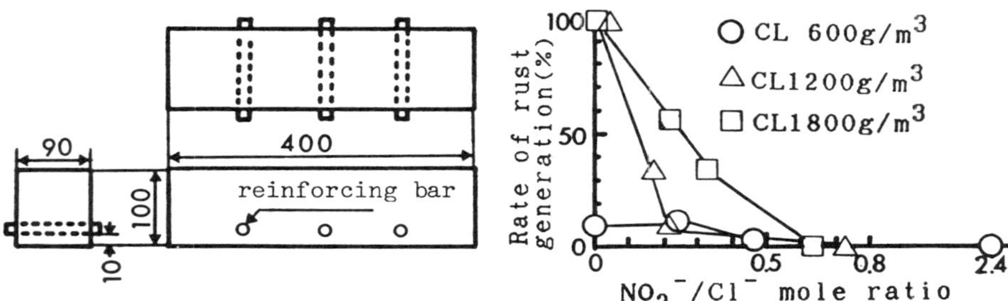

Fig. 1 Details of specimens

Fig. 2 Anti-corrosion effects

3 Anti-corrosion effects on corroded reinforcing bars

3.1 Outline of tests

All specimens are prepared and tested in the same manner as in 2.1 except reinforcing bars which already show a 32 mg/cm^2 corrosion level. Anticorrosion effects are examined through the measurement of corrosion depth and the quantitative analysis of NO_2^-.

3.2 Test results

The relation between NO_2^-/Cl^- mole ratio and decrease in corrosion is as shown in Fig. 3 and following exponential curve is fitted.

$$y = e^{6.26 - 0.31X} \qquad (1)$$

Where correlation coefficient r: 0.88
y: decrease in corrosion
x: NO_2^-/Cl^- mole ratio

NO_2^-/Cl^- mole ratio required to prevent further corrosion is in a range between 1.8 and 2.0 as shown in Fig. 4 corresponding to the decrease in corrosion of 20 mg/cm^2 measured at 5 mm from the both edges of reinforcement. Mole ratio required to prevent further corrosion on corroded reinforcing bars is far greater than the value on corrosion free reinforcing bars. This could be explained as follows.

NO_2^- ions diffuse and permeate through the corrosion sediment and reach the fresh metal surface. NO_2^- ions then act on easily isolatable Fe^{2+} forming a passive film to prevent the further corrosion.

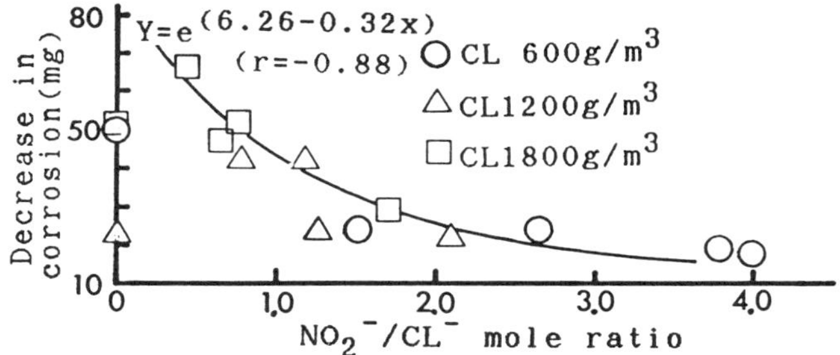

Fig. 3 NO_2^-/Cl^- mole ratio and decrease in corrosion

4 Permeability and anti-corrosion effects on cracked structure

4.1 Outline of tests

The permeability of the corrosion inhibiter into a crack is examined by using a purpose-made cracked specimen as shown in Table 1. After a crack has developed to the specimen's requirements the prescribed quantity of corrosion inhibiter is coated over the cracked surface. The coated specimen is then cured for 30 days at 20 °C room temperature and 50 % relative humidity. The corrosion test is conducted by spraying artificial sea water over the cracked surface once a week for a period of one year under the constant environment.

Table 1. Factor and Level

Factor	Level	
Cl ion in concrete	0, 1200	g/m^3
crack width	0, 0.1 0.2, 0.4 0.6, 0.8 1.2	mm
Application of corrosion inhibiter	0, 500, 1000	g/m^2

4.2 Test results

Concentration/distribution of NO_2^- ions in the crack is shown in Fig. 4. i. e. deeper distribution is achieved in a wider crack and by applying greater quantity of corrosion inhibiter.

The relationship between NO_2^-/Cl^- mole ratio and corrosion area ratio is as shown in Fig. 5. i. e. the corrosion area ratio increases sharply when the crack width exceeds 0.1 to 0.2 mm.

No significant difference is observed in various Cl^- content levels. i.e. This could be explained by the fact that the large portion of Cl^- ions are fixed as Friedel's salt.

It is therefore concluded that the corroded reinforcing bars in concrete even under cracked conditions of 0.1 to 0.2 mm in crack width could be effectively protected by the application of 1000 g/m² corrosion inhibiter.

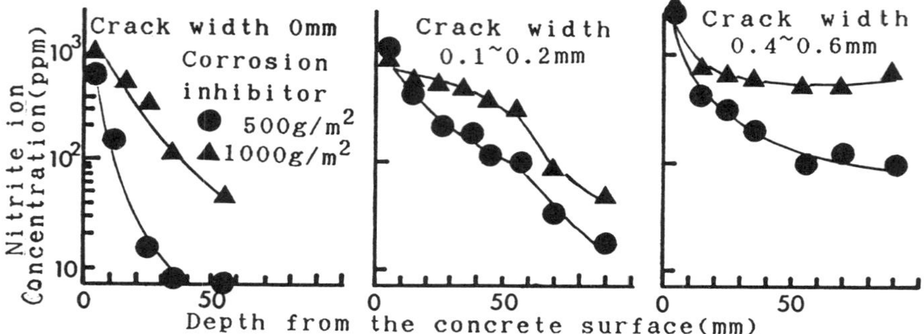

Fig. 4 Permeability of nitrite into a crack

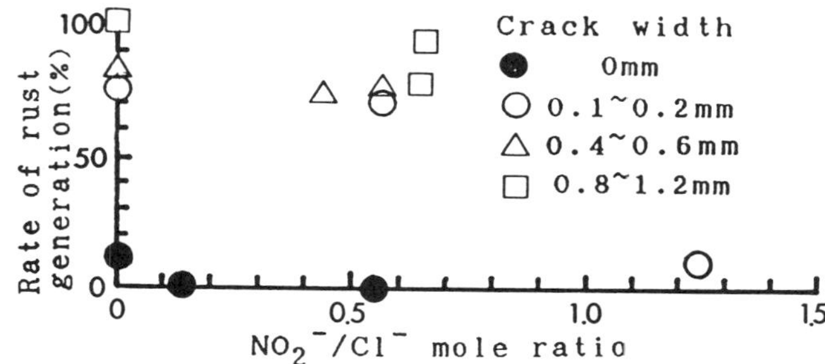

Fig. 5 Anti-corrosion effects of corrosion inhibiter on crack

5 Application of corrosion inhibiter to a large scale structure

5.1 Application examples after the one year experiment

In order to confirm the permeability of the corrosion inhibiter in an actual structure, on-site application to the deck structure of a 15 years old overbridge was conducted for an experimental purpose.

No significant cracks except local rust stains were discovered by the visual inspection and Cl^- ion content in the core-sample of the deck concrete was measured at 0.08 % which was equivalent to 1800 g/m³ and more than three times higher than the permissible salt content specified by the Japan Society of Civil Engineers.

Application of corrosion inhibiter was completed by brushing seven times working beneath the deck with the target amount of application of 1000 g/m^2 in aggregate which was equivalent to 0.5 NO_2^-/Cl^- mole ratio.

Distribution of NO_2^- ions after one year period of application is as expected and as shown in Fig. 6.

5.2 Diffusion of NO_2^- ions into the concrete

Apparent diffusion volume at each age D' (mm^3) was calculated by Apparent diffusion equation and plotted as shown in Fig. 7 by substituting NO_2^- ions concentration, apparent diffusion coefficient Dc (mm^3/day) and surface density Co (ppm). The diffusion coefficient is derived from the primary regression line and a different method of application by use of press-fitting method is also summarized for comparison purposes in Fig. 7.

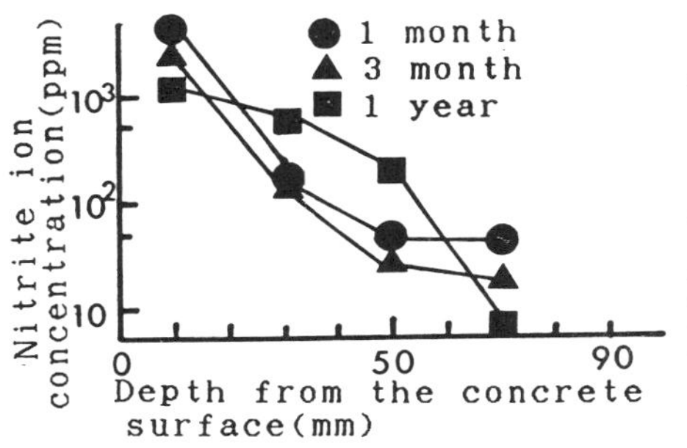

Fig. 6 NO_2^- infiltration results

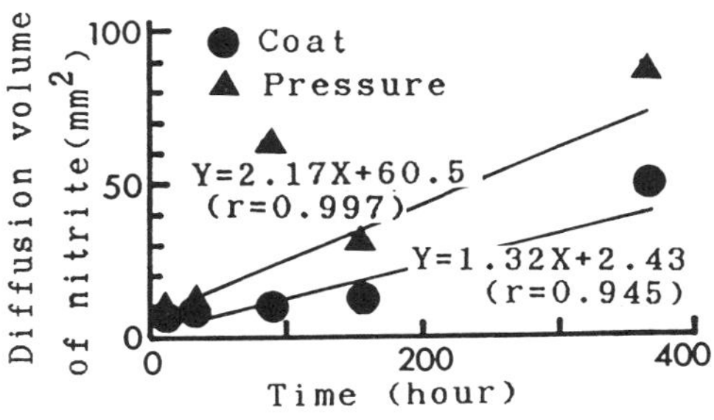

Fig. 7 Diffusion of nitrite

6 Conclusion

From the results outlined in this paper we would conclude the following points.

(1) NO_2^-/Cl^- mole ratio required for the concrete with corroded reinforcing bars is twice as high as that for concrete with rust-free reinforcing bars.

(2) Application of corrosion inhibiter to 1.82 NO_2^-/Cl^- mole ratio will prevent further corrosion effectively even for concrete with corroded reinforcing bars.

(3) Application of corrosion inhibiter over the cracked surface of 0.1 to 0.2 mm crack width is also effective by increasing the amount of application to 1.5 NO_2^-/Cl^- mole ratio.

(4) A successful degree of penetration or diffusion of corrosion inhibiter is confirmed in the actual large scale structure tested.

Electrical Half-cell Potential and Polarization Resistance Method is used for experimental purpose in order to monitor the effectiveness of corrosion inhibiter. This will be reported when the paper is presented.

7 References

Lewis, D. A. et al. (1959) Corrosion., Corrosion of Reinforcing Steel in Concrete in Marine Atmospheres, 15(7), PP. 382-388.

Losenberg, A. M. et al. (1977) ASTM STP 629., A Corrosion Inhibitor Formulated with Calcium Nitrite for Use in Reinforced Concrete, PP. 89-99.

Virmani, Y. P. Clear, K. C. and Pasko, T. J, Jr. (1983) Time To Corrosion of Reinforcing Steel in Concrete Slabs V. 5: Calcium Nitrite Admixture or Epoxy-Coated Reinforcing Bars as Corrosion Protection Systems, Report No. FHWA-RD-83-02. Federal Highway Administration, Washington, D. C..

Walitt, Arthur L. (1985) Calcium Nitrite offers Long-term Corrosion prevention. Concrete Construction.

20 CARBONIC ACID WATER ATTACK OF PORTLAND CEMENT-BASED CONCRETES

Y. BALLIM, M.G. ALEXANDER
Department of Civil Engineering, University of the Witwatersrand, South Africa

Abstract
Concretes based on portland cement (P.C.) and blends of P.C. with ground granulated blastfurnace slag (GGBS), fly ash (FA) and silica fume (SF) were subjected to carbonic acid water attack. The concrete mixes were designed on the basis of equal 28-day strength. Samples were immersed in distilled water saturated with carbon dioxide for one or two week periods after which time the water was discarded and replaced with fresh aggressive water. Measurements and observations included mass loss, leachate water composition, changes in microstructure, based on surface water absorption, and depth of attack. The results show that, on the basis of mass loss, the GGBS and FA concretes performed better than the P.C. concrete while, on the basis of surface water absorption, there was very little difference in the performance of the different mixes. The paper also presents a critique of the test procedure and emphasises the limitations of the results thus obtained. This is relevant to other durability studies and carried out using similar techniques.
Keywords: Carbonic Acid, Deterioration, Mass Loss, Absorption, Fly Ash, Blastfurnace Slag, Atomic Absorption.

1 Introduction

The deteriorating effect of carbonic acid water, ie. pure water with dissolved carbon dioxide (CO_2), on concrete is a well known phenomenon. There is, however, limited and sometimes conflicting research data regarding the influence of various cement extenders on the resistance of concrete to carbonic acid water attack, Eglinton (1975). This project was undertaken as the initial part of a broader study into the influence of South African cement extenders on the ability of concrete to resist attack by carbonic acid water. Ground granulated blastfurnace slag (GGBS), fly ash (FA) and, to a lesser extent, condensed silica fume (CSF) were investigated in this project.

Because of its 'ion hungry' nature, pure water in contact with concrete will leach calcium hydroxide ($Ca(OH)_2$) from the cement paste matrix of the concrete. This depletion of $Ca(OH)_2$ in the cement paste causes the hydrates to become unstable and decompose. In this way, pure water has the effect of undermining the physical integrity of concrete. Further, the aggressiveness of pure water is considerably increased if the water also contains dissolved CO_2. This CO_2 assists in the conversion of $Ca(OH)_2$ to the more soluble calcium bicarbonate, thereby increasing the rate at which $Ca(OH)_2$ is leached from the concrete, Lea (1970) and Basson (1989).

2 Experimental details

2.1 Concrete materials and mixes

The chemical compositions of the PC, GGBS and FA are presented in Table 1 below.

Table 1. Chemical compositions (%) of PC, GGBS and FA

Compound	PC	GGBS	FA
CaO	64,0	30,8	7,2
Al_2O_3	4,2	17,4	25,8
MgO	1,7	10,0	3,1
SiO_2	21,0	36,3	47,2
C	-	-	1,2
MnO	-	1,2	-
TiO_2	-	0,6	-
K_2O	<0,5	1,1	<0,5
Na_2O	<0,5	0,6	<0,5
S	-	0,6	0,8
Fe_2O_3	0,6	0,5	1,1
Free CaO	0,6	-	-
Glass count (%)	-	99	-

A local crushed dolerite 19 mm stone and crusher sand (fineness modulus = 3,4) were used as aggregate for all mixes. Table 2 shows the mix proportions used in this project.

Table 2. Mix proportions (kg/m^3) of concretes

	Mix number:			
Material	101	105	109	111
OPC	104 (30)*	170 (50)	312 (100)	104 (30)
GGBS	242 (70)	-	-	225 (65)
FA	-	170 (50)	-	-
CSF	-	-	-	17 (5)
Crusher sand	845	820	895	845
19 mm Stone	1200	1250	1200	1200
Water	190	180	190	190
Binder/W ratio	1,82	1,89	1,64	1,82

*Values in brackets show the % blend proportions of the various binder types

2.2 Sample preparation
Six 100 mm cubes of each mix had +/- 2 mm removed from all external faces on a high speed facing machine. This was done to remove the distorting influence of different amounts of mould oil and laitance on the surfaces. The cubes were then halved by cutting parallel to the direction of casting. This resulted in 12 'slab' samples, approximately 96 x 96 x 45 mm, which were then divided into three groups of 4 slabs each to be used as follows:

Group 1: exposed to carbonic acid water; periodically brushed.
Group 2: exposed to carbonic acid water; unbrushed.
Group 3: control group exposed to lime saturated water.

2.3 Aggressive water environment
Carbonic acid water was manufactured by bubbling CO_2 through distilled water until the pH of the water was reduced from approximately 5,9 to 4,1 +/- 0,05. Distilled water was brought into the test room at least 24 hours before bubbling to achieve thermal equilibrium.

The samples to be exposed to carbonic acid water were placed on one edge into 20 litre domestic PVC buckets, one sample group per bucket. 5 litres of carbonic acid water were added to each bucket giving a concrete surface area to aggressive water ratio of approximately 28 mm^2/ml. The test room in which all the samples in buckets as well as in lime saturated water were stored, was temperature controlled at 23 +/- 1 ^oC.

3 Test methods

3.1 Mass loss
At the time of first exposure to aggressive water, samples of mixes 101, 105 and 109 were 110 days old while the samples of mix 111 were 70 days old. Before being exposed to their respective environments, all samples were weighed in a saturated, surface dry condition to an accuracy of 0,1 g to obtain the original sample mass. Samples were weighed in this manner throughout the project.

Initially at weekly intervals and later at two-week intervals, the buckets were opened and the Group 1 samples were hand brushed using a medium stiffness brush until no further solid material could be seen to be removed. These samples were then dried and weighed. The Group 2 samples were weighed unbrushed with more care being taken in the handling and surface drying so as to avoid removal of the deteriorated surface material. The water in the buckets was discarded and replaced with fresh carbonic acid water after each weighing operation. The measured mass loss of samples in the aggressive water, was corrected for the mass gains observed for the control samples in lime saturated water.

3.2 Atomic absorption analysis
Each time the mass loss of samples was determined, a 25 ml sample of the discard water was drawn off for atomic absorption analysis of the dissolved calcium (Ca), silicon (Si) and magnesium (Mg) content.

3.3 Water absorption tests

At the end of the test programme, the unbrushed samples of mixes 101, 105 and 109 were brushed to remove the deteriorated material on the surfaces. Six consecutive slices, each 3 mm thick, were then cut parallel to a 45 x 96 mm face of one sample from each mix. The outside edges of each slice were removed so that the resulting slice measured approximately 50 x 20 mm. The blade used to cut these slices was approximately 2 mm wide.

The slices were oven dried at 105 ^{o}C for 24 hours after which they were weighed to an accuracy of 0,001 g. They were then immersed in lime saturated water such that the water surface was approximately 1 mm above the top of the slice. Each slice was weighed at 1 minute, 10 minute and 24 hour periods after immersion to determine the water absorption. The cement content of each slice was then determined using a method based on BS 1881: Part 124: 1988.

4 Results

Figures 1 and 2 show the results of the mass loss determinations of the brushed and unbrushed samples respectively. Note that the mass loss is normalised with respect to the original sample mass. Also, the low strength value for mix 111 (CSF) is due to the considerably younger age of this concrete compared to the other concretes.

Figure 3 shows the results of the atomic absorption analyses carried out on the discard aggressive water. These results are cumulative to the end of the tests and it should be noted that, for reasons of clarity, the results of the silica (Si) and magnesium (Mg) determinations have been magnified by factors of 5 and 10 respectively.

Figure 4 shows the water absorption results of the first three slices of the concretes. Very little difference was found between the results of the 3rd slice and those of the 4th to 6th slices. All water absorption results were normalised using the mass of cement in each sample. The results in Figure 4 are expressed as normalised absorption of the sample exposed to aggressive water divided by the normalised absorption of the sample exposed to lime saturated water.

5 Discussion

Figure 1 and 2 shows that the concretes with cement extenders have, on the basis of mass loss, performed better than the plain OPC concrete. In particular, the GGBS concrete showed the lowest mass loss upon brushing. This trend of better performance by the blended cements is also reflected in the results shown in Figure 3 considering the amount of calcium leached from each of the concretes. The differences, however, are not as marked as for the mass loss results. On the other hand, Figure 4 shows a complete reversal of this trend with the OPC concrete performing considerably better than the blended cement concretes. Figure 4 can be interpreted as a measure of the pore size distribution of each of the concretes; 1 minute absorption represents large, easily accessible pores while 24 hour absorption

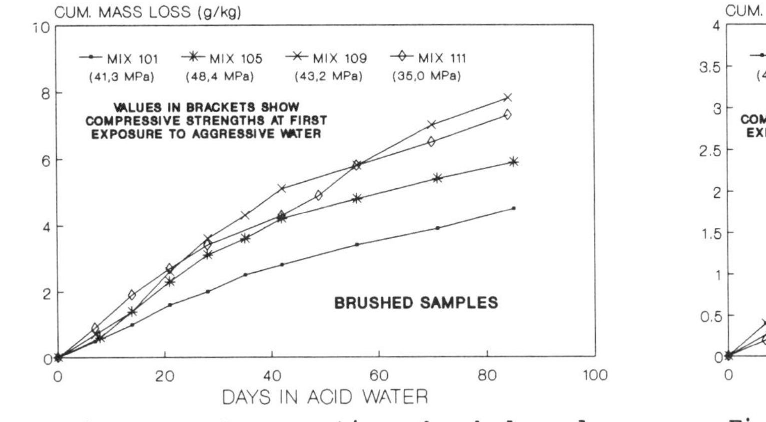

Fig.1. Mass loss vs. time - brushed samples.

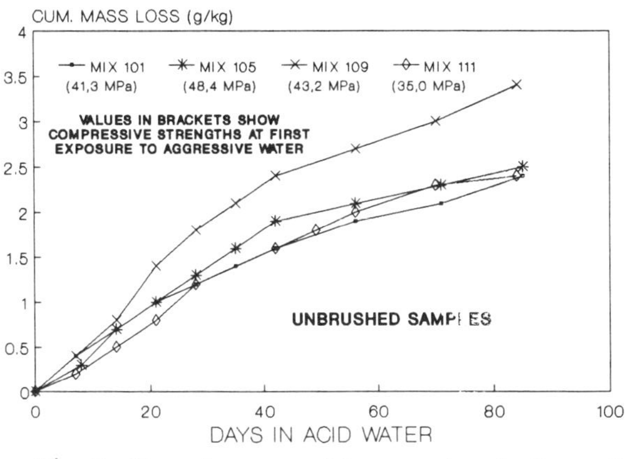

Fig.2. Mass loss vs. time - unbrushed samples.

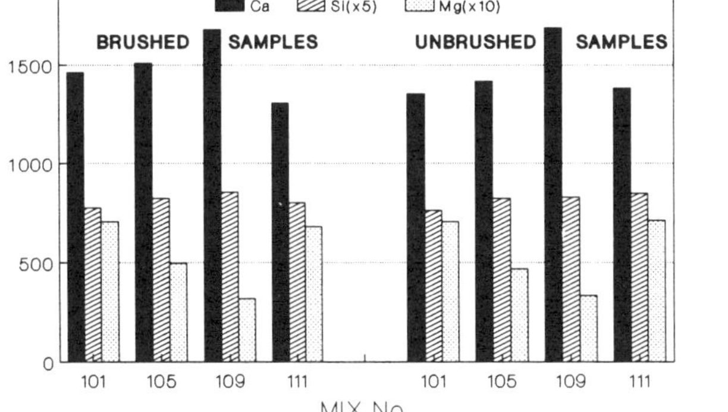

Fig.3. Results of atomic absorption analysis.

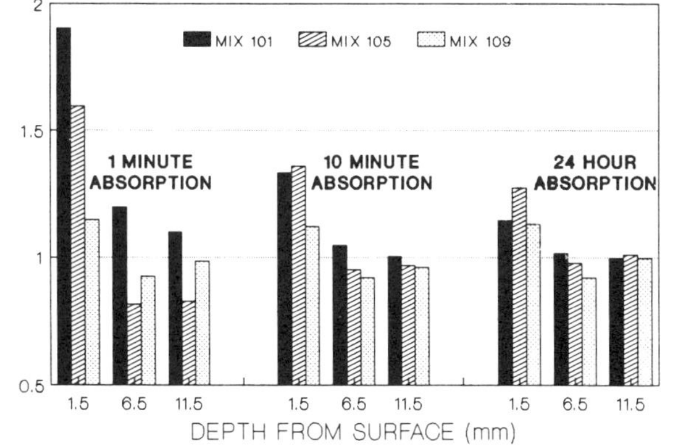

Fig.4. Thin slice water absorption results.

represents total porosity. On this basis, the exposed surfaces of the blended cement concretes have suffered an increase in the volume of large pores in comparison with their respective control samples. The 24 hour absorption results show that, at the surface, the FA mix has suffered a 55 % increase in total porosity while for the PC and GGBS mixes, total porosity has increased 28 % compared with their respective control samples. Figure 4 also shows that the depth of pore structure alteration has been approximately 5 mm except for the 1 minute absorption of the GGBS mix, where an increase was detected at a depth of 13 mm.

This project has highlighted the problem of the characterisation of deterioration. Clearly, very different conclusions can be drawn, depending on the method of characterisation used. Caution must therefore be exercised when interpreting the results of deterioration studies. The test method also presents a problem since the concretes used in this project can be expected to neutralise the aggressive water at different rates. The practice of replenishing the aggressive water at fixed time intervals therefore places in question the time that each sample group was exposed to aggressive water.

6 Conclusion

The following conclusions are drawn on the basis of the above results and discussion:

(a) For the range of concretes used, there is no relationship between strength and resistance to carbonic acid attack.

(b) On the basis of mass loss, the addition of FA and GGBS improves the resistance of PC concrete to attack. However, on the basis of changes in pore structure at the surface, the addition of FA and GGBS appears to have a negative effect. This has yet to be confirmed by further investigation on a wider sample range.

(c) The method of exposing samples to carbonic acid water used in this project requires modification to allow the performance of different concretes to be compared on an equal basis.

Acknowledgements

The authors wish to thank Slagment (Pty)Ltd. for the permission to publish the results presented in this paper.

References

Eglinton M.S. (1975) Review of concrete behaviour in acidic soils and ground waters. **CIRIA report no. 69,** September 1975.

Lea, F.M. (1970) **The Chemistry of Cement and Concrete.** 3rd Ed. Edward Arnold (Publishers) Ltd. London.

Basson , J.J. (1989) **Deterioration of Concrete in aggressive waters.** Concrete durability bureau, Portland Cement Institut Midrand, South Africa.

21 CHLOROALUMINATE AND TOTAL CHLORIDE ION AMOUNTS IN CONCRETE EXPOSED TO NATURAL SEAWATER

J.L. SAGRERA, S. GOÑI
Instituto de Ciencias de la Construcción Eduardo Torroja, CSIC, Madrid, Spain

Abstract
The behaviour of different concrete samples semi-buried in an Atlantic beach in the south of Spain has been studied. In both tidal and buried zones, the total chloride ion content as a function of the penetration depth at the age of two years has been determined by means of X-ray fluorescence spectroscopy (XRF). X-ray diffraction (XRD) has been employed to follow the evolution of the main crystalline phases.
Keywords: Chloride, Diffusion, Friedel's salt, Permeability, Concrete, Marine environment.

1 Introduction

The penetration of chloride ion into hardened cement pastes has been studied by the authors during many years (Browne (1982), Colepardi (1970), Gjörv (1979), Midgley (1984), Page (1981) and Smolczyk (1984)). The diffusion of ions play an important role in the durability of concrete marine constructions. Piers, foundations and retaining walls in sea structures are generally made of Portland ecment concrete. The physical processes of concrete deterioration are associated with chemical reactions between components of sea water and the constituents of OPC.
Under laboratory research conditions the concentration of chloride ion always has a constant value. However, it is unreasonable to apply the same conclusions to concrete structures directly exposed to sea water.

The Eduardo Torroja Institute has a marine laboratory in an Atlantic beach near Huelva to study the performance of concrete and mortar in marine environment. In this paper results up to two years are presented.

2 Experimental

Prismatic samples of 2 x 0.5 x 0.5 m made of two kinds of

cements (OPC and SRPC) and with two cement content (200 and 350 kg/m^3) were located at different situations in the beach. Each sample was buried one metre into the sand and the other metre was in the tidal zone. The chemical analysis of two cements are shown in Table 1. They have been made according to the UNE 80-215-88 standard which is equivalent to the EN 196.2 of the CEN. The mean chemical composition of the sea water is presented in Table 2 and the characterization of the different samples employed can be seen in Table 3.

To study the penetration depth of the Cl^- ion at the age of 2 years, samples were taken off either the tidal or buried zone. These samples were cut in slices to make the analyses. Cl^- ion was determined by X-ray fluorescence and Friedel's salt and the rest of crystalline phases by means of X-ray diffraction technique.

3 Results and Discussion

3.1 Chloride content in function of penetration depth

Chloride penetration profiles for different cement concretes after 2 years exposure either the tidal or the buried

Table 1. Chemical composition of cements employed (% wt)

	SiO_2	Al_2O_3	Fe_2O_3	CaO	MgO	$SO_4^=$	Ignition
OPC	19.3	6.2	3.9	61.4	1.5	4.1	0.4
SRPC	22.0	1.8	4.1	67.9	0.5	2.3	1.1

Table 2. Chemical composition of the sea water (g/l)

MgO	Cl^-	$SO_4^=$	CaO
2.4	19.9	2.9	0.5

Table 3. Characterization of samples

Concrete	Kind of cement	Content (kg/m^3)	Situation (m)
002	SRPC	200	20
017	SRPC	200	25
044	SRPC	350	20
058	SRPC	350	25
062	OPC	350	20
077	OPC	350	25
118	OPC	200	25

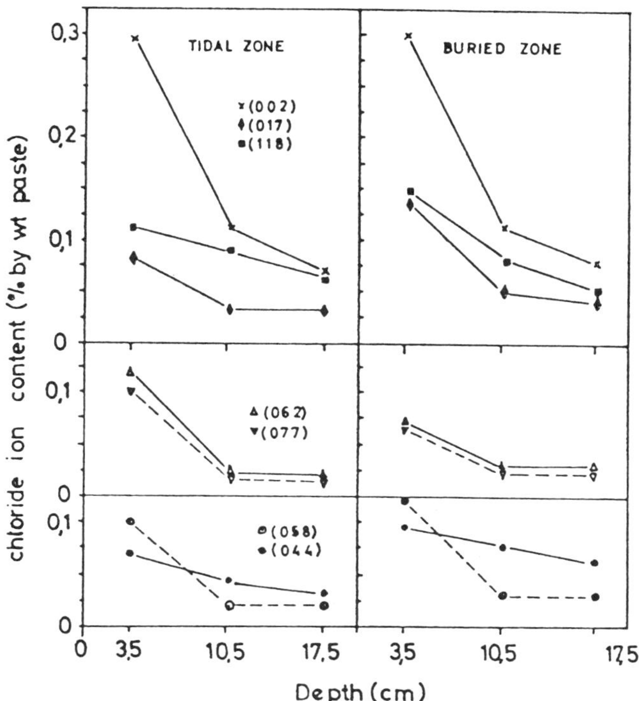

Fig.1. Cl^- ion content in function of penetration depth.

zone are shown in Figure 1. In general, the Cl^- content decreases with the penetration depth following a typical exponential function. In all cases the Cl^- content at 35 mm penetration depth (ranging from 0.075 to 0.3%) is lower than that (0.4-0.6%) needed for an active corrosion of reinforcements which normally are embedded at 40 mm from the external surface of the concrete.

With the exception of (062) and (077) concretes, it seems that the Cl^- content in the buried zone, for the same penetration depth, is higher than in the tidal zone.

In the case of concretes made of OPC, in general the Cl^- ion found is lower than that of SRPC (Compare concrete (118) and (002) in Fig.1)

As it was expected the samples made of the lower content of cement (200 kg/m^3) and therefore more porous, present the higher values of Cl^- ion (Compare concretes (002) and (118) versus (062), (077), (058) and (044) in Fig. 1).

In regard to the situation of the samples, a clear difference is found in the chloride content. So, samples located at 20 m((002)and (062))present a higher chloride content for the same penetration depth in comparison to those samples located at 25 m (017) and (077).

3.2 Partial Apparent Diffusion Coefficient

The different apparent diffusion coefficients for a determined concentration of chloride are presented in Table 4. They have been calculated from the mathematical equation:

$$X_{c=cte} = \sqrt{Dt}$$

where: X is the penetration depth for a constant chloride concentration, D is the apparent diffusion coefficient and t is the time in seconds.

This simple equation was employed by Smolczyk (1984) in a previous paper, and in fact, is very similar to the Einstein-Smoluchowsky's law (Bockris (1979)):

$$\langle x \rangle^2 = 2Dt$$

where: $\langle x \rangle^2$ is the quadratic distance reached by the ion considering a random monodimensional path:

$$\frac{x_1^2 + x_2^2 + \ldots x_n^2}{n} = \langle x \rangle^2 = 2Dt$$

The factor of 2 is introduced to consider the monodimensional path in ahead and back steps. This factor can reach a value of 6 depending on the degrees of freedom attributed to the above mentioned ion random path. Collepardi et al (1970) selected a value of 4.

As can be deduced from table 4, the more porous concretes ((118) and (002)) present the higher values of the diffusion coefficient for a constant chloride content.

Table 4. Partial diffusion coefficients. Tidal zone

Chloride content (% wt paste)	Kind of concrete	Penetration depth (cm)	D (cm^2/sg) ($\times 10^{-6}$)
0.1	(058)	3.5	0.8
"	(077)	3.5	0.8
"	(118)	6.5	2.7
"	(002)	12.5	9.9
0.075	(017)	4.4	1.2
"	(058)	5.5	1.9
"	(077)	5.5	1.9
"	(062)	6.8	2.9
"	(118)	12.7	10.3
"	(002)	16.5	17.3
0.05	(017)	8.0	4.1
"	(058)	8.0	4.1
"	(077)	8.0	4.1
"	(062)	9.0	5.1
"	(044)	9.5	5.7

On the other hand, as it is expected, the coefficient increases with the penetration depth. The probes located at 25 m from the sea water present lower values of diffusion coefficient in comparison with those of probes located at 20 m (Compare concrete (058) and (077) versus (002) and (062) in Table 4). The maximum C_3A content produces a lower diffusion coefficient (concretes (118) and (002) in Table 4).

3.3 X-ray diffraction analysis

The evolution of different crystalline phases with penetration depth for all cement concretes studied are represented in Figure 2. The non-overlapped X-ray line of higher intensity for each compound is the 100 value with respect to the rest of them, either in the tidal or buried zone.

3.3.1 Tidal zone

Calcite, portlandite and ettringite evolution is similar in four kind of cement concretes: (077), (062), (017) and (118). As it can be seen, calcite and ettringite have the higher intensity values at the more external zone, progressively decreasing towards the internal one. On the contrary, portlandite has the lower intensity value at the external zone increasing with penetration depth.

This behaviour seems to be logical since, near the sea water CO_2 is available and therefore the carbonation of portlandite is produced. On the other hand, the diffusion of sulfate ions is responsible for the higher values of ettringite reached at the external zone.

Concerning the Friedel's salt which is formed in the cement concrete with 9% C_3A (118), it seems that near the sea water, although free chloride ion is available, it nevertheless, shows its lowest value. This fact, is probably due to the lower pH value of the pore solution which is provoked by the carbonation of portlandite. Friedel's salt increases with penetration depth and ettringite decreases in an inverse relationship between the two compounds (Fig.2).

In the case of OPC cement concrete (044), (058) and (002), the evolution profiles of these compounds is not so clear. Nevertheless, the inverse relationship between calcite and portlandite is kept.

3.3.2 Buried zone

In this zone, ettringite presents the lower values at the more external zone in the case of the cement concretes (062), (017), (118), (058) and (002).
On the contrary, ettringite shows the higher value in the external zone in the case of concretes: (077) and (044). Portlandite presents the lower value at the external zone in all cases with the exception of (058)

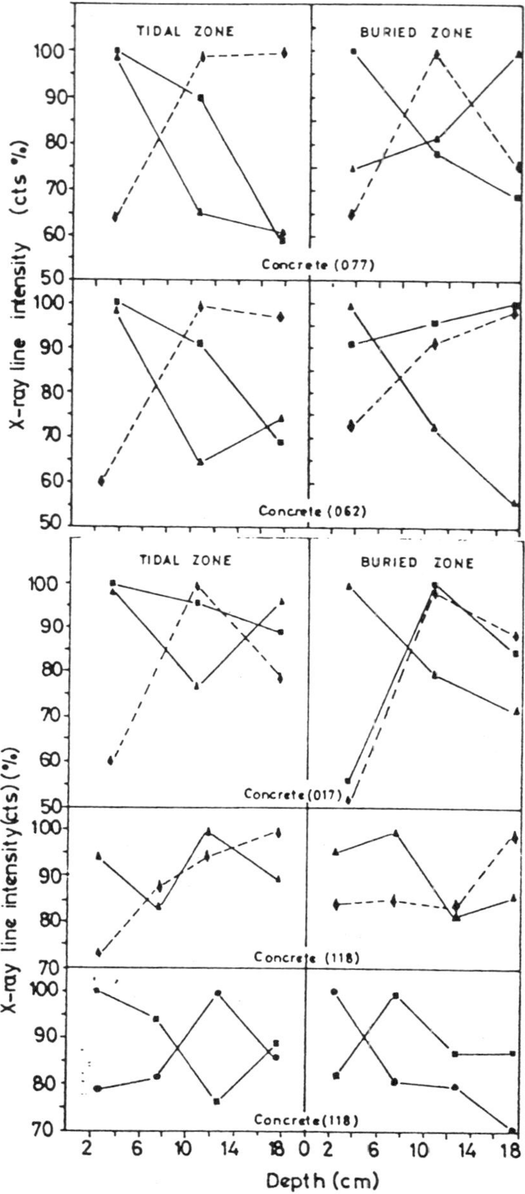

Fig.2. Relative evolution of different crystalline compounds in function of the penetration depth ▲ Calcite, ♦ Portlandite, ■ Ettringite, ○ Friedel's salt.

and (044) concretes.

Finally, calcite presents highest values at the external zone in the case of (062), (017) and (058). Friedel's salt presents the highest value at the external zone decreasing with the penetration depth to zero (See (118) concrete).

As can be deduced from the results obtained in the buried zone, the behaviour is not so clear in comparison with the tidal zone. The CO_2 available in the buried zone is lower and perhaps this was the reason of the non homogeneous behaviour at this zone.

4 References

Bockris, J.O.M. and Reddy, A.K.N. (1979) in **Modern Electrochemistry** (Plenum publishing corporation) New York, 317-319.

Browne, R.D. (1982) Design prediction of the life for reinforced concrete in marine and other chloride environments. **Durability of building materials,** 1 ,113-125.

Colepardi, M. Marcialis and Turriziani, R. (1970) La cinetica di penetrazione degli ions cloruro nel calcestruzzo. **Il Cemento,** 67, 157-164.

Gjörv, D.E. and Vennesland, D. (1979) Diffusion of chloride ions from sea water into concrete. **Cem. and Concr. Res.,** 9, 229-238.

Midgley, H.G. and Illston, J.M. (1984) The penetration of chlorides into hardened cement pastes. **Cem. and Concr. Res.,** 14, 546-558.

Page, C.L. Short, N.R. and El Tarras, A. (1981) Diffusion of chloride ions in hardened cement pastes. **Cem. and Concr. Res.,** 11, 395-406.

Smolczyk, H.G. (1984) State of knowledge on chloride diffusion in concrete. **Betonwerk-Fertigteil-Technik,** 12, 837-843.

22 THE EFFECT OF SILICA FUME TO REDUCE EXPANSION DUE TO ALKALI-SILICA REACTION IN CONCRETE

A.M. DUNSTER
Building Research Establishment, Garston, UK
H. KAWANO
Public Works Research Institute, Ibaraki-Ken, Japan
P.J. NIXON
Building Research Establishment, Garston, UK

Abstract
Concrete prism specimens containing calcined flint have been stored in humid conditions at 20°C and the influence of silica fume content, proportion of calcined flint, and alkali level on alkali silica reaction (ASR) expansion have been investigated. The addition of silica fume reduced both the rate and total amount of expansion due to ASR, and delayed the onset of expansion. Increasing the calcined flint and/or alkali level of otherwise equivalent specimens had the reverse effect.
Keywords: Silica Fume, Alkali Silica Reaction, Expansion.

1 Introduction

One of the methods of reducing the risk of alkali-silica reaction (ASR) in concrete is using partial replacement of Portland cement with pozzolanic or similar cementitious material. Silica fume (SF) has a high relative surface area with an average particle size of less than 0.1 μm. These properties give SF a high activity and offer the potential of using a lower cement replacement (5-15% by weight of cement) compared with >25% for pfa (BRE Digest 330, (1988)).

The influence of SF on ASR can be ascribed to several mechanisms. Nixon and Page (1987) reported marked reductions in hydroxyl, potassium and sodium ions in the presence of SF which equated well with its effectiveness in reducing damage from ASR. A review paper by Nixon (1987) describes an instance of increased expansion with SF of mortar bars containing aggregate with a low pessimum (Beltane opal) (Kawamura et al 1986). In these circumstances the silica/alkali ratio is a critical determinant of the amount of expansion and small additions (10-15%) of SF resulted in the worst silica/alkali ratio for expansion. 20% SF was sufficient to prevent expansion completely.

In the present work the alkali level, SF content and amount of reactive aggregate have been varied, and results on the effects of SF on ASR expansion up to 1 year age are presented.

2 Experimental

2.1 Materials

Two Ordinary Portland Cements (OPC), (total alkalis Na_2O eq. 1.0% and 0.7%), were used. Silica fume (SiO_2 content ≥ 85%, specific surface area ≥ 20,000 m^2/kg), was supplied in the form of a slurry, (solid content 50 ± 2%, total alkalis Na_2O eq ≤2.0%). The reactive aggregate Calcined Flint (CF) was manufactured by calcining crushed flint at 1400°C. CF has a greater reactivity than crushed flint due to the presence of cristobalite which reacts at a suitable rate for laboratory studies at 20°C (Lumley 1989).

2.2 Trial mixes

The proportion of reactive aggregate was varied by use of inert crushed limestone to maintain a constant ratio of cementitious material to aggregate (by volume) in contrast to the usual mix design procedure on the basis of a designed strength and workability. In practice when using SF the total content of cementitious material is usually reduced, compared with the OPC equivalent in order to achieve equal strength and workability. The chosen method of mix design consequently led to differences in workability and strength between mixes. However, adjustment of the sand/coarse aggregate ratio gave an acceptable mix design which was used for all subsequent investigations.

2.3 Mix design, casting and curing procedures

A series of specimens was cast in order to investigate the effects of SF, CF and alkali level on ASR expansion.
The mixes contained CF (as 5-15% by volume of the total aggregate), SF (as 0-15% by volume of total cementitious material), and alkali levels from 2.5-4 kg/m^3 Na_2O eq (in concrete). The cementitious contents were within the range 311-330 kg/m^3. The greatest number of SF levels were incorporated into the mixes with an alkali level 3.5 kg/m^3. This alkali level was chosen for detailed investigation because it just exceeds the maximum limit (3.0 kg/m^3) recommended by Digest 330 (1988) above which concrete is liable to ASR without protection.

Variation of the SF content necessitated a reduction in the cement proportion, and as the CF content was varied, the aggregate proportion was changed to maintain the volume ratio of cementitious material to total aggregate including CF (Table 1). The alkali level was increased where necessary by addition of potassium sulphate.
Specimens were cast, cured and measured according to draft BS 812 part 123 (1988) but were stored at 20°C.

3 Results and Discussion

3.1 The effects of silica fume on expansion at different alkali levels and CF contents

Results are illustrated in Figures 1-4. An increase in the alkali level of otherwise equivalent mixes led to an earlier onset and a greater rate and total amount of expansion. The detailed effects of alkali level on the expansion are as follows.

Table 1. Mix Proportion for prisms

CF content (vol %)		5	10	15
CF (kg/m^3)		74	147	221
Limestone sand (kg/m^3)	0.6 mm down	282	282	282
	0.6 - 2 mm	327	232	137
Limestone (kg/m^3)	5-10 mm	393	393	393
	10-20 mm	780	780	780

(all aggregate contents based on SSD condition)

Figure 1 shows data for specimens with an alkali level of 3.5 kg/m^3. No significant expansion occurred prior to 13 weeks with the exception of specimens with the highest CF (15%) which expanded slightly although subsequent expansion by up to 0.5% with associated cracking occurred in specimens with 0 or 5% SF addition. However, the addition of SF (5%) both delayed the onset of significant expansion and reduced the total expansion after 1 year by up to 0.2%.

The addition of a greater quantity of SF (10 and 15%) was sufficient to reduce the total amount of expansion to <0.2% and prevent cracking of specimens containing up to 15% CF. The highest level of SF addition (15%) completely prevented the expansion of the specimens containing 5% CF. The inclusion of SF at any level delayed the onset of expansion for a period which increased with decreasing CF content.

Significant expansion also occurred at the lower alkali levels of 2.5 and 3.0 kg/m^3 which are below the maximum limit set by BRE Digest 330 (1988) (figures 2 and 3). The effect of SF addition was similar to that at the higher alkali levels although lower total expansion and later onset of expansion were observed.

Data for the alkali level 4.0 kg/m^3 (SF additions 0-10%) are shown in figure 4. Rapid initial expansion occurred and in the absence of SF, all specimens had expanded by >0.1% within 17 weeks. The addition of SF, (10%) delayed expansion for 20 weeks. However although SF reduced the rate and total amount of expansion, SF additions of up to 10% were insufficient to prevent eventual deleterious expansion.

4 Conclusions

The addition of SF to concrete mixes, (up to 15% by weight of cement) containing calcined flint reduced both the rate and total amount of expansion due to ASR and delayed the onset of expansion.

The addition of up to 10% SF was insufficient to limit the total expansion to <0.1%. The addition of 15% SF was sufficient to prevent

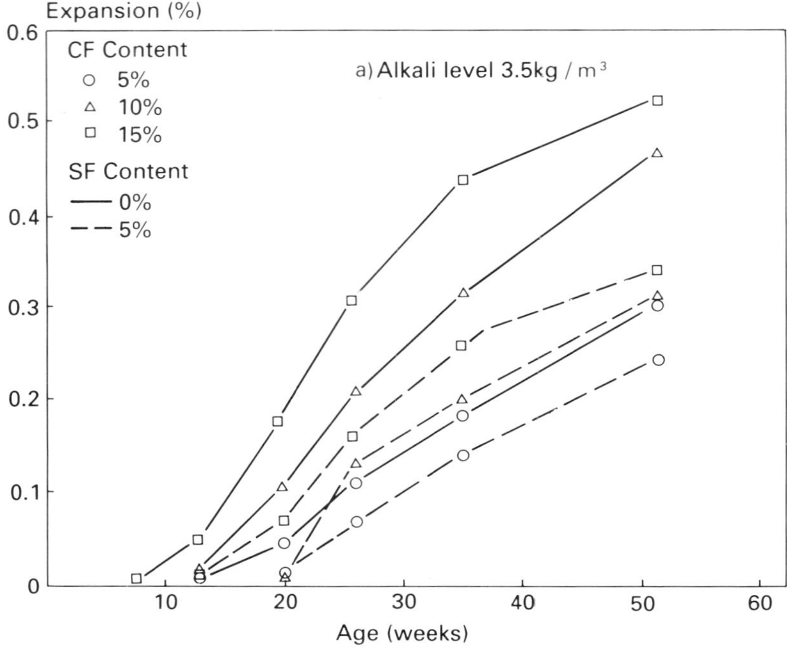

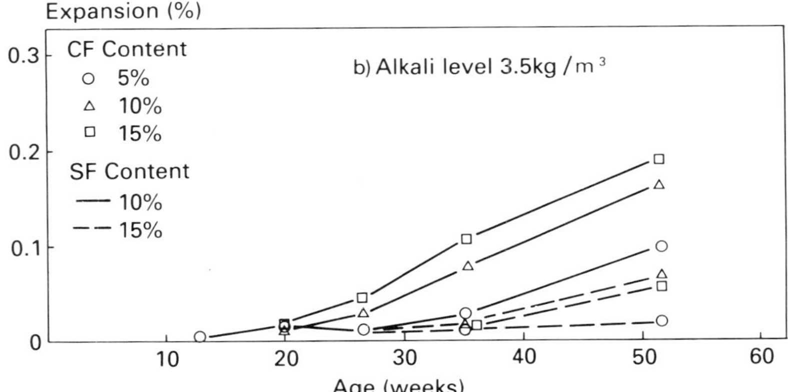

Fig 1 ASR Expansion (alkali level 3.5 kg/m³ Na_2O eq)

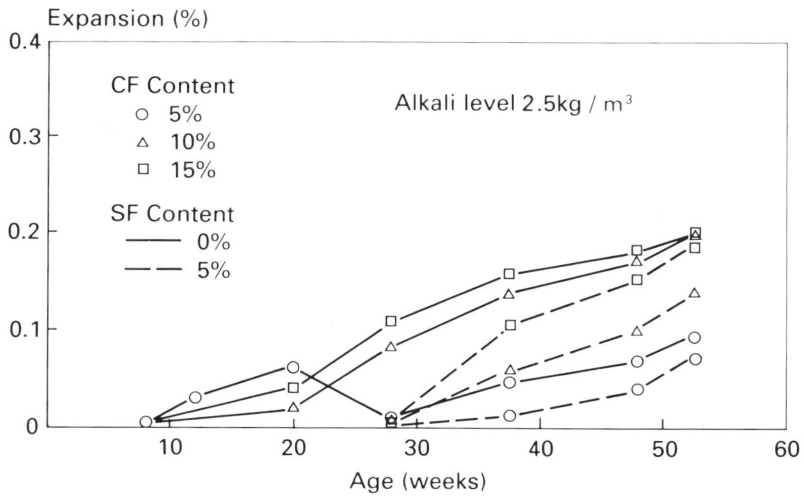

Fig 2 ASR Expansion (alkali level 2.5 kg/m³ Na_2O eq)

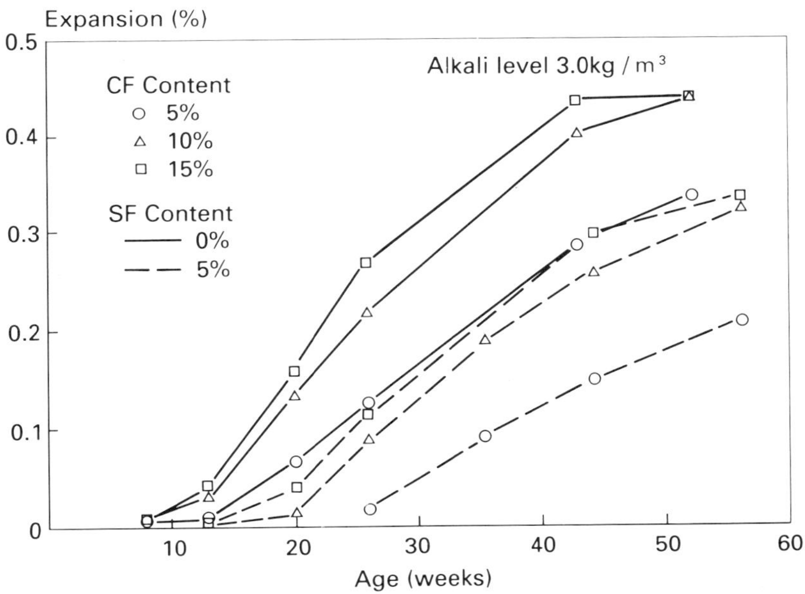

Fig 3 ASR Expansion (alkali level 3.0 kg/m³ Na_2O eq)

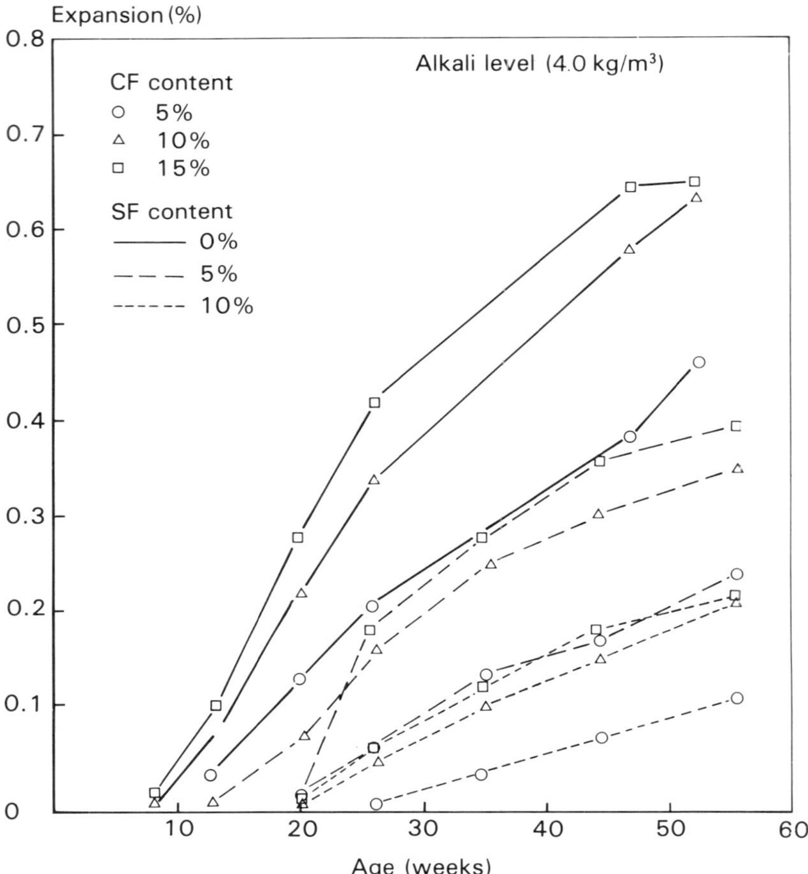

Fig 4 ASR Expansion (alkali level 4.0 kg/m^3 Na_2O eq)

deleterious expansion of specimens with an alkali level of 3.5 kg/m^3 and CF contents between 5 and 15%. However it may be necessary to increase the level of SF addition to prevent deleterious expansion at an alkali level of 4.0 kg/m^3.

This paper describes results up to 1 year. There are reports (unpublished) of very delayed expansion in concrete containing silica fume and therefore monitoring will continue. Further results will be reported in due course.

5 Copyright

This work forms part of the research programme of the Building Research Establishment and is published by permission of the Director.

6 References

British Standards Institution (1975), **Method for determination of physical properties,** BSI London, BS812 part 2.
British Standards Institution (1988), **Concrete prism method. Testing aggregates - methods for the assessment of alkali-reactivity potential.** BSI London, draft BS 812 part 123.
Building Research Establishment (1988). **Alkali aggregate reactions in concrete.** Digest 330, BRE.
Hobbs, D.W. (1988) **Alkali silica reaction in concrete.** Thomas Telford (Publishers), London.
Lumley, J.S. (1989) **Synthetic Cristobalite as a reference reactive aggregate.** 8th International Conference on Alkali Aggregate Reaction, Kyoto, Japan, 561-566.
Kawamura, M., Takemoto, K. and Hasaba, S. (1986) **Effect of silica fume on alkali-silica expansion in mortars.** Proc. 2nd Int. Conf. on Fly Ash, Silica Fume, Slag and Natural Pozzolans in Concrete, Madrid - ACI SP-91, 2 999-1012.
Nixon, P.J. (1987) **Effect of silica fume on pore solution composition.** Contribution to the Concrete Society Working Group on Microsilica. State of the Art Report 8/1-8/15.
Nixon, P.J. and Page, C.L. (1987) **Pore solution chemistry and alkali aggregate reaction.** Proc. Katherine and Bryant Mather International Conference on Concrete Durability, Atlanta ACI SP-100 1833-1862.

SESSION B

MICROSTRUCTURE AND INTERFACIAL EFFECTS

23 CONSOLIDATION AND HYDROPHOBIC TREATMENT OF NATURAL STONE

E. WENDLER, D.D. KLEMM, R. SNETHLAGE
Geological Institute/LMU – Bavarian State Conservation Office, Munich, Germany

Abstract
In several cases the treatment of natural stones with silicon-organic consolidating and hydrophobing agents reveals problems like limited durability or even subsequent damage some years after the application.
In this paper, the mechanisms of interaction between the silicon-organic molecule and the mineral surface are discussed, and the reasons of failure are demonstrated regarding the influence of the environmental situation. It is shown that silicon-organic hydrophobing chemicals do not prevent the swelling and shrinking due to changes in humidity, so that the stone material is still affected by stress-strain processes. In the case of insufficient penetration depth of the agent, two zones of highly different mechanical and hygric behaviour are joined by a sharp borderline. Periodically changing external conditions may lead to a contour scaling of the hydrophobic layer.
A pre-treatment with surfactants is useful to reduce the swelling due to humidity.
To improve the surface bond of the silicon-organic compounds, mineral-specific adhesive coupling agents which are covalently linked to the polymer backbone are tested.
Keywords: Sandstone, Consolidation, Hydrophobic Treatment, Surface Binding, Adhesive Coupling.

1 Introduction

Since almost 35 years silicon-organic compounds are used as consolidating and hydrophobing agents for natural stones. There is evidence that on buildings the durability of hydrophobing treatments is limited to 10 - 15 years (fig.1). It is well known that consolidating silica gels, formed by ethyl silicates, show a great number of shrinking fissures due to a continuous loss of structural water. However, a loss of the effectivness could not be detected (Sattler and Snethlage 1988). As a consequence of a consolidation and hydrophobation the formation of contour scales was observed in particular cases. (Wendler and Sattler 1989).

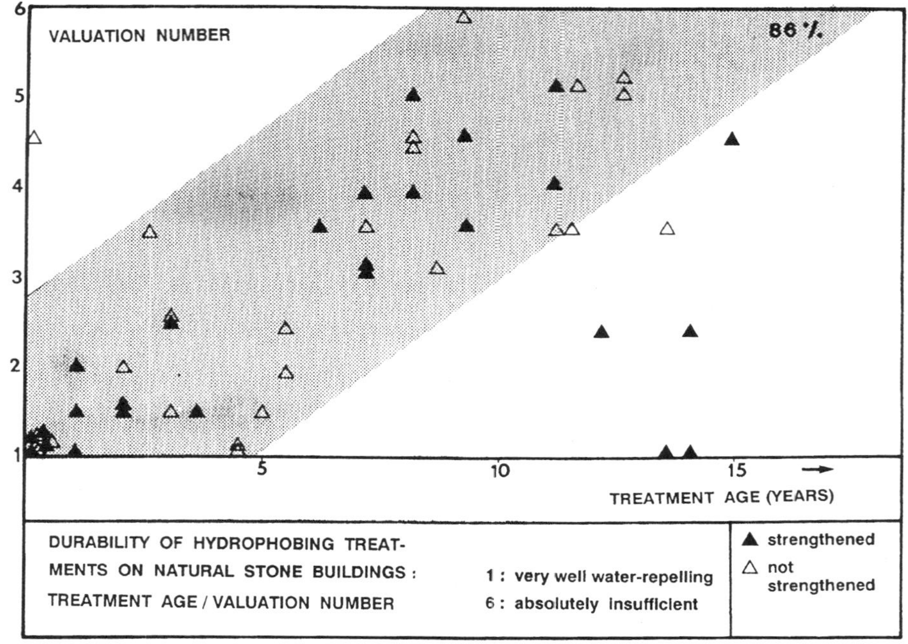

Fig.1. Durability of hydrophobing treatments.

Up to now, the mechanisms by which silica gels and polysiloxane films are fixed to the mineral surface are not clearly understood. For quartz as substrate there are two possible kinds of binding.
A covalent siloxane bond (Si-O-Si) between the mineral surface and the polysiloxane film or a weak hydrogen bridge bond between silanol groups of the substrate and the polymer. In the latter case, the bond could be simply loosened by excess rain water, while a covalent siloxane linkage should be stable in the presence of water.

Previous investigations (Wendler and Snethlage (1988)) concerning the durability of hydrophobing treatments demonstrate that the decrease in efficiency is in relation to the content of swelling clay minerals within the stone material. Therefore, it can be supposed that alternating hygric swelling and shrinking effects are responsible for the loosening of contacts between the polysiloxane film and the mineral surface.

The present paper discusses two possible solutions of this problem:
- the reduction of hygric swelling by blocking the hydration centres of the clay minerals with tensides and,
- the improvement of the surface bond of polysiloxane films by adhesive coupling agents.

2 Hygric dilatation investigations

2.1 Untreated material

Drillcores of fresh, untreated "Sander Schilfsandstone" were conditioned at 20° C/20% rel. humidity for 14 days. After weighing, the samples were fixed in a dilatometer and the rel. humidity was enhanced stepwise in 15% spaces up to 100% in time intervals of 14 days. At the end of each particular step, the dilatation and the weight increase of the samples were measured. Finally, the samples were put into distilled water. After 48 h, no further increase in weight and dilatation could be observed. In fig.2, hygric dilatation is plotted against the weight gain of the sample. Up to 100% rel. humidity, there is an almost linear increase of the dilatation with increasing weight. In the superhygric range, the dilatation is relatively decreasing.

This result is consistent with the microstructure of the sandstone: below 100% rel.humidity, the micropores, are filled with water by capillary condensation and surface

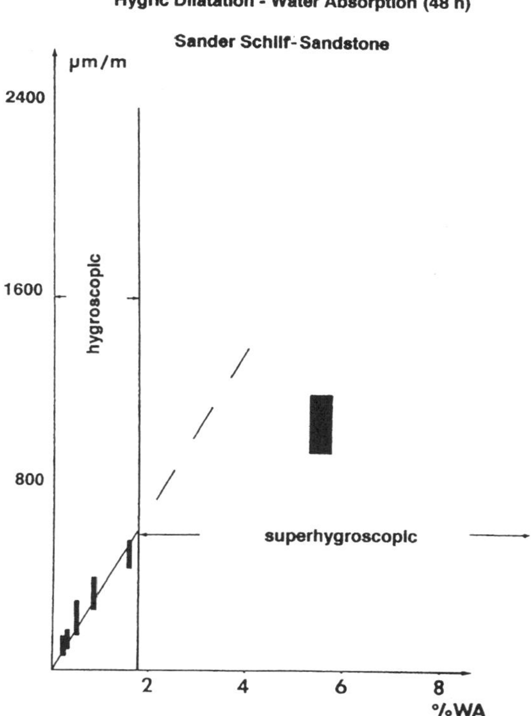

Fig.2. Hygric dilatation (μm/m) of Sander Schilfsandstone, depending on water uptake.

diffusion processes. The expansion of the clay minerals is transmitted directly to the structure of the stone. Above 100% rel. humidity, the larger pores are filled by capillarity. Within these pores, there is more space available for the expansion of the clay minerals, so that the swelling effect to the structure is reduced.

2.2 Hydrophobing treatment with commercial products
Another series of drillcore samples was conditioned at 20° C/60% rel. humidity and treated with commercial silane and polysiloxane solutions. After six weeks in 20° C/60% rel. humidity, when the condensation reaction was supposed to be finished, hygric dilatation was measured in distilled water. Fig. 3 shows the hygric dilatation as a function of time in comparison with untreated samples which were conditioned in the same way: while the untreated material reaches a constant value after some two days, the dilatation of the hydrophobed samples is delayed due to a slower penetration of liquid water. However, after some 4 to 8 days, these samples show almost the same hygric dilatation as the untreated material or even exceed it. This demonstrates that the clay component of the grain structure which is responsible for the hygric dilatation is not sufficiently coated by the hydrophobic film.

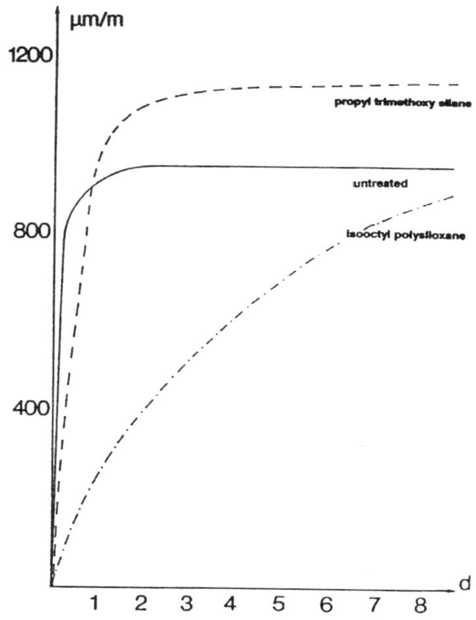

Fig.3. Hygric dilatation (µm/m) of Sander Schilfsandstone in water as a function of time.

3 Interaction of clay minerals with cationic surfactants

Cations that are fixed at the negative charge centres of clay mineral layers are able to form hydration shells whose size depends on the type of cation and the amount of water available. This process is responsible for the swelling behaviour of clay minerals. Therefore, our intention is to prevent or at least to reduce the formation of hydration shells by blocking the negative charge centres.

Previous investigations of Lagaly et al. (1970) demonstrate that cations between clay mineral layers are partially exchangeable against alkyl-ammonium ions, the amount being dependent on the location and the type of cation.

However, the immersion of sandstone samples into hexadecyl-ammonium chloride solutions leads to an almost complete destruction of the stone sample within a period of three weeks. Apparently the hydrophobic alkyl groups of the surfactant screen the contact forces between the mineral interfaces.

As a conclusion from this result an analogous experiment using bifunctional alkyl-α-ω-diammonium chlorides was carried out.

Fig.4 demonstrates the fundamental considerations concerning the contact mechanism between the bifunctional

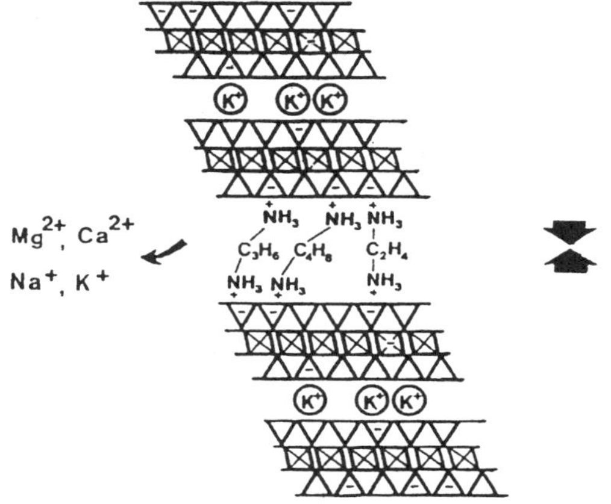

Fig.4. Model of the ionic exchange of cations against bifunctional cationic surfactants on clay mineral basal planes.

alkyl-α-ω-diammonium ions and the adjacent mineral surfaces. It is supposed that the surfactant ions replace the binding cations at the mineral interfaces. In fact, this model is confirmed by the cation exchange which can be measured when desalted sandstone powder samples are dispersed in a butyl diammonium chloride solution.

To measure the effect of the alkyl diammonium ions on the hygric swelling, drill core samples were impregnated with different concentrations of tenside solution. After that some samples were treated with a polysiloxane solution. In fig. 5, the final equilibrium values of the hygric dilatation of the different samples are compared with each other: the dilatation of the polysiloxane treated samples is almost as high as of the untreated samples. However, the dilatation is remarkably reduced by a tenside treatment. The decrease is dependent on the concentration, but higher concentrations than 0.5M are not useful because of a increasing colour change of the stone.

A combined treatment with water-repelling polysiloxanes and surfactants shows a similar decrease of the hygric dilatation, i.e., the two subsequent treatments obviously do not influence each-other.

However, a combined treatment might be particularly useful because of the following reasons:

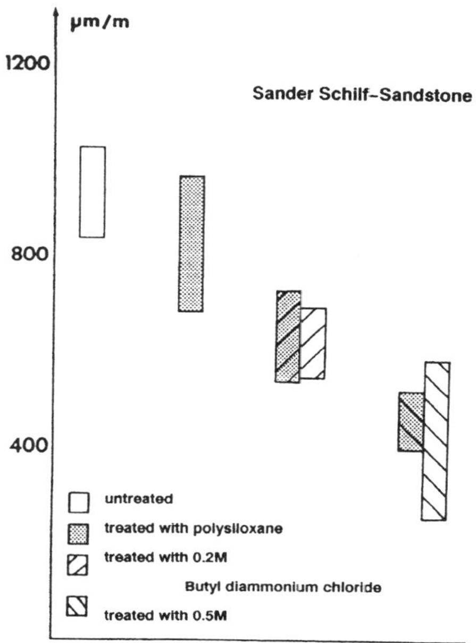

Fig.5. Equilibrium values of the hygric dilatation of differently treated Sander Schilfsandstone

on the one hand, the surfactant component reduces the swelling in the hygric region of the sorption isotherm , on the other hand, the water-repelling polysiloxane prevents the capillary uptake of water and of chemical pollutants from the atmosphere. Thus, a twofold positive effect on the weathering stability of stones can be achieved.

4 Modification of protectives by adhesive coupling

The interaction of protonated amino groups of cationic surfactants with the negative charge centres in the mineral surface can be used to strengthen the binding of silica gels and polysiloxanes to the mineral surface: if amino-groups are covalently integrated into a silicon-organic network structure, a similar interaction with the clay mineral surface should be expected as in the case of isolated alkyl ammonium ions, but additionally the polymer should also be linked to the surface (fig.6).
To confirm these considerations the following experiments were carried out.
An aqueous solution of aminopropyl silane, which was adjusted to p_H 6 by addition of HCl, was added to an ethanolic solution of propyl trimethoxy silane. The resulting mixture was poured into a glass Petri dish. After a few days, a hydrophobic film was formed which is tightly bound to the glass surface. Glass and quartz surfaces are comparable with clay minerals with respect to their negative surface potentials.

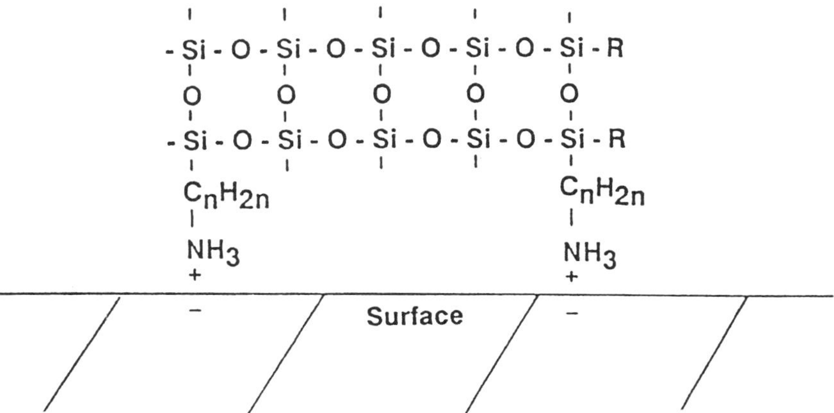

Fig.6. Model interaction of aminosilane-modified siloxanes/silica gels with negative charge centres on a mineral surface

Preliminary investigations by SIMS show that the amino groups are fixed at the surface, but further experiments have to be done to confirm the kind of chemical bond.

An analogous experiment was carried out with a mixture of aminopropyl silane and silicic acid ester. In this case, a tighter contact to the Petri dish surface was obtained than with a commercial silicic acid ester, but, due to the loss of water during the condensation reaction, shrinking gaps were formed. In contrary to the unmodified compound, the remaining fragments still had a tight contact to the glass surface.

5 Elastification of silica gel

During the condensation process of silicic acid esters, four siloxane bonds may start from one silicon atom, in contrast to alkyl silanes, which can only form three siloxane bonds to their neighbouring Si-atoms. As a consequence, silica gels are more brittle than polysiloxane networks.

To minimize the shrinkage, elastomeric bridges should be integrated into the network structure. For this purpose, dialkyl-dialkoxy silanes were used which were mixed with a stoichiometric amount of water to hydrolyze. By this way, by condensation linear oligomeric units are formed which can be added to the silicic acid ester in varying ratios (fig.7).

A 1:1 mixture yielded elastic films, which showed almost no shrinking fissures. The films, however, could be easily removed from the glass surface.

The best results are obtained by a combination of adhesive coupling by protonated aminosilanes and elastification by dimethyl siloxanes. The film had an tight contact to the glass surface; only a few shrinking fissures appeared after several weeks. The biaxial flexural strength of stone drillcore slices which were treated with these mixtures

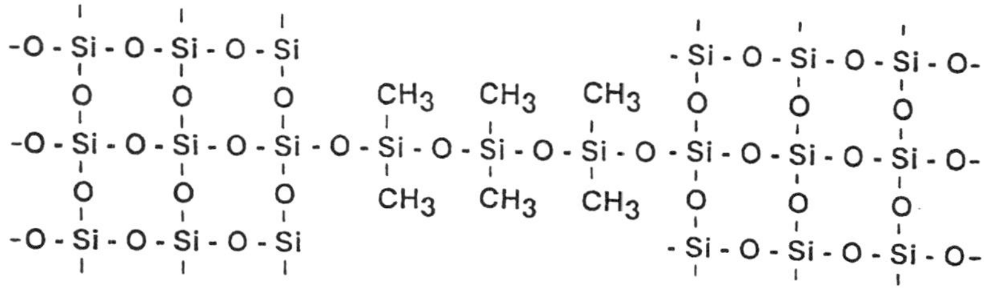

```
   |     |     |                                      |      |      |
-O-Si-O-Si-O-Si                                    -Si-O-Si-O-Si-O-
   |     |     |                                    |      |      |
   O     O     O      CH3     CH3     CH3           O      O      O
   |     |     |       |       |       |            |      |      |
-O-Si-O-Si-O-Si - O - Si - O - Si - O - Si - O -    Si-O-Si-O-Si-O-
   |     |     |       |       |       |            |      |      |
   O     O     O      CH3     CH3     CH3           O      O      O
   |     |     |                                    |      |      |
-O-Si-O-Si-O-Si                                    -Si-O-Si-O-Si-O-
   |     |     |                                    |      |      |
```

Fig.7. Model of the elastification of silica gel by dialkyl siloxane bridges

showed that the modulus of elasticity is significantly lower than in the case of a treatment with a commercial silicic acid ester, whereas the ultimate strength is the same in both cases.

6 Formation of micellar structures in aminosilane-modified solutions

It can be assumed that because of their low modulus of elasticity these polymere structures are more resistent to mechanical stresses within the grain structure of stones. When aqueous solutions of protonated aminoalkyl silanes are dropped into ethanolic solutions of alkyl silanes, the mixture becomes opaque due to a segregation, because the polarity of the solution containing the non-polar alkyl groups is enhanced by the water dipols. At a distinct amount of aminoalkyl silane, however, the mixture becomes almost clear which indicates that micellar colloids have been formed. A further addition of aminoalkyl silane causes a segregation into two separate phases.

Although the size of the micelles is not yet known, it is supposed that their shape is similar to that in fig.8. Due to the repelling forces between the positive charges, the amino groups tend to reach a maximum distance from eachother, i.e., they are arranged on the spherical surface of the micelle, while the hydrophobic alkyl chains are

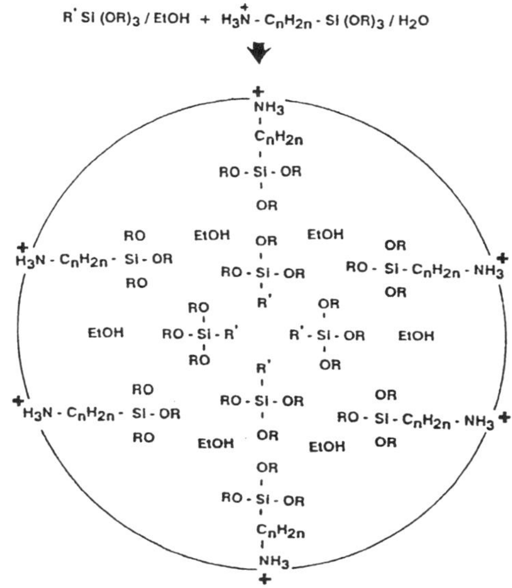

Fig.8. Supposed structure of aminoalkyl silane/alkyl silane micelles

located within its centre. Presumably, ethanol molecules are arranged between the other components. The presence of excess ethanol prevents the components from being hydrolyzed and subsequently condensed in the colloid solution even if water is present.
The micellar solutions are stable for some months. In this way, protective treatments can be carried out with ecologically compatible ethanol-water mixtures which replace poisonous solvents like ketones, toluene or hydrocarbons.

7 Conclusions

Amino-functional surfactants show strong interactions with mineral surfaces and can therefore be used in stone conservation. Bifunctional alkyl diammonium ions are able to reduce the hygric dilatation of clay rich sandstones to about a half of the original value. Amino-functional surfactants can also be used to improve particular properties of silicon organic conservation products. Aminoalkyl silanes work as primers for polysiloxanes and for SiO_2-gel on clay mineral basal planes as well as on quartz surfaces. SiO_2-gels, modified by elastomeric dialkyl siloxane bridges and primered with aminoalkyl silane, show a low modulus of elasticity and a remarkably reduced formation of shrinking fissures. Finally, on the basis of ethanolic siloxane solutions together with protonated aminosilans, it is possible to make stable conservation products which are compatible with the environment.

8 References

Lagaly, G. and Weiss, A. (1970) Formation of sequence isomers of mica type layer silicates. Z.Naturforsch. 25b, 572-576.

Sattler, L. and Snethlage, R. (1988) Durability of Consolidation Treatments with Silicic Acid Esters, in **Proc. Intern. Symp. Geol. Soc.**(eds P.G. Marinos and G.C. Koukis), 19.-23. Sept. 1988, Athens, pp. 953-956.

Wendler,E. and Sattler, L. (1989) Untersuchungen zur Dauerhaftigkeit von Steinkonservierungen mit siliziumorganischen Stoffen, **Bautenschutz und Bausanierung**, Sonderheft Bausubstanzerhaltung in der Denkmalpflege, 2. Statusseminar d. BMFT (14.-15. Dec. 1988), Wuppertal, pp. 70-75.

Wendler,E. and Snethlage, R. (1988) Durability of Hydrophobing Treatments of Natural Stone Buildings, in **Proc. Intern. Symp. Geol. Soc.**(eds P.G. Marinos and G.C. Koukis), 19.-23. Sept. 1988, Athens, pp. 945-951.

Acknowledgement:
The investigations presented in this paper were supported by the BMFT, Federal Republic of Germany.

24 ALTERATION OF MICROSTRUCTURE AND MOISTURE CHARACTERISTICS OF STONE MATERIALS DUE TO IMPREGNATION

D. HONSINGER, H.R. SASSE
University of Technology, Aachen, Germany

Abstract
The first part of this study was to elaborate performance criteria for stone treatment systems and to develop criteria for the assessment of the effectiveness of stone protecting materials. New approaches of accelerated testing were carried out in cooperation between the chemical industry and interdisciplinary collaborating scientific institutions. Agents on polymer base have been preferably developed and applied to different types of sandstones.

The assessment of the effectiveness of new impregnation-materials is carried out by comparison of physical test results and accelerated aging test results both on treated and untreated stone samples. A number of important criteria, which are to be optimized, have been quantitatively and qualitatively evaluated. The laboratory tests cover measurements on polymer properties, porous material properties and impregnated porous material properties.

Related to the new concept of natural stone protection that has been developed at the Aachen University, experimental results are exemplarily presented here. The research project is sponsored by the German Federal Research Ministry (BMFT).
Keywords: sandstone, impregnation, protecting, polymer microlayer, strengthening.

1 Introduction

Deterioration or damage of weathered structural sandstone members and sculptures in the form of crusts, surface scaling, exfoliations and granular disaggregations is most usually a result of atmospheric influences. In addition to naturally caused weathering processes anthropogenic detrimental influences have increased. Sandstone is susceptible to considerable weathering processes and the conservation of this type of rock, which is characteristic for very many old buildings and monuments in Germany and other countries, causes very serious problems.

Despite the intense efforts of curators, restorers, and architects, in many cases the actual losses of cultural and historic material cannot be prevented sufficiently by applying the available substances and procedures. At best is can be slowed down.

The present paper is meant to be an example for the techniques used for checking the effectiveness of the consolidating and protecting treatments. These techniques already existing in interdisciplinary fields are specifically modified and adapted to the field of practical stone works. In view of the fact that stone decay mainly involves mechanisms in which moisture -and more particularly liquid transport- plays a dominant role, alteration of moisture characteristics and microstructure of impregnated stone will be presented.

2 Technological requirements, concept for stone protection and consolidation

As a base for the research activities a "catalogue of requirements" for protective agents has been elaborated in interdisciplinary cooperation, Sasse (1987). The technical demands can be summarized into four main criteria:

- aesthetics of natural stone surface
- effectiveness and durability of protective means
- building site usability
- compatibility with ecological demands.

These main criteria are specified in detail by several subcriteria. The subcriteria are discussed in Sasse (1987).

The so called "supporting corset" model is understood as a protecting and strengthening polymer microlayer, coating the internal pore surfaces of the stone. The microlayer is supposed to be

- water repellent (hydrophobic) and water resistant
- impermeable against water
- restraining the water vapour diffusion
- resistant against chemical and biological agents
- rubber elastic through a wide range of temperature ($-30^{o}C$ to $80^{o}C$).

Quartz grains, loosened by loss of natural binder, are expected to be evenly coated by the microlayer and relinked to each other by polymeric "grain bridges". As a result of the planned mechanical properties the polymer microlayer displays the function of a "supporting corset". There are no significant changes in water vapour diffusion within the bulk volume of the stone, because large capillary pore-channels remain open. In simplified form this concept is explained in fig. 1, Sasse (1988). It differs from the well known hydrophobic treatment, because of the distinctive formation of a protective and impermeable microlayer, coating the inner surfaces of the stone. Compared to the strengthening by silica ester, the disadvantages of increasing the modulus drastically and/or forming a secondary pore network due to adhesive or cohesive failures do not appear.

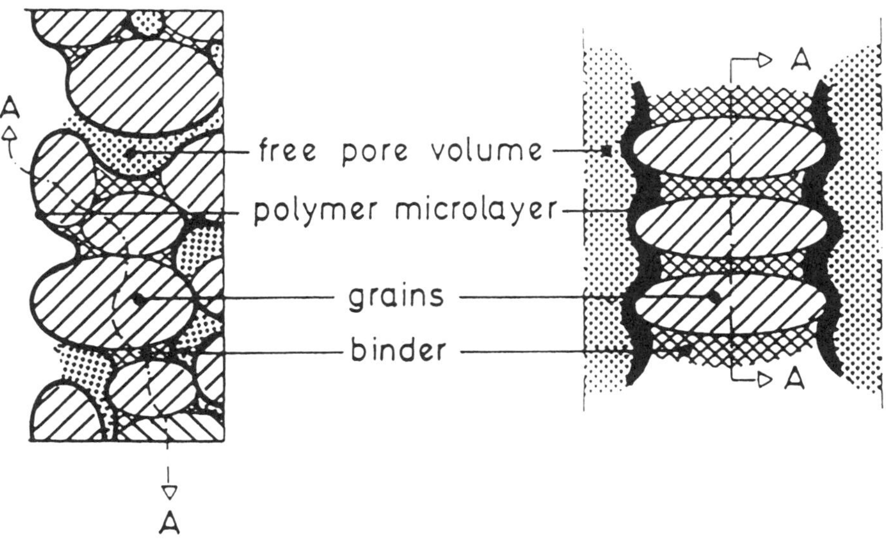

Fig. 1. The "supporting corset", protective polymer microlayer coating the pore walls

3 Experimental investigations

3.1 Aims

On the basis of the technological requirements, effectiveness tests for existing products have been carried out. In an iterative process, specifications were elaborated as a basis for developments of new polymer stone treatment systems which are able to protect natural stone monuments against harmful atmosphericals and to strengthen the microstructure. New types of polymeric materials do not only mean absolutely new materials but also developments of standard materials used for other purposes than impregnation.

3.2 Tested products, sandstone varities and application-method

About 150 cold curing polymeric products for stone treatment were studied during preliminary tests series. The products were based on solvent containing

- polyurethanes, both hardening from ambient humidity and from special chemical components (PUR)
- epoxy resins, 2-component systems (EP)
- modified acrylic resins (AY)
- silicon organic compounds, silanes, polysiloxanes (SIL)
- modified silica ester (SE)
- fluoro ethylene (FE)
- unsaturated polyesters (UP).

In addition, some preliminary tests were carried out using hydrous dispersions on the basis of polyurethane, polyesters, and acrylic resins.

Substrates were different types of sandstones, representative for German stone monuments three different quarry sandstones were selected for their typical features, such as mineral content, structure and colour. The treatments were carried out on

- Ebenheider Sandstone (EH)
- Obernkirchener Sandstone (OK)
- Sander Schilfsandstone (SS)

The samples had the dimension of 5cm x 5cm x 10cm. In order to prevent evaporation of the solvent from the stone surface during the experiment, an 1.0 mm EP sealing was applied around the sample lateral faces. The gable-ends remained free. The samples were exposed to different climates (8°C/60% R.H., 23°C/50% R.H., 28°C/95% R.H.) until reaching constant humidity. Some properties of the samples are summarised in table I.

Table I. Properties of the untreated sandstones

Stone-type	density [g/cm^3]	maximum capillary water content [wt-%]	water absorption coefficient [Kg/m^{2}s$^{1/2}$]	porosity [Vol-%]	average pore radius [µm]	specific surface [m^2/g]	water vapour diffusion factor [g/m^{2}d]
1	2	3	4	5	6	7	8
EH	2,59 - 2,73	7,99	2,439	17,3 - 20,9	3,5	2,44 - 4,10	82
OK	2,66 - 2,69	6,75	1,670	16,4 - 19,0	2,0	1,29 - 1,40	71
SS	2,64 - 2,71	7,99	1,101	19,6 - 19,9	1,9	4,49 - 7,30	91

Products were applied by capillary absorption. The sample surface of 5cm x 5cm was immersed to 0,5cm depth into the impregnation liquid under atmospheric pressure. The level was kept constant. After four hours of sucking time the samples were left to dry to a constant weight at laboratory climates 8°C/60% R.H., 23°C/50% R.H., 28°C/95% R.H. After one month the samples were weighed again to determine the residual amount of polymer left in the stone.

4 Selected test results

4.1 Penetration depth test

The penetration tests were carried out by different methods. The correlation between penetration depth and viscosity or surface tension respectively are illustrated in Honsinger (1988). In this case the different climates have no significant influence on the penetration depths of the protective liquids.

4.2 Suction of water and drying behaviour

Suction of water is understood as water absorption as a result of capillarity or adsorptive forces when wetting surfaces without external pressure. The polymer treated surface was immersed into water and suction was vertically upwards. The time related increase of weight was measured until reaching constance of weight and was reported in percent of the maximum capillary water content of the untreated specimens.

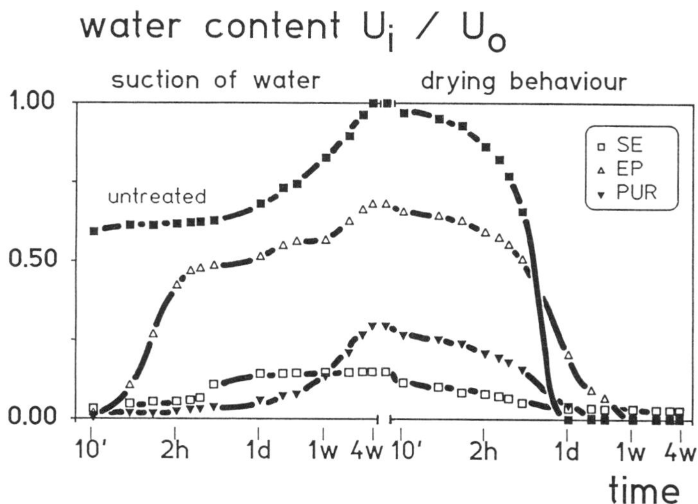

Fig. 2. Suction of water and drying behaviour on treated and on untreated sandstones

The drying behaviour is important because liquid water sometimes accumulates behind the treated zone due to condensation or moisture transport from hidden sources. Drying behaviour was determined after suction of water for 28d by measuring the weight during the drying process under 23°C/50% R.H.

The graphs in fig. 2 demonstrate the time related water absorption and -desorption processes of non treated and polymer treated sandstone samples. Compared to the untreated rock drying velocities of the successfully treated samples are considerably reduced. The samples treated with modified silica ester show a significantly hydrophobic character throughout the whole time of exposition. Samples treated with polyurethane show a moderate rise of water sorption after 7d, meaning that after that time of exposition the hydrophobic, water repellent effect is lost. As a principle it is advantageous to have as short as possible desorption times compared to adsorption times. This property was not met by the selected polymer systems.

4.3 Contact angle and laboratory weathering

Testing the contact angle a drop of water of a defined volume is deposited on a horizontally adjusted rock surface. Using a goniometer microscope the contact angle can directly be determined. Above 90° degrees, a surface can be defined as hydrophobic. Laboratory weathering tests allow a comparative evaluation of the durability of the different products. As a result of exposure the polymer microlayers at the macro surface are partly destroyed, leading to an activation of the surface energy, and thus to smaller contact angles. These surfaces can be wetted, but no suction of water into the capillary system is possible.

4.4 Contact angle and suction of water

Fig. 3 shows the effects of two different sandstones for water suction and contact angle, which were treated with modified polymers at the base of EP and PUR. As exspected the capillary absorption rises with a decreasing contact angle. Compared to the untreated sandstones the surfaces treated with epoxy resin do not show any significant hydrophobic effect. This context is no longer valid when treated surfaces are activated during service conditions. In such cases the contact angle measurements must be carried out along the cross section.

4.5 Water vapour diffusion

The polymer treated stone volume is expected not to act as a barrier towards the flow of water vapour. This is especially important for moisture exchange processes between building parts and the surrounding atmosphere.

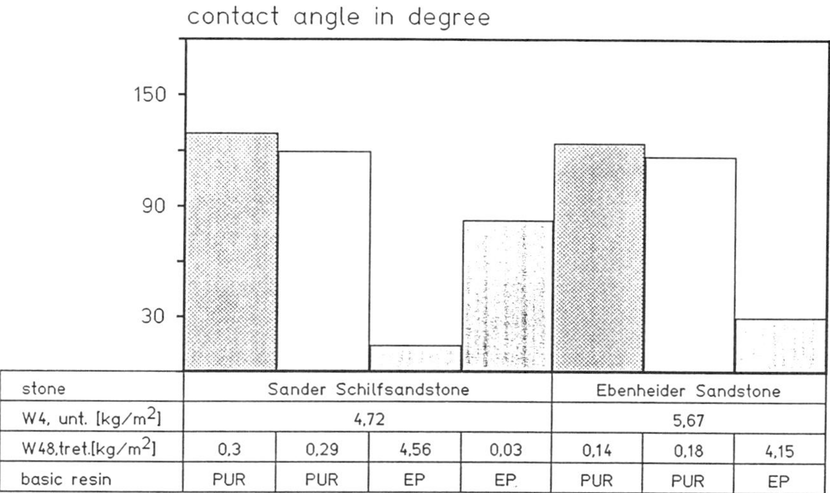

contact angle in degree
150
90
30
stone
Sander Schilfsandstone
Ebenheider Sandstone
W4, unt. [kg/m2]
4,72
5,67
W48,tret.[kg/m2]
0,3
0,29
4,56
0,03
0,14
0,18
4,15
basic resin
PUR
PUR
EP
EP
PUR
PUR
EP

Fig. 3. Contact angle and suction of water on untreated sandstones after 4h (W4) and on treated sandstones after 48h (W48)

Water vapour transmission rate (WVT) was tested according to DIN 52615 (1987). The treated samples had a cross section of 50 mm x 50 mm and a thickness of 10 mm. The test was carried out according to the so called "wet cup method".

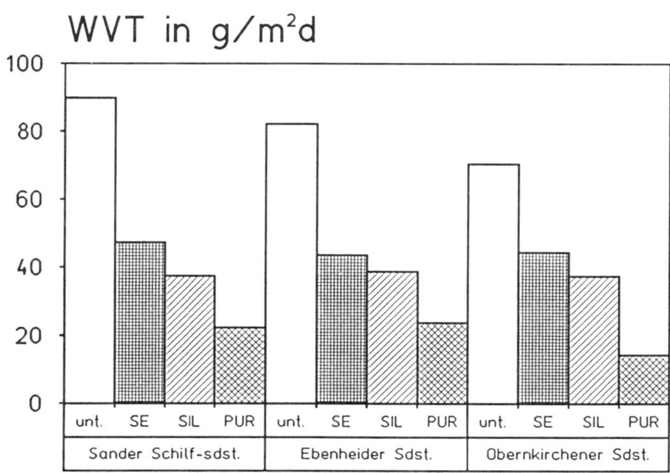

Fig. 4. Water vapour transmission rate on treated and on untreated sandstones

The applied products reduce WVT for a factor of 3 to 4, although the microlayers on the inner pore walls only affect porosity in the desired way, fig. 4. The results can be attributed to the moderate hydrophobic effects of the products, inducing a reduction of the adhesive moisture. Due to that fact transportation of water molecules within the pore system is reduced. In this case surface diffusion and capillarity only moderately attribute to the diffusion of water vapour. The results of WVT tests seem tolerable but the product's hydrophobic features should nevertheless be optimized in future.

4.6 Water vapour adsorption

The isotherm of sorption characterizes the amount of moisture being adsorbed on the specific surfaces. Hydrophobing treatments effect a reduction of water adsorbing features of the stone. This should be mainly reached by sealing pores which do not contribute essentially to the moisture transport mechanisms of the stone. Water vapour sorption was tested using 50mm x 50mm x 10mm sized samples. At 23°C relative humidities of 22%, 43%, 56%, 85% and 92% were selected.

The Sander Schilfsandstone (SS) adsorbs an essential larger amount of water compared to the Obernkirchener Sandstone (OK), fig. 5. This is caused by a larger specific surface of the SS resulting in a large surfacearea for substances to be adsorbed. Compared to the

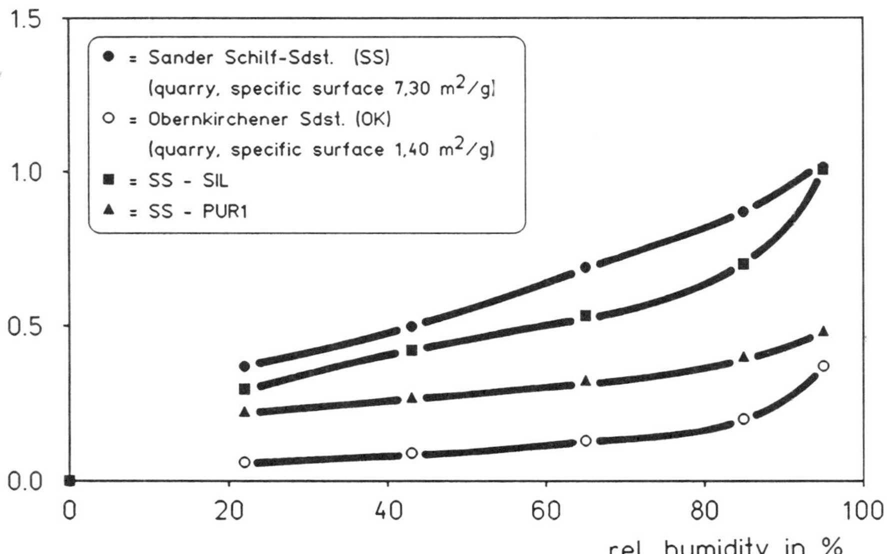

Fig. 5. Isotherms of water vapour adsorption

untreated samples the adsorption of water on the treated specimens is principally lower for all moisture grades. The larger amount of adsorbed moisture in the untreated stones can be explained by capillary condensation within pores smaller than 0.1 micron in diameter. The significantly reduced water content of the impregnated samples can be explained as a function of a smaller inner surface resulting from the polymer sealing of intercrystalline gaps. Results from SEM investigations confirm these statements. In addition, hydrophobic effects will play a role.

4.7 Investigation on the pore structure

The most important requirement for studying the formation of the polymer corset at stone surfaces and cracked sections is to recognize structure features in sufficiently small areas with the aid of the scanning electron microscope (SEM), Honsinger and Neisel (1990). In addition, an EDX-analizer equipped with an ultra thin aluminium window was used for the detection of the carbon content within the coating films. The recognized thicknesses of the polymer microlayer range between nanometers (nm) and some microns (μm) depending upon local geometric situations and type of sandstone, Honsinger (1990a). Using specially developed sample preparation methods, the topography of the inner surface features can be studied down to nanometer structures, Burchard (1987).

The visual shape of the treated samples is characterized by homogeneous continuous films bonding together the quartz crystals. The polymer does not seal the capillary pores, but only the micropores of the stone. It tends to leave a fine supporting bandage with strengthening grain-to-grain contacts. Fig. 6 shows the adhesive bond between the substratum and the polymer microlayer.

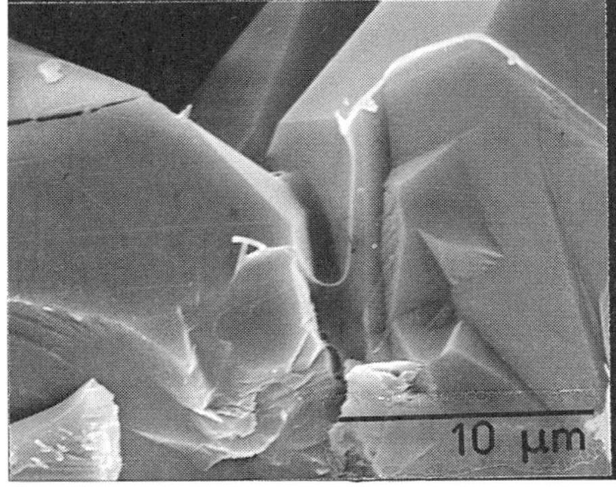

Fig. 6. Quartz crystals inside a sandstone before (left) and after treatment (right)

Fig. 7 clearly shows that the polymer is able to form a nearly continuous film covering also the crystals of clay minerals. The thickness of the polymer films mainly ranges below 1 micron. The polymer microfilms not show fracturing or blistering. Thus, the micro-mechanical stresses due to short term periods of alternating moisturing and drying of swellable clay minerals, attributing substantially to the physical decay of the stones, are reduced to a great extend.

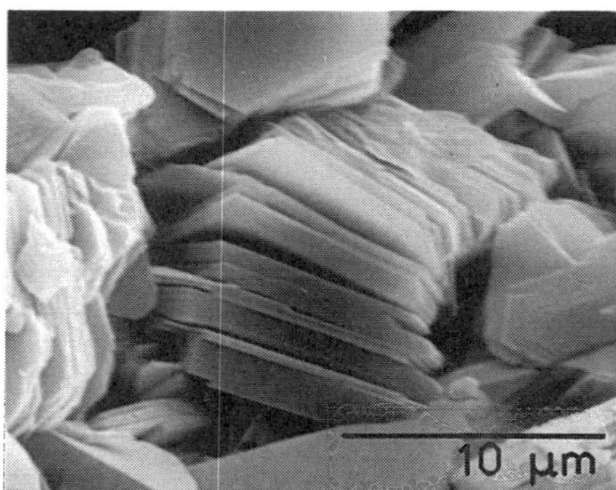

Fig. 7. Vermicular stacks, the characteristic structure for kaolinite, before (left) and after (right) treatment

SEM studies are supported by mercury-porosimetry to measure even slight changes within the distribution of pore radii as a result of the treatment. Differences in pore volume are expressed as percentage over the distribution of pore radii and show a clear mark, which can be related to the effectiveness of the treatment. Fig.8 shows that pore radii smaller than 5μm are clearly reduced, while there are no significant changes within the larger pore radii ranks. The peak of the 5μm to 10μm rank must be explained by the accidental charateristics of the measurements close to the 5μm rank boundary. The large pore channels are still available for a gaseous transport of humidity through the treated volume of the stone.

5 Evaluation of effectiveness

The effectiveness of a stone treatment can only be characterized by multiple effectiveness features. In order to carry out a computer supported, quantitative evaluation the effectiveness, criteria are differentiated into ranks. At present status only 7 effectiveness criteria of about 20 are taken into account for optimizing the stone treatment systems. The quantitative intensities of these features allow the determination of an effectiveness

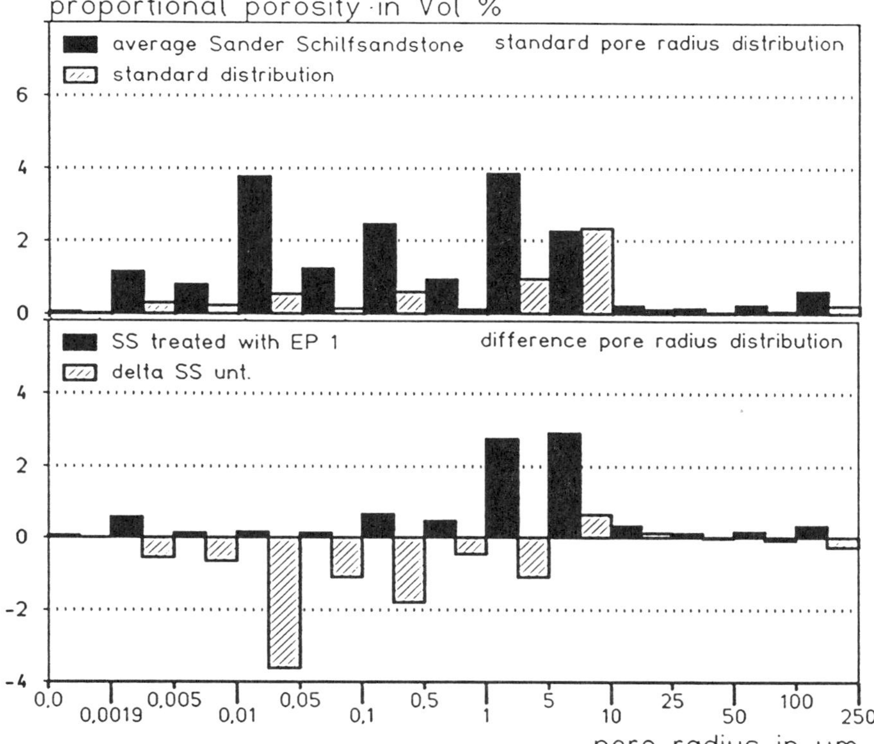

Fig. 8. Pore radius distribution

coefficient. For the most promising products further developments are discussed with the producers, Honsinger (1988).

6 Conclusions

Preliminary test series have shown that it seems promising to develop a new concept for stone strengthening and protection using microlayer forming polymers. This is a central part of a current extensive interdisciplinary research project. The research is aimed at establishing the most adequate consolidant and protective methods for different stone materials.

The results received hitherto show that several polymer systems met the technological demands which are discussed in this study. In fact these types of polymers have not yet been in use for strengthening and protecting of natural stone. Some epoxies and polyurethanes show good penetration into the stone and form nearly continuous microlayers covering the mineral pore walls. They do not seal capillary pores, but only the micropores of the stone. They tend to form a fine supporting bandage and do not drastically effect the modulus or the colour, moreover they give strength and protection including increased water repellent property. On the basis of these results a selection of the most effective products was made. Some test products seem to be worth for further development. In connection with the chemical industry these products will be developed and tested in a more sophisticated way. Of these, studying the polymer treated stone under natural weathering conditions and in short time laboratory simulation tests will be the most important future investigations.

7 References

SASSE, H. R. (1987) Polymere als Schutzstoffe für Naturstein. Bautenschutz und Bausanierung Sonderausgabe, (1) 1987, S.65-88. 1. Statusseminar des BMFT, Mainz, 17./18. Dez. 1986

SASSE, H. R. (1988) Neuartige Konzepte für den Schutz gefährdeter Natursteinbauten. Symposium über Steinzerfall. Bild der Wissenschaft, Bonn, 15. April 1988

HONSINGER, D. and SASSE, H.R. (1988) Neue Wege zum Schutz und zur Substanzerhaltung von Sandsteinoberflächen unter Verwendung von Polymeren. Bautenschutz und Bausanierung, Heft 6

DIN 52615 11.87. Wärmeschutztechnische Prüfungen. Bestimmung der Wasserdampfdurchlässigkeit von Bau- und Dämmstoffen

HONSINGER, D. (1990) Die Bedeutung von Strukturuntersuchungen im Elektronenmikroskop für die Qualitätssicherung von imprägnierten Sandsteinen. In: Tagungsband, 14.Vortragsveranstaltung des Arbeitskreises "Rasterelektronenmikroskopie in der Materialprüfung", 25.-27.04.1990. Berlin

HONSINGER, D. and NEISEL, J. (1990) Strukturuntersuchungen an imprägnierten Sandsteinen. Ibac-Kurzbericht Nr. 17. Selbstverlag Institut für Bauforschung, RWTH Aachen.

BURCHARD,W.-G. and CLOOTH, G. and SASSE, H. R. and HONSINGER, D. (1987) Rasterelektronenmikroskopische Strukturuntersuchungen an polymergetränkten Naturwerksteinen als Hilfsmittel zur Wirksamkeitsbeurteilung von Schutzmaßnahmen. In: Beiträge zur elektronenmikroskopischen Direktabbildung von Oberflächen 20 (1987). 20. Kolloquium des Arbeitskreises für elektronenmikroskopische Direktabbildung und Analysen von Oberflächen (EDO)

25 CHANGES OF SURFACE CHARACTERISTICS OF SANDSTONE CAUSED BY CLEANING METHODS APPLIED TO HISTORICAL STONE MONUMENTS

M. WERNER
Institut für Baumaschinen und Baubetrieb, Aachen Technical University, Germany

Abstract
The reasons for cleaning historical monuments made of sandstone as well as the aims of cleaning them are briefly described. The cleaning methods may be subdivided into the following groups:

cleaning methods using pure water,
cleaning methods using chemical substances and
mechanical cleaning methods.

They are described in this sequence. The effects of twelve different cleaning methods were investigated. These are basically physical and chemical investigations, most of which are standardized. The main emphases of the investigations were the changes of the capillary water absorption, chemical changes dependent on the depth profile, changes of colour and the determination of the loss of substance. The investigations were carried out on four standard varieties of sandstone, using squared stones which had to be changed in the course of a restoration. It was found that the differing state of weathering influences the result of the cleaning much more than the applied cleaning method. The effects may be assessed relatively well by simple means on test surfaces prepared on the respective test specimen. The investigations of the capillary water absorption as a measure of the opening of the pores, of the colour as a directly visible feature and of the roughness as a measure of the loss of substance are investigations which may be carried out quickly without much effort and which basically describe the result of the cleaning.
Keywords: Sandstone, Cleaning Methods, Historical Monument, Surface Characteristics, Roughness, Surface Colour, Water Absorption.

1 Problem

The physical, chemical and mineralogical characteristics of sandstone surfaces are changed by the depositing of dusts as well as gaseous and liquid pollutants from the air or from rainfall which react with the dusts or with the substrate. Measurements of immission rates of the single pollutants Danneker (1986), Kraus (1985), Krumbein (1988), Mirwald (1986), as well as investigations of crusts from different historical monuments Blaschke (1986), show that physico-chemical as well as chemical-biological processes are initiated by the concentration of pollutants which destroy the original surface of the sandstone or change it in a negative way. A layman can recognize this from the changes of colour, surface structure and the outer appearance. In order to stop the weathering process the detrimental substances have to be removed and further measures have to be undertaken.

2 Aims

The aims which have to be taken into account for the cleaning of the sandstone of historical monuments are manifold and partly contradictory, Werner (1989). On the one hand the aims comprise the improvement of the outer appearance, the opening of the capillary pores for a subsequent application of consolidants and the restoration of the physic characteristics, on the other hand the loss of substance which is caused by the cleaning is to be minimized. In order to achieve these aims different methods are used in practice, Pohle (1988).

3 Investigated cleaning methods

According to their respective types, three different groups of cleaning methods may be differentiated:

methods using water without chemical additives,
methods using water with chemical cleaners and
wet and dry mechanical methods.

Of the methods using water without chemical additives the following were investigated:

pressureless sprinkling with cold or warm water
cleaning with jets of warm water at various pressures and temperatures (60 bar/60°C, 90 bar/60°C, 150 bar/60°C, 60 bar/90°C, 90 bar/ 90°C) and
cleaning with steam jets (30 bar/150°C).

Due to the diminishing significance of the cleaning method using chemical additives, the effect of only one standard cleaner containing hydrofluoric acid was investigated. Of the mechanical methods, sandblasting using solid and dry or wet abrasives for blasting was investigated.

Abrasives for blasting were used with diameters from 0.1 mm to 0.5 mm with different grading and of different materials. All of these methods were applied to four sample sandstone varieties.

4 Natural stones

The investigations could be carried out at four historical monuments which have been undergoing large-scale restoration and from which squared stone had to be taken and changed due to the restoration measures. The squared stones are made of Sander Schilfsandstein from Schillingsfürst Castle, Werksandstein from the Residence of Würzburg, Franconian Rhaetic sandstone from the Kolb Brewery, Bayreuth and Franconian new red sandstone from Plassenburg Castle, Kulmbach. The characteristics of these sandstones will not be described in detail. The forms of pollution or of weathering are, however, characteristic of the investigated varieties of sandstone.

One can find further information on the four mentioned varieties of sandstone in Geologica Bavarica (1984) and Snethlage (1984).

5 Methods of investigation

The evaluation of the effects of different cleaning methods was based on the interpretation of physical and chemical investigations as well as on a subjective description of the microscopic and macroscopic changes of state. This included:

- water absorption capacity under normal pressure and in a vacuum according to German Industrial Standard DIN 52103,
- capillary water absorption (water absorption coefficient W in $kg/m^2 \sqrt{sec}$) according to German Industrial Standard DIN 52615,
- capillary height (water penetration coefficient B in $m/\sqrt{sec}$) according to German Industrial Standard DIN 52615,
- water penetration according to Karsten (with the Karsten testing equipment),
- water vapour permeability coefficient μ_{50-100} (WET CUP),
- water soluble components dependent on the depth profile,
- acid soluble components dependent on the depth profile,
- salt ions dependent on the depth profile,
- colour according to German Industrial Standards DIN 5033 and 55981 as well as according to CIELAB EAB*,
- roughness according to German Industrial Standard DIN 4768 and Grimm (1983),
- microscopic and SEM investigations.

The above-mentioned parameters were measured before and after cleaning. During and after a cleaning tests the following parameters were measured:

the temperature profile (methods using water),
the coordinates of the water horizon (methods using water) and
the loss of substance and organic components contained therein.

6 Results

6.1 Water absorption capacity

The water absorption under normal pressure and in a vacuum did not change significantly in any of the four investigated sandstones. The deviations of the measured values of the cleaned samples from the measured values of the quarry-fresh or weathered or polluted samples are within the usual scattering. Hence one may draw the conclusion that either no significant change of porosity occured due to weathering Schuh (1987) or cleaning, or that the measuring technique (the values were determined using samples 1 cm thick) is inappropriate for the effects of this cleaning method, these being restricted to the surface of the sandstone. Further measured values concerning the behaviour of the samples towards water lead the author to presume that the latter is more likely.

6.2 Capillary water absorption

The capillary water absorption (capillary soaking above the considered surface which watertight sides) depends very much on the substrate. A change of this parameter by the cleaning measures cannot be observed in the relatively fine-grained, dense, clayey Sander Schilfsandstein and Werksandstein, the less so since identical samples cannot be used for the investigation. The changes in the permeable varieties of sandstone are very much dependent on the degree of pollution and cleaning method used. In such a case the determination of the capillary water absorption is highly appropriate for the assessment of the effects of cleaning.

The water penetration coefficient B ($m/\sqrt{sec}$) is characteristic of every variety of stone. It is, however, almost linearly dependent on the respective capillary absorption coefficient W ($kg/m^2\sqrt{sec}$). This dependence may be explained relatively easily in physical terms Klopfer (1974). As regards the assessment of the cleaning effect, the coefficient B has only got the function of checking the coefficient W.

The determination of the Karsten water penetration may be reduced to the W-value by means of a correction-formula

Wendler (1989). The advantage of this measuring technique is (see fig. 1) that the determination may be carried out on the object before and after the treatment at the **same** measuring point without destroying it. The water content of the sample is especially important because it influences the water absorption and thus the W-value to a high degree. The measurements carried out using the Karsten equipment reconfirm the above-mentioned W-values. It may not be maintained, however, that a specification of the deviation of the W-values can be obtained for the respective cleaning method.

6.3 The water vapour permeability coefficient
The large deviations of the water vapour permeability coefficient within one variety of sandstone and the few samples (six for each investigated cleaning method) do not allow of a conclusive explanation. The only indicator is that the two fine-grained, clayey sandstones are not altered and that the coarse-grained quarzitic varieties are made more permeable by the cleaning. An exact interpretation of the slight differences seems difficult. Therefore the water vapour permeability coefficient is only relatively useful in order to assess the cleaning effect.

6.4 The water-soluble and the acid-soluble components
The determination of the water-soluble components and the acid-soluble components dependent on the depth profile is only relatively useful for the assessment of the cleaning effect for porous natural sandstone. One reason for this is, as already explained in 6.1 with regard to water absorption capacity, the little depth effect of this measure. An investigation is worthwhile if chemical cleaners are applied and their depth effects are to be assessed. The determination of the water-soluble components of a surface sample is helpful for the initial assessment of the usefulness of a sprinkling.

6.5 The ion concentration of salts
The determination of salt ions down to a depth of 3 cm in permeable sandstones and down to a depth of 1-2 cm in very dense sandstones serves to check residues of chemical cleaners after the treatment by water cleaning (fig. 2). It may clearly be seen that F^- ions from HF, Cl^- ions from HCl, SO_4^{2-} from H_2SO_4, and $HCOO^-$ from HCOOH remain to a considerable degree as residues of an acid cleaner in the surface layers. For cleaning methods without chemical additives this test does not provide any additional information which could be useful for the assessment of the cleaning effect.

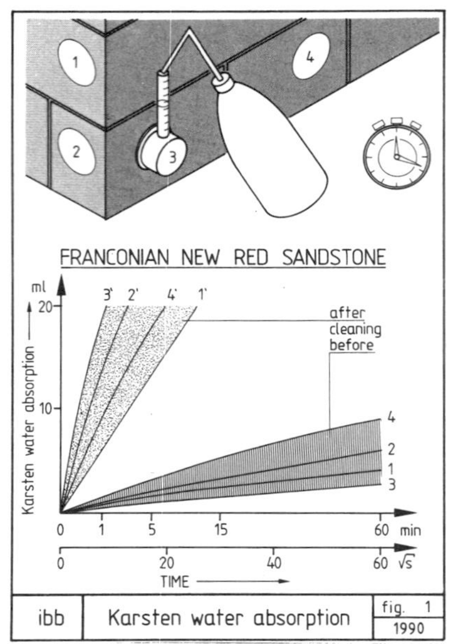

Fig. 1.

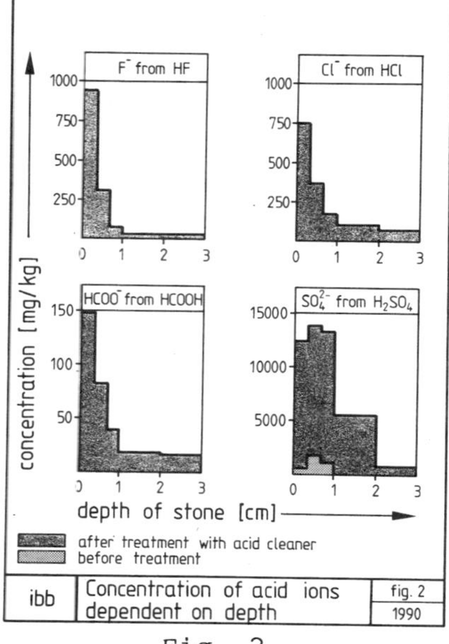

Fig. 2.

6.6 The surface colour

The determination of colour values Honsinger (1988) and Winkler (1975) is another means of describing the change of a surface. The measurement does not destroy or damage the object, it can be carried out relatively easily and quickly (fig. 3), and for many people the change of colour is the only indicator of a successful cleaning. The analysis of the measurement may be carried out by means of the x-y standardized colour table. The indication of the mean value with the respective standard deviations is a first sign of whether or not a significant colour change has been caused by the treatment (fig. 3). A similar analysis may be carried out by means of the Eab-method with the threshold value Eab* = 5 for the subjective perception. As regards the investigated sandstones, it may be said that the measured colour values correlate only sometimes with the other parameters. No interdependence can be found between the capillary water absorption and the colour difference, rather a correlation between colour change and loss of substance. This may be explained by the following notion. The enamel-like black film of the single quartz grains, which cannot be removed by cleaning without destroying the grain structure, determines the colour, but it does not, however, determine the permeability of the surface. The capillarity is increased by removing the particles which obstruct the interstices between the grains. This removal does not, however, change the colour. Only if the grain determining the colour has been removed (loss of substance) is the colour of the sample changed. The determination of the colour as an aesthetic characteristic is an appropriate means of assessing the cleaning effect.

6.7 The roughness

The determination of roughness according to German Industrial Standard DIN 4768 may either be carried out on the object itself or by means of plastic masses and the use of a scanner. The surface structure may also be visualized by means of photogrammetry Mirwald (1986). In the author's opinion, the determination of roughness according to German Industrial Standard DIN 4768 is only valid for absolute values. It is only partially useful for the determination of the roughness of sandstones for the following reasons. The grain size as the determining characteristic of the parameters of roughness is not included in this standard. Moreover, the parameters characteristic of the roughness remain the same if no grain layers, one grain layer or several grain layers are abrased parallel to the surface structure by a cleaning. In order to ensure a valid analysis many lines over a surface have to be scanned. This implies a high expenditure. The structure of a surface can be observed with the naked eye by the following trick:

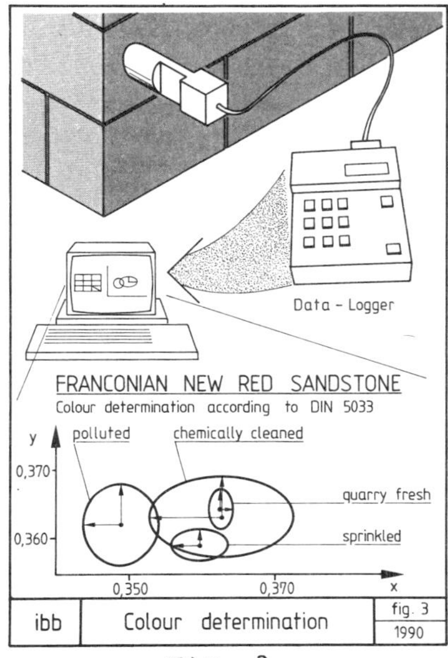

Fig. 3.

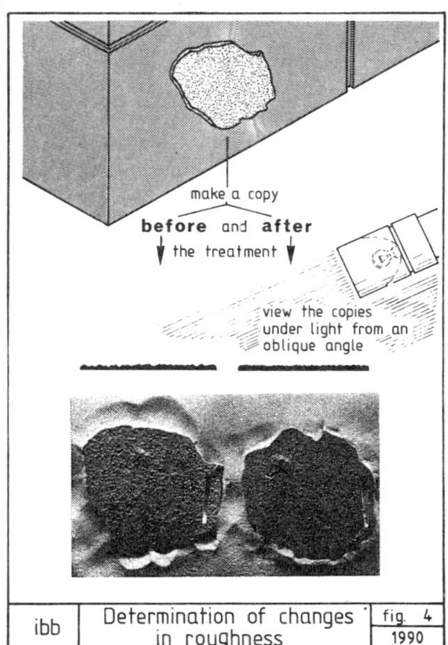

Fig. 4.

separation of roughness and colour by producing a congruent, monochromic copy made from, for example, plastic dentistry paste, and viewed in light incident on the surface at an oblique angle (fig. 4). The homogeneous colour of the copy no longer evokes the impression of roughness. Thus the two parameters colour and roughness may be evaluated separately. Hence the roughness may be determined subjectively in a visual way. In most cases this evaluation is sufficient in order to carry out a gradation of the cleaning methods as regards the change of roughness. This method is particularly successful, if the copy is made at the **identical** same point before and after the treatment. This method of determining the roughness is appropriate as an indicator of the inevitable loss of substance during a cleaning.

7 Summary

Basically, it may be said that the state of weathering at every single point of a squared stone or a statue determines the single characteristics of investigation to a much higher degree than the applied cleaning method. One exception is the pressureless sprinkling with which substrates with crusts containing gypsum can be cleaned well and which has no effect on other surfaces apart from possibly soaking them thoroughly. Another exception is the chemical cleaning method, the use of which in all absorbent porous sandstones produced residues of the cleaner which could partly be detected down to a depth of 30 mm. The methods using warm water do not differ in their effects because the impact pressure and the impact temperature depend very much on the distance valve to surface, and thus the operator can easily change the effect by a skilful handling of the jet-valve. Apart from the jet pressure, the choice of the granulate (glass, quartz sand, ground limestone, wood) and the size of the granulate are the decisive parameters of the sandblasting methods. Satisfying results can be achieved even on sensitive surfaces by an appropriate choice. In any case, the long-term damage which occurs when using chemical cleaners or thoroughly soaking the samples by sprinkling may almost be totally avoided. It is recommended that- and this may be the most significant result of this investigation- one or several test surfaces be installed for different cleaning methods at the respective object since firstly, as mentioned above, the state of weathering plays an important part and secondly the parameters can be influenced by the cleaning personel.

These standard test surfaces may be used to determine the essential parameters which describe the cleaning effect by relatively simple means, the parameters being

the capillary water absorption as a measure of the opening of the pores,
the colour as a directly visible feature and
the change of roughness as an indicator of the loss of substance.

The respective success of cleaning may be described relatively well, quickly and cheaply by these three characteristics. The capillary water absorption according to German Industrial Standard DIN 52615 may be replaced by the capillary water penetration according to Karsten. Since these methods of analysis do not destroy the objects, particularly valid results may always be obtained if the investigations of the respective **identical** point are carried out before and after the cleaning.

These investigations were carried out with the support of a financial grant from the German Ministry of Research and Technology (BMFT) as part of a joint venture project.

8 References

Blaschke, R. (1986) Schleimbildende Mikroorganismen und nitrifizierende Bakterien als Helfer der Gipsbildung in Naturstein. **Sonderheft Bautenschutz und Bausanierung**, 38-41.

Danneker, W. (1986) Einsatz eines neuartigen Passiv-Sammlersystems an Kulturbauten zur Messung sauer reagierender Schadgase in der Atmosphäre, **Sonderheft Bautenschutz Bausanierung**. 46-49.

Geologica Bavarica (1984) Oberflächennahe mineralische Rohstoffe von Bayern. **Geologica Bavarica**, 86.

Grimm, W.-D. and Völkl, J. (1983) Rauhigkeitsmessungen zur Kennzeichnung der Natursteinverwitterung. **Zeitschrift der deutschen geologischen Gesellschaft**, 134, 387-411.

Honsinger, D. and Sasse, H.R. (1988) Neue Wege zum Schutz und zur Substanzerhaltung von Sandsteinoberflächen unter Verwendung von Polymeren. **Bautenschutz und Bausanierung**, 11, 205-211.

Klopfer, H. (1974) **Wassertransport durch Diffusion in Feststoffen.** Bauverlag, Wiesbaden.

Kraus, K. (1985) **Experimente zur immissionsbedingten Verwitterung der Naturbausteine des Kölner Doms im Vergleich zu deren Verhalten am Bauwerk.** Cologne

Krumbein, W.E. (1988) Biology of stone and minerals in buildings, in **Proceedings of the VIth International Congress on Deterioration and Conservation of Stone** (eds. Nicholas Copernicus University) Nicholas Copernicus University, Torun, pp. .

Mirwald, P.W. (1986) Umweltbedingte Gesteinszerstörung untersucht mittels Freiland-Verwitterungsexperimenten. **Sonderheft Bautenschutz Bausanierung**, 24-27.

Pohle, G. (1988) Auswirkungen gängiger Reinigungsverfahren auf die Oberflächenschichten von Natursteinmauerwerk. **Sonderheft Bautenschutz und Bausanierung**, 65 - 69.
Schuh, H. (1987) **Physikalische Eigenschaften von Sandsteinen und ihren verwitterten Oberflächen**. Pfeil, Munich.
Snethlage, R. (1984) Steinkonservierung 1979-1983. Bericht für die Stiftung Volkswagenwerk. **Arbeitshefte des Bayerischen Landesamt für Denkmalpflege**, 22.
Wendler, E. and Snethlage, R. (1989) Der Wasserein-dringprüfer nach Karsten - Anwendung und Interpretation der Meßwerte. **Bautenschutz und Bausanierung**, 12, 110-115.
Werner, M. (1989) Tendenzen in der Anwendung von Reinigungsverfahren für Naturstein. in **Unpublished seminar paper for the Bauphysikalisches Colloquium**. Fraunhofer-Institut für Bauphysik, Holzkirchen.
Winkler, E.M. (1975) **Stone properties, durability in man's environment**. Springer, New York.

26 INFLUENCE OF MICROCRACKING ON AIR PERMEABILITY OF CONCRETE

S. NAGATAKI
Tokyo Institute of Technology, Tokyo, Japan
I. UJIKE
Utsunomiya University, Utsunomiya, Japan

Abstract
This study deals with the air permeability of concrete with microcrackings. Microcrackings chosen in this study are cracking induced by the differences of thermal expansion coefficients between aggregate and mortar or between aggregate and cement paste under elevated temperature as well as internal cracking formed around deformed tension bar.

The air permeability coefficient of concrete under elevated temperature above 100°C becomes more than scores of times compared with that of concrete at normal temperature. The increasing rate of air permeability coefficient of concrete by heating corresponds to the amount of microcrackings. In the case of reinforced concrete specimen subjected to sustained tensile loading, the air permeability coefficient of specimen averaged through cover has the large value compared with the specimen without tensile force. And, as the tensile stress and the diameter of steel bar become larger, the air permeability coefficient of specimen increases.
Keywords: Microcracking, Air Permeability, Elevated Temperature, Deformed Bar, Internal Cracking, Concrete Cover.

1 Introduction

Concrete cover affords the protection against the ingress of aggressive materials from the environment into the concrete. However, as concrete is porous media, such materials as chloride ions and oxygen reach the surface of reinforcing bar by diffusing through pores in concrete and corrode steel bar. The permeation characteristics of concrete is one of the most dominant factors affecting the durability of concrete structure (Lawrence,1984).

The air permeability of concrete is a rate of movement of materials in concrete and depends on the internal structure in concrete, especially, capillary pores. On the other hand, it has been also known that microcracking is induced in reinforced concrete structure by the action of external force (Goto and Ohtsuka,1980) and heat (Minami et al.,1987). Although microcrackings lower the tightness of concrete and must promote the ingress of aggressive materials, no studies are carried out on the air permeability of concrete from the viewpoint of microcracking.

The objective of this study is to make clear the effects of microcrackings on the air permeability of concrete. The generation of microcrackings was brought about in two ways, by: (1) heating the concrete, (2) applying tensile force to the reinforced concrete specimens. The effects of heating condition on the air permeability of concrete under elevated temperatures were investigated comparing with that of concrete at normal temperature. And, the magnitude of tensile stress and diameter of steel bar were chosen as the factors affecting the properties of internal cracks. The relationship between the factors and air permeability coefficient was examined.

2 Experimental procedures

Ordinary portland cement is used. River sand with specific gravity of 2.61 and fineness modulus (F.M.) of 2.85 and crushed gravel with specific gravity of 2.65 and F.M. of 6.73 from Kinu river are used as fine and coarse aggregate, respectively. Deformed bars with nominal diameter of 16, 19, 22 and 25mm are used. The mix proportions of concrete and properties of fresh concrete are given in Table 1.

the air permeability test is conducted in two ways by using an apparatus illustrated in Fig.1. On the air permeability test of concrete under elevated temperature, 15cm cubes are manufactured. After curing in water at 20°C for 28 days, the specimens are cured in air at 20°C and 60%R.H. for prescribed period. The four sides without air permeable upper and bottom surfaces of specimen are made airtight

Table 1. Mix proportion and properties of fresh concrete

W/C	s/a	Unit weight (kg/m^3)						Slump	Air cont.
(%)	(%)	W	C	S	G	WRA	AEA	(cm)	(%)
40	44	164	410	765	989	1.025	0.018	7.7	4.1

WRA:Water Reducing Agent AEA:Air-entraining Agent

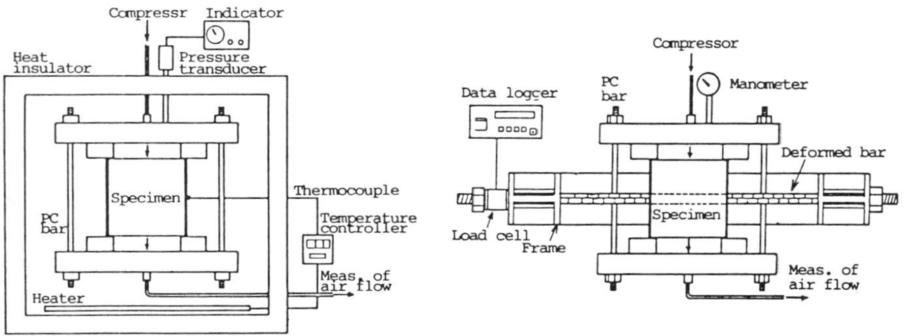

Fig.1. Air permeability apparatus

by heat-resisting adhesives. Furthermore, steel plates are glued on upper and bottom surfaces of specimen, so that the area exposed to air pressure is 13x13cm. The heating temperature is controlled by the surface temperature of specimen measured with thermocouple. The specimen is heated at a constant rate of 0.5°C per minute. The surface temperatures of specimen are kept at constant temperature of 50+2, 100+2 and 150+2°C, respectively.

On the air permeability test of tension test specimen, concrete cover of all specimen is 4cm and cross section perpendicular to the direction of air flow is 15x15cm. The specimens are cured in the moist condition for two weeks at 20°C. The specimens cured in air for prescribed period after moist curing are made airtight by the same manner as the above mentioned method using epoxy resin adhesives. The loading is applied before setting the specimen to the air permeability test apparatus. The magnitude of load applied is about 100, 150 and 200N/mm^2 in terms of reinforcement stress.

The air pressure of 0.2N/mm^2 is applied to specimen only when measuring the air flow rate. The permeated air is collected in a cylinder inverted in a water bath and its flow rate measured. The air flow rate through the heated concrete is corrected by Eq.(1), owing to remove the effects of increase of kinematic viscosity and volume expansion of air with temperature rise. The air permeability coefficient of concrete is calculated by using Eq.(2).

$$Qc=Qm\cdot(Vet/Vrt)\cdot(Tet/Trt) \quad (1)$$

$$K=[2LP_2/(P_1{}^2-P_2{}^2)]\cdot(Qc/A) \quad (2)$$

where

Qc,Qm :corrected and measured air flow rate (cm^3/s)
Vet,Vrt:kinematic viscosity at elevated and room temp. (cm^2/s)
Tet,Trt:temperature of specimen and room (K)
K :air permeability coefficient [cm^2/(s·N/cm^2)]
L :thickness of specimen (cm)
A :cross sectional area of specimen (cm^2)
P_1,P_2 :applied air pressure and atmospheric pressure (N/cm^2)

3 Experimental results and discussions

3.1 Air permeability of concrete under elevated temperature

Fig.2 shows the time-dependent changes of air permeability coefficient of concrete under three kinds of heating condition. The air permeability coefficients at the start of heating and at normal temperature are measured at room temperature of about 10°C after specimens are cured in air at 20°C and 60%R.H.. The incremental rate of air permeability coefficient in its early heating period becomes large with increase of heating temperature. However, the incremental rate of air permeability coefficient of concrete under elevated temperature above 100°C gradually decreases with the time elapsed. The air permeability coefficient at 150°C hardly increases after that rapidly increase in its early heating period.

Fig.3 shows the relationship between the air permeability

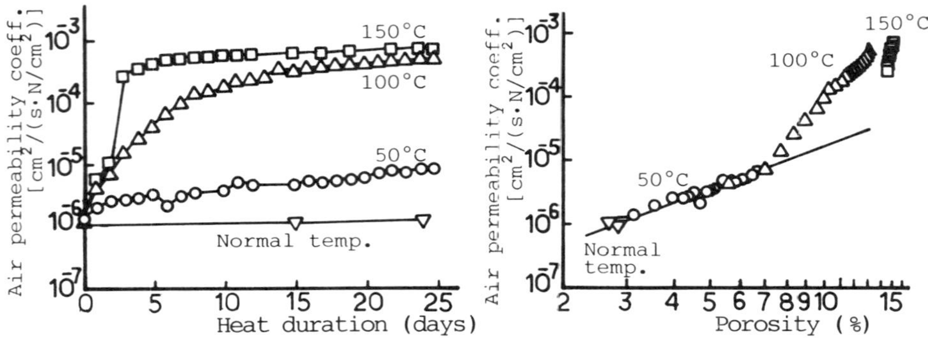

Fig.2. Time-dependent change of air permeability coefficient of concrete

Fig.3. Relationship between air permeability coefficient and porosity of concrete

coefficient and porosity of concrete, which is the volume of evaporated water from concrete divided by that of the specimen. As previously reported, it has become clear the air permeability coefficient of concrete at normal temperature is expressed by the single straight line as function of porosity (Ujike and Nagataki,1988). Although the air permeability coefficients of concrete at 50°C increase lineally with the increase of porosity lineally, the increasing rate of air permeability coefficient at 100°C becomes larger than that at 50°C.

Fig.4 shows the accumulation of AE count against the heat duration for 100 and 150°C. AE count is measured by means of acoustic emission method, which is one of detection method of microcracking. It is estimated from Fig.4 that there is more microcracking generation in concrete at 150°C than that at 100°C in its early heating period. However, the cumulative AE count of concrete at 100°C approaches to that at 150°C gradually. The cumulative AE count with the time elapsed corresponds to the time-dependent change of air permeability coefficient under elevated temperature shown in Fig.2. The amount of microcracking may be the main factor dominating the increase of air permeability coefficient of concrete by the heating.

3.2 Air permeability of tension test specimen

Fig.5 shows the air permeability coefficient of tension test specimen having the deformed and round bar of nominal diameter 22mm. A broken line in Fig.5 represents the air permeability coefficient of plain concrete. The air permeability coefficients of plain concrete and tension test specimen without tensile stress have almost the same value. The air permeability coefficient of specimen with deformed bar increases with the increase of applied stress. However, the air permeability coefficient of specimen with round bar hardly changes regardless of the magnitude of applied stress. Although specimens are manufactured under the same condition except for kinds of bar, there is difference between air permeability of tension test specimen with deformed bar and that with round bar due to the generation of

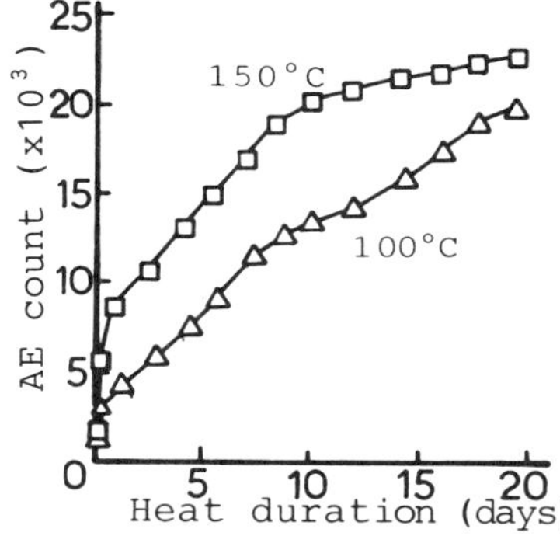

Fig.4. AE property of concrete at elevated temperature

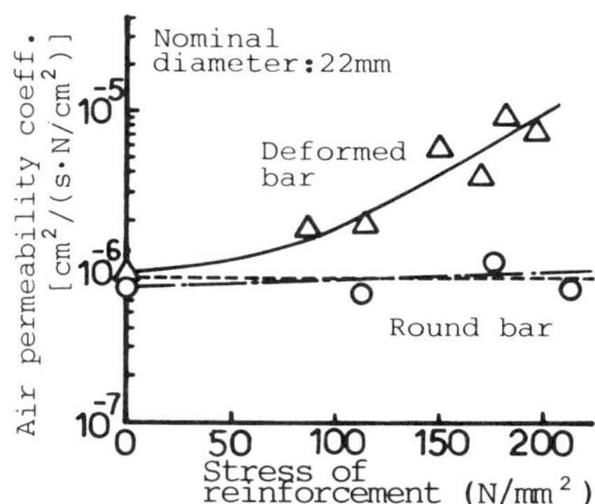

Fig.5. Air permeability of specimen having deformed and round bar

internal cracking by the action of tensile force.

Fig.6 shows the air permeability coefficients of tension test specimen using deformed bars with different diameters. As the diameter of deformed bar becomes large, the air permeability coefficient begins to increase from lower applied stress. Furthermore, the larger diameter of deformed bar is, the more considerably the air permeability coefficient increases. It was reported that internal cracking, which do not appear on the surface of reinforced concrete member, are generated at ribs on deformed bar before the applied tensile stress of deformed bar reaches 100N/mm^2. Furthermore, internal crackings at high ribs are easy to generate and are longer than that at low ribs (Goto and Ohtsuka,1980). Therefore,

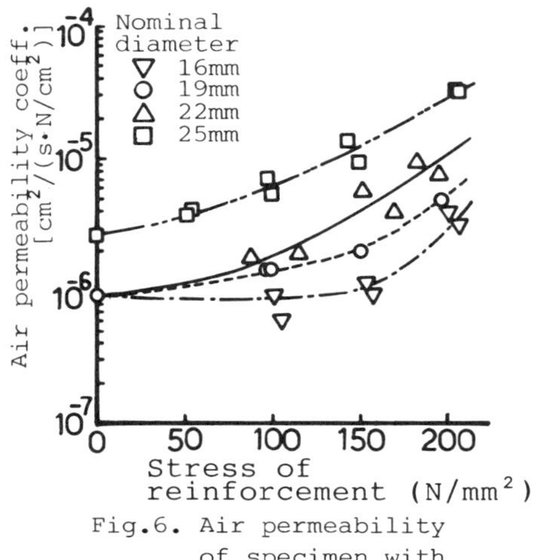

Fig.6. Air permeability of specimen with deformed bar

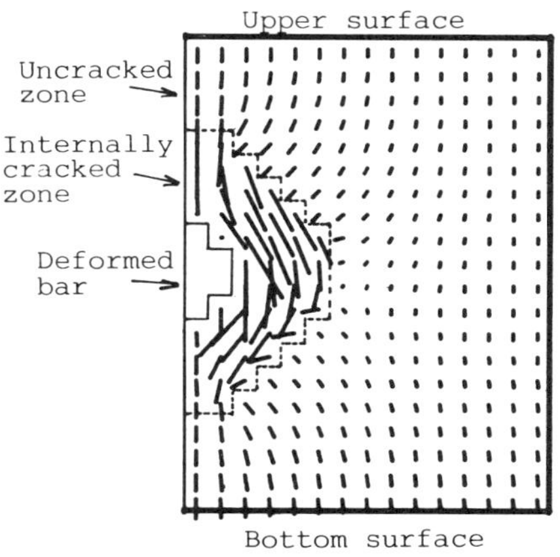

Fig.7. Air flow in tension test specimen

the difference in air permeability depending on the diameter of deformed bar is considered to be caused by the area of internally cracked zone.

Next, the air flow in concrete having more permeable zone due to internal cracking around the steel bar is shown in Fig.7. The result of calculation is obtained by solving the governing equation for air flow based on Darcy's law. A numerical solution is used the method of finite differences. Where, it is assumed that the internally cracked zone is within the area corresponding to the diameter of bar from surface of bar and its air permeability is one hundred times as large as air permeability coefficient of uncracked zone. The result of calculation indicates that the air flow gathers around the bar and air flow rate in internally cracked zone is large compared with that in uncracked zone. As mentioned above, internal crackings around the deformed tension bar make the air permeability coefficient of concrete averaged through cover increase. Furthermore, the air mainly flows internally cracked zone. These facts may mean that concrete cover becomes thin substantially.

4 Conclusions

The results obtained from this study are summarized as follows.

The air permeability coefficients of concrete at 100 and 150°C increase with the lapse of time after the start of heating. Although pore evaporated water makes path for air flow through concrete at normal temperature and 50°C, in the case of concrete at 100 and 150°C, the path of air flow is mainly composed of microcrackings caused by heating. It is become clear that air permeability coefficients of concrete under elevated temperatures of 100 and 150°C are affected by the amount of microcracking.

The air permeability coefficient averaged through cover of reinforced concrete specimen with deformed bar also increase by the action of sustained tensile force. The air permeability coefficient of specimen become large with the increase of applied tensile stress. Furthermore, the increasing rate of air permeability coefficients depends on the diameter of deformed bar when specimens have the same concrete cover.

5 References

Goto, Y. and Ohtsuka, K. (1980) Experimental studies on cracks formed in concrete around deformed tension bars. Proceedings of JSCE, No.294, 85-100.

Lawrence, C.D. (1984) Transport of oxygen through concrete. British Ceramic Society Meeting.

Minami, K., Tazawa, E., Kageyama, S. and Watanabe, Y. (1987) Effect of type of aggregate on mechanical properties of concrete subjected to high temperature. Transactions of JCI, Vol.9, 1-8.

Ujike, I. and Nagataki, S. (1988) A study on the quantitative evaluation of air permeability of concrete. Proceedings of JSCE, No.396, 79-87.

27 MECHANISM AND EVALUATION METHOD OF FROST DETERIORATION OF CELLULAR CONCRETE

O. SENBU
Building Research Institute, Ministry of Construction, Tsukuba, Japan
E. KAMADA
Hokkaido University, Sapporo, Japan

Abstract
Two types of frost deteriorations are observed in external walls made from cellular concrete in cold regions. One type of deterioration is surface **spalling** caused by freezing and thawing, the other is wide cracks caused by keeping the inner part of cellular concrete at 0° C. It is thought that frost deterioration mechanisms of cellular concrete are not same as those of ordinary concrete.

In this paper, cellular concrete deterioration caused by various test methods are analyzed, and deterioration mechanisms are investigated. From evidence gathered, an evaluation method of frost resistance in cellular concrete is proposed.
Keywords: Cellular Concrete, Frost Deterioration, Capillary Theory, Top Surface Freezing Test

1 Introduction

Cellular concrete is often used for external walls. In cold regions these walls are subject to frost deterioration. Two types of deterioration are observed: one is **spalling** of surface layers and another is wide cracks parallel to the surface. It is thought that deteriorations of cellular concrete are not same as those of ordinary concrete.

Following investigations of the deterioration of cellular concrete, it becomes clear that the frost deterioration of cellular concrete is different from that of ordinary concrete. Traditional methods can not evaluate frost resistance of cellular concrete correctly, so an evaluation method correspond to the deterioration mechanism of cellular concrete is proposed.

2 Deterioration of cellular concrete by traditional methods

2.1 Freezing and thawing method

A typical freezing and thawing method is ASTM C666. Deterioration of specimens by the method occurs in the form of **spalling.** Severe splitting which is observed under real conditions does not occur in the test. Fig.1 shows results of the freezing and thawing method in accordance with ASTM C666. It is clear that volume reduction by

spalling becames greater, as test conditions become more severe. It is thought that the **spalling** occurs as surface air voids fill with water by capillary action and freezing pressure.

2.2 Critical degree of saturation method

A typical critical degree of saturation method is RILEM Method for concrete. A rapid method consists of freezing and thawing tests to calculate the critical degree of water content sufficient to cause damage, and water absorption tests to measure water absorption. In the method, frost resistance of materials is evaluated by comparing the critical degree of saturation with water content of the water absorption test. Deterioration caused by this method is degradation of whole specimen. This type of deterioration is scarcely observed in real walls. However, critical degree of saturation is important as a measure of the quality of construction materials.

2.2 Top surface freezing test

The top surface freezing test which was proposed by Kamada et al. (1984) is a model of real external wall conditions. Fig.2 shows a test apparatus of the top surface freezing test. In the test, water is applied to the bottom surface of the specimen, the top surface is held at a constant temperature below 0°C, and the freezing position is fixed in the specimen.

As shown in Fig.3 and 4, evidence of deterioration during the test was a wide crack across the specimen located at the 0°C level. There is little correlation between the results of the freezing and thawing test and the top surface freezing test. This means that the top surface freezing test might cause deterioration by a different mechanism than that which causes deterioration in the freezing and thawing test.

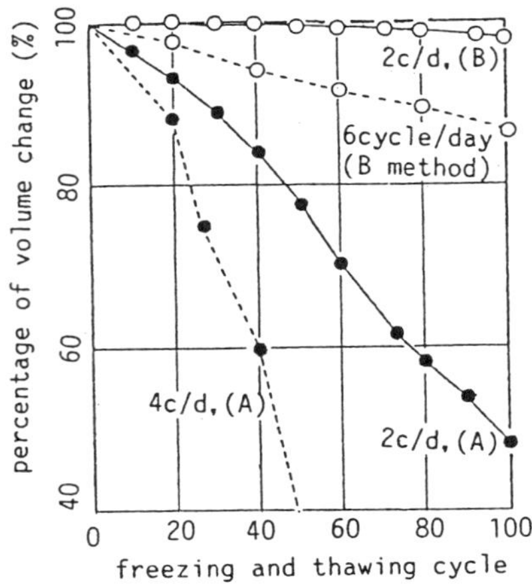

Fig.1 Result of freezing and thawing test by ASTM method (Kamada et al, 1984)

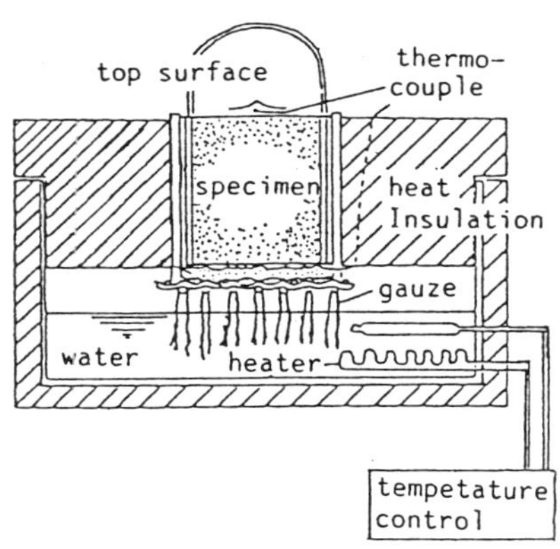

Fig.2 Test apparatus of top surface freezing test (Kamada et al, 1984)

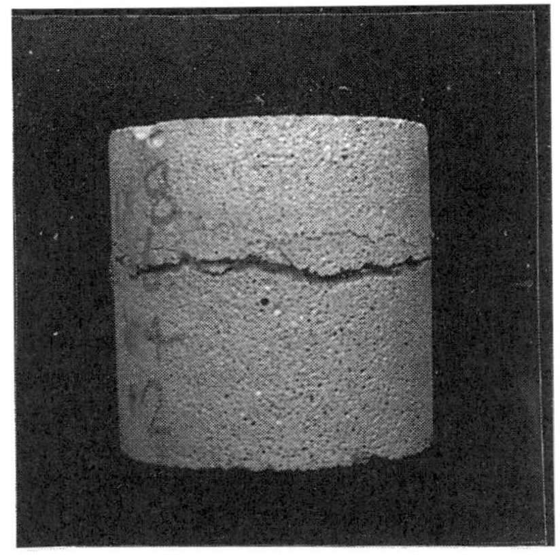

Fig.3 Cracked Specimen in top surface freezing test (Kamada et al, 1984)

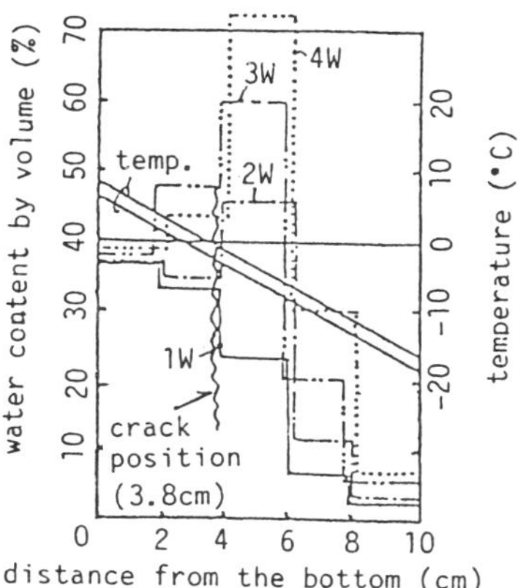

Fig.4 Changes of water content and occurance of destruction in top surface freezing test (Kamada et al, 1984)

3 Proposal of frost deterioration mechanism of cellular concrete

3.1 Pore structure of cellular concrete

Pores of cellular concrete consist of air voids and capillaries. The volumetric proportion of cellular concrete is air voids : capillaries : solid = 5:3:2. Distribution of pore size of cellular concrete and ordinaly concrete is shown in Fig.5. A decrease of the freezing point occures in the capillaries.

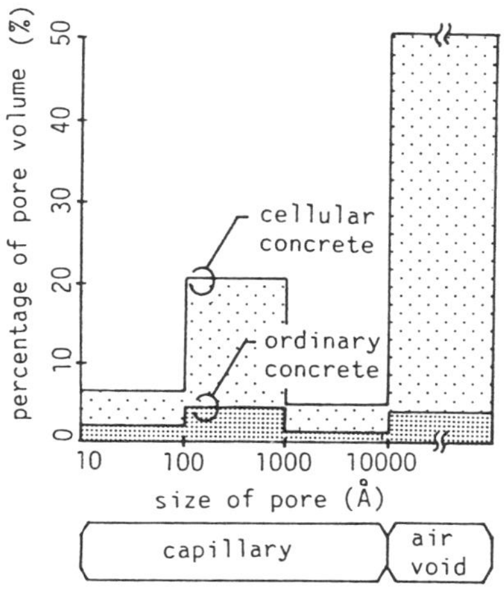

Fig.5 Distribution of pore size of cellular concrete and ordinary concrete

3.2 Application of capillary theory

In the capillary theory proposed by Everett(1961), it is assumed that ice forms in air voids and capillary water is kept unfrozen when deterioration occurs. The deterioration mechanism of cellular concrete develops as the capillary pressure differential between ice in air voids and water in capillaries increases. As shown in Fig.6, pressure (P) built up in the air void from this balance of power.

3.3 Explanation of wide cracks in the top surface freezing test

Fig.7 shows a conceptual cracked plane of a specimen in the top surface freezing test. On the plane, all the air voids are filled with ice, and pressure (P) (explained in Fig.6) builds up in all air voids. Whether or not cracks occur is determined by comparing pressure (P) per a unit area and the tensile strength of the material. Fig.8 shows the relationships between the volumetric proportion of air voids and tensile strength and/or pressure (P) per unit area. Pressure (P) is changed by the size of the capillary connected to the air void. From the data ploted in Fig.8, capillaries smaller than 200Å are thought to influence this deterioration mechanism.

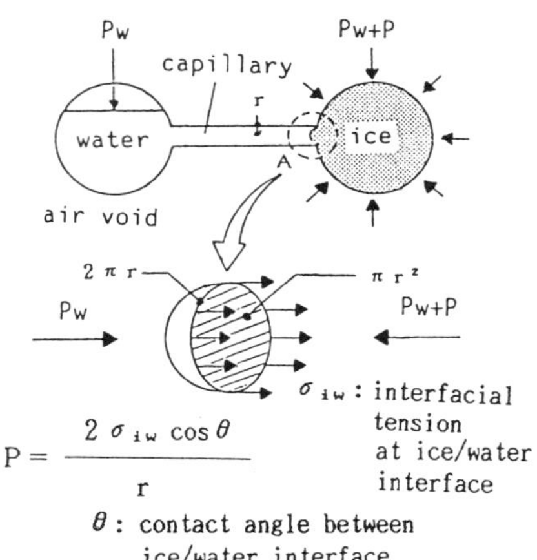

Fig.6 Pressure (P) from capillary theory

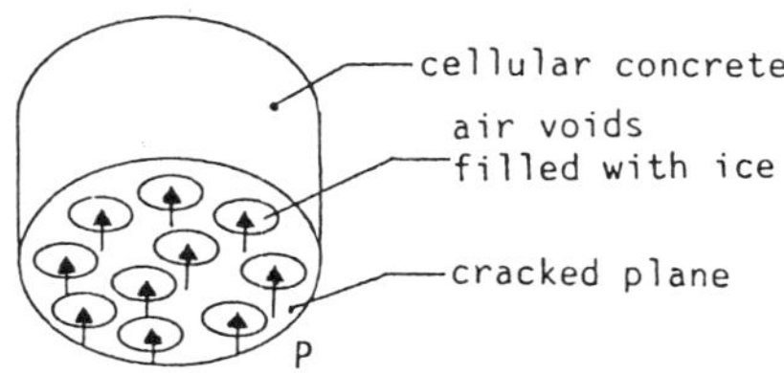

Fig.7 Model of deterioration of top surface freezing test

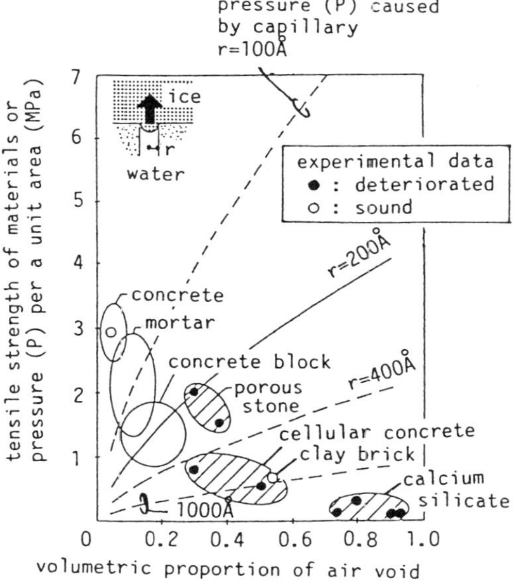

Fig.8
relationships between volumetric proportion of air void and tensile strength of materials (or pressure (P) per unit area)

3.4 Explanation of degradation of critical degree of saturation method

Although degradation is rarely observed in real external walls, **spalling** is thougth to be the degradation on infinitely small portions of the cellular concrete surface. By explaining this form of degradation, all the deterioration may become clear.

The self-consistent approximation method proposed by Baba (1978) was used to explain the degradation. In the method, the following items are assumed. Fig.9 shows a schematic presentation of the method.

(a)One air void is small enough comparing with whole composite.
(b)First, pressure (P) is occured at one air void located at the center of the spherical composite containing other air voids.
(c)The spherical composite is thought as an elastic body, and whole pressure in the composite is calculated with adding all the pressure of each air void filled with ice.

Fig.10 shows the relationships between water content and the tensile strength of materials (or pressure existing in air voids filled with ice). It is assumed that deteriorations occur when average pressure in the composite becomes greater than tensile strength. In order to cause deterioration of the composite with a critical water content of 55%vol which is calculated by Kamada et al.(1984), capillaries 600Å–1000Å in size are necessary. This result reflects the actual pore structure (Fig.5) of cellular concrete.

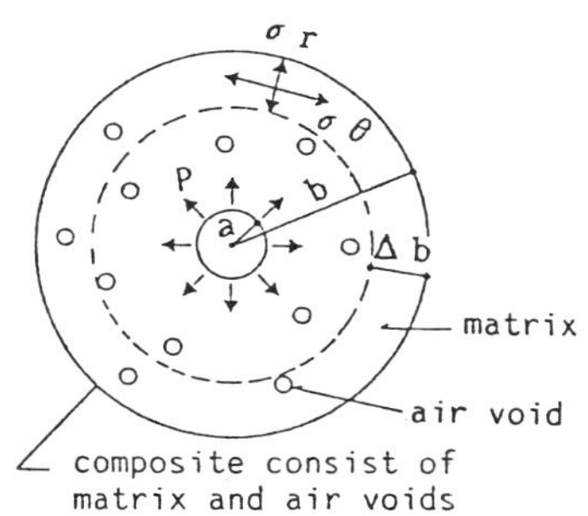

$$\sigma_m = \frac{Wp}{1 - Wp} \cdot P$$

σ_m : average pressure
Wp : volumetric ratio of air voids

Fig.9
Model of degradation based on self-consistent approximation (Baba,1978)

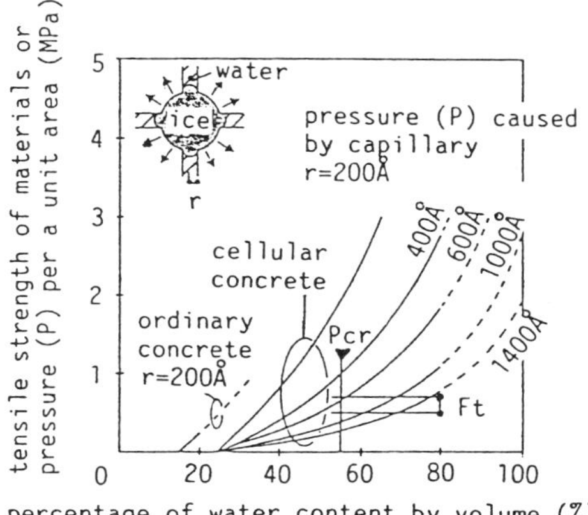

Pcr:critical water content of cellular concrete
Ft :tensile strength of cellular concrete

Fig.10
Relationships between water content and tensile strength of materials or pressure occured by air voids filled with water

4 Proposal of evaluation method

As mentioned above, frost deterioration of cellular concrete is thought to occur by the freezing of water in air voids. This is different than the deterioration mechanism of ordinary concrete which is caused by freezing water in capillaries. In the case of deterioration by freezing in air voids, it is important whether or not air voids are easily satulated with water. From the tests of section 3, it is thought that there are two effects which accelerate the filling of air voids with water. One is freezing pressure and the other is attraction of water to ice. These two effects correspond to the freezing and thawing test and the top surface freezing test respectively.

As a result, in order to evaluate the frost resistance of aerated concrete, it is necessary to apply the freezing and thawing test designated in ASTM (or its like) as well as the top surface freezing test simultaneously.

5 Concluding Remarks

When deteriorations of cellular concrete by evaluation methods are investigated it becomes clear that the frost deterioration of cellular concrete is caused by freezing water in air voids. In order to evaluate frost resistance of cellular concrete it is important to measure the two factors causing accelerated water movement to air voids. Thus, it is necessary to apply the freezing and thawing test as well as the top surface freezing test simultaneously.

6 References

Kamada,E. Koh,Y. and Tabata,M. (1984) Frost deterioration of cellular concrete, in Third international conference on the durability of building materials and components, Volume 3, pp.372-382.

Everett,D.H. (1961) The thermodynamics of frost damage to porous solids,Trans.Faraday Soc. 57, pp.1541-1551.

Baba,A. (1978) Drying shrinkage mechanism of building materials. BRI Research Paper No.77, Building Research Institute, Ministry of Construction.

SESSION C

STUDIES OF *IN-SITU* BEHAVIOUR OF MATERIALS AND COMPONENTS

28 CARBONATION. THE EFFECT OF EXPOSURE AND CONCRETE QUALITY: FIELD SURVEY RESULTS FROM SOME 400 STRUCTURES

J.H. BROWN
British Cement Association, London, UK

Abstract
This paper presents the main findings from a collection of field survey reports on 437 structures or structural elements. The data were assembled from the archives of three Test Houses, from PSA and BCA records and from some field surveys carried out specifically for this project. The data were collected by filling in a standard pro-forma, one for each structure. The primary questions concerned depth of carbonation, age, exposure and location and subsidiary questions gathered quantitive information on concrete quality, cover and other factors concerning durability where the information was available.

It was expected that the large number of data points would give well defined envelopes to the carbonation depth/age relationships in accord with generally accepted models for carbonation as a diffusion controlled process. Although such envelopes could be drawn most points showed carbonation depths well inside the boundary so in this survey the effect of age was found to be overshadowed by the influence of other factors on carbonation. Concrete quality expressed as either cement content or strength was found to be a significant controlling factor. The most surprising result, however, was that carbonation depths did not show any significant relationship with the reported exposures, which ranged from indoor, through sheltered outside to extreme outside. The lack of correlation suggests that for the structures of this survey there were factors more important than exposure controlling the carbonation process.
Keywords: Structures, Concrete, Durability, Carbonation, Cover, Exposure.

1 Introduction

This paper summarises the results of field measurements of carbonation depths on a large number of structures. It is part of a larger project, a general study of carbonation with the long term aim of improving the durability of structures by more informed design. The project is funded by the Department of the Environment (DoE) through the Building Research Establishment.

Laboratory studies of carbonation give measurements under controlled conditions. Exposure site conditions are more 'natural' but usually involve carefully prepared specimens of precise pedigree. Very old concretes are unusual in either laboratory or exposure site investigations. Field studies necessarily correspond to real structures but often the concrete mix is not well documented, the comparison of one structure with another is complicated by differences of micro and macro environment, and the results show large variability. In principle the variability could be overcome by taking large numbers of measurements but field work is expensive and this is rarely practical. Nevertheless an enormous amount of carbonation data on real structures already exists, stored in the archives of organisations concerned with testing real structures. By extraction and collation, the resulting block of data could be large enough for reliable conclusions to be drawn. Three Test Houses offered enthusiastic cooperation and, together with information from PSA and BCA archives and some tests made specifically for the project, data were collected for each of 437 structures or separate structural elements.

Essential information required was the depth of carbonation, the age and the exposure of the concrete. Other information requested, if known, included concrete quality and type, structure location and orientation, reinforcement cover and concrete deterioration.

The pattern for the analysis was to look first at the range of depths of carbonation irrespective of other factors, and then to find the effect of exposure, age and so on, so that a model of the interrelationships was built up. From the subsidiary questions relationships between cover, carbonation depth and durability were explored. Some simple comparisons - cover for in-situ compared with precast, for example - could be made, and these show the importance of factors specific to concrete practice rather than fundamental processes.

2 Measured factors

For depth of carbonation typical and maximum values were noted, with the typical depth a modal rather than a mean value of all the measurements. Similarly, the typical cover was a modal value for the more shallow steel in the area. Exposure was concerned primarily with exposure to rain and is defined as indoor; outside and completely sheltered from rain; outside with moderate shelter; and outside with no shelter. These comprised 56, 129, 172 and 76 structures respectively. A range of different assessments had been made for concrete quality by the test-houses but only compressive strength and cement content were sufficiently common to be included in the analysis.

3 Results and discussion

Considering the carbonation results alone from all 437 structures, only 5 showed typical depths greater than 50mm while 20 showed maximum depths greater than 50mm.

Age must be a factor in the carbonation process since new concrete has zero carbonation and the depth of carbonation increases with time. However, none of the 5 oldest structures (over 70 years) featured in the top 20 maximum carbonation depths, whilst 7 of the 20 were less than 20 years old. It is clear that, while age must be a fundamental factor in carbonation, other factors within the data are causing this to be obscured.

It is generally recognised that environment is an important factor in the carbonation process (many useful references in Parrott (1987)) and this follows logically from the dependence of the rate of carbonation on the moisture state of the concrete. If carbonation is controlled by a diffusion process the relationship between carbonation depth and (age)½ for a given concrete and given condition may well be linear. (Tuttii, for example, found his results fitted this form). Eight graphs are needed to show the results of the four exposures with maximum and typical depths of carbonation. Space permits only two of the graphs to be given here. Figure 1 shows the relationship between typical carbonation depth and root age for indoor exposure and Figure 2 that for maximum depth of carbonation and moderately sheltered outside exposure.

For each graph a straight line can be drawn from the origin to show a limiting depth of carbonation increasing with (age)½. However, the lines are defined by so few points they are completely unconvincing as definitions of limiting carbonation depths. The other graphs not shown here are similar to Figures 1 and 2.

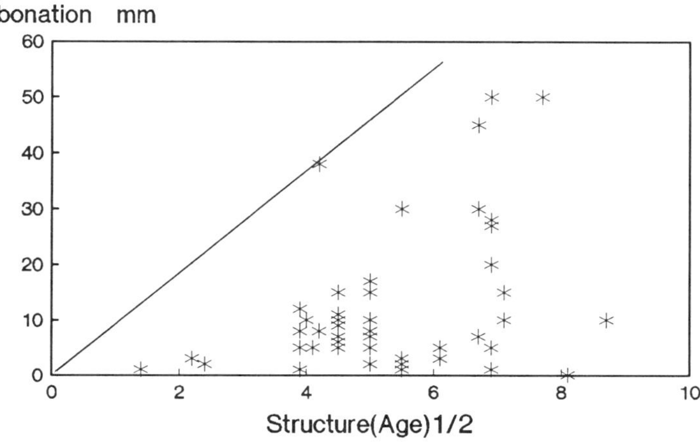

Fig.1. Typical indoor carbonation and age.

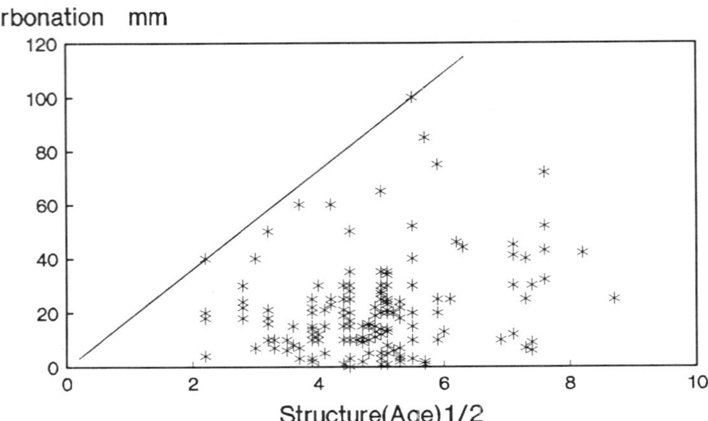

Fig.2. Maximum carbonation, moderate exposure outside, and age.

The survey included very few structures less than 10 years old which would have defined the lower part of the boundary, and the remaining points are more compatible with a plateau rather than a continuous increase of carbonation depth with age. A similar effect was noted by Martin et al (1975) with a 7 year exposure of concrete outside and sheltered from rain. However, the feature that stands out from the present survey is that very few structures show carbonation depths even approaching the limiting envelope, whatever its shape may be.

Statistical analysis brings all the measured values into account rather than simply highlighting the extremes. Table 1 shows the carbonation depths grouped by exposure and by age (older or younger than 25 years). Median rather than mean values are shown, with variability expressed as quartile values.

Looking first at the effect of age, the indoor structures showed about the same median depth of carbonation for both groups, whilst all the outside structures showed, as would be expected, the younger ones to have carbonated less. The difference is not very great - about 3mm for typical, perhaps 8mm for maximum depths - but the table also shows that the upper quartile values for all exposures are much higher for the old structures except for those outside with no shelter where there were too few old structures for quartile values to be reliably defined.

The values in Table 1 show that for the indoor structures surveyed, in general those more than 25 years old had not carbonated more deeply than those less than 25 years old. However, for outside structures those over 25 years old in general had carbonated more deeply than the younger ones. It must be remembered that any comparisons involving concretes of different ages may well be

complicated by changes of materials and methods over the years. The different depths of carbonation found do not necessarily reflect how carbonation has progressed or will progress in any particular structure.
For the effect of exposure, Table 1 shows that the depth of carbonation for outside sheltered structures is less than for those outside with more exposure. This is completely unexpected since it is generally accepted that carbonation is more rapid for sheltered than exposed sites. The indoor structures show carbonation depths not appreciably different from the moderate or extreme exposure ones, although the indoor depths are greater than those for the outside sheltered structures. If all the outside exposures are considered together then the younger ones indoor have carbonated about 2mm more than those outside, but if age is ignored there is only about 1mm more depth of carbonation for indoor than outside structures.

The small difference of carbonation depth with indoor/outside exposure and the small carbonation depths for sheltered concrete outside are unexpected observations since they do not conform to the findings of most workers in this field. Tuutti (1982), suggests that differences of perhaps 10mm more carbonation would be expected for a sheltered than exposed situation after ten years or so, and at BCA (and as part of this same project) specimens, identical except for exposure indoor or outside, showed appreciably more carbonation indoors after only 2 years.

Table 1. Carbonation, Age and Exposure

Exposure	Age (years)	CARBONATION (mm) Typical Median	Typical quartile range	Maximum Median	Maximum quartile range	Number of structures
Indoor	≤ 25	7	5 - 10	14	8 - 20	32
	> 25	7	5 - 27	15	3 - 30	24
Outside, sheltered	≤ 25	3	1 - 7	7	3 - 17	88
	> 25	6	2 - 13	15	6 - 20	41
Outside, moderate	≤ 25	7	3 - 12	15	10 - 25	113
	> 25	10	5 - 20	23	10 - 34	59
Outside, exposed	≤ 25	6	2 - 13	10	5 - 25	59
	> 25	8	3 - 13	12	5 - 19	17
all outside	≤ 25	5	2 - 4	11	5 - 21	260
	> 25	8	3 - 17	18	8 - 30	117

No satisfactory explanation of the difference between the present results and other work has yet been put forward. The assessment of environment is one source of uncertainty since environment is the least well defined factor involved, and the most subjective one. However, whilst errors between mild outdoor and severe outdoor, say, could be subjective ones it is difficult to see how indoor exposure could be noted as anything else.

A third primary factor, after age and exposure, affecting depth of carbonation is the quality of the concrete. For this survey the measure of quality is either compressive strength or cement content. Figure 3 shows the relationship between typical depth of carbonation (all outside exposures) and cement content.

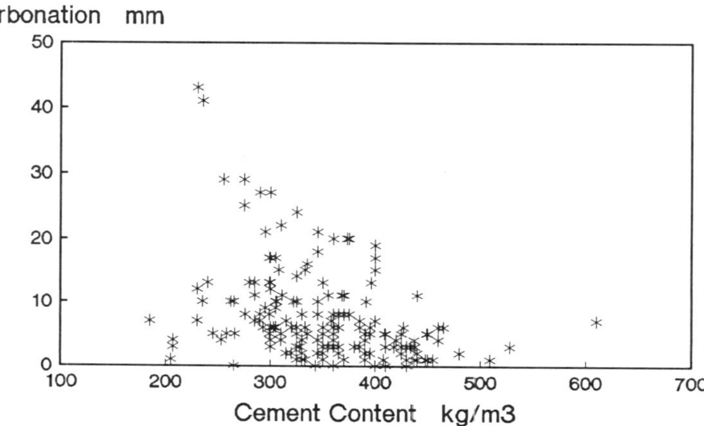

Fig.3. Typical carbonation and cement content.

The graph shows wide scatter, with most structures showing carbonation depths less than 10mm and cement contents between 260 and 450 kg/m^3. However, the extremes seem to suggest that the depths of carbonation may be limited to a depth decided by and decreasing with cement content. A plot of strength and carbonation depth shows a similar boundary envelope which extrapolates to zero carbonation for strengths over about 100 N/mm^2. Again, there are a number of references in Parrott's (1987) review which report similar trends.

It is possible to combine the age and quality parameters to try to reduce some of the variability of the depths of carbonation. The depth will increase with age of structure, possibly linearly with $(age)^{\frac{1}{2}}$, and decrease with quality. For the latter an inverse function of strength or cement content might be suitable and then the depth of carbonation (d) should relate to age (A) and strength (S) or cement content (C) by

$$d = k\,(A^{\frac{1}{2}}/S) \quad \text{or} \quad d = k(A^{\frac{1}{2}}/C) \qquad (1),(2)$$

If this is so a plot of (1) or (2) should be a straight line of slope 1/k. Figure 4 shows the strength based relationship: a plot using cement content gives a similar picture. It is clear that there is no unique relationship for all concretes in accord with equation (1) (or (2)). Whilst it is possible that the line drawn gives the limiting depth of carbonation for concrete of a given age and quality, it is clear that most structures have not carbonated to anything like this limiting depth. Even if it is assumed that the square root relationship applies only to concrete at early age (and thereafter a limiting depth is reached) most of the carbonation values found in this survey are well within the boundary that could be drawn round the extreme values even when, as in Figure 4, the concrete quality factor is included. (It may be mentioned in passing that attempts to bring in the exposure classes in more detail did not decrease the variability of the results.) The implications are that the depth of carbonation depends on a factor or factors which have not been brought into the analysis.

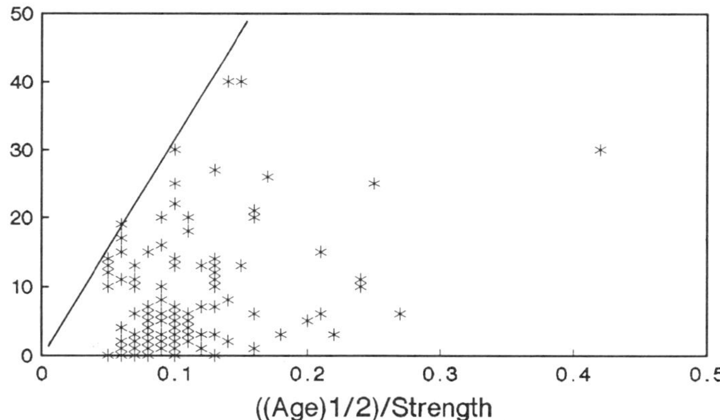

Fig.4. Typical carbonation and age/strength function.

Until these factors are unravelled, concrete design must be conservative in order to cope with the relatively few situations where the most rapid carbonation occurs. More efficient use of materials and hence design of structures will be possible only when more exact depths of carbonation for specified concretes in specified situations can be predicted.

Analysis to relate carbonation depths to general geographic area was not expected to produce conclusive results since the simple division of the country into five primary areas did not take into account local factors such as hilltop or valley, coastal or inland position. In the event, the median typical carbonation depths of the two areas of the western half of Britain are less (2mm and 3mm) than for Scotland and the North East (5mm and 6mm). S E England (8mm) showed the greatest carbonation. The carbonation depths correspond with the thesis that the Western areas are wetter and hence should carbonate less.

The orientation was noted for 94 structures and the relationship with carbonation depth was explored. The distributions North and East have median carbonation depths deeper (10, 9mm) than South or West (8,7mm). Whilst the south and West directions correspond generally with higher driving rain indices, the South face, exposed to extremes of sun and so more rapid drying, should carbonate more rapidly. Wierig (1984) found deeper carbonation on South and East faces, but his exposures do not necessarily correspond to the U.K. conditions.

A question on type of structure was included in the survey primarily to test the common belief that concrete in bridges is better than concrete in general as a consequence of better specification and perhaps more closely supervised placing. The analysis split the structures into four classes: habitable; intermediate (car parks and warehouses, for example); bridges; other structures. There were only 13 of the 272 structures in the last group. The typical carbonation depths for the habitable, intermediate and bridge structures were 10, 8 and 5mm respectively. Strength measurements were reported for about 120 structures (74 of them bridges) and these gave strengths of 37, 35 and 48 N/mm^2 for the three classes of structure respectively. The results show that bridge concrete is better quality than concretes in general, and has carbonated less. The detail figures (not given here) show also that the strength of bridge concrete has smaller variance than the others.

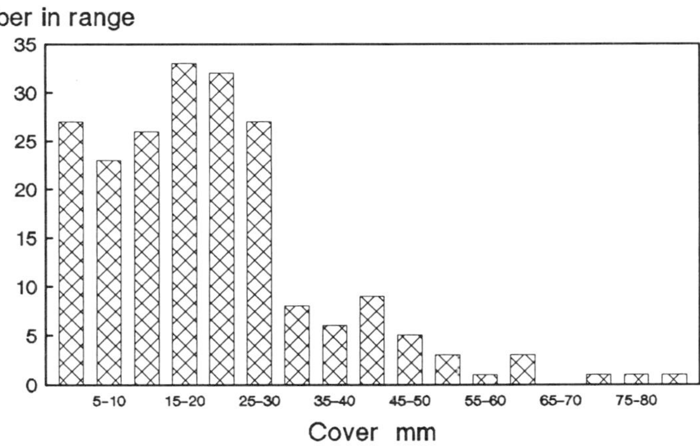

Fig.5. Worst cover distribution: all structures.

Since a large proportion of the strength data came from bridge concrete there will be some bias in the sample selection for Figure 4 but the over representation of good concrete will not affect the conclusions.

The responses on cover gave some 300 values for 'typical' and 240 for 'worst' cover. Taking all the typical values together, the distribution of cover found was near normal, with a modal value of 30 to 35mm but with 25% of all values less than 25mm cover. The 'worst' cover distribution is shown in Figure 5 and is far from normal. Here almost 25% of values show less than 10mm cover.

The data were also grouped by type of concrete and by age but full details of these results are too extensive to be included here. In summary there was little difference between cover for precast and for in-situ concrete. As far as change of cover with age or date of construction is concerned, the median typical cover for the 10-20 year olds was 35mm compared with 30mm for 20 to 30 years, and 27mm for the 30 to 40 year old structures. These depths indicate that covers are greater in more recent structures, and this conclusion is supported by the results of a much smaller study of Bridge durability (Brown, 1987).

The relationship between durability, cover and depth of carbonation was explored by comparing the cover/carbonation depth ratio with the presence or not of corrosion related damage. Only structures with low chloride content were included. The results confirm that corrosion tends to occur when the cover is completely carbonated. However, the more interesting part of the analysis is that of the 65 structures showing no evidence of corrosion, 13 had typical carbonation deeper than the worst cover. Of the 64 showing evident corrosion, 19 had typical depths less than the worst cover found. These 'odd' results emphasise the point that carbonation depths and cover vary from place to place on a structure and a coincidence of depths is difficult to predict from general measurements on either.

The effect of chloride on durability was not a major part of this investigation, but chloride levels were measured on some 130 structures and compared with the assessment of corrosion induced damage classified as 'none', 'some', and 'extensive'. To exclude damage resulting from corrosion following carbonation, only structures where the typical carbonation depth was less than the minimum cover were included in this analysis. The results are shown in Table 2.

TABLE 2. Damage and chloride level

Damage	None	Some	Extensive
Number of structures	83	35	14
Average % Cl^-/mass cement	0.36	0.74	2.13
% with Cl^- > 0.4%	33	49	79

The distribution of chloride levels within the individual groups seemed almost random, but the averages in the table confirm the common sense view that higher chloride levels are associated with more damage. However, it is difficult to set a meaningful threshold level below which chloride induced corrosion is unlikely - in this survey 21 structures (over 40%) of those showing damage had chloride levels apparently less than 0.4%/mass cement.

4 Conclusions

Although the spread of depths is very large, general carbonation depths are small compared with cover, even after many years. The survey shows that concrete quality can be related to either cement content or strength: high levels of either are associated with small depths of carbonation.

If carbonation is an active phenomenon the carbonation depth of any one structure can only increase with age. The survey does not show this with any certainty, and the implication is that other factors outweigh or hide the age effect.

In this survey carbonation depths did not show any consistent relationship to exposure. This is contrary to most established work on carbonation but no explanation of the contradiction was found.

5 References

Brown, J.H. **Exposure site measurements of the carbonation of concrete using different cements, exposures, specimen shapes and other factors.** To be published.

Brown, J.H. (1987) **The Performance of Concrete in Practice. A Field Study of Highway Bridges.** Department of Transport, Great Britain. Contractor Report 43.

Martin, H. Rausen, A. Schiessl, P. **Carbonation of Concrete Made With Different Types of Cement** in Behaviour in service of concrete structures. Colloquium, IABSE, Liege (1975), pp. 927-937.

Parrott, L.J. (1987) **A Review of Carbonation in Reinforced Concrete.** Joint publication, Building Research Establishment - Cement and Concrete Association. Report c/1-0987.

Tuutti, K. (1982) **Corrosion of Steel in Concrete.** Swedish Cement and Concrete Research Institute, Stockholm.

Wierig, H. (1984) **Longtime Studies on the Carbonation of Concrete Under Normal Outdoor Exposure.** Proc. RILEM Sem., Hannover, pp. 250-257.

Acknowledgements

I would like to thank C Broadbent (Harry Stanger Ltd), J.P.H. Frearson (Messrs.Sandberg), A.P. Keiller (Stats) and C.R.Pook (PSA) for their help in obtaining the data for this project. I would also like to thank all colleagues on the Steering Committee, and especially B.K.Marsh (BRE), for helpful discussion and support throughout.

29 EFFECTS OF OUTDOOR WEATHERING ON WITHDRAWAL AND HEAD PULL-THROUGH OF NAILS AND STAPLES IN WOOD-BASED BUILDING PANELS

P. CHOW
Department of Forestry, University of Illinois, Urbana, Illinois, USA
J.D. McNATT
US Forest Products Laboratory, Madison, Wisconsin, USA
L. ZHAO
Beijing Forestry University, Beijing, China

Abstract
No information is available on the effects of outdoor weathering on direct fastener withdrawal and head pull-through performance in new wood-based panel products. The suitability and safety of the wall and roof system of a house made of wood-based material depend on the type of wood material and nails or staples used. So this study was started in 1983 to obtain fastener withdrawal and head pull-through resistance data on the suitability of new structural wood composite sheathing and siding panel products for exterior use. Six kinds of materials including the conventional C-D grade Douglas-fir, plywood, waferboard, two oriented strandboards, veneer composite, and embossed hardboard siding were used in this study. The performance of direct nails and staple withdrawal, and nailhead or staple crown pull-through in the wood composites during five years' outdoor weathering was determined. The test also involved several laboratory accelerated exposure conditions (vacuum-pressure-soak-dry, ASTM 6-cycle accelerated aging, American Plywood Association 6-cycle aging, and 24-hour water soak).
Keywords: Composites, Density, Durability, Fastener, Laboratory accelerated aging, Nailhead pull-through, Oriented strandboard, Outdoor weathering, Sheathing, Staple, Thickness swell, Waferboard, Wood-based panel.

1 Introduction

The application of wood sheathing materials in housing construction often involves the use of nails, and staples. The stability and safety of the wall and roof system depends on the type of wood material, and fasteners used (Chow, 1974). As a result of rapidly increasing engineering use of newly developed wood-base panels, the performance and proper use of mechanical fasteners is of utmost importance. New information on the performance of fasteners in these new sheathing and siding products is needed for comparative purposes. To date, little information is available in the literature on the withdrawal or holding power resistance of nails and staples in new structural wood-base sheathings and siding panel products, especially in actual outdoor weathered or aged condition, compared to conventional wood sheathing materials.

Direct withdrawal resistance of nails and staples is the force required to pull embedded fasteners from the wood sheathing. The holding power of fasteners is especially important when applying shingles or siding materials to roof or wall sheathing. A standard American Society of Testing Materials (ASTM) nailhead pull-through test has been used to determine the resistance of a wood material by pulling the head of a nail through the thickness of the specimen. These tests are meant to simulate the conditions encountered when wind or earthquake forces tend to pull sheathing or siding from a wall (Chow et al., 1988).

Previous study determined the effects of outdoor weathering on lateral fastener resistance in wood-base products (Chow, 1985, 1989). However, little information is available on the effects of outdoor exposure on the fastener direct withdrawal and head pull-through resistance performance of new structural wood-base sheathing panels. The objectives of this study were to determine the effects of wood-based panel type, outdoor exposure (5 years) and accelerated aging tests (vacuum-pressure-soak-dry, ASTM 6-cycle accelerated aging, American Plywood Association 6-cycle aging, and 24-hour water soak) on the maximum direct nail and staple withdrawal and head pull-through performance of six commercial structural wood-base sheathing and siding panel materials.

2 Experimental Procedure

2.1 Materials

This investigation used eight sheets (1.22-by 2.44-meter) each of Aspen (Populus) waferboard, Aspen (Populus) oriented strandboard, five-ply, Douglas-fir, plywood, embossed pine hardboard siding, pine oriented strandboard, and larch veneered flakeboard. All specimens were approximately 12.7 mm thick (with the exception of the 9.5 mm embossed hardboard) and were cut randomly from commercial panels. Twenty 76 mm by 152.4 mm specimens of each panel type (cut parallel to the long direction of the original sheet) were used for each type of laboratory exposure treatments.

2.2 Outdoor weathering test

Four 1.22 by 1.22-meter specimens of each panel type were used for the long-term out-door weather exposure test. A total of 20 specimens (1.22 by 1.22-meter) were exposed and were hung vertically on posts, facing south at the Illini Plantation Site, Department of Forestry, University of Illinois, Urbana, Illinois, USA. No protection was given to the back or edges. Specimens (76 by 152-mm) were cut from each exposed panel at five years, weighted, measured, and equilibriated at 20°C/65% humidity before testing.

2.3 Laboratory accelerated aging tests

The coded specimens were subjected to various exposure tests. A list of these tests appears below along with a summary of the procedure for each test.

2.3.1 ASTM D1037-78 (1981), 6 cycles, 12 days to complete: a) Soaked in water at 49°C for 1 hour, b) Steamed at 93°C for 3 hours, c) Frozen at 12° for 20 hours, d) Dried at 99°C for 3 hours, e) Steamed again at 93°C for 3 hours, and f) Dried at 99°C for 18 hours.

2.3.2 Vacuum-pressure-soak-dry, 6 cycles, 6 days to complete: a) Soaked in water at 736 mm Hg vacuum for 30 minutes, b) Soaked in water at 413 kPa pressure for 30 minutes, and c) Dried at 71° for 23 hours (Baker, 1978).

2.3.3 American Plywood Association, 6 cycles, 3 days to complete: a) Soaked in water at 381 mm. Hg vacuum for 30 minutes, b) Soaked in water for 30 minutes, c) Dried in oven at 71°C for 6 hours, d) Repeat (a) and (b), and e) Dried in the oven at 71°C for 15 hours. This completes two cycles (APA, 1980).

2.3.4 24-hour water soak test: a) Specimens were submerged in water at room temperature, approximately 21°C, and b) Specimens were removed after 24 hours and were tested at wet conditions. (ASTM, 1981).

2.3.5 Dry (65% relative humidity) at 20°C temperature for 3 weeks.

2.4 Test Conditions

Six groups of randomly selected specimens were subjected to six different exposure condition tests. Specimens for accelerated aging were nailed and stapled directly before treatment. The dry specimens and the outdoor exposed specimens did not receive their fasteners until immediately before testing.

2.4.1 Direct Withdrawal Resistance Test

Each of the 6-penny (2.87 mm. shank diameter and 50.8 mm long) common wire nails, and 12.7 mm crown, 16 gauge power-driven staples (1.42 mm diameter and 50.8 mm long) was driven through the thickness of the specimen and had at least 12.7 mm of the fastener top protruding above the surface of each specimen using special jigs. The common wire nail tests were performed using the apparatus specified by the ASTM standard. A newly designed test jig was made to fit the universal testing machine for conducting the staple withdrawal tests.

A heavy-duty power driven staple gun was used to drive the 12.7 mm crown staple into the specimens. It has been commonly used for wood-base wall sheathing attached to wood members by the factory-built house and prefabricated panel industries.

2.4.2 Fastener Head Pull-Through Test Methods

A standard ASTM nailhead pull-through test was used to determine the resistance of specimens to having the head of a 6-penny common nail pull-through the thickness of a specimen.

The staple crown pull-through resistance test was performed using a newly designed jig. The legs of the 12.7 mm crown staple are placed in 16-gauge-size holes in steel block which has two Allen screws on two sides. The four Allen screws are then tightened to hold two legs in place while the staple crown is pulled through the specimen.

3 Results and Discussions

Table 1 shows the effects of the five years outdoor exposure and four laboratory accelerated aging tests on the thickness and density of six types of wood-based materials. At dry condition, plywood had the lowest average density value at 0.50 g/cm^3. After exposure to different conditions, excessive thickness swelling occurred to both waferboard and oriented strandboard. Their average density values were greatly reduced except after the 24 hour water soaked condition. The reason for the increased density after the water-soaked condition is that all density values were calculated from the wet weight and volume of all soaked specimens. On the other hand both hardboard and plywood showed good dimensional stability, and had the least change in average thickness and density.

Table 2 lists the average values of direct nail and staple withdrawal of six wood-base materials after subjection to five years of outdoor exposure and four kinds of accelerated aging. Both at control condition and after exposure to the five years weathering condition, average direct nail withdrawal values of each type of wood material are lower than those of the staple performance. In some cases, the nail withdrawal test results of the accelerated aging treatment were significantly higher than those of the control specimens. One possible reason could be corrosion of the nail. Nail corrosion could lead to a closer bond between the nail shank and wood fibers. Table

2 shows that the reductions of average direct nail and staple withdrawal during the five years of the weathering ranged from 38% to 75%. The nail withdrawal value of specimens exposed to five-year outdoor weather probably would have been higher due to nail corrosion if the nails had been driven into the specimens prior to the outdoor exposure.

Table 3 indicates the exposure effects on the value of nailhead and staple crown pull-through resistance of different wood composite panel products. The reductions of average nailhead and staple crown pull-through values for all specimens ranged from 5% to 40%. On the average the nailhead pull-through is higher than the staple crown pull-through except for a few instances in composite plywood and OSB specimens. In most of the tests, waferboard and OSB products usually performed as well as plywood under various exposures.

The analyses of variance (SAS, 1988 and Steel, 1960) shows that the average nail and staple withdrawal resitance values and nailhead and staple crown pull-through resistance values for all specimens were significantly (5% level) affected by two main factors, the outdoor and laboratory accelerated aging exposure and the wood-based panel material type.

4 References

American Plywood Association. (1980): Performance standard and policies for APA rated sheathing panels. APA Report No. 445, Tacoma, WA 98411.

American Society for Testing and Materials. (1981): Standard methods of evaluating the properties of wood-base fiber and particle panel materials. ASTM Design D-1037-78, and D2277-75, ASTM, Philadelphia, PA.

American Society for Testing and Materials. (1979): Standard methods of testing mechanical fasteners in wood. ASTUM Design D-1761-77, ASTM, Philadelphia, PA.

Baker, A. J. and Gillespie, R. H. (1978): Accelerated aging of phenolic-bonded flakeboards. Proceedings of USDA Forest Service Symposium. General Tech. Report Wo-5. USDA, Forest Service, Washington, D. C. pp. 93-100.

Chow, P. (1974). Three tests on nailhead pull-through performance in wood-base panels. For. Prod. Jour. 24(1): 41-44.

Chow, P., McNatt, J. D. (1988). Direct withdrawal and head pull-through performance of nails and staples in structural wood-based panel materials. Forest Products Journal 38(6):19-25.

Chow, P., McNatt, J. D. (1985). Effects of test methods and exposure conditions on lateral nail and staple resistance of wood-base panel materials. Forest Products Journal 35(9):13-19.

Chow, P. and J. D. McNatt (1989). Performance of lateral nails and staples resistance in wood composite panels during five years exposure to weather. Proceedings of the Second Pacific Timber Engineering Conference, Auckland, New Zealand pp. 201-204.

Elias, Edward G. (1981): Performance-based testing for fastener holding. Report No. PT 82-5, American Plywood Association, Tacoma, Washington.

Statistical Analysis System (SAS) User's Guide: Statistics, (1982): edition, SAS Institute Inc., Cary, NC.

Steel, R. G. D. and J. H. Torrie. (1960): Principles and procedures of statistics. McGraw-Hill Book Co., Inc., New York.

Table 1. Effects of outdoor exposure and accelerated aging on the thickness and density of the wood-based materials.

Materials	Fastener Type	Control 0 Year	Outdoor Exposure 5 Years	Accelerated Aging APA[c]	VPSD[d]	24H[e]	ASTM[f]
Plywood Sheathing C-D grade	Thick. (mm)	12.1[a]	12.9 (+6)[b]	12.4 (+2)	12.4 (+2)	12.9 (+6)	12.7 (+4)
	Dens. (g/cm^3)	0.50	0.49 (-2)	0.49 (-2)	0.47 (-6)	0.69 (+38)	0.50 (0)
Aspen Wafer-Board	Thick. (mm)	12.9	15.7 (+22)	18.0 (+39)	16.2 (+26)	14.7 (+14)	17.5 (+35)
	Dens. (g/cm^3)	0.69	0.49 (-29)	0.54 (-22)	0.56 (-19)	0.88 (+28)	0.56 (-19)
Hard-board Siding	Thick. (mm)	9.6	10.7 (+11)	10.2 (+5)	10.4 (+8)	9.9 (+3)	10.4 (+8)
	Dens. (g/cm^3)	0.67	0.59 (-12)	0.63 (-6)	0.61 (-9)	0.73 (+9)	0.63 (-6)
Larch Com-posite Plywood	Thick. (mm)	12.4	14.2 (+14)	13.7 (+10)	14.7 (+18)	13.7 (+10)	14.5 (+17)
	Dens. (g/cm^3)	0.66	0.53 (-20)	0.65 (-2)	0.57 (-14)	0.88 (+33)	0.56 (-15)
OB-1 (Pine)	Thick. (mm)	13.2	14.0 (+6)	14.7 (+11)	16.0 (+21)	15.7 (+19)	16.0 (+21)
	Dens. (g/cm^3	0.66	0.54 (-18)	0.57 (-14)	0.57 (-14)	0.80 (+21)	0.58 (-12)
OSB-2 (Aspen)	Thick. (mm)	12.7	15.0 (+12)	14.2 (+12)	14.2 (+12)	15.2 (+20)	15.7 (+24)
	Dens. (g/cm^3)	0.67	0.52 (-22)	0.51 (-24)	0.56 (-16)	0.80 (+19)	0.57 (-15)

Note: [a]Mean value, each value is an average for 16 measurements (1 inch = 25.4 mm.)
[b]Change in percent.
[c]APA-American Plywood Association 6 cycles test.
[d]VPSD-Vacuum-Pressure-Soak-Dry exposure.
[e]24H-24 hours water soak. Density is based on wet weight and volume.
[f]ASTM-ASTM D1037 6 cycles test.

Table 2. Average direct nail and staple withdrawal (Newtons) of wood-based materials after five years of outdoor exposure.

Materials	Fastener Type	Control 0 Year	Outdoor Exposure 5 Years	Accelerated Aging APA[c]	VPSD[d]	24H[e]	ASTM[f]
Hardboard Siding	Nail	142.2[a]	35.3 (0.25)[b]	106.9 (0.75)	92.2 (0.62)	152.0 (1.06)	169.7 (1.19)
	Staple 1.27cm.	267.7	98.1 (0.37)	133.4 (0.52)	142.2 (0.53)	152.0 (0.57)	53.0 (0.20)
Plywood Sheathing (C-D grade)	Nail	294.2	111.8 (0.38)	476.6 (1.62)	338.3 (1.15)	155.9 (0.53)	111.8 (0.38)
	Staple 1.27cm.	418.7	254.0 (0.61)	275.6 (0.65)	271.6 (0.65)	240.3 (0.57)	84.3 (0.30)
Aspen Waferboard	Nail	325.6	133.4 (0.41)	352.1 (1.07)	280.5 (0.86)	231.4 (0.71)	276.5 (0.85)
	Staple 1.27cm.	753.2	254.0 (0.34)	396.2 (0.53)	285.4 (0.38)	441.3 (0.59)	71.6 (0.10)
Larch Veneered Composite Plywood	Nail	289.3	178.5 (0.62)	329.5 (0.77)	294.2 (0.68)	183.3 (0.48)	200.1 (0.69)
	Staple 1.27cm.	597.2	245.2 (0.41)	365.8 (0.61)	245.2 (0.41)	249.1 (0.42)	53.9 (0.10)
Pine (OSB-1)	Nail	320.7	98.1 (0.31)	521.7 (1.63)	476.6 (1.48)	155.9 (0.49)	401.1 (1.25)
	Staple 1.27cm.	561.9	315.8 (0.56)	369.7 (0.66)	374.6 (0.66)	280.5 (0.50)	98.1 (0.18)
Aspen (OSB-2)	Nail	280.5	106.9 (0.38)	213.8 (0.76)	276.5 (0.98)	222.6 (0.79)	356.0 (1.27)
	Staple 1.27cm.	744.3	217.7 (0.29)	245.2 (0.33)	303.0 (0.41)	409.9 (0.55)	102.0 (0.55)

[a]Each value is an average for 20 tests.
[b]Ratio retained from control condition.
[c]APA: American Plywood Association 6 cycles test.
[d]VPSD: Vacuum-Pressure-Soak-Dry exposure (6 cycles).
[e]24H: 24 hours water soak.
[f]ASTM: ASTM D1037 6 cycles test.

Table 3. Comparisons of Nail Head and Staple Crown Pull-through (Newtons) between accelerated aging cycles and 5-year outdoor exposure.

Materials	Fastener Type	Control 0 Year	Outdoor Exposure 5 Years	Accelerated Aging[f] APA[c]	VPSD[d]	24H[e]	ASTM[f]
Plywood	Nail	1461[a]	1236 (0.85)[b]	1520 (1.04)	1167 (0.80)	1481 (0.77)	1128 (0.77)
Sheathing C-D Grade	Staple 1.27cm.	1187	1030 (0.87)	1216 (1.02	1030 (0.86)	1069 (0.90)	1177 (0.99)
Aspen Waferboard	Nail	1638	1138 (0.70)	1255 (0.77)	1295 (0.79)	1481 (0.90)	1059 (0.65)
	Staple 1.27cm.	1579	961 (0.61)	1334 (0.84)	1236 (0.78)	1520 (0.96)	1059 (0.67)
Veneered Composite Plywood	Nail	1353	932 (0.69)	1108 (0.81)	1079 (0.80)	922 (0.68)	1000 (0.74)
	Staple 1.27cm.	1353	736 (0.54)	1245 (0.92)	1206 (0.89)	912 (0.67)	1118 (0.82)
Hardboard Siding	Nail	500	412 (0.82)	422 (0.84)	461 (0.92)	451 (0.91)	432 (0.86)
	Staple 1.27cm.	451	275 (0.60)	373 (0.82)	382 (0.84)	432 (0.96)	324 (0.71)
Pine (OSB-1)	Nail	1549	1167 (0.75)	1569 (1.01)	1618 (1.04)	941 (0.61)	1265 (0.82)
	Staple	1245	1040 (0.83)	1412 (1.13)	1383 (1.11)	922 (0.74)	1196 (0.97)
Aspen (OSB-2)	Nail	1481	1236 (0.83)	1069 (0.72)	1089 (0.73)	1294 (0.87)	1275 (0.86)
	Staple 1.27cm.	1471	961 (0.65)	667 (0.72)	1167 (0.79)	1236 (0.84)	1255 (0.85)

[a]Each value is an average for 20 tests.
[b]Ratio retained from control condition.
[c]APA: American Plywood Association 6 cycles test.
[d]VPSD: Vacuum-Pressure-Soak-Dry exposure (6 cycles).
[e]24H: 24 hours water soak.
[f]ASTM: ASTM D1037 6 cycles test.

30 CATHODIC PROTECTION OF PORTCULLIS HOUSE: PSA 5-YEAR TRIAL PROVES THE SYSTEM

C.C. MORTON
PSA Specialist Services, DCES, Croydon, UK

Abstract
Portcullis House, Southend is a 14 storey office block which was completed in 1965 and occupied by the Customs and Excise organisation. The structure comprises a mix of insitu and precast construction of generally good quality dense concrete. However the building has a history of cracking and spalling as a result of reinforcement corrosion. High chloride levels were found in the mullions and capstone beams arising from use of an accelerating admixture during manufacture of the precast components. The structure as one would expect has needed to be repaired several times and in 1984 PSA's Directorate of Civil Engineering Services were called in to assess the condition and advise.

It was again decided to conventially repair the entire structure where necessary but also to cathodically protect one storey height (level 2 to 3) around the building and monitor performance for a minimum of five years and compare against the unprotected areas. A cathodic protection system with a surface coating anode was installed.

Although this paper describes the system and method of monitoring, its main purposes is to record the success of the cathodic protection.

Deterioration is clearly evident in the form of cracking and spalling both above and below the trial whilst the CP area remains in a sound condition throughout.

1 Introduction

Concrete has not proved to be as durable as we all thought and its deterioration to a varying degree has become a universal problem. Consequently it was inevitable that resources would need to be committed to studying the subject and co-ordinating policy on how one should best repair and protect reinforced concrete structures.

Cathodic protection has been used on steel structures for many years but its application to reinforced concrete is a comparatively recent innovation. Doubts prevail as to whether it really works and, if so, how effective the treatment is. It has been suggested that the costs of installation and maintenance are prohibitive and that the exercise becomes unrealistic when applied to complete buildings.

Against this background of uncertainty DCES decided to conduct a 5 year trial on an area over one storey height around the perimeter of an office building, corrosion damaged by an accelerating admixture containing calcium chloride use in the concrete manufacture.

2 Background

Portcullis House, Southend is a 14 storey block which was completed in 1965 and is occupied by HM Customs and Excise. The structure comprises a mix of insitu and precast construction, as can be seen in Figures 1 and 2. The concrete was generally good quality dense concrete but had suffered cracking and spalling as a result of reinforcement corrosion. Chloride was found present in the concrete and its concentration was highest in the mullions and capstone beams between the second and fifth floors with typical percentages of chloride by weight of cement ranging between 1% and 3%.

Cracking and spalling developed during the 70s and in 1979/80 repairs were carried out using an epoxy resin mortar. By 1984 fairly widespread deterioration was again evident both in new areas and in the existing patch repairs. It was decided that further repair work had to be carried out but that this time a cathodic protection system would be installed over part of the structure simultaneously and in this way the new repairs would serve as a control to compare with the cathodic protection.

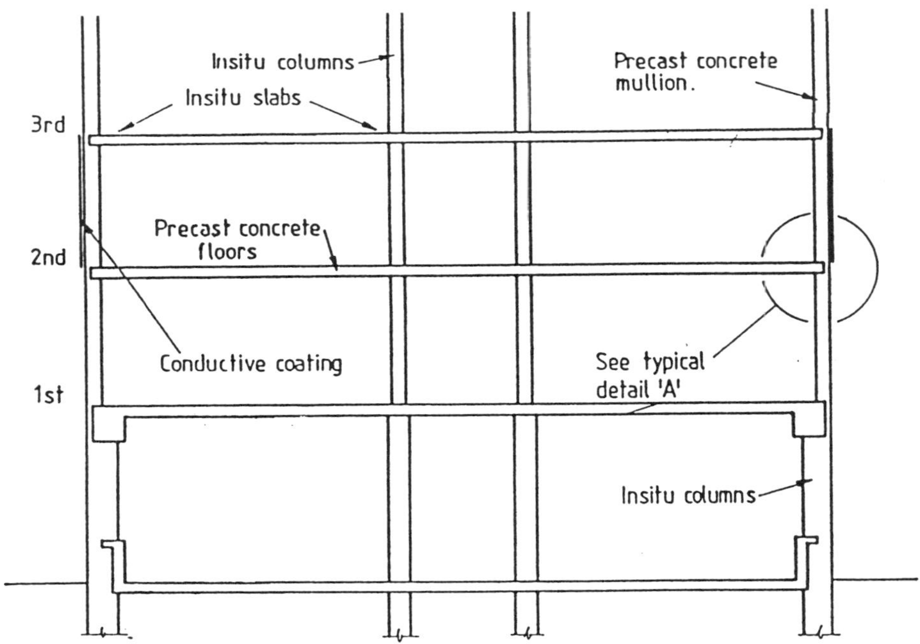

Fig.1. Typical Section

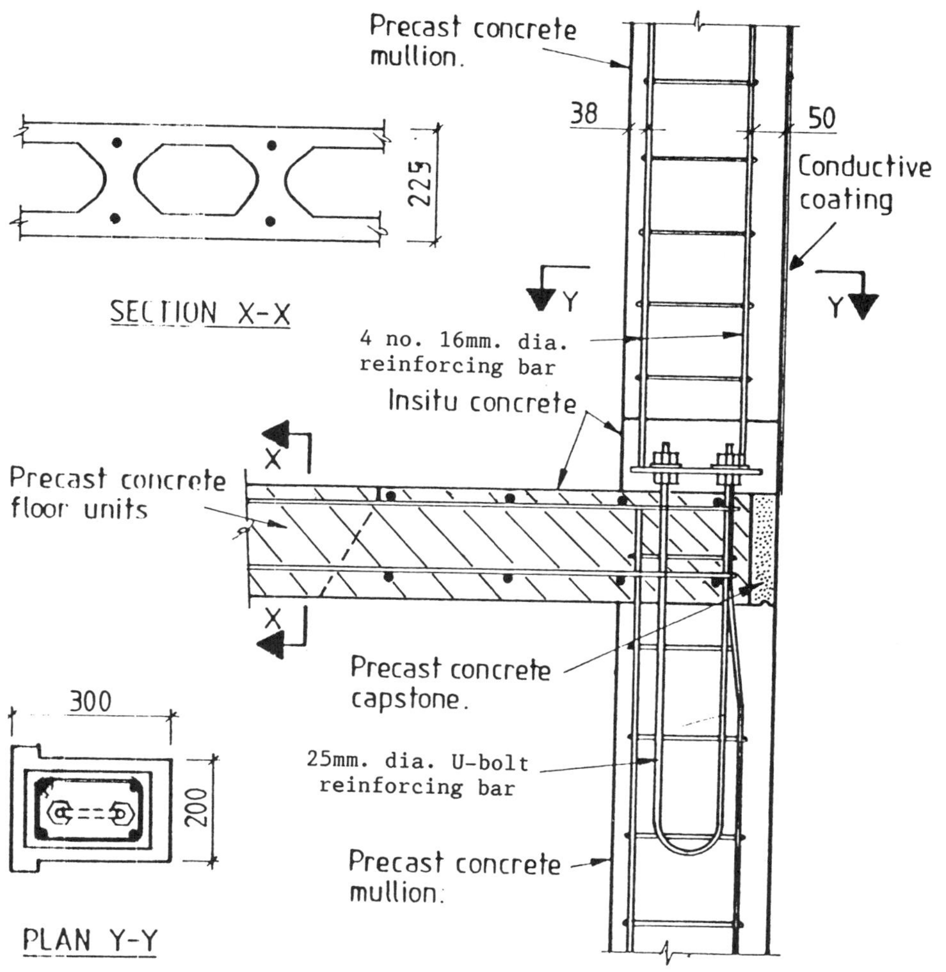

Fig. 2. Typical Detail 'A'

3 Principles of Anode Cathode Reaction

Let us consider the simple electrochemical principles of what is happening in reinforced concrete. Reinforcement installed in normal uncarbonated low-chloride concrete is protected by the alkaline environment which stimulates the formation of a passivation layer around the bar. This layer comprises a thin film of hydrated iron oxide which remains stable in conditions of high alkalinity and so provides the required protection for as long as the concrete environment remains alkaline.

CONCRETE (ELECTROLYTE)

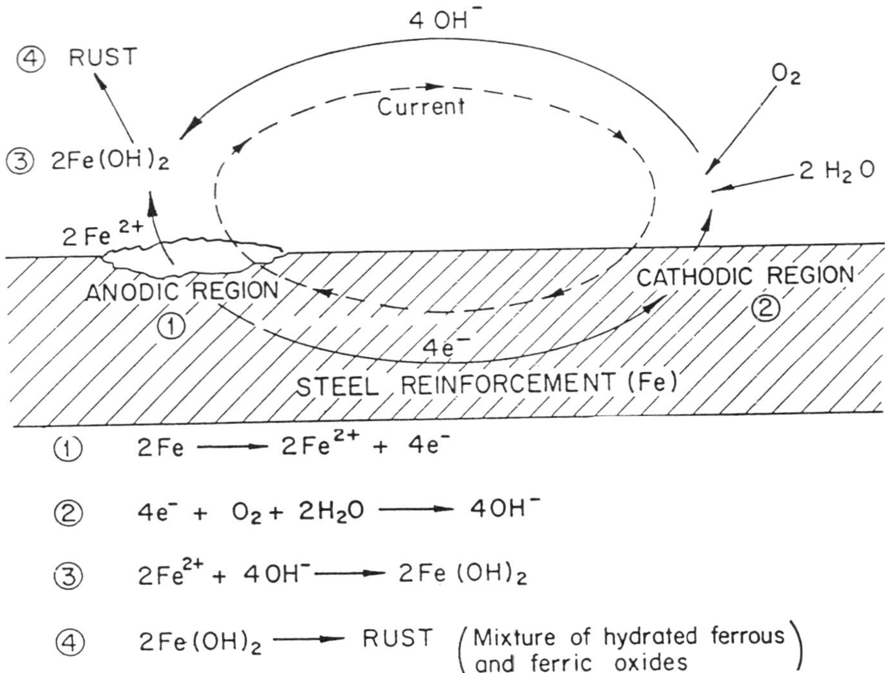

Figure 3. Idealised Corrosion Reaction

In Figure 3 the passive layer has been destroyed locally through carbonation of the concrete. Provided that there is sufficient moisture and oxygen an anode: cathode reaction will form and the reinforcement will corrode as shown. Electrons pass from the anode along the bar to the cathode., Here they combine with moisture and oxygen to form hydroxyl ions. These travel through the concrete electrolyte back to the anode where they combine with iron ions to form ferric and ferrous oxides commonly identified as expansive rust. If chloride ions are present then ferrous chloride is formed initially. In the presence of moisture this product is hydrolysed forming iron hydroxide and liberating hydrogen and chloride. Pitting corrosion occurs where large cathodic areas support corrosion at small anodes give rise to very high rates of local attack. The corrosion product may not be expansive unless oxygen can penetrate to the anodic sites where further oxidation proceeds giving rise to a volumetric expansion of the corrosion product. Figure 4 gives an indication of the chemical reactions when there is an abundance of chloride ions.

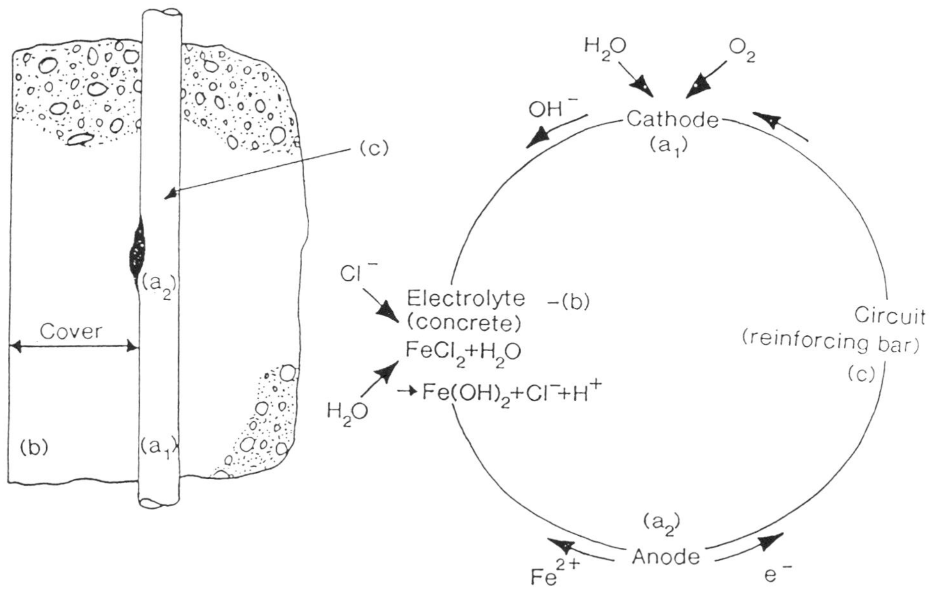

Figure 4. Schematic diagram of chloride induced corrosion of steel in concrete

4 Principles of Cathodic Protection

If an external current of sufficient magnitude and direction is applied to the reinforcement via a circuit set up through a conductive material applied to the surface of the concrete the corrosion reaction can be arrested. In practice the positive terminal of a DC power supply is connected to a conductive surface material which becomes the anode, the negative terminal is connected to the reinforcement (the cathode) and a small DC current is applied. This causes a flow of negative charge from the reinforcement to the conductive membrane. The applied current is then increased to a level which will oppose the negatively charged electron flow from the most active corrosion sites, thus rendering the reinforcement cathodic to the surface anode allowing control of corrosion. The corrosion reaction now occurs at the external anode which can be designed to resist attack. Figure 5 indicates the electrochemistry after the CP system has been installed and switched on.

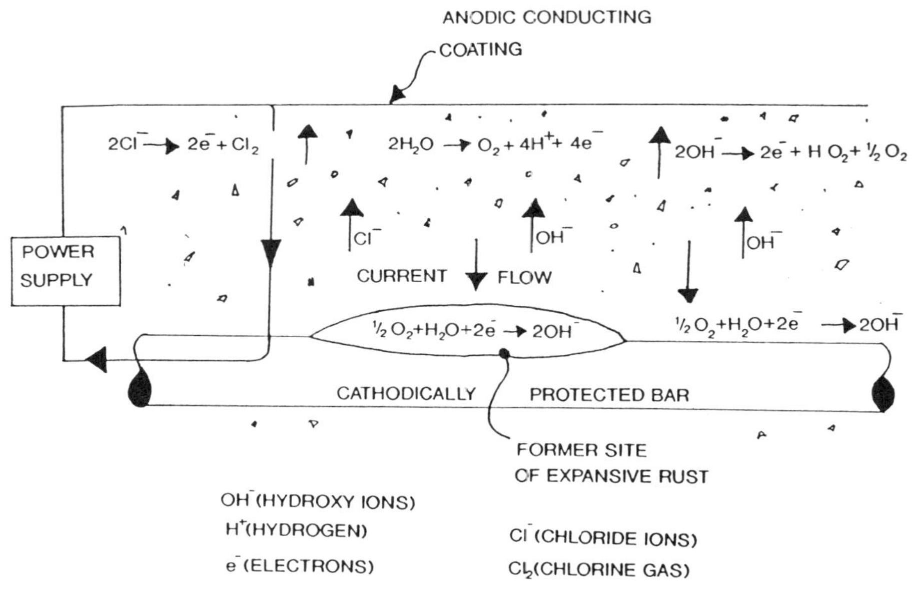

Figure 5. Electrochemistry of cathodically protected steel in concrete

5 Practical application of CP to Portcullis House

In 1985 the PSA Directorate of Civil Engineering Services let a commission with Taywood Engineering Ltd to design, install, maintain and monitor the trial on a shared cost basis. BRE would maintain an active monitoring role and PSA would appraise the performance and develop a policy as to its feasibility for more widespread use according to the findings of the trial.

5.1 Repair and potential mapping

After patch repairs of damaged areas had been completed as part of the main contract reinforcement was exposed to allow electrical continuity checks. Where necessary additional cable connections were made between reinforcement cages for continuity and in each ground bed area. Electro-chemical potential mapping of the beams using portable half cells and mullions of the proposed site (level 2 to level 3) was then carried out to identify areas of corrosion activity at the reinforcement surface.

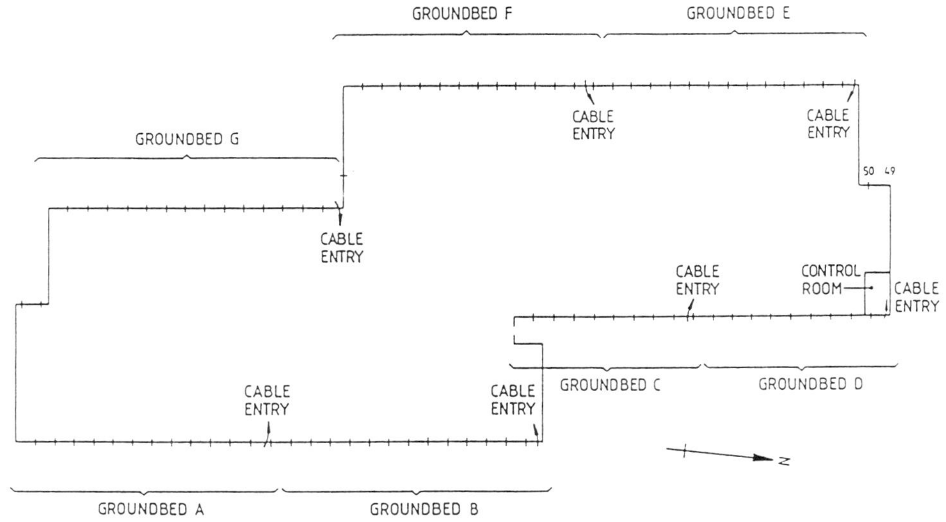

Figure 6. Plan of Portcullis House.

5.2 Ground beds

Because of the length of the site the CP system was divided up into 7 ground beds, each to operate independently of the next (Figure 6) in order to maintain continuity of the circuit, reinforcement was cross connected at beams and mullions throughout the ground bed.

5.3 Reference electrodes

Between four and six reference electrodes per ground bed were embedded in the concrete between the steel and the surface in order to measure the electro-chemical potential of the reinforcement in different parts of each ground bed. Locations were selected to cover a range of activity indicated by the potential mapping. For example, one or two were positioned in areas of high negative potential and close potential contours whilst others were placed in areas of lower potential and with one in a repair. A well tested silver/silver chloride type reference half cell was chosen as the embedded electrode. These were similar in operation to the devices used for surface potential mapping but designed to be sufficiently durable to withstand embedment in concrete.

5.4 Anode system

The anode system comprises primary and secondary anodes. The primary anodes carry the current to secondary anodes which then distribute the current evenly over the concrete surface. After repair and surface preparation by grit blasting, one of two types of primary anode was applied. Platinised titanium wire was affixed to the concrete in ground bed D using an insulating epoxy glue whilst in the remaining

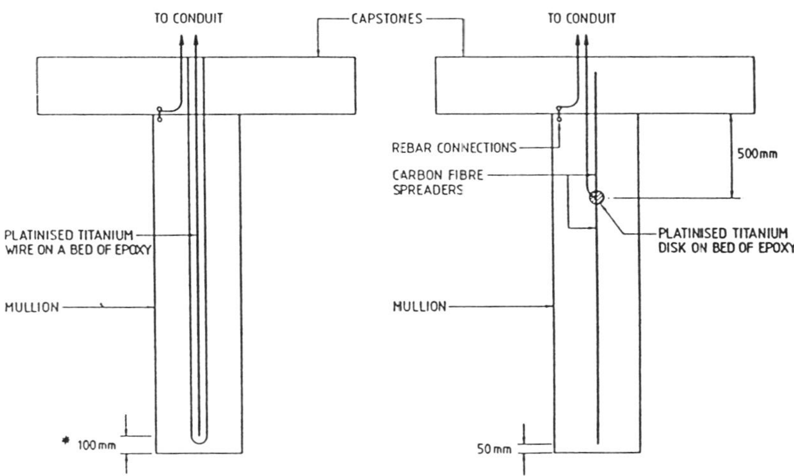

Figure 7. Primary Anode Systems

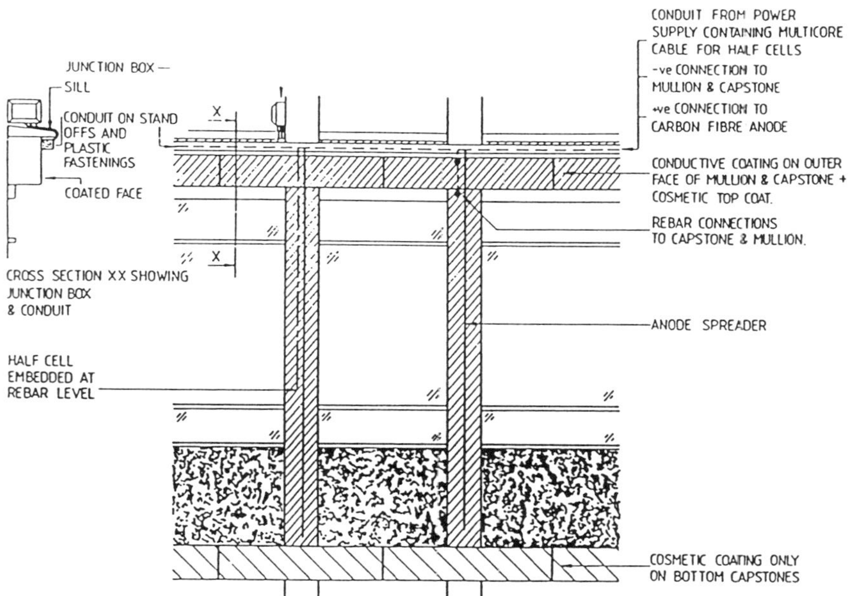

Figure 8. Schematic diagram of CP system applied to a typical pair of mullions and capstones.

ground beds a platinised titanium disc and carbon fibre wire was similarly attached. Wiring from anodes, reinforcement connections and embedded reference electrodes were carried in trunking to the power and control units (see Figures 7 and 8). Following installation of the primary anode circuits and instrumentation the conductive coating system was applied to complete the anode system.

5.5 Anode requirements

As the coating forming the secondary anode was required to distribute current evenly from the primary anode to the concrete surface, it had to meet a number of requirements. Essentially it should exhibit good adhesion, sufficient conductivity, durability, requisite vapour transmission performance and resistance to acid and other aggressive evolutions at the anode/concrete interface. A conductive coating was developed and tested to meet these requirements.

The coating system comprises a primer, 2 coats of a modified chlorinated rubber conducting paint and finally 2 cosmetic coats of Micatex masonry paint all applied to the appropriate specification.

5.6 Power and control units

The power and control units are located in a room adjacent to the anode installations together with the modem unit which provides a telecom. data link to Taywood Engineering in London.

Power is supplied from the mains to a programmable transformer and rectifier to provide DC voltage. The supply is designed to provide up to 500 milliamps to a maximum of 8 ground beds with output voltages varying between 0 and 15 volts.

Constant current modules are provided for each ground bed. These enable the required current density to be set and maintained up to a limiting voltage of 15 volts. This drive voltage is automatically adjusted to provide the desired current density. The drive voltage constantly changes due to changes in the restivity of the system. Resistivity changes are a function of the weather conditions, eg an increase in concrete moisture content due to the environment causes a decrease in resistivity. The system drive voltage increased to the maximum require to maintain the selected current density. The system is said to be voltage limiting when drive voltage reaches a maximum of 15 volts.

The power transformer and constant current modules are located in one instrument housing. Although capable of manual operation by setting and re-setting at a constant current the system is controlled by a microprocessor at Portcullis House. The microprocessor which is contained in a separate instrument housing provides signal conditioning and remote control via a model link.

6 Criteria of adequate CP

There is a lack of universal agreement on methods and criteria for monitoring the effectiveness of CP for reinforced concrete. there are two background points to the criteria tests which should be noted:

(1) Embedded half cells are widely accepted as monitoring devices for cathodic protection, there being no direct way of showing

that corrosion rates have been reduced sufficiently by the action of the impressed current.

(2) Since the potential measured by the half cell is affected by the passage of current in the cathodic protection system, the potential is measured at 'instant off', usually within one second of switching off the current. This removes the interference effect of the impressed current but gives the polarised potential before the cathodic protection has decayed. All potentials are therefore either 'instant off', 'decay potentials' (after a period of inactivity) or 'base potentials' prior to the application of cathodic protection.

Figure 9 illustrates the criteria used to assess the performance of the CP trials, namely:

(a) 150-200 mV potential shift between the 'initial' (the base) potential prior to application of CP and 'the instant off' potential after CP.

(b) 100 mV decay between 'instant off' potential and the potential measured 4 hours after de-energising the system.

6.1 Assessment Tests

The range of tests, both electrochemical and non-electrochemical, available for assessment of the CP system are as follows:

(1) Electrochemical data
  - (a) Compliance with the potential shift criteria
  - (b) Compliance with the 4 hour decay criteria.

(2) Coating performance
  - (a) Pull-off tests
  - (b) Visual inspection
  - (c) Down coating voltages

(3) Reinforcement condition
  - (a) Visual inspection of reinforcement exposed by coring.

7 Performance to Date

7.1 Conventionally repaired areas

Inspection $4^1/_2$ years after installation indicated that the building was still in reasonably good condition. A small number of repaired areas outside the CP zone exhibited map cracking and a few individual cracks. Investigation by coring indicated that the reinforcement was in reasonable condition and was not the cause of the cracking which was attributed to shrinkage in the repair mortar. Elsewhere two out of four cores indicated corrosion described as moderate and slight respectively.

Mullions and capstone beams have started to crack and spall; seven new areas having been identified. The time taken for new anodes to become established is longer than previously observed, probably because earlier repairs removed the most active anodes and subsequent fresh sites corroded at slower rates.

7.2 Cathodic protection area Level 2 to Level 3

The CP area was devoid of any cracks or spalls. Recent pull off tests indicated that the conducting coating was adhering effectively to the concrete surface with mean results well above 5 kg/cm^2. Both the concrete and the coating appeared to be in very good condition.

7.3 Electrochemical criteria

Readings for one of three of the ground beds (A) are shown in Figures 10-15 recording electrical data against time. It can be seen that data from the various cells are widely scattered although the majority are above or near the designed 150 mV potential shift and 100 mV 4 hour decay potential. Another feature of the results indicates that the criterion of setting the level of applied current and allowing the drive voltage to adjust according to the resistivity of the concrete has not been totally successful. This is because the voltage has

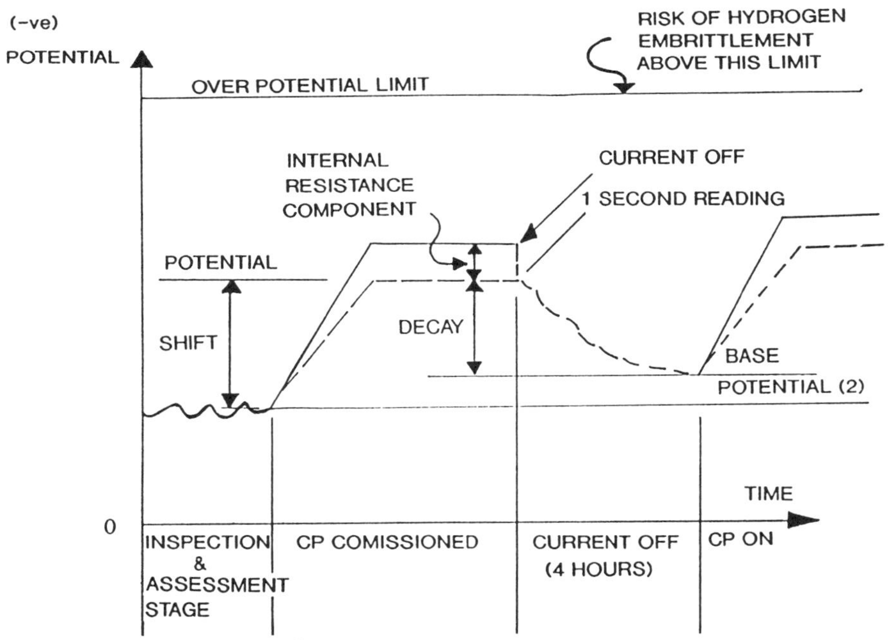

Decay Criterion	100 mV in 4 hours
Shift Criterion	150 mV
Potential Criterion	Measure within 1 second of circuit interruption
Over Potential Criterion	Hydrogen evolution potential

Figure 9. Protection Criteria

reached the maximum fairly quickly and was unable to sustain the set level of applied current due to the high resistivity of the concrete. Nevertheless, absence of further deterioration in this area is taken to indicate that the system is operating satisfactorily.

The protection levels needed, however, should reduce with time as chlorides migrate away from the steel/concrete interface under the influence of the applied voltage and the concrete surrounding the rebar tends to repassivate and thereby requires less potential shift.

Ground bed 'D' has not achieved the designated criteria and therefore has been depolarised and repolarised again after several weeks.

It is known that the concrete in ground bed 'D' is particularly resistive and this could be a reason for failure to reach criteria. Other factors may also be important, but, as with 'A' there is no evidence of corrosion induced cracking and the CP system is believed to be providing the necessary protection level.

Summary and Conclusions

The conventionally repaired parts of the structure have begun to deteriorate again as expected. The comparative dryness of the external concrete has delayed the inevitable damage by a year or so. Since this is now taking place entirely outside the CP area there is a growing confidence in the system. Clearly in time the conventional repair process will need to be repeated if the building is to remain serviceable and provided the present promising performance continues then application of cathodic protection throughout the building should be seriously considered.

The conducting coating appears to have worked well according to all criteria and should have at least several years of life remaining.

High concrete resistivity and the low limit on drive voltage have not so far been detrimental to the system.

The expertise of those at Taywood Engineering and BRE involved in the development and application of the CP System is acknowledged and accredited by the PSA.

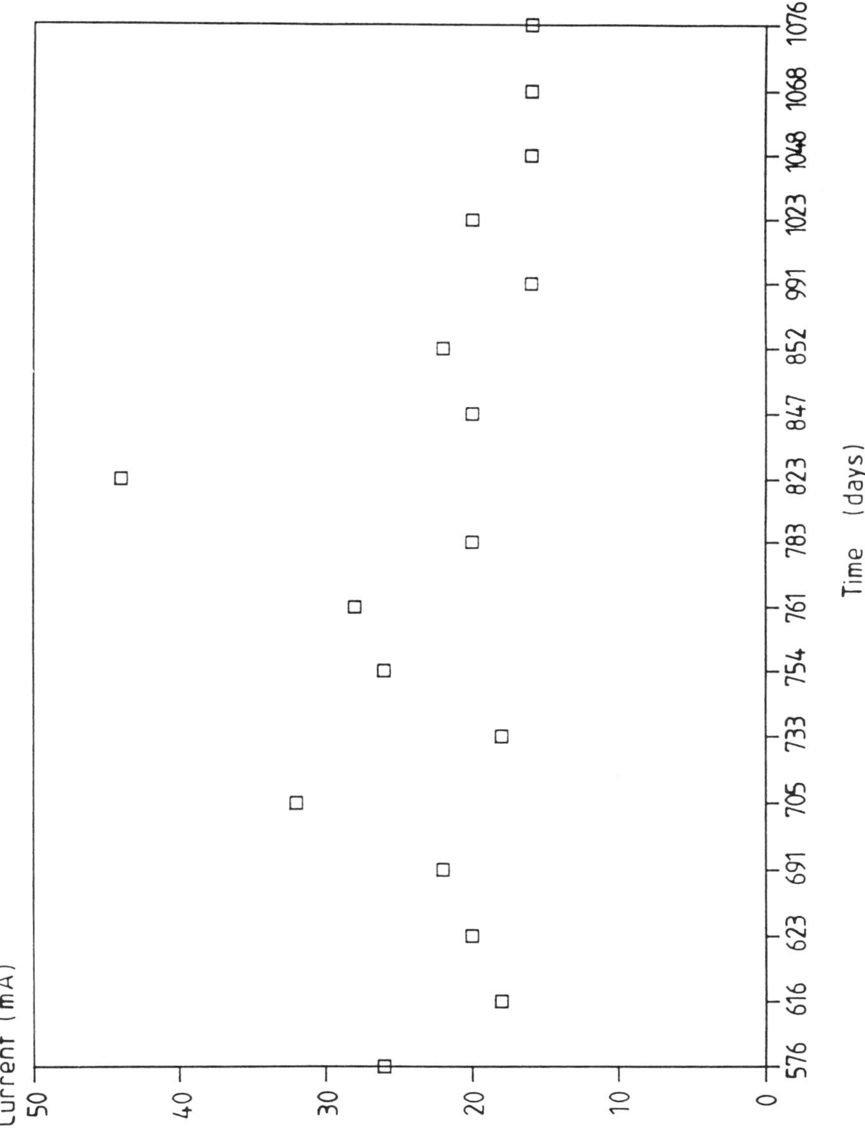

Current vs. Time - Groundbed A

Figure 10

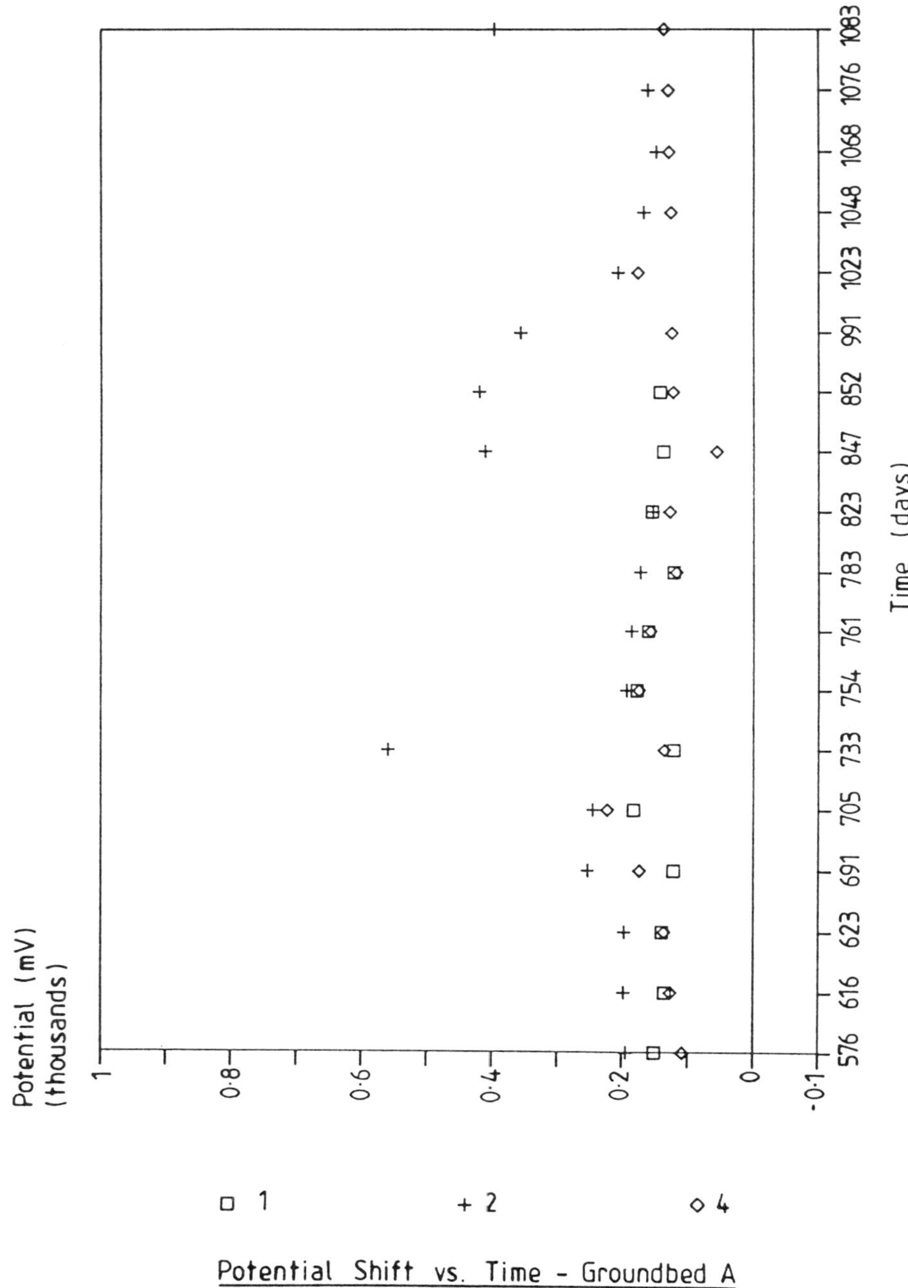

Potential Shift vs. Time - Groundbed A

Figure 11

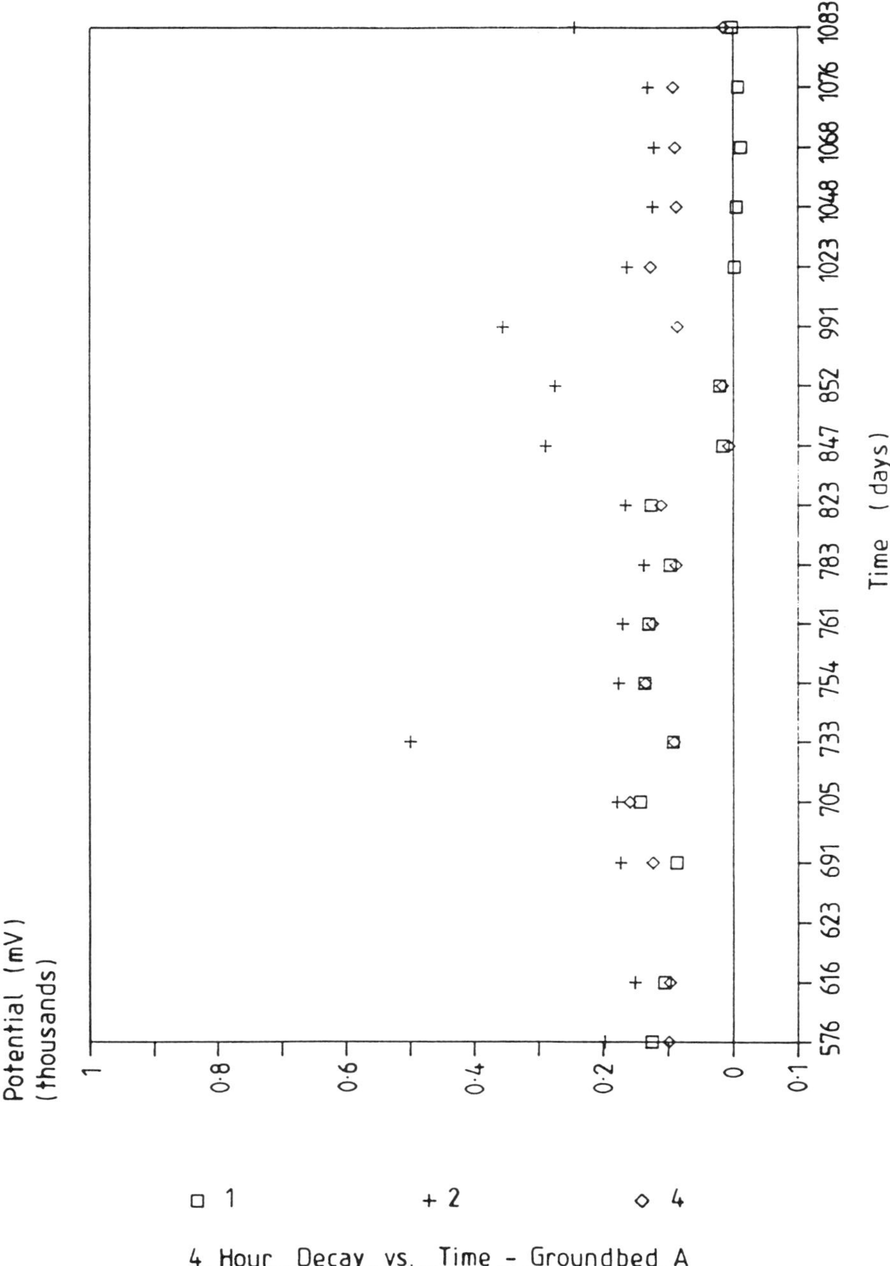

4 Hour Decay vs. Time - Groundbed A

Figure 12

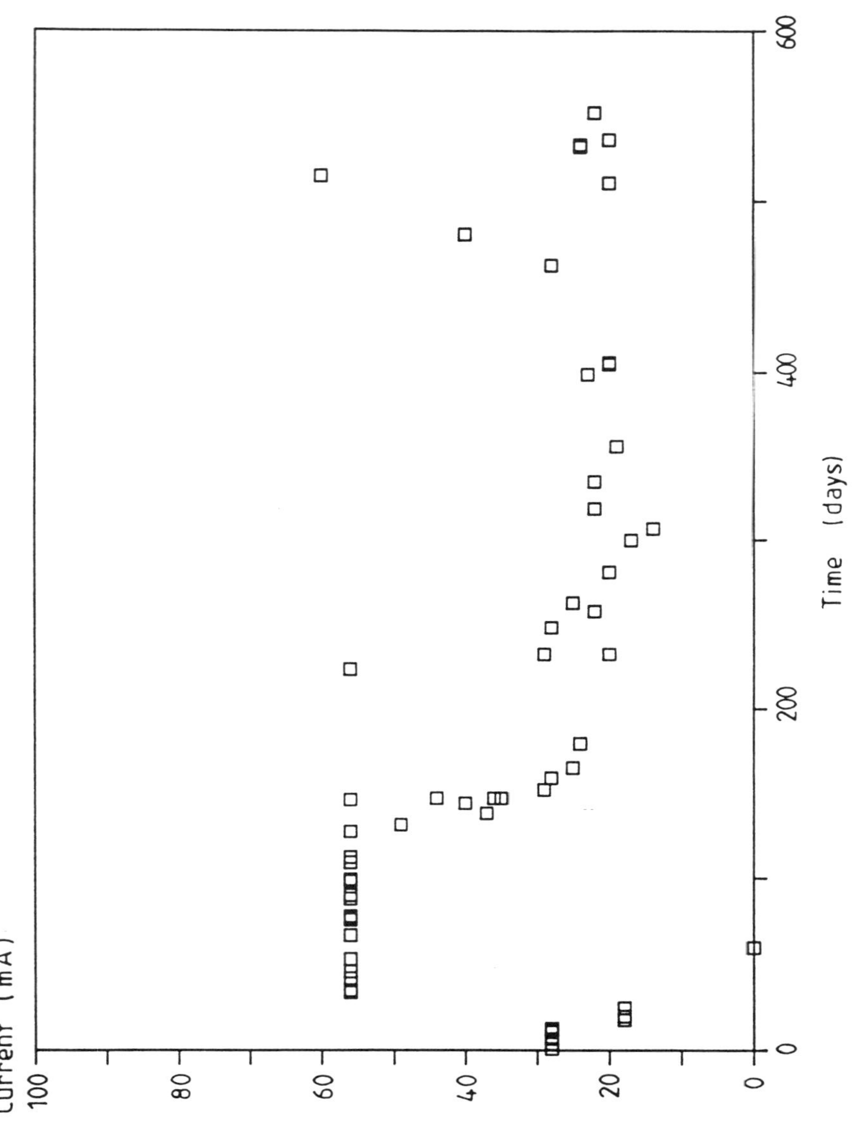

Current vs. Time - Area A

Figure 13

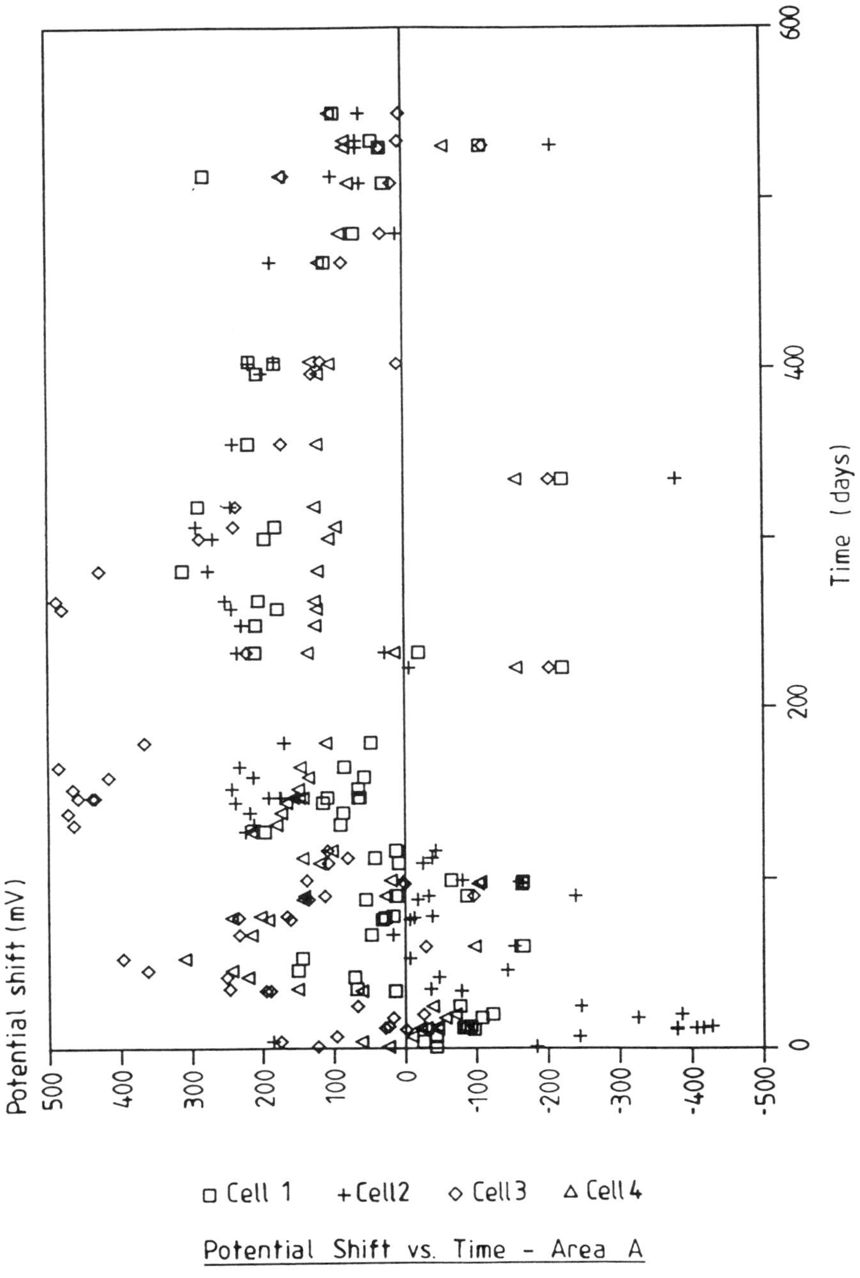

Potential Shift vs. Time - Area A

Figure 14

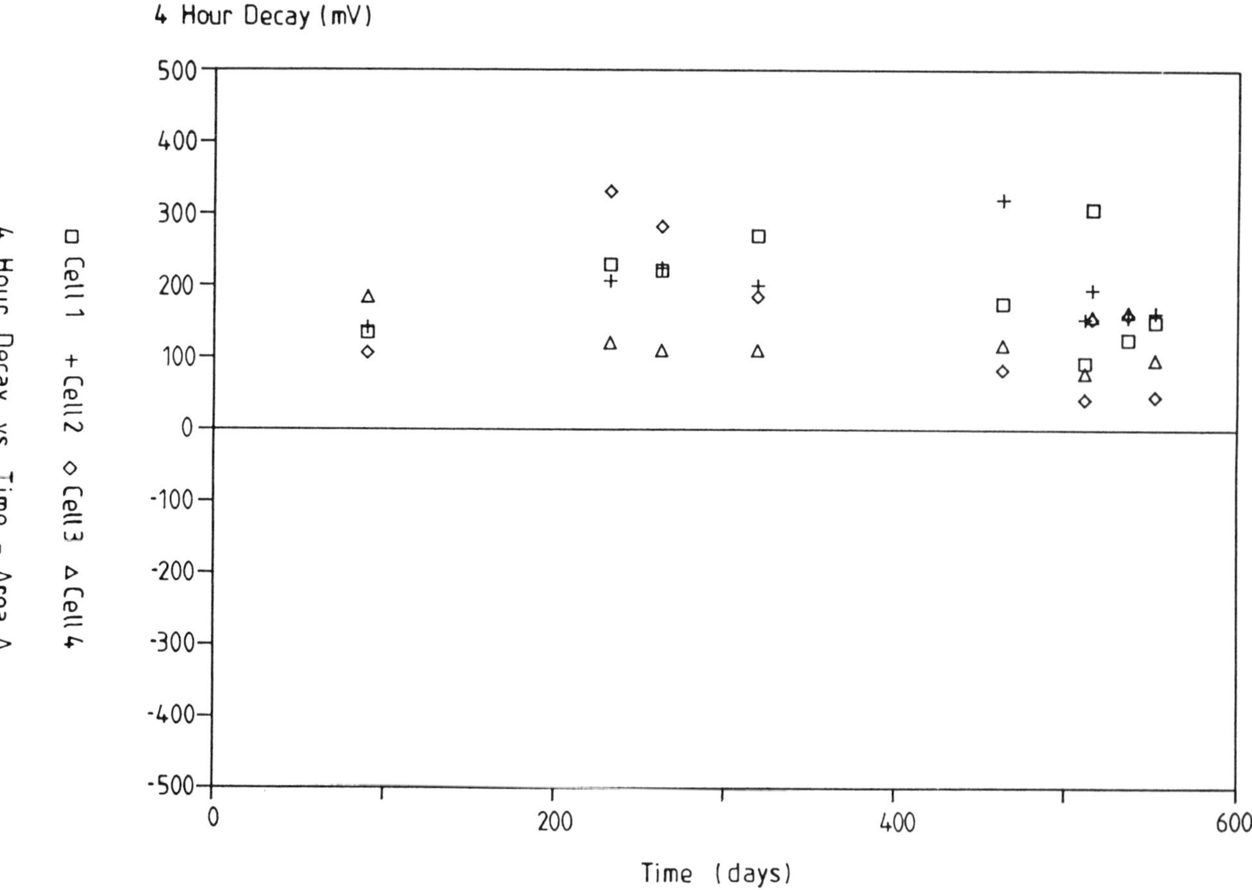

4 Hour Decay vs. Time - Area A

Figure 15

31 COLLECTION OF IN-SERVICE PERFORMANCE DATA: STATE OF THE ART AND APPROACH BY CIB W80/RILEM 100-TSL

C. SJÖSTRÖM
Materials and Structures, National Swedish Institute for Building Research, Gävle, Sweden
E. BRANDT
Building Physics, Danish Building Research Institute, Hørsholm, Denmark

Abstract
The paper reports the work performed by the joint CIB and RILEM committee W80/100-TSL on the Prediction of Service Life. The work is a continuation of an earlier joint activity (CIB W80/RILEM 71-PSL), with a concentration on generic methodologies for Feedback from Practice of Durability Data by in specific Inspection of Buildings.

Different concepts to establish data on the in-service performance of building materials and components are presented. An approach to the design of field inspection surveys of buildings, when the purpose is to establish durability data, is reviewed and discussed.

How to consider the role of the exposure environment when performing feedback investigations is shortly commented.
Keywords: Durability Data, Feedback, Field Inspections, Inspection Routines, In-service Conditions, Exposure Environment.

1 Introduction

In 1982 a joint CIB and RILEM committee (CIB W80/RILEM 71-PSL) on service life prediction of building materials and components was established. In accordance with the plans the initial programme was completed in 1985 and the results of the work was reported through CIB and by Masters (1986). A main result of the work, done by this committee, is a systematic methodology for the prediction of service life of building materials and components. This methodology was recently accepted as a RILEM recommendation, Masters, L.W. and Brandt E. (1989).

In 1987 a new RILEM committee was established to, in cooperation with CIB W80, continue the work. The scope of this new committee is to

- encourage and aid in the implementation of the generic methodology for service life prediction earlier developed
- further develop and detail that methodology
- provide a focal point for service life research and contribute to advance the knowledge base of service life prediction.

It was decided to concentrate the committee´s work on general methodologies concerning feedback on the performance of materials and building products in service. This paper is based on the report of W80/100-TSL which is published by CIB, Sjöström, Ch. ed, (1990).

Data on service life are available from experience and from testing. In spite of the fact that reliable service life predictions to a great extent have to rely on experience based on use of the products in actual buildings, there is a lack of agreed and approved methodologies for the gathering of such in-use data. Furthermore only after careful examination of the degradation mechanisms obtained in service life testing and after ensuring that the mechanisms are the same as in the in-service situation, can the data from testing be used in a prediction.

2 Approaches to long term ageing data

Figure 1 shows in a condensed form the generic methodology for service life prediction expanded and detailed to include different ways of generating data from long term ageing. Testing at field exposure sites, in-use testing and testing by use of experimental buildings may be described as more pure experimental techniques, in comparison with the evaluation of building products through inspection of buildings.

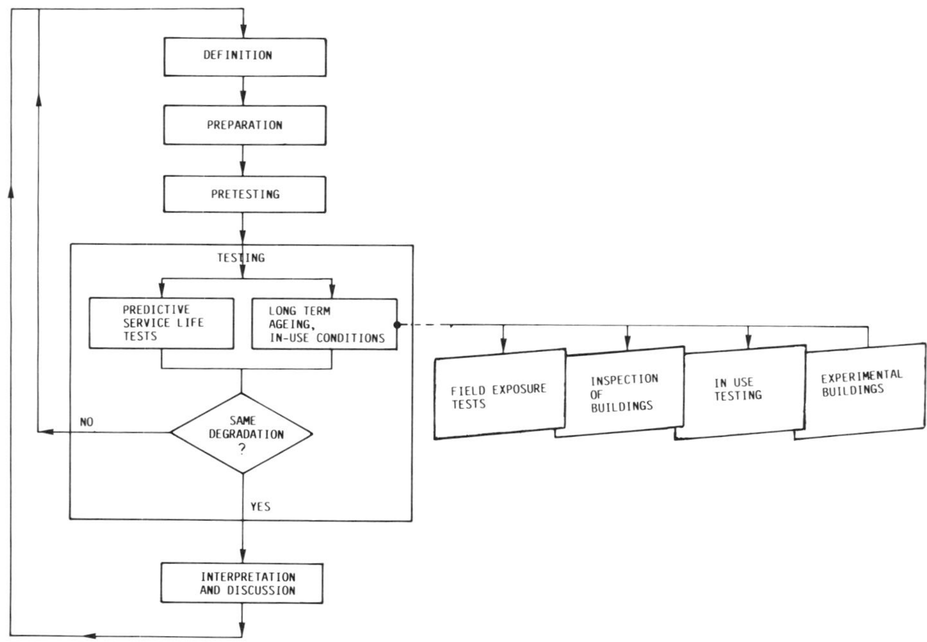

Fig. 1. The Service Life Prediction Methodology expanded and detailed to include different ways of generating data from long-term ageing

For all four approaches - Field Exposure Tests Inspection of Buildings, Experimental Buildings and In-Use Testing - there are basically the same demands and needs regarding evaluation techniques, i.e. the methods and equipments to be used when evaluating the degradation indicators.

In the following a short general overview of the different concepts is given, and the build up of field inspection investigations is presented in chapter three.

2.1 Testing at field exposure sites

Standardized ways of performing field exposure tests are since long in operation and there exists a deep and wide spread knowledge among materials researchers on the applicability and usefulness of field exposure testing to different types of materials.

It is essential to bear in mind that:

1. results from field exposure testing relate to the specific test site; it may cause great difficulties to transfer the data to another geographic location,
2. as weather does not repeat itself one has to be cautious in drawing conclusions from one exposure period to another, and
3. exposing material samples to the environment, e.g. on exposure racks with the inclination 45^0 and with a direction towards the sun, may be regarded as an accelerated ageing test; the degree of acceleration varying with the type of material under exposure.

2.2 Inspection of buildings

The purpose of inspection of buildings may be to describe the in service state of materials or components for single buildings or for a whole well defined population of buildings.

With the limited scope of drawing conclusions regarding only those objects that have been studied, the procedure of sampling buildings may be non statistical. If the goal is to generalize the observations to a population of buildings, an inspection of a statistical sample of objects from the population is an effective way of performing the investigation.

Evaluating the service life of building products through inspection of buildings involves several difficulties that have to be accounted for:

1. it may be difficult to obtain data on the history of inspected materials or components, i.e. data on the original performance values, information on performed maintenance etc.
2. the in-use environment normally cannot be controlled, and may be difficult to measure and describe.

Feedback from practice by inspection of buildings has the possible advantage, when the investigations are properly designed, of giving direct correlation between the state of materials, exposure environment and the building use. When the investigation is based on inspection of statistical samples it is also possible to reach reliability data for the studied materials.

2.3 Experimental buildings

Durability evaluation of building products may be carried out by testing the product in special experimental test houses. If the test procedure and the test house are properly designed, with regard to the questions to be answered, this approach for establishing data from long term ageing under conditions close to an in-use situation has several advantages. Among the most important positive assets are

1. that the building product may be tested in full scale, and
2. that if the scale of the experiment is sufficiently large it is possible to collect data on reliability.

No standard procedures exist for the use of experimental buildings in durability evaluation of building materials and components, and this fact puts a special demand on describing of the used test procedure when reporting the investigation. The difficulties of generalizing the results that were pointed out for field exposure testing may also apply to investigation involving the use of experimental houses.

2.4 In-use testing

By in-use testing is meant an intentional use of a material or a building product in a full scale building or structure under normal use, in order to evaluate the service life of the product. The aim of the approach is to create an experimental situation as controlled as possible, when the test material or the component is under the influence of the full range of degradation factors of the in-service situation.

It is easy to recognize several similarities between the use of experimental buildings and the in-use testing approach. The limitation of the in-use testing approach is basically the same as for experimental buildings. The in service environment affecting the test product is normally more difficult or not at all possible to control and may also be difficult to measure and describe when adopting the in-use testing concept.

3 Feedback by inspection of buildings

3.1 Sampling of buildings

The sampling procedure is the decisive step determining the possibilities to generalize the results. If one wants statistically to generalize the data, the objects to be inspected must be sampled at random. If the inspected buildings are not randomly sampled each of the buildings must be conceived as a case study, and a generalization of the results has to be based on other considerations than statistical.

One method of sampling building is to divide maps or areal photographs into grids with squares of the same size, and to sample by random some of the squares. The buildings situated in the sampled squares are then inspected. The method has been used in several studies aiming at estimating the amount of certain building materials.

If suitable registers of real estates or buildings are available these registers can be used for statistical sampling. This technique has been developed and used in several investigations of the technical status of the building stock performed by the National Swedish Institute for Building Research. The different options for statistical sampling are treated more in detail by Sjöström and Tolstoy (1989) and in the CIB W80/RILEM 100-TSL Report, CIB Report, Sjöström, ed., (1990).

3.2 Routines for inspection

Inspections should be carried out in accordance with an inspection routine. In the following a generic approach to the build up of such a routine is outlined (see fig. 2).

An inspection routine must aid the user in assuring that all important aspects of the investigation are treated. This can be done by designing the routine as a checklist of the topics to be examined. The inspection should be carried out in a logical order, thereby revealing possible defects with the smallest possible efforts. Finally the results of the building inspection should be recorded in a way that allows comparison with results of similar investigations and also allows comparison with laboratory studies.

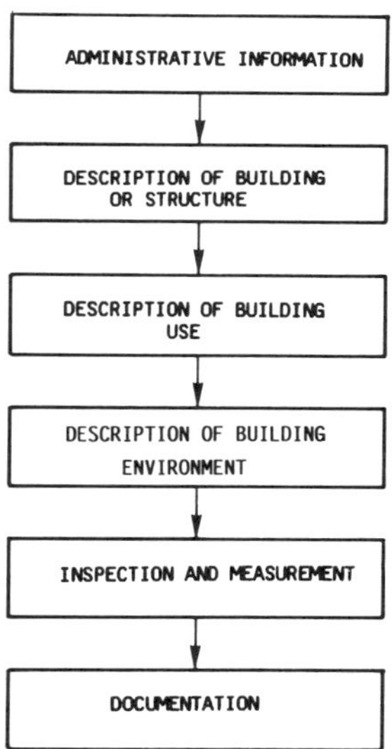

Fig. 2. The build up of an inspection routine

The information sought in a specific investigation depends on the purpose of the investigation, but often the following knowledge is required:

- condition of materials or components
- extent of defects

- cause of defects
- necessary repair works

The boxes of fig. 2 may be examplified as follows:

Administrative information
State the necessary administrative information; e.g. location or address of the building or structure, names and addresses of involved persons (contact person at building, investigator).
Description of the inspected building or structure
E.g. year of construction and of important reconstruction, used building technology etc.
Description of the use of the building
Describe the normal use of the building, indicate known deviations from normal use, account for maintenance procedures etc.
Description of the building environment
Present all relevant data on the local environment of the building important for the degradation of the materials that are studied.
Inspection guides and measurement techniques
Guide the evaluation of specific materials by inspection schemes that also indicates recommended evaluation techniques.

The level of the inspection must be chosen in accordance with the purpose of the investigation. Lower level, for instance visual inspection, may be quite satisfactory for some purposes whereas a more thorough investigation may be necessary in others demanding the use of sophisticated equipment. The cost of an inspection will very much depend on the chosen level of inspection.

When assessing the condition of materials or components it is practical to distinguish between different levels. These levels can be described in different ways depending on the purpose of the investigation. An example of four condition levels may be as follows (see also fig. 3)

Condition 0: Intact; no maintenance necessary
Condition 1: Minor damages; some maintenance is suggested
Condition 2: Malfunction; maintenance needed as soon as possible
Condition 3: Out of order; replace or repair immediately

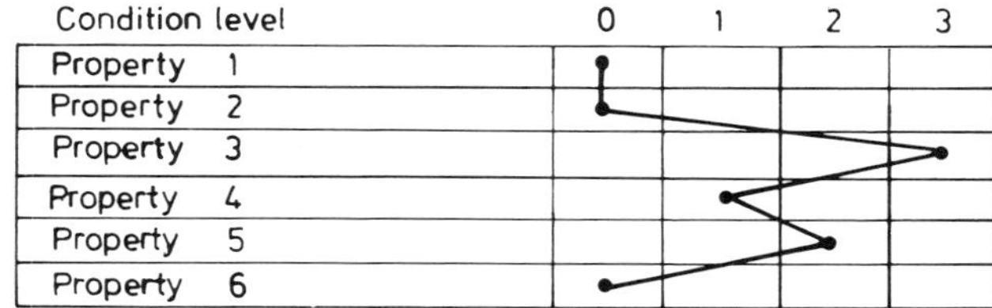

Fig.3. Presentation of results. An example of an investigation with four condition levels

The conditions levels may be summarized in inspection schemes, as exemplified in table 1 showing an inspection scheme for evaluation of coated sheet metal.

Table 1. An example of an inspection scheme for evaluation of sheet metal and coating

COATING / METAL SUBSTRATE	00 **No damages** Blistering ≥8 F Cracking ≥8 Chalking ≥8 Flaking ≥8	10 **Minor damages** 6 F Blistering <8 F 4< Cracking <8 4< Chalking <8 4< Flaking <8	20 **Repair need** Blistering ≤6 F Cracking ≤4 Chalking ≤4 Flaking ≤4
00 **No damages** No damages type scratches, buckling or split open sheets Corrosion ≥8	00 No damages to fasteners, no mechanical damages Corrosion ≥8 Blistering ≥8 F Cracking ≥8 Chalking ≥8 Flaking ≥8	10 No damages to fasteners, no mechanical damages Corrosion ≥8 6 F< Blistering <8 F 4< Cracking 4< Chalking <8 4< Flaking <8	No damages to fasteners, no mechanical damages Corrosion ≥8 Blistering ≤6 F Cracking ≤4 Chalking ≤4 Flaking ≤4
01 **Minor damages** Minor and few scratches (not down to metal), minor buckling and split open sheets 5< Corrosion <8	01 Minor mechanical damages or damages to fastenes 5< Corrosion <8 Blistering ≥8 F Cracking ≥8 Chalking ≥8 Flaking ≥8	11 Minor mechanical damages or damages to fasteners 5< Corrosion <8 6 F<Blistering <8 F 4< Cracking <8 4< Chalking <8 4< Flaking <8	21 Minor mechanical damages or damages to fasteners 5< Corrosion <8 Blistering ≤6 F Cracking ≤4 Chalking ≤4 Flaking ≤4
02 **Repair need** Serious damages type buckling, split open sheets; scratches down to metal, damages to fasteners Corrosion ≤5	02 Serious damages to fasteners and mechanical damages Corrosion ≤5 Blistering ≥8 F Cracking ≥8 Chalking ≥8 Flaking ≥8	12 Serious damages to fasteners and mechanical damages Corrosion ≤5 6 F< Blistering <8 F 4< Cracking <8 4< Chalking <8 4< Flaking <8	22 Serious damages to fasteners and mechanical damages Corrosion ≤5 Blistering ≤6 F Cracking ≤4 Chalking ≤4 Flaking ≤4

<u>Standard Methods of Evaluation</u>

Corrosion (Rusting)	ASTM D610-68
Blistering	ASTM D714-56
Cracking	ASTM D661-44
Chalking	ASTM D659-80
Flaking (scaling)	ASTM D772-47

3.3 Field applicable evaluation techniques

It has been found that the majority of ordinary inspections, for example in connection with maintenance, can be performed with good results with limited resources, i.e. simple or no instruments and at low cost. To ensure uniform results it is an advantage to use inspection guides, e.g. in the form of catalogues with pictures describing various defects and their importance, i.e. the level of condition.

In some cases, however, it is necessary to profit from more thorough evaluation techniques including more or less advanced measuring methods. A great number of measuring methods are available, most of them demanding experienced technicians. The technician must have a general knowledge of the conditions to be examined, and on the basis of a technical and economic evaluation be able to select the best suited method of evaluation, i.e. which measurements or tests to be carried out.

In the CIB W80/RILEM 100-TSL appendix report, CIB Report, Brandt, E, ed, (1990), a compilation of field applicable evaluation techniques is presented.

The purpose of most measurements is to give a numeric description of the object to be measured. Normally, the measurements will primarily be connected to the properties of the products.

The measurements to be carried out can be direct or indirect. As an example of direct measurements, the evaluation of the thickness of layers of paint can be mentioned. By indirect measurements a certain property is measured to be used for an opinion about another wanted property. A measurement of this kind is for example thermography, whereby the infra-red radiation of an object is applied in connection with the evaluation of insulation conditions.

In some cases the desired properties can also be determined indirectly if the physical behavior can be described by a mathematical formula. The loadbearing capacity of wooden joints can for example be determined after measurement and evaluation of the wood quality.

In certain instances it may be necessary - or more expedient - to select testing objects for a closer inspection in a laboratory. Laboratory examinations are primarily used for analyzing the composition of materials, but in certain instances it might be more purposeful to examine components or building parts by means of a laboratory test.

During all measurements various errors are likely to occur. The reason and the magnitude of these errors may vary depending on the specific task. Indication of test results should therefore always be provided with a "declaration of quality" in the form of an indication of the uncertainty of the result.

For the practical use in connection with measurements in the field it is especially important to distinguish between the uncertainty of the test object and the uncertainty of the test method (apparatus, surrounding etc).

It is normal to distinguish between stochastic errors and systematic errors:

Stochastic errors (or uncertainty in measurement) occur as random deviations between the test results from repeated measurements under similar conditions.

Systematic errors are characterized by the fact that they occur in a predetermined way and cause a systematic deviation from the true value. In as far as the systematic errors are known their influence on the test result may be corrected by calibration. Effects pertaining to other systematic errors might often occur too; errors that are known, but the magnitude of which is unknown. The influence of such errors has to be evaluated subjectively.

Traditionally, the influence of stochastic errors is determined by repeated measurements, the results of which are used for the calculation of the average value and the standard deviation.

The total uncertainty of the test result should contain the contribution from the systematic errors as well as from the stochastic ones. The tendency is to treat the remaining systematic errors as equal to the stochastic errors. That means that the law of accumulation for indefiniteness for stochastic errors can be applied also when estimating the total uncertainty.

Measuring method and instrument should be chosen with due consideration to the uncertainty allowed. Ideally, the uncertainty of the measuring should be considerably below the uncertainty of the object to be measured. In practice an uncertainty of the method of 1/2 - 1/3 of the uncertainty of the subject to be measured is sufficient. (Too strict demands to the uncertainty may easily result in a considerable increase in the cost of the test.)

There are, however, other factors to take into consideration when choosing a test method e.g. the method should

- as far as possible, be objective, i.e. independent of the person
- be reproducible i.e. tests on the same object with the same method but, for example, with another instrument should give the "same results"
- be handy
- be easy to calibrate.

3.4 Description of exposure environment

The exposure environment may be considered in categories of meteorological, biological, chemical and physical degradation factors. For a specific material only a part of the environmental factors identified by the building context are degrading factors. In the same way only a part of all the material characterizing properties are performance or function characteristics in that specific context. For instance, optical transmission of plastics glazing may involve a particular spectral range of visible light when defining performance, but the degradation of this sensitive property may involve only the ultra violet component of solar radiation.

The methods for measurement and description of the exposure environment are similar irrespective as to whether functional or degrading factors are being measured. There is a tendency, however, that the measurement of many functional factors are done according to standard test methods whereas the measurement of many degrading factors are of an investigative nature developing as the further understanding of material degradation mechanisms is proceeding. An often used ground for classification of climate is the division into macro, meso and micro climate. This division means a definition of different scales for the description of the variations in the meteorological variables. There exist no exact and common definitions of the different scales.

With macro climate is normally meant the gross climate described in terms like polar climate, subtropical climate and tropical climate. The descriptions are based on measurement of meteorological factors such as air temperature, precipitation etc.

When describing meso climate the effects of the terrain and of the build environment are taken into account. The climatological description is still based on the standard meteorological measurements.

The micro climate describes the meteorological variables in the absolute proximity of a material surface. Decisive for materials degradation is the micro climate or the micro environment. The most important meteorological variables describing micro climate are relative humidity, surface moisture, surface temperature, irradiation and deposition of air pollutants. Measurement and description of the micro climate or the micro environment calls for specially developed measurement devices and measurement techniques. The need for research and development work in this area is great.

In practice, when performing feedback of durability data by inspection of buildings one has, for the description of the exposure environment, to rely upon available standard meteorological data and on direct and indirect measurement of the important degradation factors. Direct measurements may be examplified by the measurement of surface moisture by use of miniature sensors or by the measurement of ultra violet radiation by UV-dosimeters applied directly on the building. Indirect measurement of the degradation environment is done by the exposure of material specimens mounted on the buildings; the degradation of the material specimens after a certain time of exposure is evaluated in laboratory. To define corrosivity categories and thus classify the corrosivity of the atmosphere by the corrosion effect on standard specimens is for example one basis for the ISO/DIS Standards 9223-9226.

3.5 Reporting of in-service durability data

The report of a field inspection of buildings should, presented by stating the most essential head lines, contain the following information:

Purpose of field investigation
Present the aim of the complete service life study, and in particular the purpose of the field investigation.

Performer of investigation
Present the institute or company responsible for the investigation; name and affiliation of investigator.

Inspected object or objects
Give conclusive descriptions of the inspected type of structures, buildings, components, materials, etc.

Sampling of objects to inspect
Describe the method used.

Inspection routine used for the field investigation
Describe the used inspection routine by presenting the head lines of the main steps of the routine.

Inspection guides and methods for measuring degradation
Present a compilation of the used inspection guides and measurement methods.

Description of exposure environment
Present conclusive results from performed measurements of exposure environment account for measurement methods.

Compilation of the results
Account for number of measurements, calculations made, uncertainties of results etc.

Presentation and discussion of results
Present the investigation results in for instance report schemes, graphs etc. Discuss reliability and validity. Compare the field investigation results with experimental data.

4 Concluding remarks

There are great needs for research and development connected with the generating and gathering of data from long term ageing under in-use conditions. Among the most vital needs are:

a development of generic routines for evaluating materials and components durability by use of experimental buildings and by the in-use testing approach

a further development of non-destructive field applicable assessment and evaluation techniques. This is an area in rapid progress and there are promising techniques waiting for a more common use. An example is the condition assessment of materials in the building envelope by the use of computer image processing of video-films.

a development of the knowledge base regarding the measurement and description of the environmental factors causing materials degradation. Of specific importance is the development of direct and indirect techniques for the measuring of degradation factors on the building or structure.

5 References

CIB report (1986), Masters, L.W. (ed) **Prediction of service life of building materials and components.** Publication 96, International Council for Building Research Studies and Documentation (CIB), Rotterdam.

CIB report (1990), Sjöström, Ch (ed) **Feedback from practice of durability data - Inspection of buildings.** Under publication by International Council for Building Research Studies and Documentation (CIB), Rotterdam.

CIB Report (1990), Brandt, E. (ed) **Feedback from practice of durability data - Appendix -** Examples of field investigations of buildings and structures. Under publication by International Council for Building Research Studies and Documentation (CIB), Rotterdam.

Masters, L.W. (1986) Prediction of service life of building materials and components. **Matériaux et Constructions/Materials and Structures**, vol 19, No 114.

Masters, L.W. and Brandt, E. (1989) Systematic methodology for service life prediction of building materials and components. **Matériaux et Constructions/Materials and Structures**, 22, pp. 385-392.

Sjöström, Ch. and Tolstoy, N. (1989) **Feedback from practice of durability data. Proceedings of XIth CIB Congress**, International Council for Building Research Studies and Documentation (CIB), Rotterdam.

32 *IN-SITU* BEHAVIOUR OF GALVANIZED REINFORCEMENT

R.N. SWAMY
University of Sheffield, Sheffield, UK

Abstract
Although galvanized steel has many inherent qualities to provide high resistance to corrosive environments, there is surprisingly considerable discrepancies on its performance reported from some laboratory studies and field tests. This paper makes a critical review of these test results to fathom the reasons for this apparent anomaly. It is shown that results obtained from tests on liquid solutions simulating concrete environment cannot be extrapolated to predict corrosion behaviour of galvanized steel in concrete and chlorides. The threshold potential and critical chloride concentration affecting the electrochemical stability of galvanized steel in concrete are shown to be influenced by cover to steel, concrete quality and cracking. To evaluate the relative significance of these parameters, tests on concrete prisms containing uncoated or galvanized/chromated bars and exposed under stress to accelerated exposure regime and natural exposure in a tidal zone are reported. These results are related to field data reported in literature. Cover to steel is shown to be the most critical factor affecting corrosion resistance of any type of steel. With adequate cover, galvanized steel can give excellent corrosion resistance in the most severe corrosive environments.
Keywords: Reinforcing Steels, Corrosion Resistance, Galvanized Steel, Accelerated Tests, Field Tests, Marine Exposure, Cover, Concrete Durability.

1 Introduction

Concrete has the most unique property: of all construction materials, it provides a safe and protective alkaline environment for embedded steel reinforcement. Further, it has the inherent potential to be durable, but its ability to resist external aggressive environmental agencies and sources of internal deterioration very much depends on the care and control exercised in proportioning, placing, compacting and curing the material. There are innumerable examples in practice of concrete structures that have successfully withstood the effects of time and environment; but equally, there are many instances where human factors, which are very much involved in moulding and fabricating concrete, have led to lack of quality control and subsequent reduced durability. Protection of embedded steel has thus become an essential component of structural design if reinforced concrete structures are to remain serviceable for the period of their design life.

Like concrete, galvanizing or zinc coatings have certain unique advantages in providing protection to steel in a corrosive environment. The most important fact is that the coating is metallurgically bonded to the base steel, thus forming a tightly adherent and integral barrier preventing contact between any aggressive agent and the steel surface. In addition, the zinc also provides galvanic or sacrificial protection to the steel so that a galvanized rebar will not grossly corrode in any one area until all of the zinc coating is destroyed in the area. There are other advantages also. The oxidation products of iron are highly insoluble in concrete with the result that the steel corrosion products, viz., iron oxides, hydroxides and hydrates, occupy several times the volume of the original steel. The zinc corrosion products are more soluble in the concrete environment, so that they may diffuse some distance from the metal-concrete interface; and zinc oxide, the usual corrosion product of zinc, occupies only about 50 per cent more space than the parent metal. The net volume increase and splitting pressures due to corrosion are thus very much reduced with galvanized steel than with uncoated steel. Zinc corrosion products are also generally greyish white and do not therefore cause the unsightly brown staining usually observed with steel.

In spite of these inherent advantages of galvanizing as a means of corrosion protection of steel in concrete, there is considerable confusion in the minds of engineers as to the long term stability and durability of galvanized rebars in concrete, because of the conflicting and contradictory data in published literature. There are worrying discrepancies in the data reported from laboratory tests and field exposure studies on the performance of galvanized reinforcing bars: whilst some studies show substantially improved durability over uncoated steel bars, others appear to indicate slightly better, equal, or sometimes even worse performance. Data acquired from real structures in service, on the other hand, show excellent long term stability of galvanized steel in concrete structures exposed to aggressive environments. Since resistance to chlorides of steel bars is fundamentally related to other factors such as cover to steel, quality of concrete, nature and severity of exposure conditions used in the tests, and the stress state of the test specimens, discussion of and conclusions derived from test results in isolation without reference to these significant influential parameters can be misleading, and very unrealistic.

The aim of this paper is to identify the reasons for the reported pessimistic evaluations while field experience contradicts these, and to synthesize the findings of three sets of studies. The first is concerned with accelerated laboratory tests on concrete prisms containing uncoated bars or galvanized and chromated reinforcement. The second study is related to natural exposure tests in a tidal zone of similar reinforcement. The tests were carried out on concrete prisms with embedded steel, the main variable being the cover to steel. To simulate real life situations, the tests were carried out with the prisms in a loaded condition, i.e., with the bars under stress and the concrete cracked. The results from these two series of tests are then co-related to other published data and to the long term performance characteristics of zinc-coated reinforcing bars in real structures exposed to high chloride environments. The paper aims to evaluate the role of the zinc coating in chloride-contaminated environments, the development of corrosion rates compared to uncoated bars, and the influence of zinc-corrosion products on the integrity and stability of reinforced concrete structures.

2 Corrosion of zinc in concrete

Many of the laboratory electrochemical studies reported in literature on the chemical reactions and corrosion of zinc have been carried out in solutions representing the chemical environment in concrete Roetheli et al (1932), Rehm and Laemmke (1974), Duval and Arliguie (1974) and Unz (1978). These studies are highly significant as they clarify the nature of the basic chemical reactions of pure zinc, but aqueous hydroxide environments are unrealistic representations of the real nature of concrete. Concrete remains in the plastic state for only a few hours. Once it has hardened, the amount of free moisture is progressively and drastically reduced by cement hydration and drying. Ionic diffusivity in concrete will thus be fundamentally different to that of zinc in liquid solutions. Extreme caution should therefore be exercised in extrapolating the corrosion performance of zinc in saturated calcium hydroxide solutions or in pure acids and alkalies, and particularly in aerated agitated solutions. Concrete also contains many compounds whose action on zinc is not clearly established. For all these reasons, the behaviour of galvanized steel in concrete cannot be evaluated by tests in liquid solutions. This also explains why in real life situations galvanized steel shows a far superior performance than that indicated by tests in liquid solutions.

3 Dependence on pH

The corrosion rate of zinc is highly dependent on the pH, and tests show that in dilute NaOH the minimum corrosion rate occurs at a pH of about 12.5, Roetheli et al (1932). Beyond this value the corrosion rate increases almost exponentially; increases in pH of about 0.1 may cause substantial increases in the dissolution of zinc. Concrete has generally an initial pH of 12.4 to 12.6, similar to that of saturated calcium hydroxide solution; however, the presence of small amounts of sodium and potassium oxides in cement can increase this initial alkalinity, and pH values higher than 13.2-13.3 may occur in concrete, Verbeck (1975). The corrosion rate of pure zinc may therefore be very much enhanced in liquid solutions of high pH.

More recent tests in saturated $Ca(OH)_2$ solutions, however, show that there is a threshold pH of 13.35±0.10 which defines the onset of active corrosion in galvanized steel, Macias and Andrade (1983). This implies that active corrosion will only occur when the Na^+, K^+ and SO_4^{2-} contents in the cement are very high. Cements with low alkali contents would thus provide a stable environment for galvanized steel. Even with high alkali content cements, the pH values of concretes exposed to real environments are unlikely to reach or exceed this critical threshold value.

This is confirmed by both laboratory tests and tests on existing concrete structures. Figure 1 shows the measured pH values in concrete prisms containing one 19mm diameter high tensile galvanized chromated bar (zinc coating thickness about 70 μm) with steel covers of 20, 40 and 70mm and subjected to 36 months of marine exposure in a highly corrosive tidal zone, Swamy et al (1988). The specimens in these tests had been cracked and kept under load during exposure with a steel stress of about 200 MPa. The results show that for all cover depths, the pH of concrete in the vicinity of the reinforcing steel remained in the range of 12.6 to 12.8, well within the safe limits for pure zinc.

Figures 2 and 3 show the measured pH values from two bridges reported by Stark and Perenchio (1975). Figure 2 refers to dual three span bridges on I35 carrying traffic over Long Dick Creek near Ames, Iowa, and the bridge decks were exposed to deicer salts. The concrete in the bridges had a cement content of 420 kg/m^3, water-cement ratio of 0.40-0.41 and 5.2 to 6.2% air entrainment. Figure 3

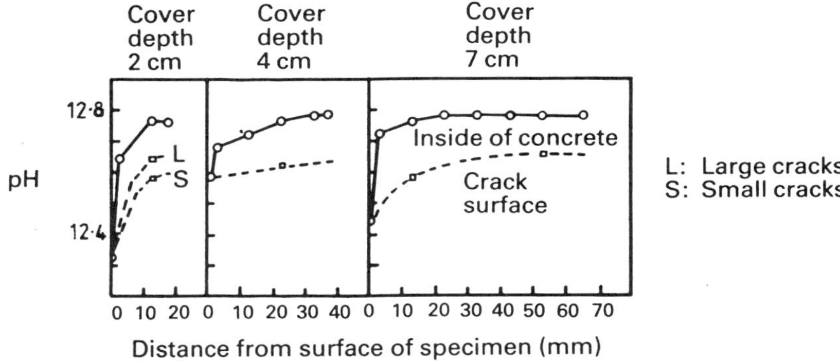

Fig. 1. pH variation in concrete.

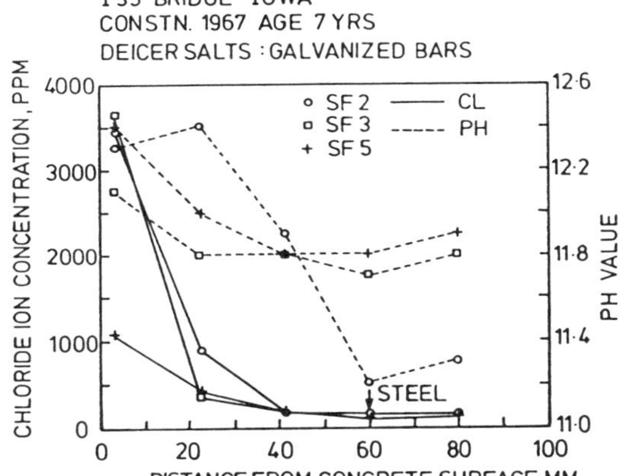

Fig. 2. pH values in bridge deck.

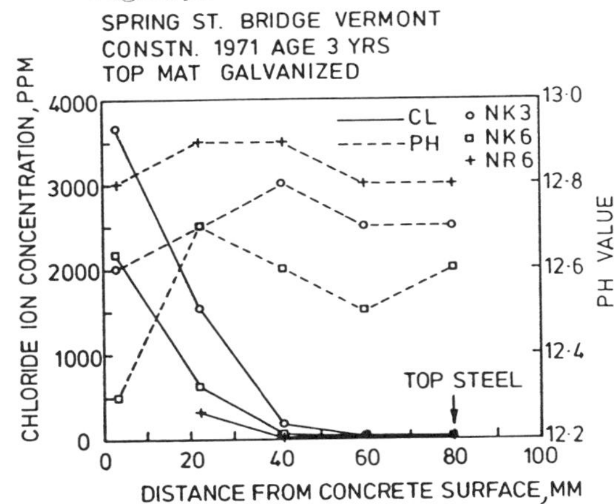

Fig. 3. pH values in bridge deck.

(Figs 2 and 3 derived from Stark and Perenchio, 1975)

refers to a composite steel-concrete bridge in Montpelier, Vermont. The concrete in the deck had cement contents of 360 to 390 kg/m^3, water-cement ratio of 0.44, and 6% air entrainment. The results show pH values of 11.2 to 12.4 in the Iowa Bridge after 7 years of exposure; and values of 12.4 to 12.7 in the Vermont Bridge after 3 years of construction. These data confirm that pH itself is never critical to the stability of galvanized rebars in concrete, although many laboratory tests on saturated solutions show otherwise. Many concrete structures in service show similar pH values less than critical thresholds.

There is also the added factor that pH values in concrete structures exposed to real environments will only decrease with time and not increase. In real structures carbonation will proceed, albeit very slowly. Galvanized steels have the advantage that they will remain passive in carbonated concrete, and the corrosion rate is generally of the same order of magnitude as that in non-carbonated concrete. The corrosion rate of plain carbon steel, on the other hand, is more than a factor of 10 greater in carbonated concrete, Maahn and Sorensen (1986). This superior performance of galvanized steel is also confirmed by tests on concrete prisms, Martin and Rauen (1974).

4 Effect of chlorides on galvanized steel in concrete

Galvanized steel, like untreated steel, is normally passive in the highly alkaline environment of concrete. The penetration of chloride ions to the metal surface can, however, break down this passivity, and initiate the sacrificial corrosion of zinc. The critical pitting potential and the critical chloride concentration which will promote pitting corrosion have again been investigated for pure zinc and galvanized steel by a number of investigators, Ishikawa, Cornet and Bresler (1969), Duval and Arliguie (1974), Clifton (1977), Unz (1978), but these studies are again in saturated $Ca(OH)_2$ solutions without or with added chlorides. The results are somewhat variable, but generally show enhanced critical levels for zinc compared to untreated steel. A significant result from these studies is the formation of the reaction product calcium hydroxy-zincate, $Ca(Zn(OH)_3)_2.2H_2O$ at a very early age. This zincate is soluble at pH values of about 12.5 and above. At pH values of 12.5 and less, zincates form a tightly adherent diffusion barrier which has a definite passivating effect on the zinc surface, Cornet and Bresler (1981). It is to be emphasized that these studies were carried out in liquid solutions, often with pure zinc, and it would be interesting to know whether galvanized steel gets passivated by $Ca(OH)_2$ in concrete. These studies also indicate that sodium dichromate has some beneficial influence in delaying the appearance of corrosion nodules on the galvanized steel. Some of these conclusions may have a positive bearing on the superior performance of galvanized bars observed in actual concrete structures.

5 Threshold potential and chloride resistance

In spite of the inherent qualities of galvanized steel to effectively resist atmospheric and chloride-induced corrosion, there is considerable conflicting and contradictory test results on its performance characteristics in concrete. The situation is confusing and intriguing, to say the least. However, many of the discrepancies can be explained when it is realised that corrosion (rusting) and the concomitant cracking are very much influenced by cover to steel and concrete quality; further, the rate of corrosion initiation is also dependent on whether the concrete is cracked or not at the time of exposure. It has already been shown that tests in aqueous solutions are unlikely to yield valid and relevant data that can either explain the discrepancies of these test results in concrete or be used as a basis for design. Results of tests based on carefully proportioned concrete mixes with low water-cement ratios and adequate cover to steel show superior performance of galvanized steel, Sopler (1973), Duffaut et al (1973), Bernhardt and Sopler (1974); tests, on the other hand, derived from studies involving poor concrete quality with water-cement ratios of 0.6 to 0.9, and very low cover to steel of 10 to 20mm, Griffin (1969), Treadaway et al (1980), Treadaway et al (1989), have shown reasonable or poor performance of galvanized steel against atmospheric and chloride-induced corrosion. This should not come as a surprise: one of the main factors influencing chloride ion intrusion into concrete is the degree of concrete permeability. At w/c ratios of 0.5 and above even 28 day water curing is inadequate to block the formation of continuous capillary pore system, Powers et al (1959). With concretes of w/c ratios of 0.6 and above, it is only a matter of time before corrosive agents penetrate to the steel, the benefit of service life depending on the cover to steel and severity of exposure.

5.1 Cover to steel

Cover to steel is the other important paramter determining the stability of the reinforcing steel. Laboratory accelerated tests and field exposure studies on concretes with a w/c ratio of 0.55 show that cover is the major factor controlling the penetration of chloride ions into concrete and the rate of corrosion of the steel bars, Swamy (1990). Table 1 shows the rate of chloride ion penetration into concrete whilst Tables 2 and 3 show the rate of corrosion of plain and galvanized/chromated bars in severe chloride contaminated exposure conditions. The data confirm that galvanized/chromated bars have a high chloride tolerance level and excellent corrosion resistance in severe chloride contaminated environments.

Table 1 indicates that with a cover of 40mm and normal quality concrete, galvanized/chromated bars should be able to resist effectively chloride ion concentrations of about 4000 ppm. This implies that, for example, concrete containing about 225 kg/m^3 cement content (or 4 US bags/ yd^3) cover should be able to resist the effects of the equivalent of over 10% anhydrons $CaCl_2$ by weight of cement or over 25 lbs. Cl^- per cu. yd. of concrete. These amounts are of course far in excess of the chlorides that are likely to enter concrete even in the severest corrosive environments. Whilst these data cannot obviously be literally transferred to practice, what they imply is that galvanized steel has a very high tolerance limit, much higher than what one would deduce from published research.

Tables 1 to 3 also show that with the inevitable high chloride ion intrusion and consequent high corrosion rates in cover depths of 10 to 20mm, cover to steel is the most critical factor influencing the electrochemical stability of steel in concrete. At these cover depths even the best corrosion protection will eventually begin to detriorate. Ferritic stainless steels with low carbon and medium chromium content have not survived in such environments; even the austenitic, chromium-nickel steel (300 series) have shown signs of pitting corrosion in concrete with high levels of chloride, Treadaway et al (1989).

5.2 Corrosion resistance

The superior performance of galvanized steel in these environments is confirmed by the field investigation of four structures reported by Stark (1978). Seven samples were taken from these structures varying in age from 7 to 23 years, containing galvanized reinforcement in concrete and exposed to severe marine environment. The results shown in Table 4 confirm that galvanized steel can resist chloride induced corrosion effectively, and that it can tolerate very high levels of chloride ions in normal quality concrete. The average corrosion layer measured in these structures varied from 2 to 8 μm with one value of about 13 μm. Over 90% of the zinc coating was considered to be remaining on the bars. The depth of corrosion of the coating was thus extremely low; and even though the period of exposure of the reinforcing bars to the chloride levels shown in Table 4 was not known, it is clear that galvanized steel is able to tolerate chloride concentrations well above the 275-325 ppm (0.66 to 0.78 kg Cl^- ions per cu.m of concrete) considered necessary to corrode untreated steel.

Table 1. Chloride ion penetration into concrete

Distance from surface, mm	Chloride conc. ppm at cover in mm		
	20	40	70
	Tidal zone exposure: 3 years		
10	8,500	7,500	6,000
20	10,000	7,000	5,000
40	-	7,500	4,000
	Accelerated exposure : 1 year		
10	8,200	8,000	5,000
20	9,000	6,000	3,500
40	-	< 4,000	2,000

Table 2. Rate of corrosion - tidal zone exposure

Type of bar	Cover mm	Frequency of rust, %	
		1 yr.	3 yrs.
Uncoated	20	95	46
	40	10	38
	70	2	8
Galvanized	20	Negligible	1.5
	40	Negligible	1.0
	70	Negligible	Negligible

Table 3. Rate of corrosion - accelerated exposure

Type of bar	Cover mm	Frequency of rust, %	
		1 yr.	3 yrs.
Uncoated	20	80	100
	40	25	98
	70	8	75
Galvanized	20	15	72
	40	7	8
	70	5	6

Table 4. Field performance of galvanized steel

Structure	Age yrs.	Chloride in concrete at level of steel (kg/m^3)	Average thickness of corrosion layer (µm)	Zinc coating (%)
Penno's wharf - St. George (SG17) (1969)	7	3.0	2.54	98
Jetty - Bermuda Yacht Club (BYC 3)	8	3.6	0.00	100
Hamilton dock (H 22)	10	1.9	5.08	95
Hamilton dock (H 26)	10	3.6	7.62	96
Penno's wharf - St. George (SG 10) (1966)	10	4.6	5.08	99
Penno's wharf - St. George (SG 9) (1964)	12	6.4	12.7	92
Longbird bridge (LB 20)	23	4.3	5.08	98

(adapted from Stark, 1978)

Figures 4 to 7 show typical chloride profiles in concrete in structures exposed to marine spray - two in Bermuda and two in the United States reported by Stark (1975, 1978), all of which have shown excellent performance in service. The Longbird bridge (Fig.4) and Penno's wharf (Fig.5) in Bermuda have very high concentrations of chloride at the steel level. The galvanized coating was only slightly affected by corrosion in the bridge, and retained, after 23 years service, an appreciable layer of zinc over the intermetallic layers. In the Penno's wharf, the average corrosion depth varied with age (Table 4) and was limited to about 2.5 to 12.7μm, although more severe attack was observed in some local areas.

In the Boca Chica Bridge in Florida (Fig.6) it was thought that some considerable chloride was introduced during mixing and placing. The span with uncoated steel showed high levels of chloride, and there was evidence of minor corrosion. The span with galvanized steel had very much higher levels of chloride ions, but there was no evidence of corrosion. The cores also revealed some evidence of a frothy texture at the steel-concrete interface. pH values were about 12.2 to 12.5 in the two spans. In the Seven Mile Bridge, Florida (Fig.7), corrosion had occurred in the uncoated steel, and the concrete showed signs of distress. No corrosion or distress was found in the panel with galvanized steel.

All these data show that galvanized steel is capable of superior performance in highly corrosive environments. In many instances, the average depth of corrosion of the zinc coating was extremely low after several years of exposure to marine spray or deicer salts. Even where there was evidence of hydrogen evolution at the steel-concrete interface, the integrity of the concrete and of the structure as well as the corrosion resistance of galvanized steel remained unimpaired. In the four structures in Bermuda exposed to severe seawater environment, more than 90% of the original coating thickness remained intact, and there was no evidence of cracking in the concrete due to zinc corrosion (Table 4), Stark (1978).

6 Conclusions

Galvanized steel has excellent inherent qualities to provide high corrosion resistance in concrete contaminated by chloride ions. Nevertheless some laboratory and field studies report poor performance while field experience contradicts such pessimistic evaluations. This paper examines critically the reasons for these reported discrepancies, and presents laboratory and field data to show excellent corrosion resistance of galvanized steel in concrete containing high levels of chloride.

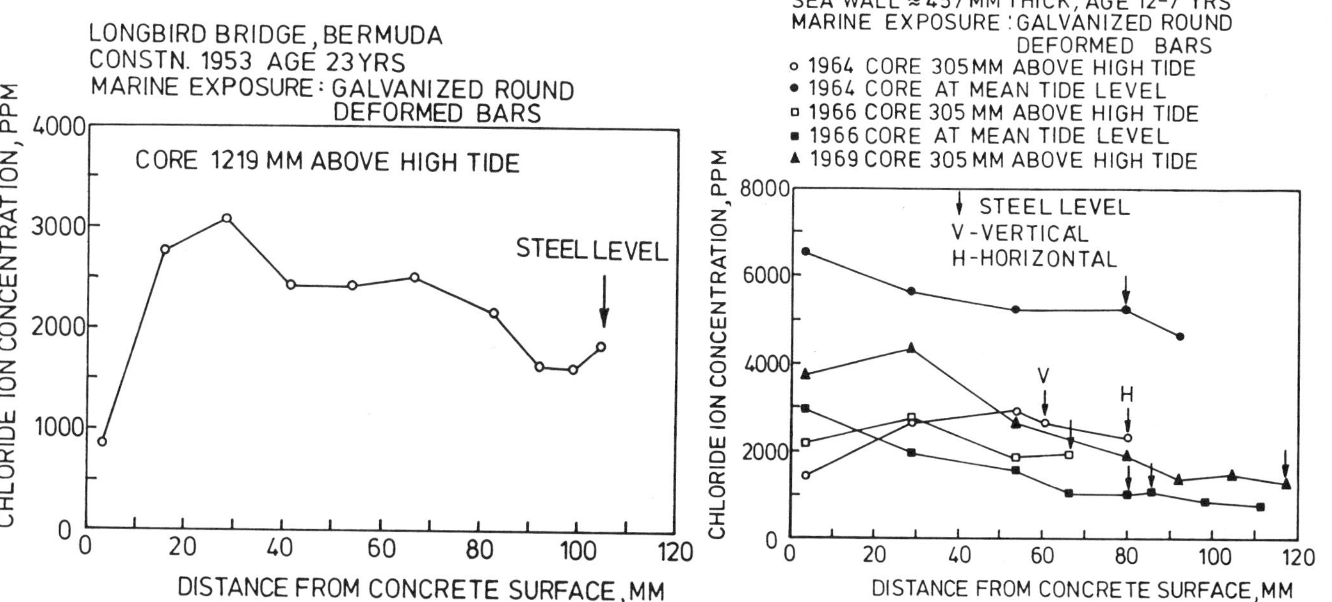

Fig.4. Chloride profile: Longbird bridge.

(derived from Stark, 1978)

Fig.5. Chloride Profile: Penno's wharf.

(derived from Stark, 1978)

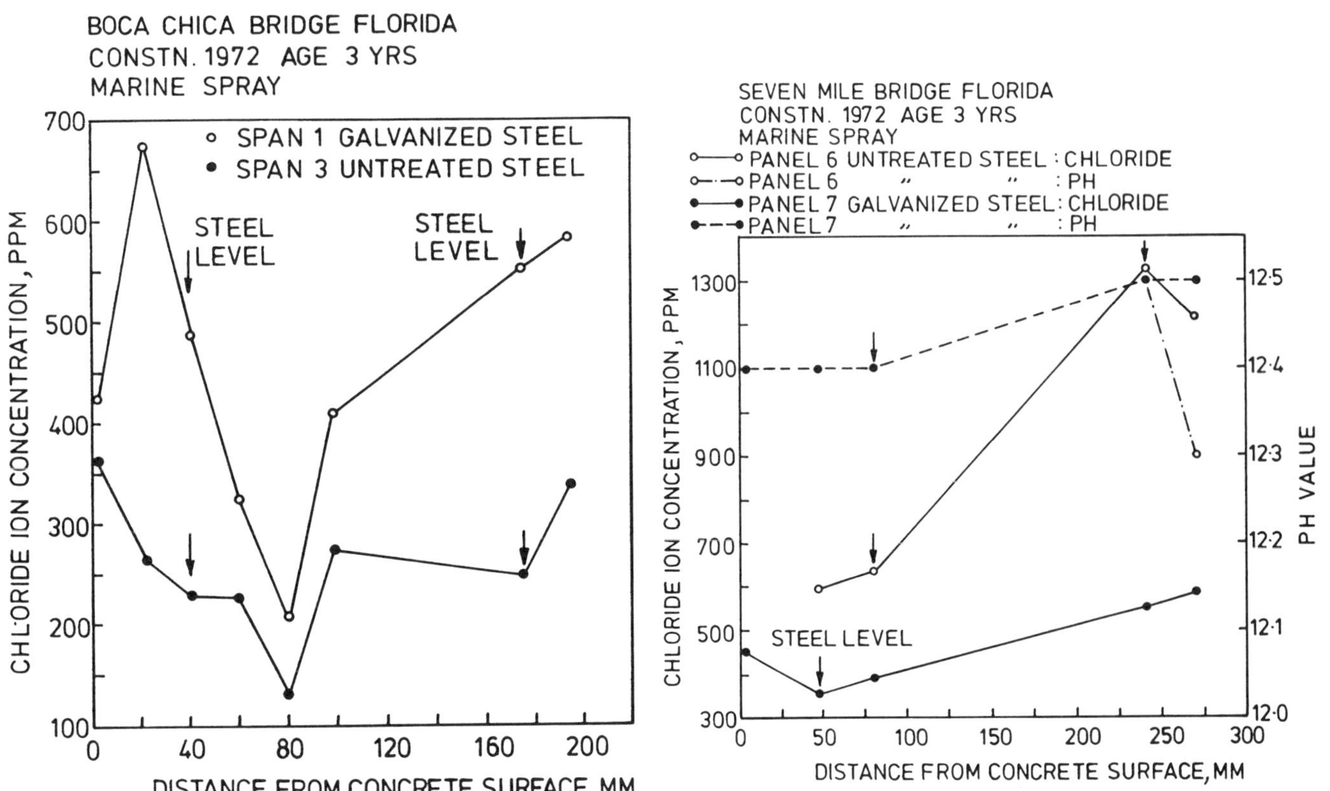

Fig. 6. Chloride profile: Boca Chica bridge.

Fig. 7. Chloride profile: Seven Mile bridge.

(derived from Stark and Perenchio, 1975)

Electrochemical studies of pure zinc or galvanized steel in solutions simulating the chemical environment of concrete are unrealistic to represent the true behaviour of galvanized steel in concrete. Although very high pH values in excess of 13.0 may be observed in pore solution studies, cores taken from structrues exposed to real environments show pH values less than the critical threshold value, providing a stable environment for galvanized steel. This is confirmed by tests of concrete prisms exposed to marine environments. There is the added benefit that galvanized steel, unlike plain steel, has similar corrosion rates in carbonated concrete as in non-carbonated concrete. There is also some evidence that calcium hydroxyzincate provides a passivating layer on the zinc surface.

It is shown that cover to steel, concrete quality and cracking are the three major factors controlling chloride intrusion into concrete and determining the stability of steel against corrosion. At water-cement ratios in excess of 0.6, continuous capillary pores can never be blocked, and it is simply a question of time before chloride ions penetrate into concrete. At cover depths of 10 to 20mm, even the best corrosion protection will begin to deteriorate with time. Laboratory and field test data are then correlated to show that with adequate cover and normal quality concrete, galvanized/chromated bars have excellent corrosion resistance in concrete with high levels of chloride.

7 References

Bernhardt, C.J. and Sopler, B. (1974) An experimental study of the corrosion of steel in reinforced concrete in marine environment, Nordisk Betang, 2.

Clifton, J.R. (1977) Corrosion of galvanized reinforcing bars, National Bureau of Standards, USA.

Cornet, I. and Bresler, B. (1981) Galvanized steel in concrete: Literature Review and Assessment of Performance, Galvanized Reinforcement for Concrete - II, Int. Lead Zinc Res. Org. Inc., 1-56

Duffaut, P., Dahoux, L. and Heuze, B. (1973) Corrosion of steel in reinforced concrete. Tests conducted in the Rance Estuary from 1959 to 1971, Annales de d'Institut Tech. du Bat. et des Tr. Pub., Supplement No.305.

Duval, R. and Arliguie, G. (1974) Memoires Scientifiques Rev. Metallurg, LXXI, No.11.

Griffin, D.F. (1969) Effectiveness of zinc coating on reinforcing steel in concrete exposed to a marine enviornment, Tech. Note N-1032, Naval Civil Eng. Lab., USA.

Ishikawa, T., Cornet, I. and Bresler, B. (1972) Electrochemical study of the corrosion behaviour of galvanized steel in concrete, Proc. of the Fourth Int. Congress on Met. Corr., 556-559.

Maahn, E. and Sorensen, B. (1986) The influence of microstructure on the corrosion properties of hot-dip galvanized reinforcement in concrete, Corrosion, 42, 187-196.

Macias, A. and Andrade, C. (1983) Corrosion rate of galvanized steel immersed in saturated solutions of $Ca(OH)_2$ in the pH range 12-13.8, Br. Corr. J., 18, 82-87.

Powers, T.C., Copeland, L.E. and Mann, H.M. (1959) Capillary continuity or discontinuity in cement pastes, J. Portl. Cem. Ass. Res. and Dev. Labs., 1, 38-48.

Rehm, G. and Laemmke, A. (1974) Corrosion behaviour of galvanized steel in cement mortars and concrete, Deutscher Ausschuss fur Stahlbeton, Heft 242, Berlin, 45-60.

Roetheli, B.E., Cox, G.L. and Littreal, W.B. (1932) Effect of pH on the corrosion products and corrosion rate of zinc in oxygenated aqueous solutions, Metals and Alloys, 3, 73-76.

Sopler, B. (1973) Corrosion of reinforcement in concrete, FCB Report 73-4, The Norwegian Inst. of Tech.

Swamy, R.N. (1990) Resistance to chlorides of galvanized rebars (to be published).

Swamy, R.N., Koyama, S., Arai, T. and Mikami, N. (1988) Durability of steel reinforcement in marine environment, Performance of Concrete in Marine Environment, ACI Publn. SP-109, 147-161.

Stark, D. (1978) Galvanized reinforcement in concrete containing chlorides, Project No. 2E-247, Constn. Tech. Labs., 35 pp.

Stark, D. and Perenchio, W. (1975) The performance of galvanized reinforcement in concrete bridge decks, Project No. 2E-206, Constn. Tech. Labs., 80 pp.

Treadaway, K.W.J., Brown, B.L. and Cox, R.N. (1980) Durability of galvanized steel in concrete, Corrosion of Reinforcing Steel in Concrete, ASTM STP 713, 102-131.

Treadaway, K.W.J., Cox, R.N. and Brown, B.L. (1989) Durability of corrosion resisting steels in concrete, Proc. Instn. Civ. Engrs., 86, 305-331.

Unz, M. (1978) Performance of galvanized reinforcement in calcium hydroxide solution, ACI Journal, 91-99.

Verbeck, G.J. (1975) Mechanisms of corrosion of steel in concrete, Corrosion of Metals in Concrete, ACI SP-49.

33 HYGROSCOPIC SALTS – INFLUENCE ON THE MOISTURE BEHAVIOUR OF STRUCTURAL ELEMENTS

H. GARRECHT, H.K. HILSDORF, J. KROPP
University of Karlsruhe, Institut für Massivbau und Baustofftechnologie, Karlsruhe, Germany

Abstract
High moisture concentrations in building elements occur due to capillary rise of water out of ground water, wet soil or driving rain. In this way dissolved salts may be transported into the structural element. Then these salts influence the moisture behaviour of the building element. Therefore, the hygroscopic effect of salts in sorption isotherms as well as the influence of salts on capillary flow are presented and the consequences on the moisture balance of a salt contaminated structural element are discussed.

1 Introduction

In the repair and conservation of masonry structures the protection of building elements against moisture is of particular importance because most types of degradation of materials such as natural stones, clay bricks, mortars and plasters occur in the presence of moisture.
Aside from rain, in many cases excess moisture is absorbed by the foundations of the building due to capillary rise of water. Then, soluble salts may be transported into the pore system of the building materials by capillary flow. Despite of the fact that normally the adsorption of moisture from the atmosphere results in a low moisture content, a drastic increase of moisture concentration can be observed due to the hygroscopic effects if the material is contaminated with soluble salts. At the evaporation zone these salts may cause damage mechanisms like efflorescence and salt crystallization depending on the temperature and the relative humidity of the ambient air.
To estimate the influence of these salts on the moisture behaviour of structural elements it is necessary to determine the hygroscopic effects on the water repellant properties of building materials.
Then the record of the climate conditions inside and outside the building, the analysis of the salt concentration and the hygric boundary conditions prevailing on the site will indi-

cate, whether the hygroscopic effect of salts, the condensation of water vapour or the capillary rise of water causes the high moisture content of the particular structural element.

2 High moisture content - a consequence of water vapour adsorption

2.1 Sorption Isotherms

The relation between the moisture content of porous materials in equilibrium with the relative humidity of the ambient air is given by sorption isotherms. The up-take or loss of moisture occurs through adsorption of water molecules at the interior surfaces of the pores in an initially dry material or through desorption processes of a water-saturated material. In Fig. 1 the desorption isotherms of some porous building materials are presented. To determine such relations, thin discs with a height of 10 mm and a diameter of 30 mm were conditioned in 12 climates with different relative humidities at 20 °C for more than 3 months. Subsequently, the moisture concentration gained through adsorption or desorption were determined gravimetrically.
The experimental results demonstrate that at relative humidities below 70% the moisture content is low, caused by the small specific surface area of most building materials. Because the percentage of pores with diameters $d < 10^{-8}$m is small, also the moisture content due to capillary condensation is low even at high relative humidities. Consequently the well known hysteresis effect between de- and adsorption is not very pronounced [1].

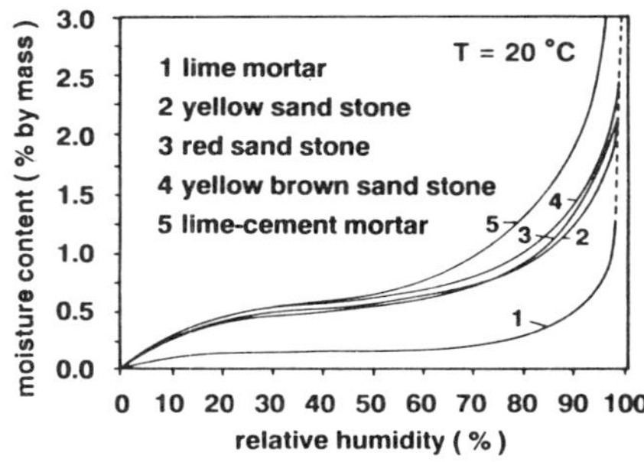

Fig.1. Sorption isotherms for different building materials

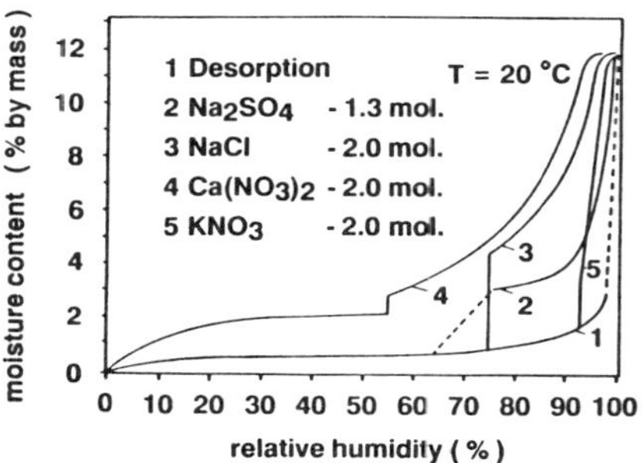

Fig.2. Influence of salts on sorption isotherms

2.2 Influence of Soluble Salts

In some instances salt solutions may penetrate a structure. The sources of such salts may be a natural salt content of surrounding soils, fertilizer, industrial contamination, or salts in some building materials.

In order to illustrate the effect of such salts on the moisture behaviour of building materials, samples were charged with defined concentrations of different salts frequently encountered in buildings.

To ensure a defined salt content the thin discs were initially water saturated. Then they were placed in one of the specific salt solutions until the dissolved salt ions were distributed unifirmly throughout the entire pore system. After two weeks the samples were conditioned in the boxes with the different relative humidities at 20°C until the equilibrium moisture concentration of the material was reached. The desorption isotherms shown in in Fig. 2 and Fig. 3 were determined gravimetrically.

The isotherms demonstrate that for ambient relative humidities higher than a critical value the equilibrium moisture content of the material rises drastically. This critical relative humidity is governed by the partial vapour pressure above the surface of a saturated salt solution.

For high salt concentrations and high relative humidities even saturation of the pore system may be obtained.

To illustrate the hygroscopic effect of salts the results obtained with a yellow sand stone charged with different concentrations of sodium chloride (NaCl) presented in Fig. 3 will be discussed. For comparison, curve No. 1 represents the desorption isotherm of the uncharged material.

Contrary to the salt-free sand stone a contamination with a 0.4 molar NaCl-solution and a 2.0 molar NaCl-solution leads to the hygroscopic effect at 75% relative humidity. At this moisture concentration of the ambient air water vapour will be absorbed from the atmosphere until the saturated solution becomes a diluted solution.

At higher relative humidities more water molecules will be absorbed, and a saturation of the pore system may be obtained for high salt concentrations.

Bellow 75% r.H. the desorption isotherms are identical to those of the uncharged material. In this range the salt only exists in the crystal form [1,2].

In contrast to the desorption isotherms for NaCl and KNO_3 solutions the corresponding curves for the Na_2SO_4 or $Ca(NO_3)_2$ solutions shown in Fig. 2 show a higher moisture content than the reference curve 1 below the critical relative humidity (55%r.H. for $Ca(NO_3)_2$, 75%r.H. for Na_2SO_4). These salts readily hydrate and dehydrate in response to changes in temperature and relative humidity of the atmosphere; adjustment to the new environment takes place as the salt changes to a more stable hydrate. The absorption of water in the crystal structure of such salts leads to an increase of volume of the salt. Thus a high pressure developes, which may cause damage of a porous material.

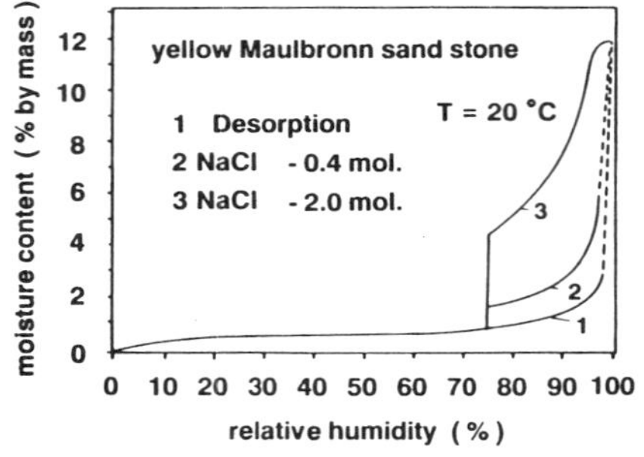

Fig.3. Influence of salts concentration on sorption isotherms

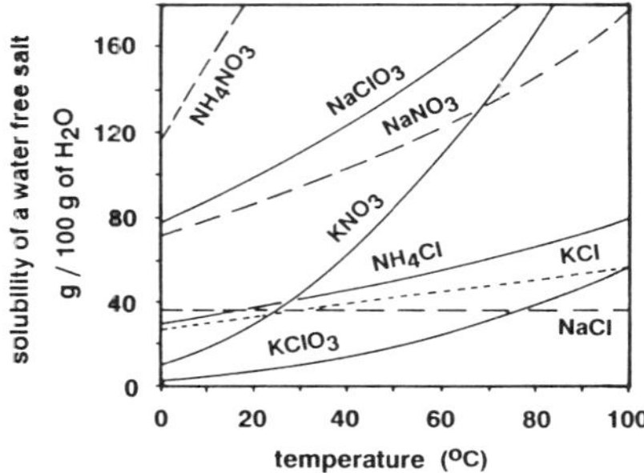

Fig.4. Solubility of salts

2.3 Theoratical estimates of the influence of salts on sorption isotherms

The precipitous rise in moisture content at the critical relative humidity can be calculated if the salt concentration inside the pores of a material and the solubility of the salt are known.

Fig. 4 shows the solubility of some salts as a function of temperature. For NaCl a maximum of 38 g is dissolved in 100 g of water [3].

One cm^3 of yellow sand stone which has a porosity of 25% contains 0.25 cm^3 of a salt solution. For a 2 molar NaCl-solution this volume contains 0.029 g of NaCl. At such a salt content the solubility of NaCl leads to an absorption of moisture of approximately 0.076 g so that the saturated solution becomes diluted. With a density of 1.97 g/cm^3 for the yellow sand stone this quantity is equivalent to a moisture content of 3.85% by mass. This value agrees very well with the experimental results presented in Fig. 3.

At relative humidities higher than this critical value additional water molecules are absorbed in the diluted solution until an equilibrium state is obtained.

Since soluble salts restrict the evaporation of water molecules from the liquid phase, the partial pressure above a salt solution is lower than the partial pressure above pure water.

From Raoults law the relation between partial pressure above a salt solution p_1 and above distilled water p_{10} can be calculated:

$$\frac{p_1}{p_{10}} = \frac{n_1}{n_1 + n_2} \qquad (1)$$

where : n_1 number of mole of water
n_2 number of mole of salt ions

On one hand eq.(1) allows to determine the critical value of the relative humidity at which the salts cause an up-take of moisture. According to Fig. 4 no more than 38 g of NaCl can dissolve in 100 g of water which is equivalent to n_1=5.555 moles of H_2O and n_2=0.65 moles of NaCl.
Dissolved salts only exist as ions. Therefore the number of ions f of a salt molecule has to be considered in calculating n_2. For NaCl n_2 results in n_2=2*0.65 because the ions Na^+ and Cl^- are dissolved in the solution.
For the 2 molar NaCl-solution a ratio of the partial pressures p_1/p_{10} of 0.8 can be calculated from eq.(1). This value is close to the critical value of 75% r.H..
On the other hand eq.(1) allows an estimate of the relative humidity at which the entire pore volume is filled with the diluted solution.
Since 1 cm^3 of the yellow sand stone charged with a 2 molar NaCl-solution contains 0.029g NaCl the number of moles of salt ions can be calculated as follows:

$$n_2 = f * m_2 / u_2 = 2*0.029/58 = 0.001 \text{ mole} \qquad (2)$$

where : f = number of soluble ions of one salt molecule
m_2 = mass of salt inside the pores of the material
u_2 = molecular weight of the salt

If the entire pore volume is filled with the diluted solution 0.25 g of water per cm^3 of the yellow sand stone is absorbed. Then, with n_1 = 0.014 mole of water and n_2 =0.001 mole of salt ions a ratio of the partial pressure above the solution to the partial pressure above pure water of p_1/p_{10} =0.93, can be calculated, i.e. if the yellow sand stone is charged with a 2 molar NaCl-solution it will become saturated at 93% r.H.. This is in aggreement with the experimental observations shown in Figs.2 and 3 [2].

3 Capillary flow of water and salt solutions

The take-up of moisture in building materials may occur either by permeation of liquid water under an external pressure, by capillary suction of water or by diffusion of water vapor.

3.1 Capillary flow of water in porous materials

At higher moisture concentrations transport of water in the liquid phase occurs in porous materials as unsaturated or as saturated capillary flow [2,4,5]. With the help of Buckingham-Darcy's law the flow of water may be expressed as follows:

$$\frac{\partial c}{\partial t} = \nabla K_l(c,\vartheta) (\nabla p_c + \nabla z) \tag{3}$$

where
c = moisture concentration
t = time
ϑ = temperature
z = height of the water front

$K_l(c,\vartheta)$ is the moisture concentration dependent hydraulic conductivity. If a relation between the capillary pressure p_c and the moisture concentration c for porous materials exists it is possible to calculate the water diffusivity D_l from:

$$D_l(c,\vartheta) = K(c,\vartheta) \cdot \frac{d\, p_c}{d\, c} \tag{4}$$

Then, equation (5) can be formulated as the fundamental diffential equation of capillary flow in unsaturated porous media:

$$\frac{\partial c}{\partial t} = \nabla(D_l(c,\vartheta) \nabla c) + \frac{d\, K_l(c,\vartheta)}{d\, c} \cdot \frac{d\, c}{d\, z} \tag{5}$$

The capillary diffusivity $D_l(c)$, the capillary conductivity $K_l(c)$ and the capillary pressure p_c may strongly vary with c, and only few measurements have been made of these parameters for building materials [2,4,5,6].
To evaluate the respective transport coefficient D_L for sand stone drilled cores with a diameter of 25 mm and a length of more then 200 mm were sealed along their circumference. One end of the cylinders was exposed to liquid water, and the opposite end was kept at 0% r.h.. For various time steps the concentration profiles of the absorbed water were measured nondestructively by a Nuclear Magnetic Resonance (NMR) Spectrometer. A personal computer coordinates the position of the magnet unit and triggers the measurements of the NMR in steps of 2 mm beginning at the wet end of the cylinder. From these concentration profiles the capillary diffusivity can be calculated as a function of the moisture concentration [2,6].
Fig.5 gives the results obtained for a red sand stone.

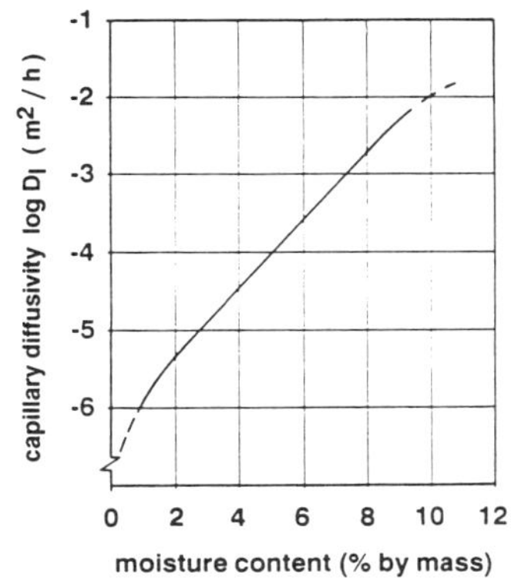

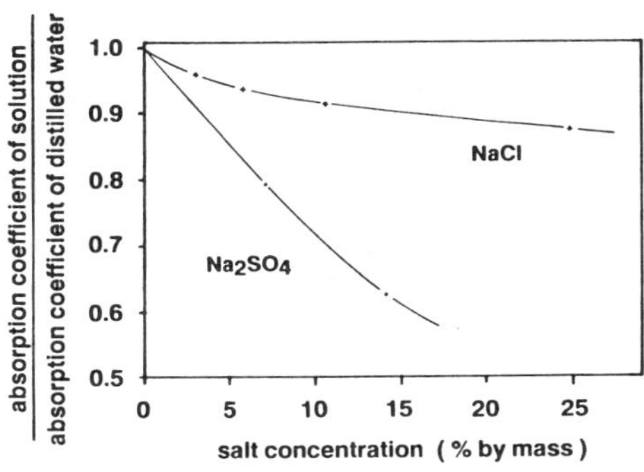

Fig.5. Capillary diffusivity of a red sand stone

Fig.6. Influence of salts on the absorption coefficient

3.2 Influence of soluble salts

To analyze the influence of salts on the capillary flow in porous materials, the capillary suction of salt solutions was determined from drilled cores with a diameter of 30 mm and a length of 100mm. The specimens were sealed along their circumference. Then the bottom of the specimens was brought into contact with the salt solution, and the up-take of the liquid was measured. The absorption m of the liquid at time t can be described approximately by :

$$m = A * t^{n} \qquad (6)$$

where : A = absorption of the solution after 1 h
n = exponent (for most building materials $n \approx 0.5$)

As an example Fig. 6 gives the experimental results for the yellow sand stone exposed to NaCl and Na_2SO_4 solutions. The diagram demonstrates that the ratio of the absorption coefficients between the solution and distilled water decreases as the concentration of salts increases.

This effect is caused by the change in physical properties of the water by the dissolved salts, mainly the surface tension σ, the viscosity η, the density ρ and the contact angle δ [2,3].

In a salt solution normally the surface tension σ increases with increasing salt concentration. In Fig 7 the surface tension of NaCl-solutions is shown as a function of the salt concentration. It demonstrates the wide range of scatter of the results obtained in different investigations.

To approach a state of minimum energy, the salt ions which have a higher potential than the water molecules will move

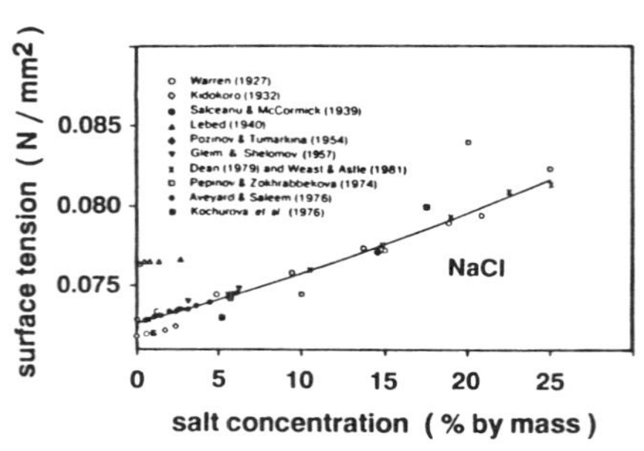

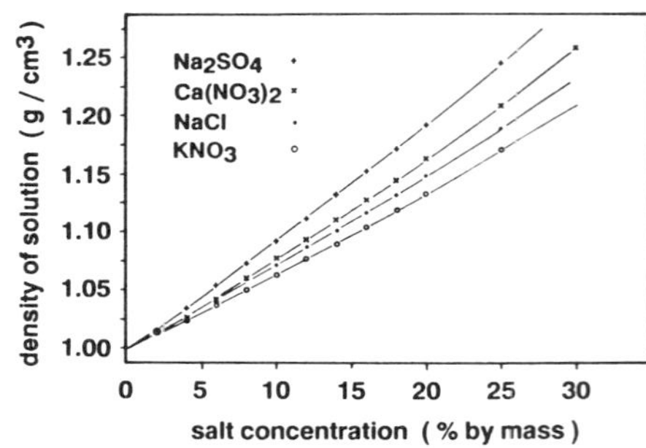

Fig.7. Surface tension of NaCl-solutions

Fig.8. Density of salt solutions

towards the interior of the liquid. In contrast, the osmotic effect causes a transport of ions to the surface to gain an equilibrium distribution of the ions in the solution. The concentration of salt ions at the surface as a consequence of these two counteracting effects determines the magnitude of surface tension.

Figs. 8 and 9 demonstrate that also the viscosity and the density of the solution increase with increasing salt concentration. The contact angle will show the same effect.

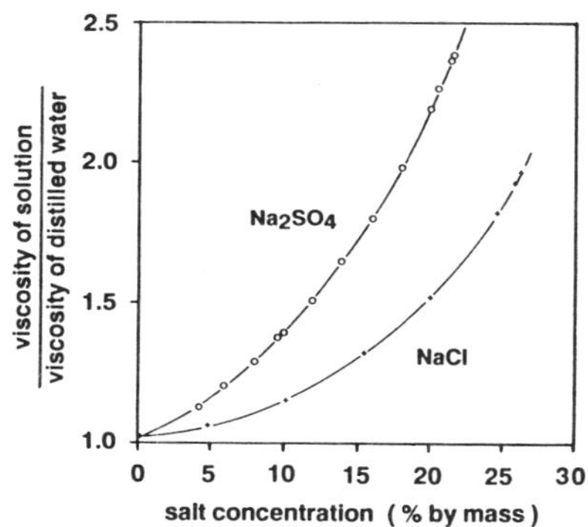

Fig.9 Viscosity of salt solutions

Capillary flow in cylindrical capillaries can be described by the Washburn equation :

$$\frac{dz}{dt} = \frac{r^2}{8 \eta z} \cdot \left(\frac{2 \gamma \cos \vartheta}{r} - \rho g z \right) \qquad (7)$$

where z = height (length) of the column of liquid in the capillary
r = radius of the capillary

From eq.(7) the rate of capillary flow for salt solutions in a single capillary can be calculated to study the influence of the salt concentration. As a result of these calculations Fig. 10 shows that the rate of capillary flow decreases as the salt concentration increases. For these calculations a radius of the capillary of 10^{-6} m and a height of the meniscus of the liquid in the capillary of 20 mm was assumed. Also the maximum of the capillary rise in this capillary may be obtained from eq.(7) as shown in Fig. 11. Similar to Fig. 10 the higher the salt concentration of the liquid the lower is the maximum height of capillary rise.

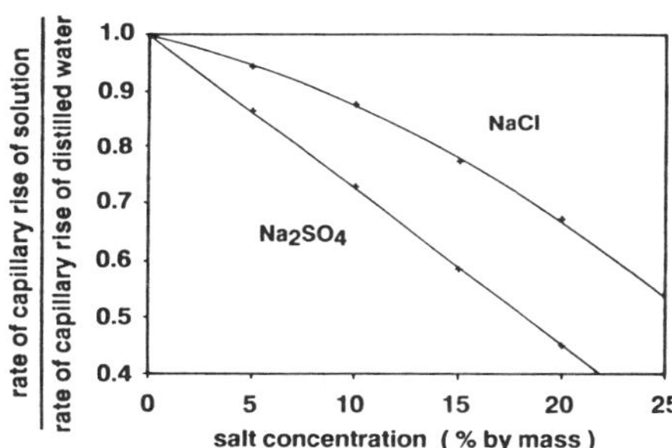

Fig.10. Rate of capillary rise of solutions in cylindrical capillaries

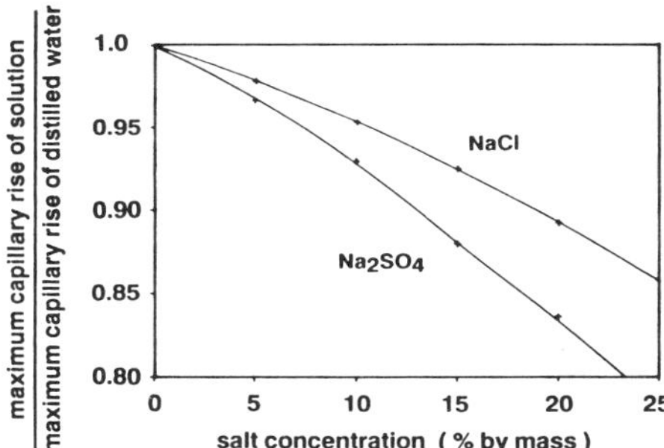

Fig.11. Capillary rise of the meniscus of solutions in cylindrical capillaries

To calculate the influence of salts on transport phenomena in porous materials described by eq.(5), Gummerson et al. [7] defined an intrinsic diffusivity D_l' given by:

$$D_l' = D_l \cdot \frac{\eta}{\sigma} \tag{8}$$

To consider the influence of changes of density and contact angle, equation 8 has to be expanded to:

$$D_l'' = D_l \cdot \frac{\eta \cdot \rho}{\cos\gamma \cdot \sigma} \tag{9}$$

Here, the intrinsic diffusivity D_l'' is independent of the properties of the liquid phase.

4 Moisture balance of a salt contaminated structural element

The wetting and drying behaviour of structural elements in buildings is determined by the moisture transport and storage capacity inside the structure as well as by the boundary conditions prevailing on the site.
Water in the liquid phase will be absorbed by the building element and transported through the pore system by capillary suction. At the evaporation zone the desorption of water is governed by the boundary conditions. The higher the temperature and the lower the relative humidity of the climate in the ambient atmosphere the better the drying conditions of a wetted building element.
The quantity of water m which may evaporate from the surface of a liquid in time t can be calculated from:

$$\frac{dm}{dt} = \beta \, (p_l - p_a) \qquad (10)$$

where β = coefficient describing evaporation resistance
p_a = partial water vapour pressure of the atmosphere
p_l = partial water vapour pressure above the liquid

If the pores of a building material are contaminated with salts, hygroscopic effects as described in section 2 occur. At relative humidities higher than the critical value absorption of water vapour from the atmosphere will take place until the equilibrium state between the vapour pressure of the atmosphere and the vapour pressure above the newly formed diluted solution is reached.
Because the partial pressure above the salt solution p_l is lower than the one above distilled water, a salt contamination of the building material reduces the difference of the partial pressures given in eq.(10), and a smaller quantity of water will evaporate. As a consequence of these less favorable drying conditions the capillary rise of water may be higher so that salts are transported to initially uncontaminated regions.
Therefore, for analyses of the moisture behaviour of buildings charged with soluble salts not only the salt concentration and associated hygric material properties are required but also the climatic conditions inside and outside of the building have to be studied. A record of the climatic conditions for a longer period of time then will show whether relative humidity of the ambient air inside and outside the structure will allow an evaporation of water. Measurements in historic buildings of southern Germany, particulary churches showed that in unheated parts such as the like nave , church choir etc. the relative humidity never falls below 80% r.H during the cold season, i.e. September through

April. The atmosphere outside it may also vary between 75 to 90% RH. Under such climatic conditions no drying occurs primarily because of the influence of salts as described in section 2.
If the building is contaminated with salts, eq.(3) has to be expanded so that the moisture balance of a structural element may be formulated by equation 11 [4,5].

$$\frac{\partial c}{\partial t} = \nabla (D_v \nabla p_v + K_p \nabla p + D_T \nabla \vartheta + K_l \nabla z + D_l \nabla c) \quad (11)$$

In eq.(11) the material parameter D_v characterizes the flow of moisture by diffusion due to a gradient of water vapour pressure ∇p_v. Moisture movement due to the influence of a temperature gradient is determined by the thermal diffusivity D_T. Moisture movement caused by a hydraulic pressure is charaterized by the hydraulic conductivity K_p. The last two terms of eq.(11) represent the moisture transport due to capillary effects in an unsaturated porous material (see chapter 3) [2]. All of these transport parameters are functions of the local moisture concentration c and the ambient temperature and have to be determined experimentally.
The transient field problem expressed by eq.(11) was solved by the finite element method for the two dimensional flow of water. Thus the moisture distribution in masonry was calculated taking into account the boundary conditions prevailing on a particular site. Also case studies of possible protection methods are analysed. In this way different methods may be compared as to cost, effectiveness and minimizing changes in masonry structures [1,4].
In future analyses also the influence of soluble salts will be taken into account. For this the distribution of the salts in the structural element has to be known. Such information can be obtained only from measurements on existing structures. Therefore, the results of laboratory tests to determine the influence of these salts on the hygric material properties as well as records of the microclimate on the site are needed as input data for the solution of eq.(11).
As shown by the sorption isotherms in Figs. 2 and 3 the moisture content increases drastically if a porous building material is contaminated with soluble salts. It is absolutely necessary to consider this effect in eq.(11) in future investigations. Another problem will be to consider the diffusion of dissolved salts in the pore system. This is essential to characterize the hygroscopic effects as well as the rate of absorption of water molecules out of the atmosphere in the diluted solution. Then the moisture behaviour of real building elements may be simulated numerically, a process which is possible so for only under the assumption that the effects of salts are negligible [4].

5 Conclusions

Though high moisture concentrations in building elements occur due to capillary rise of water out of ground water, wet soil or driving rain sections only exposed to the free atmosphere will not exhibit high moisture concentrations over a prolonged period of time. The equilibrium conditions between capillary rise and evaporation limit the level of the moisture front, and the capillary rise does not always proceed to a considerable height in elements exposed to the atmosphere.
From the sorption isotherms presented in Figs.2 and 3 it may be concluded that high moisture concentrations found in elements of some masonry buildings may also be caused by a contamination of the building materials with hygroscopic salts which will absorb large amounts of water, so that high moisture concentrations in the material are in equilibrium with the moisture content of the surrounding air. Although in this case capillary rise per se is not the immediate cause of a high moisture content dissolved salts are transported into the building materials by this mechanism. Then, protective measures for the foundations or sections in contact with soil will not reduce the high moisture content, however, an impervious moisture barrier applied to the foundation will stop further ingress of salts and a further enrichment of salts in the evaporation zone.Presently, we attempt to consider the influence of salts on the moisture behaviour of structural elements. Then the moisture balance may be modelled numerically taking into account the prevailing boundary conditions and the salt contamination.

6 References

1. Hilsdorf,H.K.; Kropp,J.; Garrecht,H.; Ursachen und Wege der Feuchtigkeit in Baukonstruktionen; Jahrbuch SFB 315 - Erhalten historisch bedeutsamer Bauwerke, 1986, Verlag Ernst u. Sohn, Berlin, (1987)
2. Garrecht,H.; Kropp,J.; Hilsdorf,H.K.; Mauerfeuchte als Folge bauschädlicher Salze; Jahrbuch SFB 315 - Erhalten historisch bedeutsamer Bauwerke, 1987, Berlin, (1988)
3. Moore,W.;Hummel,D.; Physikalische Chemie. Berlin (1973)
4. Garrecht,H.; Kropp,J.; Hilsdorf,H.K.; Computer Modelling of Moisture Protection Measures in Historic Buildings; First Int. Conf. on Structural Studies, STREMA 1989 Birkhäuser Verlag Basel Boston Berlin (1989)
5. De Vries,D.A.; Simultanious Transfer of Heat and Moisture in Porous Media; American Geophysical Union 39(1958)
6. Krischer,O; Kroll,K; Wissenschaftliche Grundlagen der Trocknungstechnik; Springer Verlag, Berlin (1963)
7. Gummerson,R.J.;Hall,C;Hoff,W.D.; Water Movement in Porous Building Materials-II. Hydraulic Suction and Sorptivity of Brick and other masonry Materials; Building and Environment (1980) Vol.17, S. 101 -108

34 LONG-TERM BEHAVIOR OF PRESTRESSED GIRDER SLABS IN CATTLE STABLES

M. HERGENROEDER
Technische Universitaet Muenchen, Lehrstuhl fuer Massivbau, Munich, Germany

Abstract
Considerations on the assessment of the structural state of about 5000 prefabricated prestressed girder slabs in cattle stables are described. The collapse of two such slabs in the early eighties led to extensive research activities. Material investigations showed that the damages were caused by many brittle fractures of the quenched prestressing wires due to stress corrosion cracking. The stress corrosion was induced by superficial corrosion at areas where the concrete cover was too small or not sufficiently consolidated.

A statistical evaluation of the achieved results gave evidence that other collapses were to be expected. Therefore immediate supporting measures were suggested for about 5000 slabs produced by three different manufacturers in the period from 1954–1963. The Bavarian authorities followed this recommendation.

As damaged slabs can not be identified by simple test methods research has been done to establish non–destructive test methods. Two of them namely magnetic field measurements to detect wire fractures and infra–red thermography to detect defects in the concrete cover are presented in this paper.
Keywords: Assessment of Existing Structures, Case Study, Prestressing, Stress Corrosion Cracking

1 Introduction

In recent years the assessment of the long–term behavior of existing structures has become increasingly important as damages due to time–dependent deterioration processes have occurred in large numbers. Although structural reliability considerations based on modern probabilistic methods have been applied in structural engineering in the last decades, general approaches to assess the reliability of existing structures are not yet available (see e.g. CEB, 1989, DABI, 1988). Furthermore the knowledge about the deteriorating processes is not yet sufficiently developed.

In this paper a case study is given which describes a specific assessment problem on prestressed girder slabs in cattle stables. Due to the aggressive deterioration mechanism it was possible to separate the time dependency of the damage process from the assessment decisions. Material investigations to define the deterioration process, statistical considerations to estimate the structural state of the numerous slabs in use and research to establish non–destructive test methods are presented.

2 Construction

The construction consists of prefabricated beams (I or ⊥ –cross section, pretensioning) hollow blocks and a reinforced cast in–situ layer. Figure 1 shows an example for a typical slab cross section. The girders were prestressed by quenched wires ($A_z = 20 / 30\ mm^2$, St 145/160 kg/mm^2). Following DIN 1045 (1953) the minimum concrete cover of the prestressing wires was 14–15 mm, which is 10 mm plus the stirrup diameter. The girders have usually been produced using parallel installed sliding forms. At the bottom side of the beams mostly a heat insulation layer (brickwork, bituminum saturated felt, polystyrene) was inserted in the prestressing mold before casting the concrete. The investigations have included the product from three manufacturers covering a total amount of about 5000 slabs in the period from 1954 to 1963.

3 The Assessment process

3.1 Assessment of damage

In 1980 and 1984 prestressed girder slabs collapsed in two cattle stables due to many brittle wire fractures of the prestressing steel. According to the owners no warnings such as deflections or cracks were visually observed before. In the case of the first collapse only some parts of broken beams could be investigated. Specific results such as an unacceptable high prestressing and longitudinal microcracks in the wires indicated an unusually high sensitivity to corrosion of the prestressing steel. However, the examination of the second collapse and destructive investigations on a random sample of such slabs showed the same type of wire fracture without these observations.

In addition the important fact was established that fractures always occur near to superficially corroded areas. Examining fractured cross sections by a scanning electron microscope, characteristics of a hydrogen induced stress corrosion mechanism were found. The hydrogen originates from the cathodic reaction of the superficial corrosion reaction and diffuses into the steel. In principle the mechanism was demonstrated in carbonated $Ca(OH)_2$ solutions to occur at pH–values of 9 to 10, Stoll (1982). Two main reasons for this superficial corrosion were found as

- incorrectly located heat insulation on the bottom side of the beams, which reduced the concrete cover and
- concrete cover, which was partially not sufficiently consolidated.

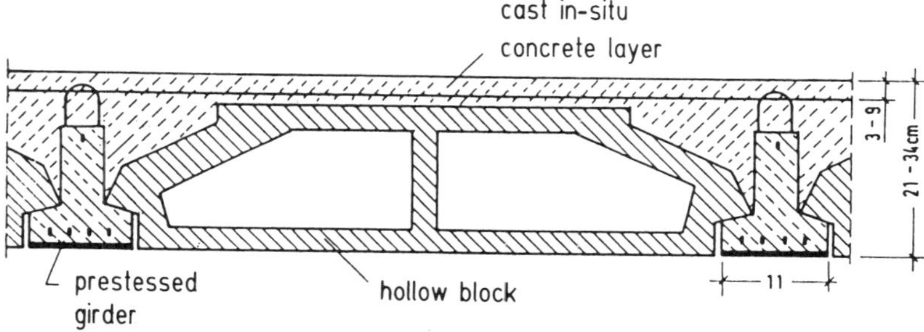

Fig. 1: Typical slab cross section

It was shown that voids and cavities could not be reliably detected by a visual inspection of the uncovered bottom side of a beam only. Furthermore a very strict correlation between such defects in the concrete cover, superficial corrosion and stress corrosion cracking was found due to the permanent humid exposure conditions in the stables. Therefore small defective areas can cause major damages. In dense concrete the mean carbonation depth has been found to be 0–2 mm only. Once fractures have occurred in such randomly distributed areas, local overloading of other wires next to the broken ones and of neighbouring beams can lead to concrete cracks and consequently new corrosion spots. Thus a continuing damage process can induce the collapse of the slab.

Such a collapse will normally happen without any major deflections or wide cracks due to the prestressing of the beams and the good bond performance of the ribbed wires which can induce splice effects between wire fractures.

Considering all these facts it was concluded that due to defects (too small or unsufficiently consolidated concrete cover) a considerable risk of a sudden unindicated collapse is given for these specific constructions after a service life of about 30 years. As a consequence of the aggressiveness of the deterioration mechanism it was sufficient and necessary to neglect the influence of time on general considerations about the actual structural state of the slabs.

3.2 Statistical evaluations

The potential danger of a sudden collapse for girder slabs with partially defective areas in the concrete cover raised the question of how many of all existing slabs show such damage. Applying Bayes decision theory the amount of damaged slabs was estimated on the basis of the results from only a few investigated objects. For the statistical evaluation it was assumed that the beams from one slab can be considered as one unit. During the investigations it was found that in the case of major damage always a bigger part of the beams had been affected. Thus a good–bad decision for each slab was assumed to be reasonable. The examined slabs were considered as randomly chosen because no visual damages had been observed before the destructive investigation.

The amount of damaged slabs was estimated using equation 1, Benjamin and Cornell (1970). The problem was formulated so as to determine the likelihoods of future events solely on the basis of a–posteriori data, without assumptions about the form of the underlying distribution. Equation 1 represents the density function of the "Bayesian" distribution

$$p(y) = \frac{n+1}{s+y+1} \binom{m}{y} \cdot \binom{n}{s} \cdot \left[\binom{m+n+1}{s+y+1}\right]^{-1} \tag{1}$$

where y is the number of damaged elements included in the amount of potential damaged elements, which here is the amount of existing but not yet examined slabs, s the amount of damaged elements in the random sample and

$$m = N-d-n \tag{2}$$

here N is the amount of elements in the population, d the amount of collapsed slabs, n the amount of examined slabs.

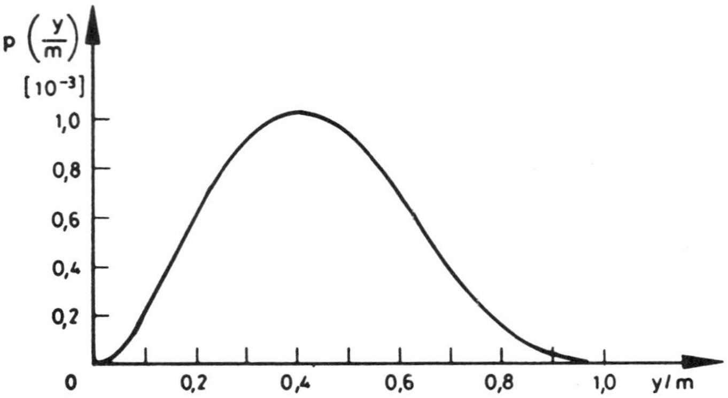

Fig. 2: Bayesian density function for the estimation of the relative frequency of damaged slabs

Figure 2 gives an example of a density function. Although only five slabs from this manufacturer were investigated, finding damage in two of them, an estimation could be made with high probability ($P = 0.90$) that more than 20 % of all slabs ($N \simeq 2000$) are damaged. The expected value was about 40%. Later test results of two manufacturers were summarized ($n = 14$). Thus the expected value for the relative frequency of damaged slabs was still about 40% and with a very high probability ($P = 0.98$) more than 20% of all slabs were estimated to be damaged.

3.3 Conclusions

On the basis of the illustrations above the following conclusions were drawn:

a) The risk of a sudden failure of slabs in the future was high. It was considered to exceed commonly accepted failure probabilities of structures by orders of magnitude.
b) As visible damage normally will not occur before the collapse it was suggested to support all slabs of the three manufactures immediately.
c) Each slab has to be entirely investigated by non–destructive methods.
d) If no damage is found the supports can be removed.

The Bavarian authorities followed these recommendations.

4 Non–destructive test methods

4.1 Demands for test methods

The specific type of the damage processes (see 3.1) made the following separation of two demands reasonable:

- Assessment of the actual structural state of a slab by investigation of the actual risk of failure due to corrosion damage
  → detection of wire fractures
- To prove, that there is no structural risk due to wire fractures which will occur in the future
  → detection of defects in the concrete cover of the beams and of areas with too small concrete cover

Test methods should be both reliable and inexpensive relative to the measured area. Therefore the following methods were found to be suitable:

Detection of wire fractures	:	Magnetic field measurements
Detection of defects in the concrete cover	:	Infra–red thermography

For the detection of areas with too small concrete cover instruments are available. The ultrasonic–impulse–echo–method was tested to detect defects, Hergenroeder and Kupfer (1989). It was found that it could only be successfully applied in such rare cases where the bottom side of the beams is very plain.

4.2 Magnetic field measurements

4.2.1 The method

The method will be described very briefly. If a prestressing steel is exposed to a magnetic field it will react as a magnet itself due to its remanence. At the ends of the wires but also at wire fractures poles originate. At wire fractures north and south pole are located next to each other forming a magnetic dipole. These dipoles can be found by a sensing element. Applying the developed device it is also possible to detect such dipoles by a flux leakage examination simultaneously to the magnetization. Differential magnet resistant sensors were used as sensing elements which proved to be a robust, reliable and economic solution.

Magnetization by an electromagnet and measurements are carried out by a slide on a rack which is installed up to 80 millimeters beneath and parallel to the wires. Wire fractures can easily be identified and reliably be distinguished from stirrups which induce a considerably weaker signal. Figure 3 shows an example of measured data. More details are included in Gerling and Hergenroeder (1990).

In principle the same method was simultaneously developed and is applied now by the Hochtief AG, Frankfurt, and the FMPA Stuttgart. However, different sensing elements (Hall sensors) are used.

4.2.2 Test results and first experiences

During laboratory tests on 20 beams out of slabs in cattle stables the instruments were calibrated and improved. All beams were destructively tested to check the measurements. The reliability of the system thus was increased to a high level. The fast and variable device has been applied in several cattle stables. Slabs up to 120 m^2 can be investigated in one day.

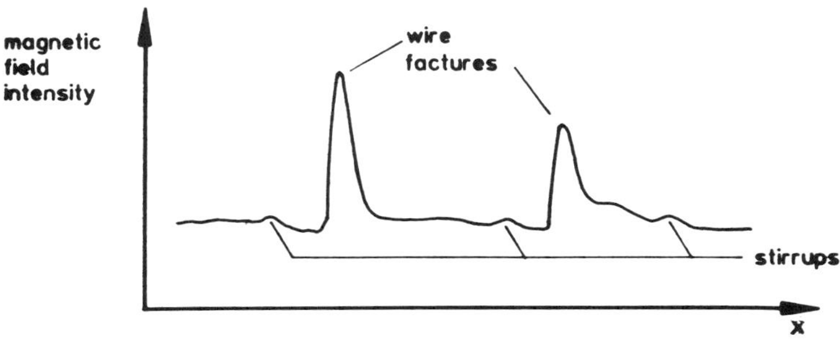

Fig. 3: Measurement of the magnetic field intensity indicates wire fractures, data from one of up to five parallel measuring tracks

4.3 Infra–red thermography

4.3.1 The method

The detection of defects (voids, cavities) in the concrete cover behind a thin surface layer is based on their influence on heat transportation phenomena. The air volume in the defects provides a heat insulating effect when the bottom side of the beam is exposed to infrared radiation and heat dissipation into the beam takes place. The heat accumulation in the covering layer induces temperatures on the beam surface, which are locally relatively higher than those of the surrounding dense concrete. Such spots can be measured on the surface using an infra–red thermography device. A system by AGEMA was applied with a measuring range down to temperature differences of $\Delta T = 0{,}07\ ^{o}C$ at $30^{o}C$. Figure 4 shows an example for such temperature differences depending on the time after the end of the infrared radiation. The end of the measuring time was characterized when spots over minor defects disappeared and outlines of the spots became indistinct.

Before applying the method plaster and heat insulation have to be carefully removed under the beams. Furthermore it must be considered that open voids will cause no heat insulating effect and local lobes at the bottom side of the beams will get relatively warmer during radiation.

4.3.2 Test results and first experiences

During a research program the parameters of the radiation and the period which is usable for the measurement were established. Performing tests on beams from cattle stables it was found that the best results can be achieved by applying a high related wattage of the heat lamp (2.7 W/cm^2) for one minute.

After radiation the allowable inspection period lasts from ten seconds to about four minutes. The reliability of the method proved to be high. Having investigated 28 beams from four different slabs all defects bigger than 6 mm in diameter were detected. Assuming sufficient concrete cover only such defects have been considered to provide a significant risk of corrosion for the prestressing wires.

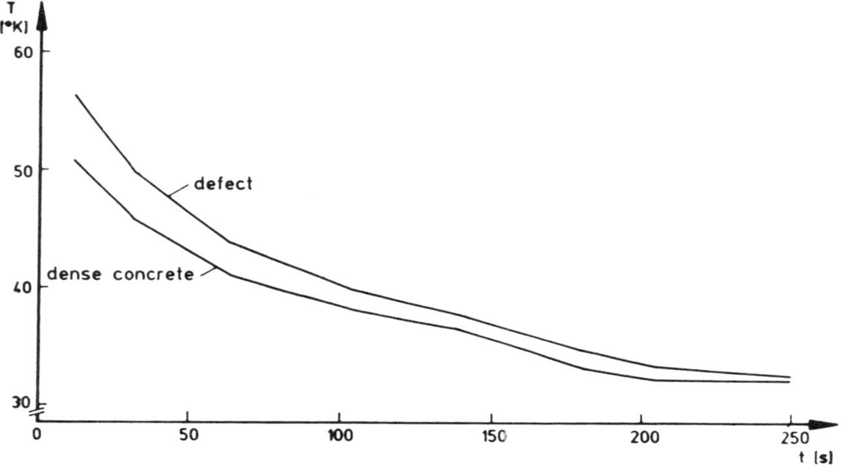

Fig. 4: Infra–red thermography, time–depending surface temperatures T(t) at the bottom side of a girder, comparison of an area with a defect and an area of dense concrete nearby

Having taken proper account of the preparations listed (see. 4.3.1) the amount of registered spots which did not indicate voids was in fact very small. It was found to be considerably below 1 % related to the examined total length of the beams. Here zones of structural decay are included which have not yet caused corrosion of the wires. Slabs up to 120 m^2 can be investigated in one day.

5 Actual State (Feb. 90)

The investigations on the potentially damaged slabs started in autumn 1989. Until February this year a total of 23 slabs have been examined by magnetic field measurements (Hochtief AG, TU Muenchen). It was found that ten slabs had wire fractures. Thus the estimates have been confirmed (see 3.2). The assessment criteria were recommended by a committee of the Institut für Bautechnik Berlin, an Institution of the German States. This committee stated in preliminary findings that concrete quality and cover should be investigated at a few randomly chosen points. Further research on representative slabs is considered to justify this assumption which was substantiated by assuming that a significant number of wire fractures will not occur after a service life of about 30 years. In principle it was determined that slabs with dense and adequate concrete cover are durable. Investigations by infra–red thermography have been accepted as a possibility to obtain much more information about each slab. However, this method will be used only in those cases where such additional information is regarded to be attended with major economic advantages.

6 References

Benjamin, J.R. and Cornell, C.A. (1980) **Probability, Statistics and Decision for Civil Engineers**. McGraw–Hill Book Company

CEB Bulletin N°192 (1989) **Diagnosis and Assessment of Concrete Structures**

DABI Danish Concrete Institut, Int. Symposium (1988) **Re–evaluation of Concrete Structures**

Gerling, H. and Hergenroeder, M. (1990) Magnetfeldmessungen und Infrarotthermothermographie zur Untersuchung der Standsicherheit von Spannbetondecken. in **Qualitaet und Zuverlaessigkeit durch Materialprüfung im Bauwesen und Maschinenbau**, VMPA–Tagung 1990, Muenchen

Hergenroeder, M. and Kupfer, H. (1989) Die Untersuchung von Spannbeton–deckentraegern mit der Ultraschall–Impuls–Echo–Methode. **Baustoff-Forschung, Anwendung, Bewaehrung** (Festschrift Prof. Springenschmid), edited by TU Muenchen Baustoffinstitut

Stoll, F. **Elektrochemische Untersuchungen zur Passivität und Spannungsriß–korrosion von Spannstählen in alkalischen und karbonatisierten Calcium–hydroxidlösungen**. Dissertation Universität Erlangen, 1982

35 THE EFFECT OF ADMIXTURES ON THE WATER TIGHTNESS OF CONCRETE

T. AL-KADHIMI, I. AL-KURWI, H.S. JAMIL,
J.A. AL-MULLA
Building Research Centre, Baghdad, Iraq

Abstract
Concrete cubes of low-medium workability (3-10 cm slump) incorporating various proportions of industrial waste were prepared, cured, and stored in a temperature of 25 ±2°C for a period of 28 days. Durability evaluations in terms of permeability, compressive strength and initial surface absorption were conducted. Four types of waste products were used as admixtures for concrete to study their effect on the water tightness of the concrete. Two of them show a clear improvement on permeability with no effect or very little reduction in compressive strength, while the other two manifest adverse effect on the permeability but act as water repellants.

1 Introduction

Permeability is that property of a material which permits the passage of liquid through its internal structure. The water tightness of concrete may be of greater significance than compressive strength, particularly in a water retaining structure. The movement of water through porous materials has many important consequences in building construction. In inhabited buildings the migration of water through porous materials influences the performance of the individual members and their fabric and underlies such practical problems as water penetration and rising damp. The capillary absorption of a liquid by a porous solid depends partly on the properties of the liquid and partly on the micro structure of the solid.

The controlling properties of the liquid rely on its surface tension and viscosity , Gummerson (1980). It was found that changes in permeability depend on changes in w/c ratio, hydration course, porosity and surface area of the permeable media, Nyame (1987).

Other factors were mix proportion, fineness of aggregates and bonding materials (cement, pfa, ground blast furnace slag ... etc.), curing condition and sulphate content of the mix, Al-Kadhimi (1987). Nevertheless good materials and workmanship are the first consideration in any structure and are essential for watertight concrete. The use of admixtures such as water proofers or damp proofers, either integral or superficial, should not be considered as being compensation for poor workmanship, lean mixes, or deficient materials, Waddell (1962). However, two extreme examples of admixture requirements can clearly be seen, first; low cement content of mass concrete may well be porous and yet only show minor leakage due to low pressure gradient, second; thin sections, such as renders, may be dense but still leak due to high pressure gradient. Several methods of testing were devised and carried out to measure the water tightness and permeability of concrete and a number of related specifications were established, Hope (1984), Vourine (1987) and Levitt (1971).

This paper presents results showing the effect of certain waste materials used as admixtures for concrete, in terms of permeability, initial surface absorption, and compressive strength.

2 Experimental

Ordinary portland cement satisfying Iraqi Standard No.5 was used. The chemical composition and physical properties are shown in Table 1. The fine aggregate was natural sand obtained from Karballa area, and the coarse aggregate was graded natural gravel with a maximum grain size of 20 mm, collected from Nabai area. Aggregate properties are given in Table 2. Four types of admixtures were used through the whole programme and listed below.

2.1 Basic Sodium Silicate (Na_2SiO_3)

This material is produced by melting sodium silicate rock in a chamber at 160°C and 5 bar which is currently used by the Vegetable Oil State Company, for producing detergents. Its pH value is 11 and concentration 38 - 40%.

2.2 Soap Powder

This is a by-product produced during the spray drying of soaps.

Soap produced by hydrolysis of fats (saponification) originates from a mixture of palm oil, stearic oil, caprylic oil and oleic oil. It is a mixture of sodium salts of long-chain of fatty acids.

$$
\begin{array}{lllll}
CH_2 - O - C - R & & CH2OH & & RCOO^- \;\; Na^+ \\
\qquad\qquad\;\; O & & & & \\
CH \;\; - O - C - R' & NaOH & CHOH & + & RCOO^- \;\; Na^+ \\
\qquad\qquad\;\; O & & & & \\
CH_2 - O - C - R'' & & CH_2OH & & RCOO^- \;\; Na^+ \\
\qquad\qquad\;\; O & & & & \\
\text{A glyceride} & & & & \text{Soap} \\
\text{(A Fat)} & & & &
\end{array}
$$

2.3 Soap Stock

This is a liquid produced from chemical treatment of fat with caustic soda to remove fatty acid.

$$
RCOOH + NAOH \qquad \text{Soap stock} + \underset{\text{Soap}}{RCOONa} + H_2O
$$

This liquid contains 10% fat, 10% soap, and 80% water with pH value of 8 - 9.

2.4 Oil Grade 60
An extract of reduced crude oil, it contains 80% aromatic compounds, 10% paraffinic compounds and 80% sulphonate and nitrogen compounds.

One concrete mix (cement:sand:gravel, 1:2:4) of low to medium workability (3-10 cm slump) and various proportions of admixtures (by weight of cement) were used to prepare concrete cubes. The specimens were cured and stored in water of 25 ±2°C temperature for a period of 28 days.

Concrete cubes of 100x100x100 mm were used for compressive strength, initial surface absorption (according to BS 1881:Part 4: 1970), and 200x200x200 mm size concrete cubes were used for permeability (according to DIN 1048-1971). All tests were conducted 28 days after casting.

Table 1. Cement characteristics

Chemical analysis		Hypothetical compounds		Physical properties		
Oxide	%	%				
CaO	61.40	C_3S	35.25	Fineness (Blaine)(cm2/g)		3286
SiO_2	21.90					
Al_2O_3	5.53	C_2S	36.27	Initial setting time (min)		95
Fe_2O_3	2.54					
MgO	4.40	C_3A	10.36	Final setting time (min)		145
SO_3	2.60					
L.O.I	0.57	C_4AF	7.72	Compressive strength (N/mm2)		
Insoluble					3 days	13.7
residue	1.06				7 days	21.5

Table 2. Aggregate properties

Sieve size (mm)	Gradation (% passing)			
	Gravel Typical Sample	BS 882-73	Sand Typical Sample	BS 882-73 (Zone 2)
38.10		100		
19.00	98	95 - 100		
12.70	70			
9.52	45	25 - 55		
4.76	5	0 - 10		
2.36			100	77 - 100
1.18			85	55 - 90
0.60			55	35 - 59
0.30			20	8 - 30
0.15			5	0 - 10
0.07			0.0	

3 Results and discussion

Figure 1 shows that soap stock and soap powder have adverse effect on permeability, which reflects the action of foaming agents under pressure as those additives may emulsify or disperse through the cement slurry, forming open or closed fragile voids of various sizes. These voids soon will open up with hydraulic pressure, thereby changing the skeleton of concrete into a more permeable mass. Contrarily, sodium silicate and oil grade 60 show an improvement in permeability resistance at dosage of 0.5% by weight of cement. Beyond that, sodium silicate manifests no further improvement while oil grade 60 indicates more stability and reduces the permeability of concrete by actively reacting with either or both of calcium hydroxide and lime, thereby releasing more alkali and depositing excess calcium silicate hydrate in the pores of concrete. Oil grade 60 inhibits the movement of water through the concrete by coating cement particles with a hydrophobic layer and acting internally as a water repellant.

In Figure 2 the various proportions (0.5-1.5% by weight of cement) of soap stock and soap powder incorporated in concrete cause an average reduction in 28 days compressive strength of 30-35% due to the occurrence of excess voids created by the foaming action of the additives. Sodium silicate however, despite acting as an active pore filler causes an average drop in 28 days compressive strength of 18%. This may be due to the release of alkalis from the reaction of sodium silicate with some cement compounds which increase the rate of hydration of silicate phases (C_2S and C3S) in cement, and consequently cause an increase in early strength and a decrease in late strength (28 days), Lea (1970). The presence of oil grade 60 in concrete seems to have very little or no adverse effect on the 28 days compressive strength of concrete. This indicates that the adsorbed oil by the cement and the aggregate surface particles has not influenced the cement hydration or its bonding with aggregate.

Figure 3 shows the results of initial surface absorption test (ISAT) on the admixture concrete specimens under investigation. The presented data indicate higher fluctuations at the early time of testing (10 min.), as this represents the rate of suction at the upper layer of the specimen surface. Apart from the common effect of the contact angle and number, orientation and tortuosity of the continuous capillaries, this is governed by the surface conditions of the concrete; such as surface irregularity and the variations of moisture content. However, all concrete samples incorporating the admixtures under consideration manifested an improvement in ISAT values and they were within the suggested tentative ISAT limits, Levitt (1971), for weathered concrete, except for specimens containing various percentages of sodium silicate and low percentage (0.5%) of stock soap. The adverse effect of sodium silicate on ISAT may be due to the interference or alteration of the capillary structure as a result of its chemical activity with the silicate phases of cement. Addition of oil (grade 60) as integral admixture to the concrete shows a considerable improvement on ISAT at 0.5% (by weight of cement) dosage, though higher percentages of the additive seem to reduce the improvement due to the reverse changes of the contact

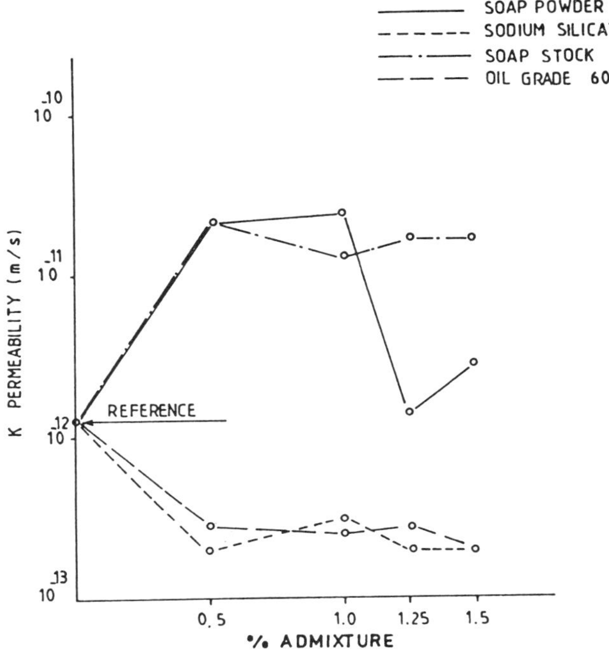

FIG. 1. THE EFFECT OF DIFFERENT PROPORTIONS OF ADMIXTURES ON THE PERMEABILITY OF CONCRETE AT 28 DAYS.

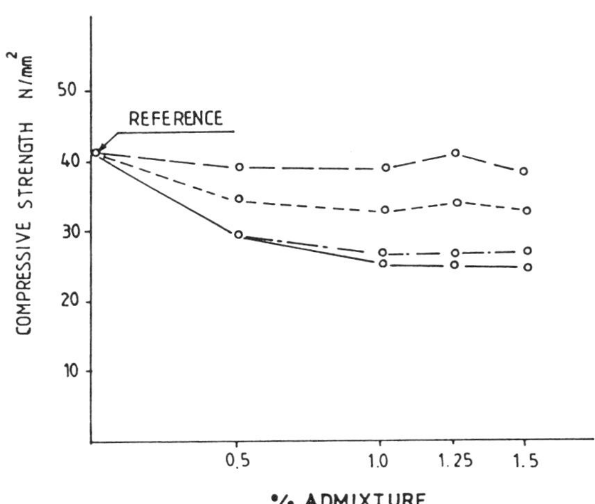

FIG. 2. THE EFFECT OF DIFFERENT PROPORTIONS OF ADMIXTURES ON COMPRESSIVE STRENGTH OF CONCRETE AT 28 DAYS.

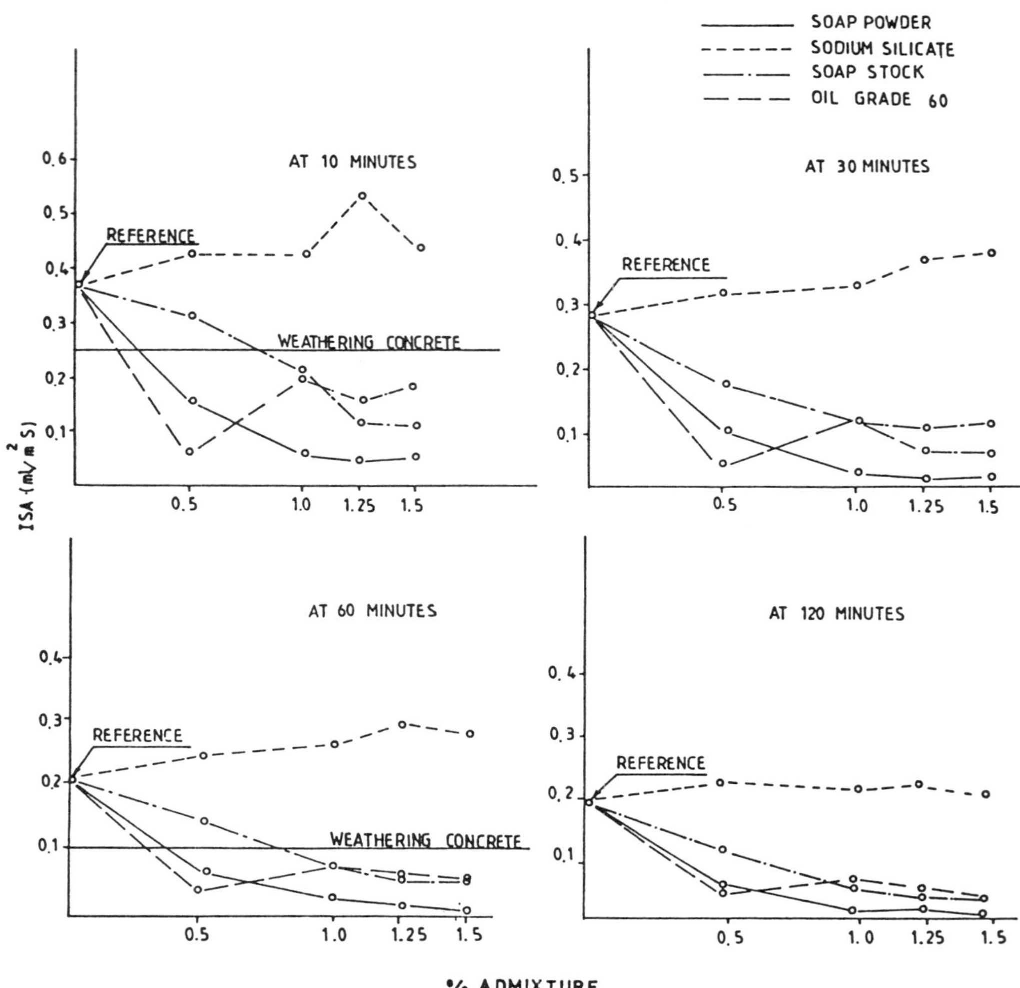

FIG. 3. THE EFFECT OF DIFFERENT PROPORTIONS OF ADMIXTURES ON THE INITIAL SURFACE ABSORPTION OF CONCRETE AT 28 DAYS.

angle at the water/concrete interface by the excess coverage of the large oil molecules on the inner wall of the capillaries. Nevertheless, oil bearing concrete, generally manifested water repellent characteristics below the suggested required limits, Levitt (1971) for weathering concrete. For soap stock and soap powder, although they expected to produce air-bubbles which may interfere with the capillary structure when incorporated in concrete in percentages higher than 1% (by weight of cement), appear to improve the ISAT considerably and produce a concrete of higher water repellent characteristics.

4 Conclusions

The quality of concrete that ensures durability can be assessed by its water tightness. Four industrial by-products were incorporated as additives to improve the water tightness of concrete. For the materials used in this study, the following conclusions have been drawn:

1 Petroleum extract mainly composed of aromatic compounds (oil grade 60) appear to improve the water tightness of concrete without any mechanical repercussion.
2 Alkaline sodium silicate from the vegetable oil industry can improve the resistance of concrete to water permeability, though a precautionary measure should be taken regarding the required strength and the damp proofing of concrete.
3 Soap stock and powder resulting from the hydrolysis of fats in vegetable oils decrease the rate of water penetration of dry concrete, when they are added in percentages higher than 1% (by weight of cement). The addition of both materials can be detrimental to concrete strength and may reduce its resistance to water permeability.

5 References

Al-Kaadhimi, T. Jamil, H.S. and Philip, S. (1987) Permeability testing for concrete quality - effect of curing conditions, mix proportions, and sulphate content (1987) **Symposium quality and performance of cement and concrete produced in the Arab world proceedings**, Baghdad, Iraq, 120-136, Nov.
Gummerson, R.J. Hall, C. and Hoff, W.D. (1980) Water movement in porous building materials - II - hydraulic suction and sorptivity of brick and other masonry materials. **Bldg. Envir.**, 15, Vol. No.2, 101-108.
Hope, B.B. and Malhotra, V.M. (1984) The measurement of concrete permeability, **Can. J. Civ. Eng.**, Vol. 11, 287-292.
Lea, F.M. (1970) The chemistry of cement and concrete, 3rd edition, pp 303, 547.
Levitt, M. (1971), The ISAT, A non destructive test for the durability of concrete, **British Journal of Non Destructive Testing**, 106-112, July.

Nyame, B.K. and Illston, J.M. (1981) Relationships between permeability and pore structure of hardened cement paste. **Magazine of Concrete Research**, Vol. 33, No. 116, September, 139-146.
Vuorinen, J. (1985) Application of diffusion theory to permeability tests on concrete, Part II: Pressure-saturation test on concrete and coefficient of permeability, **Magazine of Concrete Research**, Vol. 37, No. 132, 153-161, Sept.
Waddell, J.J. (1962) Water tightness, in **Practical quality control for concrete**, pp 80-86.

36 STUDY ON THE INFLUENCE OF SOME PARAMETERS OF MIX DESIGN ON THE PERMEABILITY OF HARDENED CONCRETE

J.M. GÁLLIGO, F. RODRÍGUEZ
Laboratorio Central de Estructura y Materiales, MOPU, Madrid, Spain

Abstract

A study on the different available techniques to test the concrete permeability has been carried out in the Laboratorio Central de Estructuras y Materiales in Madrid (Spain). This research programme has included several test methods to apply to specimens in laboratory, and also to obtain in-situ measurements. One of them has been the water penetration method. Some operational problems can appear when the test is carried out according to ISO 7031: 1.983. Some modifications were developed to solve them, in order to develop the Spanish Standard UNE 83.309, which deals with this method.

Besides, the research work has included a test programme to assess the influence of some parameters of the dosage in the permeability of concrete. The water/cement ratio, the maximum size of aggregate and the curing duration were taken into account.

Five test methods were used in the experimental work: water penetration method, initial surface absorption test, the Figg methods to evaluate the air and water permeability of concrete and the Hansen, Ottosen and Petersen's method for permeability of concrete to gases under low pressures. About one hundred and fifty specimens with ten differents dosages were tested. The first experimental results obtained with the two first methods mentioned above, are presented in this work.
Keywords: Permeability, Absorption, Testing Methods, Dosage Parameters.

1 Introduction

The recent statistical studies on the expenditures for the maintenance, repairs and strengthening of road bridges, show the importance , from an economic viewpoint of the problems related to the durability of concrete. The studies that have been carried out by the OECD are very significant in this point. The expenditures origined in this concept during 1.985-1.990, have increased more than 65 per cent. [1]

Regarding above, a great interest has appeared about all the problems related to the concrete durability. Thus it is necessary to reach an adequate Knowledgement to obtain a good quality of concrete during its design and making. This is the only way to get a concrete with a good performance concerning durability.

The aggresive agents of the environment where structurals element are situated introduce into concrete through its pore structure. The mechanisms of transport are numerous. Phenomena such as diffusion, absorption and , sometimes, permeability join their effects to cause the penetration of the aggresive elements into the concrete. Then, it is neccesary to carry out a study on the influence of some of the parameters of the mix design in the performance of concrete related to the mechanisms of transport of fluids into it.

2 Experimental programme

Regarding the above, an experimental study was carried out in the Laboratorio Central de Estructuras y Materiales. Several methods were included to determine the permeability, not only in strict sense as the result of a pressure gradient, but also in a broad sense those phenomena that are originated by another types or gradients, such as the absorption.

2.1 Objectives

The objectives of the research programme were to study the influence of the following parameters:

- Influence of the water/cement ratio
- Influence of the curing
- Influence of the maximum size of aggregate
- Relation between the results obtained for each method and the strength resistance of concrete.

2.2 Test methods

The test methods were the following:

- Method for the determination of depth of water penetration under pressure
- Initial surface absorption test (I.S.A.T.)

2.3 Mix designs

In Table 1 the parameters more significants in the dosages used are shown. A total of 150 specimens ϕ15x30 cm with ten different mix designs were made.

The type of cement was the same for all mixes because it can introduce modifications in the results , and its study would increase too much the number of tests. It was chosen P-450-ARI as the cement to be used. The reason was the absence of slag and fly ash and its high specific surface, which provides a fast development of the hydration processes, with the consequent evolution of its pore structure.

The w/c ratios were varied between 0.4 and 0.7. This range was considered as a representative sample of the common water/cement ratios usually employed in concretes for road bridges structures, as well as building elements.

The maximum sizes of aggregate used were 10, 20 and 40 mm. The curing was carried out at 20 $\pm$2 ºC and 90 $\pm$ 5 % relative humidity, varying its duration from 3 to 14 days. After that period, specimens were conserved at 50 $\pm$ 5 % R.H. and 20 $\pm$ 2 ºC until the realization of the tests.

The type of aggregates used were the same in every dosage. They were rounded siliceous aggregates, and the gradings were adjusted to the Fuller's curve to obtain the maximum compactness.

Table 1. Mix designs

DOSAGE Nº	1	4	5	6	2	3	7	8	9	10
TYPE OF CEMENT	P-450-ARI									
W/C RATIO	0.53	0.60	0.70	0.40	0.46	0.41	0.53	0.53	0.53	0.53
CEMENT CONTENT (Kg/m^3)	300	300	300	300	350	400	300	300	300	300
MAX. SIZE AGGREGATE	20	20	20	20	20	20	40	10	20	20
CURING DURATION (days)	7	7	7	7	7	7	7	7	3	14
WATER CONTENT (l/m^3)	160	180	210	120	160	165	160	160	160	160

3 The water penetration method

The Standard ISO/DIS 7.031: 1.983 deals with the operative method for the determination of the depth of water penetration under pressure in concrete specimens. The direct use of that method can offer some problems. In our study, they appeared in the initial phases, when we carried out the first tests. If the concrete has some internal humidity , it is not always possible to obtain the sufficient contrast of greys to distinguish the saturated concrete in the test from the other areas where the front of penetration has not arisen. That was the reason why it was neccesary to develop some modifications in the operative method in order to solve the mentioned problems.

First, it was tried to introduce some colouring substance into the water to obtain a colored front of penetration that was easily distinguishable from the rest of the concrete. Various products were used (rhodamine, metilen blue, fuchsine,...). The results were the same for all of them. The colouring particules settled in the pore surfaces thus decreasing their sections, with the consequent "filter effect". Therefore, it appeared a displacement between the colouring front and the penetration front. [2]

In consequence, it was decided to carry out a previous drying of the specimens. This process should not introduce any modification into the pore structure of concrete. Thus, it was adopted a drying at 45 ± 2 ºC during 24 hours. This period was considered enough to solve the above problem, in according to the first achieved trials.

All this modifications are included in the new Spanish Standard UNE 83.309 (next to be published).

4 Results and discussion

The specimens were tested at age of 28 days. In the ISAT, three measurements were obtained: at 10, 30 and 60 minutes. The average measurements obtained for each dosage, are shown in Table 2.

4.1 Influence of the water/cement ratio

The Figures 1 to 3 show the results obtained in relation to the w/c ratio.

The important role of the w/c ratio in the permeable capacity of concrete is extensively known,

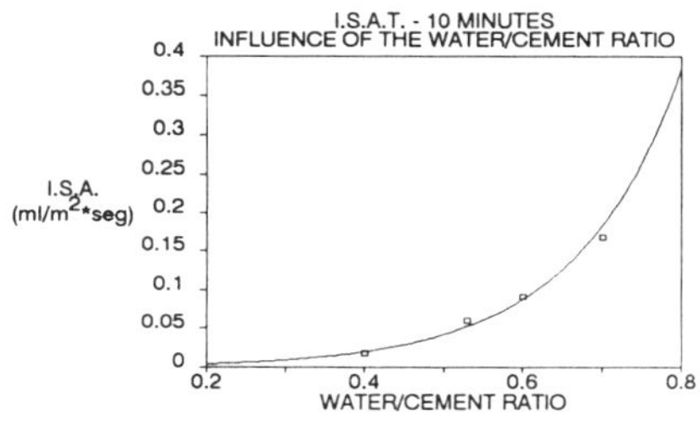

Figure 1.

Table 2. Average results obtained for each dosage.

Dosage nº	Pmax(mm)	Pmed(mm)	ISAT (ml/m^2 seg) 10 min	30 min	60 min
1	47,67	37,00	0,0602	0,0450	0,0157
2	51,67	36,72	0,0802	0,0170	0,0108
3	47,33	36,69	0,1427	0,0295	0,0138
4	104,33	95,22	0,0903	0,0512	0,0340
5	149,67	141,97	0,1684	0,0986	0,0651
6	86,00	20,67	0,0181	0,0086	0,0046
7	74,67	62,39	0,1647	0,0511	0,0294
8	54,67	40,89	0,0752	0,0510	0,0166
9	57,00	37,06	0,0737	0,0253	0,0166
10	44,00	25,39	0,1183	0,0383	0,0232

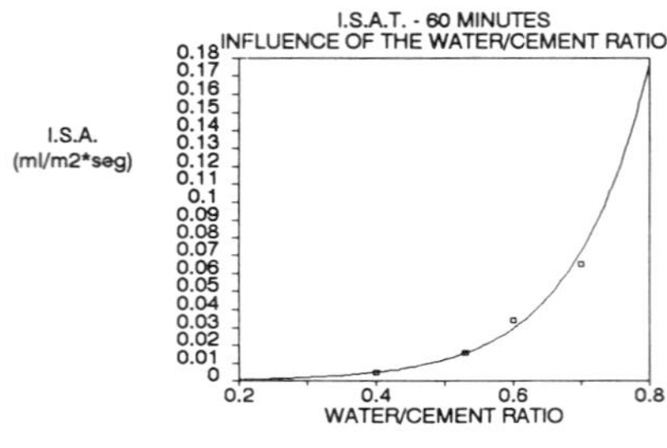

Figure 2.

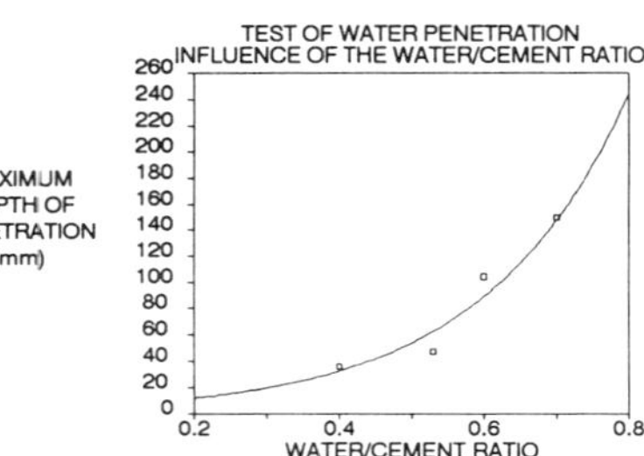

Figure 3.

with the existence of a range of w/c ratios between 0.40 and 0.60 where the minimum of permeability is obtained.

The results confirm the excellent behaviour of measurements at 10, 30 and 60 minutes when the ISAT was employed. The correlation coefficients were 0.993, 0.963 and 0.995. If we consider the good sensitivity and speed of the method, we can deduce its utility to obtain results, at least of quality character, on the absorption capacity in a concrete, and its consequent resistance to the environmental agressive substances.

Also, the water penetration method is very sensitive to the influence of the w/c ratio. The correlation coeficients obtained are 0.984 and 0.974 for the maximum and average depths.

4.2 Influence of the type of curing

The duration and the type of curing have a decisive influence on the internal structure of concrete. This process controls the surface water losses that can affect the adequate formation of the products of hydration during the early stages of hardening. In this work, the type of curing was a constant, but its duration was one of the following ones: 3, 7 or 14 days. The results obtained for the water penetration method are shown in Figures 4 and 5.

This method is one of the best in order to reproduce the transport of water into concrete. It explains its sensitivity to the conditions in which the curing is carried out because this can alter the pore structure and vary the capillar distribution through which the water penetrates. The correlation coeficients obtained in this case were 0.931 and 0.934.

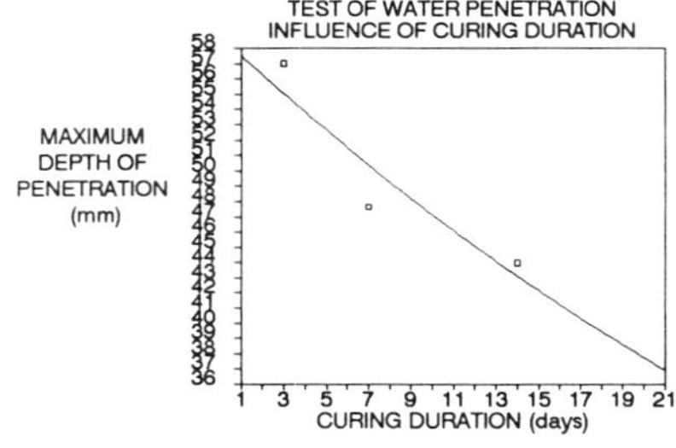

Figure 4.

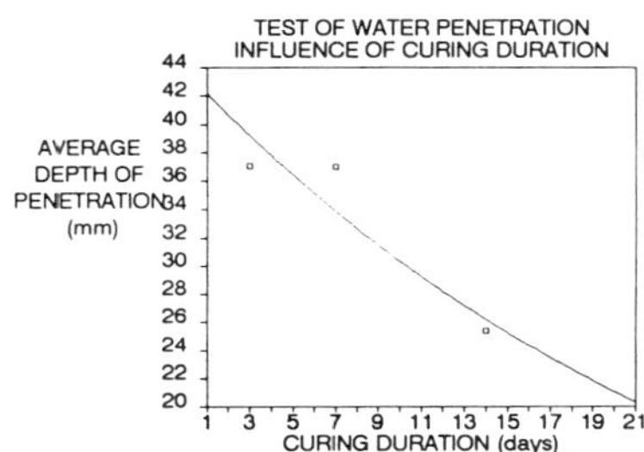

Figure 5.

4.3 Influence of the maximum size of aggregate

In Figures 6, 7, 8 and 9, the results obtained are shown. The measurements at 10 and 60 minutes in the Initial Surface Absoption Test show a coherent behaviour for the different sizes of aggregate used in the mix designs. In this case, like in the water penetration one, it can be observed the increment of water transport when the maximum size of aggregate is increased. The values obtained for ISAT-60 min. and water penetration are particularly significants in this point. It can be explained because of the major frecuency in the arising of lacks of continuity and voids in the aggregate-paste interphase when the size of the aggregate is increased.

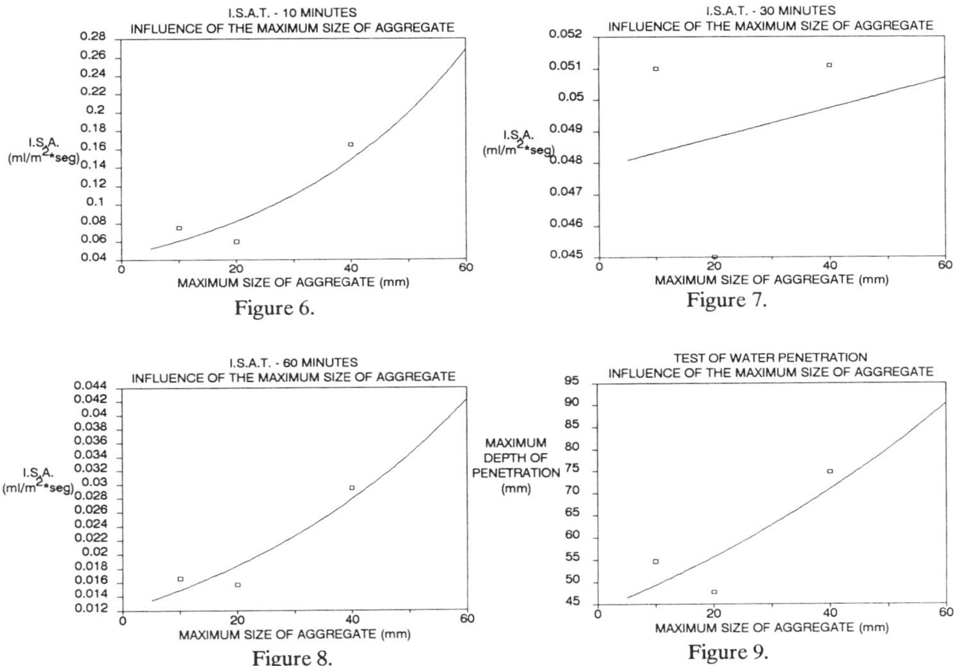

Figure 6.

Figure 7.

Figure 8.

Figure 9.

Nevertheless, the results obtained for ISAT-30 min. show a great dispersions and a behaviour which is not in agreement with the other ones. The elapsed time from the beginning of the test can explain this effect. It is possible that the absorpted water has reached the interphase between the surface cement layer and the concrete. Then, the different layers which Kreigjer calls "skin of concrete" [3] can have a relevant role. A specific study on these effects would be necessary to evaluate their influences on the mentioned behaviour.

4.4 Relation between the strength and the absorption or water penetration

Both the fluid transport and the strength of concrete are strongly related to the characteristics of its internal structure. This was the reason why the evaluation of the possible relationship existing between those parameters were considered interesting. In Figures 10 and 11 , the results are shown.

It can be observed the exponential behaviour of the relations obtained, with correlation coeficients of 0.905 and 0.912. That confirms the existence of a strong dependence between a low permeability and a high strength for concrete having the same type of cement and aggregates.

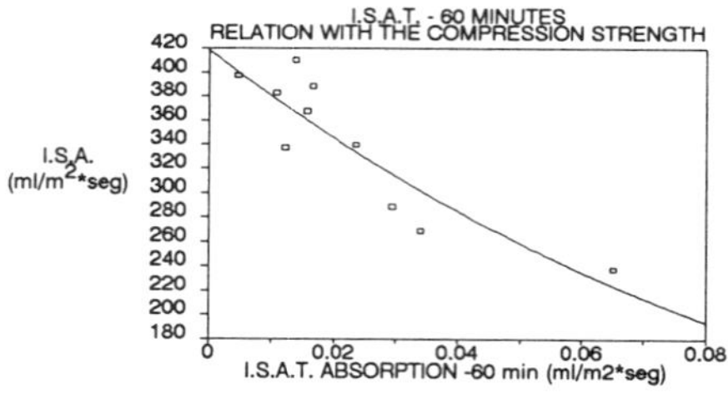

Figure 10.

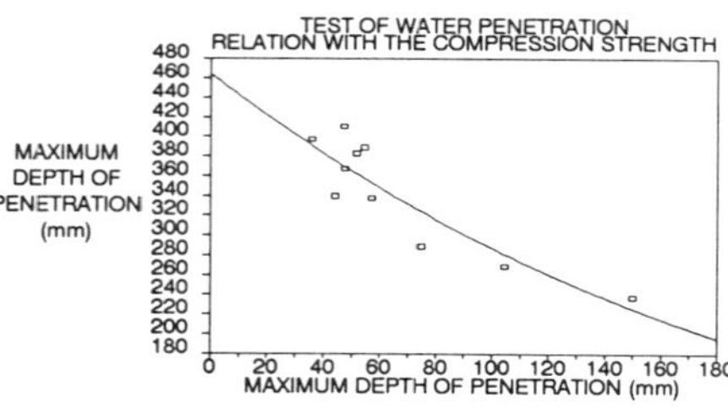

Figure 11.

5 The project of European Standard ENV-206: permeability requirements

The present document of ENV-206, in the epigraph 7.3.1.5. ("resistance to water penetration") [4], specifies maximum values of depths of water penetration , in order to get the adequate low permeability of concrete. Those values are 50 and 20 mm. for maximum and average depths obtained in the tests.

From the results of the tests which were carried out in the experimental programme, a linear relation between both of those depths was obtained, with a correlation coefficient of 0.970. For the maximum depth of water penetration, the value of 50 mm can be adequate, but in the case of the average depth, the 20 mm is considered as very restrictive, and disconnected with the other measurement. Probably, it would be better to limit the respective depths to 50 and 30 mm.

6 Conclusions

- It is neccesary to dry previously the specimen for carrying out the water penetration method, in order to distinguish clearly the front of penetration.
- The ISAT and water penetration methods are adequate to study the influence of the parameters considered (w/c ratio, duration of curing, maximum size of aggregate). The correlation coeficients obtained are very close to 1.
- The water/cement ratio is the most sensitive parameter in relation with the fluid transport mechanims. When the w/c ratio is greater than 0.53, an increase of the permeability is observed.
- Some effects related to the existent interphases between the surface cement layer and the concrete can appear. The ISAT values are sensitive to these effects.
- It is very restrictive to limit the average depth of water penetration to 20 mm. In accordance to the experimental results, the requirements of the ENV-206 for the low permeability of concrete would be 50 and 30 mm for each one of the measured depths (maximum and average depths).

7 References

[1] Organization for Economic Co-operation and Development (OECD) (1.989). Durability of Concrete Road Bridges, pag 29.

[2] Gálligo, J.M.; Rodríguez, F. (1.989). El ensayo de penetración de agua como método para el control de la durabilidad de un hormigón utilizado en puentes de carretera. Hormigón y Acero, 171, pag. 143-152.

[3] Kreigjer, P.C. (1.984). The skin of concrete. Composition and properties. Materials and Structures, 100, pag. 275-283.

[4] European Committee for Standardisation. Technical Committee CEN TC/104 (1.988). Concrete: Performance, Production, Placing and Compliance Criteria. pr-ENV-206.

37 DEVELOPMENT OF THERMAL DEGRADATION LOAD MAP FOR ELASTOMERIC ROOFING SHEETS IN JAPAN

K. TANAKA, M. KOIKE
Tokyo Institute of Technology, Yokohama, Japan

Abstract
A load map for predicting thermal degradation of elastomeric roofing materials has been developed. First, being based on the meteorological data for ten years, thermal degradation loads at 66 points in Japan were calculated, and a load map was drawn so as to divide a whole area into several zones according to the level of them. Second, a field test of two types of rubber sheet was carried out at three cities in Japan for two years to verify the applicability of the map. The degraded values of elongation at break in the outdoor exposure were compared with the predicted ones, and it might be concluded that the proposed method with the thermal degradation map is applicable for prediction of degradation of materials which were mainly affected by heat.
Keywords: Thermal Degradation, Degradation Load Map, Roofing Sheet, Weather Element, Prediction, Elongation at Break, Japan

1 Introduction

We have experienced varying thermal degradation of roofing materials in Japan, because it is situated from the subarctic to subtropical zones. Then, much difference in degradation is expected among the places where roofing materials are used. In order to make durable roof systems or maintain them in a satisfactory condition for a long period, it is necessary to know how roofing materials degrade outdoors and to predict it exactly at the stage of specifying materials.

Reliable information on durability of them is basically obtained from an outdoor exposure test, but it is practically limited to a very few places because of economical reasons or the like. Therefore, it is convenient if we can make a system to quantitatively predict degradation of roofing materials in all areas without exposure tests at the site. The information of degradation can be obtained by an artificial weathering test such as a heat aging test. If the relation between the results obtained by an artificial weathering test and those by an exposure test is made clear, the degradation of a material can be predicted from the former test alone. For this purpose, we intended to propose the concept which quantitatively connects both test results, and proposed a practical

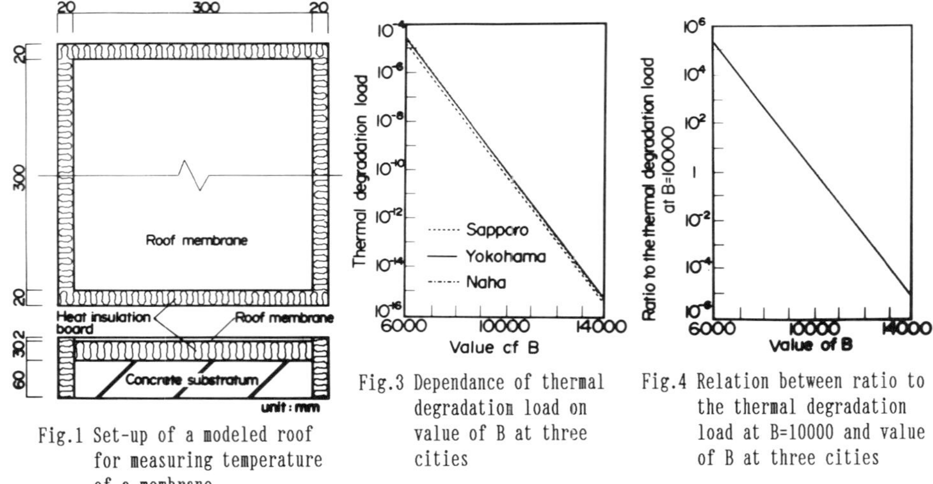

Fig.1 Set-up of a modeled roof for measuring temperature of a membrane

Fig.3 Dependance of thermal degradation load on value of B at three cities

Fig.4 Relation between ratio to the thermal degradation load at B=10000 and value of B at three cities

factor such as thermal degradation load. If thermal degradation loads were obtained at many places and expressed on a map, we can easily calculate the degradation of roofing materials at any places concerned.

Elastomeric roofing sheets were selected as samples to predict degradation, because the quantitative relation between the degradation by an artificial test and an exposure test had been already made clear by the authors.

2 Temperature of roofing membrane

A polyurethane membrane with blackly coated surface on a thermal insulation board shown in Fig.1, in which a thermocouple for measuring temperature was embedded, was exposed in Sapporo, Yokohama and Naha. The weather elements, such as ambient air temperature, solar radiation and wind velocity, which affect the temperature of the membrane, were also measured at Yokohama.

Thermal degradation load can be theoretically calculated with temperature of the membrane and the length of time. So, the

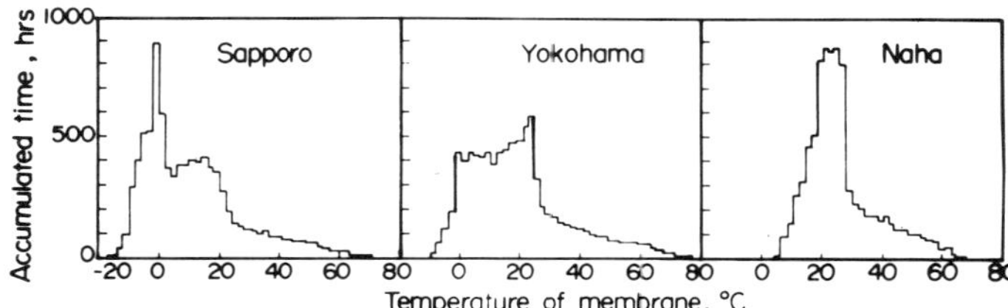

Fig.2 Histogram of temperature of roof membranes at three cities

temperature was measured at an interval of 10 minutes. Fig.2 shows the histograms of accumulated time at every 2°C increment at three cities.

3 Thermal degradation load

A change in physical property of a elastomeric roofing sheet by heat can be expressed by the following equation,

$$\frac{1}{n-1}\left(\frac{1}{y^{n-1}} - \frac{1}{y_o^{n-1}}\right) = \Sigma \exp(-B/T) \cdot t \qquad (1)$$

where n = apparent reaction order(-)

A, B = material constants related to thermal degradation

y = value of physical property of a material

y_o = initial value of y

T = Temperature of material(K)

t = time(hrs).

$\Sigma(-B/T) \cdot t$ in the right side in equation (1) is the part relating the degree of degradation, then it is named "thermal degradation load", of which abbreviation is T.D.L.,in this paper. It expresses the load to degrade a material when it is exposed at T during t. If T.D.L. is obtained, we can calculate a change in property from the equation(1). Elongation at break is adopted as physical property in this paper.

T.D.L. varies in accordance with value of B. As the range of the value of B for most rubber sheets is from 6000 to 14000 in our study, we calculate T.D.L.s within these range of B. Fig. 3 shows the relation between T.D.L. and value of B at three cities. The slopes of three lines in the figure can be regarded almost same, then, the difference among cities can be expressed as parallel vertical shifting of a line.

Now, we assume a material of which material constant B is 10000 as a standard one. The value of the T.D.L. for this material is called here as "standard thermal degradation load". Ratios of T.D.L. of materials with different B values to the standard one are calculated at three cities, and shown in Fig.4 as a over-lapped line. This means that there is no difference in the ratios among cities, and we can obtain coefficients for other materials from this figure.

4 Standard thermal degradation load map

4.1 Equation for expressing daily equivalent temperature

To make a map, measurement of temperatures of roofing membranes is basically needed in all areas. They were carried out, however, at only three cities. It is not enough for drawing the map, then the data by the meteorological stations were used. The daily equivalent temperature, which is defined as the constant temperature causing the same degradation during one day, was used in this paper. The temperature of the membrane mainly depends on ambient air temperature, solar irradiation and wind velocity. So, we considered that the daily equivalent temperature of a roofing membrane could be expressed by these three weather elements, and formulated the following equation for expressing the daily equivalent temperature with the record of temperature and the weather data at Yokohama.

$$T_{ed} = 1.00 \cdot \theta_d + \frac{7.72 + \dfrac{11.7}{5.8 + 4.0 \cdot W_d}}{2.58 \cdot W_d + 5.79} \cdot J_d - 1.79 \qquad (2)$$

where T_{ed} = daily equivalent temperature(°C)
θ_d = daily averaged ambient air temperature(°C)
W_d = daily averaged wind velocity(m/s)
J_d = daily accumulated solar radiation(MJ/m^2).

4.2 Drawing of standard thermal degradation load map

Based on meteorological data from January 1979 to December 1988 at 66 points in Japan, the daily equivalent temperatures were obtained, and the standard T.D.L.s for one year were calculated. They are shown on a map dividing a whole area into several zones. We call it the standard thermal degradation load map and show it in Fig.5.

4.3 Procedure for predicting degradation by the map

The procedure for predicting degradation of elastomeric roofing sheets by applying the map is as follows.

Step 1; Obtain apparent reaction order,n, and material constants, A and B, from heat aging tests at several temperature,
Step 2; Read a ratio from Fig.4 in accordance with the B value,

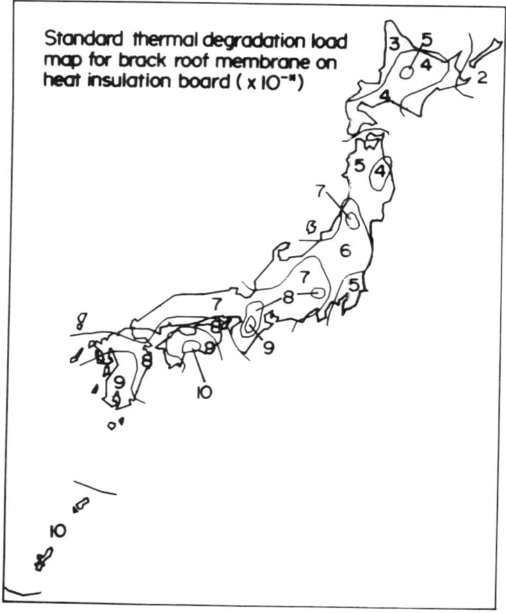

Fig.5 Standard thermal degradation load map for black roof membrane on heat insulation board

Table 1 Equation for expressing thermal degradation

Sample	Elongation at break(%)	Equation for expressing thermal degradation
IIR+EPDM-1	443	$\frac{1}{6}\cdot(\frac{1}{y^6}-1)=1.1\times10^{11}\cdot\Sigma\exp(-\frac{11400}{T})\cdot t$
IIR+EPDM-2	468	$\frac{1}{8}\cdot(\frac{1}{y^8}-1)=8.8\times10^{11}\cdot\Sigma\exp(-\frac{12700}{T})\cdot t$

Step 3; Read a value of the standard T.D.L. at the place concerned on the map shown in Fig.5,

Step 4; Calculate a total T.D.L. by the next equation, (Total T.D.L.)=(Ratio)・(Standard T.D.L. value)・(Expected service period),

Step 5; Put the total T.D.L. into equation(1), and calculate the change of the property of the material.

5 Verification of procedure for predicting degradation

5.1 Prediction of degradation

Two kinds of commercial IIR/EPDM sheets, of which abbreviations are IIR+EPDM-1 and IIR+EPDM-2, were exposed in an air oven at 70°C for

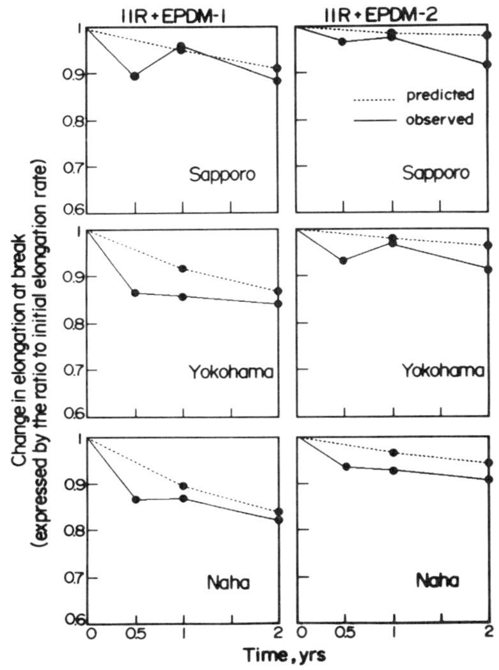

Fig.6 Comparison between predicted values and observed ones

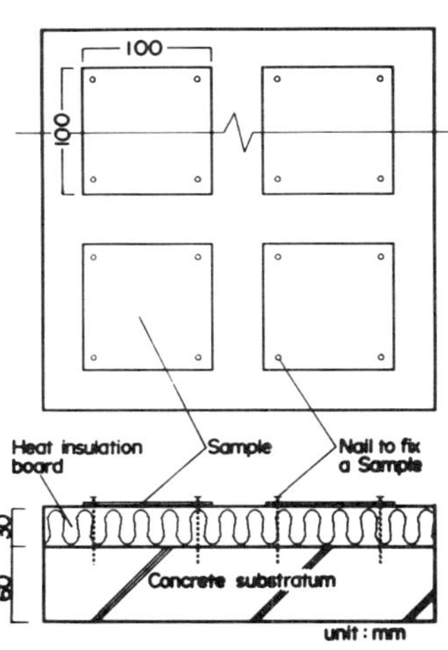

Fig.7 Specimen of roofing sheet for a field test

1500hrs,90°C for 800hrs and 110°C for 400hrs. Dumbbell specimens were cut from the sheets at every scheduled time, and the elongation of them were measured. The apparent reaction order and material constants were taken from the results and equations expressing thermal degradation were obtained as shown in Table 1. According to the aforementioned procedure, the degradation of the two rubber sheets were calculated. They are shown with the dotted lines in Fig.6.

5.2 Field tests and comparison with the predicted degradation
Four pieces(100mmx100mm) of each roofing sheet were exposed on a heat insulation board as shown in Fig.7. The outdoor exposure was carried out at three cities for two years from May 1987 to April 1989. A piece of them was taken at six months, one year and two years, and the tensile test was made. The test results thus obtained are shown with solid line in Fig.6.

The changes of elongation of the two samples in the field test are somewhat larger than the value predicted by the map. The difference between two results are quite understandable because the materials in the field were affected by the other elements such as ultra-violet radiation, rain water and so on, except thermal effect induced by ambient air temperature and solar irradiation.

7 Summary and conclusions

We proposed the thermal degradation load map to calculate changes in property of elastomeric roofing sheets. And we showed the procedure how to use it. If an apparent reaction order and material constants related to thermal degradation are obtained from heat aging tests, we can roughly predict a change in elongation of a sheet on a heat insulation board at any place in our country.

It is concluded that the map must be useful for materials which mainly degrade by heat effect or are used in a condition to be affected by it. However, there are many other factors to degrade materials, and it should be noticed, that the values predicted by this method are always somewhat less than the those degraded by all weather elements on roof.

8 References

Michio Koike and Kyoji Tanaka (1976) The Estimation of Thermal Aging for Rubber and Plastic Sheets under Varying Temperature, Report of RLEMTIT, 1, 177-184.

Michio Koike and Kyoji Tanaka (1977) Thermal Effect on Rubber and Plastic Sheets in Outdoor Exposure, Report of RLEMTIT, 2, 167-178.

Michio Koike and Kyoji Tanaka (1977) Weatherability of Elastomeric Roofing membranes(part 3), Transactions of Architectural Institute of Japan, 255, 9-15.

Takashi Tomiita (1989) Thermal Degradation of Outdoor Thermal Degradation Environment(part 1), Journal of Structural and Construction Engineering, 395, 13-20.

38 SERVICE LIFE ESTIMATION FOR MULTI-PLY FLAT ROOF MEMBRANES

F. AKÖZ
Yildiz University, Istanbul, Turkey
M.S. AKMAN
Technical University, Istanbul, Turkey

Abstract
A method to predict the service life of bituminous multi-ply roofing elements has been developed. The damage factors have been wetting-drying, wetting-partial drying, freezing-thawing, surface heating-cooling and ultraviolet radiation. Tensile tests have been carried out on specimens aged in laboratory and atmosphere. The frequencies of the damage factors have been obtained hypothetically from meteorological survey records. A total correcting factor has been assessed by means of the specimens exposed to open atmosphere. The annual total damage ratio is the corrected sum of products of damage ratios by relevant frequencies. The service life is the ratio of the critical damage level to the annual total damage ratio. Assuming that the means and standard deviations of tensile strengths are linear functions of time, confidence limits may also be estimated.
Keywords: Service life, Multi-ply, Meteorologic Survey, Stochastic.

1 Introduction

A mathematical model has been developed in order to estimate the service life of a bituminous glass fibre reinforced multi-ply waterproof roofing element. Water permeability and strain controlled tensile tests have been carried out on damaged specimens.

The serviceability of these elements is in fact the low pressure water impermeability. However the time to obtain significant results is too long, and the evaluation has been based on tensile strength and toughness.

Wetting-drying, wetting-partial drying, freezing-thawing, surface heating-cooling and ultraviolet radiation have been considered as damage factors. (Knöfel et al 1987)

2 Experimental studies

Ageing tests were carried out on 30x50 cm size specimens shown in fig.1. 12 specimens were cured under laboratory conditions for a year and 12 specimens laid down in open atmosphere for twenty five months as control specimens, and five groups of 12 specimens were subjected to accelerated tests in laboratory. The details of these processes have

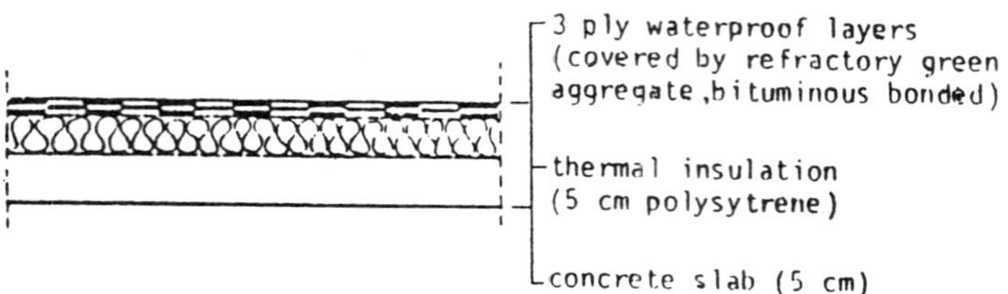

fig.1. Cross section of specimens.

been: 1) Wetting in 20°C water, drying in 65°C oven, 2) Wetting in water, drying at 70% RH and 65°C in humidity oven, 3) Freezing at -15°C, thawing at 20°C in water, 4) Heating at 70°C on the surface for 8 hours, then letting to cool for the subsequent 16 hours, 5) exposure to ultraviolet radiation for the 8 hours per day.

After repeating every damage process cycles up to certain number, tensile tests of constant strain rate have been made on strip specimens sawn from the elements. Water permeability tests have also been carried out by fixing 2 m long tubes on damaged elements. In this way, tensile strength and toughness of the elements have been studied as functions of number of cycles or duration of exposure to damage factors (fig.2,3). Damage ratios (e_f, e_t) for one cycle or one hour of exposure have been determined by regression analysis (table 1). Total damage ratios for tensile strength (16.6 %) and for toughness (51.6 %) have also been determined for the specimens in the open atmosphere.

The toughness is the necessary energy for breaking a material per unit volume and equal to the area under stress-strain curve.

Table 1. Damage ratios of the damage factors

Damage ratios	Wetting -drying	Wet.-partial drying	Freezing -thawing	Surface heating	Ultraviolet radiation
e_f T.strength	4.31×10^{-3}	1.08×10^{-3}	4.99×10^{-3}	0.06×10^{-3}	0.12×10^{-3}
e_t Toughness	7.53×10^{-3}	10.57×10^{-3}	12.12×10^{-3}	0.14×10^{-3}	0.34×10^{-3}

3 Mathematical model

The mathematical model proposed consists of the assessment of the annual total damage ratio and the estimation of the service life.

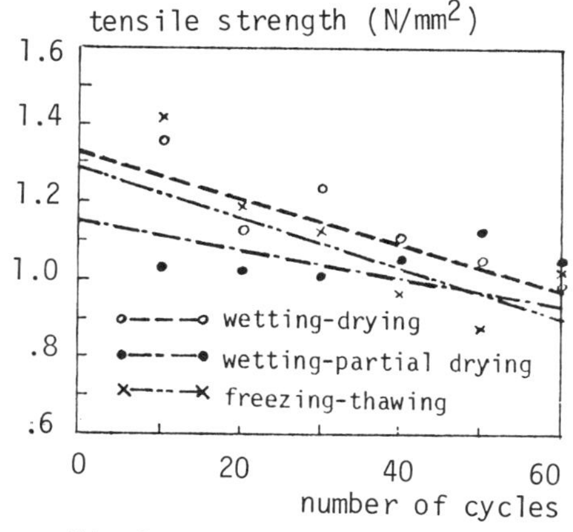

fig.2. Tensile strength-versus number of cycles

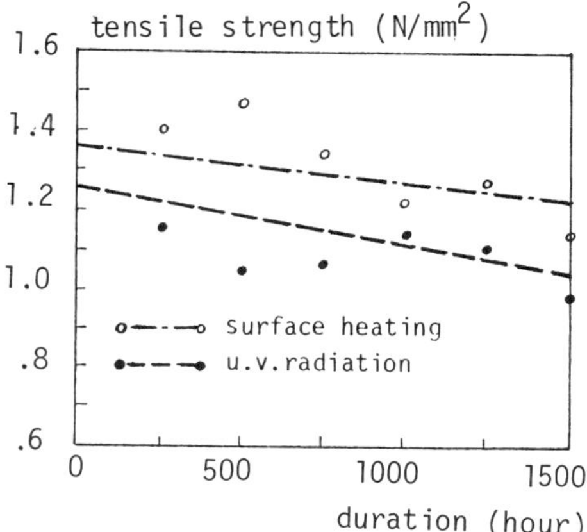

fig.3. Tensile strength-versus duration of exposure

3.1 Computation of damage ratio

The annual total damage ratio was taken as the sum of products of each damage ratio by the annual number of cycles or the duration of exposure. As meteorological phenomena are interactive, their effects can not be studied separately and it was necessary to correct the theoretical annual damage ratio computed by a correcting factor (d), to obtain the real annual damage ratio (Dg_r). Then the first step of the model has been formulated as :

$$Dg_r = d\left(\sum_{i=1}^{3} e_i . n_i + \sum_{j=4}^{5} e_j . h_j\right) \quad (1)$$

Where e_i, e_j are the damage ratio for each damage and n_i, h_j are their annual cycles or durations.

The following hypotheses were adopted to obtain n_i and h_j from the meteorological survey records:

1) The rainfall rate has to exceed 0.1 mm for wetting, for drying the max. temperature should be more than 25°C and the RH less than 90 %, or the max. temperature must be more than 10°C and the RH less than 40 %. (ACI, 305 R-77, 1982)

2) For partial drying, the max. temperature should be more than 25°C and the RH between 70 and 80 %.

3) For freezing, the temperature must fall below -0,1°C, for thawing the temperature must be over 5°C and the rainfall more than 0.1 mm.

4) To heat the element on the surface the max. temperatures has to exceed 25°C. The exposure duration was assumed to be 70 % of the daytime and the daytime was assumed as a function of zenith angle and latitude.

5) The ultraviolet radiation may be efficient during 70 % of the daytime, the efficiency is 80 % for sunny, 45 % for cloudy and zero for overcast sky.

The meteorological records give the number of days during which the phenomena described took place. If more than one meteorological condition were necessary for incurring the damage mentioned, the minimum among the frequencies cited should be taken. Besides, the number of cycles in a month is limited by 15. Frequencies and durations in table 2 have been obtained from the means of meteorological observations of 40 years.

3.2 Prediction of service life

To predict the service life, it is necessary to fix the critical damage ratio generally based on economic considerations. It has been defined as the ratio of the permeable area to the total area (Wp_c). In this study, the critical damage ratios have been defined in terms of tensile strength (Dgf_c) or toughness (Dgt_c), the relations among Dgf_c, Dgt_c and Wp_c could be established in the case of freezing and thawing tests: Dgf_c=0,47 Wp_c, Dgt_c=1,14 Wp_c. In numerical evaluation of the model, Wp_c has been fixed as 40 %.

One of the methods proposed to predict the service life is purely deterministic and similar to "value analysis I" of Pihlajavaara. The average theoretical service life is:(Pihlajavaara, 1984).

$$Kf_m = \frac{\sum_{i=1}^{n} Kf_i . P_i}{\sum_{i=1}^{n} P_i} \qquad (2)$$

where Kf_i is the theoretical service life according to a single damage factor i and equal to $Dgf_c/Dgfa_i$, $Dgfa_i$ is the annual damage ratio for damage i. The significance factors P_i are obtained from the expert judgement, but in this study they were determined as :

$$P_i = \frac{Dgfa_i}{(Dgfa_i)\ max} \qquad (3)$$

To predict the real service life, Kf_m should be divided by the correcting factor.

The second method proposed to predict the service life enables also stochastic estimation. In this method, numbers of cycles or durations should be first expressed in real time scale. This transformation is quite sensitive and valid only for the region. The arithmetic means (f) and standard deviations (s) of the tensile strengths measured were assumed to be linear functions of the time (t) and written in the following forms :

$$f = at + f_o \qquad (4)$$

$$s = pt + q \qquad (5)$$

In these equations, f_o and q are the mean and standard deviation of the tensile strengths of the intact specimens, a and p can be obtained by statistical evaluation of the test results.

It is possible to estimate the average service life from the equation (6) :

$$Kf_m = \frac{-Dgf_c.f_o}{a} \tag{6}$$

By dividing this theoretical value by the correcting factor, the real average service life may be predicted.

The confidence limits for the service life are in the case of normal Gaussion distribution :

$$(Kf_m)_{max} = -\frac{Dgf_c.f_o+Zq}{a+Zp} \quad , \quad (Kf_m)_{min} = -\frac{Dgf_c.f_o-Zq}{a-Zp} \tag{7}$$

Z is the tolerance parameter corresponding to the confidence level chosen.

Table 2. Frequencies and durations of damage factors in the regions of Turkey

Region (city)	$n_1^{(*)}$	n_2	n_3	h_4	h_5
Marmara Region (Istanbul)	29	29	15	680	1210
Agean Region (İzmir)	21	1	7	1450	1660
Mediterranean Region (Antalya)	23	0	0	1500	1700
South East Anatolia (Urfa)	23	0	0	1740	1700
East Anatolia (Erzurum)	23	0	30	525	1330
Black Sea Region (Trabzon)	39	39	0	645	1020
Central Anatolia (Ankara)	34	34	50	1090	1410

$n_1^{(*)}$: number of cycles of wetting-drying
n_2 : number of cycles of wetting-partial drying
n_3 : number of cycles of freezing-thawing
h_4 : surface heating duration (hour)
h_5 : ultraviolet radiation exposure (hour)

4 Conclusions

The service life of a building element damaged by the atmosphere will be predicted by comparing the critical damage level adopted with the annual total damage ratio estimated.

The annual total damage ratio will be estimated as follows :

a. Distinction of damage factors,

b. Expression of serviceability by a measurable property,

c. Determination of the damage ratios by accelerated laboratory tests,

d. Obtaining regional numbers of cycles from meteorological records,

e. Summing up the products of damage ratios by the relevant number of cycles.

It is necessary to correct this total ratio calculated, by a unique factor obtained from tests on specimens laid down in the open atmosphere. The agents in nature are interactive, therefore, a separate correcting factor for each agent can not be assessed.

The critical damage ratio will be fixed according to economic and technical data. The theoretical service life is the ratio of the critical damage ratio to the annual total damage ratio.

The average service life of the multi-ply waterproof element has been estimated as 11 years and 9 months, the maximum life as 17 years and 6 months, the minimum life as 5 years and 10 months for 90 % confidence level in Marmara region.

5 References

Fagerlund,G. (1979) Service life of structures, Int.Sym. on Quality control of concrete structures, Stockholm, Proc., pp. 199-218

Aköz,F. (1989) The prediction of service life of waterproof layers on flat roofs, Ph.D. thesis, Technical University of Istanbul (in Turkish)

Knöfel,D.K., Hoffmann,D. and Snethlage,R. (1987) Physico-chemical weathering reactions as a formulary for time-lapsing ageing tests, Materials and Structures, vol.20, no.116, pp. 127-145

ACI 305 R-77 (1982) Hot weather concreting, ACI manual of concrete practice, part 2-1986, American Concrete Institute, Redford Station Detroit, Michigan

Rakhra,A.S. (1984) Durability of building materials, market structure and the economic development of the construction industry, Third international conference on the durability of building materials and components, vol.1, pp. 178-189

Pihlajavaara,S.E., (1984) The prediction of service life with the aid of multiple testing, reference materials, experience data, and value analysis, Third international conference on the durability of building materials and components, vol.1, pp. 37-64

SESSION D

MANUFACTURING PRACTICE AND DURABILITY

39 METHODS FOR PREDICTING THE SERVICE LIFE OF CONCRETE

J.R. CLIFTON
National Institute of Standards and Technology,
Gaithersburg, USA

Abstract
At present the design of concrete is usually based on empirical relationships between materials and properties, and experience with its performance. Another approach for selecting concrete is based on predictions of service life. While this approach is not yet often used, it is likely to have an increasingly important role in designing concrete.

There are several methods for predicting the service lives of construction materials. They include i) estimates based on experience, ii) deductions from performance of similar materials, iii) accelerated testing, iv) applications of reliability and stochastic concepts, and v) mathematical and simulation modeling based on the chemistry and physics of degradation processes. These methods are discussed in this paper, along with examples of their applications.
Keywords: Accelerated Testing, Concrete, Mathematical Modeling, Reliability Methods, Service Life Predictions, Simulation Modeling, Stochastic Methods.

1 Introduction

The design of concrete is usually based on i) empirical relationships between materials and the physicochemical and mechanical properties of concrete; and ii) experience with the effects of service environments on their performances. These approaches are based on the premise that concrete will have the desired life and there is no need to predict its service life. Usually concrete performs adequately for its design life; if not, premature degradation is often attributed to factors as poor quality control, improper characterization of (or unanticipated changes in) the service environment, or unanticipated changes in the use of the structure.

Another approach to selecting concrete is based on prediction of service life. While this approach is not often used, it is likely to have an increasingly important role in designing concrete because of i) applications that require significantly increased service lives, ii) increased use of concrete in harsh environments, iii) the high cost of rebuilding and maintaining national infrastructures, and iv) the development of innovative concretes for which a record of performance is not available. In addition, improved understanding of the factors controlling the life of concrete can contribute to the development of more durable concretes.

The purpose of this paper is to present an overview of methods for predicting the service life of new concrete, with examples of their applications.

2 Approaches for Making Service Life Predictions

Several methods are used for predicting the service lives of construction materials. They include i) estimates based on experience, ii) deductions from performance of similar materials, iii) accelerated testing, iv) applications of reliability and stochastic concepts, and v) mathematical and simulation modeling based on the chemistry and physics of degradation processes. Although these approaches are discussed separately in this paper, they may be used in combination.

2.1 Estimates Based on Experience.

A common method of making semi-quantitative predictions of the service life of concrete is expert judgement based on the accumulated knowledge from laboratory and field testing, and experience. This body of accumulated knowledge contains both empirical knowledge and heuristics; collectively, these provide the largest contribution to the basis for standards for concrete. It is assumed that, if concrete is made following the standard guidelines and practices, it will have the required life. This approach gives an "assumed service life" prediction. The concrete may perform adequately for its design life, especially if the design life is fairly short and the service conditions are not too severe. This approach breaks down when it becomes necessary to predict the service life of concrete which is required to be durable for a time which exceeds our experience with concrete, when new environments are encountered, or when new concrete materials are to be used. For example, an approach being considered in the United States for disposing of low-level radioactive waste is to place it in underground concrete vaults (Clifton 1989). A life of 500 years would be required for the vaults. Our knowledge of the durability of underground concrete is limited and we have less than 150 years total experience with portland cement concretes, and much less with concretes containing mineral admixtures. Fagerlund (1985) analyzed several examples of using this approach and concluded that experience or qualitative assessments of durability do not form a reliable basis for service life prediction.

2.2 Predictions Based on Comparison of Performance

The comparative approach has not been commonly used for concrete but, with a growing population of aging concrete structures, we can expect that its use will increase. In this approach, it is assumed that, if a concrete has been durable for a certain time, a similar concrete exposed to a similar environment will have the same life. A problem with this approach is each concrete structure has a certain uniqueness because of variability in materials, geometry, and construction practices. Also, over the years, the properties of concrete materials have changed. For example, portland cements are ground finer today than they were 40 years ago, to achieve increased early age strength. The smaller particle size may be detrimental to the durability of concrete (Neville 1987). In contrast, advances in chemical and mineral admixtures have led to the development of concrete with potentially improved performance and durability. Another problem with the comparison approach is differences in the microclimates to which concrete structures are exposed may have unanticipated effects on the concrete durability. Therefore, comparisons between the durabilities of old and new concretes is not straightforward.

2.3 Accelerated Testing

Most durability tests for concrete involve the use of elevated stresses (concentration of reactants, temperature, humidity, etc.) to accelerate degradation. Accelerated testing programs, if properly designed, performed, and interpreted, should provide a sound basis for predicting the performance and service life of concrete. Use of accelerated testing for predicting the service life of several types of building materials was discussed by Frohnsdorff et al. (1980). An important requirement for using accelerated testing is that the degradation mechanism in the accelerated test should be the same as that responsible for the in-service deterioration. Often, accelerated tests are developed and used without properly identifying the degradation mechanism activated by the test conditions (Frohnsdorff et al., 1980).

RILEM recently published a systematic methodology for predicting the service life of building materials and components (RILEM, 1989). This document, which is similar to ASTM E 632 (ASTM 1981), addresses the identification of needed information, the selection or development of tests, the interpretation of data, and the reporting of results. One of the earliest descriptions of the type of methodology described in the RILEM and ASTM documents was given by Plum et al. (1965). The purpose of their paper was to describe how the service life of each element of a concrete structure might be estimated. They emphasized the understanding of the degradation mechanism and the development and application of mathematical models. The procedures used were similar to the steps prescribed by the RILEM and ASTM methodologies.

Both the RILEM and ASTM documents describe the following methods for quantifying the effects of an accelerated test. If the degradation proceeds at a proportional rate by the same mechanism in both accelerated aging and long-term in-service tests, an acceleration factor, K, can be obtained, from:

$$K = R_{AT}/R_{LT} \tag{1}$$

where R_{AT} is the rate of change in accelerated tests, and R_{LT} is the rate of change in long-term in-service testing. If, as is usually the case, the relationship between the rates is non-linear, then mathematical modeling of the degradation mechanism is recommended to establish the relationship.

2.3.1 Applications of Accelerated Testing

The following study illustrates an approach for using accelerated tests for estimating service lives, while also indicating possible sources of problems in designing accelerated tests. The US Bureau of Reclamation estimated the service life of concrete exposed to sulfates by combining the results of accelerated tests and field tests (Kalousek et al. 1972). In this study, a set of concrete specimens were continuously immersed in a 2.1% Na_2SO_4 (sodium sulfate) solution until failure (defined as an expansion of 0.5%) occurred or until the investigation was completed. The age of specimens at the completion of the continuous-immersion study was between 18 to 24 years. Companion specimens were subjected to an accelerated test in which the specimens were exposed to repeated cycles of immersion in a 2.1% Na_2SO_4 solution for 16 hours and forced air drying at 54°C (130°F) for 8 hours. From a comparison of the times for specimens to reach an expansion of 0.5% in the accelerated test and the continuous immersion test, it was estimated that 1 year of accelerated testing equalled 8 years of continuous immersion. A 2.1% solution of Na_2SO_4 is a severe environment and if concrete was exposed to a lower concentration of sulfate the life expectancy would be expected to be longer. This method could be used for predicting the service life of concretes continuously immersed in a different concentration of sulfate ions, if the acceleration factor of 1:8, obtained using the 2.1% sulfate solution, holds for other sulfate concentrations.

In reality, it is likely that the degradation mechanisms associated with the accelerated tests were not the same as those occurring in the continuous immersion tests. During the continuous immersion, sulfate ions would migrate into the concrete by diffusion alone while, in the immersion and drying tests, the sulfate would likely be transported by convection, diffusion, and capillary forces. Also, in the immersion and drying test, the concentration of sulfate ions would likely be higher near the surfaces of the test specimens than in the immersion solution, because the evaporation of the solute would tend to draw the sodium sulfate near the specimens surfaces, where it could precipitate. While the severity of sulfate attack increases with concentration, the reaction mechanism also changes. For example, with low concentrations of sulfate (less than 830 mg/l), ettringite formation is considered (Biczok 1967) to be the main cause of deterioration while, at higher concentrations, gypsum formation tends to become the dominant deterioration process. A 2.1% sodium sulfate solution gives a concentration of 21000 mg/l, which means that it is likely that both ettringite and gypsum formed in the test.

2.4 Stochastic Methods

The use of stochastic concepts in making service life predictions of construction materials has been explored by several researchers, most notably by Sentler (1984) and Martin (1985). Service life models using stochastic methods are based on the premise that service life cannot be precisely predicted (Vesikari 1988). A large number of factors affect the service life of concrete and their interactions are not well known. These factors include the extent of adherence to design specifications, variability in the properties of hardened concrete, randomness of the in-service environment, and materials response to microclimates. Two stochastic approaches are the reliability method and the combination of statistical and deterministic models.

2.4.1 Reliability Method

The reliability method combines the principles of accelerated degradation testing with probabilistic concepts in predicting service life. This method has been discussed by Martin (1985) and applied by him to coatings (Martin 1989) and roofing materials (Martin and Embree 1989). Application of the method is described below by considering concrete subjected to a hypothetical laboratory durability test.

As is typical of any engineering material, supposedly identical concrete specimens exposed to the same conditions will have a broad distribution of times to failure (Fig. 1). The reliability method takes into account the time-to-failure distributions. For example, the solid curve in Fig. 1 gives the results of a hypothetical laboratory durability test of concrete specimens in

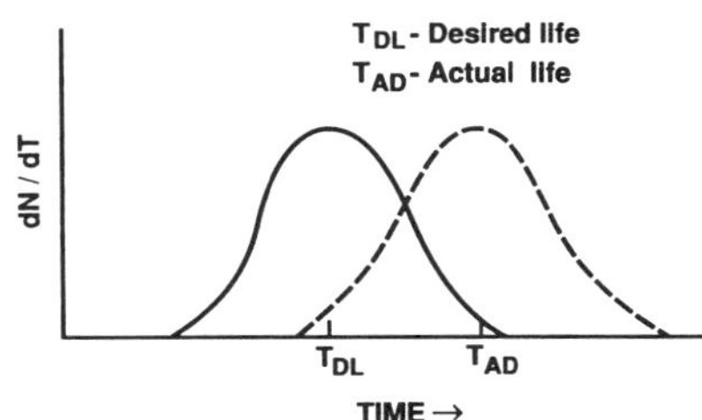

Figure 1. Time-to-failure distribution.

which the passing criterion is based on the length of time that the value of a certain property is equalled or exceeded. According to the curve, the criterion for passing, indicated as the desired life, T_{DL}, is reached when around 50% (depending on the form of the distribution curve) of the concrete specimens have failed. If a lower probability of failure at T_{DL} is desired, then the distribution needs to be shifted to longer lifetimes as indicated by the broken-line curve, with an actual average life denoted by T_{AD}. The data represented by the solid line in Fig. 1, is from one set of hypothetical test conditions. By elevating the stresses which have the major effect on accelerating the failure rate, similar sets of data can be obtained, and can be plotted as shown in Fig 2. These plots are based on the premise that data follows a Weibull distribution (Martin 1985). If the failure rate increases as the stress level increases, the life distribution at in-service stresses can be related to the life distribution at elevated stress by the time transformation function, $p_i(t)$, which is expressed as (Martin 1985):

$$F_i(t) = F_o(p_i(t)) \tag{2}$$

where t is time, $F_i(t)$ is the life distribution at the i'th elevated stress level, and $F_o(t)$ is the life distribution at the in-service stress level. From Equation 2, a probability-of-failure stress time-to-failure (P-S-T) diagram can be prepared as shown in Fig 3. The curves in a P-S-T diagram, e.g., the $F(t) = 0.10$ curve, are iso-probability lines. The iso-probability lines give, for each stress level, the time at which a given fraction of a group of specimens can be expected to have failed (Martin 1985). The P-S-T diagram gives a basis for predicting the service life of concrete if the in-service conditions are in the range covered by the diagram and are not anticipated to change significantly. If significant changes are anticipated, then a cumulative damage function model known as Miner's rule (Martin 1989) can be applied to predict the service life.

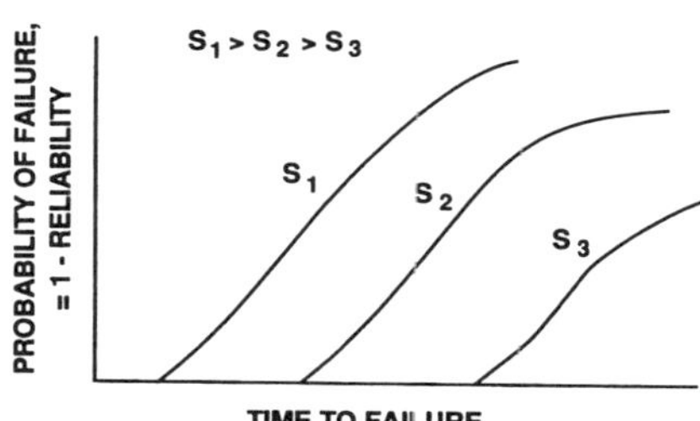

Figure 2. Probability of failure at different stress levels (S).

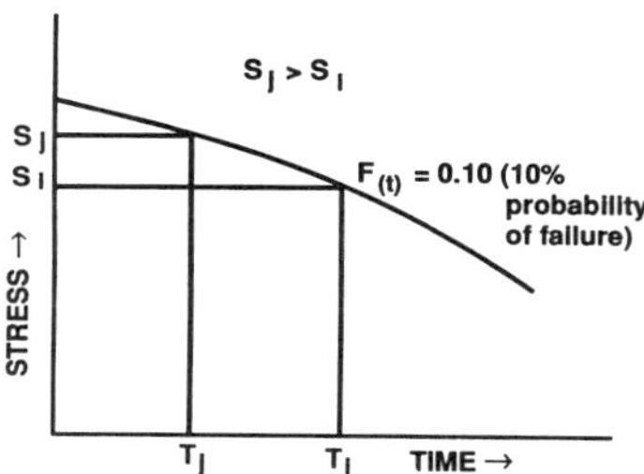

Figure 3. P-S-T diagram showing the 10% probability of failure curve.

The time transformation function approach is applicable if i) the deterioration mechanism under all tested stress levels is the same as that under in-service conditions, ii) deterioration begins at the instant of stress application, and iii) deterioration is an irreversible cumulative process (Martin 1982).

Synergistic effects may develop between deterioration processes. The importance of synergistic effects can be assessed by the combination of time transformation functions for different processes with a time synergistic function, using a multivariate analysis (Martin 1982).

2.4.2 Combination of Statistical and Deterministic Models

Often statistical models are combined with deterministic models. For example, Siemes et al. (1985) predict the mean service life of buildings by using mean values for the parameters in deterministic models. The standard deviation of the service life was calculated using the expression:

$$\sigma^2(t_1) = \sum_{j=1}^{n} \left\{\frac{\partial t_1}{\partial x_j} \cdot \sigma(x_j)\right\}^2 \qquad (3)$$

where $\sigma(t_1)$ is the standard deviation of service life,

$\sigma(x_j)$ the standard deviation of the variable x_j,

$\frac{\partial t_1}{\partial x_j}$ the partial derivative of t_1 with respect to x_j, and

n is the number of variables.

The partial derivatives $\partial t_1/\partial x_j$ are calculated for the mean values of the stochastic variables.

Instead of normal distributions, log-normal distributions are recommended for representing the service life distributions (Siemes et al., 1985).

2.5 Mathematical and Simulation Modeling

2.5.1 Mathematical Modeling

Several models have been developed for predicting the service life of concrete subjected to degradation processes such as frost damage (Fagerlund 1981; Bryant and Mlaker 1989), corrosion of reinforcing steel (Tuutti 1982; Bazant 1979a and 1979b), and leaching (Atkinson 1985). The feasibility of using mathematical models for predicting service lives of concrete has been discussed by Pommersheim and Clifton (1985). Many of the degradation processes of concrete, excluding those caused by excessive mechanical loading, are associated with the intrusion of concrete by one or more of water, ions and gases. For such processes, mathematical and simulation models for predicting service life can be developed by considering i) the rate of penetration of aggressive media into concrete, and ii) the rate of chemical reactions and physical processes. Mathematical models of degradation processes controlled by the ingress of water, salts, and gases into concrete by convection and diffusion have been described by Pommersheim and Clifton (1990). A model based on a two-step mechanism has been developed by Tuuti (1982) for corrosion of reinforcing steel in concrete. A proposed conceptual model for sulfate attack is discussed in the following to illustrate the factors which should be considered in developing reliable models.

At present, only empirical models have been developed to predict service life of concrete exposed to sulfate. Several studies are currently being performed to develop models based on a mechanistic understanding of sulfate attack (Atkinson et al. 1989; Odler and Gasser 1988). One approach for modeling sulfate attack is to mathematically describe the rate of intrusion of sulfate ions in concrete, the reaction process, and the effect of cracking on the intrusion rate. The range of usefulness of any mathematical model of sulfate attack depends upon the adequacy of the conceptual models upon which it is based. For example, if the zone of sulfate attack cracks or delaminates, fresh exterior surfaces are created which are exposed to the aggressive environment. Therefore, the rate at which the sulfate front moves through concrete could exceed that predicted from consideration of convection and diffusion processes only.

A conceptual model for degradation due to ettringite formation is presented in Fig. 4. It consists of intrusion of sulfate ions into the concrete and expansion of the concrete caused by ettringite filling the pore space. It is described by two steps: Step I, movement of sulfate ions to a unit element in a concrete component; and Step II, the expansion and cracking of the unit element of concrete due to the formation of ettringite. Step I is considered to be completed when the threshold concentration (concentration of sulfate ions required to initiate concrete expansion through ettringite formation) is reached at the unit element of concrete. Thus, in Step I, the ettringite is formed in pre-existing pore space. When no further ettringite crystals

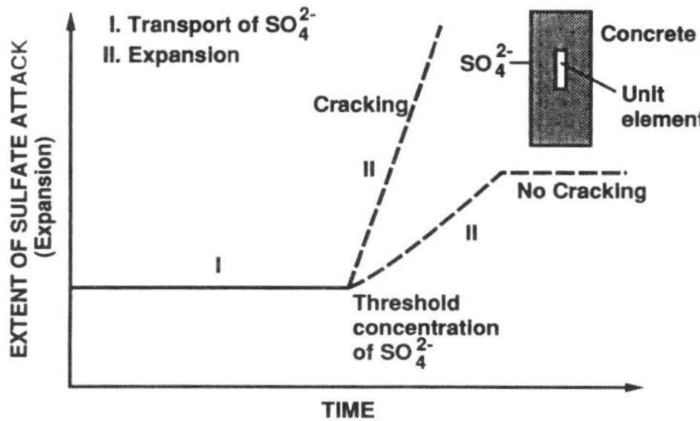

Figure 4. Schematic of expansion rate of concrete caused by ettringite formation.

can be accommodated in the pore space, expansion will start (beginning of Step II). Cracking will occur when the internal stresses exerted by ettringite formation exceed the tensile strength of the concrete. Failure is considered to occur when the unit element cracks. If the time for the completion of Step I is much greater than that for Step II, then the prediction of service life is reduced to only considering Step I. Even if only Step I needs to be modeled, determination or estimation of penetration rates is not sufficient, by itself, to predict the service life. The model must also be capable of predicting if a sufficient amount of ettringite will be formed by the sulfate and tricalcium aluminate reaction to cause cracking. Otherwise the model may falsely predict failure when the tricalcium aluminate content of the cement is too low to induce cracking. Based on considering transport processes only, the model will predict that increasing the tricalcium aluminate content of the cement will reduce the diffusion coefficient of sulfate ions. This is a correct prediction but the model must also consider that cracking will occur when a critical amount of ettringite is formed. A comprehensive model should consider factors controlling the rate of sulfate intrusion (e.g., cracking of concrete, reactions with tricalcium aluminate hydrates and monosulfate, and whether diffusion, capillary effects or convection is rate controlling), thermodynamic factors affecting the formation and stability of ettringite, and the reaction between sulfate ions and calcium hydroxide.

2.5.2 Simulation Modeling

A major difficulty in solving mathematical models of degradation processes is obtaining data on the rate of ingress of water or the transport of dissolved salts and gases in concrete. The obtaining of such data is complicated for sulfate attack. Portland cement contains an appreciable amount of calcium sulfate added to control set time. Also, the transport rate of sulfate ions in concrete will be low. The diffusion coefficient for sulfate ions in a mortar was determined by tracer studies to be around $6x10^{-14}$ m^2/s (Spinks et al. 1952).

Another approach to mathematical modeling based on differential equations is simulation modeling. Advances in applications of simulation modeling to cementitious materials being pursued at the National Institute of Standards and Technology include the simulation of the microstructure of portland cement (Garboczi and Bentz 1990), computation of effective diffusivities, and the use of percolation-based-equations for calculating the permeability of porous materials (Garboczi, 1990). To illustrate the power of simulation modeling the computation of effective diffusivities is described.

The approach for computing effective diffusivities is based on random walk simulations (Garboczi and Bentz 1990). The simulation is performed by using "blind ants" (random walkers) at sites scattered throughout the pore space of a simulated, or real, microstructure of a porous material (Banavar and Schwartz 1989). The ants are allowed to move an unit distance in a random direction. If the projected landing point lies within the pore space, the move is completed. If the point lies within the solid space, then the move is not made. By such simulations the effective diffusivity can be obtained. The effect of adsorption or reaction on diffusion can be treated by placing "sticky" sites on the surface which can capture the ants. The "blind ant" algorithm can be adapted to work on digitized images, where microstructural information is stored in terms of pixels. The images can be of actual microstructures of porous materials or computed microstructures developed using simulation models of microstructure development during cement hydration.

Development of expansive stresses, e.g. caused by ettringite formation, can be estimated by simulation modeling of the effect of expansive inclusions in lattices (Thorpe and Garboczi 1990). If the tensile strength is exceeded, then cracking of the matrix can be assumed to occur.

In the view of the author, simulation is likely to become an important tool for obtaining information needed to predict the service life of concretes.

3 Concluding Remarks

Service life predictions of concrete are in an embryonic stage. At present we can predict with reasonable confidence the relative service lives of concretes, which is useful in the service life design of structures. However, we cannot be assured that the concrete will last as long as predicted, largely because of a lack of verification of the predictions. The large population of aging concrete structure provides an attractive opportunity to test the different service life approaches.

Many of the major needs for improving service life predictions have been reported by Tassios (1985) and Borges (1985). They include improved understanding of degradation mechanisms, improved characterization of service environments, development of advanced models, and development of standards and guidelines for making service life predictions. Also, the development of integrated knowledge system for service life predictions should be undertaken. The proposed integrated knowledge system would include mathematical models, expert systems, and databases on the performance and service lives of concrete.

4 Acknowledgements

The authors appreciates the many helpful suggestions by and discussions with Drs. Jon Martin and Geoffrey Frohnsdorff, National Institute of Standards and Technology (NIST) during the preparation of this paper. Also, the support by the U.S. Nuclear Regulatory Commission and NIST for the author's studies on service life predictions is appreciated.

5 References

ASTM E 632 (1981) Standard practice for developing accelerated tests to aid prediction of service life of building components and materials. ASTM, Philadelphia.

Atkinson, A. (1985) The time dependence of pH within a repository for radioactive waste disposal. Report AERE-R1177, Harwell Lab., Oxfordshire.

Atkinson, A., Haxaby, A., and Hearne, J.A. (1988) The chemistry and expansion of limestone-portland cement mortars exposed to sulphate containing solutions. Report NSS/Ri27, Harwell Lab., Oxfordshire.

Banavar, J.R. and Schwartz, L.M. (1989) Transport properties of disordered continuum systems. Phys. Rev. B, 39, p. 11965.

Bazant, Z.P. (1979) Physical model for steel corrosion in concrete: sea structures-theory. ASCE J. Structures Div., 105, pp 1137-1153.

Bazant, Z.P. (1979) Physical model for steel corrosion in concrete: sea structures-application. ASCE J. Structures Div., 105, pp. 1155-1166.

Biczok, E. (1967). Concrete Corrosion and Concrete Protection. Chemical Publishing Co., New York.

Borges, J.F. (1985) Report of discussion group on inorganic materials, in Part VI of Problems in Service Life Prediction of Building and Construction Materials (ed. L.W. Masters), Martinus Nijhoff Publishers, Dordrecht, pp. 267-270.

Bryant, L.M. and Mlaker, P.F. (1989) Estimation of concrete service life. U.S. Army Corps of Engineers Report No. J650-89-002-1420, Vicksburg, MS.

Clifton, J. R. (1989) Service life of concrete. National Institute of Standards and Technology, NISTIR 89-4086.

Garboczi, E.J. (1990) Permeability, diffusivity, and microstructural parameters: a critical review. Cem. Concr. Res., in press.

Garboczi, E.J. and Bentz, D.P. (1990) Analytical and numerical methods of transport in porous cementitous materials, in MRS Symposium on Scientific Basis for Nuclear Waste Management, in press.

Fagerlund, G. (1985) Essential data for service life prediction, in Problems in Service Life Prediction of Building and Construction Materials (ed. L.W. Masters), Martinus Nijhoff Publishers, Dordrecht, pp. 113-138.

Fagerlund, G. (1978) Prediction of the service life of concrete exposed to frost action, in Studies on Concrete Technology, Swedish Cement and Concrete Research Institute.

Frohnsdorff, G., Masters, L.W., and Martin, J.W. (1980) An approach to improved durability tests for building materials and components. National Bureau of Standards, NBS Technical Note 1120.

Kalousek, G.L., Porter, L.C., and Benton, E.J. (1972) Concrete for long-term service in sulfate environment, Cem. Conc. Res., 2, pp 79-90.

Martin, J. (1985) Service life predictions from accelerated aging tests using reliability theory and life testing analysis, in Problems in Service Life Prediction of Building and Construction Materials (ed. L.W. Masters), Martinus Nijhoff Publishers, Dordrecht, pp. 191-212.

Martin, J. (1989) Accelerated aging test design for coating systems, in Proc. 15th Inter. Conference on Organic Coatings Science and Technology, Athens, Greece, pp. 237-253.

Martin, J. and Embree, E. (1989) Effect of contaminants and cure time on EPDM single-ply joint strength. J. Materials in Civil Engineering, 1, pp. 151-168.

Neville, A. (1987) Why we have concrete durability problems, in Concrete Durability: Katherine and Bryant Mather International Conference (ed. J.M. Scanlon), American Concrete Institute, pp. 21-30.

Odler, I. and Gasser, M. (1988) Mechanism of sulfate expansion in hydrated portland cement. J. Am. Ceram. Soc., 71, pp. 1015-1020.

Plum, N.M, Jessing, J., and Bredsdorff, P. (1965) A new approach to testing of building materials, RILEM Bulletin, No. 30, p. 123.

Pommersheim, J. and Clifton, R. (1985) Prediction of concrete service-life. Materials and Construction, 18, pp. 21-30.

Pommersheim, J. and Clifton, R. (1990) Models of transport processes in concrete. National Institute of Standards and Technology, in press.

RILEM (1989) Systematic methodology for service life prediction of building materials and components. Materials and Structures, 22, 385-392.

Sentler, L., (1984) Stochastic characterization of carbonation of concrete, in 3rd International Conference on the Durability of Building Materials and Components, Technical Research Centre of Finland, pp. 569-580.
Siemes, A., Vrouwenvelder, A., Beukel, A. (1985) Durability of buildings: a reliability analysis. Heron, 30, pp. 3-48.
Spinks, J.W.T., Baldwin, H.W., and Thorvaldson, T. (1952) Tracer studies in set portland cement. Canadian J. of Technology, 30, pp. 20-28.
Tassios, T.P. (1985) Report of discussion group on inorganic materials, in Part III of Problems in Service Life Prediction of Building and Construction Materials (ed. L.W. Masters), Martinus Nijhoff Publishers, Dordrecht, pp. 139-144.
Thorpe, M.F. and Garboczi, E.J. Elastic properties of central-force networks with bond-length mismatches. Submitted to Phys. Rev. B.
Tuutti, K. (1982) Corrosion of Steel in Concrete. Swedish Cement and Concrete Research Institute, Stockholm.
Vesikari, E. (1988) Service life of concrete structures with regard to corrosion of reinforcement. Technical Research Centre of Finland, Research Report 553.

40 EVALUATION OF METAL FASTENER PERFORMANCE IN CCA TREATED TIMBER

J.N. CROSS
CAPCIS, University of Manchester Institute of Science and Technology (UMIST), Manchester, UK

Abstract
This paper describes a recent two-year research project at CAPCIS that evaluated the corrosion performance of metal fasteners in constructional timbers treated with copper - chromium - arsenic (CCA) preservatives. The background to the work and the experimental methodology are briefly summarised, and followed by presentation and discussion of the most important results.
Keywords: CCA Preservatives, Timber, Metal Fasteners, Corrosion, Electrochemical Monitoring.

1 Introduction

Metal fasteners are widely used for joining timber. Mild steel and galvanised mild steel are the most common materials employed and over the last few years doubt has been raised regarding the integrity of some joints made with these metals, such as those used in roof trusses[1-4]. The concern centred around the possibility of enhanced corrosion of the fasteners when copper-chromium-arsenic (CCA) preservatives were used to treat the timber.

Evidence for enhanced corrosion was based on observations of the corrosivity of fresh (un-fixed) CCA treated timber and treated timber exposed to very humid environments. However, timbers are rarely used in such conditions. A minimum fixation time is specified within which freshly treated timber should not be used, during which most of the preservative salts become fixed. The moisture content of timber in very humid environments is 22% and upwards, tending to be greater in treated than untreated timber. Roof trusses experience much lower moisture contents of, perhaps, 10-18%.

In view of these observations, a need was identified for the corrosion rate determination of metal fasteners in contact with treated timber at typical "service" moisture contents[5]. In this range, 10-22% moisture content, the electrical conductivity of timber is too low to enable conventional (d.c.) corrosion measurement and monitoring techniques, such as polarisation resistance, to be used. However, UMIST have developed and are using electrochemical noise and impedance techniques which are applicable to low conductivity media.

A collaborative research project was therefore instigated at CAPCIS, funded by the Department of the Environment (through the Building Research Establishment) and two industrial sponsors, Hickson World Timber Limited and Rentokil Limited. The main objective of the project was to determine, using modern electrochemical corrosion monitoring techniques, the corrosion performance of metal fasteners in CCA treated timber with particular emphasis on normal timber moisture contents.

This paper presents the essential findings of the investigation, while the full results are to be published by the Building Research Establishment.

2 Programme of work

In January 1986, a two-year experimental programme was initiated involving the exposure of timber blocks (dimensions 20cm x 10cm x 2.5cm) in which various metals were embedded. The blocks were exposed in artifical laboratory and natural roof space environments. The variables which were studied included:-

(a)	Timber type	- Baltic Redwood	- Pinus Sylvestris
		- European Spruce	- Picea Abies
(b)	Treatment type	- CCA salt formulation	- Tanalith C
		- CCA oxide formulation	- Celcure A
(c)	Laboratory temperature	- 10°, 25° and 35°C	
(d)	Relative humidity	- 87% to 100% RH	
(e)	Timber moisture content	- 17% to 26%	
(f)	Metals	- Mild steel, zinc electroplated steel, galvanised steel, stainless steel (type 304) and an aluminium alloy (BS 6063).	

Site exposures were made at four locations in the North West of England (Liverpool, Flixton, Sale and Ramsbottom); these were chosen to represent the spectrum of roof design, usage pattern and hence micro-climates present in dwelling houses. The measured moisture contents of timber samples were from 10% to 18.5%.

The corrosion monitoring techniques used to determine corrosion rates and corrosion mechanisms for the various metals embedded in timber were A.C. impedance, zero resistance ammetry (ZRA) and electrochemical noise (ENC). Information relating to these techniques can be found in refs 6 to 11. The high resistivity of dry and medium-moisture content timber precluded the use of conventional corrosion monitoring techniques such as linear polarisation resistance measurement (LPRM). A novel electrode configuration was developed to allow monitoring in the timber; the 'comb electrode' as shown in Figure 1.

Fig. 1. CAPCIS comb electrodes. A three electrode arrangement is inserted in the timber.

3 Results

Selected results are presented here to illustrate the most important effects that were observed, although additional confirmatory and proving experiments were conducted. Most of the experiments were carried out using untreated and salt formulation CCA treated timbers.

3.1 The effect of moisture content

Table 1 illustrates the strong dependance of corrosion rate of mild steel on timber moisture content.

Table 1. Corrosion rate of mild steel (μm/year)

	Moisture content/temperature		
	17%/35°C	**21%/25°C**	**24.5%/25°C**
Untreated redwood	1	18	26
	19%/35°C	**26%/25°C**	**30%/25°C**
Treated redwood	2	250	500

(a) The corrosion rates were estimated by electrochemical impedance measurements and checked by visual examination.
(b) The moisture contents were determined by moisture meter. The values for treated redwood are not corrected for the effects of salts.
(c) Preservative loading was 14.5Kg/m^3, after fixing.

The data indicate that there exists a 'critical' moisture content below which the rate of corrosion is negligible, and above which the rate of corrosion increases measurably in both treated and untreated timber, especially the former. This effect can probably be explanined by the increase in conductivity of the timber at higher moisture contents, enabling corrosion by acetic acid and/or differential aeration to proceed. In treated timber, the presence of hygroscopic salts of sodium sulphate are believed to further increase the conductivity.

3.2 Differences between metals

Table 2 compares the typical corrosion rates obtained after approximately 1 years' exposure at high moisture content in treated (30% moisture content) and untreated (24.5% moisture content) redwood. **N.B.** The data for electroplated zinc were determined using pressure saturated timber (sodium sulphate saturated **untreated** timber and distilled water saturated **treated** timber) after 5 months exposure.

The results confirm that more corrosion resitant materials such as stainless steel can be used in very damp timber, both treated and untreated.

Table 2. Materials and estimated general corrosion rates (μm/yr) at 25°C and with high moisture contents

	Untreated redwood (24.5% MC)	Treated redwood (30% MC)
Mild steel	26	500
Electroplated zinc[b]	39	88
Galvanised steel[c]	13	43
Stainless steel	0.22	0.26
Aluminium[d]	0.36	4.7

(a) The corrosion rates were estimated by A.C. electrochemical measurements and checked by visual examination.
(b) The zinc coating had corroded away completely (15-30μm) from the underlying steel in both treated and untreated redwood. Attack to the steel was very slight in the sodium sulphate saturated untreated timber and in the range 15-30μm in the distilled water saturated treated timber.
(c) The zinc coating had corroded away completely from the nail plate surfaces in contact with the treated timber and localized attack of the mild steel had progressed to estimated depths of up to 100μm (0.1mm). There was evidence of slight zinc corrosion on the

contact surfaces of the nail plate in untreated timber.

(d) Pitting attack had progressed to a depth of approximately 10μm (0.01mm) on combs in untreated timber and approximately 50μm (0.05mm) on combs in treated timber. Furthermore, intergranular corrosion had occurred to the extent of approximately 500μm (0.5mm) on combs in the treated timber.

3.3 Site exposure testing

Continuous temperature and relative humidity recording, coupled with periodic moisture content determination was undertaken at the four exposure sites over the ten months between February and November 1987. Figure 2 shows timber blocks in the loft space at Liverpool. The average monthly temperatures ranged from 5°C (November) to 17°C (June), the average monthly relative humidities ranged from 55% (June) to 85% (November) and the mositure contents ranged from 10% to 14.5% for the untreated timber and from 12.5% to 18.5% for the treated timber.

Corrosion monitoring was carried out on mild steel, zinc electroplated steel and aluminium electrodes embedded in treated and untreated redwood. Corrosion rates were generally immeasurably low, the highest rate indicated being some 0.02μm/yr for zinc electroplated steel in treated timber. The results of corrosion monitoring under service conditions are in accordance with those obtained in the laboratory studies.

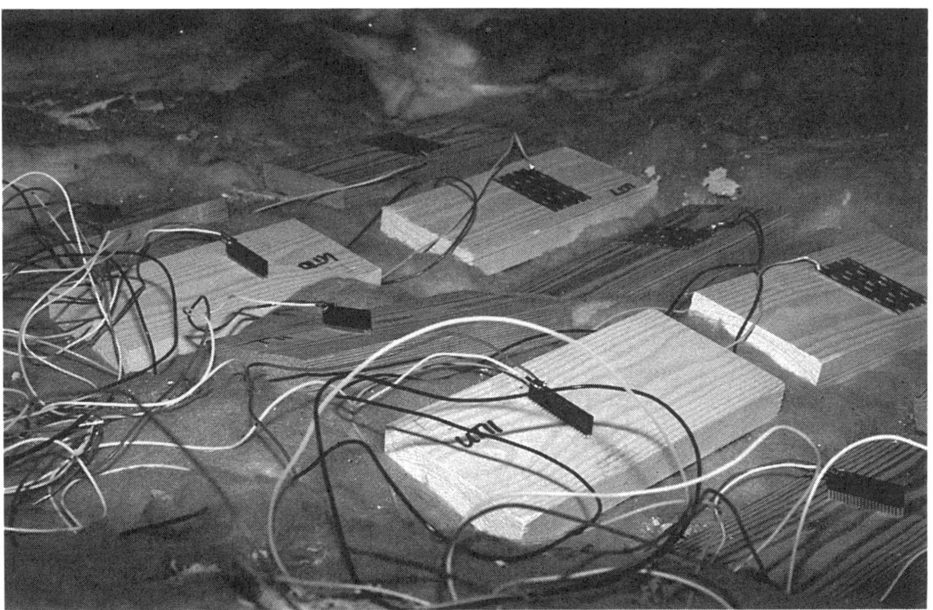

Figure 2. Timber blocks/electrodes exposed in the loft space at Liverpool.

3.4 General

A number of additional experiments were carried out, the results of which are summarised below. Some of these experiments were based on small numbers of observations; this, coupled with the fact that the corrosion monitoring techniques being employed were often close to their lower limits of detection, demands that care should be taken in the interpretation of the results.

The effect of temperature on corrosion rate at a relative humidity of 100% was variable. At 10°C, the corrosion rate of mild steel in both treated and untreated redwood was similar. At 25°C, the corrosion rate of mild steel in treated redwood had increased fifty fold over that at 10°C, while in untreated redwood the increase at 25°C was very slight (see Table 1, right hand column, for rates at 25°C).

Redwood and spruce were compared in their corrosivity towards mild steel. Redwood was found to be more corrosive than spruce by a factor of approximately 5½ in the untreated condition and approximately 8½ in the treated condition, despite the apparent higher (3%) moisture content of the spruce samples.

At moisture contents of approximately 30% and 26% respectively, salt formulation CCA treatement (loading of 14.5kg/m^3) was found to be some five times more corrosive towards mild steel than oxide formulation CCA treatment (loading of 5.2kg/m^3).

In treated redwood of approximately 26% moisture content, the corrosion rate of mild steel combs sampling along the grain was approximately 50% greater than combs sampling across the grain, this probably being due to increased ionic conductivity along the grain.

4 Conclusions

The essential conclusion of this two year research programme is that at normal moisture contents i.e. below approximately 19%, the corrosion rates of mild steel, electroplated/galvanised steel, aluminium alloy and stainless steel are extremely low in both CCA treated and untreated timbers, and that the enhanced corrosion due to CCA preservatives is insignificant. Corrosion rates of several microns per year are low enough to be ignored.

At moisture contents of approximately 26%, rates of some 0.2mm/yr are experienced by steel and zinc (in the form of galvanising and zinc electroplate) in treated timber. At very high moisture contents of approximately 30%, steel in treated timber corrodes at up to 0.5mm/yr. Stainless steel (type 304) and aluminium alloy (BS 6063) exhibited negligbly low corrosion rates under all test conditions except for pitting and severe intergranular corrosion of the aluminium in treated timber at very high moisture levels.

It is believed that the results of this work should significantly alleviate the concerns of builders, architects and local authorities regarding any corrosion risks associated with CCA preservatives.

5 References

1. Greater London Council **Development and Materials Bulletin,** nos. 87,91,135.
2. Building Research Advisory Service, Cl. Sf. B. Xt. 6(S2).
3. Building Research Establishment Information Sheet ISII/77 (August 1977).
4. TRADA Seminar **Integrity of Trussed Rafter Roofs,** 18th May, 1979.
5. Bailey, G and Schofield, M.J. (1984) Corrosion of Metal Fasteners in CCA Treated Timber. **J. Inst Wood Science.,** 14-18.
6. Callow L.M., Richardson J.A. and Dawson J.L., British Council Journal, **11**, 123, (1976).
7. Callow L.M. Richardson J.A. and Dawson J.L., British Corrosion Jornal, **11**, 132, (1976).
8. Hladky K., Callow L.M., and Dawson J.L., Br. Corr. J. **15**, 20, (1980).
9. Haruyama S., and Tsuru T., ASTM STP 727, P 167 (1981).
10. Dawson, J.L. Gill, J.S. Al-Zankin, I.A. Woollam, R.C. (1986) **Electrochemical Corrosion Testing using Electrochemical Noise, Impedance and Harnomic Analysis.** Dechema-Monographs vol.-VCH Verlagsgesellschaft.
11. Dawson J.C., Farrell D.M., Aylott P.J. and Hladky K. Paper No. 31, Corrosion '89 New Orleans, Pub. NACE 1989.

Acknowledgements

The author wishes to thank the sponsors of this research programme for permission to publish the work.

41 A COMPARISON OF THE PROPERTIES OF OPC AND PFA CONCRETES IN 30-YEAR-OLD MASS CONCRETE STRUCTURES

M.D.A. THOMAS
Building Research Establishment, Garston, UK

Abstract
This paper presents the results of an investigation of two mass concrete structures, a buttress dam and a sea defence wall,. both constructed over 30 years ago using concrete with and without pulverized-fuel ash (pfa). The pfa used in both cases was unclassified ash from local power stations. A range of laboratory tests were carried out on concrete cores taken from these structures in order to compare the durability properties of opc and pfa concrete within the same structure. The compressive and tensile strength of cores were slightly greater for the pfa concrete in both structures. Average carbonation depths were less than 10mm for all cores tested and the results from initial surface absorption tests (ISAT) and gas permeability tests for all the concretes were typical of concretes of low to average permeability. Determinations of chloride content in powder samples taken from the seaward face of the sea wall show the pfa concrete to have a significantly increased resistance to the ingress of chloride ions compared with the opc concrete in the same structure.

The findings from this study show that the use of unclassified pfa as a partial replacement in these concretes to achieve reductions in hydration temperatures did not compromise the strength or durability of the hardened concrete.
Keywords:Concrete structures, durability, permeability, puverized-fuel ash

1 Introduction

Recent comparisons of the performance of opc and pfa concrete in real structures (Thomas, 1989; Thomas et al. 1990) have confirmed the benefits of incorporating good quality pfa in structural concrete, with increased strength and reduced permeability being observed for pfa concrete compared with equal grade opc concrete. In both these studies the pfa used complied with BS 3892: Part 1 'Specification for pulverized-fuel ash for use as a cementitious component in structural concrete' (British Standards Institution, 1982) and the comparable opc and pfa concrete mixes were designed on the basis of equal strength grade and workability.

This paper describes two mass concrete structures more than thirty-years old where unclassified, 'run-of-station' pfa was incorporated into the concrete mix as a partial replacement for opc. Both structures also contained a comparable opc concrete (same cement content) with no pfa replacement. The properties of concrete cores from these structures were measured in a laboratory study in order to assess the contribution of pfa to the long-term performance of the concrete.

2 Structures examined

2.1 Lednock Dam, Scotland

The Lednock Dam is a mass concrete buttress dam which was constructed between 1954 and 1957 using both opc concrete and concrete containing 20% pfa from Braehead power station. The construction of the dam has been well documented (Allen, 1959) and details of the pfa and the concrete mixes used in the buttresses are given in Tables 1 and 2. Pfa was utilised as it was readily available, would reduce the demand on opc (then in short supply in Scotland) and would reduce the overall cost of construction. However, sufficient information on the suitability of Braehead pfa for use in concrete was not available at the commencement of construction and consequently opc concrete was used for concreting during the first eighteen months.

Fluctuations in the carbon content of the pfa supplied (due to the wide variety of coals used at Braehead power station) produced a wide variation in the strength of works cubes. Consequently, the minimum 28-day strength of the pfa concrete was reduced to 2000 lb/sq.in. (13.8 MN/m^2) compared with the minimum strength of 2800 lb/sq.in. (19.3 MN/m^2) specified for the opc concrete.

Temperature rises were monitored during construction and the peak temperature rise in the pfa concrete during hydration was reduced by up to 20% compared with the opc concrete. It was concluded that the lower heat rise measured in the pfa concrete decreased the tendency for thermal cracking (Allen, 1959).

2.2 Aberthaw Sea Wall, Wales

The second structure examined was a sea wall at Aberthaw 'A' Power Station constructed in the 1950's. Although no documentation of the mix details was available, concrete from this structure was sampled and tested as part of an earlier survey of pfa concrete carried out between 1978 and 1980 (Taylor Woodrow Research Laboratories, 1990).

The results from this survey reported the aggregate/cement ratio to be 4.1 with a fine aggregate proportion of 34% (wt.% of total aggregate). Records available at the time of this earlier investigation indicated that both opc and pfa concrete were used in the construction. Although tests to locate the pfa concrete were inconclusive, chemical analysis showed the concrete from one particular location to have a relatively low $Ca(OH)_2$ content and high alumina and titania contents. Such results may be indicative of the presence of pfa, and calculations based on alumina and titania contents suggested levels of pfa replacement of 25 to 30%. Common practice in CEGB construction at this time was to use 1:1½:3 mix

Table 1 Chemical composition of Braehead pfa

Oxide (%)	Braehead pfa
SiO_2	42-51
Al_2O_3	28-41
Fe_2O_3	5-17
CaO	1.3-4.4
MgO	0.6-2.1
Na_2O)) K_2O)	0.4-5.1
SO_3	0.3-0.8
Carbon	1.5-9.2

Table 2 Details of concretes sampled from Lednock Dam

Concrete Mix	PFA Cont. %	Agg/ Cem	W/C Ratio	C.F.*	28 day strength (MN/m^2)	
					Trial Mixes	Works Cubes
F	0	9	0.64	0.804	32	26
F/FA	20	9	0.60	0.806	26	20

* Compacting factor

proportions for general concreting, with replacement levels of 20% pfa where used. The likely source of pfa for this construction would have been Portishead 'B' Power Station.

3 Experimental details

Engineering drawings of the Lednock Dam gave details of the actual concrete mix used in each pour for the whole construction. Two adjacent buttresses were selected where the concrete in the lifts at ground level was opc and pfa concrete respectively. Six 225mm diameter and three 150mm diameter cores were cut from the south-west face of each buttress. The taking of such large diameter cores was necessitated by the large maximum aggregate size (75mm) used in the construction of the dam.

In a preliminary investigation to locate the pfa concrete in the sea wall at Aberthaw, small concrete samples were removed by hammer and chisel from every concrete pour in the section of wall running from the most easterly point for 500m. In all, some 30 samples were taken from 20 different sections of the wall. The location of these samples included the location of the cores taken during the previous survey (Taylor Woodrow Research Laboratories, 1990). These samples were washed with dilute hydrochloric acid prior to examination by optical and electron microscopy. In contrast to earlier findings, spherical pfa particles were easily identified in abundance in one particular section of the wall. This sample coincided with a sample taken in the previous study which was diagnosed as containing pfa on the basis of the $Ca(OH)_2$, alumina and titania contents with estimated pfa levels of 25 to 30% being reported.

In the present study, six 100mm diameter cores were taken from the landward side of this section of the wall. In addition, samples of the concrete were also obtained by drilling the concrete on the seaward face with a rotary hammer drill, the powdered samples being collected in depths up to 30mm in increments of 5mm. The sampling procedure was repeated for the adjacent concrete section which was found to contain no pfa.

The concrete cores were later cut into test specimens for the following schedule of tests;(1) density, (2) water absorption, (3) compressive strength, (4) indirect tensile strength, (5) initial surface absorption (ISAT), (6) original water content, (7) cement content, (8) depth of carbonation, (9) oxygen permeability, (10) preparation of petrographic sections and (11) $Ca(OH)_2$ content. Tests (1) to (7) were carried out in accordance with the relevant part of BS 1881 'Methods of testing concrete (British Standards Institution, 1983). The depth of carbonation was determined by spraying freshly broken concrete surfaces with phenolphthalein indicator and measuring the depth of the uncoloured zone. $Ca(OH)_2$ determinations were carried out using thermogravimetric analysis. The permeability to oxygen of 100mm diameter x 50mm slices from the sea wall was determined using a permeability cell and test method developed at the British Cement Association (Lawrence, 1986) The large aggregate (up to 75mm) used in the Lednock Dam necessitated the testing of much larger cores which precluded the use of the permeability cell. A large scale triaxial cell was used to measure the permeability of 225mm diameter x 150mm concrete cores from this structure. The principle of the test is the same as that used in the BCA test. Porous plates were placed on the two flat surfaces of the test specimen and the curved surface was sealed using a neoprene membrane. The whole assembly was placed in the triaxial cell which was filled with water at a pressure of 7 bar. Oxygen was then applied at a pressure between 2 and 5 bar to one of flat surfaces via the porous plate and the gas flow was measured from

Table 3 Properties of Concrete Cores

Structure	Mix	Density Kg/m^3	Water Absorp. (%)	Orig. Water (%)	Cem. Cont. (%)	$Ca(OH)_2$ Content (%)
Lednock	opc	2399	2.09	7.1	10.9	1.55
Dam	pfa	2442	1.90	6.6	-	0.94
Aberthaw	opc	2473	3.86	7.9	16.0	3.68
Sea Wall	pfa	2450	3.95	7.1	-	0.77

the other surface. The coefficient of permeability was then calculated using a combination of Darcy's law and the Poiseuille equation.

The chloride content of the powdered concrete samples from the sea wall was determined using x-ray fluorescence spectrometry.

4 Results

Visual examination of cores showed all the concretes sampled to be well-compacted, with little excess voidage and with well-graded and evenly distributed aggregate. Petrographic examination of thin sections by optical microscopy confirmed these findings and successfully identified abundant pfa spheres in the concretes designated as 'pfa concrete'. The pfa concrete sections contained lower quantities of portlandite than the opc concrete sections and there was evidence of a relatively high proportion of carbonaceous matter in the pfa concretes.

Density, porosity and water absorption values are given in Table 3 together with results from the original water, cement and $Ca(OH)_2$ content determinations. The pfa concretes showed reduced $Ca(OH)_2$ content and original water content compared with the opc concrete from the same structure; otherwise no consistent difference was observed between the opc and pfa concretes from the same source. The concretes from the sea wall have a higher measured cement content and water absorption than the concretes from the dam.

Average compressive and tensile strength and carbonation data are given in Table 4, and each of these results represents the average of 3 test specimens. Figure 1 shows the average compressive strength of the concrete cores together with the strength results for laboratory trials and works cubes for the concrete mixes used at the Lednock Dam. Both 30-year-old opc and pfa concrete cores have considerably greater strength than the laboratory and works cubes, the proportional strength of the cores to 28-day works cubes being 238% for pfa and 160% for opc concrete. The concretes from the sea wall have higher strength and lower depths of carbonation than concretes from the dam and within either location the pfa concrete exhibits higher strength

Table 4 Strength and carbonation data from concrete cores

Structure	Mix	Av. Strength (MN/m²)		Carbonation Depths (mm)	
		Comp.	Tensile	Range	Mean
Lednock Dam	opc	42.5	3.86	2.5-10.0	5.0
	pfa	48.0	3.94	5.5-12.5	8.5
Aberthaw Sea Wall	opc	59.2	4.66	0-0.5	0.5
	pfa	63.5	4.71	1.0-1.5	1.5

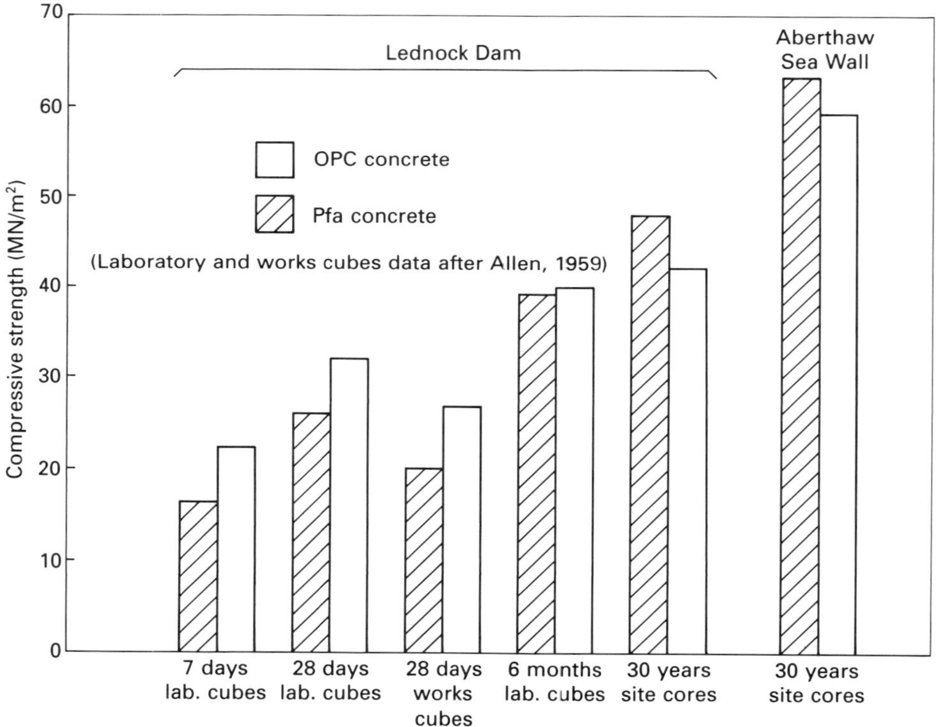

Fig. 1 Strength of Concretes

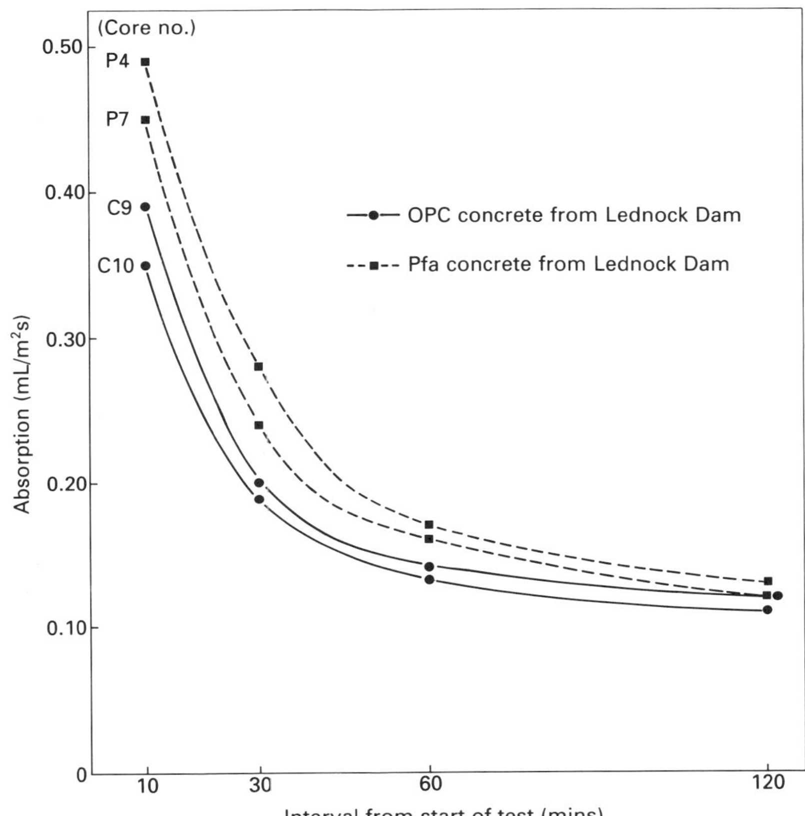

Fig. 2 ISAT results for concrete cores from the Lednock Dam

and depth of carbonation than the opc concrete.

ISAT and oxygen permeability results for individual specimens are given in Figures 2 and 3. The ISAT results for the concrete cores from the Lednock Dam and permeability results for all the concrete specimens are typical of concretes of average permeability. There is little significant difference between the ISAT values for opc and pfa concretes (opc concretes have slightly lower absorption values) although the permeability results show the opc concrete from both structures to be considerably more permeable than the pfa concrete from the same structure. The average permeability of the opc concrete is between 70% and 105% higher than the average for the pfa concrete. No consistent variation in the permeability of opc concrete with depth of sample was observed. However, the results in Figure 3 show that for all the pfa concrete cores the section including the surface is consistently more permeable compared with sections taken from the same concrete at greater depth. The concretes from the dam are on average 2.5 times more permeable than the concretes from the sea wall.

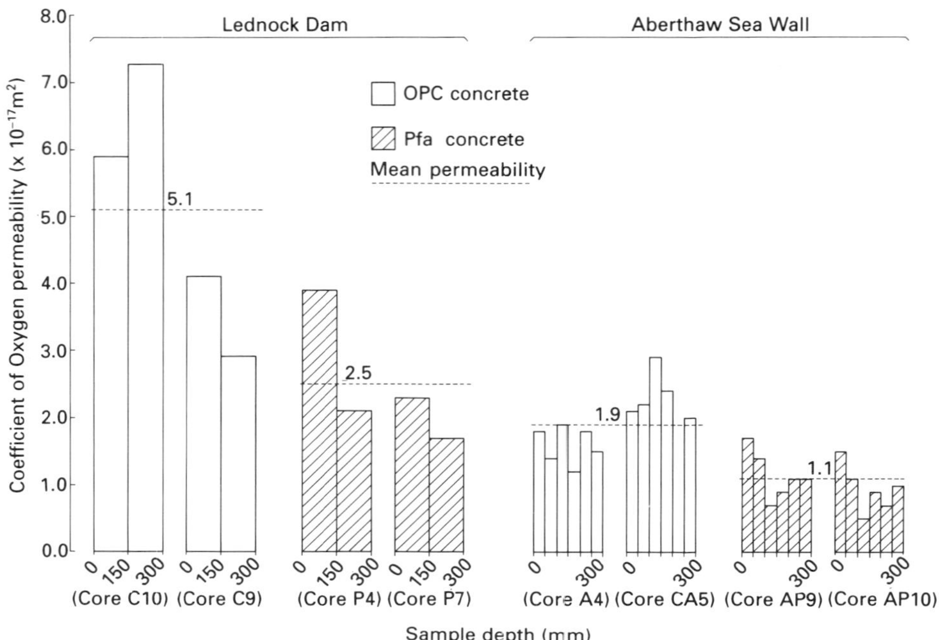

Fig. 3 Oxygen permeability results

Figure 4 shows the variation in chloride concentration with depth for the concrete in the sea wall. The chloride ion concentration reduces with depth in both concretes although the reduction is considerably more marked in the pfa concrete. Consequently, in spite of similar concentrations of chloride at the surface, at depths below about 15mm the chloride concentration in the opc concrete is approximately twice that in the pfa concrete.

5 Discussion

The two structures examined in this study were constructed using both opc and pfa concrete and these concretes were successfully located and sampled. The concretes examined from the Lednock Dam were known to be of equal total cement (opc + pfa) content, the level of replacement in the pfa concrete being 20%. The pfa used in this concrete was unclassified, run-of-station pfa and was subjected to very limited quality control, a maximum carbon content of 10% being the only specification. The low quality of the pfa is reflected in the strength results of the works cubes, the pfa concrete only achieving 71% of the strength of the opc concrete at 28 days. The proportion of strength would be expected to be higher for a good quality pfa Matthews and Gutt, 1978). Although no precise information on the concrete mixes and constituents used in the sea wall is available,

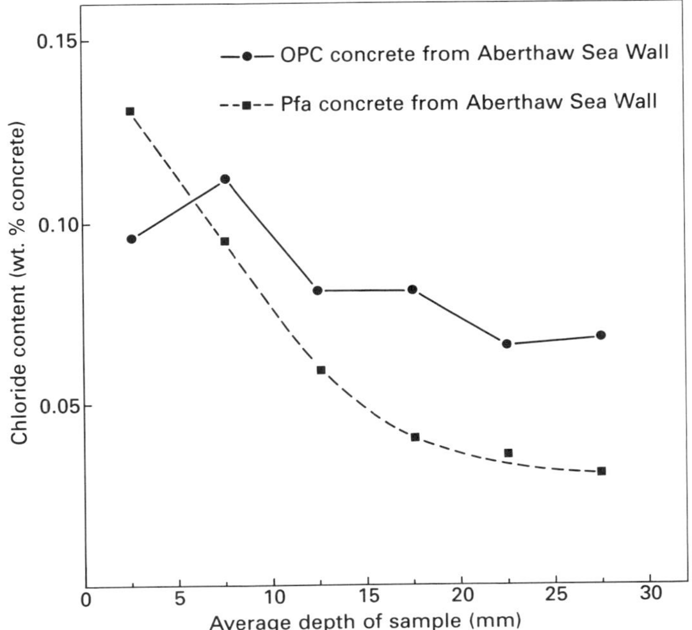

Fig. 4 Chloride concentration profiles for concretes from Aberthaw Sea Wall

general practice and the results of a previous investigation would suggest that the opc and pfa concretes would have equal cementitious content with a replacement level of between 20-30% pfa being used.

The analyses of cores from these thirty-year-old structures showed that the opc and pfa concretes used in the sea wall had higher cement contents and lower water/cement ratios compared with Lednock Dam concretes. This is reflected in the higher strength, reduced depths of carbonation and lower permeability observed for the concrete samples taken from the sea wall.

Both opc and pfa concrete cores from the Lednock Dam showed considerable strength increase when compared with the results from 28-day works cubes. The improvement in strength was much more marked for the pfa concrete (in spite of the poor quality of the pfa) with the compressive strength of pfa concrete cores being 13% greater than the opc concrete cores. Similarly, for the concrete sea wall, the pfa concrete cores had a compressive strength 7% greater than the opc concrete cores.

The increased strength of the thirty-year-old pfa concrete cores in these structures is the result of pozzolanic reaction, evidence of which is provided by the depletion of $Ca(OH)_2$ in these concretes compared with the corresponding opc concrete.

In addition to having increased strength the pfa concretes were less permeable than the opc concrete from the same structure. The reduced permeability of the pfa concretes can be variously ascribed to; (i) the reduced water content of the pfa concretes (see Tables 2 and 3), (ii) pozzolanic reaction between pfa and $Ca(OH)_2$ resulting in the precipitation of hydration products which may block pores leading to a finer pore structure (Manmohan and Mehta, 1981) and (iii) temperature rises during hydration which may enhance the pozzolanic reaction thereby decreasing the permeability of pfa concrete but lead to increased permeability in opc concrete due to poorer dispersion of hydration products (Owens, 1985; Bakker, 1983). This last process may partially explain the reduction in permeability with depth observed for the pfa concretes (ie increased temperature rises with depth during hydration).

The permeability of concrete is an important parameter in determining the durability as it provides a measure of the resistance of the concrete to the penetration of deleterious agents. In reinforced concrete, the resistance of the concrete cover zone to the penetration of carbon dioxide and chloride ions is of particular importance in terms of protecting the steel from corrosion. Although the concretes examined were not reinforced, the permeability data suggest that the pfa concrete could be expected to provide at least equal protection to steel reinforcement as the opc concrete.

Carbonation and ISAT results from the Lednock Dam concretes are at variance with the permeability data, with the pfa concrete exhibiting increased carbonation and surface absorption. The contrast between the permeability and carbonation data may be partially the result of chemical differences between opc and pfa concrete, the lower $Ca(OH)_2$ of the latter providing less resistance to carbonation. In addition, these differences are thought to emanate from changes in the microstructure of the concretes with depth. It has been shown (Patel et al., 1985) that the pore structure of cement pastes becomes more refined with depth from the drying surface and an earlier examination of field concretes (Thomas et al, 1990) showed that the gradient of the change in pore structure with depth is more marked for pfa concrete. In the outer surface zone (0 to 10mm approx) the pore structure of pfa concrete was found to be coarser than opc concrete but at greater depths the situation was reversed. This may result in a more rapid carbonation and higher initial surface absorption of the outer zone of the pfa concrete despite the overall lower permeability of the pfa concrete cover. At greater depths, however, the more refined pore structure of pfa concrete might be expected to provide greater resistance to carbonation compared with opc concrete.

In the structures examined in this study carbonation had not penetrated to any significant depths and could be considered to be inconsequential if the concretes had contained steel reinforcement. The chloride concentration profiles for the concrete sea wall show a marked increase in the diffusion of chlorides in the opc concrete. Existing guidelines (Everett and Treadaway, 1980) state that there is a 'medium risk' of reinforcement corrosion when the chloride level at the location of the steel approaches 0.4% (by weight of cement). The cement content of the concrete sea wall was determined to be 16% (by weight) and consequently if steel reinforcement had been present at locations where the chloride level was above 0.064% (by weight of concrete), it would be at risk of corrosion. The results in Figure 4 suggest that steel reinforcement with less than 30mm cover in the opc concrete would be at risk from corrosion, whereas for the pfa concrete, the steel reinforcement would probably be protected from corrosion provided the depth of cover was at least 15mm.

Overall, the results from this study support the decision to use pfa in these structures to reduce cost, demand on cement and hydration temperatures even though the pfa was unclassified. Generally, the properties of thirty-year-old opc and pfa concrete cores were similar and the concretes were in good condition. Improvements in the strength and permeability of the pfa concretes compared with opc concrete were observed and were attributed to the pozzolanic reaction. Similar findings have been reported (Cabrera and Woolley, 1985) from a study of twenty-five year old foundation structures constructed using plain opc concrete and a comparable concrete with 20% of the cement replaced by a pfa which did not meet the present British Standard specification. In addition, two recently published studies (Thomas, 1989; Thomas et al, 1990) of concrete structures constructed with equal grade opc and pfa concretes using quality assured pfa have confirmed the considerable improvements in concrete durability that can be achieved thorugh the use of pfa.

6 Conclusions

Two mass concrete structures, each constructed using both opc and pfa concrete, were sampled in order to compare the performance of the two types of concrete. Both structures were in good condition and the results from the laboratory tests on cores show all the concretes to be of good quality. The pfa concretes from both structures had higher strengths and lower permeabilities compared with the opc concrete from the same structure and these improvements are attributed to the pozzolanic reaction. Slightly higher carbonation depths were observed for the pfa concretes, but carbonation was generally low (less than 10mm) for all the concretes tested. The diffusion of chlorides was considerably reduced by the presence of pfa in the concretes in the Aberthaw sea wall. The results from these tests show that low quality pfa can be used in mass concrete structures to reduce cost and temperature rise during hydration without detriment to the performance of the hardened concrete.

7 Acknowledgement

The work described has been carried out at the Building Research Establishment in collaboration with a National Power/Imperial College Research Fellowship. The work forms part of the research programme of the Building Research Establishment and is published by permission of the Director. The author would like to thank Mr C Beak of the North of Scotland Hydro-Electric Board for his co-operation and Mr C Haynes for his assistance with both the site and laboratory investigation.

8 References

Allen. A.C. (1959) Features of Lednock Dam, including the use of fly ash. **Proc. Instn. Civ. Engineers** 13, 179-196.

Bakker, R.F.M. (1983) Permeability of blended cement concretes. **Proc 1st Int. Conf. on the use of Fly Ash, Silica Fume, Slag and other Mineral By-Products in Concrete, ACI, SP-79,** Vol 1, American Concrete Institute, Detroit 415-433.

British Standards Institution (1982) **Specification for pulverized-fuel ash for use as a cementitious component in structural concrete.** BS3892: Part 1, BSI, London.

British Standards Institution (1983) **Testing Concrete.** BS1881. BSI, London.

Cabrera, J.G. and Woolley, G.R. (1985) A study of twenty-five year old pulverized fuel ash concrete used in foundation structures. **Proc. Instn. Civ. Engrs**, Part 2, March, 149-165.

Everett, L.H. and Treadaway, K.W.J. (1980) Deterioration due to corrosion in reinforced concrete. **BRE Information Paper IP 12/80,** Building Research Establishment, Garston.

Lawrence, C.D. (1986) Measurements of permeability. **Proc. 8th Int. Cong. Chemistry of Cement**, Volume V, Rio de Janeiro, 29-34.

Manmohan, D. and Mehta, P.K. (1981) Influence of pozzolanic, slag and chemical admixtures on pore size distribution and permeability of hardened cement pastes. **Cement, Concrete and Aggregates,** 3(1), 63-67.

Matthews, J.D. and Gutt, W.H. (1978) Studies of fly ash as a cementitious material. **Proc. 1st Int. Ash Marketing and Technology Conference,** London, 284-302.

Owens, P.L. (1985) Effect of temperature rise and fall on the strength and permeability of concrete made with and without fly ash. **Temperature Effects on Concrete, ASTM STP858** (ed T.R. Naik) American Society for Testing Materials, Philadelphia, 134-149.

Patel, R.G. Parrott, L.J. Martin, J.A. and Killoh, D.C. (1985) Gradients of microstructure and diffusion properties in cement paste caused by drying. **Cement and Concrete Research,** 15, 343-356.

Taylor Woodrow Research Laboratories (1990). Surveys of existing concrete marine structures. **Concrete in the Oceans Technical Report No. 21. Offshore Technology Report OTH 87/244.** HMSO, London/

Thomas, M.D.A. (1989) An investigation of conventional ordinary Portland cement and pulverized-fuel ash concretes in 10-year-old concrete bridges. **Proc. Instn. Civ. Engrs,** Part 1, Dec, 1111-1128.

Thomas, M.D.A. Osborne, G.J. Matthews, J.D. and Cripwell, J.B. (1990). A comparison of the properties of opc, pfa and ggbs concretes in reinforced concrete tank walls. **Proc. Concrete for the Nineties,** Leura, NSW, Australia.

42 DURABILITY TESTS ON PRECAST CONCRETE WALL PANELS WITH EXPOSED AGGREGATE FINISHES

S.L. LEE, C.T. TAM, W.J. NG, S.K. TING, Y.H. LOO
Department of Civil Engineering, National University of Singapore, Singapore

Abstract
Exposed aggregate finishes provide an aesthetic alternative to bricks, tiles and painted concrete for building facades. Besides creating identities through variation, its main advantage over the painted concrete finishing is the reduction of long-term maintenance costs. To evaluate the long-term performance and durability of exposed aggregates as surface finishes in precast concrete wall panels, an investigation was carried out involving four types of specimens and a number of tests. Included in these tests are shear bond strength tests of the exposed aggregate layer and initial surface absorption tests, both before and after simulated weathering, carbonation tests, sulphuric acid tests and tests for microbial growth. Weathering effects were simulated by cyclic compressions as well as wetting and drying cycles. It was observed that the quality of the face mix was generally lower than that of the body mix even though they were of the same water/cement ratio. In order that the exposed aggregate layer may contribute to effective cover for reinforcement, the water/cement ratio ought to be reduced while increasing the cement content of the face mix so that the rate of water penetration, carbonation and microbial growth will be effectively reduced. Also, exposed aggregate surface made with graded aggregates is preferred over single-sized aggregates as the former provides higher shear bond strength and better resistance against simulated temperature effects.
Keywords: Exposed Aggregates, Surface Finish, Carbonation, Microbial Growth, Simulated Weathering, Durability, Initial Surface Absorption.

1 Introduction

1.1 Scope of the study

The present investigation deals with precast panels with exposed aggregate surface finish. Four types of precast panels each with a different system of reinforcement or different exposed aggregate were tested. The performance of the exposed aggregates as surface finish was evaluated through a study on the base concrete to exposed aggregate bond. Direct shear test on small areas of exposed aggregate surface was conducted on specimens which had been exposed to various degrees of simulated weathering as well as on unexposed specimens. Effects of temperature induced stresses was simulated in a cyclic eccentric compression test while alternate climatic changes of sunshine and rain was simulated in the laboratory by using a weatherometer.

The rate of penetration by carbon dioxide was studied by exposing the specimens to carbon dioxide at various concentrations in sealed chambers. The extent of discoloration by sulphur oxides in the atmosphere was studied by applying sulphuric acid of different concentrations to the exposed aggregate surface.

1.2 Microbial growth

Various micro-organisms ranging from viruses, bacteria, fungi (moulds) and algae, sporadically appear on building surfaces resulting in their unsightly discoloration. This is inevitable as they are abundant in the air and are easily transferred. Generally the dusty spots and discoloration on surfaces of buildings are caused by fungi and algae. To deal successfully with this menace, we need to determine the predominant species that are habitually attacking the exposed concrete surfaces. This can be achieved firstly by propagating the organisms in laboratory media to get more material for study and secondly to intentionally inoculate or apply the predominant microbes on concrete slabs under laboratory conditions but simulating outdoor conditions.

2 Test program

2.1 Test specimens

All specimens were 300 mm wide, 400 mm long and 100 mm in depth. Four types of specimens were cast. The cube strengths and other specimen details are presented in Table 1. The mix proportion of the concrete (Cement:Water:Fine Aggregate:Coarse Aggregate) for the exposed aggregate layer (face mix) was 370:185:370:1480 (kg/m^3) and for the precast element (body mix) was 360:180:775:1025 (kg/m^3). The exposed aggregate layer was first cast against formwork coated with a retarder. The body mix was then added. After 24 hours the specimens were demoulded and the aggregates exposed by washing and brushing away the paste in between them.

Table 1. Specimen details

Specimen Type	Reinforcement (Layer x BRC Mesh)	Cover (mm)	Mix Type	28 Day Cube Strength (MPa)	Aggregate Grading
A	2 x A6	25	Face	35.3	20mm Graded
			Body	44.0	
B	1 x A6	50	Face	35.5	20mm Graded
			Body	41.3	
C	2 x A6	25	Face	23.3	Single-size
			Body	38.8	12.5 to 20 mm
D	2 x A4	25	Face	39.3	Single-size
			Body	43.3	12.5 to 20 mm

2.2 Shear bond tests

To provide an indication of the bond strength between the exposed aggregate layer and the base concrete, shear bond tests were carried out after grooves were cut on the exposed aggregate surface. A shear area of 40 mm x 60 mm was used for all the shear bond tests (Fig. 1) as trial tests showed that lengths greater than 60 mm often led to partial failure of the shear area.

2.3 Initial surface absorption tests

The initial surface absorption test (ISAT) is described in BS 1881 : Part 5 : 1970. It was carried out directly on the exposed aggregate surface. The test serves to indicate the rate of flow of water into the concrete through unit area of the surface at various intervals of time since the start of the test and under a constant applied head (200±20 mm of water) and temperature (27±2°C). The diameter of the circular contact area is 80 mm. The tests were carried out at two locations for each specimen.

The ISAT results have been correlated to the performance under normal weathering conditions and changes in mix parameters by Levitt (1971). He estimated that the head of 200 mm used in the test is equivalent to a combined wind and rain speed of about 190 kph (120 mph). Based on laboratory and field observations, he has recommended tentative initial surface absorption limits for various applications. Levitt (1985) presented the use of the initial surface absorption test values to relate performance against onset of corrosion of embedded reinforcement.

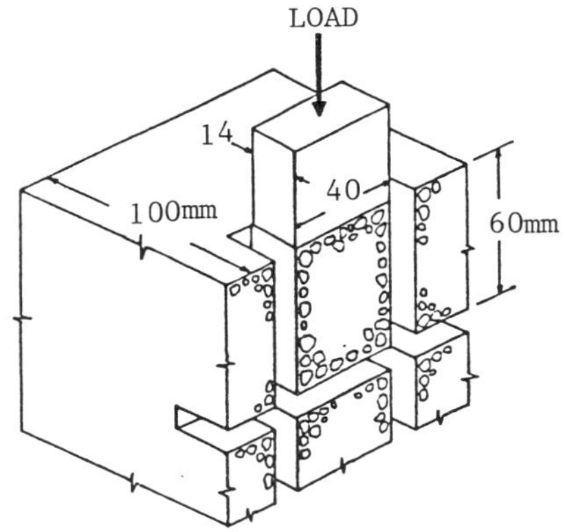

Fig. 1 Shear bond test

2.4 Simulation of temperature effects

The absorption of solar radiation by exposed building surfaces is accompanied by a rise in temperature. In Singapore, the temperature on directly exposed surfaces is likely to reach 70°C during the hottest part of the day. Building surfaces can also lose heat by emitting long wavelength radiation. The drop in temperature on clear nights can cause surface temperature of dark surfaces to fall below shade air temperature. Thus a daily cycle of temperature change of 40°-45°C is not unusual. Also, the internal surface is kept cool by natural ventilation or air-conditioning giving rise to a temperature gradient between the outer and inner surfaces. This difference of temperature depends on the conductivity of the material.

Temperature changes cause dimensional changes in materials. These changes cause stress which if not accommodated, can exceed the strength of the components causing distortion and rupture. Temperature changes can be quite fast. Sunlight breaking through clouds can heat up a surface especially a dark surface very rapidly. Rain falling on a sun heated surface applies a severe quenching shock. The bond between aggregate and cement matrix can thus undergo the first initial breakdown leading to subsequent deterioration.

Due to differential thermal expansion, a wall panel will assume the shape of a shallow spherical shell. If the edges of the panel are not restrained there would be no stress induced. However, in addition to mortar grout filling the gaps between panels and supporting members, the panels are usually restrained by a number of dowel bars. Such restraints would prevent the panel from assuming a spherical shape and produce two-way bending action in the panel. This causes compressive stress on the exterior surface of the panel and in the surrounding members. There is thus a need to study the effects of these cyclic compressive stress on the exposed aggregate surface.

The effects of thermal stresses were simulated by applying compressive stresses on the exposed aggregate side of the specimens. Previous studies on tiled wall panels have shown that the probable stress level under specific conditions (Lee et al 1985) is in the region of one-third the wall material strength. It was judged appropriate to employ a maximum compressive stress level of one third the static strength of the base material. This is probably more severe than what the panel will experience in practice.

The specimens required for this test were of dimensions 400x300x100 mm. They were subjected to eccentric compression. An eccentricity of 8 mm was employed to ensure that no tension would occur in the specimens. This also reduced the amount of load needed to produce the required compressive stress at the exposed aggregate surface, and set up a stress gradient that would be in line with that which may occur in practice.

An Instron testing machine was used to apply 18,250 cycles as well as 36,500 cycles of compression with the load oscillating between 50 kN and 250 kN at 10 Hz. These cycles were used to represent daily cycles over about 50 and 100 years respectively. After completion of the loading cycles, the initial surface absorption tests were carried out for each specimens before they were subjected to the shear bond test.

2.5 Simulation of weathering effects

This test was devised to study the durability of the panels subjected to wetting and drying cycles similar to those caused by climatic changes. At present, no recommended standard test exists in either the American or British Standards. However, a few researchers, e.g. Beeby (1979) and Treadaway (1979), have carried out weathering tests on concrete materials. Similar in nature to these tests, a weatherometer was developed.

The weathering effects were simulated by subjecting the specimens to wetting and drying cycles each consisting of 3 hours of wetting under shower roses followed by 1 hour of drying under strong direct lighting at an air temperature of about 60°C. The rate of spraying was high in order to spread the water over the entire surface of the specimen. The specimens (400x300x100 mm) were placed, exposed aggregate side up, at a slight inclination in the weatherometer to allow easy runoff of water from the top surface. They were subjected to 100 and 200 cycles of wetting and drying. After the specified cycles, the specimens were subjected to shear bond tests.

2.6 Carbonation tests

Concrete quality and concrete cover are important factors in ensuring durability and protection against reinforcement corrosion. One process which influences the durability of reinforced concrete is carbonation. There are a number of aspects of carbonation which influences the properties of concrete, the most important of which are shrinkage and the neutralization of the alkaline conditions of the hydrated cement paste. In the presence of an alkaline environment, corrosion of reinforcement is inhibited by the rapid formation of a thin protective film of iron oxide on the metal surface which renders it passive. If the alkaline environment is destroyed and, air and moisture permeate to the reinforcement, the conditions are right for corrosion to take place.

The depth of carbonation is determined by treating a freshly broken surface with a phenolphthalein indicator solution. Where the concrete is still highly alkaline, a purple-red coloration is obtained. If no coloration occurs, carbonation has taken place and thus the depth of the carbonated surface layer can be measured.

The specimens were exposed to 4 different environment. The carbon dioxide concentrations in the environment were 0.03%, 0.3%, 3% and 30%. For the Australian environment, Ho and Lewis (1987) showed that the depth of carbonation after a week of testing at 4% carbon dioxide concentration is approximately equivalent to one year of laboratory storage.

2.7 Sulphuric acid tests

The sulphur oxide tests were carried out by applying sulphuric acid at a concentration of 0.1N to 6N (4 to 240 g/l in terms of SO_3). This was to determine the susceptibility to staining of the exposed aggregate surface due to the sulphur oxides present in the atmosphere. The average concentration of sulphur dioxide in the air in 1984 was 26 $\mu g/m^3$ (26×10^{-9} g/l) in the industrial zone of Singapore (Ministry of Environment, 1984).

2.8 Fungal/algal growth

2.8.1 Sampling and isolation

Altogether, three samples were obtained from discoloured/stained exposed concrete walls and bricks. Direct sampling was carried out by simply transferring the coloured mass on the wall to pure agar culture plates, using a sterilized scalpel or needle. In addition, scrapings from the wall were transferred to sterile distilled water and kept in glass bijou bottles for further isolation in the laboratory.

2.8.2 Cultivation of isolated samples

Czapek-Dox agar, Potato Dextrose agar, Sabouraud's agar and Sabouraud's agar containing concrete chips and sand were used for the selection of fungal species. For the enrichment of algal species, the samples were inoculated into several inorganic algal culture medium including Bold's medium.

In the case of isolation on solid medium, the agar plates were incubated for a few days at 28°C. Broth cultures were incubated in a 28°C psychrotherm with agitation up to 7 days for fungal cultures and up to several weeks under fluorescent illumination for algal cultures. Thus prepared, the bacteria, fungi and algae species will appear on the plates invariably in mixed populations. It is necessary to obtain pure cultures of each isolate by direct transfer techniques so that each isolated species can be examined microscopically and macroscopically for gross morphology. Hence serial platings have to be carried out to obtain pure cultures.

2.8.3 Microscopic examination

Each isolated species was mounted in water or lacto-phenol blue on glass slides. They were examined under x10, x40 and x100 magnifications.

The fungal species isolated were putatively identified by their gross morphology on agar media and also by their microscopic features. The common genera found were <u>Cladosporium sp, Monilia sp</u> and <u>Aspersillus</u>.

Most of the fungal isolates obtained were species which could sporulate profusely and were pigmented. Spores produced by fungi are highly resistant to harsh conditions. These germinate to produce new mould colonies when they land in suitable places. Species predominance is highly dependent on the type of substrate and the growth environment. Their basic growth requirements are water, a carbon and a nitrogen source.

Algae, on the other hand, requires mainly water, inorganic compounds and carbon dioxide and are highly sensitive to moisture levels (e.g. <u>Trentepohlia</u>

thrives better in low moisture conditions) and strong sunlight (e.g. Chlorella grows better under illumination).

2.8.4 Preparation of inocula for microbial challenges on the specimens

The individual species were each grown in larger volumes (300 ml in one litre flasks) of Sabouraud's Broth wherein pellets of mycelial growth can be seen. These were then mixed so that the concrete slabs can be challenged with the major species isolated previously from contaminated walls and bricks. Five such cultures were prepared and mixed in equal proportions to produce the mixed culture inocula. No algal cultures were included in the inocula because none predominated in the samples obtained from the contaminated walls and bricks.

Concrete specimens with an exposed aggregate surface measuring 100 mm by 100 mm were prepared by first cutting a groove approximately 3 mm deep. This divided the exposed aggregate surface into two equal halves each measuring 50 mm by 100 mm. Three samples were drawn from each of the four types of concrete specimens (A, B, C and D) and labelled 1, 2 and 3. Samples labelled 1 were left uncarbonated while samples labelled 2 had their exposed aggregate surfaces carbonated to a depth of 3 mm. Samples labelled 3 were carbonated to a depth of 5 mm.

Prior to subjecting the specimens to the first microbial challenge, they were rinsed with potable water and then completely immersed in the water for 24 hours. The specimens were then withdrawn from the water and left to stand under room conditions for 1 hour to allow excess water to drain away.

Four culture boxes each measuring 84x25x30 cm (LxWxD) were fabricated from plexiglass. The concrete specimens were placed into these boxes and sand was packed around these specimens to a height such that the surface of the sand was approximately 4 cm from the top of the specimens. The sand was saturated with water. The boxes were covered with a plexiglass sheet.

Each specimen was challenged on only one half of its exposed aggregate surface with 6 ml of the inocula. Subsequent challenges would always be performed on this half of the surface. The specimens were exposed to continuous illumination (fluorescent) of approximately 1,500 lux.

Specimen Types A and C were challenged at intervals of two weeks whereas Types B and D were challenged once every four weeks. The specimens were visually inspected at regular intervals and an estimate would be made as to the percentage of the area on the challenged half which was covered by microbial growth. No growth would be assigned 0 while complete coverage would be 100.

3 Test results and discussions

3.1 Compressive strength

The compressive strengths of the face and body mixes for each of the four types are shown in Table 1. It can be observed that the mixes for Type C are weaker than the corresponding mixes for the other samples. In particular the face mix for this sample is very much lower than the others. Since the mix proportions for all face mixes are intended to be the same, this indicates that a tighter quality control is needed. Since this mix is made with single-sized aggregates, there is a high tendency to segregate as no fine aggregates are included. On the other hand, all the body mixes show similar strengths, but that for Type C is still about 7% below the average.

3.2 Shear bond strength

Table 2 shows that Type D specimen has the lowest and Type A specimen has the highest shear bond strength. The variability of the test is high, with the average standard deviation slightly above 1 MPa. For an average shear bond strength within the range of 2.88 to 3.67 MPa, the average coefficient of variation is 32%. However, Types A and B (graded aggregate mix) specimens show higher strength than Types C and D (single-sized mix) specimens.

3.3 Initial surface absorption

The results of initial surface absorption tests correlated well with the exposed depth measurements. Types A and B specimens showed lower absorption rates than Types C and D specimens (Table 3). Type C specimen had the highest absorption rate. Only Type A specimen had absorption rates after 10, 30 and 60 minutes from start of test within limits recommended by Levitt (1985) for reinforced concrete. However, they were still higher than the limits for 50 years of severe weathering proposed by Levitt (1971). As the nominal cover of 25 mm was reduced by the exposed aggregate depth (about 9 mm) the face layer of exposed aggregate should not be counted upon as cover for the protection of reinforcement unless the mix is redesigned to achieve acceptable values of initial surface absorption.

3.4 Effects of cyclic compression

Two specimens each of the four types of specimens were subjected to each of the two loading cycles: 18,250 cycles of eccentric compression, representing 50 years of daily temperature effects and 36,500 cycles, representing 100 years. The stress applied at the exposed-aggregate face of the specimens was 13 MPa in every case. After the application of the eccentric compression, the specimens were tested for the shear bond strengths between the exposed aggregate and the body concrete.

From the results (Table 2), shear bond strengths of all four types of specimens decreased by about 14% to 44% after the application of eccentric compression. When these results were tested statistically with t-distribution at 95% confidence interval, all except Type B showed significant decreases in shear-bond strength. Under the same treatment, however , it was found that the differences between the shear bond strengths after 18,250 cycles and those after 36,500 cycles were statistically not significant. Thus, while Type B specimens seemed little affected by cyclic compression, even after 36,500 cycles, the other three types of specimens were weakened by about 20% to 40% after 18,250 cycles, but did not deteriorate significantly beyond this stage.

3.5 Effects of accelerated weathering

After the specified wetting and drying cycles, the specimens were subjected to the shear bond tests (Table 2). After 100 cycles of wetting and drying, Types A and B specimens have an average shear bond strength of 81% and 85% compared to that of the original unweathered specimens. Type C specimens actually showed an increase in strength to 110%. On the other hand, Type D specimens showed only a strength of 74% of original. When these results were tested using a t-distribution and at 95% confidence interval, only Type D showed a statistically significant decrease in shear bond strength. Although Types A and B showed a decrease in shear-bond strength, these were found to be statistically not significant at 95% confidence interval. Even at 90% confidence interval, Types B and C still showed no significant difference in

Table 2. Shear bond strength

Conditions of Specimens	Specimen Type	A	B	C	D
Without simulated weathering	No. of data	12	11	10	11
	Ave. shear strength f_{s0} (MPa)	3.67	3.39	3.12	2.88
	Std. dev. (σ_0)	1.01	1.09	1.26	0.75
	$f_{s0} - 1.64\ \sigma_0$	2.01	1.60	1.05	1.65
After 100 cycles of wetting and drying	No. of data	12	12	12	12
	Ave. shear strength f_{s1} (MPa)	3.00	2.88	3.45	2.12
	Std. dev. (σ_1)	0.77	0.68	0.81	0.68
	f_{s1}/f_{s0} (%)	81.7	85.0	110.6	73.6
	$f_{s1} - 1.64\ \sigma_1$	1.74	1.76	2.12	1.01
After 200 cycles of wetting and drying	No. of data	12	12	12	12
	Ave. shear strength f_{s2} (MPa)	3.18	3.12	3.32	2.25
	Std. dev. (σ_2)	0.59	0.45	0.62	0.47
	f_{s2}/f_{s0} (%)	86.6	92.0	106.4	78.1
	$f_{s2} - 1.64\ \sigma_2$	2.21	2.38	2.30	1.48
After 18250 cycles σ_{max} = 13 MPa	No. of data	25	29	28	23
	Ave. shear strength f_{s3} (MPa)	2.86	2.91	2.20	1.62
	Std. dev. (σ_3)	0.74	0.76	0.83	0.60
	f_{s3}/f_{s0} (%)	77.9	85.8	70.5	56.3
	$f_{s3} - 1.64\ \sigma_3$	1.65	1.66	0.84	0.64
After 36500 cycles σ_{max} = 13 MPa	No. of data	23	23	26	23
	Ave. shear strength f_{s4} (MPa)	2.51	2.84	2.02	1.82
	Std. dev. (σ_4)	0.66	0.80	0.81	0.40
	f_{s4}/f_{s0} (%)	68.4	83.8	64.7	63.2
	$f_{s4} - 1.64\ \sigma_4$	1.43	1.53	0.69	1.16

Table 3. Initial surface absorption

Condition of specimens	Type	Absorption Rate* (ml/m²/s) at 10 min	30 mins	1 hr	2 hrs
Without simulated weathering	A	0.30	0.17	0.11	0.08
	B	0.95	0.55	0.35	0.20
	C	1.22	0.70	0.49	0.33
	D	1.10	0.56	0.38	0.26
After 100 cycles of wetting and drying	A	0.06	0.02	0.01	0.01
	B	0.10	0.06	0.04	0.03
	C	0.19	0.10	0.07	0.05
	D	>3.6	-	-	-
After 200 cycles of wetting and drying	A	0.06	0.03	0.02	0.01
	B	0.025	0.012	0.01	0.005
	C	0.12	0.07	0.04	0.02
	D	0.26	0.12	0.08	0.06
After 18250 cycles σ_{max} = 13 MPa	A	0.58	0.28	0.18	0.11
	B	1.21	0.67	0.37	0.28
	C	0.80	0.42	0.31	0.20
	D	1.77	0.98	0.60	0.27

* Average of two tests

shear bond strength. There is no significant difference in the shear bond strength between the 100 and 200 cycles of weathering. For Type D, most of the reduction in shear bond strength occurred during the first 100 cycles of weathering. The larger decrease in strength in Type D specimens could possibly be due to the prevalence of honeycombing in these specimens which allowed greater depth of water penetration.

3.6 Carbonation tests

Carbonation depths had been measured for the four levels of carbon dioxide environment (0.03%, 0.3%, 3% and 30%) after 14 and 26 weeks of exposure. The data for the side faces, which were either moulded or cut, showed that the cut surfaces had very shallow depths of carbonation compared to the moulded surfaces (Fig. 2).

After 14 weeks, the carbonation depth for the exposed aggregate face reached nearly the full layer depth even for the lowest carbon dioxide level which is similar to that of natural environment. After 26 weeks of exposure, there was no further increase in carbonation depth beyond that of the exposed aggregate layer. The opposite face (finished top surface as cast) of all types of samples shows only 2 mm carbonation depth at the 0.03% carbon dioxide level. This is because the strengths of the body mixes were similar. At higher carbon dioxide levels greater depths were measured. The depths did not always show a consistent trend as different locations along the length of each specimen had to be used for determining the carbonation depth after each period of exposure. Generally, the exposed aggregate

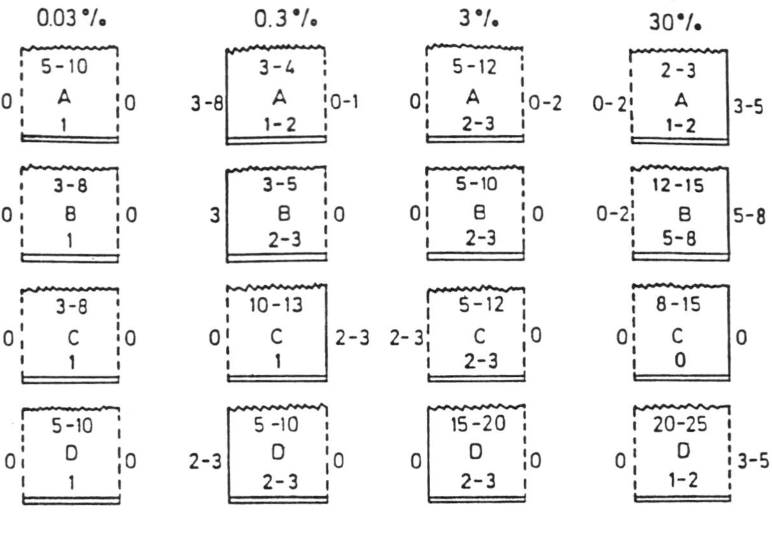

(a) Carbonation Depth after 14 weeks

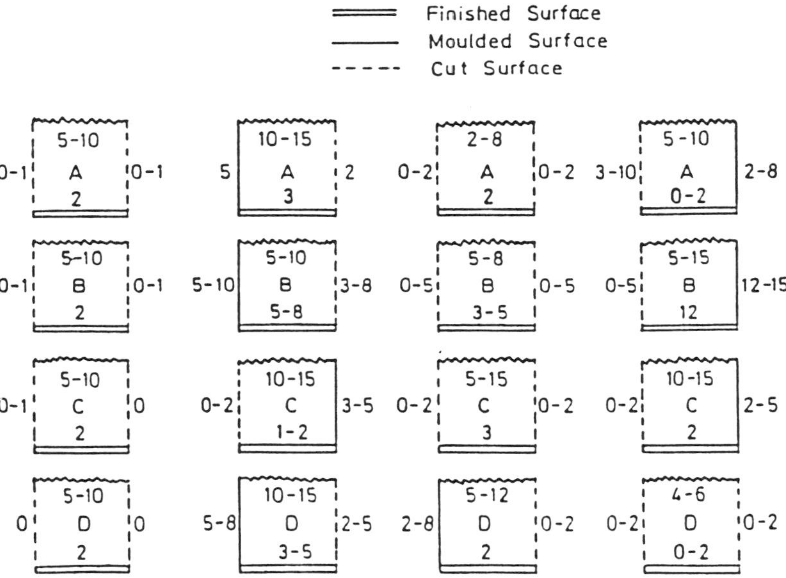

(b) Carbonation Depth after 26 weeks

Fig. 2 Carbonation depths (mm)

surface of Types A and B made with graded aggregate showed lower carbonation depths than those of Types C and D made with single-sized aggregate.

3.7 Sulphuric acid tests

The result of painting sulphuric acid with a concentration of 0.1N up to 6N (4 to 240 g/l in terms of SO_3) had shown no staining of the aggregates. However, at a concentration of above 1N, an increasing amount of white calcium sulphate was noticed on the surface as a result of chemical action on the cement paste matrix,.

3.8 Fungal/algal growth

All concrete specimens showed some microbial growth eventually. Fungal growth, indicated by white patches on the concrete surface, occurred first. This was followed by algal growth about one to two weeks later. Algal growth was indicated by the presence of green patches. It was observed that algae grew primarily on the concrete matrix while fungi grew on both the concrete matrix and the exposed aggregates. The results are shown in Table 4. Note that for ratings purposes, no attempt was made to separate fungal from algal growth. Instead all growth was grouped together as microbial growth.

Uncarbonated specimens resisted microbial growth the longest. Specimens A1 and C1 which were challenged once every 2 weeks resisted growth for about 6 weeks while B1 and D1 which were challenged once every 4 weeks resisted for about 7 weeks. For all specimen types, the uncarbonated samples showed the least microbial growth suggesting that the concrete's alkalinity is a good defence against microbial attack. The amount of microbial growth increased as the carbonated depth increased for all specimen types. A3, B3, C3 and D3 therefore showed the heaviest microbial growth. In addition, microbial growth spreaded to the unchallenged half in these specimens. Specimens carbonated to 5 mm depth appeared to lose their resistance to microbial attack. All the uncarbonated specimens, A1, B1, C1 and D1, showed an absence of microbial growth on the unchallenged half and the presence of numerous spores on the challenged half suggesting that uncarbonated concrete surface was not conducive to fungal growth. After 41 weeks the effect of microbial growth on the carbonated specimens becomes increasingly obvious. The uncarbonated specimens (A1, B1, C1 & D1) resisted microbial growth to a large extent even after 41 weeks.

Microbial growth on the concrete specimens would generally show more spread following each challenge but would recede after about a week. The spread would have been contributed by the fresh introduction of nutrients and microbial mass following each challenge. It can reasonably be concluded that any concrete specimen used in an environment where microbial challenges and nutrient addition occur frequently enough would show microbial growth.

Of the 4 types of concrete specimens, Type B indicated the highest resistance to microbial growth. Note that the nominal cover for Type B specimens was 50 mm as opposed to 25 mm for the remaining types. It is known that the presence of iron may favour the growth of certain micro-organisms and in this context particularly algae. Soluble iron may have been leached from the reinforcement and migrated upward to the surface of the specimens. Type B could be more resistant to such migration than the other types due to the larger cover. A comparison of Types D and A or C suggested that types of reinforcement mesh resulted in little difference in the microbial growth.

No correlation was observed between intensity of microbial growth and surface absorption. In general Type D showed the highest surface absorption

Table 4. Microbial growth intensity on concrete specimens

Sample No.	Week No. 1	4	5	6	7	8	9	10	11	12	13	14	15
A1	0*	0*	0	0*	10	20*	25	25*	30	20*	15	5*	5
A2		0*	0	5*	20	40*	45	45*	45	35*	30	5*	5
A3						5*	5	5*	60+	45+*	50+	50+*	60+
B1	0*	0*	0	0	0	10*	10	10	10	0*	0	0	0
B2		0*	0	0	5	10*	10	10	10	0*	0	0	0
B3						5*	5	5	20	10*	10	10	15
C1	0*	0*	0	0*	10	10*	10	15*	15	5*	0	0*	0
C2		0*	10	10*	20	30*	30	35*	35	25*	25	0*	10
C3						15*	15	20+*	45+	45+*	50+	50+*	50
D1	0*	0*	0	0	0	10*	10	10	10	5*	0	0	0
D2		0*	10	10	10	20*	20	25+	35+	35+*	30+	20+	20+
D3						5*	5	10+	10+	50+	50+	50+	55+

+ Spread of microbial growth on to unchallenged half.
* A challenge on that week
- Figures represent the portion of the challenged half which was covered by microbial growth expressed in terms of 0 for no observable growth and 100 for complete microbial coverage. This area would not include the area covered by microbial growth on the unchallenged half.
- A1, B1, C1 & D1 are uncarbonated concrete specimens.
- A2, B2, C2 & D2 are concrete specimens carbonated to 3 mm depth.
- A3, B3, C3 & D3 are concrete specimens carbonated to 5 mm depth.

followed by C, B and A. In the natural environment some correlation might be expected since moisture is a prerequisite for life. However, in this study, correlation was not expected because moisture had been made a non-limiting parameter.

The intensity of microbial growth on Types A, C and D indicated that there was only marginal difference between specimens with graded or single-sized aggregate. However, microbial growth appeared to be affected by the exposed aggregate area. The exposed aggregate area of Type B specimen was 74%, Type C 71%, Type D 70% and Type A 68%. When the specimen types were arranged based on increasing intensity of microbial growth, the sequence was also B, C, D and A. So it is not whether the aggregates are graded or not but how much aggregate is exposed that is important to microbial growth resistance.

4 Summary and concluding remarks

4.1 Summary of results

a) The compressive strength of the face mix was lower than that of the body mix even though they were of the same water/cement ratio.

b) The initial surface absorption values of the exposed aggregate layer for all types of specimens were high, particularly for those with single-size aggregate mix (Types C and D). The values were further increased after 18,250 simulated temperature cycles.

c) There was a general reduction of the initial surface absorption values after 100 cycles of wetting and drying and a further reduction for those after 200 cycles.

d) The shear bond strength was higher for face mixes with graded aggregates (Types A and B) than those with single-sized aggregates (Types C and D). Although Type B specimens seemed little affected by simulated temperature cycles, even after 36,500 cycles, the other three types of specimens were weakened by about 20% to 40% after 18,250 cycles, but did not deteriorate significantly beyond this stage.

e) There was a general reduction of shear bond strength after 100 cycles of wetting and drying, except for Type C specimens. However, there was a general trend of higher shear bond strength for specimens after 200 cycles. This may be attributed to the effect of additional curing.

f) Due to the high variability of shear bond strength testing, the differences due to 100 or 200 wetting and drying cycles were not statistically significant even at 90% confidence interval.

g) Uncarbonated specimens showed an absence of microbial growth on the unchallenged half and the presence of numerous spores on the challenged half. Thus uncarbonated concrete surface are not conducive to fungal growth.

h) Concrete specimens carbonated to 5 mm depth appeared to have lost their resistance to microbial attack.

i) Type B specimens with a nominal cover of 50 mm, as opposed to 25 mm for the other types, indicated the highest resistance to microbial growth.

j) There was only marginal difference in microbial growth between specimens with graded or single-sized aggregates. Microbial growth appeared to be more intensive with decreasing exposed aggregate area.

k) After 14 weeks of exposure at 0.03% carbon dioxide concentration, carbonation of the exposed aggregate face reached nearly the full layer depth.

l) In general, the exposed aggregate surface with graded aggregates showed lower carbonation depths than those made with single-sized aggregates.

m) Painting the exposed surface with sulphuric acid with a concentration of 0.1N to 6N showed no staining of the aggregates. However, at a concentration of above 1N, an increasing amount of white calcium sulphate was noticed on the surface as a result of chemical action on the cement paste matrix.

4.2 Recommendations

Based on the test results, the following recommendations are proposed:

a) In order that the exposed aggregate layer may contribute as an effective cover, the quality of the face mix has to be comparable or even superior to that of the body mix so as to reduce water and carbon dioxide penetration. This can be achieved by a reduction in water content and an increase in cement content of the face mix, which will effectively reduce the rate of carbonation and hence the rate of microbial growth.

b) Exposed aggregate surface made with graded aggregates is preferred over single-sized aggregates as the former provides higher shear bond strength and better resistance against simulated temperature effects. However, it should be noted that no aggregate spalling was observed in any of the specimens tested even after 36,500 cycles of eccentric compression.

5 References

Beeby, A.W. (1979) Cracking and Design against Corrosion in Reinforced Concrete, in **Corrosion of Steel Reinforcements in Concrete Construction,** Society of Chemical Industry, London, 15-27.

Ho, D.W.S. and Lewis, R.K. (1987) Carbonation of Concrete and Its Prediction. **Cement and Concrete Research,** Vol. 17, No. 3, 489-504.

Lee, S.L., Tam, C.T., Mansur, M.A., Ong, G.K.C. and Loo, Y. H (1985) Performance Tests on Bond between Ceramic Split Tiles and Precast Wall Panels for HDB Flats. A report submitted to Lee Kim Tah (Pte) Ltd.

Levitt, M. (1971) The ISAT - A Non-Destructive Test for the Durability of Concrete. **British Journal of N.D.T.,** July, 106-112.

Levitt, M. (1985) The ISAT for Limit State Design of Durability of Concrete. **Journal of the Concrete Society,** London, Vol. 19, No. 7, July pp 29.

Ministry of the Environment, Singapore, Anti-Pollution Unit, Annual Report 1984, 49 pp.

Treadaway, K.W. (1979) Durability of Steel in Concrete, in **Corrosion of Steel Reinforcements in Concrete Construction,** Society of Chemical Industry, London, 1-11.

43 COMPARATIVE SULPHATE RESISTANCE OF SRPC AND PORTLAND SLAG CEMENTS

J.J. KOLLEK, J.S. LUMLEY
Blue Circle Technical Services Division, Greenhithe, Kent, UK

Abstract
This paper reports on an investigation into the sulphate resistance of Portland cements and Portland blastfurnace slag cements having a range of slag levels. Mortar bars were immersed in $MgSO_4$ and Na_2SO_4 solutions with SO_4^{--} concentrations from 0.6 to 4.2%. Some of the slag cements performed relatively well in Na_2SO_4 solutions but were particularly vulnerable in $MgSO_4$ solutions. None of the slag cements performed as well as the low C_3A Portland cement. The performance of the composites was dependent on the properties of the base cements, the properties of the slags and their substitution levels. Strength results ranging from 3 to 910 days indicated that slag levels which give a relatively good sulphate performance suffer from poor early strength. The alumina content of the slags appears to be a major factor influencing the strength development and sulphate performance. The sulphate resistance of slag cements can be improved significantly by targeted SO_3 additions.
Keywords: Sulphate Resistance, Mortar Bar Test, Portland Cement, Composite Cement, Slag, Performance, Strength, SO_3 Addition.

1 Introduction

It is only a comparatively small amount of concrete (estimated to be about 4% of total) that is designed to withstand the attack of sulphates. However, although the volume is small, such concrete is used in critical parts of major structures and as such justifies our more than average attention. The consequences of sulphate attack are expansion, cracking, loss of strength and with time, under extreme conditions, complete disintegration.

The phenomenon of sulphate attack has been known for many years yet the detailed chemical and physical processes involved are still largely a matter of intense debate. The wealth of literature on the subject, extending now over a period of more than 100 years or so, is enormous and this has recently been reviewed by Lawrence (1990).

The practice oriented approach to this problem tends to revolve around the question of the choice of the primary materials and particularly that of the binder in an effort to achieve a concrete of improved sulphate resisting properties. Portland cement to BS 12 with a high C_3A content is known to be particularly vulnerable to sulphate attack although its performance is known to be greatly influenced by mix design parameters, compaction and curing.

Sulphate resisting Portland cement (SRPC) was developed in the 1930's. Manufacture and large scale usage in the UK started in 1949 since when sulphate damage has been reasonably successfully kept at bay. The first BS 4027 appeared in 1966. More recently, composite cements based on combinations of Portland cement and secondary binders such as ground granulated blastfurnace slag (ggbs or slag) and fly ash and in some countries also natural pozzolana and silica fume have been approved and used in situations where greater sulphate resistance was required. In some countries these secondary binders are not combined into factory produced cements but are added to the concrete mixer. An important difference between a factory produced slag cement and a site blend is that the SO_3 level in the factory cement can be controlled. In the UK the SO_3 level of a 70% slag blend will normally be lower than 1%. This difference is often overlooked when field experience in continental Europe with factory produced slag cements is attributed to site blends.

The present use of Portland cements (BS 12 and BS 4027) and slag or fly ash in sulphate bearing ground conditions is regulated by the BRE Digest 250 and Table 6.1 of BS 8110 Part 1 which reflect the current technical consensus. The adoption of European standardisation may require a reassessment of the nature of sulphate resisting concrete.

The need to be informed and to inform one's clients requires the primary concrete material manufacturer and supplier to generate experimental and/or other evidence pertaining to the performance of his products under the various conditions of usage. This does not limit him to the performance of his material in isolation but inevitably encompasses the use of his material in combination with others if such probability exists. This poses the question of interactions.

The case for a virtually continuous generation of experimental data will be further justified if one considers that cements and secondary binders which are used today may be significantly different from those which were used in the formulation of current perceptions and regulations. If these are to be changed, be it for reasons of the creation of a common European approach or for any other reasons, then it appears important that these be based on sufficient experimental evidence and sound understanding of the factors involved.

Portland cements, even from the same works, may undergo a change with time with regard to both chemical composition and physical properties (Concrete Society Technical Report No 29, 1987) and which, although small, may be significant in terms of performance. This may extend to sulphate resistance.

In the case of slags, the probability of change with time in the chemical and physical properties may also be significant and even greater than those of Portland cements. They are dictated by the requirements of an efficient running of a blastfurnace rather than the requirements of the optimum performance of its by-product in terms of its cementitious properties.

The above-mentioned reasons indicate that it is prudent if not essential to check the sulphate resistance of all binders but particularly those which are based on a combination of a given Portland cement with a given slag in a given market situation. The fact that the sulphate resistance of a given binder requires years, even in an accelerated test, to be established, demonstrates the need for such a programme of tests.

This paper is to be viewed with the above points in mind.

2 The test method for sulphate resistance

Difficulties always arise with the choice of the experimental method which should be adopted in the assessment of cements or binder combinations. Many different methods based on various criteria have been tried and advocated over the years (Hansen, 1966). No method has emerged which is free from criticism or has gained universal application. The main lack of concordance of views is not least related to the fact that different methods lead to a different ranking order of binder types.

The method adopted in this investigation is the mortar bar expansion method which is similar to the one adopted subsequently in ASTM C 1012 "Length change of hydraulic mortar exposed to sulphate solution". It embodies the advantages of being simple, and undemanding in terms of equipment, operator's time and numbers of specimens. The recorded parameter (expansion) has a strong although not unique relationship to the phenomena associated with sulphate attack. However, mortar specimens of slag cements often disintegrate after only a comparatively small degree of expansion. In these cases, the age at which integrity is lost is taken as a measure of sulphate resistance. Since the objective of this work was to compare different binders, ie determine their relative performance, the authors consider that any lack of strict correlation of the method with the performance of real structures under field conditions would be small and possibly insignificant.

The specimens used were bars of dimensions 25x25x285 mm provided with stainless steel studs for length change measurement using a comparator. The mortar was 1:4.0:0.6 with the quartzitic sand of a narrow particle size distribution in the range -850+600 μm. The bars were cast and cured at 20±2°C under humid conditions and then stored under water at the same temperature up to the age of 7 days. A length comparator reading was then taken.

From a batch of nine prisms, eight were immersed in sulphate solutions and one in water as control. Four of the solutions were magnesium sulphate of concentration 0.6, 1.2, 2.4 and 4.2% SO_4^{--} by mass and four were sodium sulphate of similar concentration. Comparator readings were again taken after 28 days of immersion and after further various periods dictated largely by the rate of visual deterioration. The "sulphate attack expansion" was calculated by subtracting the apparent expansion of the control prism from each of the test prisms. This practice removed the possibility of error due to the moisture movement of the specimens. Failure of the bar (expressed as the survival time of the bar in months and years) was recorded when its expansion reached 0.5% or the bar lost its integrity.

3 Materials

Three UK works produced cements ('A'-sulphate resisting Portland cement, SRPC, 'B'- and 'C'-ordinary Portland cements, OPC) were selected. The chemical composition and the usual physical properties of the three cements are given in Table 1. Cement 'A' is characterised by a low potential C_3A content. Cement 'B' is characterised by a higher than average alkali content (0.92% eq Na_2O) and a SO_3 content of 3.0% which is fairly average for present day OPC production. The potential C_3A content is 8.6%. The specific surface area at 448 m^2kg^{-1} is more typical of rapid hardening Portland cement production. Cement 'C' is characterised by a lower than average alkali content (0.34% eq Na_2O) and a SO_3 content of 2.3% which is low for present day OPC production. The potential C_3A content is 9.6%. The specific surface area of cement 'C' at 375 m^2kg^{-1} is within the range of OPC production but somewhat higher than average. In all other respects the last two cements were fairly similar to each other and typical of UK production.

The selection of the cements 'B' and 'C' was made on the expectation that they would have a different activation capacity of the ggbs's which would reflect in a different strength growth and possibly also in a different sulphate resistance of the various binder combinations. This was indeed justified on both counts.

Three slags ('X', 'Y', 'Z') typical of current use in the UK were selected. Their chemical compositions and the usual physical properties are given in Table 1. Slag 'X' was pelletised while the other two were water-quenched by the granulation process. The Al_2O_3 content of slag 'X' is relatively low (11.2%) while that of slag 'Z' is unusually high (16.0%). Slag 'Z' exhibits a relatively high alkali content (0.99% eq Na_2O) which together with a high ratio of Al_2O_3 to SiO_2 confirmed in a high Hydraulic Index (= 2.0) points to a high reactivity slag and, therefore, to a high early strength contribution in concrete.

4 Experimental and results

In the main series of tests, 24 different composite cements were produced by blending each of the 3 slags with the two Portland cements 'B' and 'C' in the ratios of:-

slag/Portland cements: 23/77, 46/54, 69/31 and 92/8

Compressive strength tests were carried out according to BS 4550: Part 3:Section 3.4: 1978 on concrete of 315 kg m^{-3} of binder, W/C=0.6 using specified coarse and fine aggregate mixed by machine and compacted manually with a compacting bar into 100 mm cubes. The slumps were determined with each mix but within the range of mixes they did not vary significantly. The earliest strength was determined at 3 days, followed by a determination at 7 days (coinciding with the immersion of the expansion bars in sulphate solutions), then at various periods up to 910 days (2.5 years). The strength results (mean of 3 cubes) are given in Table 2.

Table 1. Chemical and Physical Data of Binders

%	Portland cements			Ground granulated blastfurnace slags		
	A	B	C	Z	X	Y
SiO_2	20.0	20.2	19.8	31.0	36.8	33.6
IR	(0.20)	(0.91)	(0.50)	(0.40)	(0.31)	(0.30)
Al_2O_3	3.9	5.4	5.8	16.0	11.2	13.4
Fe_2O_3	5.5	3.4	3.4	1.5	0.4	1.5
CaO	64.6	63.9	63.8	36.3	40.0	40.4
MgO	1.1	1.4	3.0	11.0	7.7	8.3
SO_3	2.2	3.0	2.3	0.16	0.13	0.05
S				1.1	0.97	0.82
LOI	1.4	1.4	0.8	0.4*	0.6*	0.4*
K_2O	0.30	0.93	0.43	0.65	0.47	0.52
Na_2O	0.15	0.31	0.06	0.56	0.35	0.33
$(Na_2O)_e$	(0.35)	(0.92)	(0.34)	(0.99)	(0.66)	(0.67)
Free lime	1.8	1.2	1.2			
Cl				0.06	0.01	0.01
C_3S	61.9	52.0	54.0			
C_2S	11.1	18.7	16.1			
C_3A	1.0	8.6	9.6			
C_4AF	16.7	10.4	10.4			
Specific surface $m^2\ kg^{-1}$	462	448	375	384	397	394
45 µm residue	8.0	9.6	13.6	2.1	6.2	6.5
Apparent particle density $kg\ m^{-3}$	3160	3130	3150	2910	2920	2930
Crystallinity Index**				0	11	0
Hydraulic Index***				2.0	1.6	1.8

IR - Insoluble residue
LOI - Loss on ignition
Value in parentheses not included in total
* in nitrogen
** by x-ray diffraction

$$CI = \frac{\text{Area under peaks} \times 100}{\text{Total area}}$$ in range 12-38° 2θ, Cu K_α radiation

*** $$\text{Hydraulic Index} = \frac{C + M + A}{S}$$

Table 2. Strength of Portland and Portland - ggbs composite cements

cement	ggbs	subst. level %	Strength in ($N\ mm^{-2}$) at age in days 3	7	28	91	365	910
'B'		0	26.1	35.6	44.3	47.1	49.7	50.9
	'X'	23	18.4	25.4	38.3	50.1	54.3	59.7
		46	10.7	16.0	33.0	47.3	58.8	65.7
		69	5.2	10.2	26.6	41.1	55.0	62.6
		92	2.3	7.3	14.4	19.1	22.9	26.4
	'Y'	23	19.0	29.7	48.8	52.5	53.0	56.2
		46	11.8	23.4	48.5	59.0	60.1	65.3
		69	8.3	22.4	42.5	57.5	62.6	65.7
		92	5.0	12.8	28.0	35.4	40.7	47.5
	'Z'	23	18.9	30.3	46.1	50.7	50.8	53.0
		46	14.1	27.7	48.7	55.5	57.0	61.2
		69	10.6	27.1	44.2	53.8	59.8	65.7
		92	6.9	14.9	28.5	35.1	43.0	50.5
'C'		0	20.2	31.5	48.0	51.8	52.7	53.6
	'X'	23	14.2	22.4	41.0	55.4	62.4	59.0
		46	9.0	14.6	33.1	52.5	66.6	68.2
		69	4.4	9.0	25.6	43.4	59.3	65.6
		92	2.2	6.1	13.6	20.3	24.8	26.9
	'Y'	23	14.3	24.7	47.0	55.9	57.5	53.6
		46	10.6	20.8	45.9	60.7	63.0	61.0
		69	7.7	19.4	41.2	58.2	65.2	65.2
		92	5.2	12.4	27.3	37.1	43.7	46.0
	'Z'	23	14.5	27.0	48.0	53.0	51.3	53.7
		46	11.8	26.1	51.7	59.3	59.7	57.8
		69	9.4	26.4	46.3	55.2	57.7	61.5
		92	6.4	15.0	28.7	35.5	40.9	47.6

Two series of sulphate resistance tests were conducted: main and subsidiary.

In the main series the test bars were prepared (using the method and procedure given in Section 2 above) from the 3 "classical" cements 'A', 'B' and 'C' and the 24 composite cements. 243 bars were involved in this series. The results are presented in graphical form in Fig 1, 2 and 3.

In the subsidiary series of tests - effect of SO_3 addition by blending ground gypsum into the composite cement - only slag 'Y' and cements 'B' and 'C' were involved. The composite cements were in the ratios:-

slag/Portland cement: 46/54 and 69/31

and the sulphate solution exposure type and SO_4^{--} levels as well as the expansion results were those indicated in Fig 4. 28 bars were involved in this series.

5 Discussion

The results confirm that Portland cements with a moderate to high C_3A content (cements 'B' and 'C' in this investigation) exhibit a poor sulphate resistance as determined by the test method of this investigation. Cement 'B' (C_3A content 8.6%) exhibited a better sulphate resistance than cement 'C' (C_3A content 9.6%). It is possible, however, that the relatively small difference in the C_3A content does not entirely account for the difference in performance and that the higher specific surface of cement 'B' (448 $m^2 kg^{-1}$ versus 375 $m^2 kg^{-1}$ of cement 'C') and the higher early strength properties of cement 'B' (35.6 N mm^{-2} versus 31.5 N mm^{-2} of cement 'C' at 7 days) made significant contributions. The higher alkali of cement 'B' could also have made a contribution. It should, however, be noted that the late strength of concrete made with cement 'C' and cured in water is higher than that of cement 'B'.

The Portland cement with a low C_3A content (cement 'A' in this investigation) exhibited a high sulphate resistance as shown in Fig 1. This figure shows that bars exposed to 4.2% SO_4^{--} (as $MgSO_4$) failed the expansion limit of 0.5% at 3 years while all the bars exposed to lower strength sulphate solutions have survived the period of 8 years. The performance of the bars in Na_2SO_4 was only slightly better at corresponding strengths of solution.

The slag composite cements, even at high slag substitution levels, have not performed as well as cement 'A' as can be seen from Fig 2 and 3.

This is at variance with recent findings on similar materials by Frearson (1986), who concluded that slag blends with more than 70% slag content gave a superior sulphate resistance to a neat Portland cement with 0% C_3A. However, he worked with a different type of specimen, and a different procedure and used only Na_2SO_4 solutions.

The results also indicate that the performance of slag cements is influenced by the properties of the base Portland cement, the chemistry and the mineralogy of the slag as well as the slag substitution level in the composite.

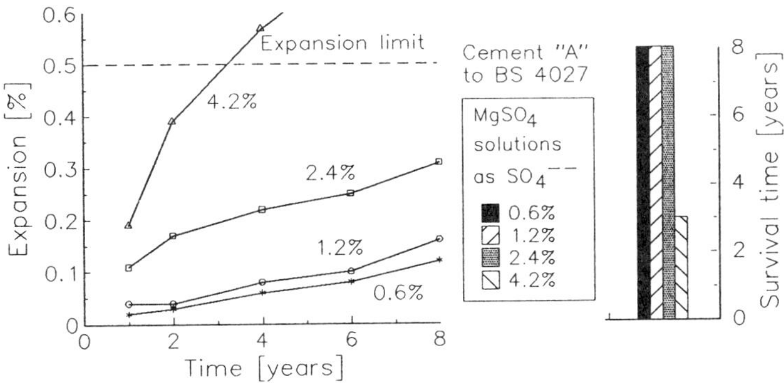

Fig 1. Performance of a sulphate resisting Portland cement (cement 'A') presented as bar expansion percentage and in survival time in years

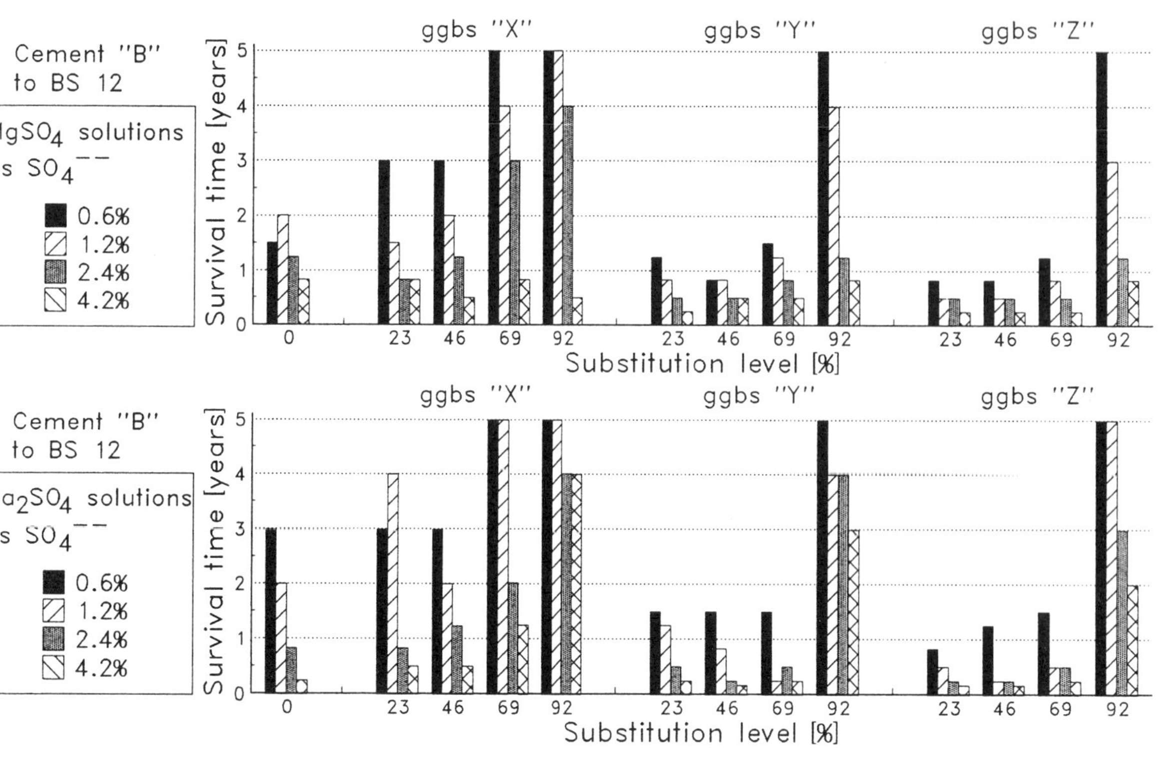

Fig 2. Performance of slag cements based on cement 'B' and slags 'X', 'Y', 'Z' in various sulphate solutions

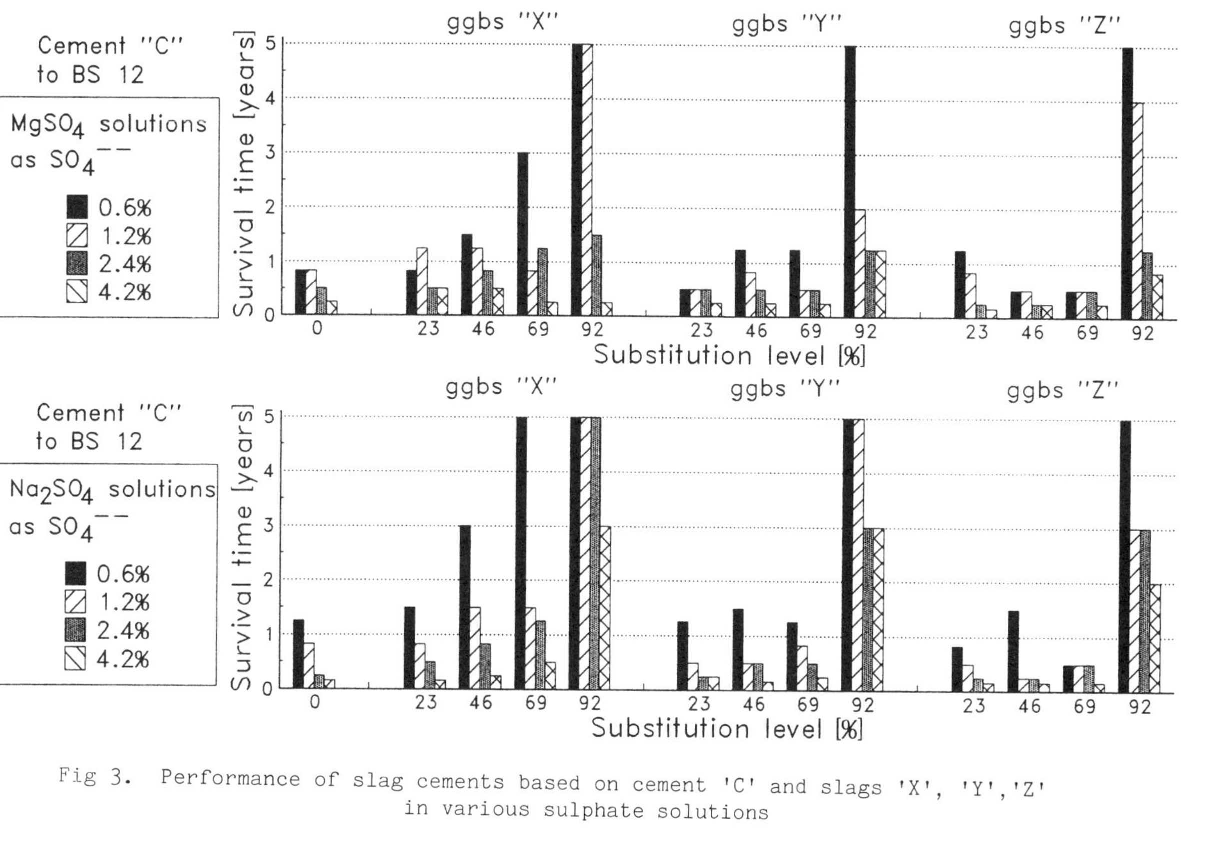

Fig 3. Performance of slag cements based on cement 'C' and slags 'X', 'Y','Z' in various sulphate solutions

Fig 2 indicates that cement 'B' gives a generally better performance with all slag combinations than cement 'C'.

Slag 'X' gives the best performance with both base cements and the improvement appears to increase fairly monotonically with the slag substitution level. Slags 'Y' and 'Z', on the other hand, appear to show a deterioration in performance with respect to the base cement at low to medium substitution levels, rising to a high performance at high substitution level.

The results show clearly that the aggressivity of the $MgSO_4$ solution is distinctly more severe and that slag cements are more vulnerable to $MgSO_4$ than to Na_2SO_4 solutions and that this applies more at higher SO_4^{--} concentrations. At high substitution levels, the performance of the slag cements in Na_2SO_4 is distinctly better than that of either of the base cements (see Fig 2 and 3) while that in $MgSO_4$ is dependent on the base cement - that of cement 'B' combined with slag 'Y' and 'Z' being no different even at a high substitution level.

The poor performance of slag cements in $MgSO_4$ solutions has been known for some time and the phenomena associated with this are understood. It is legitimate to argue that in the case of $MgSO_4$ we experience the superposition of two attacks: that of the SO_4^{--} and of the Mg^{++} ions. This, however, does not alter the fact that this combination of ions occurs frequently in ground waters and soils and practical concrete needs to be designed around it.

The poor performance of slags 'Z' and 'Y' is probably mostly related to their high Al_2O_3 levels of 16.0 and 13.4% respectively compared to that of slag 'X' of 11.2%. Locher (1966), Osborne (1989), working with similar materials but on smaller specimens tested in bending reached similar conclusions. A high alumina content in a slag is not entirely negative in its effect on the properties of a composite cement, its contribution to an early reactivity and, therefore, early strength contribution being highly desirable. Table 2 indicates that the strength at one year of a 92% slag 'X' combination with base cement 'B' gives 22.9 N mm^{-2} while a 92% slag 'Z' combination with the same base cement gives 43.0 N mm^{-2}. It is unfortunate that the same parameter should offset an improved strength performance with a diminished sulphate resistance.

The subsidiary investigation was aimed at establishing whether the addition of sulphate to slag, as advocated earlier by Kondo (1960) and others, has a positive effect on the sulphate resistance.

Fig 4 demonstrates clearly that this is so with slag 'Y' when combined with either of the two base cements 'C' and 'B', irrespective of the type and strength of sulphate solution and of the level of slag substitution. This presumably is so because the higher SO_3 leads to the formation of ettringite in larger quantities in the early stages of hydration and reduction of its subsequent disruptive expansion due to the ingress of external sulphates. It would appear that a correctly targeted sulphate content of a composite cement or a slag to be blended-in in a concrete mixer is highly desirable and should be adopted in practice where sulphate resistance is a design requirement.

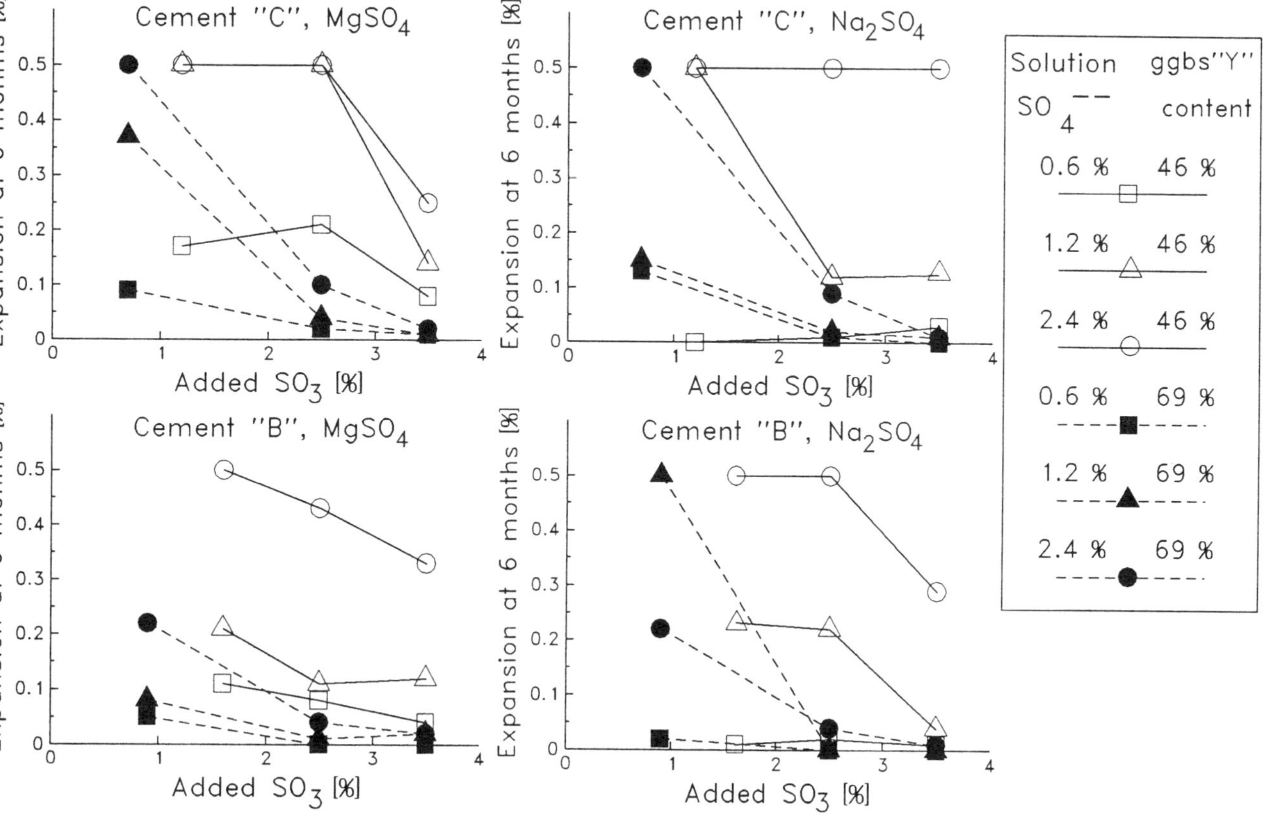

Fig 4. Effect of added SO_3 to ground granulated blastfurnace slag on the sulphate resistance of a composite cement

6 Conclusions

The mortar bar expansion method for the determination of sulphate resistance of cements is convenient and meaningful in terms of the phenomena associated with sulphate attack.

Composite cements based on the replacement of part of the base Portland cement with ground granulated blastfurnace slag can give an improved sulphate resistance compared to the base cement. This performance is, however, dependent on the properties of the base cement, the properties of the slag and on the slag substitution level. The slag substitution level which is likely to give a good sulphate resistance is also likely to be one which gives poor early strength. High alumina slags give excellent reactivity and hence early strength but are a major cause of relatively poor sulphate performance.

Slag cements are more vulnerable to $MgSO_4$ than to Na_2SO_4 solutions. Both species of cation are encountered in soils and ground waters.

The sulphate resistance of slag composite cements and of site blends is significantly improved by targeted SO_3 additions.

Sulphate resisting Portland cements with a controlled low level of C_3A give a superior performance to slag composite cements or site blends in both $MgSO_4$ and Na_2SO_4 solutions.

7 References

Concrete Society Technical Report No 29 (1987) London, UK. Changes in the properties of Ordinary Portland Cement and their effects on concrete.

Frearson, P.H. (1986) The sulphate resistance of combinations of Portland cement and ground granulated blast furnace slag. Fly Ash, Silica Fume, Slag and Natural Pozzolans in Concrete, Proceedings Second International Conference, Madrid, Spain (ed V.M. Malhotra), Vol 2, SP 91-74, pp. 1495-1524.

Hansen, W.C. (1966) Attack of Portland cement concrete by alkali soils and waters - a critical review. Highway Research Record 113, pp. 1-32, Highway Research Board, National Academy of Sciences - National Research Council, Washington D.C., USA.

Kondo, R. (1960) Chemical resistivities of various types of cements. 4th International Symposium on the Chemistry of Cement, Washington D.C., USA, Vol II, pp. 881-888.

Lawrence, C.D. (1990) The sulphate attack of concrete. To be published in Magazine of Concrete Research, Vol 42, No 152.

Locher, F.W. (1966) Zur Frage des Sulfatwiderstandes von Hüttenzementen. Zement-Kalk-Gips, Vol 19, No 9, pp. 395-401.

Osborne, G.J. (1989) Determination of the sulphate resistance of blastfurnace slag cements using small-scale accelerated methods of test. Advances in Cement Research, Vol 2, No 5, pp. 21-27.

44 PERFORMANCE OF PARKING GARAGE DECKS CONSTRUCTED WITH EPOXY COATED REINFORCING STEEL

G.G. LITVAN
Institute for Research in Construction, National Research Council, Ottawa, Canada

Abstract
Excessive cracking was discovered in a large parking garage built in 1981 with epoxy coated reinforcing steel. Because the use of epoxy coated steel is not only approved by the relevant standard but has been made mandatory in Ontario for the construction of garages, an investigation has been launched to determine whether the experienced excessive cracking is typical of slabs constructed with epoxy coated steel and whether the cause of this deficiency is corrosion. It has been established that suspended garage decks constructed with epoxy coated steel appear to be prone to excessive cracking. This deficiency, however, does not seem to be caused by corrosion of the reinforcing steel. It is, therefore, necessary to install a waterproof membrane over the concrete decks to prevent leakage through the slab and to avoid infiltration of salt solution if the lower mat consists of bare steel.
Keywords: Parking Garages, Epoxy Coated Steel, Corrosion of Reinforcing Steel, Cracking of Concrete.

1 Introduction

Deterioration of parking garages has become a serious problem in North America in the last decade. Rusting of the reinforcing steel, promoted by the chloride ions of the sodium or calcium chloride used for deicing purposes, is the cause of the problem.

In order to impart corrosion resistance to new garages, the use of epoxy coated steel appears to be an attractive solution; the high dielectric coating almost completely isolates the steel from the aggressive environment, thus preventing the corrosion from taking place.

In the last few years since 1981, epoxy coated reinforcing steel has been utilized for the construction of new garages at an increasing rate. The common practice is to use epoxy coated steel for the top mat of the reinforcement in garage decks while the bottom mat is constructed with bare steel.

The Canadian Standards Association (1987) standard for parking structures acknowledged the merits of using epoxy coated rebars and required its use in the top 100 mm zone of the deck in "heavy-use areas" such as entrance ramps. In addition, the installation of a waterproofing elastomeric membrane over the concrete deck is also specified. Conformance with the standard is compulsory in the Province of Ontario, while elsewhere in Canada compliance is voluntary.

Excessive cracking discovered in the course of routine inspection of a large parking facility built in 1981 with epoxy coated reinforcing steel was, therefore, of considerable interest and concern.

Consequently, it was decided to launch an investigation, aiming to determine whether suspended concrete slabs containing epoxy coated steel of parking garages are prone to excessive cracking or the noted deficiency is due to causes unrelated to the use of epoxy coated steel.

2 Case Histories

2.1 Garage No. 46

An underground, enclosed, three level parkade (225,000 sq. ft.) of a highrise office tower (900,000 sq. ft.) and three levels of retail shops, was built in 1981.

The top course of the steel reinforcement in the slab is epoxy coated with a specified concrete cover of 1 inch thickness. The deck has no slope to drains. A concrete sealer was applied at the time of construction.

A few months after construction additional drains were installed to alleviate ponding problems.

A large number of cracks have also developed, primarily around the columns (Plates 1-4). These and those developed during the five year warranty period were repaired by the consultant. Typically, cracks were grouted with Isoflex 800. The repairs have not proven to be satisfactory; many of the old cracks reopened and new ones developed at a rate of 10,000 lineal ft. per year.

2.2 Garage No. E-1

Office tower in Toronto, one of the first garages (1981) constructed with epoxy coated bars in the top mat of the suspended slab. The specified cover is 1.5 inch. The concrete had low slump, low w/c and high cement content.

The cracks appeared a few months after construction requiring the installation of a membrane. Since the provision of the waterproofing overlay the garage has performed satisfactorily.

2.3 Garage No. 50

An exposed open parking structure with one suspended level (293 by 540 ft. in size) serving a shopping centre in Central Canada.

The construction took place in the summer of 1987 between June and late September.

The structure comprises six sections each supported at the centre, toward which the shrinkage of the slabs is unimpeded.

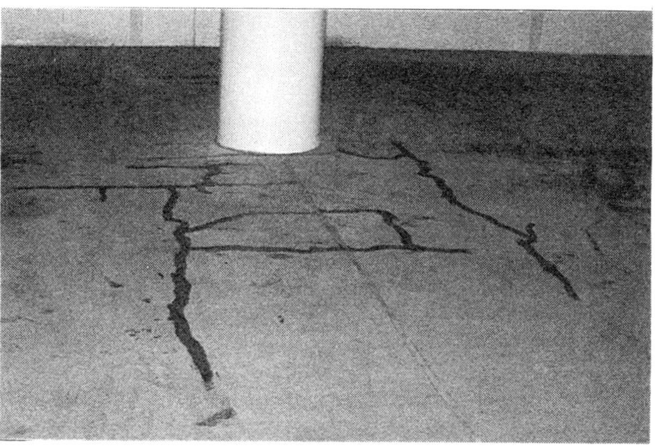

Plate 1

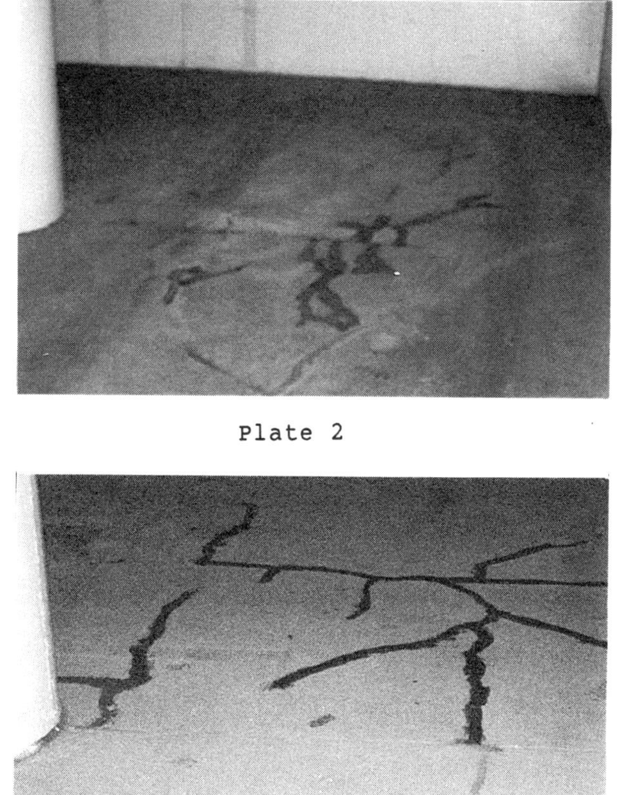

Plate 2

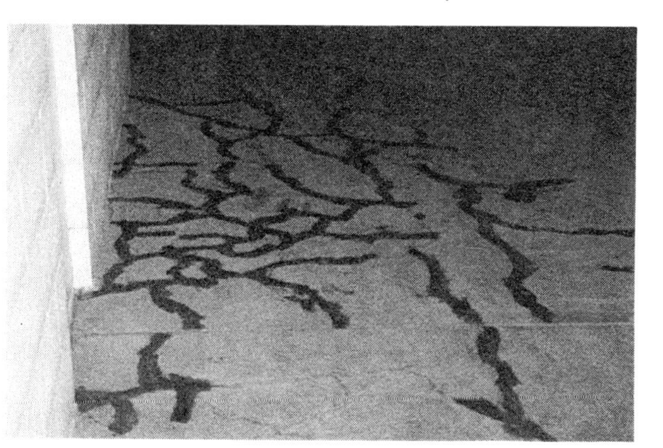

Plate 3

Plate 4

The slab is 10.5 in. thick and the top mat of reinforcement is epoxy coated, while the bottom mat is black steel. The concrete cover over the steel is 1.5 inch thick.

The concrete, which conformed to the requirements of the pertinent Canadian Standards Association (1977) for class A exposure, contained 460 kg cement per cubic metre of concrete, 5 to 8% entrained air and a superplasticizer admixture. The water:cement ratio was 0.33. Concrete placement was fully supervised on the job site; slump (1.5-2"), air temperature, concrete temperature and air content were monitored and the shipment was rejected if it did not meet the specified values. No construction joints were allowed.

There were altogether six concrete pours, four during a very hot summer period (17 June, 8 July, 5 August, 13 August) and the last two in September (the 16th and 25th).

The concrete was wet cured for seven days.

The designed 28 day compressive strength was 35 MPa, and the actual average compresive strength ranged between 36.3 and 47.8 MPa.

No waterproofing membrane was installed because of concern that it would be damaged by the snow plowing operation. A waterproofing concrete sealer was applied to the surface.

In November, two months after the last concrete pour, excessive cracking was noted.

The total length of the leaking cracks of the various sections (each representing a single pour) at 40, 68, and 205 days after the last pour is shown in Figure 1.

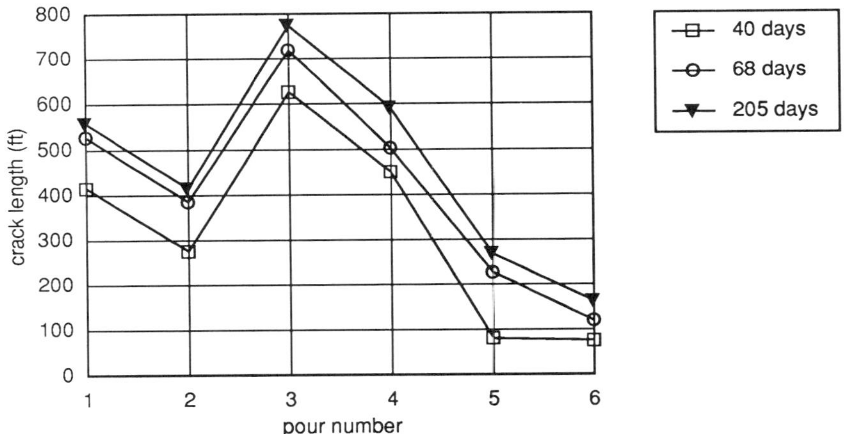

Fig. 1 Length of leaking cracks of sections at indicated age.

It is apparent that the extent of cracking is the greatest in those sections which were poured during the summer months (#3 and #4 pours) despite all the recommended precautions being taken for hot weather concreting. Pour numbered 5 and 6, which were effected in September, have suffered the least amount of cracking.

Whereas most of the cracking occurred in the first six months the rate of crack growth is still significant (Figures 2 and 3).

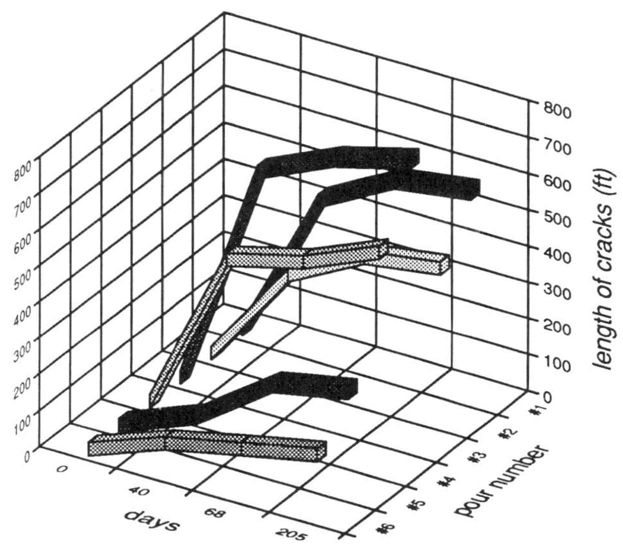

Fig. 2 Increase with time of the length of leaking cracks.

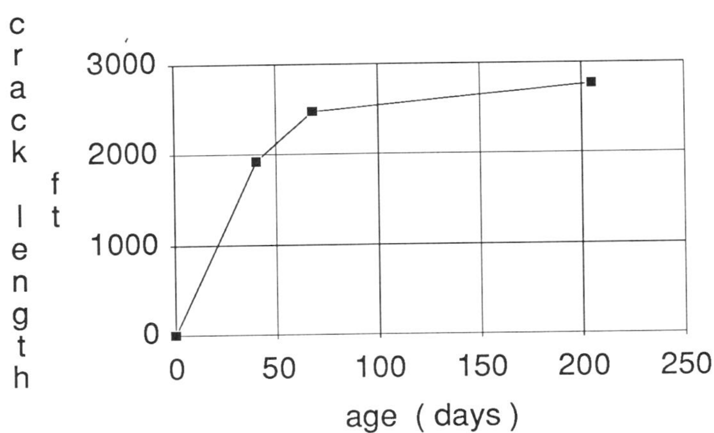

Fig. 3 Increase in total length of leaking cracks.

Eighteen months after completion of the construction of the structure a visual survey was undertaken and the following observations were made:

a) Cracks visible from the soffit side of the suspended slab run along the column line, column to column. Many of the cracks leak, and stalactites have grown (Plates 5-6).
b) The top surface of the suspended slab shows numerous smaller cracks (Plates 7-8).
c) The outline of the reinforcing steel mat is visible in several areas of the underside of the deck. This can be taken as indication for corrosion of the bottom mat of the reinforcing steel.
d) Cracks radiate outwards from shear walls and stair wells.
e) At two locations in the top surface of the slab fairly deep spalling occurred (Plate 9), apparently due to the action of freezing and thawing. As the air content and spacing factor of the spalled concrete has been found to meet the specified requirements (Table 1) (6% air and <0.2 mm) it has to be concluded that the damage was caused by freezing of the water that accumulated in some voids. Obviously, the cracks made water infiltration possible.

Table 1. Air void parameters of spalled concrete

Specimen	Air Content (%)	Spacing Factor (mm)
5	5.66	0.16
6	5.41	0.21

2.4 Garage No. E-2

This parking structure is open and exposed to the elements. It was built in 1988 and has one suspended slab, that is 250 mm thick. The upper mat of reinforcing steel is epoxy coated and has a 20 mm specified cover. The bottom course is black steel. No membrane has been installed because of concern about its vulnerability to the snow plowing operation.

At the time of the site visit snow covered the top surface of the deck and only the underside of the suspended slab could be inspected. It should be noted that no car has yet parked on the upper level because the facility the garage is part of has been only partially opened for business.

Plate 5

Plate 6

Plate 7

Plate 8

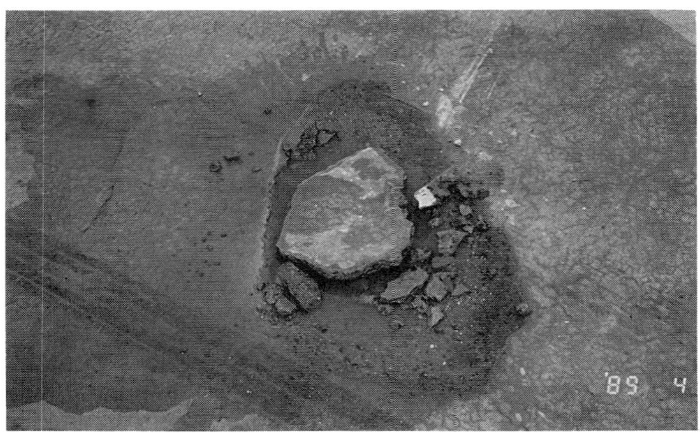

Plate 9

The underside of the slab showed excessive cracking of various types; longitudinal, map and radiating. The outline of the steel is visible in several areas of the bottom side of the suspended slab indicating a high probability of corrosion.

3 Historical Review of the Use of Epoxy Coated Reinforcing Steel

In response to the increasing incidences of corrosion of highway bridge decks induced by de-icing salt, the Federal Highway Administration (FHWA) of the USA initiated research on epoxy coatings in the 1960s.

A major research project was carried out by the U.S. National Bureau of Standards to evaluate the different types of protective coatings for reinforcing steel. The conclusion of the study was that the only impervious and tough/bendable coatings were the epoxy powder coatings applied by the electrostatic spray fusion-bonding process originally developed by the pipeline industry for the coating of steel pipes [Clifton 1974].

The development of the use of epoxy coated reinforcement progressed through further laboratory testing, manufacturing process development, powder formulation, construction specification and experimental use in construction.

The first major field application of epoxy coated reinforcing bars was in a Pennsylvania bridge deck over the Schuykill River near Philadelphia in 1973. Fusion-bonded epoxy coated reinforcement reached the commercial market in 1976.

In 1981 the American Society for Testing and Materials (ASTM) Standard Specification for Epoxy-Coated Reinforcing Steel Bars was issued permitting a range of epoxy thickness between 5 and 12 mils (0.13 and 0.30 mm). Suggested provisions for Epoxy-Coated Reinforcing Bars for Inclusion in Project Specifications were issued by the Concrete Reinforcing Steel Institute in 1984.

Epoxy-coated reinforcing steel developed for bridges and other highway structures were utilized also for the construction of parking garages since 1981.

Although originally epoxy-coated steel was used only in the top layer of reinforcement in bridge decks soon the trend resulted in the use of only epoxy coated steel throughout the bridge deck. In parking garage decks, however, in most cases still only the top mat is specified to be epoxy-coated.

4 Discussion

4.1 Field Exposure

On the whole, the corrosion performance of epoxy-coated reinforcing steel, if properly fabricated and placed under field conditions, appears to be satisfactory. The most notable exception is the case of the corrosion of steel in the substructures of bridges in the Florida Keys [Kessler and Powers (1987), Powers (1988)].

The bridges affected were built between six and nine years ago for service in an environment with high yearly mean temperatures and exposure to sea water containing typically 2.6% chlorides [Zayed et al. (1989)]. Not only the bent bars in the deck but also the straight vertical bars were found to corrode. After cutting through the epoxy with a knife the coating could be easily peeled off revealing the steel surface of the exposed bars to be darkened.

Some members of the epoxy-coated steel industry are of the opinion that the problems experienced in Florida are caused by insufficient quality control in the plant of the supplier. Whether this explanation is correct or not cannot be ascertained but, whatever the reason for the failure of the coating to protect the steel, it seems to be an isolated occurrence.

Both researchers and field engineers reported, however, that new bridge decks with epoxy-coated reinforcing bars have been developing an excessive number of deep cracks during the early stages of curing [American Concrete Institute (1988)].

It was concluded that this may be an interactive effect resulting from:

a) Greater shrinkage of the concrete (with higher cement contents and certain admixtures), along with higher concrete strength,
b) Greater cover over the reinforcement (steel not as effective in restraining cracking at the surface), and
c) Lower "in and out" bond strength (transfer of tensile thrust into the reinforcing bar at cracks and out away from cracks) of epoxy-coated bars.

A need was stated to determine the optimum combination of parameters to minimize cracking while still using epoxy-coated reinforcement, especially since eventual corrosion would concentrate at cracks.

Strength of bond between epoxy-coated reinforcing steel and the concrete matrix is less than that existing between grey steel and the concrete. The reduction of bond strength in quantitative terms is the subject of controversy. In stiff slabs the decrease in bond strength is probably not more than 15%.

4.2 Cause of Cracking

The documented case histories leave little doubt that suspended slabs of parking garages are prone to excessive cracking if constructed with epoxy-coated steel.

Epoxy-coated reinforcement has been used only recently to avoid the durability problems experienced with the former type of garage construction. In addition to the coating of the steel several other measures have also been taken as a protection, such as lower water:cement ratio of the concrete, higher cement content of the mix, greater concrete coverage over the steel. Still, cracking of the concrete is attributed to the epoxy-coating because no adverse occurrences of this type have been observed in recently constructed garages, in which uncoated black steel was used as reinforcement as well as changes in concrete mix design have been adopted.

The cause of the excessive cracking is not known. It appears to be reasonable to accept the hypothesis that the defect is related to the decreased adhesion between the epoxy-coated steel and the concrete matrix compared to that existing between bare steel and concrete. Whereas the achieved bond between the steel and the concrete varies in the range of 10 to 15 percent, in the two systems, great differences exist in the nature of the interaction: in the bare steel system strong adhesive bond exists; in the epoxy-coated steel system, however, adhesion plays a lesser role and the mechanical component of the bond, is important.

The cured concrete decks have essentially similar mechanical properties whether reinforced with bare or coated steel. But there may well be great differences in the nature of the responses of the system during setting when considerable shrinkage takes place.

There is no indication that the cracking of concrete is caused by the corrosion of the steel or that epoxy coating will fail to protect the steel from corrosion in the future. Excessive cracking, however, does permit the ingress of water and chlorides into the deck and if appropriate measures are not implemented in time, corrosion of the lower mat of steel is to be expected. In addition, structures exposed to the elements may suffer damages due to freezing and thawing because of the accumulated water in the crevices and cracks.

4.3 Practical Consequences

The unavoidable consequence is that, in conformity with the CAN/CSA S413 Standard requirement, the installation of a waterproofing membrane is essential .

In existing structures the installation of a waterproof membrane creates not only a financial problem but also poses a technical difficulty if snow clearing with steel bladed equipment is deemed necessary. In these instances a solution has to be found by careful consideration of the condition of the structure, its reserve load carrying capacity, the level of snow clearing required and the available financial resources.

In new structures designers must consider the installation of waterproofing membrane a necessity. In cases of exposed structures some means for the protection of the membrane from the snow plows have to be found. Possibly, the exposed deck will have to be covered with an asphalt wearing course, like bridges, which would however, require a structure with increased load capacity and thus escalate the cost.

Alternately, or simultaneously, consideration may be given to use epoxy coated steel not only for the top but also for the bottom mat. By doing so the danger of corrosion is eliminated even if ingress of salt occurs in the lifetime of the structure. Although the price of epoxy coated steel is about $0.10/lbs. or about 25% more than that of uncoated steel typically this represents only 3.6% of the total cost of the typical garage and is a very small cost in comparison to that of the repairs.

5 Conclusions

Suspended decks constructed with epoxy-coated steel appear to be prone to excessive cracking;
Cracking of decks containing epoxy-coated steel does not seem to be caused by corrosion of the reinforcing steel;
The installation of a waterproofing membrane over concrete decks reinforced with epoxy-coated steel is necessary to avoid serious distress (corrosion of the lower mat of steel, frost action), and leakage through the slab;
If snow clearing with rubber edged blades is not acceptable for a garage exposed to the elements, provisions - such as an asphalt wearing course with increased dead load capacity of the structure if needed - for the protection of the membrane have to be made;
Epoxy-coated steel reinforcement in both top and bottom mats seems to be warranted for adequate protection.

6 Acknowledgement

We are indebited to numerous individuals and organizations who assisted this investigation. In particular, we appreciate the contributions of the following experts (in alphabetical order) Tony Alexander, Construction Control Limited, J.R. Clifton, U.S. National Institute of Standards and Technology, A. Czumachenko, M.S. Yolles Partners, G.C. Hoff, Mobile

Research & Development Corp., D.G. Manning, Ministry of Transport, Province of Ontario, J. Prosser, VSL Canada Ltd., Mr. Schupack, Schupack Suarez Engineers, H. Vander Velde, Vanco Structural Services Ltd., Professor P.Z. Zia, North Carolina State University. Especially appreciated is the cooperation of owners of garages and their technical personnel who provided all the required information.

7 References

American Concrete Institute (1988) ACI Workshop on epoxy coated reinforcement, **Concrete International**, 10, 80-84.

Canadian Standards Association (1977) Concrete Materials and Methods of Concrete Construction Standard CAN3-A23.1-M77.

Canadian Standards Association (1987) Parking Structures Standard CAN/CSA-S413-87.

Clifton, J.R., Beeghley, H.F., and Mathey, R.G. (1974) Non-metallic coatings of concrete reinforcing bars, U.S. National Bureau of Standards Report No. FHWA-RD-74-18, PB 236424.

Kessler, R.J., and Powers, R.G. (1987) Corrosion evaluation of substructure, Long Key Bridge Corrosion Report No. 87-9A, Florida Department of Transportation, Gainsville, Fla.

Powers, R. (1988) Corrosion of epoxy coated rebar, Keys segmented bridges, Corrosion Report No. 88-8A, Florida Department of Transportation, Gainsville, Florida.

Zayed, A.M., Sagues, A.A., and Powers, R.G. (1989) Corrosion of epoxy coated reinforcing steel in concrete, in **Corrosion 89**, National Association of Corrosion Engineers, Paper No. 379.

45 DURABILITY OF CONCRETE ORNAMENTS ON HISTORIC STRUCTURES

S.E. THOMASEN, C.L. SEARLS
Wiss, Janney, Elstner Associates, Inc., Emeryville, California, USA

Abstract
Historic concrete structures typically have ornamental elements which constitute important architectural features. These elements often experience environmental deterioration because of their large surface area, thin cross sections and bold exposure. These design features make the concrete vulnerable to temperature changes, moisture infiltration and attack by harmful environmental agents. The examination of the performance of such concrete construction provides insight into the durability of not only the original material but also information on the success or failure of design features and construction techniques.

The performance of concrete under a variety of exposure conditions is illustrated by several case studies of historic concrete structures in Southern California. These include a large 75 year old multiple arch concrete bridge which has received only minimum maintenance, a facade on an art museum with highly sculptured cast-stone ornaments, and a 100 year old, well maintained concrete/masonry dam.
Keywords: Concrete, Deterioration, Historic.

1 Introduction

Concrete structures are designed to provide long-term serviceability, safety and acceptable appearance. Designers and builders often assume that concrete is a maintenance free material and that its performance is constant with time. These assumptions have been challenged by failures observed in historic concrete structures. Examination of these structures shows that deterioration affects not only the surface but that decay sometimes reaches the inner mass of the concrete.

The study of historic structures provides information on the long-term durability performance of concrete, as illustrated in the following three case studies.

2 Case study - The Colorado Street Bridge

2.1 Introduction

The Colorado Street Bridge in Pasadena near Los Angeles, California, is an open spandrel, eleven-span arched reinforced concrete structure built in 1912-13 (Fig. 1). At 447 m long and 46 m above the bed of the arroyo, it was the longest and highest concrete bridge of its time built in the United States. The bridge features extensive decorative detailing and picturesque refuge bays set into the side railings on each pier. The bridge was designed by Joseph Alexander Low Waddell, one of the nation's foremost bridge engineers, and built by John Drake Mercereau, a well known California builder.

Fig. 1. Colorado Street Bridge

The two-lane bridge served as the major connection between Pasadena and Los Angeles until the 1950s, when an adjacent bridge was built. Since then only minimal maintenance has been provided for the old bridge. Approximately ten years ago, the State of California Department of Transportation observed serious deterioration of the bridge and undertook a series of studies to evaluate its condition and make recommendations for its repair.

2.2 Environmental exposure

The climate in Pasadena is mild with only occasional frost. The average yearly rainfall is 350 mm. The acid content in the rain is low and the salt content in the rain is low to moderate. Snow is seldom seen and the bridge is not exposed to deicing salts.

The general environment might be mild but the design of the bridge has created severe microenvironments. Expansion joints fill up with wet soil and drains at the bridge deck are easily clogged causing ponding conditions and the deck drains empty at top of the arches creating prolonged wetness of the concrete arch surfaces.

2.3 Materials

Petrographic examination of samples from the bridge showed the concrete had a cement content of 270 kg/m^3. The cement paste was moderately soft. The large aggregate was sound and very hard. Chemical analysis of concrete cores found a chloride content well below the threshold limit for corrosion of reinforcing steel. Compressive strength tests were made of selected concrete cores and results were correlated with Schmidt Hammer tests conducted in the field. The compressive strength varied from 17 to 27 MPa. Concrete cores removed from the bridge structure had a depth of carbonation of from 50 to 75 mm, a depth greater than the typical concrete cover over the reinforcing bars.

Fig. 2. Soffit of bridge deck shows evidence of extensive leakage

2.4 Observed conditions

The bridge deck was inspected visually and with a vehicle-mounted sound recording device to detect subsurface deterioration. Delaminations within the concrete slab at the level of the top reinforcing mat were recorded. The amount of delamination ranged from 33% to 93% of the surveyed deck area.

Fig. 3. Spalling and corrosion at arches. Large sections of reinforcing steel are exposed

The piers and arches were surveyed from the bridge using a truck-mounted "snooper scaffold" and from below using a high-lift crane. Deterioration was observed throughout the structure. The underside of the bridge deck had severe cracking, softening of the cement paste and residue of efflorescence as a result of leakage through the concrete (Fig. 2). The concrete arches exhibited extensive spalling and concrete delamination, particularly in areas where the concrete had been exposed to constant leakage. Reinforcing steel was exposed in many locations where corrosion had caused the concrete to spall (Fig. 3). Many concrete surfaces were covered with organic growth.

3 Case study - Museum of Art, San Diego

3.1 Introduction

The Museum of Art in San Diego was built in 1926 at the site of the Panama - California Exposition in Balboa Park. The building design was inspired by Spanish renaissance structures such as the University of Salamanca and Siguanza Cathedral but the design was executed in cast stone rather than in carved, natural stone (Fig. 4). The two-story building has a concrete frame with clay tile infill walls. The exterior walls are covered with a thin layer of integral colored cement plaster rendering. The many decorative elements are precast concrete sections -- also called 'cast stone' -- which are set into the structure. The cast stone includes the roof cornice with dentils and motifs of carved scallop shells and the frontispiece at the main entrance, which has an elaborate plateresque design embodying pilasters infilled with arasbesque, arms and candelabras, portraits of El Greco and De Ribera and statues of Velasquez, Murillo and Zurbaran.

Fig. 4. Museum of Art, San Diego

3.2 Environmental exposure

The climate in San Diego is mild with no frost and with the surface temperature at the facade varying from 4 to 50°C. The average yearly rainfall is 400 mm and the prevailing wind is from the ocean about 10 km away. The content of acid in the rain is low but the salt content is moderate to high.

3.3 Materials

Cast stone ornaments were a popular building material in the 1920's. They are fabricated by pressing a 5 cm layer of colored sand mortar into a plaster form and then filling up the form with a regular concrete mix. The cast stone blocks are reinforced and they can be solidly cast or hollow with internal web stiffeners. The blocks are installed in the wall with mortar joints and anchored to the substrate with steel fasteners similar to those used in the setting of marble and stone ornaments.

Fig. 5. Crispness of design retained

Fig. 6. Vertical cracks in cast stone blocks from corrosion of embedded steel

3.4 Condition assessment

The surfaces of the cast stone blocks are in excellent condition. The texture shows the grain of the original rich sand mix with only minor surface erosion to depths of up to 4 mm. The crispness of the design has been retained (Fig. 5). The cast stone blocks are placed in the wall with mortar joints. Moisture infiltration through deteriorated joints and at unprotected horizontal surfaces has caused corrosion of the embedded reinforcing bars and steel anchors, resulting in cracking of the cast stone blocks from rust volume expansion (Fig. 6).

4 Case study - Sweetwater Dam, Chula Vista

4.1 Introduction

The Sweetwater Dam is located north of the Mexican border about 12 km inland from the Pacific ocean. The original masonry dam was erected in 1888 with stone blocks cut from nearby hills. In 1911 the structure was enlarged by using the masonry as a core to create a concrete gravity dam that was 150 m long and 36 m high (Fig. 7). The concrete was mixed on-site using cement imported from Belgium. The dam was finally enlarged in 1916 to a height of 40 m and a length of 210 m. The structure has been kept in continuous service with only a minimum of maintenance. It functions as a drinking water reservoir for the nearby communities and the water is collected partly as runoff from surrounding mountains and partly pumped in from the Colorado River.

Fig. 7. Sweetwater Dam, Chula Vista

4.2 Environmental exposure

The climate in Southern California is mild with no frost. The average yearly rainfall is 300 mm. The acid content in the rain is low but the salt content in the prevailing wind from the nearby Pacific Ocean is moderate to high. Most of the upstream dam face is under water. The downstream face is well drained and it receives only occasional runoff. The masonry core of the dam is to some degree permeable and seepage occurs through the dam. The water typically evaporates at the downstream face but in some areas it appears as flowing leakage, keeping the concrete surfaces constantly wet.

Fig. 8. Form patterns are well preserved

4.3 Materials

Petrographic examination of concrete samples from the dam face found that the concrete had a cement content of 300 kg/m^3. The cement, which was extremely hydrated, was a Type IV. The paste was firm but with some chalky areas. Both the fine and the large aggregate were sound but the large aggregate had a weathered appearance, which was judged to be from exposure prior to being placed in the concrete mix. The concrete showed no evidence of silicate attack or alkali silika reaction.

4.4 Condition assessment

The upstream face of the dam is structurally sound with no evidence of distress. The concrete surfaces were originally board formed and the form patterns are preserved (Fig. 8). The concrete at the pour lift lines was badly bonded and the segregated aggregate at these lines now shows up as open joints. Penetration and evaporation of water at these pour lines has further deteriorated the joint edges. The concrete surface at the downstream dam face is peeling off in about 20 mm thick layers, exposing the large aggregate. The exfoliation affects the total lower dam face and occurs as the water that seeps through the dam evaporates at the surface. This water has dissolved calcium hydroxide and other cement binders on its way and the volume expansion from the recrystallization causes the exfoliation. The prolonged leakage in some sections has also resulted in surface deposits of calcium carbonate (Fig. 9).

Fig. 9. Surface exfoliation and calcium deposits

5 Durability failures

5.1 Introduction

Types of concrete deterioration most often found in historic structures are surface erosion, environmental deterioration, water related deterioration, expansion and contraction failures, corrosion damage, material and workmanship failures, and deterioration from lack of maintenance.

5.2 Surface erosion

Surface erosion is the weathering of the concrete surface by sun, water, or wind-borne sand. Mechanical and chemical erosion attack the softer cement paste, first at the surface layer which exposes the aggregate and then the paste between the aggregate. Surface erosion to a depth of several millimeters was seen at the three case study structures. The erosion exposes more porous concrete, and allows more water infiltration which can impair the general durability of the structures.

5.3 Environmental deterioration

Carbon dioxide reacts in the presence of moisture with the cement paste at the concrete surface. The reaction neutralizes the alkalinity of concrete and reduces the protection of embedded metal, permitting corrosion of reinforcing steel. The high level of concrete carbonation at the Colorado Street Bridge has been a contributing factor to its excessive corrosion damage.

5.4 Water related deterioration

Moisture is the environmental agent most harmful to concrete. Water exerts pressure when it freezes, supports biological growth, and is a carrier for chemicals and salts.

The presence of biological organisms such as algae, moss, lichen, and plants, prevents the concrete from drying out, and the chemical products of their metabolism disintegrate the cement binder. The formation of biological growth in a crack can exert considerable pressure which ultimately fractures the concrete. Biological growth is seen at both Colorado Street Bridge and at Sweetwater Dam. It has caused minor decay in form of surface pitting of the cement paste and extensive spalling at edges of cracks containing organic growth.

Infiltrated water is a carrier of chemicals, which dissolve cementitious binders as the water permeates through the concrete. As the water evaporates, salt particles expand and cause erosion and surface delamination, as observed on the downstream face of the Sweetwater Dam.

5.5 Expansion and contraction failures

Volume changes can be one-way or cyclic and they cause distress when the structure has no provisions for differential movement. Creep and shrinkage of concrete and thermal expansion and contraction can result in cracking and crushing. Good design and workmanship will reduce most detrimental effects. The Sweetwater Dam was built without expansion and contraction joints but the concrete was reportedly placed slowly, with time for the initial shrinkage to take place before the next pour.

5.6 Corrosion damage

Corrosion causes strength loss of reinforcing steel, and volume expansion from rust products results in cracking, spalling and delamination of concrete. Factors contributing to corrosion are insufficient concrete cover, high permeability of the surface layers, loss of protective alkalinity and high percentage of soluble chloride ions in the concrete.

The Colorado Street Bridge concrete has a low chloride content but it has insufficient concrete cover, highly carbonated concrete and high porosity due to a low cement content and high water/cement ratio. As a result corrosion has occurred causing severe cracking and delamination of the concrete. The structural strength is diminished by the loss of steel section.

The reinforcing in the Sweetwater Dam has a concrete cover of more than 70 mm and this has prevented any corrosion damage despite the extensive exfoliation of the concrete surface.

5.7 Material and workmanship

Early concrete structures were often constructed with soft aggregates such as crushed bricks which expand when wet. Early concrete was also poorly consolidated when placed in the forms, and this tended to leave voids at congested areas and 'honeycombs'--aggregate without cement paste--at the surface. Problems of workmanship, some of which are not unknown in modern concrete construction, are cold joints where one layer of deposited concrete is allowed to harden before the next layer is poured, as seen at the Sweetwater Dam, and insufficient cover over reinforcing steel, as seen at the Colorado Street Bridge.

5.8 Deterioration from lack of maintenance

Regular maintenance is critical to durability. When maintenance is neglected the concrete often deteriorates at an accelerated rate, as seen on the Colorado Street Bridge. The maintenance must begin before the deterioration has progressed to the point of requiring major repair.

6 Conclusions

Concrete structures depend on the durability of the material to provide serviceability, safety and acceptable appearance. Evaluation of the performance of concrete in historic structures provides information on durability of the material under long-term exposure. It is found that while some durability failures only concern the surface appearance others affect the safety and serviceability of the structures. Most failures are associated with moisture infiltration indicating that to achieve maximum durability the most important material property is low porosity of the concrete.

46 IMPROVING DURABILITY BY ENVIRONMENTAL CONTROL

P. LAMBERT, J.G.M. WOOD
Mott MacDonald Limited, Croydon, UK

Abstract
The life of a component is controlled by the material properties and the environmental conditions. Where the environment is particularly aggressive it may be necessary to either enhance the material properties or alter the environment local to the component. With building materials such as concrete it is more common to concentrate on the material characteristics such as strength and permeability.

The purpose of this paper is to examine the environmental factors which influence the service life of building materials and show how they may be controlled by the use of techniques such as coating and cladding, so enhancing durability. In addition, an examination will be made of the potential risks associated with incorrect use of such techniques.
Keywords: Durability, Concrete, Reinforcement, Corrosion, Environment, Cladding, Coating.

1 Introduction

Many surviving traditional structures have mellowed with age and remain serviceable as a tribute to their builders' understanding of the materials and their owners' prudent maintenance. The record of durability of reinforced concrete structures inspires less confidence. The mistakes of our forefathers fell down long ago and have been forgotten, but from them the classic traditional styles of masonry, brick and timber construction developed which gave long, useful lives with only simple maintenance. The rapid development of concrete during the twentieth century has coincided with the fragmentation of education in the professions, resulting in a failure to appreciate the characteristics of the material that relate to overall durability rather than just strength. The product of this lack of understanding has been ugly structures, suffering from premature staining, cracking and spalling, marring the reputation of reinforced concrete.

The service life of a structural component is ultimately controlled by the interaction of the materials (from which it is constructed), with the environment to which it is exposed. For many structural materials there is a wealth of past experience which may be drawn upon. With more recently introduced materials such as polymers or new combinations of materials, such experience may not exist and reliance must be placed upon laboratory evaluation and accelerated testing. For both traditional and new materials we must quantify the physical and chemical characteristics and examine their compatibility.

The purpose of this paper is to examine some of the environmental factors which can influence the service life of building materials and how they can be controlled to be compatible with the material properties by the use of techniques such as coating and cladding. The rapid deterioration of basically good materials in adverse environments is also highlighted.

2 Improving the Performance of Materials

Where a structure is to be exposed to a particularly aggressive environment, there is a choice between enhancing the material properties or adjusting the environment local to the component so as to achieve the design life. The full structural design life may be achieved by using replaceable short life units, moderate quality units with additional protection or regular maintenance, or high quality units made from 'immortal' materials.

When specifying construction materials such as concrete it is usual to concentrate on the improvement of material characteristics such as strength and permeability. However, while strong and impermeable bricks are recognised to resist frost damage, the introduction of a damp proof course or application of a water-repellant coating can impart similar properties on weaker, more permeable brickwork. Similarly, the use of pitched roofs with good overhangs, gutters and hanging tiles has enabled such inherently poor materials as wattle and daub, timber and half timbered buildings to last for hundreds of years.

3 The Durability of Reinforced Concrete

The environment provided by good quality concrete for the embedded steel reinforcement is one of high alkalinity (generally >pH 13), produced by the hydroxides of sodium, potassium and calcium released during the various hydration reactions, Barneyback and Diamond (1981). In addition, the bulk of surrounding concrete acts as a physical barrier to most of the substances that may lead to degradation of the

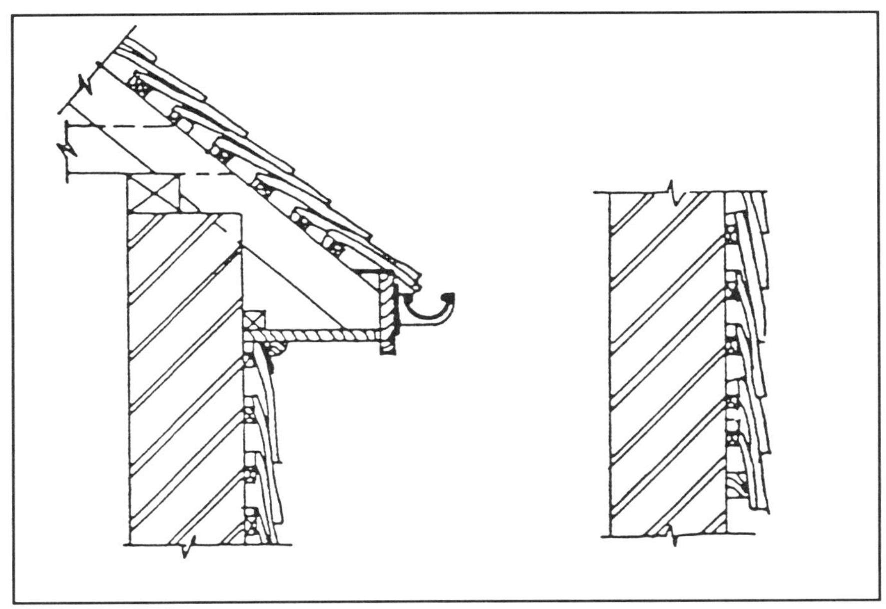

Fig.1 Traditional details for dry, durable walls

reinforcement. Provided this environment is maintained, the steel remains passive and any small breaks in the stable protective oxide film are soon repaired. However, if the alkalinity of the surroundings is reduced, for example by reaction with atmospheric carbon dioxide (carbonation), or if depassivating chloride ions are made available at the surface of the steel then corrosion may be initiated, resulting in loss of steel section and spalling of cover concrete, Page and Treadaway (1982).

The first route to improved durability must therefore be to resist the ingress of these aggressive substances by using a 'high quality' concrete. A thick cover depth of low permeability concrete will greatly assist in preventing depassivation of the reinforcement, however, structural factors may limit the maximum cover to 50mm. The composition of cement is clearly important, while the use of cement replacement materials such as fly ash (pfa) and blast furnace slag (ggbs) can also be beneficial, particularly when improving the resistance to chloride ions, Page et al. (1986).

The corrosion rate of steel in concrete is dependant upon the presence of an ionically conductive aqueous phase in contact with the steel, the existence of anodic and cathodic sites on the metal surface in contact with this electrolyte and the availability of oxygen to enable the

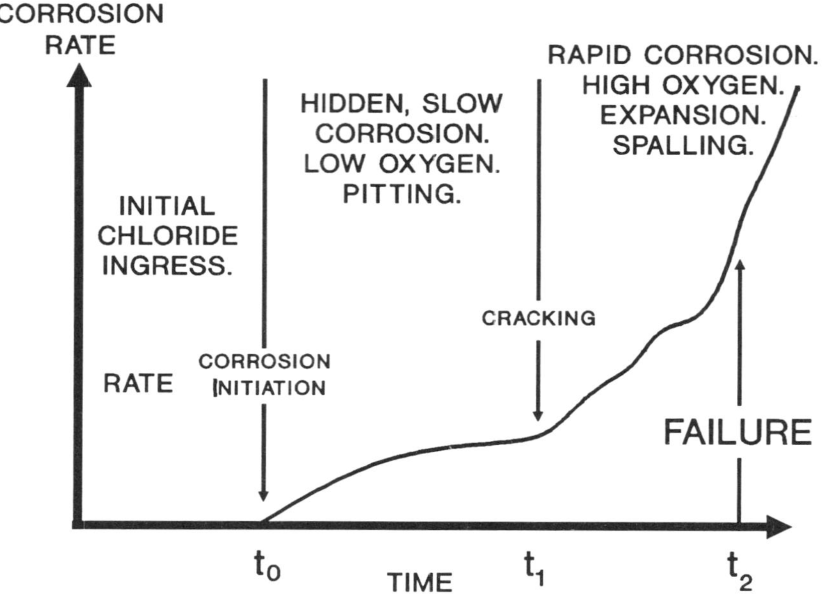

Fig.2 The development of chloride induced corrosion

reactions to proceed. This indicates a second route to restricting corrosion and hence improving durability through control of the environment, Wood (1985). It is certainly possible to stifle corrosion by restricting oxygen availability, although the amount of oxygen required to maintain corrosion is very small and may be difficult to limit other than by complete immersion in water, Page and Lambert (1987). Alternatively, reinforcement corrosion may be controlled by limiting water access and allowing the concrete to 'dry-out', the associated increase in electrical resistance limiting the magnitude of corrosion currents, Tuutti (1982). The influence of defects is likely to be of great importance with regard to the success of such techniques.

For the consideration of how design or modification of reinforced concrete structures to create a benign environment can influence durability, it is essential to understand how exposure conditions influence the following factors:-

The rate of breakdown of protection and initiation of corrosion or other deterioration.
The rate of corrosion or deterioration.
Secondary deterioration processes.

For example, carbonation develops much more in dry (60-70% RH), internal conditions than in wet, near saturated conditions, Parrott (1987). Once carbonation has reached the steel, corrosion rates are negligible at 50% RH but most rapid at 95% RH or under conditions of wet and dry cycling, Wood (1985). Once corrosion has developed sufficiently to crack the cover, corrosion is accelerated. Spalling of the cracked cover by frost action further accelerates the deterioration.

In the case of chloride induced corrosion, chloride ion ingress can only occur through the liquid phase of the concrete. The maximum rate of ingress occurs during cycles of wetting and drying. Chloride induced corrosion may also be separated into different stages, as shown in figure 2. It is a distinct function of temperature, humidity and cycles of wetting and drying, Tuutti (1882).

For alkali aggregate reaction (AAR), the most damaging condition is from 95% RH to near saturation. Total immersion may leach alkali and ameliorate the reaction, while at 80% RH AAR proceeds very slowly until re-wetting occurs. AAR is effectively dormant below 75% RH, Wood and Johnson (1989).

4 Controlling the Environment

Continuing with the example of steel reinforced concrete, and taking the widest possible definition of environmental control, there are a number of different regions where the environment can be altered or controlled in order to improve the durability. The most important areas are shown in figure 3. Further sub-division of these regions is possible, for example the environment within a single gel pore, but in this case it is sufficient to concentrate on the areas indicated.

4.1 The environment surrounding the re-bar (A)

Normally protective to the re-bar, it may be enhanced, eg. by the addition of inhibitors, Short et al. (1989), or excluded, eg. by the use of a high quality epoxy coating, Rostam et al. (1990).

4.2 The environment provided by the bulk concrete (B)

Greatly influenced by material factors such as mix design, cement type, cement replacement materials, mixing water and admixtures.

4.3 The environment at the surface of the concrete (C)

The importance of the concrete surface tends to be underestimated when considering the performance of the

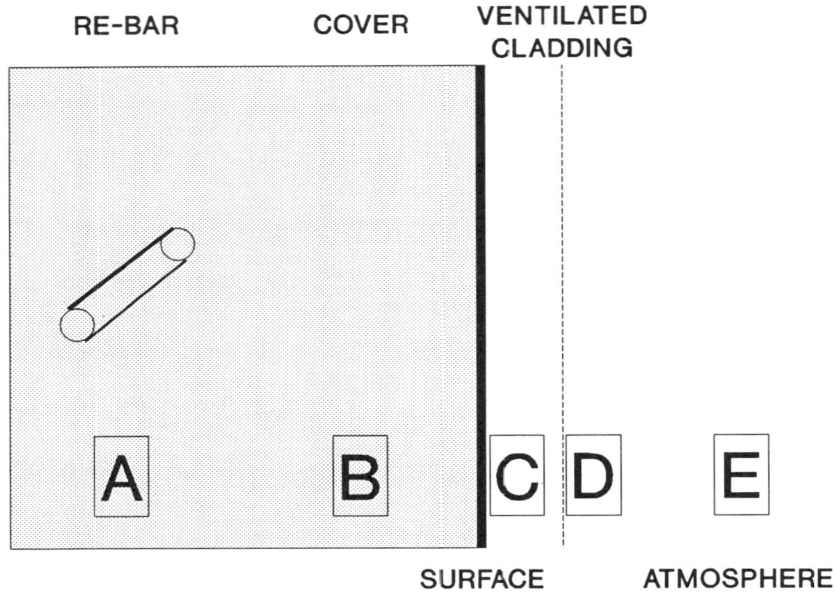

Fig.3 Regions of possible environmental control for reinforced concrete

structure. Most of the mobile species that affect the durability of both the concrete and re-bar (eg. water, oxygen, carbon dioxide, chloride, sulphate), must cross this boundary. It is also the easiest to influence, for example, by the application of a coating or, more recently, a surface treatment such as silane.

4.4 The environment provided by ventilated cladding (D)

Carefully designed ventilated cladding is more durable and more effective than any coating in reducing moisture ingress and shielding from chlorides, Anderson and Gill (1988).

4.5 The external atmospheric environment (E)

Full control of the external environment is clearly impractical in all but the most demanding applications, such as certain nuclear storage facilities. Partial or local control is possible by careful drainage design.

5 Possible Risks

While environmental control can be seen to be a powerful technique in the improvement of durability, it is always important to ensure that the introduction of new products does not replace one set of problems by new, possibly unexpected ones. For example, while sheltering concrete from the elements may be beneficial in limiting existing corrosion and slowing AAR, it can also accelerate the rate of carbonation of the concrete cover leading to the destruction of the protective alkaline environment, Parrott (1987). A further example is the use of anodic corrosion inhibitors as concrete additives, where the level of addition is critical; too much inhibitor may weaken the concrete, too little may result in enhanced localised corrosion of the reinforcement, Short et al. (1989). A major limitation of some of these products is that they are leached from the concrete faster than the chlorides enter. Consequently, when the chloride ions reach the steel, there is little or no inhibitor to resist their action.

The use of surface treatments and coatings on concrete may also be of mixed value. Impermeable coatings can limit carbonation or chloride ingress but they can obscure and disguise areas of damage and by trapping moisture can aggravate AAR. They may also prevent the use of non-destructive monitoring techniques such as half-cell potential surveys, or more seriously, distort the results to give a false impression of the condition of the structure. There is particular concern with the use of surface treatments such as silanes which can interfere with the electrolyte path between the surface of the concrete and the reinforcement. Being vapour permeable, silanes are likely to be of little use with AAR, except on rain-swept vertical faces, and ineffective in resisting the ingress of carbon dioxide. These are clearly areas requiring further work.

6 Conclusions

As enhanced durability through improved material performance becomes more difficult and expensive to achieve, there is an increasing need to control or alter the exposure environment so that the capabilities of existing materials are not exceeded. It is important to fully understand the nature of both internal and external environments and the precise way in which they are to be changed.

Where new or enhanced performance materials are required, laboratory evaluation and accelerated testing is essential so that the full range of physical and chemical properties are quantified. There may be advantages in increasing the use of numerical modelling techniques to predict the durability of both new and traditional materials in the increasingly complex and changing environments to which they are exposed.

7 References

Anderson, J.M. and Gill, J.R. (1988) **Rainscreen cladding - a guide to design principles and practice.** CIRIA.

Barneyback Jr.,R.S. and Diamond, S. (1981) Expression and analysis of pore fluids from hardened cement pastes and mortars. **Cement and Concrete Research, 11**, pp. 279-285.

Page, C.L. and Lambert, P. (1987) Kinetics of oxygen diffusion in hardened cement pastes. **Journal of Materials Science, 22**, pp. 942-946.

Page, C.L. Short, N.R. and Holden, W.R. (1986) The influence of different cements on chloride-induced corrosion of reinforcing steel. **Cement and Concrete Research, 16**, pp. 79-86.

Page, C.L. and Treadaway, K.W.J. (1982) Aspects of the electro-chemistry of steel in concrete. **Nature, 297**, No.5862, pp. 109-115.

Parrott, L.J. (1987) **A review of carbonation in reinforced concrete.** C&CA/BRE, DoE, Building Research Station, Garston.

Rostam, S. Ecob, C.R. and King, E.S. (1990) Epoxy-coated reinforcement cages in precast concrete segmental tunnel linings - durability. **Third International Symposium on Corrosion of Reinforcement in Concrete Construction**, Society of Chemical Industry, UK.

Short, N.R. Lambert, P. and Page, C.L. (1989) Effect of corrosion inhibitors on pore solution chemistry of hardened cement pastes. **Durability of Concrete - Second International Seminar**, Gothenburg, pp. 218-228.

Tuutti, K. (1982) **Corrosion of steel in concrete.** Swedish Cement and Concrete Institute.

Wood, J.G.M. and Johnson, R.A. (1989) An engineers perspective on UK experience with alkali aggregate reaction. **Eighth International Conference on AAR**, Kyoto.

Wood, J.G.M. (1985) Methods of control of active corrosion in concrete. **First International Conference on Deterioration and Repair of Reinforced Concrete in the Arabian Gulf**, Bahrain, pp. 139-150.

47 DESIGN CHARTS FOR CRACK CONTROL IN CONCRETE MEMBERS REINFORCED WITH WELDED WIRE FABRIC

K.H. TAN, M.A. MANSUR, K. KASIRAJU, S.L. LEE
Department of Civil Engineering, National University of Singapore, Singapore

Abstract
Welded wire fabric (WWF) has been found to be effective in reducing crack widths in concrete members. In this paper, the basis for the development of design charts for crack control in tension members, beams and one-way slabs reinforced with WWF is explained. Typical charts covering the normal range of sectional and material properties are presented and examples are given to illustrate the use of these charts.
Keywords: Design Charts, Crack Control, Tension Members, Beams, One-Way Slabs, Welded Wire Fabric (WWF).

1 Introduction

In reinforced concrete structures, cracking is almost inevitable. This is because concrete has a low tensile strain capacity compared to the usual working strains in the steel reinforcement. However, crack control in reinforced concrete structures is of concern due to three main reasons: appearance, leakage and durability.

Wide cracks are not only aesthetically unpleasant; they may lead to unnecessary public alarm. In liquid-retaining structures, leakage, the prevention of which is of utmost importance, is a function of crack width. Cracking enables the carbonation of the concrete or the penetration of chlorides to reach the steel bar surface quickly. The time required for an electrolytic cell to be established and corrosion of steel reinforcement to start depends on the crack width

Crack widths can be reduced by improving the bond characteristics between the steel bars and concrete, by dispersing the reinforcement uniformly or by using welded wire fabric (WWF) as reinforcement. The use of WWF is of significance since it leads to greater speed and economy in construction. The authors have earlier shown that, with proper spacings for the welded transverse wires of the WWF, substantial reduction in crack widths can be achieved in both tension as well as flexural members (Lee et al. 1987, 1989a, 1989b). A simple approach to calulate the crack widths in such members has also been recently proposed by the authors (Mansur et al. 1990).

To expedite the design process, however, graphical charts are useful. In this paper, design charts for crack control in tension members, one-way slabs and beams reinforced with WWF are presented. The basis for the development of the design charts is explained and examples are given in Appendix to illustrate the use of these charts.

2 Basis for development of design charts

According to the crack spacing model developed earlier by the authors (Lee et al. 1987, 1989a 1989b) for concrete members reinforced with WWF, the spacing of cracks for various spacings S_t of welded transverse wires are governed by one of the following two parameters: (a) the critical transfer length L_t'; and (b) the primary (flexural) crack height h_c. The critical transfer length is defined as

$$L_t' = K_p \frac{A_{ce} f_t}{\Sigma_o} \tag{1}$$

where K_p is the bond coefficient; f_t is the tensile strength of concrete; Σ_o is the total perimeter of tension reinforcement and A_{ce} is the effective concrete area [see Fig. 1] defined by (Desayi 1976):

$$A_{ce} = \left(\frac{2b}{D+b} \right)^{1/2} \left(\frac{\pi}{4} D^2 - A_s \right) \quad \text{for tension members} \tag{2a}$$

$$A_{ce} = 0.8 \left(2(D - d)b - A_s \right) \quad \text{for beams} \tag{2b}$$

$$A_{ce} = 0.5 \left(2(D - d)b - A_s \right) \quad \text{for one-way slabs} \tag{2c}$$

in which D = overall depth; d = effective depth to tension reinforcement; b = width and A_s = area of tension reinforcement.

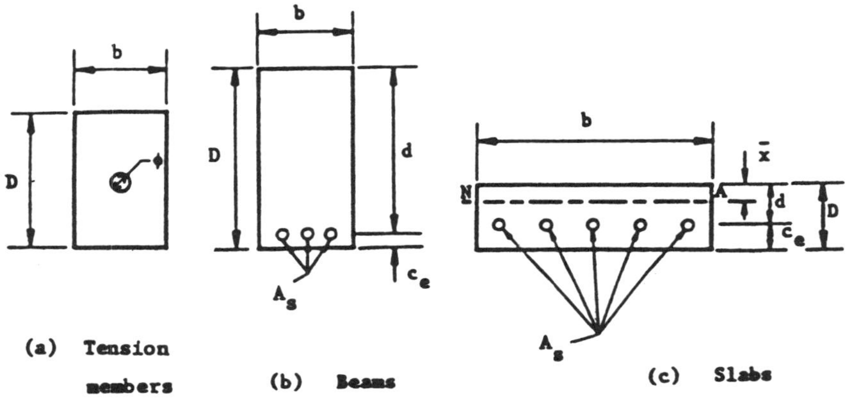

Fig. 1. Effective concrete area in tension.

The primary crack height can be obtained from elastic analysis and is given by (neglecting the presence of compression steel):

$$h_c = D[1 + m\rho\alpha\ (1 - \sqrt{1 + (2/m\rho)})] \tag{3}$$

where $\alpha = d/D$, $\rho = A_s/bd$ and $m = E_s/E_c$; E_s and E_c being the modulus of elasticity for steel and concrete respectively.

In general, the minimum and maximum crack spacings, a_{min} and a_{max}, are:

$$\left.\begin{array}{llll} \text{when } S_t < X; & a_{min} = S_t & \text{and} & a_{max} = 2S_t \\ \text{when } X < S_t < 2X; & a_{min} = (S_t - X) & \text{and} & a_{max} = S_t \\ \text{when } 2X < S_t < 3X; & a_{min} = (S_t - 2X) & \text{and} & a_{max} = 2X \\ \quad\vdots & \quad\vdots & & \quad\vdots \\ \text{when } nX < S_t < (n+1)X; & a_{min} = (S_t - nX) & \text{and} & a_{max} = 2X \end{array}\right\} \tag{4}$$

where $X = L_t'$ for tension members and beams and $X = h_c$ for one-way slabs of usual thickness and amount of longitudinal reinforcement.

Fig. 2 shows the variation of minimum and maximum crack spacings with transverse wire spacing. The use of WWF leads to smaller crack spacings and hence smaller crack widths when the spacing of transverse wires is restricted to certain values. Two zones for the spacing of the welded transverse wires have been recommended. Zone 1 refers to spacings of less than 0.75X, which if not practical, may be replaced by those of Zone 2, that is, spacings of between X and 1.5X.

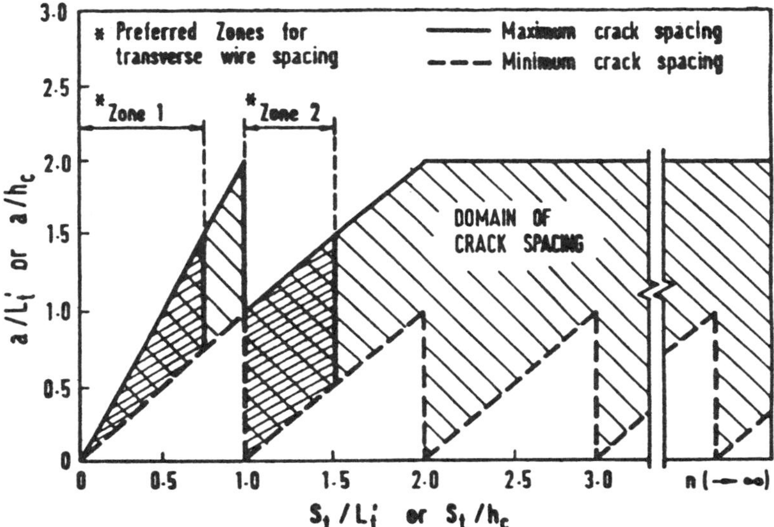

Fig.2. Domain of crack spacing.

Once the values of minimum and maximum crack spacings are determined, the corresponding crack widths for a particular stage of loading can be calculated by appropriately assuming the distribution of slip along the steel reinforcement between cracks and using an idealized bond-slip relationship for the reinforcement. The authors have considered two variations in the slip distribution. In the rigorous method (Lee et al. 1987, 1989a, 1989b), slip between steel and concrete is assumed to occur only within a short distance from the crack whereas in the simplified method (Mansur et al. 1990), slip is assumed to occur throughout the entire length of the segment between two consecutive cracks. In both methods, the maximum slip at the face of a crack is first calculated. The maximum crack width, which is of particular interest in serviceability design, is then taken as twice the value of this maximum slip. For beams and slabs, the maximum crack width occurs at the extreme tensile fibre and is taken as twice the maximum slip multiplied by the cover ratio R, where R = h_c/h_s and h_s is the distance of the centroid of tension reinforcement from the neutral axis. The rigorous method was found to give accurate predictions of the maximum crack widths while the simplified method leads to conservative but reasonably accurate estimates (Mansur et al. 1990). In this paper, the rigorous method is used for developing the design charts.

In general, the maximum crack width w_{max} at a given stress level in steel f_s, depends on the maximum crack spacing a_{max}, the bond properties of the reinforcement as represented by slip modulus k, critical slip u_c and ultimate bond stress f_{bu}, modular ratio m, diameter of the longitudinal bars ϕ_ℓ and the effective reinforcement ratio ρ_e (= A_s/A_{ce}). Assuming a bi- linear idealization of bond-slip relationship, $u_c = f_{bu}/k$. Hence,

$$w_{max} = f(a_{max}, k, f_{bu}, m, \phi_\ell, A_s/A_{ce}) \qquad (5)$$

A systematic study carried out to determine the effects of various parameters on maximum crack width indicated that the influences of k, m and A_{ce} are very small, and hence can be neglected. Therefore,

$$w_{max} = f(a_{max}, f_{bu}, \phi_\ell, A_s) \qquad (6)$$

2.1 Tension members

In order to minimize the value of w_{max}, a_{max} should be as small as possible. For Zone 2 of the preferred transverse wire spacings for tension members [see Fig. 2], the minimum possible value for a_{max} is obtained when the transverse wires are located at a spacing of L_t'. However, this spacing is highly sensitive. A slight reduction in S_t below L_t' would result in a maximum crack spacing a_{max} equal to $2L_t$'.

In this case, WWF is not at all effective in crack control, and it will behave in a manner similar to conventional reinforcement. To avoid this and to account for the uncertainties in material properties, the optimum transverse wire spacing $(S_t)_o$ is taken as 1.1 L_t' and this equals the minimum feasible value of a_{max}. In other words,

$$(S_t)_o = 1.1\ L_t' = a_{max} \tag{7}$$

However, if the spacing given by Eq. (7) exceeds the maximum spacing specified by codes or is not possible due to minimum transverse steel requirement, then exactly half the value of the above spacing, which falls in Zone 1, can be used, that is,

$$(S_t)_o = 0.55\ L_t' = \frac{a_{max}}{2} \tag{8}$$

Combining Eqs. (7) and (8) and substituting the value of L_t' from Eq. (1), the following relationship is obtained

$$(S_t)_o = (0.825 \pm 0.275)\ K_p\ \frac{A_{ce}\ f_t}{\Sigma_o} \tag{9}$$

Inserting the values of A_{ce} from Eq. (2a) and Σ_o in terms of ϕ_ℓ, Eq. (9) can be rewritten in non-dimensional form as

$$\frac{(S_t)_o}{\phi_\ell} = (3 \pm 1)\left(\frac{1.1\ K_p f_t}{4}\right)\left(\frac{\pi}{4\rho_t} - \gamma\right)\left(\frac{1}{\gamma(1+\gamma)}\right) \tag{10}$$

where $\rho_t = A_s/bD$ and $\gamma = b/D$.

Eq. (10) forms the basis for the development of design charts to satisfy the serviceability requirement of maximum crack width in tension members. For a given value of γ, a set of curves, each corresponding to a particular value of $K_p f_t$ may be plotted with $(S_t)_o/\phi_\ell$ or $2(S_t)_o/\phi_\ell$ as ordinate and the reinforcement ratio ρ_t as abscissa. Fig. 3 shows two series of such curves, one for $\gamma = 1.5$ and the other for $\gamma = 0.8$.

Knowing the value of $(S_t)_o/\phi_\ell$, the value of w_{max}/ϕ_ℓ can be calculated from Eq. (6) provided the values of f_{bu} and A_s (or ϕ_ℓ) are known. Fig. 3 shows the relationship between $(S_t)_o/\phi_\ell$ and w_{max}/ϕ_ℓ for f_{bu} = 2.5 MPa and 5 MPa and ϕ_ℓ = 8 mm and 12 mm. In the development of these charts, the value of k is taken as 45 N/mm^3, the average value as obtained from pull-out tests on both deformed and smooth bars (Lee et al. 1987). The value of m is taken as 8, the average value for normal range of structural concrete (f_{cu} ranging from 25 to 40 MPa), and the stress level in steel at service load is assumed as 0.58 f_y.

Using part (a) of these charts, the recommended spacing of the

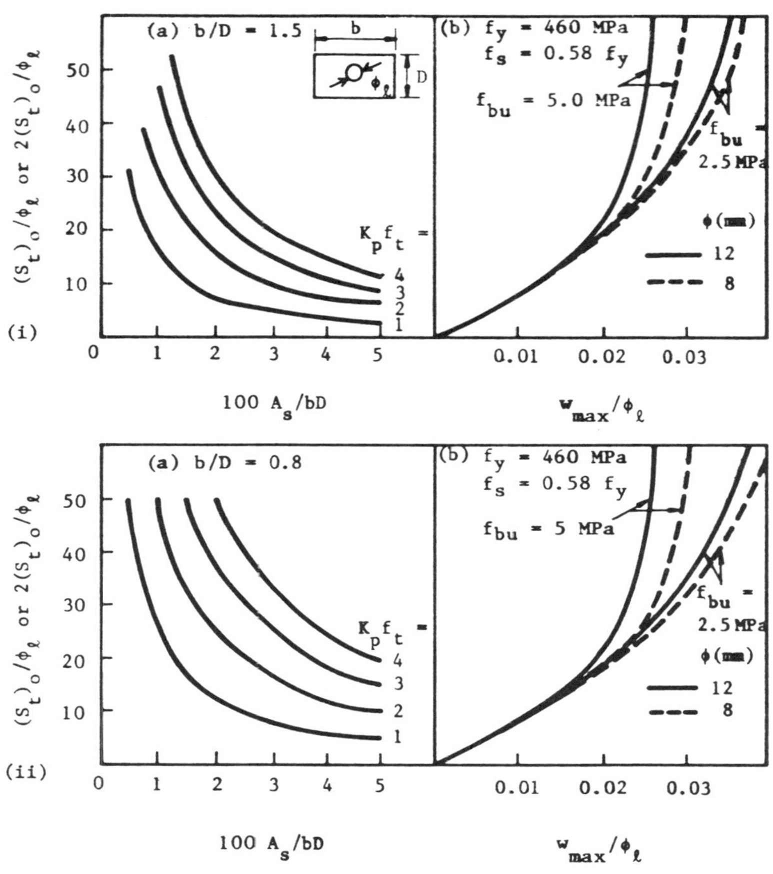

Fig. 3. Design charts for crack control in tension members.

welded transverse wires can be determined for a given section; the material properties of which have been selected from ultimate strength consideration. The corresponding value of w_{max} can be read off from part (b) of the chart and check against the permissible crack width.

2.2 Beams

Similar to tension members, the optimal spacing of transverse wires is given by Eq. (9). Substituting the value of A_{ce} from Eq. (2b), Eq. (9) can be written as

$$\frac{(S_t)_o}{\phi_\ell} = 0.11\ (3 \pm 1)\ K_p\ f_t\ \left(\frac{1}{\rho} \left(\frac{1}{\alpha} - 1 \right) - \frac{1}{2} \right) \tag{11}$$

Thus, for a given value of α, a set of curves, each corresponding to a particular value of $K_p f_t$ may be plotted with $(S_t)_o/\phi_\ell$ or $2(S_t)_o/\phi_\ell$ as ordinate and the tension reinforcement ratio ρ as abscissa [Figs. 4(a) and 5(a)]. Knowing the value of $(S_t)_o/\phi_\ell$, the value of w_{max}/ϕ_ℓ can be calculated for different area of longitudinal reinforcement A_s, provided the values of f_{bu} and ϕ_ℓ are known [see Eq. (6)].

The relationship between $(S_t)_o/\phi_\ell$ and w_{max}/ϕ_ℓ can be obtained for

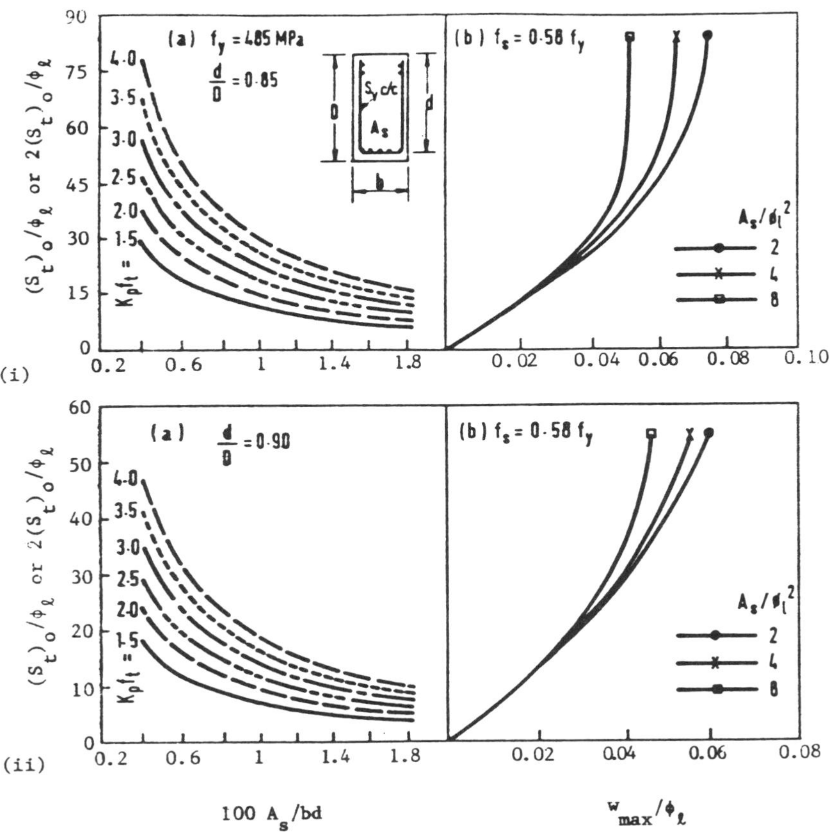

Fig. 4. Design charts for crack control in beams (f_{bu} = 1.5 MPa).

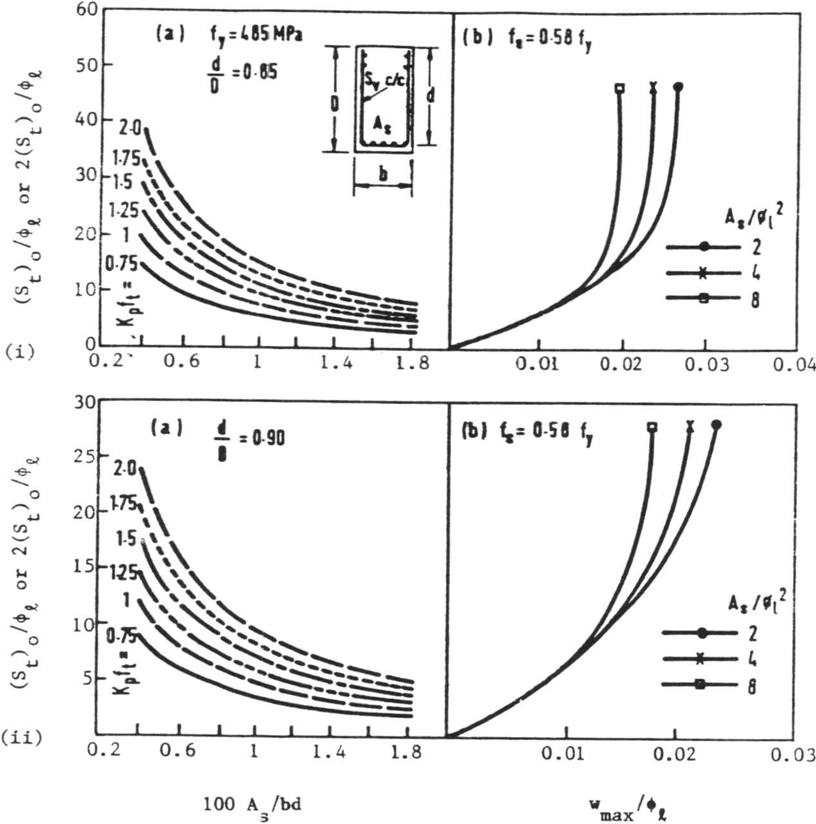

Fig. 5. Design charts for crack control in beams (f_{bu} = 5.0 MPa).

selected values of A_s/ϕ_ℓ^2, for a specified value of f_{bu}. Such plots are shown in Figs. 4(b) and 5(b). The curves have been drawn for ϕ_ℓ equal to 12 mm, the maximum bar diameter for which welding is currently feasible. The values of k and m are taken to be 45 N/mm^3 and 8 respectively.

In reinforced concrete design, checking for serviceability with respect to maximum crack width is usually preceded by ultimate strength design from which the section and material properties are selected. Knowing these, the recommended value of $(S_t)_o/\phi_\ell$ can be determined from part (a) of the chart. The corresponding value of w_{max}/ϕ_ℓ can then be obtained from part (b) of the same chart.

2.3 One-way slabs

For one-way slabs, the preferred transverse wire spacing of Zone 1 [see Fig. 2] may not be economical. In Zone 2, the minimum possible value for a_{max} is obtained when the transverse wires are located at a spacing of h_c. However, this spacing is sensitive and hence the optimum transverse wire spacing $(S_t)_o$ may be taken as 1.1 h_c. Thus, equating

$$a_{max} = (S_t)_o = 1.1\ h_c \tag{12}$$

and substituting the value of h_c from Eq. (3) and assuming m = 8, Eq. (12) can be expressed in non-dimensional form as

$$\frac{a_{max}}{D} = \frac{(S_t)_o}{D} = 1.1\ \{1 + 8\rho\alpha\ (1 - \sqrt{1 + (1/4\rho)})\} \tag{13}$$

Eq. (13) forms the basis for the development of serviceability design charts for one-way slabs. A series of curves, each corresponding to a particular value of α, may be plotted with $(S_t)_o/D$ as ordinate and tension reinforcement ratio ρ as abscissa [Fig. 6]. Thus, for known values of ρ (= A_s/bd) and α (= d/D), the optimum spacing of transverse wires $(S_t)_o/D$ can be read from these curves.

Once the value of $(S_t)_o/D$ is known, the corresponding w_{max}/D can be calculated for different slab depths, by using Eq. (6). Therefore, for a particular value of f_{bu}, the relationship between w_{max}/D and $(S_t)_o/D$ can be obtained for selected values of ϕ_ℓ/D and D. Typical charts for f_{bu} ranging from 2 MPa to 6 MPa are shown in Fig. 6. These charts have been plotted for ϕ_ℓ/D ranging from 0.025 to 0.200, slab depths of 75, 150 and 225 mm, and d/D ranging from 0.7 to 0.9. The value of k has been taken as 45 N/mm^3 and the stress level in steel f_s as 0.58 f_y.

Knowing the values of A_s, b, D, d and the material properties from ultimate strength design, the recommended $(S_t)_o/D$ can be determined from part (a) of the chart. The corresponding w_{max}/D can then be obtained from part (b) provided the value of ϕ_ℓ/D is known. If the above w_{max}/D exceeds the permissible limit, then w_{max} can be reduced by adjusting ϕ_ℓ/D. It is obvious that a smaller diameter of longitudinal wires will provide smaller crack widths, and according to test results reported earlier (Lee et al. 1989a), transverse wires should have a diameter of at least 0.3 times the diameter of longitudinal wires. If w_{max} is still greater than the permissible limit, then the slab should be redesigned with a higher ρ.

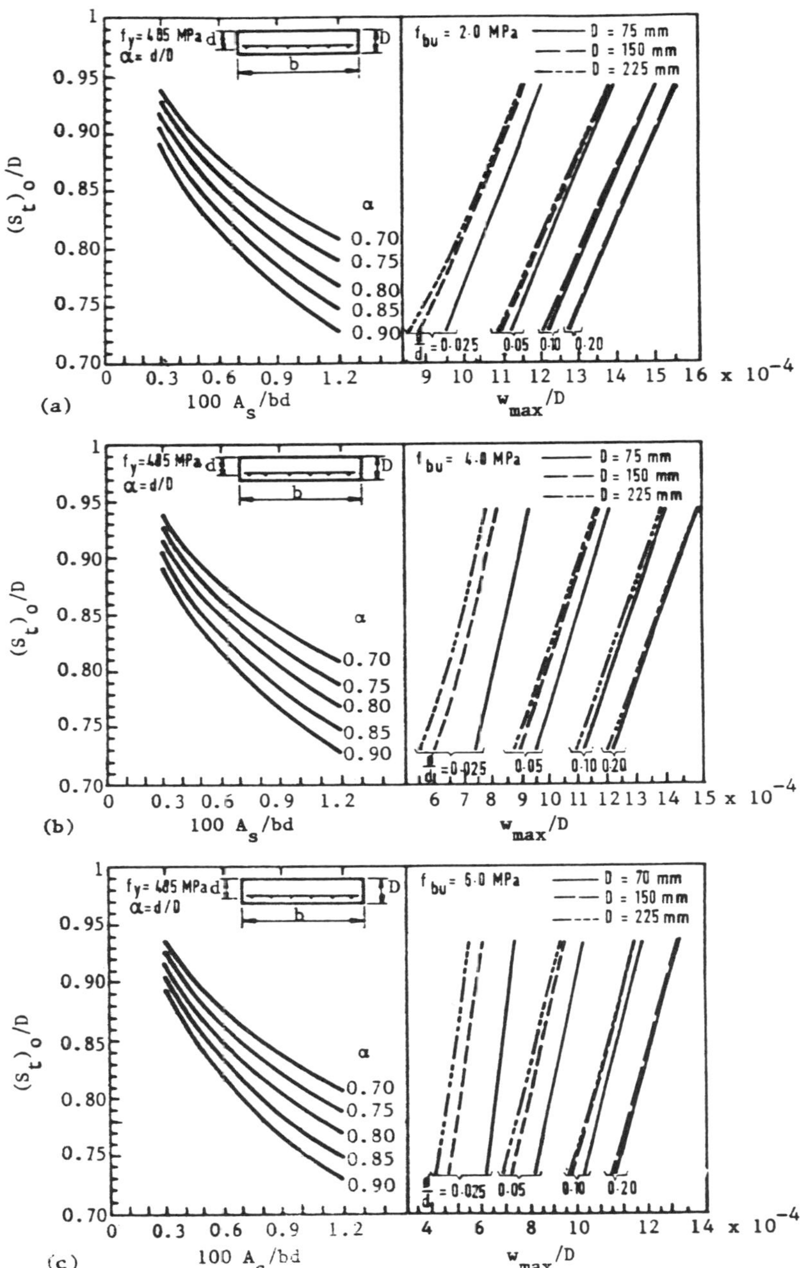

Fig. 6. Design charts for crack control in one-way slabs.

3 Summary and Conclusion

The use of welded wire fabric in concrete members leads to substantial reduction in crack width, especially when the spacing of the transverse wires is set to certain values. Based on the crack spacing model and a rigorous method proposed earlier by the authors (Lee et al. 1987, 1989a, 1989b) to calculate crack widths, graphical charts suitable for the serviceability design of tension members, beams and one-way slabs reinforced with WWF are developed and presented in this paper. The use of the design charts are illustrated by three examples in the Appendix.

4 References

Desayi, P. (1976) Determination of the maximum crack width in reinforced concrete members. **J. Amer. Concr. Inst.**, 73(8), 473-477.

Lee, S.L., Mansur, M.A., Tan, K.H. and Kasiraju, K. (1987). Cracking behavior of concrete tension members reinforced with welded wire fabric. **ACI Struct. J.**, 84(6), 481-491.

Lee, S.L., Mansur, M.A., Tan, K.H. and Kasiraju, K. (1989a). Cracking behavior of one-way slabs reinforced with welded wire fabric. **ACI Struct. J.**, 86(2), 208-216.

Lee, S.L., Mansur, M.A., Tan, K.H. and Kasiraju, K. (1989b). Crack control in beams using deformed wire fabric. **J. Struct. Engrg.**, 115(10), 2645-2660.

Mansur, M.A., Tan, K.H., Lee, S.L. and Kasiraju, K. (1990). Crack width in concrete members reinforced with welded wire fabric. (Submitted for publication in **ACI Struct. J.**).

5 Appendix. Examples

(a) A tension member measuring 120 mm by 75 mm, is reinforced with two nos. of 12 mm diameter bars. Determine the spacing of welded transverse wires in order to satisfy the serviceability criterion of maximum crack width. Given: f_y = 460 MPa; K_p = 0.6 mm^2/N; f_t = 2.8 MPa; f_{bu} = 5.0 MPa and permissible w_{max} = 0.2 mm.

Solution: Since $\rho_t = 2x(\pi x12^2/4)/(120x75) = 2.51$ %, b/D = (120/2)/75 = 0.8, $K_p f_t$ = 0.6x2.8 = 1.68, then from Fig. 3(i),

$(S_t)_o/\phi_\ell$ = 16.5 or 8.25 or $(S_t)_o$ = 198 mm or 99 mm;

w_{max}/ϕ_ℓ = 0.017 or $w_{max} = 0.017x12 \cong 0.20$ mm. O.K.

Therefore, use welded transverse wires at 200 mm spacing.

(b) A rectangular beam exposed to a marine environment has a cross-section 200 mm by 400 mm. If tension reinforcement ratio required from the ultimate strength consideration is 0.95 percent, determine the required stirrup spacing of WWF in order to satisfy the serviceability criterion of maximum crack width. Given: $K_p = 0.65\ mm^2/N$; $f_t = 2.5$ MPa; $f_{bu} = 5.0$ MPa; $f_y = 485$ MPa; and permissible $w_{max} =$ 0.15 mm.

Solution: Assuming 3 pairs of 12 mm diameter longitudinal wires and a clear concrete cover of 40 mm, effective depth d = 400 - 40 - 6 = 354 mm and reinforcement ratio $\rho = A_s/bd = 679/(200 \times 354) = 0.96\%$.

For $\rho = 0.96$, $K_p f_t = 0.65 \times 2.5 = 1.63$ and d/D = 0.885, linear interpolation between Figs. 5(i)(a) and 5(ii)(a) gives

$$(S_t)_o/\phi_\ell = 9.44 \text{ or } 4.77 \quad \text{or} \quad (S_t)_o = 113 \text{ mm or } 57 \text{ mm.}$$

Assuming $(S_t)_{max} = d/2 = 177$ mm, adopt $S_t = 113$ mm. For $\rho = 0.96$, d/D = 0.885, $K_p f_t = 1.63$ and $A_s/\phi_\ell^2 = 4.7$, interpolation between Figs. 5(i)(b) and 5(i)(b) gives

$$w_{max}/\phi_\ell = 0.0125 \quad \text{or} \quad w_{max} = 0.15 \text{ mm.} \qquad \text{Hence O. K.}$$

Therefore, use bent-up deformed WWF cage with 3 pairs of 12 mm diameter main longitudinal wires and welded stirrups at a spacing of 115 c/c.

(c) A one-way slab exposed to a wet environment has an overall depth of 150 mm. If tension reinforcement ratio required from the ultimate strength consideration is 0.91%, determine the required transverse wire spacing of WWF in order to satisfy the serviceability criterion of maximum crack width. Given: $f_{bu} = 2.5$ MPa and permissible $w_{max} =$ 0.2 mm.

Solution: Assuming 12 mm diameter smooth longitudinal (main) wirews at 100 mm c/c and a clear concrete cover of 20 mm, effective slab depth d = 150 - 20 - 6 = 124 mm, $\alpha = d/D = 124/150 = 0.83$, $\phi_\ell/D =$ 12/150 = 0.08 and $\rho = 0.91$. From design chart for slabs [Fig. 6], linear interpolation gives

$$(S_t)/D = 0.792 \quad \text{or} \quad (S_t) = 0.792 \times 150 = 119 \text{ mm, say 120 mm; and}$$

$$w_{max} = 12.4 \times 10^{-4} \times 150 = 0.186 \text{ mm} < 0.2 \text{ mm.}$$

Use WWF with a grid size at 100 mm x 120 mm.

48 INFLUENCE OF CEMENT C_3S LEVEL ON CONCRETE DURABILITY

S. KELHAM, G.K. MOIR
Blue Circle Industries PLC, Greenhithe, UK

Abstract
The effect of cement C_3S level on the properties of concrete has been studied using five specially prepared cements differing only in the relative proportions of C_3S and C_2S. In well cured concrete the C_3S level of the cement was found to have little effect on the concrete properties examined. However, when poorly cured, the high C_3S cements produced more durable concrete as a result of the faster hydration. The greater rate of reaction leads to higher temperatures in large pours but more work would be needed to determine whether this leads to an increased risk of thermal cracking.
Keywords: C_3S, Strength, Durability, Curing, Carbonation, Freeze/thaw, Porosity, Permeability.

1 Introduction

In a recent Concrete Society technical report (No. 29, 1987) it was pointed out that changes in cement properties had resulted in the strength of standard concrete mixes tested at the BRE increasing from ~30 MPa in 1950 to ~45 MPa in 1980. This implies that concretes designed to specified strengths will currently have lower cement contents and higher water/cement ratios than they did in 1950. It has been argued that the improvements in cement strengths may thus have led to the production of less durable concrete. The Concrete Society report identified an increase in cement C_3S content as the most notable change in the physical and chemical properties over the same period. This historical link between increasing cement C_3S content and a perceived decrease in concrete durability has resulted in the development of the idea that high cement C_3S content leads directly to reduced concrete durability.

The effect of cement C_3S content on concrete properties was studied by Ben-Bassat, Nixon and Hardcastle (to be published) who used three production cements with widely differing C_3S contents. They concluded that with good curing the differences were insignificant but with poor curing the faster hydration of the high C_3S cement produced more durable concrete. However, these cements also differed in other important respects e.g. C_3A content, SO_3 content, alkali content, which would be expected to influence the concrete properties.

2 Materials and test methods

The cements were prepared from five clinkers produced in a batch rotary kiln with target C_3S contents of 25%, 35%, 45%, 55% and 65%. The physical and chemical properties of the cements are given in Table 1.

Most tests were carried out on a concrete mix of type C1, BS 4550 Part 3.4 (1:2.5:3.5, 0.6 w/c) using aggregates to BS 4550 Parts 4,5 (crushed rock and graded sand), referred to as Mix C. The same aggregates were also used in a 0.5 w/c ratio mix, 1:2.15:3.08, which is referred to as Mix E. All concrete was cured for 24 hrs at 20°C before demoulding except one set of mix C concretes which was prepared and cured at 5°C.

Compressive strengths of concrete cubes were measured at ages up to one year. The well cured samples were cured in water at 20°C until tested. The poorly cured samples were exposed to air at 20°C, 50% rh from demoulding until 24 hrs before testing and then immersed in water to resaturate. The mixes prepared at 5°C were cured in water at 5°C to 28 days and then in water at 20°C. Flexural and compressive mortar strengths were measured to EN 196-1.

The carbonation, porosity and permeability measurements were carried out at one year on cast cylinders, 150 mm. diameter, 50 mm. length, which were exposed to air at 20°C, 50% rh after 24 hours or 28 days of curing. The permeability to oxygen and effective capillary porosity were measured using published methods (Kollek, 1989; Kelham, 1988) and the cylinders were then split and the carbonation depth measured by spraying the fracture surfaces with a solution of phenolphthalein. The freeze/thaw resistance was measured using the method given in ASTM C666. Temperature

Table 1. Physical and chemical properties of cements

Target	Bogue analysis (%)				EqNa	SO_3	SSA	+45μm
C_3S (%)	C_3S	C_2S	C_3A	C_4AF	(%)	(%)	(m^2/kg)	(%)
25	24	50	9	12	0.59	2.8	354	18.8
35	34	42	9	11	0.57	2.6	352	12.5
45	42	34	8	11	0.54	2.9	351	13.2
55	54	23	8	11	0.66	2.9	355	12.7
65	68	9	8	10	0.64	2.8	347	9.1

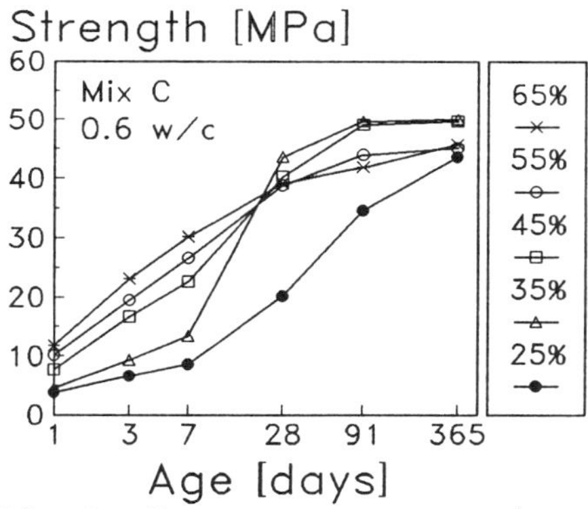

Fig.1. Concrete compressive strength development.

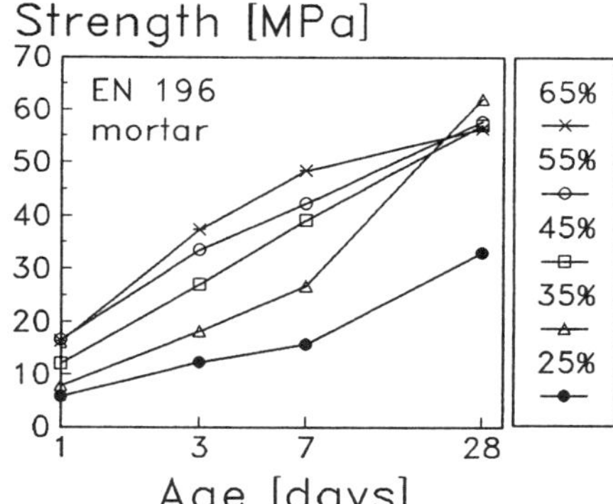

Fig.2. Mortar compressive strength development

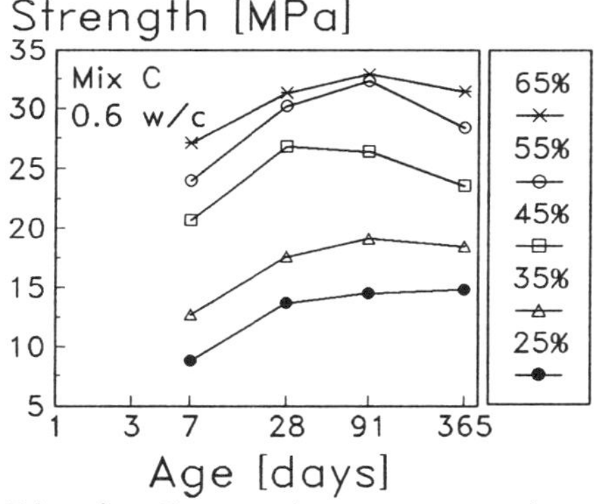

Fig.3. Concrete compressive strength development in air.

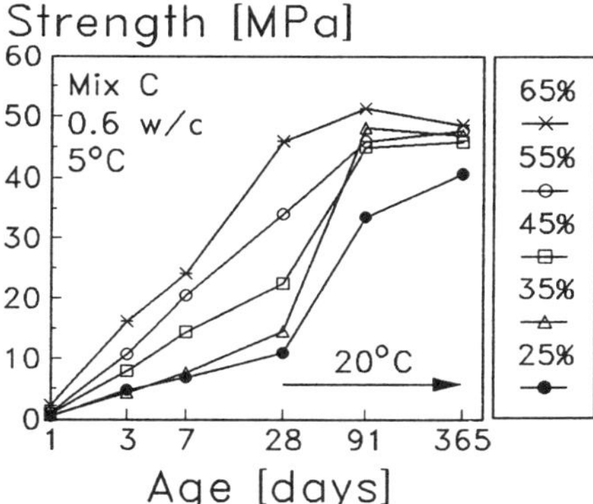

Fig.4. Concrete compressive strength development at 5°C

rises in 3m. pours were obtained in a controlled thermal leakage semi-adiabatic calorimeter. The freeze/thaw and temperature rise tests were carried out on mix C concretes.

3 Experimental results

3.1 Strength results

The compressive strength development results for water cured concrete (mix C) and mortar are shown in Fig. 1 and Fig. 2. Up to seven days the strengths increase with increasing C_3S content. At 28 days C_3S level in the range 35% to 65% had little influence on compressive strength. The highest strengths at one year were given by the cements

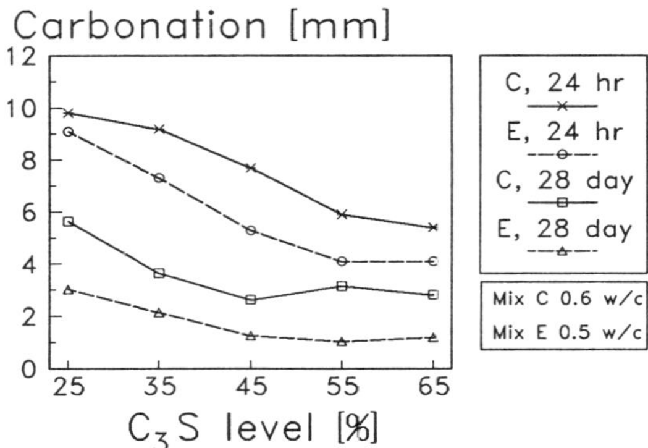

Fig.5. Carbonation after 1 year at 20°C, 50% rh.

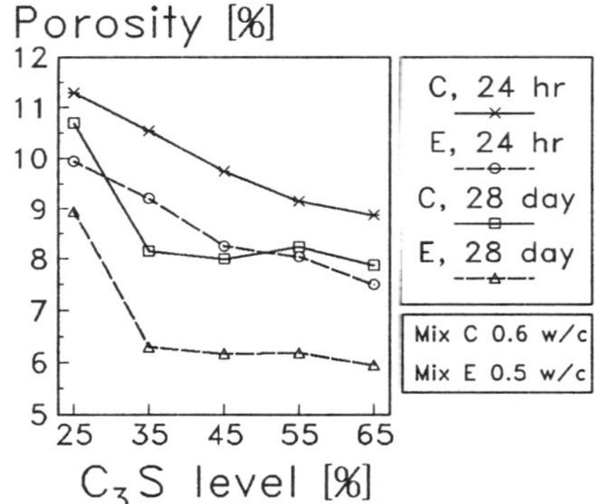

Fig.6. Effective capillary porosity.

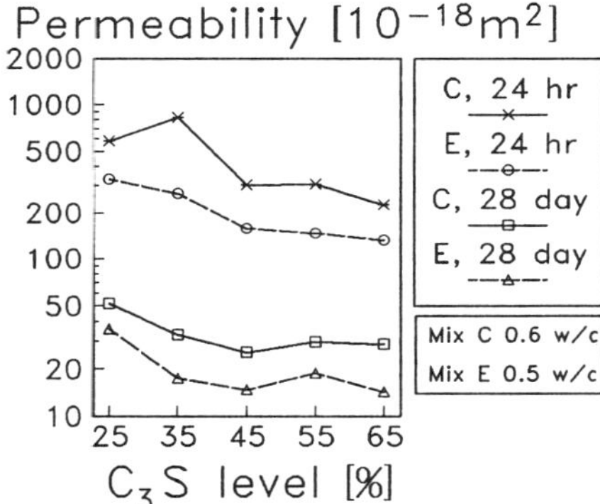

Fig.7. Permeability to oxygen.

with 35% and 45% C_3S. Exposure to air at 24 hours prevents the later gains in strength, Fig. 3, and the strengths remain directly related to C_3S content. Curing at 5°C greatly reduces the 28 day strengths of the low C_3S cements but increases the strength of the 65% C_3S cement, Fig. 4. Subsequent 20°C curing leads to a rapid increase in strength in the low C_3S cements to the same values as the high C_3S cements.

3.2 Carbonation, porosity and permeability results

The carbonation depths after one year of exposure are plotted in Fig. 5 against cement C_3S level. The carbonation depths decreased with increasing C_3S content with a reduced

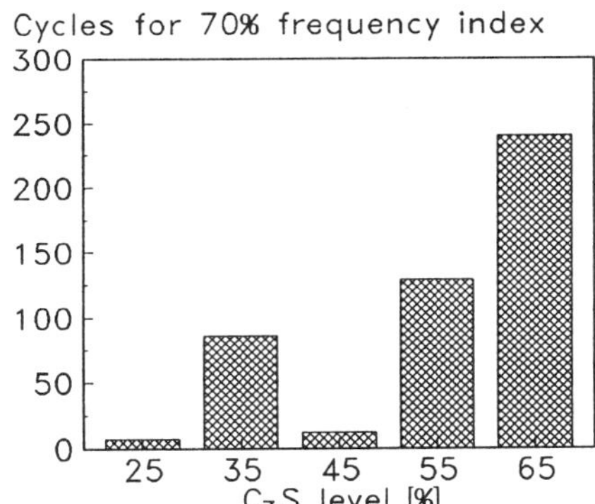

Fig.8. Freeze/thaw resistance.

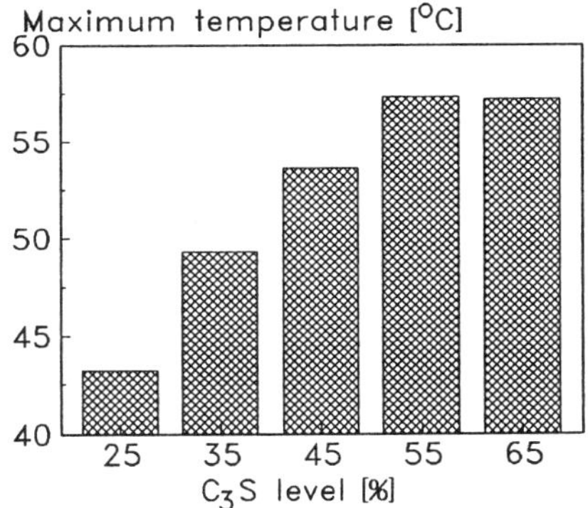

Fig.9. Temperatures in simulated large pours.

dependence on C_3S level in the well cured concretes. This difference between the 24 hour and 28 day cures is also clear in the porosity results, Fig. 6. Following 24 hour curing the porosity decreases steadily with increasing C_3S content. After 28 days curing the porosity is independent of C_3S level between 35% and 65%. This agrees well with the strength results. The permeability results, Fig. 7, are similar except that the effect of curing is much more important than w/c or cement type.

3.3 Freeze/thaw resistance results

Fig. 8 shows the number of freeze/thaw cycles required to reduce the elastic modulus of the concretes to 70% of their initial values. The trend is for the high C_3S cements to survive longer than low C_3S cements. There is no obvious explanation for the anomalous relative performance of the 35% and 45% C_3S cements.

3.4 Large pour simulation results

The maximum temperatures obtained in simulated 3m. pours using mix C concretes (310 kg/m^3) are shown in Fig. 9. The temperatures increase with increasing C_3S content. The maxima occured at between 60 and 108 minutes after the start of mixing.

4 Discussion

When cured in water under standard laboratory conditions the cements with nominal C_3S contents of 35% and 45% gave 28 day compressive strengths which were similar to those of cements with C_3S levels of 55% and 65%. Thus, in designed concrete mixes similar cement contents and w/c ratios would be permitted.

It has been demonstrated, however, that the strength development and the porosity, permeability and carbonation rate of concretes prepared from these low C_3S cements are very sensitive to curing conditions. Inadequate curing, particularly under low temperature conditions, would result in concretes which may be unable to give adequate protection to reinforcing steel or resist freeze-thaw cycling.

The low heat characteristics of low C_3S cements have been confirmed. The concrete strengths, in particular the low strength associated with the 25% C_3S cement, need to be considered when assessing these results. Further tests are in progress to determine the compressive and tensile strength development under matched temperature curing conditions.

It can be observed that in many respects the low C_3S cements have yielded concretes with similar characteristics to those of Portland Blastfurnace Slag Cement (PBFC).

5 Conclusions

Within the range giving acceptable cement properties, an increase in cement C_3S content leads to :-

Higher early strengths
An almost unchanged 28 day strength when tested under standard laboratory conditions
No change in the potential durability of well cured concrete
Improved potential durability of poorly cured concrete
Increased temperatures in large pours

6 References

Ben-Bassat, M.,Nixon, P.J. and Hardcastle J. To be published

Concrete Society Technical Report, No 29 (1987) Changes in the properties of ordinary Portland cement and their effects on concrete.

Kelham, S. (1988) A water absorption test for concrete. **Mag. Conc. Res.**, 40.143, 106-110.

Kollek, J.J. (1989) The determination of the permeability of concrete to oxygen by the Cembureau method - a recommendation. **Materials and Structures**, 22, 225-230

49 DURABILITY OF PVC-COATED FABRICS FOR MEMBRANE STRUCTURES

T. NIREKI
Building Research Institute, Ministry of Construction, Tsukuba, Japan
H. TOYODA
Taiyo Kogyo Corporation, Osaka, Japan
Y. ONUMA
Membrane Structures Association of Japan

Abstract
PVC-coated polyester fabrics are widely applied for membrane structures providing relatively low-cost and reasonable performance. This paper deals with the results of seven year outdoor exposure tests under both at material level and at full scale level of commercial fabrics, intended to obtain more practical data for the prediction of service life time of the fabrics. The results show a significant decrease of tensile or tear strength, surface appearance,water resistance, including flame resistance.
Keywords; Durability, PVC-coated fabric, Membrane Structures, Polyester fibre, Outdoor Exposure, Deterioration.

1 Introduction

Performance over time should be the most important factor in the use of PVC (polyvinyl chloride) composite sheets for pneumatic, suspension or steel-framed membrane structures as a front line of the defence against the external environments.

After the revision of Codes in 1987, the membrane structures which meet the certified conditions have been treated as the permanent structures, thereby, leading to the requirement of durability as in the ordinary building constructions.

View point from the available material in the enforced code at present, PTFE-(polytetrafluoethylene) coated glass fibre fabric (Class A), PVC-coated glass fibre fabric (Class B) and PVC-coated polyester fibre fabric (Class C) can be applicable. Among above materials,PVC-coated polyester plain-weave fabric has relatively short term of service life, whereas, much attention should be taken into account of serviceabilitiy and maintainability of the membrane structures of PVC fabrics. This provided the motivation to start a joint research project on the durability of PVC fabrics between Building Research Institute (BRI) and Membrane Structures Association of Japan (MSAJ) in 1982.

2 Specimen

Three different commercial PVC-coated polyester fibre fabric (called A, B and C) were selected and Table 1 compares manufacturer's data, and their general structure is shown in Fig.1.

Table 1. Properties of PVC-coated plain-weave fabric (warp×fill)

Specimen	Yarn	Density (count/cm)	Composition of the PVC compound(unit:parts by weight)					Thickneness	Weight
			PVC	plasticizer	stabilizer	flame retardant	pigment	(mm)	(g/m^2)
A	polyester ECC10/1 × ECC10/1	52×46	100	DOP:75	6.5	Sb_2O_3 : 20	25	0.53	642
B					6.5	Sb_2O_3 : 20	25	0.47	562
C					6.0	Sb_2O_3 : 25	23	0.49	586

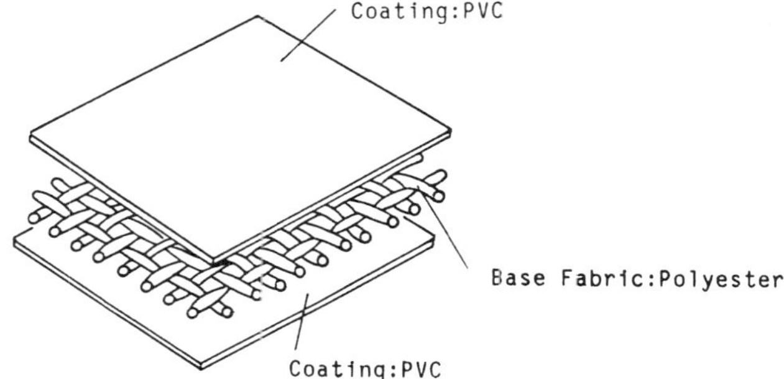

Fig.1. Composition of PVC-coated polyester plain-weave fabric

3 Test Method

3.1 Outdoor Exposure

Two levels(AT MATERIAL AND FULL-SCALE LEVEL) of outdoor exposuretests are carried out. Outlines of test methods are shown in Fig.2 and Fig.3. Both outdoor exposure tests started in November 1982 and specimens were sampled in 1985 and in 1989 for evaluations.After the removing of samples from the full scaled specimen, the resulting holes were repaired by the same type fabric for evaluation of effectiveness of repairing.

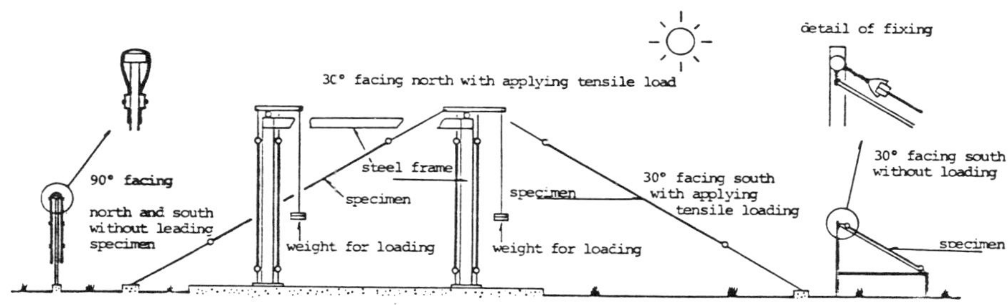

Fig.2. Outline of outdoor exposure test (AT MATERIAL LEVEL)

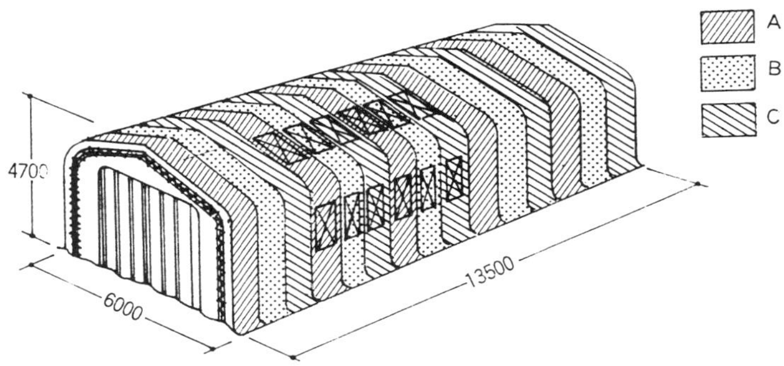

Fig.3. Outline of outdoor exposure test (AT FULL-SCALE LEVEL)

3.2 Evaluation Methods

Evaluation methods and their details are shown in Table 2 and in Fig.4.

Table 2. Details of test items

items	test method and condition		size of specimen
mass	JIS K 6328		10cm × 10cm
weight	JIS K 6328		10cm × 10cm
Tensile strength	JIS L 1096.6.12.1 strip Method,	Tension speed:20cm/min	3cm × 30cm
Elongation		Clamling distance:20cm	
Tear strength	JIS L 1096.6.15.1 single tongne method,	Tension speed:20cm/min	10cm × 25cm
Stiffness	JIS L 1096.6.19.2 Slide method,	see Fig.4	2cm × 15cm
Discolouration	JIS Z 8722		
Water resistance	JIS L 1092 5.1.1.A		15cm × 15cm
Flame resistarce	JIS A 1322.B	Heating time :2 minutes	30cm × 20cm

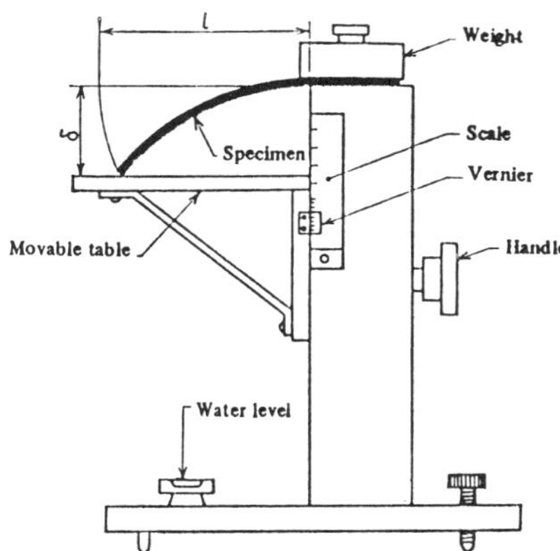

Fig.4. Slide type tester for stiffness (JISL1096)

4 Results and Discussion

4.1 Tensile strength

Change of tensile strength over time is one of the dominant factors for selecting the fabric materials. The exposure condition clearly effects the reduction of tensile strength as in Fig.5. Facing south at an angle of 30° , a conventional evaluation method for ordinary building materials, or simulates a sloping part of an actual building which is a more severe condition than that of facing north at an angle of 90° (vertical), and facing south at an angle of 90° is intermediate between these two exposure conditions.

The uniformly distributed loading set at 5% of the initial tensile breaking strength than was effect expected. Among three different commercial fabrics, specimen C behaved well regardless of exposure conditions. Fig.6 shows the results at full scale level. All specimens are extracted from the specified part of the full scale mock-up, and the extracted parts were repaired by using the same type of fabric for the evaluation of maintainability, especially, the performance of the welded part on site. It can be said that the residual strength of roof covering in facing south is decreased in comparison with that of at material level.

Generally, the above tendency of changing tensile strength is similar to the trend of loss of weight (in g/m^2) and thickness (in mm) of the fabrics under the various exposure conditions; maximum weight loss is measured when the roof faces south and exposure at 30° gives the second highest loss.

4.2 Tear strength

Tear strength is one of the vital performance requirement for structures. As shown in Fig.7 and 8, the loss of tear strength is relatively high under various exposure conditions.

4.3 Stiffness

Extensive increase of stiffness can be observed starting after three years exposure in the whole specimens, especially the specimens extracted from the full scale mock-up shows too stiff to evaluate according to the specified test method as in Fig.4.

4.4 Discolouration and visual observation

Discolouration was measured as "colour difference (ΔE)". All specimens of a greenish colour show high values of colour difference, specimen facing south turned to pale green, due to fading and soiling, excessive chalking is observed. Moss or mildew is also observed on the surface of fabrics, especially the specimens facing north due to the longer wet ambient atmosphere.

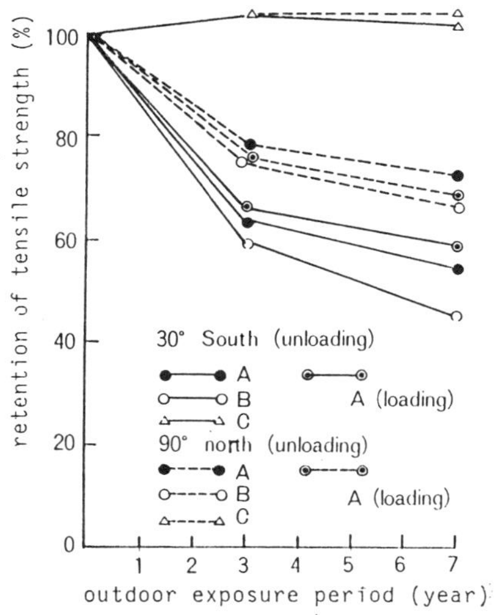

Fig.5.Result of tensile strength retention under various conditions (AT METERIAL LEVEL)

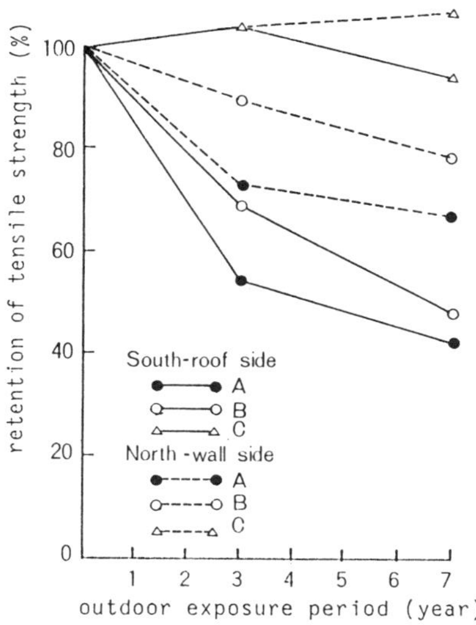

Fig.6.Result of tensile strength retention (AT FULL-SCALE LEVEL)

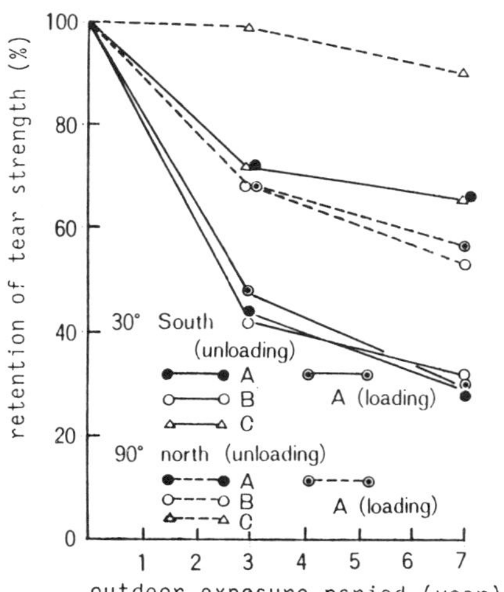

Fig.7.Result of tear strength retention under various conditions (AT METERIAL LEVEL)

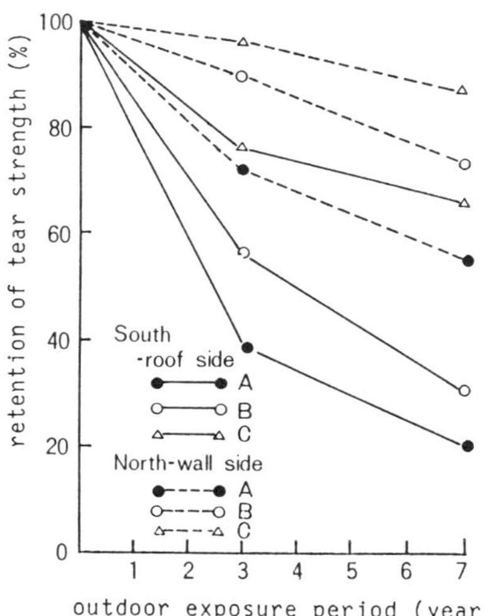

Fig.8.Result of tear strength retention (AT FULL-SCALE LEVEL)

Growth of mildew within the fabric layer is clearly observed for specimen C in the full scale test but not at the material level. Cracks in the PVC coating layer were also observed in most specimens.

4.5 Water resistance

Water resistance after seven years exposure fell to less than one third to one fifth of initial value, however, no water leakage is recognized yet in the mock-up.

4.6 Flame resistance

Originally the PVC-coated fabrics have to meet "Class 2" flame resistance as a minimum quality standard, however, some specimens are decreased as "Class 3" after seven years exposure.

4.7 Service life of PVC-coated fabric

Service life time of PVC-coated fabric has said to be some ten years [Ansell(1980)]. Mambrane Structures Association of Japan (MSAJ) has issued a guide for maintaining membrane structure, and the service life time of fabrics is determined on the basis of several criteria such as; tensile strength, tear strength, water resistance, cracks, peeling, etc.. On the basis of the above criteria, the deterioration state of the full scale mock-up at present can be outlined as follows;

1) No critical functional failure was observed.
2) From the aspect of residual strength of tensile or tear, they approach to the stage of whole replacement is required, even though the water resistance keeps satisfactory performance.
3) As to the partical repair of roof or wall of PVC-fabrics aged after 7 or 8 years, the welding by hot jet air on site might not be able to give adequate bonding of new fabric due to surface deterioration (loss of polymeric component in the coating and embrittlement).

5 References

Ansell, M.P., and Harris, B.(1980), "Fabrics: charactaristics and Testing" proceedirgs on symposium, Air-Supported Structures : The State of the Art,Institution of Structured Engineers, London, p.124.

50 AN APPROACH TO DESIGN FOR DURABILITY OF THE BUILDING TECHNOLOGICAL SYSTEM

A. LUCCHINI
Dipartimento di Ingegneria dei Sistemi Edilizi e Territoriali, Politecnico di Milano, Milan, Italy

Abstract

The aim of the design for durability or, more exactly, of quality over time of the building technological system is to achieve a real duration which equals the expected duration.

This is possible only if the real over time performance, and therefore the service life of the single components and of the whole, meet design expectations.

This means that performance decay has to follow the planned trends, i.e. that over the whole building process (design, production and management of the technological system) there have to be no anomalies, or better disturbance factors, interfering.

The paper introduces an approach to design for durability of the building technological system based on the theory of disturbance factors.
Key words: quality over time, design for durability, disturbance factors.

1 Introduction

In order to achieve a real duration which equals the design duration it is essential that the over time performances and therefore the service life of the single building components and of the whole meet design expectations.

This means that the decay of performances has to follow the planned trend, that is, there have to be no anomalies, or better disturbance factors, interfering.

2 Propaedeutic definitions

The definition of disturbance factor requires the introduction of a few propaedutic definitions.

It has to be underlined that the behaviour over time of any kind of building component evolves in compliance to the laws of chemistry and physics imposed by its material nature and the environment of exposition. Therefore disturbance factors should be tackled by referring to a precise schedule of quality which expresses the design expectations towards the behaviour over time of the given component.

We therefore define the:

complex of duration conditions:as the complex of hypotheses and design logics for the behaviour over time of the whole technological system.
It determines the schedule of over time quality of the single elements and of the whole, and provides information for the definition of duration requirements and specifications of each building component. Must be defined by the designer when laying down the technological design;
design deterioration: as the evolution of the behaviour over time of the single building components or of the whole technological system, as forecasted at design level.
It sums up the logics and hypotheses defined at design level and expresses the nature of the design dynamic of quality;
quality schedule: as the complex of conditions, duration requirements and specifications, design deteriorations of the building technological system.
It has to be the reference point for the building object's design, production, commissioning and management. For a correct setting up it must be borne in mind that a building object is a system of components, each following the logics of behaviour and functioning over time imposed by its material constitution, objectual complexity, and by the role it has been assigned within the system, i.e. the relation complexity.
If this is not the case the schedule of quality over time of the single components and of parts of the system to which it is related will not be kept.

On the basis of the previous definitions and considerations we define as:

pathological condition: any anomaly, concerning the quality schedule, in the functioning and evolution of the behaviour over time of a building component affecting its performance over time and service life. Such a loss in performance may have negative consequences also for the parts of the system related to the component in case. A pathological condition invalidates the congruence of the component with the designed system;
early deterioration of pathological nature: deterioration due to a pathological condition, linked to

a shortening of the design service life;
disturbance factor: a factor triggering anomalies in the functioning or evolution of the over time behaviour of a building component leading to a pathological condition.

To sum up a pathological condition is an anomalous situation, since it is undesired and unforeseen, involving an early deterioration linked to a shortening of the design service life, i.e a change in the design quality schedule.

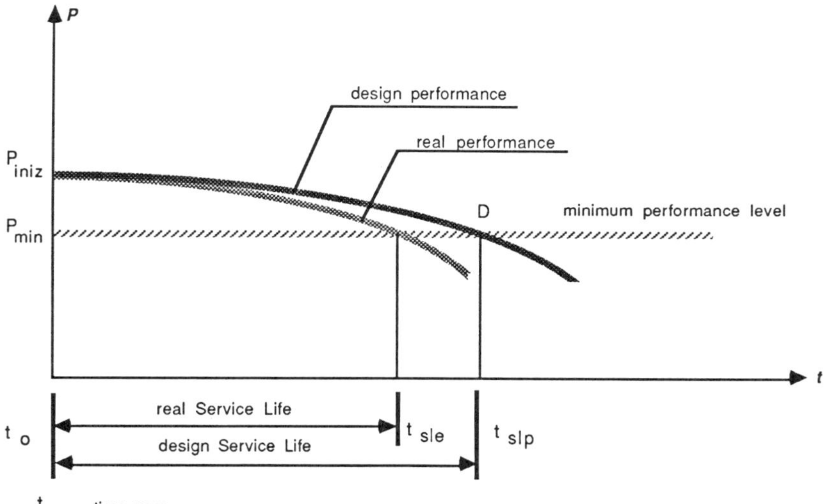

Fig.1

3 Disturbance factors and building process

The study of disturbance factors must be tackled in consistency with the organization of the decisional and operational stages of the building process.

As to the causes leading to disturbance factors, it has been established that they are unnoticed errors, done in any of the design and production phases of the building process having an influence on the functioning and evolution of the over time behaviour of building components.

The following main stages which can cause errors leading to disturbance factors have been detected:

- technological design
- operational design
- management design
- manufacturing of components
  . manufacturing processes involving factory precasting and on site assembly
  . manufacturing processes involving on site production
- maintenance of components

4 Tackling of errors leading to disturbance factors

The following criteria should be applied for the recognition and elimination of the errors leading to disturbance factors.

- Fundamental criteria for the methodological control approach:

. errors in the technological, operational and management design (the latter only for the part concerning the fulfilment of the maintenance requirement), and in the manufacturing, exist already from the commissioning of the building.
The presence of one or more of them, if not eliminated will affect the evaluation of the over time performance and the service life prediction for the building object.
Such evaluations will prove unreliable for the components and parts of the technological system, affected by such errors;

. management design (except the part concerning the fulfilment of the maintenance requirement) and maintenance errors can become active only after the commissioning of the building, with the application of the generating phase.
Again the presence of one or more of them, if not eliminated will affect the evaluation of the over time performance and the service life prediction for the building object.
Such evaluations will indeed prove unreliable for the components and parts of the technological system, affected by such errors;

. the reliability of the prediction of the over time performances and service life of building objects requires:
correctness of design phases;
correctness of predictions;
congruency between design and product, achievable only through correct production and management phases.
Each of these 'correctnesses' may be achieved only through an adequate quality control by which possible errors leading to disturbance factors can be detected and eliminated.

- **Classification criteria:** the decay of the over time behaviour of a building component follows qualitatively one of the following two models:
. sudden decay or on-off functioning model: characterized by a virtually constant performance and by a virtually istantaneous loss of performance (collapse of the performance) (see Fig.2.a)

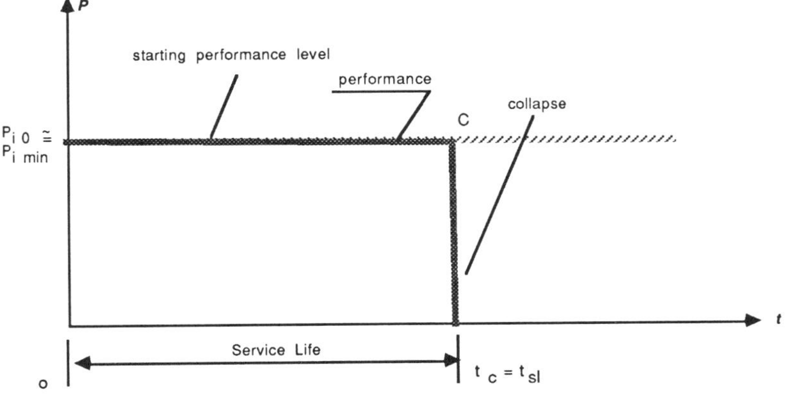

Fig.2.a

. progressive decay model: characterized by a progressive loss of performance according to a decay gradient (see fig.2.b)

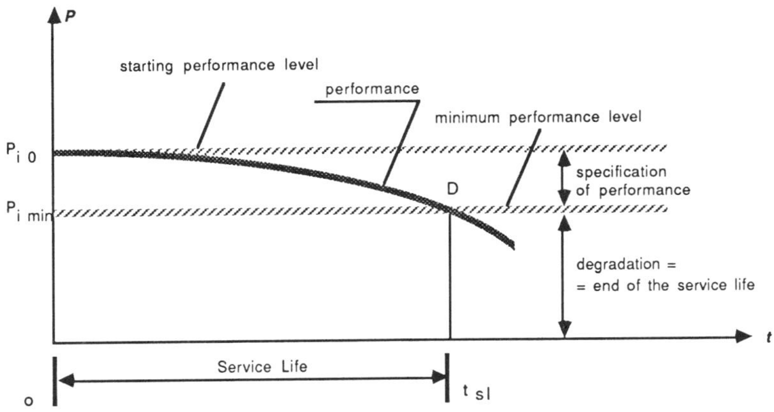

Fig.2.b

Consequently the errors generating disturbance factors

can be classified as follows:
. **first level errors**, since they lead, in one or more components, considered individually or in their whole, to the lack of performance levels required at time zero (insufficient initial performance);
. **second level errors**, since they lead, in one or more components, considered individually or in their whole, to the lack of the performance levels required for the expected duration (insufficient over time performances and service life);
. both a first level and second level error may be identified as **critical** or **not critical** error, depending on whether the lack of required performance levels to which they lead in a component, does or not affect the performance levels otherwise expressed by other components, belonging to the same typological category or to another one.
- **General control criteria**: for a correct methodological approach to the control of errors generating disturbance factors (i.e. recognition and elimination) the following general control criteria have been defined:
. priority general criteria
. procedure general criteria
It has been noticed, that in the control of over time quality a substantial difference exists between first and second level control, indeed:
in the case of first level control the objects have to be examined at time zero and therefore characteristic properties have to be dealt with no regard to the modifications taking place over time;
in the case of second level control, objects have to be examined over their useful service life. Characteristic properties are therefore tackled with regard to their over time modifications.
As a consequence performances assumed at time zero and verified with first level control procedures, make up the basic reference for the prediction of over time performances.
Such prediction can be done by mediating the values of the parameters of performances at time zero, with factors obtained through models believed to be reliable for the prediction of the real over time evolution, in the expected service conditions.
- **Specific competence criteria**: it has been established how the quality control should be organized in each of the processual stages mentioned in 3, according to an interdisciplinary approach reaching over the fields of Durability, Quality Control, Planned Maintenance and Building Pathology. Such an approach should enable a more efficient application of the experiences made in each field as well as the implementation of the level of knowledge as a

consequence of practice.
In particular the definition of competence criteria specific to the control of primary and secondary errors and of regulations prescribing the action to be taken in case that such errors have not been recognized it has been worked out for each of the processual stages, design (technological, operational and management), production and maintenance.

5 References

Lucchini, A. (1989) **Metodologia per la valutazione dell'affidabilita' di elementi tecnici edilizi fuori sistema e problematica della durabilita,** Doctoral Thesis, Milan and Turin Polytechnics

Lucchini, A. (1990) **Organizzazione di un approccio sistematico interdisciplinare e modelli del processo per la previsione della service life.** Consiglio Nazionale delle Ricerche, Comitato di Ingegneria e Architettura, Roma

Lucchini, A. (1989) Situazione internazionale delle aree di ricerca della Durabilita' e della Patologia Edilizia - Conoscenze di base - Ricerche - Attivita' complementari, **Politecnico di Milano, Programma di Istruzione Permanente della Facolta' di Ingegneria, I° Corso di Aggiornamento in Patologia Edilizia**

Lucchini, A. (1989) An approach to Building Pathology, in **Seminar on Building Failures",** The National Swedish Institute for Building Research, Gävle, Sweden

Maggi, P. N.- Croce, S.- Gottfried, A.- Morra, L. and Lucchini, A. (1989) Evaluation of Building Components Long-Term Quality, in **Quality for Building Users Throughout the World,** XIth CIB Congress, Paris

51 THE DURABILITY OF POLYMER CONCRETE

P.J. KOBLISCHEK
Interacryl GmbH, Frankfurt, Germany

Abstract
Polymer concrete (PC) is a composite in which graded aggregates are bonded by a polymer. This polymer is normally formed through polymerization or polyaddition of reactive resins without addition of heat and at room-temperature. In contrast to cement, concrete PC contains no inter-connected pore system and therefore no harmful substances can be absorbed. In addition to this it will be reported about porous PC, which has the same pore structure as natural stone, used in the renovation of buildings and monuments built of natural stone.
Keyword: Polymer concrete, PC, Reactive resins, Application, Facades, Slabs, Feeding troughs, Machine bases, Synthetic marble.

1 Introduction

The following reactive resins are applied in the building industry and are also used as coatings and repair mortar for steel and concrete as well as for prefabricated parts. These can be anything from elastic to hard, and are clearly defined under DIN 1694: Epoxy resins (EP), Methacrylic resins (MMA), Polyester resins (unsaturated) (UP), and Polyurethane resins (PUR). If one needs a material with resistance against UV-exposure, acids, alkalis, and with thermal stability of over 100^0C, there is only one PC, based on methylmethacrylic resins, available. This material has been applied in practice since 1968. These weather exposed parts, eg. facades, kerbstones, railway platform edges, feeding troughs, balcony parapets, electrically heated angular stairs and balustrade coverings do not show any signs of efflorescence and therefore proved the durability. In contrast to cement, concrete PC based on MMA has a constant reverse bending strength.

2 Polymer concrete, PC

PC contains only organic bonding agents. The durability depends on the quality of PC and requires knowledge of the characteristics of the resins and the hardeners.

2.1 Curing systems

The curing is started by adding the hardening agent appropriately specified for the respective resin; once initiated the reaction cannot be stopped.

Every organic chemical reaction is dependent upon temperature. This holds good for polyaddition (EP, PUR) as well as for radical polymerization (MMA, UP) curing systems.

In polyaddition two components in stoichiometric amounts react with each other.

In radical polymerization, the polymerization is triggered off by radicals derived mainly from peroxide. Should hardening at room temperature be required, then it will be necessary to employ a redox reaction. The redoxsystem comprises hardening agents; peroxide, and accelerator, tertiary aromatic amines or organic metal salts. These form reactive molecular fragments, i.e. radicals, which then react with the double bonding elements causing them to form polymer chains and networks.. For radical polymerization only very small quantities (1-3%) of the redox system will be required.

Only a PMMA polymerized by heat or ultra-violet rays is absolutely non-yellowing, whereas there is yellowing of the aromatic decomposition products of the radical. As a result, PMMA resins are only non-yellowing if dosing is perfectly balanced. For special applications like synthetic white marble, we have a redoxsystem without any yellowing.

A major disadvantage - which does, however, also apply to all the other reaction resin systems which harden by radical polymerization - is the fact that some inhibition comes through oxygen. It is therefore not possible to manufacture a porous acrylic concrete, as this would not harden completely.

2.2 Principle

As with cement binder, so with PC, every particle of filler and each grain must be wetted. On grounds of expense, and also in order to make the most of the mechanical and thermal properties of the PC, it is absolutely essential to fill the reactive resin with additional mineral material, even of flour-grain particle size.

This requirement presupposes a lower viscosity of the resin used as the binding agent. The only type of reactive resin that makes this possible is that of methacrylate resins, which is available at a viscosity of their base monomer: methylmethacrylate (MMA). Neither the epoxy resins (EP) nor the unsaturated polyester resins (UP) show this low viscosity.

By polymerization from the monomer MMA, polymethylmethacrylate (PMMA) is formed. This has been known for many years under the name "acrylic glass" and under trade names such as PLEXIGLAS, PERSPEX etc. The weatherproofing properties have been proved and not for nothing has it been called the "King of Plastics". PMMA is absolutely permeable to UV-rays.

2.3 Aggregates

The quality of the aggregates determine the quality of PC. This applies to both the mechanical and chemical characteristics. Where structural components of PC are to be subjected to static or dynamic loadings, only unbroken and screened aggregates may be used. The smallest moisture content in the aggregates reduces the mechanical characteristics of the PC. Where exposure to chemical influence is expected, the specific characteristics of the hardened bonding agent and of the aggregates and additives must be taken into consideration. The concentration of bonding agent is dependent upon its composition and grain size distribution. An optimum density of the dry mixture should be sought. This requires the addition of fine filling material, 0 - 0.2mm, as a prerequisite. The proportion, of filler is dependent upon the viscosity of the applied resin: high viscosity, smaller portion of fine; low viscosity, larger proportions of fine. The maximum grain size of the aggregates is dependent upon the wall thickness of the PC-structural component, whereby as a rule the largest grain size shall not exceed 1/3 of the wall thickness; the proportion of resin declines with increasing grain size. For the composition of the filling material mixture, the screening curves according to Fuller or Rothfuchs may be applied, as these also take the various grain forms into consideration.

2.4 **Methylmethacrylic and Polymethylmethacrylic resins (MMA & PMMA)**
Through polymerization of MMA or dissolution of PMMA in MMA we can produce resins with increased viscosity.

Table 1. The influence of the viscosity to the principal material properties:

Filler:	Quartz 0 - 8 mm Particle size			
	Unit	2	10	150 mPa.s
Resin content	%	8	11	14
Specific gravity	g/cm^3	2.3	2.2	2.1
Compressive strength	N/mm^2	138.5	133.7	120.0
Bending tensile strength	N/mm^2	29.1	29.6	29.8
Thermal dimensional-stability	^{0}C	108	103	95

In all test breakages of the aggregate regularly occurred, i.e. the adhesion of the MMA/PMMA resins to the grain was greater than the strength of the aggregate itself. Since there is no compelling reason for working with resins of higher viscosity, we use only resins as binder for PC with a viscosity between 2 and 10 mPa.s. For gelcoats only we use resins with a higher viscosity.

A further advantage of MMA resins is their high reactivity, even at temperatures of 0°C and below. This, too, is demonstrated by the results of tests:

Test sample: 40 x 40 x 160 mm
Mixture : Resin to filler 1:6
Granulation: 0 - 8 mm,
"Fuller curve", 6 grain fractions.
Manufactured and stored at 0°C and +20°C.

Table 2. Hardening at 0°C and +20°C.

	Compressive strength in N / mm^2 after:				
Temp.	1 hour	2 hours	4 hours	1 day	3 days
0°C	75	80	85	87	120
+20°C	82	94	90	101	115

2.5 Unsaturated polyester resins (UP)
The cheap standard resins are not alkali-resistant, and are not sufficiently UV - and weatherproof. As their long-term performance has also produced unsatisfactory results, this class of material is not used by us.

2.6 Polyurethane resins (PUR)
These types form normally elastic polymers. The hardening is strongly catalyzed through moisture. Using aliphatic isocyanates in conjunction with aliphatic polyalcohols, light - and weatherproof polyurethane systems are produced. The resistance to UV has be improved by adding UV-inhibitors.

2.7 Epoxy resins (EP)
Harden by separation of the epoxide groups and by addition of hardener with active hydrogen atoms. The hardening reaction is dependent of the temperature and the type of hardener:

Test sample: 40 x 40 x 160 mm
Mixture : Resin to filler 1:7
Granulation: 0 - 2.0 mm,
Manufactured and stored at +5°C and +23°C.

Table 3. Hardening at + 5°C and + 23°C

Compressive strength N/mm^2 after:		1	2	3	7 days(s)
Standard EP-hardener:	+ 5°C	3	40	54	68
	+23°C	80	87	85	83
Special EP-hardener	+ 5°C	29	38	47	59

It is vital that attention must be paid to the form retention under heat of cold-hardening EP-systems.

The "HDT" Test (heat distortion temperature test), i.e. form retention under heat, amounts to only 40-45 °C for low-yellowing EP systems under DIN 53461 or ISO R 75 - HDT/FA, with load A (bending stress 1.85 N/mm^2). This figure only applies for systems containing no reaction dilutant.

4.0 Application

PC will never become a material such as, for example, normal concrete. The advantages are its high chemical constancy, its dimensional stability due to its impermeability to moisture and pollutants and its ability, also when unreinforced, to absorb tension. The comparatively quick curing and therefore prompt availability of the product result in ever new application possibilities.

The durability of PC can be documented by the application done in the past years.

4.1 Electrically heated angle-type steps, MMA-PC

A considerable number of such steps have been installed at Munich, Cologne and at first 1969 in Frankfurt/M. Since then these units have remained fully functional and shown acceptable wear of the non-slip surface.

4.2 Railway platform edge units, MMA-PC

Since 1974 numerous platform edges have been constructed with these slip-proof and highly wear-resistance units in Hamburg and Frankfurt/M.

4.3 Facades, MMA-PC

Since 1974 also a considerable number of facades in different decors were produced. One of the oldest is in Sprendlingen near Frankfurt/M. and in Saarbrucken, without any visible damage on the surface.

4.4 Slabs, MMA-PC

A thickness of only 8 mm has proved very satisfactory for highly wear-resistant industrial floorings. A non-skid surface is provided by suitable texture. These slabs have been manufactured industrially since 1975 and are used for preference in locations where extremely severe service conditions exist.

4.5 Feeding troughs. MMA-PC

The increasing use of silage fodder and the modern concepts of cattle-breeding require new shapes for troughs and mangers, which cannot be achieved with stoneware as a resistant alternative to conventional concrete. Smooth pore-free surfaces, good accuracy of fit of the moulded components with resulting narrow joints and ease of assembly are advantages securing these products an ever increasing share of the market. Since 1977 these troughs were manufactured and until today no failure due to the quality of this product is known.

4.6 Synthetic sandstone, EP-PC

The selected EP-system is a mortar with a content of only 7% binder. This mortar is not poured but pounded by hand in the prepared mould. Another type allows free-modelling. the surface can be reworked like natural stone. With this technique status and monuments can be copied. The original goes to the museum and the copy takes its place. This system was used first in Switzerland in 1977 and is still used today for restoration e.g. in Metaponto, Prov. Matera, Italy.

Missing parts and pieces can be supplied in order to make the reconstruction with the ancient pieces possible. This EP-sandstone has the same water vapour permeability as the natural stone but nearly total resistance against the harmful substances in the air and rain.

4.7 Synthetic marble, PMMA-PC and PUR-PC

Using aliphatic isocyanates in conjunction with aliphatic poly-alcohols, light- and weatherproof polyurethane systems are produced. The resistance to UV has to be improved by adding UV-inhibitors.

By PMMA is the modified redoxsystem necessary if a non yellowing white marble is required. Both systems were used in the renovation of antique monuments e.g. the Corinthian capitals at the ETH in Zurich, or white marble columns in the south of Turkey.

4.8 Machine bases, MMA-PC

Since 1980 this modified PC has been used an an alternative material for grey cast iron and welded steel construction in the manufacture of machine tools. The degree of damping in comparison with grey cast iron is 10 times, and with steel 20 times higher. The constant reverse bending strength was tested on a rotation bending test stand with a test frequency at 5 Hz and 2 million load cycles. MOTEMA-AC showed a constant reverse bending strength of 3.2 N/mm^2. This was the first proof of the dynamic stressability and long service life of a PC in the world.

5 Summary

After more than 20 years of experience in the application of PC and observance of the different carried out samples of the praxis, the durability has to be granted to these components. The basic condition is the exact knowledge of the specific characteristics of the reactive resins.

There is no bad PC, only its wrong application.

SESSION E

ASSESSMENT METHODS

52 ASSESSMENT OF RAPID CHLORIDE PERMEABILITY TEST OF CONCRETE WITH AND WITHOUT MINERAL ADMIXTURES

M. GEIKER, N. THAULOW, P.J. ANDERSEN
G.M. Idorn Consult A/S, Birkeroed, Denmark

Abstract
A study of permeability by means of the AASHTO T 271-831, Rapid Determination of the Chloride Permeability of Concrete, was made on ten concrete mixes differing in cementitious material (w/c from 0.31 to 0.36, 16.5 weeks and 36 weeks old). The mixes included one, two and three powder mixtures of either low alkali sulphate resistant Portland cement or rapid hardening Portland cement with and without silica fume and fly ash.

The repeatability, expressed as the coefficient of variation within specimens from the same mix varied from 5% to 18%.

For unblended cements a linear correlation between electrical charge passed (coulomb) and porosity was found. For blended cements with either fly ash, silica fume or both much lower values of coulomb without a similar change in the amount of porosity were obtained.

The measurements of electrical charge passed correlate well with resistivity measurements, especially for dense concretes. The increase in electrical current passed during the six hour long test period experienced for porous concretes tested according to the AASHTO T 277-831 method is a temperature effect.
Keywords: Permeability, Porosity, Chloride, Test, Concrete, Silica Fume, Fly Ash.

1 Introduction

During the last decades increasing awareness has been given to the durability of concrete structures. Low permeability of concrete is perhaps the most decisive parameter for its optimal durability in aggressive environments. Because of this, the specification and testing of the permeability of concrete is most relevant.

One test method for measuring the permeability of concrete receiving increasing interest is the AASHTO T 277-831, Rapid Determination of the Chloride Permeability of Concrete. This method was originally developed by Whiting (1981) and is used extensively throughout the USA. The method consists of monitoring the amount of electrical current passing through water saturated concrete cores over a period of six hours while the specimen is being subjected to an electrical potential of 60 volt D.C. The total electrical charge passed (coulomb) is generally related to the chloride permeability of the concrete.

The name: 'Rapid Determination of the Chloride Permeability of Concrete' is somewhat deceptive since the method only indirectly measures the sensitivity towards the ingress of chloride ions. The ionic transport under an applied field is dependent on the physical as well as the chemical nature of the specimen. The determining factors in concrete are the properties of the pore system and the conductivity of the pore fluid.

The effects of mineral admixtures (silica fume and fly ash) on the correlation between the electrical charge passed and the porosity of the concrete will be discussed.

2 Experimental

2.1 Materials

This paper is primarily based on experiments performed on ten concrete trial mixes for the Danish Great Belt Tunnel. The compositon of these concretes (designated A to J) are shown in Table 1.

Table 1. Composition of concretes tested

		Mix									
		A	B	C	D	E	F	G	H	I	J
SRPC	kg	310	405	370	345	370	345	0	0	0	0
RPC	kg	0	0	0	0	0	0	310	405	370	345
PFA	kg	70	0	70	0	70	0	70	0	70	0
SF	kg	30	0	0	30	0	30	30	0	0	30
Water	kg	139	134	130	135	125	130	144	139	135	140
w/c		0.34	0.33	0.32	0.33	0.31	0.32	0.36	0.34	0.33	0.35
Air	%	4.7	3.9	6.1	4.3	4.1	6	6	4.4	5.3	4

SRPC: Low alkali, sulphate resistant Portland cement.
RPC: Rapid hardening Portland cement. PFA: Fly ash. SF: Silica fume.
Air: The amount of entrained air measured by microscopy on hardened concrete.

The concretes to be tested were cast in blocks of 400 mm x 400 mm x 200 mm. After approximately 16 weeks two cores (diameter 100 mm, length 200 mm) were taken from each test block. From each core three slices (diameter 100 mm, thickness 51 mm, marked 1, 2 and 3, starting from the surface) were cut, dried for one day in the laboratory and coated with a watertight paint on the cylindrical surface. The paint was allowed to cure for one day whereafter the slices were water saturated and tested as described below. After the first test the slices were kept under laboratory conditions for approximately one month and then stored in water filled plastic bags until the second test was carried out. Thereafter the slices were kept under laboratory conditions.

2.2 Tests

The permeability of the concrete (age 16.5 weeks and 36 weeks) has been measured indirectly by means of the AASHTO T 277-831, Rapid Determination of the Chloride Permeability of Concrete. For this purpose equipment that allows easy assembling of the test cell, automatic collection of data as well as print-outs of data reports has been developed by G.M. Idorn Consult A/S (Holm, 1989).

This equipment measures the permeability of the concrete by positioning a water saturated slice (diameter 100 mm, thickness 51 mm) of a concrete core or test cylinder into a cell containing on either side an electrolyte as prescribed in the AASHTO T 277-831 (anode chamber 0.3 N sodium hydroxide solution, cathode chamber 3% w/w sodium chloride solution). A constant voltage of 60 volt D.C. is applied to the cell and the electrical current passing through the concrete slice is recorded at five minute intervals for a total period of 6 hours. The total electrical charge passed is calculated by Simpson integration of the current vs. time. The electrical charge passed (coulomb) is corrected for smaller deviations in the dimensions of the concrete slices from the dimensions required according to the standard.

The porosity of the concretes (age 1 year) was determined by means of the ASTM C 642-82, Standard Test Method for Specific Gravity, Absorption, and Voids in Hardened Concrete. This method is based on weighing the samples oven dried (110°C) and water saturated in air and submersed.

3 Results and Discussion

The AASHTO T 277-831 test is basically identical to a conductivity measurement. The current passing through the concrete slice is according to Ohm's first law equal to the electrical potential across the slice (60 volt (applied) - 2 volt (polarisation potential) = 58 volt) divided by the resistance. For dense concretes the measured correlation between the electrical charge passed and the resistivity appears to be according to the theory (Fig. 1). This figure also includes data from Burke (1989). Over a period of six hours the temperature in the cell increases due to Ohm's second law causing a decreased resistivity and thereby an increased current flow through the relatively porous concretes. Due to this temperature effect, a deviation from the theoretical relationship between electrical charge passed and the resistivity is observed for these concretes (Fig. 1).

During the test period the ions in the pore solution (K^+, Na^+, Ca^{2+}, OH^-) will migrate out of the slice and ions from the electolyte (Na^+, Cl^-) will migrate into the concrete. As the conductance of the ions migrating out is higher than the conductance of the ions migrating into the slice, the conductivity of the pore solution will decrease during the test (Castellan, 1964). The conductivity of the pore solution in a non-pozzolanic Portland cement can be twice the conductivity of a 3% sodium chloride solution (Arup, 1989). Repetitive testing on a concrete (approx. 1500 coulomb, no pozzolans) showed no measurable change in the electrical charge passed during the first and the second test (the second test was performed after cooling the

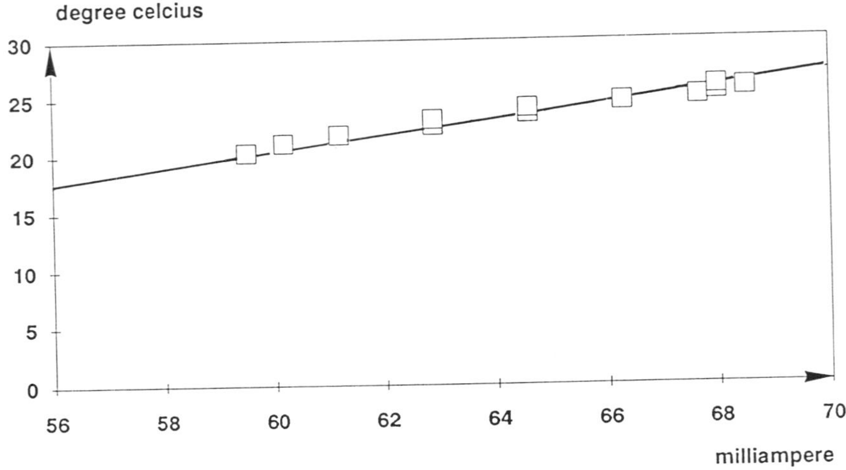

Fig. 2.
Temperature increase (^{o}C) of electrolyte versus electrical current. The curve shown illustrates the temperature dependence of the ionic equivalent conductivities (Castellan, 1964).

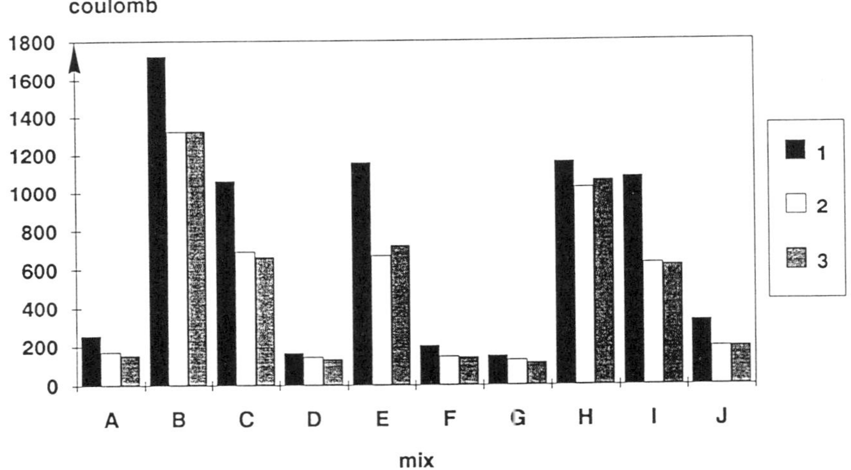

Fig. 3.
Electrical charge (coulomb) passed through 16.5 week old concrete slices with one cast and one cut surface (average for slices marked 1) and two cut surfaces (average for slices marked 2 & 3).

A distinct difference in the electrical charge passed can be observed for several of the mixes both after 16.5 weeks and 36 weeks of curing (Fig. 3 and Fig. 4). This is further illustrated in Fig. 5.

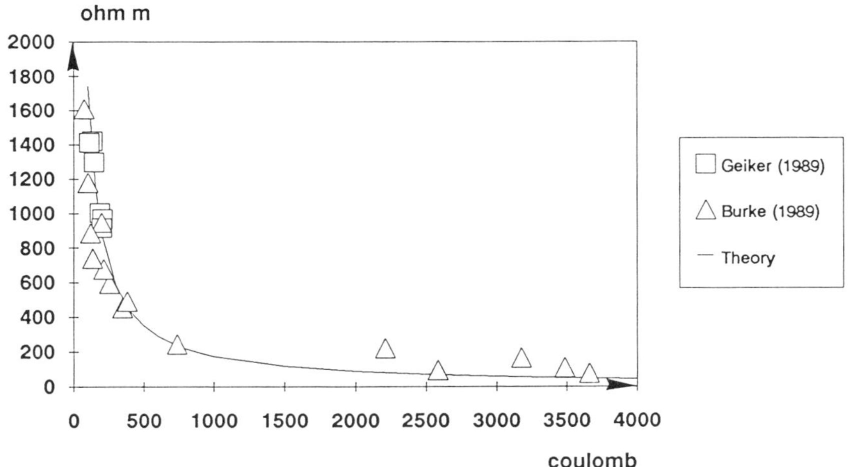

Fig. 1.
Resistivity (ohm m) versus electrical charge passed (coulomb) (Geiker, 1989, Burke, 1989). The curve drawn is the theoretical relationship.

sample and the cell to ambient temperature). This indicates that only a small amount of chloride ions migrated into the slice during the test period.

As mentioned above, measuring over a period of six hours gives rise to a temperature increase especially in the more porous concretes. To illustrate this the current through a slice of concrete (w/c=0.47, 375 kg/m^3 cement, no admixtures) and the temperature in one of the electrode chambers of the permeability cell have been monitored during the test period, Fig. 2. For most ions in water the conductivity increases with temperature by about 2% per degree celcius (Castellan, 1964). The data obtained correlates reasonably well with this relationship.

The electrical charge passed through concrete slices of the compositions given in Table 1 after 16.5 weeks and 36 weeks of curing (same slices) is shown in Fig. 3 and Fig. 4.

Several features can be observed in these figures. During the first test (after 16.5 weeks of curing) the slices with one cast and one cut surface (marked 1) have generally allowed a larger electrical charge to pass than the slices with two cut surfaces (marked 2 and 3) (Fig. 3). This effect was not observed at the later test (after 36 weeks of curing) (Fig. 4) and is therefore believed to be of chemical rather than physical nature. One explanation for the effect measured at the first test could be that evaporation from the surface during the period of curing has caused an increased concentration of alkalies in the pore liquid and thereby a higher conductivity than in the rest of the concrete. After soaking in the same water for approximately three months, the alkali concentration will be nearly the same in all three slices (marked 1, 2 and 3) and as observed, the electrical charge passed through the three slices during the later test will therefore be similar.

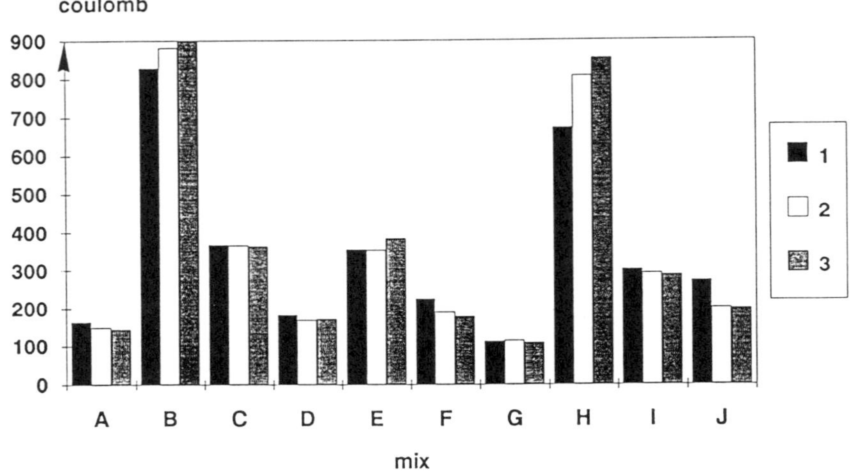

Fig. 4.
Electrical charge passed (coulomb) through 36 week old concrete slices with one cast and one cut surface (average for slices marked 1) and two cut surfaces (average for slices marked 2 & 3).

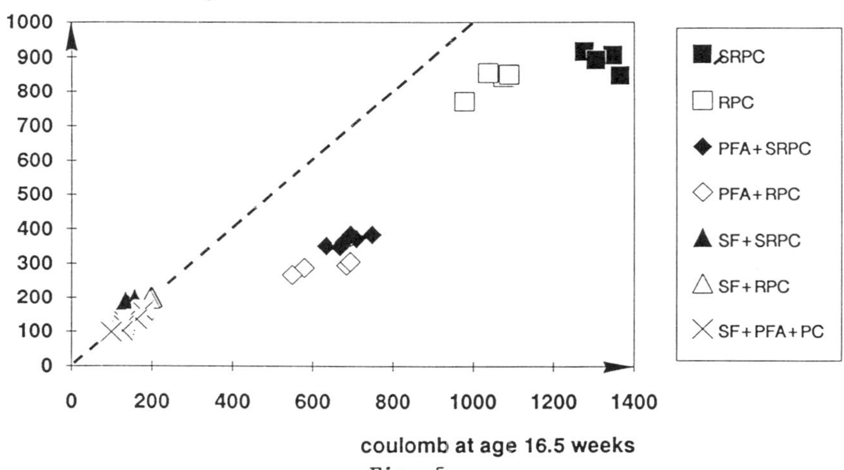

Fig. 5.
Electrical charge passed (coulomb) after 36 weeks of curing versus the charge passed (coulomb) after 16.5 weeks of curing. The dotted line is the line of direct proportionality.

In this figure the electrical charge passed through the slices with two cut surfaces after 16.5 weeks of curing is correlated with the charge passed after 36 weeks of curing. To facilitate the reading of the figure, the line of direct proportionality is drawn.

It is observed that the composition of the concretes is decisive for the amount of electrical charge passed. A grouping of the data is seen, dependent firstly on whether or not the concrete contains silica fume, secondly on the content of fly ash and finally on the type of cement. The concretes containing silica fume show the lowest values of electrical charge passed. All the concretes were manufactured with similar effective w/c. The activity factors used were 0.5 for fly ash and 2 for silica fume.

The concretes containing silica fume do not exhibit a measurable change in the electrical charge passing at 16.5 weeks and at 36 weeks. Concretes without silica fume are seen to have allowed a decreased electrical charge to pass after prolonged hydration. The effect is most pronounced for the concretes containing the low alkali sulphate resistant cement.

The determining factors for the electrical charge passed through a concrete specimen are the properties of the pore system, the amount and tortuosity of the porosity and the conductivity of the pore liquid. The porosity in question is the amount of entrapped air, entrained air and capillary porosity.

The relationship between electrical charge passed and porosity measured after 36 weeks of curing is shown in Fig. 6. In this figure data from Andersen (1990) for concretes (w/c=0.5, no chemical or mineral admixtures) varying in amount of entrapped air is included. There appears to be no correlation between the electrical charge passed and the total porosity.

When testing according to the AASHTO T 277-831 method the part of the porosity due to air entrainment was not water filled. The water filled porosity has been calculated by subtracting the amount of entrained air measured by optical microscopy on the hardened concrete (Table 1) from the total porosity. The relationship between the water filled porosity and the electrical charge passed is shown in Fig. 7.

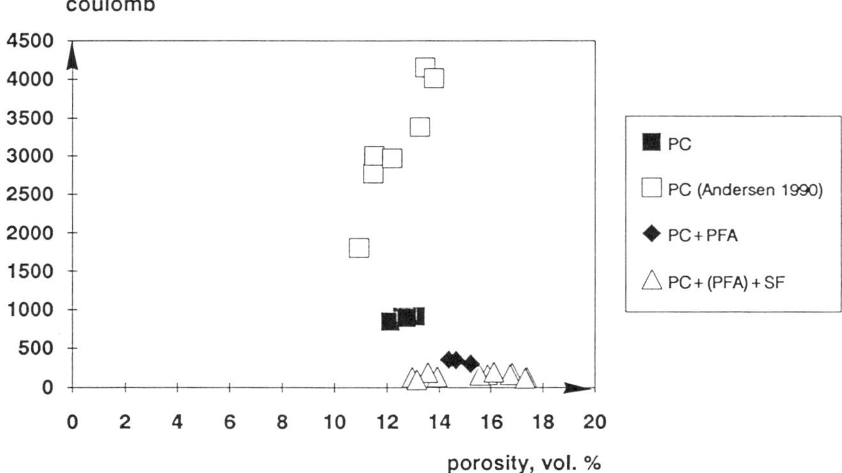

Fig. 6.
Electrical charge passed (coulomb) versus total porosity (vol. %).

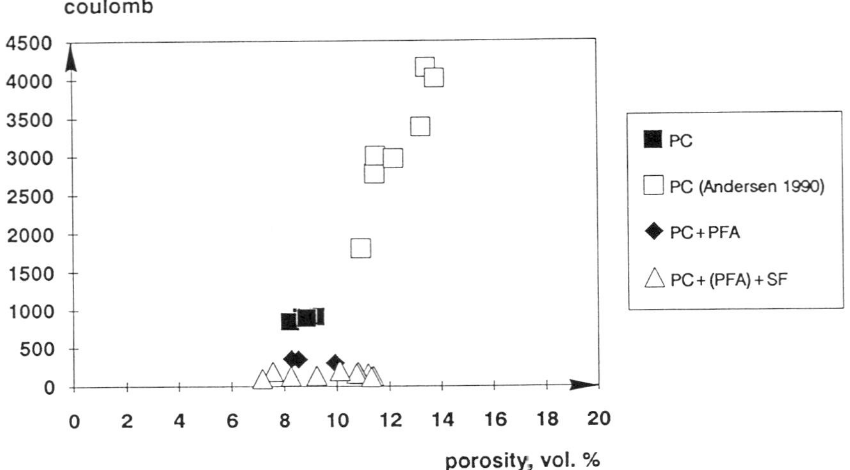

Fig. 7.
Electrical charge passed (coulomb) versus water filled porosity (vol. %).

From Fig. 7 it appears that the electrical charge passed is directly proportional to the porosity of concretes without mineral admixtures and that macro porosity due to entrapped air and capillary porosity have the same effect on the relationship.

The effect of the composition of the cementitious material is illustrated in Fig. 8, in which a part of Fig. 7 has been enlarged. The data was obtained on concretes with similar effective w/c.

The effect of the capillary porosity on the electrical charge passed seems to be influenced by the composition of the cementitious material. This is probably caused by a combined effect of the fineness of the pore system, the tortuosity, and the conductivity of the pore liquid. The effect of silica fume is more pronounced than that of fly ash.

Based on pore solution composition data (Glasser, 1988, Marr, 1983) the conductivity of the pore solution in a concrete without silica fume will be approximately four times the conductivity of the pore solution in a concrete containing silica fume. Concretes containing fly ash will, especially after prolonged hydration, exhibit the same trend, but to a smaller extent.

The concretes investigated were very strong and dense. Due to this it was not possible to squeeze out the pore liquid for chemical or electrical analyses.

The repeatability of the permeability measurements have been calculated from the data obtained on the 36 week old concrete slices. The coefficient of variation within slices from the same mix (same operator) varies from 5% to 18% and appears to be independent of the electrical charge passed, Fig. 9. Whiting (1981) reported values of repeatability from 6% to 7%.

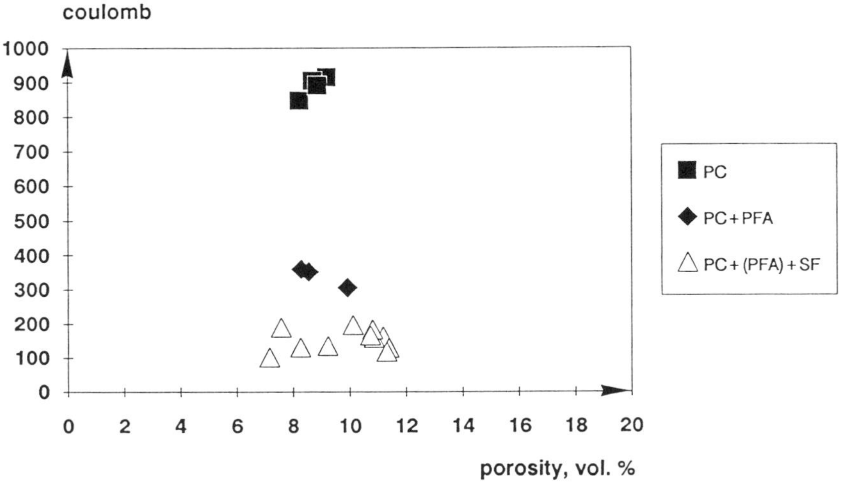

Fig. 8.
Electrical charge passed (coulomb) versus water filled porosity (vol. %).

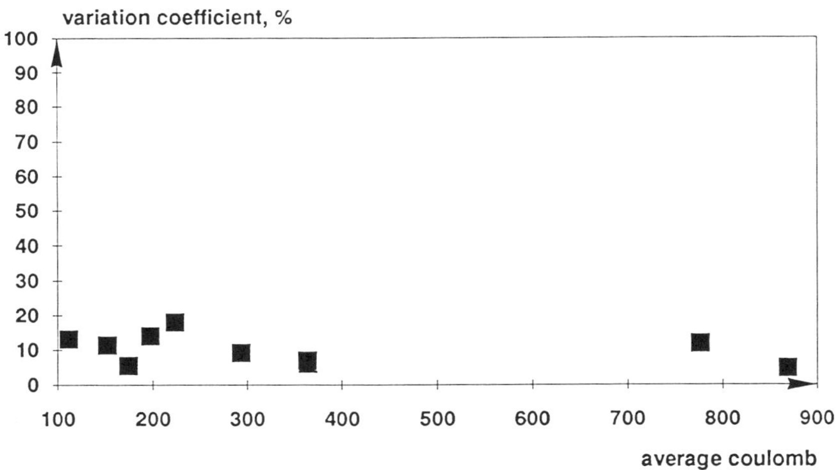

Fig. 9.
Coefficient of variation within slices from the same mix versus electrical charge passed (coulomb).

4 Conclusion

For AASHTO T 277-831 a repeatability of 5% to 18% was found. The method has, in contrast to measurements of porosity, allowed distinction between concretes (similar effective w/c) with and without mineral admixtures.

For unblended cements a linear correlation between electrical charge passed and porosity was found, whereas for blended cements, with either fly ash, silica fume or both, much lower values of electrical charge passed without a similar change in the amount of porosity were obtained.

For blended cements the effect of the mineral admixtures on the conductivity of the pore liquid should be taken into account when interpreting permeability data obtained by the AASHTO T 277-831 method.

The increase in electrical current passed during the six hour long test period experienced for porous concretes tested according to the AASHTO T 277-831 method is a temperature effect.

5 Acknowledgement

The authors are indebted to Great Belt A.S., Denmark for their permission to publish relevant data.

6 References

Andersen, P.J.(1990) **Thesis**. The Danish Academy of Technical Sciences.
Arup, H. (1989) **Dansk Beton**. Vol. 6, p.41.
Burke, N.S. (1989) 3rd Int. Conf. Fly Ash, Silica Fume, Slag and Natural Pozzolans in Concrete, p.861. Trondheim.
Castellan, G.W.(1964) **Physical Chemistry**. Addison-Wesley Publishing Company.
Geiker, M. and Sørensen, B. (1989) Report to Great Belt A.S.
Glasser, F.P. et al (1988) **Cement and Concrete Research**. Vol. 18, p.165.
Holm, J. and Andersen, P.J. (1989) **Dansk Beton**. Vol. 6, p.35.
Marr, J. and Glasser, F.P. (1983) Proc. 6th Symp. on Alkali Aggregate Reactions, Copenhagen, p.239.
Whiting, D. (1981) Report No. FHWA/RD-81/119. Federal Highway Administration, Washington D.C.

53 ASSESSMENT OF CONCRETE DURABILITY BY INTRINSIC PERMEABILITY*

R.K. DHIR
Department of Civil Engineering, University of Dundee, Scotland, UK
P.C. HEWLETT
British Board of Agrément, Garston, Watford, UK
Y.N. CHAN
Ove Arup & Partners, London, UK

Abstract

The measurement of the intrinsic permeability of concrete is described. The tests developed are rapid and reliable. It is shown that the intrinsic permeability of concrete can be characterised using the air permeability test. Hydraulic permeability data can also be derived if required. The permeability of concrete in the surface layer (that comprising the cover to reinforcement) is greatly affected by both the water/cement ratio and curing. Concrete quality should not be estimated from strength data alone. The link between measured insitu permeability and estimated durability is the subject of current research.

1 Introduction

Durable concrete should retain its required quality and serviceability with a minimum of maintenance throughout its design life. The permeation characteristics of the near surface of concrete, which affords both the chemical and physical resistance against the ingress of deleterious elements from the environment, have a major influence on the long term performance of the concrete structure.([2])

This paper deals with the permeability of the cover concrete to steel reinforcement in particular the intrinsic permeability (k) and its inferred quality. It does not at this stage deal with the relationship between the measured k value of insitu concrete that is of unknown composition or history. However the intention is to extend these procedures to that objective.

The intrinsic permeability of concrete (k) is a geometric factor and relates to the internal structure of concrete and is independent of the properties of the migrating fluid; a sample tested with any liquid or gas should yield all other factors being equal, the same k value. As such it is a useful and more absolute basis for comparing one concrete with another.

*Paper originally published in Magazine of Concrete Research ([1])

The main objectives of this study were to develop test methods that can overcome or minimise the technical difficulties associated with the measurement of the intrinsic permeability of concrete.

2 Permeability measuring apparatus

The apparatus developed for the measurement of intrinsic permeability using air and water has been published elsewhere ([1]).

2.1 Air permeability

Air permeability measurements can be very sensitive to the level of moisture within concrete([3]) and for the various reasons discussed in a previous publication by the authors([4]) it was decided to adopt the 105°C oven drying method for conditioning specimens prior to testing.

To establish the duration of drying required for the test specimens, 50mm and 100mm diameter by 50mm thick, a series of tests were carried out, an example of data from which is illustrated in Figure 1. On this basis it was concluded that a suitable criterion for judging the required period to dry a specimen was when a weight change of less than 0.1% over 24 hours was observed.

The coefficient of variation of the permeability measurements was found to be 9% and 14% for the 100mm and 50mm diameter specimens respectively.

2.2 Water permeability

The water permeability pressure cell was essentially similar to the air permeability cell, except that it had an additional water inlet to fill the system completely at the outlet side of the cell before testing. The specimens were also vacuum saturated before testing.

Once the flow rate at the outlet equalled that of the inlet the steady state flow and the corresponding pressure were recorded. Tests were then repeated with reduced water pressure P_1 of 4.8 N/mm^2, 3.5 N/mm^2 and 2.1 N/mm^2. A plot of pressure against flow rate was drawn as the measurements were made as a means of controlling the accuracy of the results.

The apparatus allowed the permeability to be measured at the first pressure within 15 minutes. Shorter times were required for the rest of the test pressures and normally a maximum of up to 1 hour was needed to complete the test of a specimen.

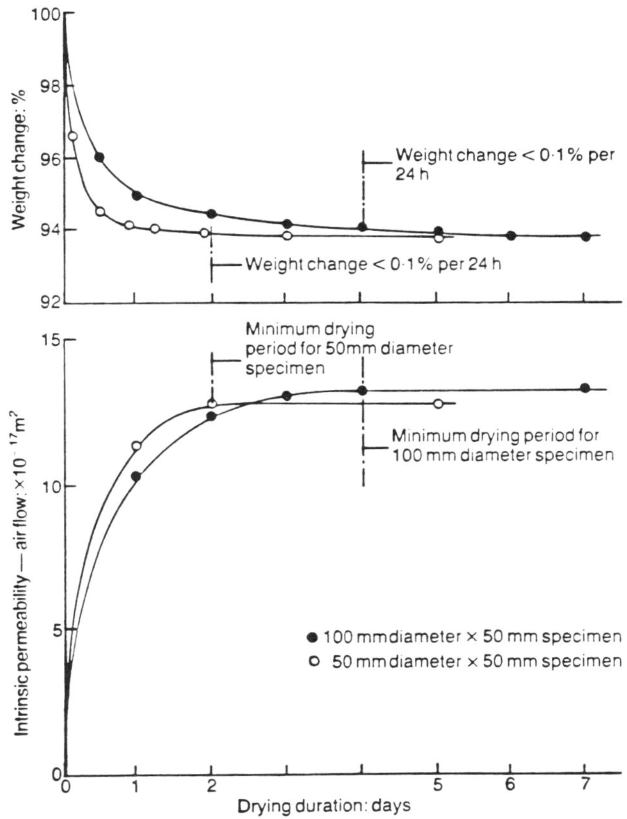

FIG.1. MINIMUM DRYING PERIOD FOR AIR PERMEABILITY TEST; WATER / CEMENT RATIO = 0·55. SPECIMENS CURED 3 DAYS WATER, FOLLOWED BY AIR AT 20°C, 55% RH. SOAKED IN WATER FOR TWO DAYS AND DRIED AT 105°C PRIOR TO TEST.

3 Calculation of intrinsic permeability

Flow rates in concrete are normally very low, resulting in laminar rather than turbulent flow. Laminar viscous flow is described by Darcy's Empirical law, which is dependent upon the properties of the fluid, i.e. viscosity and density, as well as the characteristics of the porous medium. A more rational concept of permeability may be expressed by a coefficient which is independent of the fluid properties as

$$\frac{Q}{A} = -\frac{K}{\mu} \cdot \frac{dp}{dl} \qquad (1)$$

Where Q is the volume rate of flow, A is the cross sectional area perpendicular to the flow direction, k is the intrinsic permeability, μ is the viscosity and dp/dl is the pressure gradient in the direction of flow.

For a non compressible fluid (for example, water) the integration of equation (1) can be performed directly to obtain

$$k = \frac{Q\mu L}{A(P_1 - P_2)} \qquad (2)$$

Where P_1 is the inlet pressure, P_2 the outlet pressure and L is the length of specimen.

For a compressible fluid, (for example, a gas) the quantity Q varies with change of pressure P, according to the relationship PQ = a constant = PmQm

Where $P_m = \frac{P_1 + P_2}{2}$ and Q_m is the flow at the mean pressure P_m

Substituting for Q in equation (1)

$$\frac{P_m Q_m}{PA} = -\frac{K}{\mu} \cdot \frac{dp}{dl}$$

Integrating,

$$\frac{Qm}{A} \int_0^l = -\frac{K}{\mu Pm} \int_{P_1}^{P_2} Pdp \qquad (3)$$

Therefore $K = \frac{2\ \mu l P_2 Q_2}{A\ (P_1{}^2 - P_2{}^2)}$

Where Q_2 is the outlet volume flow rate

In theory it should be possible to use any liquid or gas to obtain the intrinsic permeability of a porous material using equations (2) or (3) While this is the case for liquids, provided there is no chemical or physical interaction between the fluid and the material that changes the characteristics of either or both, the flow capacity for a gas depends on its nature and the mean flow pressure and therefore a correction factor is introduced to take these into account when calculating permeability ([5]). One of the main factors that causes the measured permeability to a gas to be greater than that to a liquid is gas slippage.

In this study to overcome the gas slippage effect we used the following procedure.

For each specimen a series of tests is carried out at various mean pressures and the straight line plot of k' against $\frac{1}{Pm}$ is extrapolated to infinite mean pressure to deduce the intrinsic permeability. (k' = apparent intrinsic permeability and P_m = mean pressure of flow). By this method the errors involved with using an empirical constant are eliminated. This method is suitable for all flow gases and porous media.

A similar correction was applied for visco-intertial or flow effects([1]).

4 Experimental results

Details of mixes used, curing and preconditioning have already been given ([1]). Four series of concrete mixes were tested after 28 days namely,

Series 1 - variable water/cement ratio
Series 2 - variable workability
Series 3 - variable maximum aggregate size
Series 4 - special constituent materials

The test specimens were of diameter 100mm or 50mm and 50mm thick. The maximum diameter of 100mm was chosen to minimise the possibility of cutting reinforcement when cores are taken from in-situ structures. Since the quality of cover concrete has a dominant influence on the durability of concrete, it was decided to limit the thickness to the outside 50mm of specimens cored from 150m cubes, subject to different curing conditions.

4.1 Factors affecting permeability of concrete

The effect of water/cement ratio (Series 1) and initial moist curing on the air permeability of cover concrete is illustrated in Figure 2. It can be seen that permeability increases almost exponentially with increasing water/cement ratio. The duration of moist curing also has a very significant effect. The results show that to produce a low permeability concrete it is equally essential to ensure adequate curing as to specify a low water/cement ratio. For a given water/cement ratio or curing, air permeability has a close relationship with concrete compressive strength (Figure 3). However the results clearly demonstrate that it is misleading to suggest that strength generally be used to predict permeability.

The workability (Series 2) except when approaching a collapse slump, was found to have no signficant effect on air permeability.

In general there was no significant difference in air permeability of concrete made with various maximum aggregate sizes up to 20mm. (Series 3).

The influence on the air permeability of the inclusion of either an air entraining agent (AEA), rapid hardening Portland cement (RHPC), pulverised fuel ash (PFA), or microsilica (MS) in concrete (Series 4) was compared with that of corresponding ordinary Portland cement (OPC) concrete, as shown in Figure 4. The RHPC, PFA and microsilica concretes perform marginally better than the OPC concrete.

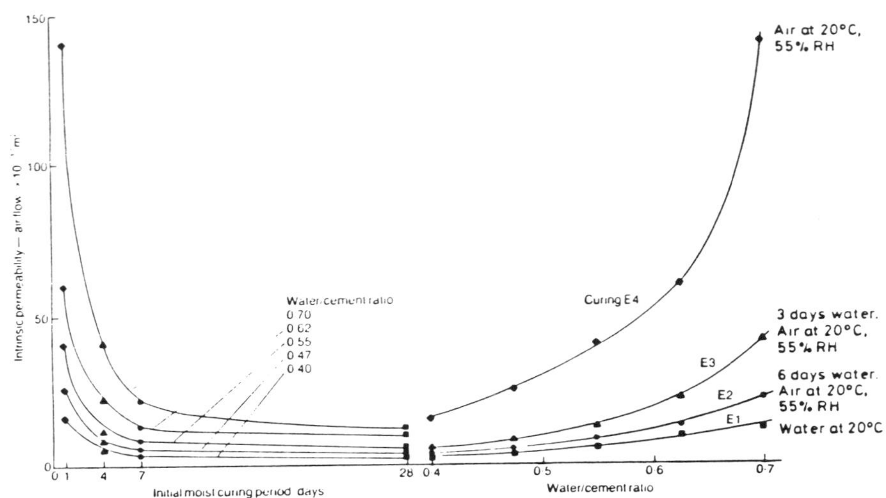

FIG 2 EFFECT OF INITIAL MOIST CURING PERIOD AND WATER / CEMENT RATIO ON INTRINSIC PERMEABILITY OF CONCRETE ; FLOW FLUID WAS AIR, SPECIMEN OVEN DRIED AT 105°C

4.2 Water permeability

To study the water permeability, tests were carried out using the same four series of concrete mixes used previously for the air permeability measurement.

Two types of pre conditioning were used,

1. Oven dried at 105°C then vacuum saturated (C1)
2. Laboratory dried at 20°C (55%RH) the vacuum saturated (C2)

The full results have been published elsewhere.([1]) The general trends in the water permeability results were found to be essentially similar to those observed for the air permeability.

The difference in the two sets of water permeability results show the signficance of the pre-conditioning of test specimens. Thus, until a standard test procedure is accepted, it is important to state the type of preconditioning used when presenting water permeability data.

The relationship between the two sets of water permeability data obtained with different preconditionings, is shown in Figure 5 and can be expressed as

$$k_{w_1} = \mathrm{EXP}\ (0.56 + 0.75 \log k_{w_2})$$

where k_{w_1} is the intrinsic permeability of the C1 preconditioned specimen and k_{w_2} is the intrinsic permeability of the C2 preconditioned specimen.

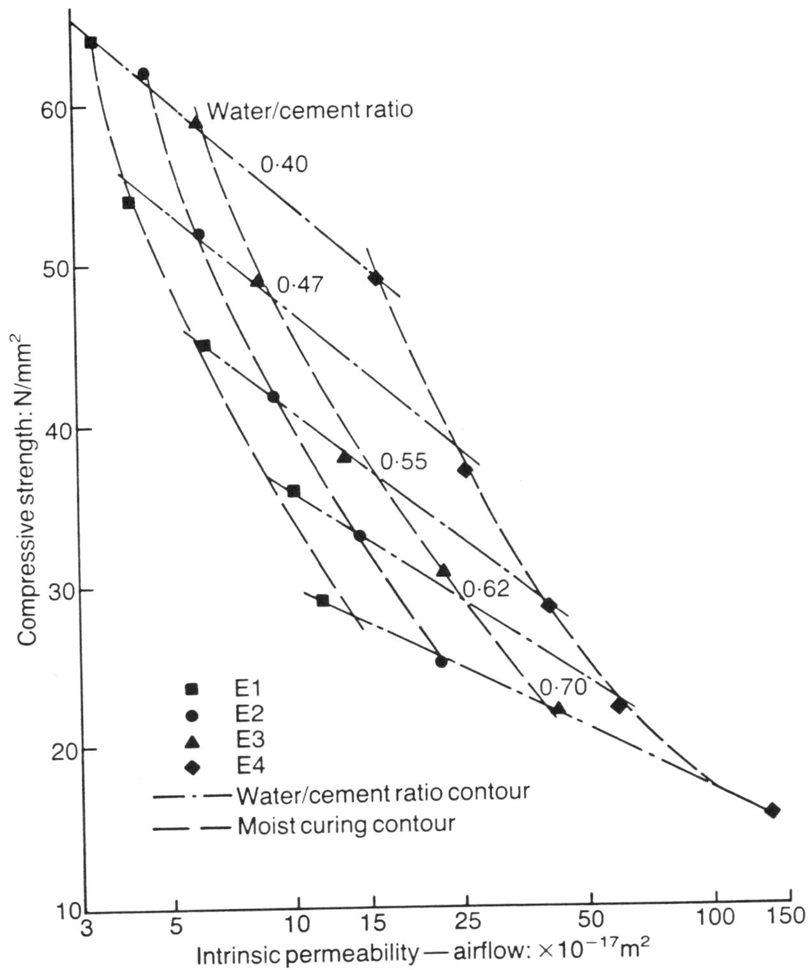

FIG.3. RELATIONSHIP BETWEEN COMPRESSIVE STRENGTH AND PERMEABILITY OF CONCRETE; FLOW FLUID WAS AIR, SPECIMEN OVEN DRIED AT 105° C

4.3 Relationship between air and water permeability

It has been shown that even when the same test specimen is used, the duration of test is less than one hour and the gas slippage and temperature effects are considered in the calculation, the derived intrinsic permeability of concrete can still vary, depending on whether air or water is used for its measurements. This departure from the concept of a single absolute value is thought to be due to the dilation (swelling) of the cement hydrates when in contact with water.

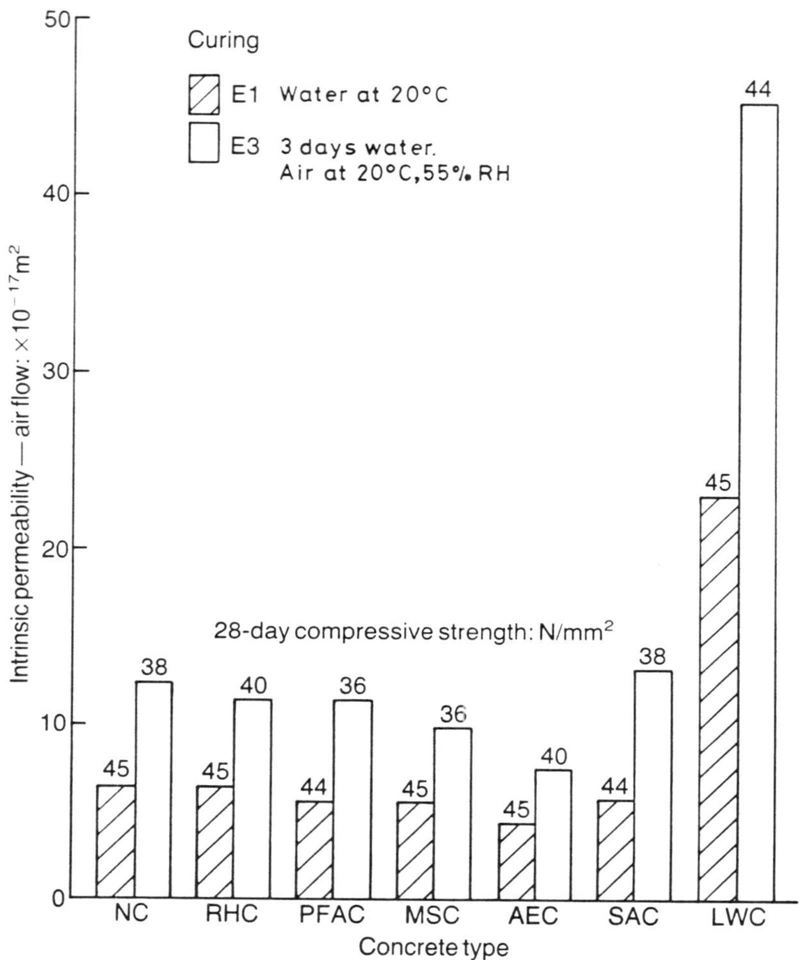

FIG. 4. EFFECT OF CONSTITUENT MATERIALS ON INTRINSIC PERMEABILITY (AIR FLOW) OF CONCRETE WITH EQUIVALENT STRENGTH

The permeability results obtained using air or water for all the four series of concrete mixes are shown plotted in Figure 6. It is shown that when the water permeability is known intrinsic permeability k of concrete can be obtained.

The most important point to emerge from this study, however, is that the air permeability test provides directly a measure of the instrinsic permeability of concrete, and therefore it can be used to characterise concrete in this regard. Moreover, when

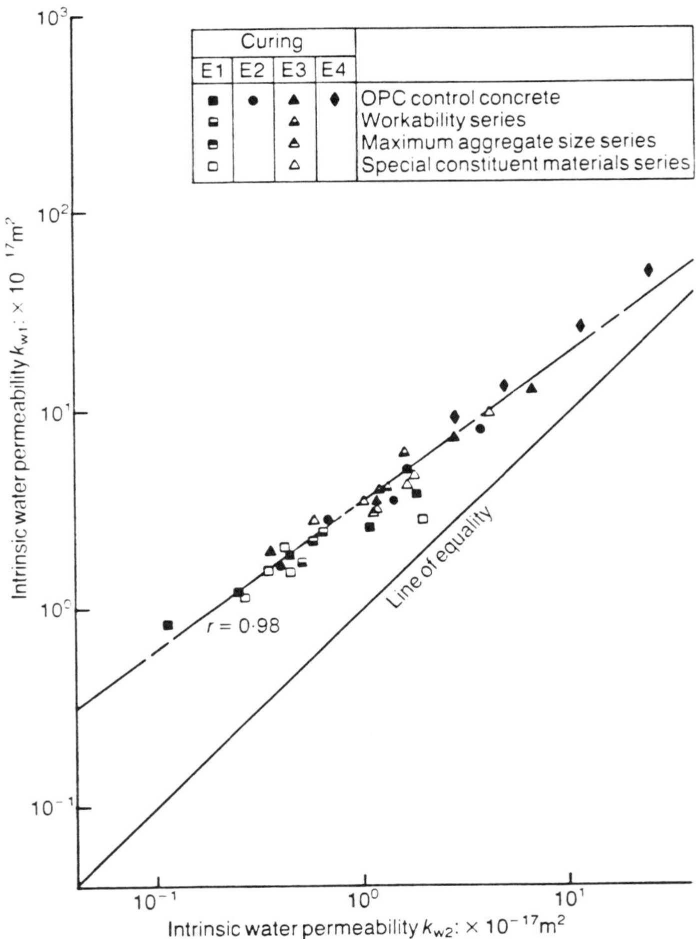

FIG 5. RELATIONSHIP BETWEEN k_{W1} AND k_{W2}

$k_{W1} = \exp(0{\cdot}75 \log k_{W2} + 0{\cdot}56)$

required the test can also be used to derive the corresponding hydraulic permeability of concrete with the aid of relationships such as those shown in Figure 6.

At the present time it is not possible to use the measured intrinsic permeability of an "unknown" concrete and make a prognosis on durability other than in a general way. That work is the subject of a current research programme and will be reported at a later date.

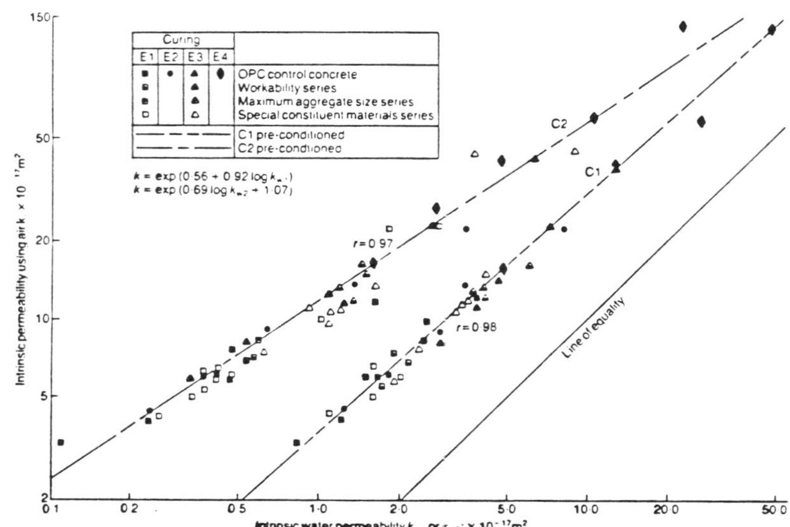

FIG 6 RELATIONSHIP BETWEEN INTRINSIC AIR PERMEABILITY AND INTRINSIC WATER PERMEABILITY; $k = \exp(0{\cdot}56 + 0{\cdot}92 \log k_{W1}) = \exp(0{\cdot}69 \log k_{W2} + 1{\cdot}07)$

5 Conclusions

Apparatus for the measurement, at steady state flow, of permeability of concrete to air and water has been developed. The test methods are rapid and repeatable and they overcome many of the limitations associated with most of the methods previously reported.

The gas slippage and viscoinertial flow should be taken into account in deriving the intrinsic permeability of concrete using air. The swelling of cement hydrates in concrete is believed to be the main factor causing the measured intrinsic permeability using water to deviate from that obtained using air. Empirical relationships have been derived between the intrinsic air permeability and the intrinsic water permeability of concrete.

It has been shown that concrete can be characterised in terms of intrinsic permeability using the air test; where hydraulic permeability is required that may also be derived. Alternatively, hydraulic permeability of in situ concrete can be directly measured using the developed test method.

The permeability of covered concrete has been shown to be very sensitive to small changes in the water:cement ratio and the degree of initial moist curing. It is also shown that the compressive strength of concrete alone should not be used as a means of assessing its permeability potential.

REFERENCES

1. Dhir RK, Hewlett PC and Chan YN "Near surface characteristics of concrete: intrinsic permeability" 1989, Magazine of Concrete Research, Volume 41, No 147 June 87-97

2. Dhir, RK, Hewlett PC and Chan YN "Near surface characteristics and durability of concrete: an intial appraisal" 1986, Magazine of Concrete Research, Volume 38 No 134 March, 54-56

3. Loughborough MT "Permeability of concrete to air" Ontario Hydro Res Quart 1966, 1st quarter 14-25

4. Dhir RK, Hewlett PC and Chan YN "Near surface characteristics of concrete: assessment and development of in situ test methods" 1987 Magazine of Concrete Research, volume 39 No 141, Dec 183-195: Discussion Magazine of Concrete Research 1988,Vol 40, no 145, Dec 240-250

5. Klinkenberg LJ, "The permeability of porous media to liquids and gases" Drilling and Production Practice 1941, American Petroleum Institute, Chicago 200-202.

54 X-RAY PHOTOELECTRON SPECTROSCOPY STUDIES OF COATINGS ON SHEET METAL

P. JERNBERG, D. LALA, C. SJÖSTRÖM
Materials and Structures, National Swedish Institute for Building Research, Gävle, Sweden

Abstract
In general degradation is an environment governed process taking place at the surface of a material. By means of X-ray Photoelectron Spectroscopy (XPS) one can reveal elemental and chemical conditions in the few outermost atomic layers of a material and, accordingly, XPS stands out as an excellent tool in the investigations of degradation processes.

In this study we have monitored the degradation of some of the most frequently used binders for coil coated sheet metals. The panels analysed were unexposed and natural aged in periods ranging from 3 to 8 years at six Nordic exposure stations representing different environments. In addition, laboratory SO_2 and weatherometer aged panels were studied aiming to evaluate the significance of these types of accelerated testing.

The discussion is mainly focussed upon the results obtained in the studies of the PVF_2 coating while the PVC-plastisol and acrylic-latex results are treated more briefly. The degradation levels of the coatings were shown to depend on both the exposure time and the exposure site to a varying extent for the three types of coatings. By comparing the results from the studies of the natural and accelerated aged PVF_2 and PVC panels respectively, one can conclude that the used procedure used for testing in SO_2 polluted atmospheres introduces degradation mechanisms not present in natural ageing whereas this discrepancy was not found for the weatherometer testing.
Keywords: XPS, Coatings degradation, PVF_2, PVC-plastisol, Acrylic latex, Natural ageing, Accelerated ageing.

1 Introduction

X-ray Photoelectron Spectroscopy (XPS) has developed to become a very useful method when studying solid materials, surfaces and thin films. XPS gives information about which elements are present in a sample and, furthermore, as the actual photoelectron energies depend upon the chemical states of the atoms involved, chemical information is also obtained. The measured signals arise from a very thin surface layer, about 50 Å in thickness, and therefore the method is especially well suited for studying the properties of solid surfaces.

In this report results are presented from XPS studies of coatings on coil coated sheet metal. The degradation of three of the most important coating binders, PVF_2, PVC-plastisol and acrylic latex respectively, have been monitored.

The XPS-studies contribute to a deep knowledge of the chemical composition of the coatings and on the current deterioration processes. The usefulness of XPS (also called ESCA) for this type of studies was earlier reported and discussed by Sjöström (1987). However, as our research is at an early stage, degradation models still have to be established. Therefore most of the results presented are based on elemental analysis and the detailed chemical information is only used to a minor extent.

2 The coatings and the coating process

In the coil coating process the paint is roll applied on to the sheet metal, normally in a 25 um thick layer for PVF_2 and acrylic latex whereas PVC-plastisol is applied with a thickness of 150-200 um. The metal band then passes a drying compartment (an oven) and after that the sheet metal is roll shaped and finally the band is cut into the desired lengths.

The studied PVF_2-coating is a dispersion of polyvinylidine fluoride in a solvent. As bearer of the pigment is used another type of binder, normally an acrylic, which is also the case for the coating type investigated. The amount of polyvinylidine fluoride is about 80 % and the remaining is acrylic.

The PVC coating is also a dispersion, consisting mainly of polyvinyl chloride , softerners and stabilizers in a minor part of solvent. The mixing weight ratio of PVC: softener is in the range 0.4 - 1.2.

These two coatings are of the thermo-plastic types meaning that no chemical reaction is taking place after the application. The binders, used as film formers, consist already from the beginning of high molecular polymers. During the evaporation of the solvent the polymer chains build up a coating film through a sintering process.

The acrylic latex coating consists from the beginning mainly of a dispersion of acrylic binders in water and some organic solvent. When the water and the solvents evaporate a cross binding process takes place, thus building up larger molecules.

3 Exposure and ageing of the coatings

Test panels from coated galvanized steel and aluminium sheet metal were subjected to exposure at field exposure stations in the Nordic countries and to ageing tests in laboratory. Fig.1 shows the location of the exposure stations and table 1 gives a qualitative description of the exposure site climate.

The laboratory ageing tests evaluated by use of XPS were accelerated weathering in UV-radiation according to ASTM G26 - 70, and a non-standardised test in an SO_2-contaminated atmosphere.

For the accelerated weathering in UV-radiation an ATLAS Xenon

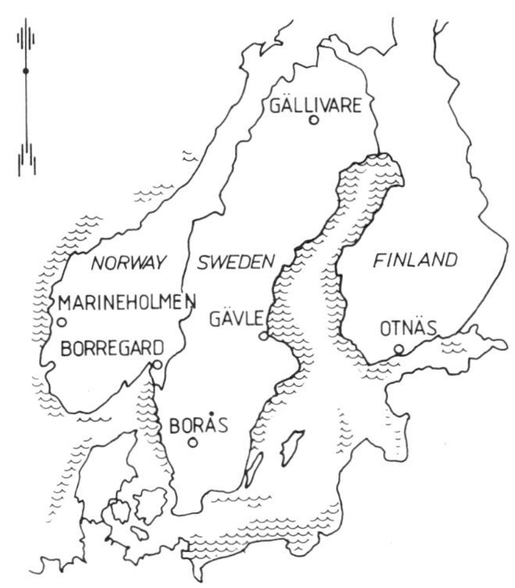

Fig. 1. Location of exposure stations

Table 1. Qualitative description of exposure site climate

Exposure site		Climate
Borregaard,	Norway	Industrial
Marineholmen,	Norway	Marine
Gävle,	Sweden	City, coastal
Otnäs,	Finland	City, coastal
Borås,	Sweden	City, inland
Gällivare,	Sweden	Urban, arctic

weatherometer of the type 600 DMC - WRC was used. The weatherometer was equipped with a water cooled xenon-arc lamp. The exposure was performed for 4000 hours with the cycles of 102 minutes radiation followed by 18 minutes of combined radiation and water spray.

The non-standardised testing in SO_2-contaminated atmosphere was performed according to an experimental climate test design simulating industrial atmosphere. The climate test programme, as shown in fig.2, is developed by the Norwegian Institute for Air Research. The complete coating test was designed as follows:

Some of the test panels were pre-aged in an ATLAS Weatherometer (ASTM G26 - 70) for 1000 hours and then exposed to the SO_2-atmosphere test programme. The whole SO_2 test cycle takes 84 hours and was repeated 10 times with the SO_2 concentration 1 ppm, where after the test panels were visually inspected. The test was then continued for 12 cycles with an SO_2 concentration of 3.4 ppm.

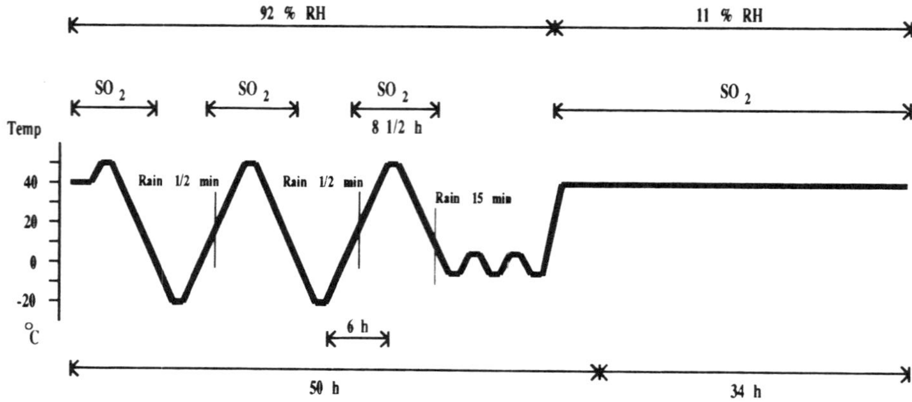

Fig.2. Climate testing in SO_2-contaminated atmosphere

4 XPS experimental

Samples were cut from the unexposed and exposed coil coated steel and aluminium panels. The sample size was about one square-centimeter.

The samples were carefully cleaned by submerging them in an isopropanol filled beaker immersed in an ultra-sonic bath. After the cleaning most of the samples were out-gassed for at least 5 hours in the chamber of a turbopump.

The spectrometer used is a modified HP5950B, which utilizes Al K-alpha radiation. The main chamber pressure was in all cases less than 10^{-7} torr. An overview spectrum running from 0 to 1000 eV was recorded for each sample (see fig.3).

Each element present substantially was recorded separately in detail using an energy range of 20 - 30 eV with the most intense electron peak centered. Elements recorded in detail were:

C, O, F	for PVF_2
Cl, C, O, Sn	for PVC
C, O	for acrylic latex

The surface sensitivity depends on the spectrometer acceptance angle of the emitted electrons, where the lowest surface sensitivity is obtained for electrons emitted perpendicular to the plane of the sample. Normally the angle used in this study was 52.0^o. For some of the PVF_2 samples, series of elemental spectra were recorded where the acceptance angle was varied by rotating the sample into 8 different positions ranging from 90.0^o down to 17.5^o. From the analysis of the relative intensity variations in such a series, chemical depth profiles are obtained, Sjöström et.al. (to be published).

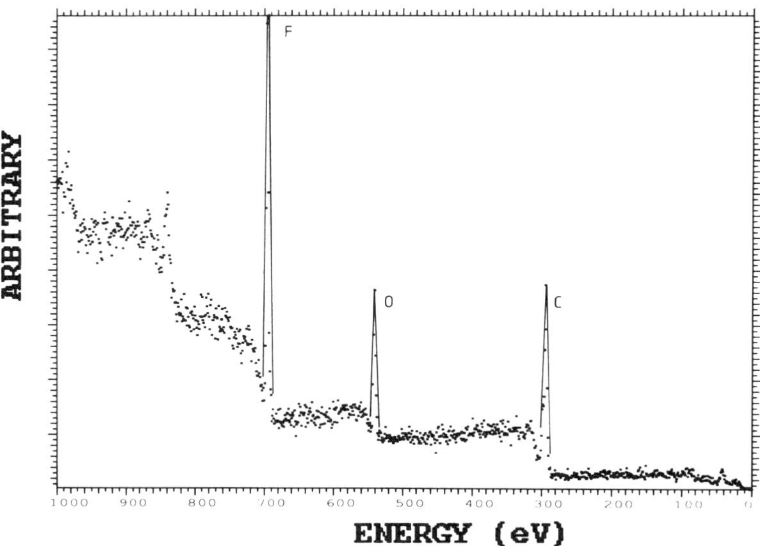

Fig.3. Overview spectrum of an unexposed PVF_2-coating

5 Analysis methodology

5.1 Conditions and presumptions

For the analysis a software package named "Crunch", Vasquez et al. (1981), was used. The software was slightly modified in order to realise a mixed Gaussian/Lorentzian lineshape, Ansell et al. (1979), in the non-linear least squares curve fitting procedure. Prior to the fittings the spectral backgrounds were removed using a method proposed by Shirley (1972) followed by a simple linear background subtraction.

Since most of the constituents of the polymers under study are well documented by other researchers such as Clark (1973), Pijpers and Donners (1985) and Clark and Thomas (1978), we in principle adopted their values of the binding energies as a standard and for calibration purposes. To allow for unknown constituents and degradation products typical binding energies of expected functional groups were used, Peeling and Clark (1981). In most cases the full width at half maximum (FWHM) of the carbon peaks were estimated to be about 1.5 eV.

Elemental and chemical information can be revealed by comparing the peak areas at the different energy positions (chemical shifts).

5.2 Spectral interpretation

The XPS sensitivity varies from element to element due to both physical and technical factors. To make a quantitative elemental analysis possible it is therefore necessary to find the relative sensitivity factors for the elements involved. In this work, for PVF_2 we used an empirical approach to determine the factors dependent on the sample material. These factors vary smoothly from element to element and this variation is similar for all polymers. Thus we could without making significant errors, but somewhat improper, from the empirical model approximately calculate the factors of the PVC and acrylic latex. The spectrometer dependent factors were adopted from Elliot et al. (1983).

So, the obtained normalised elemental intensities can be used to monitor the degradation by comparing samples of different exposure periods and from different sites. Most interesting is to study the oxygen-carbon relations to obtain the oxidation rates. For PVF_2 and acrylic latex the oxygen to carbon intensity ratios were taken. However, for PVF_2 only the acrylic carbon intensity was considered as it has been shown by Peeling and Clark (1981) that the oxidation can be expected to be much more extensive in the acrylic constituent in the PVF_2 coating and because the occurrence at the surface of the two constituents varied strongly between different samples. From the ratios also one third was subtracted as this is roughly the inherent ratio for a typical acrylic mixing used in coatings. For PVC instead the relative oxygen occurrence was calculated since the relative tin occurrence was seen to be appreciable and to vary strongly between the samples, and since tin can also be subject to oxidation. The tin occurrence is attributed to the thermostabilizers used which seem to be enriched at some parts of the surface.

6 Results and discussions

6.1 PVF_2-coating

6.1.1 The unexposed case

Fig.4 show the carbon, oxygen and fluorine peaks of an unexposed PVF_2-coated galvanized steel sheet test sample. The sample, cut from the same sheet metal as those samples subjected to the different ageing tests and exposed at the exposure stations, was stored in darkness and at room temperature. As a control of this unexposed reference sample another reference was produced by painting a galvanized sheet metal with the same type of PVF_2-coating, baking the painted sample at 220 C in an oven and then immediately analysing the coating in the spectrometer. This latter unexposed reference is named "fresh" in the diagrams shown in fig.5-6.

The analysis has shown relatively great differencies in the PVF_2 to acrylate mixing ratio between different spots on a coated sheet metal. The inhomogeneous mixing of the coating film results in a measured variation of the fluorine to acrylate-carbon intensity ratio of 0.31 - 0.70. Theoretically this ratio should be about 2.50 for a coating consisting of 80 % PVF_2 and 20 % acrylic. The reason for the low intensity ratios observed is that the coating surface is enriched with acrylic, resulting in a reduced fluorine signal. This is revealed by the analysis of the depth profile obtained from varying acceptance angle measurements, Sjöström et.al. (to be published).

(a)

(b)

(c)

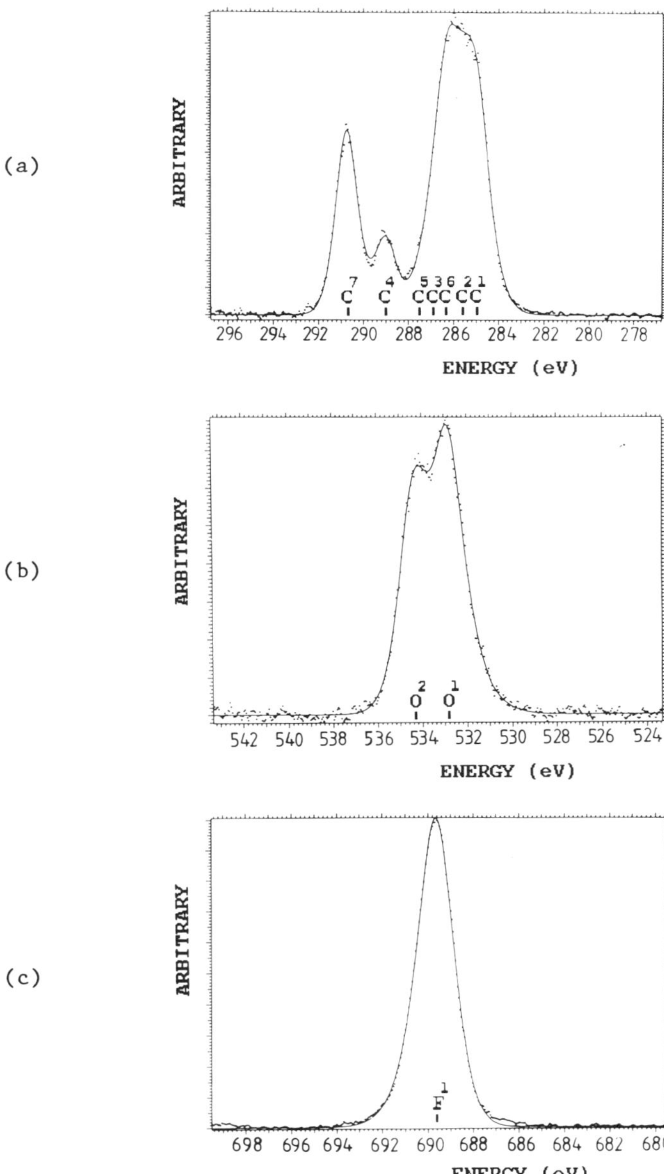

Fig.4. Recorded and fitted spectra from an unexposed PVF_2 panel. Energy positions of peaks attributed to different chemical states are indicated. The elements recorded are carbon (a), oxygen (b) and fluorine (c).

6.1.2 Oxidation of the coating

Fig.5 shows the ratio excess oxygen to acrylic carbon intensity for different exposure modes. For a close description of the parameter, see paragraph 5.2 Spectral interpretation.

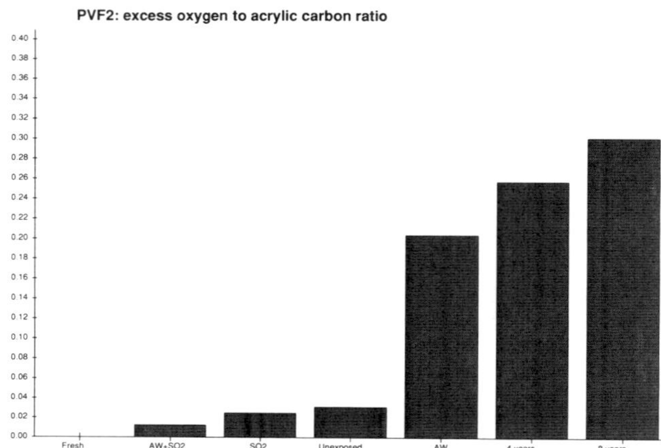

Fig.5. PVF_2 excess oxygen to acrylic carbon ratio for fresh, unexposed and exposed test specimens of different modes.

An oxidation of the coating occurs both when exposing samples in the weatherometer and at the exposure stations. Ratios of the latter modes are mean values of the different stations. The testing at the exposure sites reveals that this oxidation happens during the first years of exposure. Exposure for 4000 hours in the weatherometer results in approximately the same level of oxidation as 4 years at the Nordic exposure sites.

The samples tested in the SO_2 contaminated atmosphere show practically no oxidation; the ratio excess oxygen to acrylic carbon is approximately the same for both an unexposed and SO_2 exposed coating.

In fig. 6 is shown the ratio excess oxygen to acrylic carbon intensity for the different exposure sites. As the major part of the oxidation obviously occurs within the first 4 years the data from samples exposed both 4 and 8 years are averaged. It is also important to note that the ratio excess oxygen to acrylic carbon does not depend on the variation in the PVF_2 to acrylate mixing ratio of a specific sample.

6.2 PVC-plastisol coating

Qualitatively the results for PVC are similar as for PVF_2 which can be seen by comparing fig.5-6 to fig.7-8. Again the SO_2 exposure test method results in a negligible oxidation whereas the weather-ometer testing results in an oxidation level comparable to 4 years of field exposure. A difference is that here gives the combined weather-ometer and SO_2 testing method also a significant oxidation level, while for

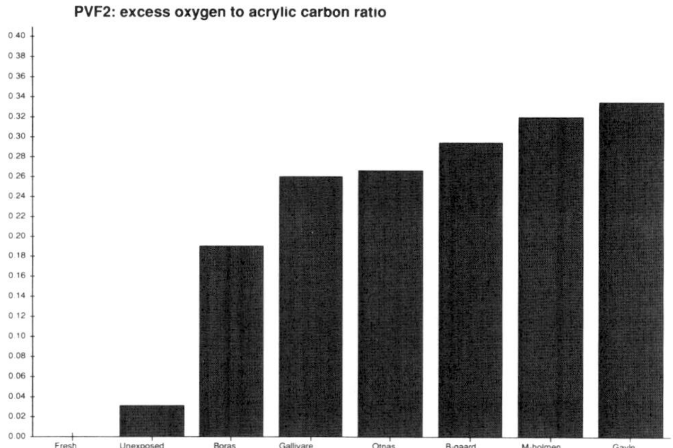

Fig.6. PVF_2 excess oxygen to acrylic carbon ratio for fresh, unexposed and exposed test specimens from each site.

PVF_2 it did not. The order of the different exposure sites is the same as for PVF_2 with Gävle as the toughest site. The results of Marineholmen and Otnäs are lacking in fig.8 since at those sites no 8 year exposures were made. Therefore a proper average comparison can not be made to the other sites since for PVC there is a significant difference between 4 and 8 years of exposure. For convinience the 3 years exposure at Borås is treated as a 4 years exposure.

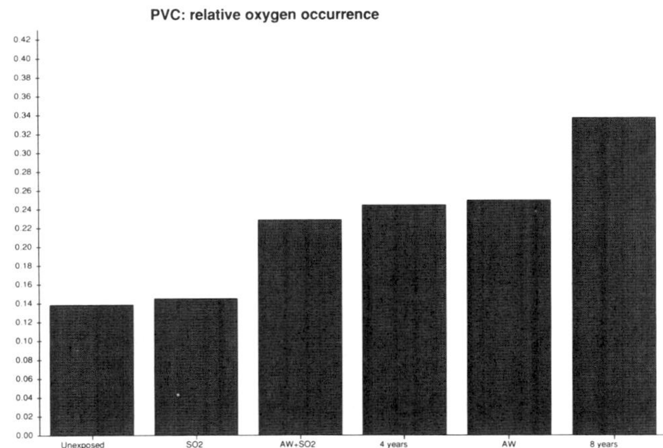

Fig.7. PVC relative oxygen occurrence for unexposed and exposed test specimens of different modes.

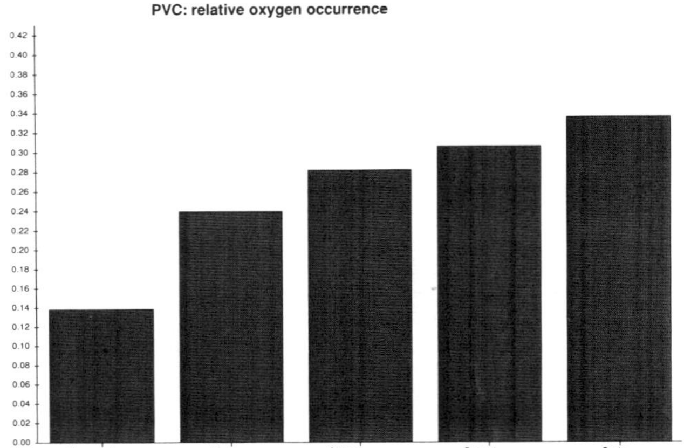

Fig.8. PVC relative oxygen occurrence for unexposed and exposed test specimens from four sites.

6.3 Acrylic latex

Since for this coating type the analysis has not been completed and also because of a hidden failure of the XPS spectromter we have to resort to comparisons of unexposed and 8 years exposed samples. The results are given in fig.9. Again Gävle shows to be the toughest site, while there is a change in order of the other three sites. Some results of the 4 years exposures, not accounted for here, indicate significantly less oxidation compared to the 8 years exposed samples.

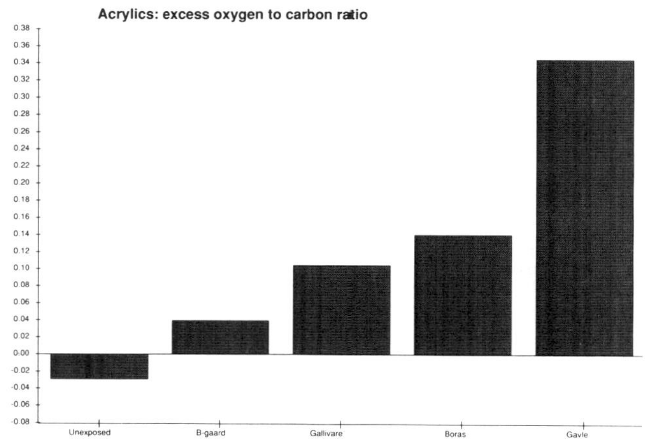

Fig.9. Acrylic latex excess oxygen to carbon ratio for unexposed and exposed test specimens from 8 years exposure at four sites.

7 Conclusions

The studied coatings are durable products that undergo minor chemical changes during exposure to normal environments. The chemical alterations of the coating are evidenced by an oxidation, which can be observed through XPS-analysis.

In Nordic environments, for PVF_2, the major part of the degradation occurs within the first 2 to 4 years, after which the deterioration processes slows down. For PVC-plastisol and acrylic latex the oxidation seems to proceed at a constant rate over a period of 8 years.

All three types of coatings were observed to have the greatest oxidation rate at the exposure station in Gävle. The reason for this is not clear till now but it may be explained by a fairly high insolation together with the presence of air pollutants as SO_2, NO_x and poly aromatic hydrocarbons.

A significant result of the analysis is that the ageing of the PVF_2 and PVC coating in an artificial SO_2-contaminated atmosphere results in almost no oxidations in contradiction to the field exposures. On the other hand, the ATLAS weatherometer , with the used test procedure, seems to simulate the chemical degradation in an appropriate way. Exposure for 4000 hours in the weatherometer results in approximately the same level of oxidation as 4 years at the Nordic exposure sites. Analysis of the defluorination of PVF_2, Sjöström et.al. (to be published), gives a result in line to that of the oxidation analysis and thus support a conclusion that the used weather-ometer exposure is a relevant accelerated test whereas the SO_2 exposure is not. This conclusion is also based on the similarity between recorded carbon spectra from the weatherometer and the naturally aged samples.

The depth profile analysis, Sjöström et.al. (to be published), reveals that the outer most surface of the PVF_2-coating is enriched with acrylate. The surface composition of the unexposed coating is in the range of a likely acrylic mixture, while the interior composition shows an approximate PVF_2 to acrylate mixing ratio of 80/20. This depends on a segregation process taking place after the coating is roll applied in the coil coating process.

8 References

Ansell, R.O. et al. (1979), **J. Electroanal Chem.**, 98, 79.

Clark, D.T., Feast, W.J., Kilcast, D. and Musgrave, W.K.R. (1973), **Pol. Sci.**, 11, 389-411.

Clark, D.T. and Thomas, H.R. (1978), **J. Pol. Sci.**, 16, 791-820.

Elliot, I., Doyle, C. and Andrade, J.D. (1983), **J. Elec. Spec. Rel. Phen.**, 28, 303-316.

Peeling, J. and Clark, D.T. (1981), **Pol. Deg. Stab.**, 3, 177-185.

Pijpers, A.P. and Donners, A.B. (1985), **J. Pol. Sci.**, 23, 453-462.

Shirley, D.A. (1972), **Phys. Rev.**, B5, 4709.

Sjöström, Ch., Jernberg, P and Lala, D (submitted for publication in J. Mat. Constr.).

Sjöström, Ch. (1987) On the use of ESCA for durability evaluation of coil coated sheet metal. **Reprint series of National Swedish Institute for Building Research**, 34.
Vasquez, R.P. et al. (1981), **Relat. Phenom.**, 23, 63-81.

55 MEASUREMENT AND EVALUATION OF ABRASION RESISTANCE OF CONCRETE SURFACES

R. KUNTERDING, H.K. HILSDORF
Institut für Massivbau und Baustofftechnologie, University of Karlsruhe, Karlsruhe, Germany

Abstract
In an experimental and theoretical investigation an attempt is made to develop a better understanding into the basic mechanisms resulting in the abrasion of concrete surfaces. Furthermore, experimental procedures are being developed which may result in better correlation between test results and actual performance. An analysis of the stress state around a sphere acting on a concrete surface either through impact at various angles or through friction shows that the critical stress state for concrete is tension-compression and occurs directly on the concrete surface. In the experimental investigation 4 test procedures are employed to simulate different local fracture mechanisms. The experiments give some insight into differences in the behavior of paste, aggregate and interface.
Keywords: Abrasion, Abrasion resistance of concrete, Abrasion test methods, Stress analysis, Fracture mechanisms.

1 Introduction

The surfaces of some concrete structures such as pavements, industrial floors or some interior parts of silos may be exposed to mechanical, mechanical-chemical or mechanical-thermal attack. As a consequence the surface regions of such concrete structures may be damaged particularly by abrasion. In extreme cases abrasion may even lead to the complete exposure of the reinforcement. Thus, not only the serviceability but also the structural integrity of such elements is endangered.

In this paper various test methods to determine the abrasion resistance of concrete are evaluated and compared to the actual conditions prevailing in concrete surfaces exposed to abrasion. Experimental results are presented which allow an evaluation of the effectiveness of technological methods to improve the abrasion resistance of concrete. Based on these results failure mechanisms are described which cause the deterioration of concrete surfaces exposed to abrasion.

Test method 1	Test method 2	Test method 3	Test method 4
Abrasive wear tester	Microhardness	Sandblasting	Grinding wheel acc. to Böhme
—	—	ASTM-C-418-81	DIN 52 108

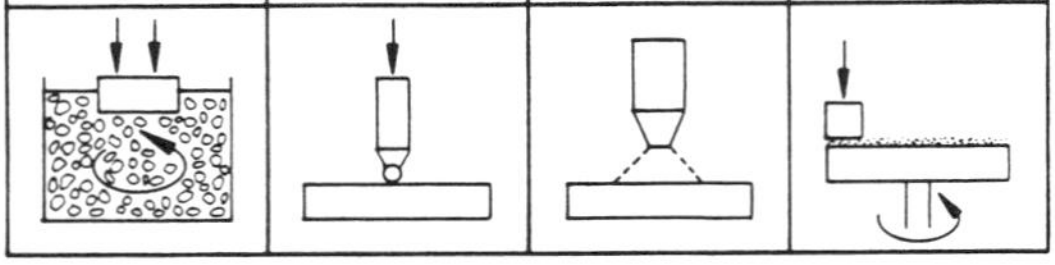

Fig. 1. Test methods

2 Test methods

To simulate the various conditions to which a concrete surface may be exposed realistic test methods are required. In Fig. 1 four different methods are presented which result in different states of exposure of a concrete surface. Test method 1, exposure to bulk material, may be useful to estimate the abrasion resistance of concrete surfaces to hard particles such as in the interior of concrete silos where bulk material e.g. gravel, crushed aggregates or cement clinker slide along the concrete surface. Test method 2, microhardness testing, simulates the penetration of a single, hard particle into a concrete surface. Exposure of concrete surfaces to the impact of particles under various angles can be simulated by means of sandblasting e.g. according to ASTM-C-418-81 (test method 3). A grinding exposure of a concrete surface can be simulated by the grinding wheel according to Böhme (test method 4).

For the test methods 1, 3 and 4 the amount of abraded material is determined by weighing, for the microhardness testing the depth of penetration due to a constant load is measured.

3 Stress analysis

The test methods described in section 2 generate different states of stress in the exposed concrete element and, therefore, result in different types and mechanisms of failure. Therefore, some of the states of stress which may occur in a concrete surface exposed to abrasion were analysed.

Fig. 2 summarizes the various exposure conditions of a concrete surface and the stress components resulting from such an exposure. An abrading particle exerts a vertical force F_V and a horizontal force F_H on a surface. In addition, the particles may move horizontally at a rate v_H or they may rotate at a rate ω . The various loading conditions can be separated into four basic types expressed by F_V; F_H; v_H and ω : impact, rolling, sliding or combinations thereof.

For an estimate of the magnitude and distribution of stresses in the surface region of a material exposed to a spherical particle closed solutions are available e.g. [2, 3, 4, 8]. For our work the solution developed by Hamilton and Goodman [3] has been used and tested with a finite element programme ADINA. Since the closed solution

Loading conditions	F_V, F_H, V_H			
Basic types	Impact	Rolling	Sliding	Combinations
	F, α	F_V, ω	F_V, $F_H = \mu \cdot F_V$	F_V, ω, F_H
Stress components	$\sigma = f(\alpha)$ $\int_A \tau \, dA = \mu \cdot F(\alpha)$	σ	σ $\int_A \tau \, dA = \mu \cdot F_V$	σ $\int_A \tau \, dA = \mu \cdot F_V$

Fig. 2. Types of exposure of a concrete surface

is valid only for a linear-elastic material a non-linear FE-analysis has been carried out to study progressive fracture in a concrete surface incorporating the concrete model proposed by Bathe and Ramaswamy [1] which can be implemented into the ADINA programme.

3.1 Stresses due to normal forces

If a spherical particle subjected to a vertical force F_V exerts pressure on a surface, stresses σ_x; σ_y and σ_z as shown in Fig. 3 are generated. In Fig. 3 the solution acc. to [3] (solid lines) is compared to the results of the FE-analysis. The interface between particle and surface is circular. Therefore, the distribution of normal stresses σ_z follows a half-sphere. The stresses σ_x, σ_y and σ_z in the surface are rotationally symmetric with respect to the z-axis. In the center of the contact surface (x/a = 0, y/a = 0, z/a = 0) a triaxial state of compressive stresses prevails. At the interface between loaded and unloaded surface (x/a = +/- 1,0, y/a = 0, z/a = 0) a critical biaxial state of compression-tension can be observed. At these interfaces $|\sigma_x| = |\sigma_y|$, and both components are approximately 1/5 of the maximum compressive stress acting in the center. In the unloaded surface the stresses σ_x and σ_y rapidly decrease.

The results of the finite element analysis represented by individual data points in Fig. 3 are in close agreement with the solution acc. to Hamilton and Goodman [3].

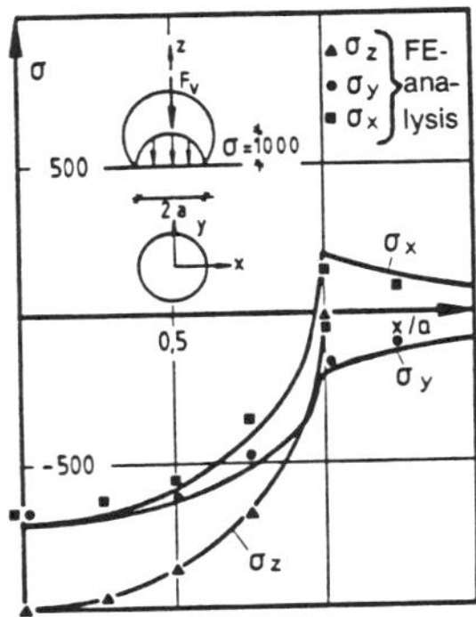

Fig. 3. Distribution of stresses in a concrete surface acc. to Hamilton/Goodman [3] and results of a FE-analysis

3.2 Stresses due to normal and tangential forces

If the spherical particle is subjected to a normal force F_V as well as to a tangential force F_H normal stresses σ_z are generated due to F_V as shown in Fig. 3. The force F_H is introduced into the surface following Coulomb's law of friction. It results in shear stresses which are distributed according to a half-ellipsoid. The distribution of the normal stresses σ_x in the surface can be taken from Fig. 4. It depends on the coefficient of friction between the surface and the particle. At x/a = -1,0 the tensile stresses caused by the normal force, $\mu = 0$, are increased. At x/a = +1,0, however, the tensile stresses due to the normal force, F_V, $\mu = 0$, are reduced or they change in sign. The σ_y-stresses are little affected by the tangential force F_H.

3.3 Non-linear stress analysis

Based upon the non-linear finite element analysis using the concrete model of Bathe and Ramaswamy [1] the critical regions in a surface exposed to normal forces as shown in Fig. 5 have been determined. Initially tensile cracks are developed close to the surface in the transition region between loaded and unloaded surface (x/a = 1,0). If the load is increased further the region in which tensile cracks occur increases. Subsequently, in the center below the surface (x/a = 0, y/a = 0, z/a = 1,25) an additional region develops where tensile fracture occurs. This is the result of tensile splitting stresses. In the center of the contact surface above the second region of tensile failure a state of triaxial compression prevails (x/a = 0, y/a = 0, z/a = 0,75 < 1,25).

No non-linear analysis has been carried out for the case of combined normal and tangential forces since it would require the solution of a three-dimensional problem. However, it is to be expected that tensile fracture due to biaxial compression-tension at the line x/a = 1 will occur at lower values of F_V whereas the regions of tensile splitting will be little affected by F_H.

The actual stress state generated by an abrasive bulk material will be influenced by a number of additional parameter not taken into account in this analysis e.g. the interaction between several partic-

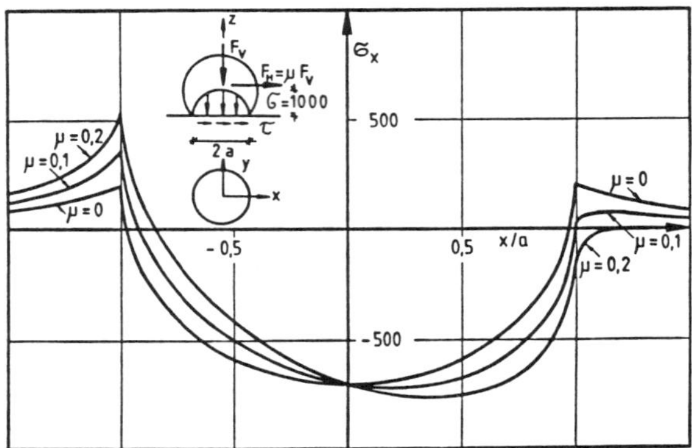

Fig. 4. Distribution of stresses in a concrete surface to normal and tangential forces

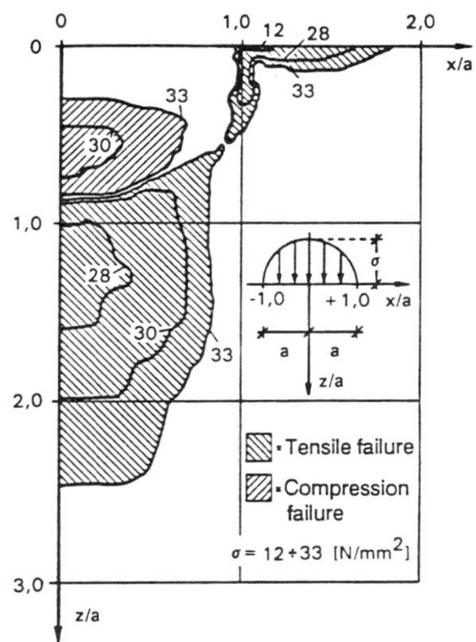

Fig. 5. Theoretical crack configuration

les, the effect of impact or of aggregates. Nevertheless, the analysis clearly shows that the resistance of a concrete surface to a biaxial stress state of compression-tension is a decisive parameter for the abrasion resistance of concrete.

4 Experiments

4.1 Experimental program

In an extensive experimental program various types of concrete were exposed to abrasion according to test methods 1 to 4. Water/cement-ratio, type of cement and duration of curing of the concrete were some of the major parameters. In addition type and size distribution of aggregates as well as additions, in particular silica-fume and steel fibres, have also been investigated.

4.2 Test procedures

For test method 1, 60 dm^3 of bulk material are filled into a rotating container. Within this container the bulk material is rotated with respect to the container by means of a helix. A stationary concrete specimen, Ø 150 mm, depth 50 mm, is placed on top of the container and subjected to a constant compressive force of 100 N. After 30 min. the bulk material is replaced by new material. The specimens are subjected to abrasion for 3 x 30 min. Test method 2 - microhardness testing - has been used to test the mortar matrix of the concrete. A Vickers - pyramide with a constant force of 60 N has been chosen in these experiments. Sandblasting experiments (test method 3) were carried out according to ASTM-C-418-81. For this, granulated steel, 0,30 - 0,70 mm, has been used. The angle between specimen and sandblasting beam was varied between 15 and 90°. The experiments with the grinding wheel (test method 4) have been conducted according to DIN 52108. Further details with regard to the test set-ups as well as the execution of the experiments can be found in [5, 6, 7].

4.3 Experimental results

In Figs. 6 through 14 some of the experimental results are summarized. In all of these figures the ordinate gives the relative abrasion value w* (%) as a function of the particular parameter investigated. The abrasion value of the reference concrete corresponds to 100 %. Fig. 6 shows the effect of the duration of curing and type of cement on abrasion resistance. The influence of type of cement can also be seen from Fig. 8. Fig. 7 demonstrates the effect of water/cement-ratio. The results obtained for concretes made of an ordinary Type 1 Portland cement with a water/cement-ratio of 0.5, size distribution of aggregates acc. to A16, DIN 1045 and cured for 7 days are summarized. Along the abscissa the various parameters studied which may improve concrete abrasion resistance are given, whereas the ordinates show the relative abrasion value w* (%).

4.4 Discussion of experimental results

Fig. 6 clearly demonstrates the expected tendency that the relative abrasion value decreases as the duration of curing is increased. This tendency can be observed irrespective of the particular test method employed. However, which concrete is particularly sensitive to curing depends both on the type of cement as well as on the particular test method. Test methods which result in a severe exposure of the matrix such as test method 1, 2 and 3 show the effect of curing more clearly

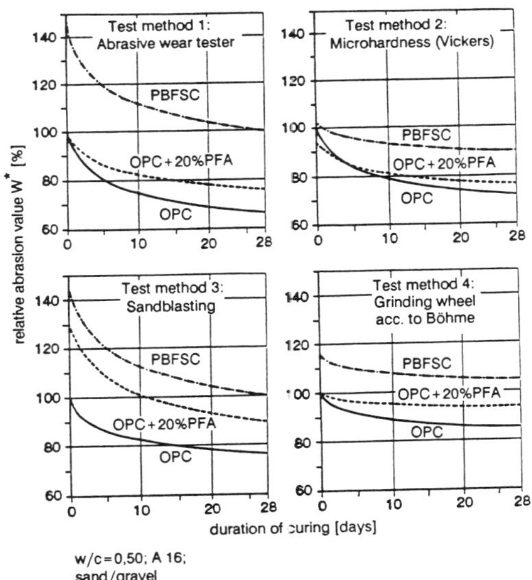

Fig. 6. Effect of duration of curing on abrasion resistance for different types of cement based on different test methods

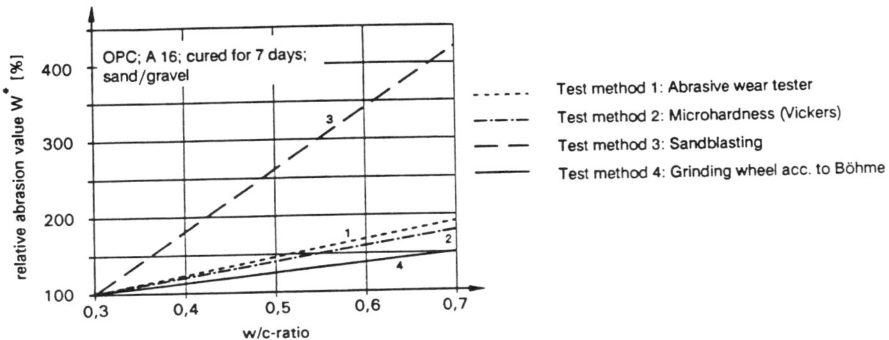

Fig. 7. Effect of water/cement-ratio on abrasion resistance based on different test methods

than test method 4. Among the three types of cement blast furnace slag cements are more affected by duration of curing than the concretes made of other types of cement. Accord. to Fig. 7 the increase of the relative abrasion value w* with increasing water/cement-ratio becomes most pronounced if sandblasting is used as a test method: In this case an increase of the water/cement-ratio from o.3 to 0.7 results in an increase of a relative abrasion value by a factor of 4, whereas only an increase by a factor of 1.5 is observed if the grinding wheel is used as a test method. A similar trend can be found when evaluating the influence of type of cement on the relative abrasion value w*. Again, sandblasting is the most sensitive, whereas the grinding wheel is the least sensitive test method.

The effect of various additions and modifications of aggregates on the relative abrasion value follows from Figs. 9 through 12. In these diagrams the results obtained for a particular test method are given individually. It becomes evident from these figures that the ranking order of various measures to improve abrasion resistance depends on

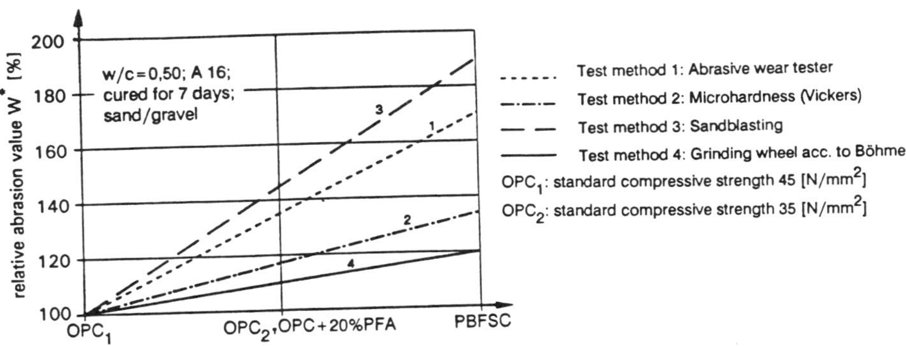

Fig. 8. Effect of type of cement on abrasion resistance based on different test methods

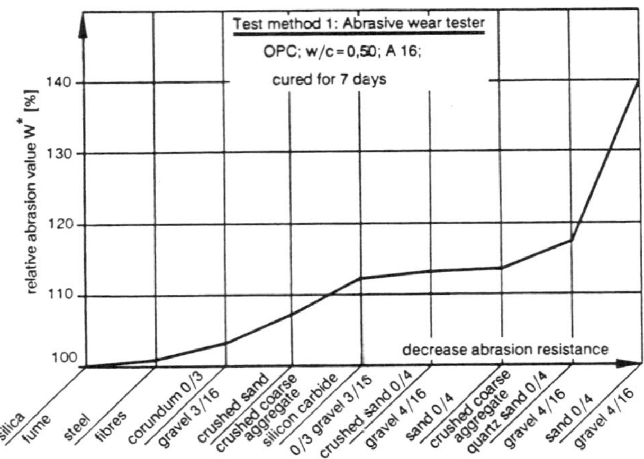

Fig. 9. Optimization of abrasion resistance - Test method 1: Exposure to bulk material

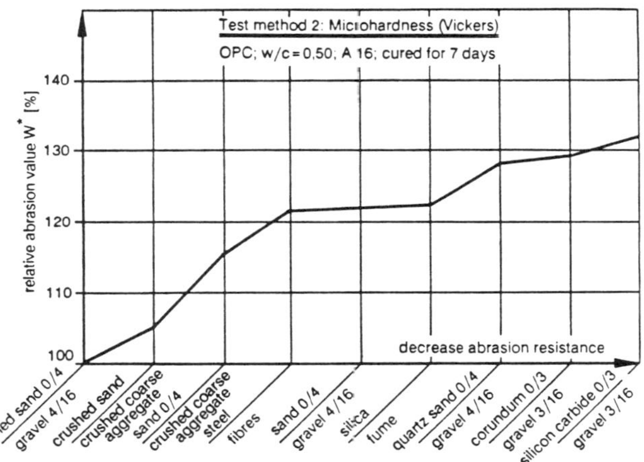

Fig. 10. Optimization of abrasion resistance - Test method 2: Microhardness testing

the type of test method employed. For the case of exposure to bulk material (test method 1) the concretes containing silica-fume or steel fibres had the highest abrasion resistance. For the case of microhardness testing (test method 2) and for sandblasting (test method 3) the highest resistance to penetration was observed for con cretes containing crushed sand. The resisting to grinding exposure (test method 4) appeared to be the highest for concretes containing hard particles such as siliconcarbide or corundum.

In Fig. 13 the effect of the orientation of the sandblasting beam with respect to the exposed surface on the relative abrasion value is

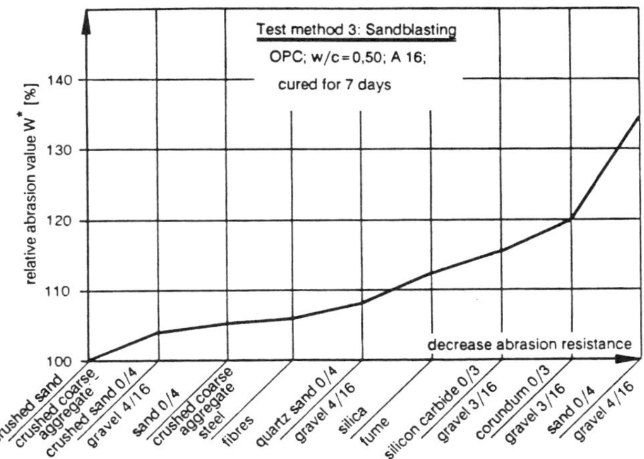

Fig. 11. Optimization of abrasion resistance - Test method 3: Sandblasting

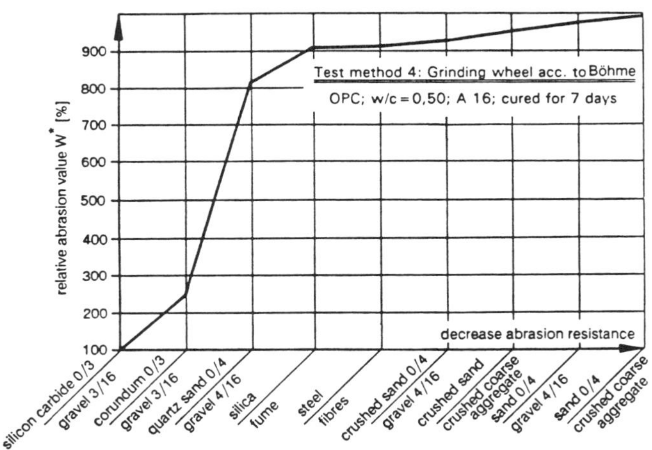

Fig. 12. Optimization of abrasion resistance - Test method 4: grinding wheel

given for concrete together with the results obtained for structural steel (St37) and for basalt. It can be seen that the relative abrasion value does indeed depend on the angle α . For concrete the most critical value is approx. 75°, whereas it is about 30° for structural steel and 90° for the basalt. The behaviour of concrete is in close agreement with the results of the stress analysis given in section 3. For a coefficient of friction of $\mu = 0.20$ the most critical state of biaxial tension-compression occurs at an angle $\alpha = 75°$.

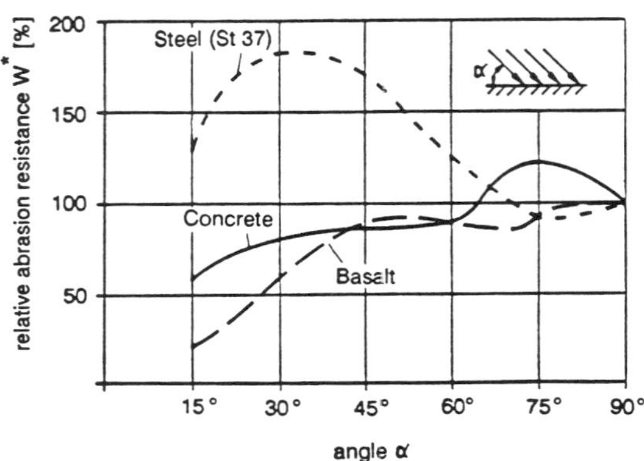

Fig. 13. Effect of sandblasting angle on abrasion resistance of various materials

As a summary it may be concluded that the highest abrasion resistance of concrete has to be expected if the resistance of the matrix, of the aggregates and of the aggregate-matrix interface is optimized. This can be achieved by the choice of very low water/cement-ratios such as w/c = 0.3, sufficient curing and the use of silica fume as an addition together with well-graded, hard aggregate particles. This leads to a new and promissing application of high strength concrete as proposed e.g. in [9, 10].

5 Failure mechanisms

Depending on the particular type of exposure various types of failure mechanisms are responsible for the abrasion of concrete surfaces as summarized in Fig. 14.

In the case of exposure to bulk material (test method 1) failure of the matrix, of the aggregates and of matrix aggregate bond are responsible for the failure of the surface. As a consequence of such an exposure even larger aggregate particles are gradually removed from the concrete due to debonding.

When employing test method 2 the stress states calculated and evaluated in the theoretical stress analysis are closely simulated. When an individual particle penetrates a surface a state of compression-tension occurs at the border between loaded and unloaded surface. Therefore, cracks along the circumference of the loaded area are to be expected. They could be found in experiments employing microhardness testing according to Brinell. Such exposure also leads to weakening of a region below the surface due to splitting tension as has been demonstrated in Fig. 3. Such cracks coalesce and result in the formation of cone-shaped fracture surfaces and seperation of individual cone-shaped particles (Fig. 14; 2. Microhardness).

A particularly severe exposure of the matrix occurs in the case of sandblasting (test method 3). For a sandblasting beam perpendicular to the surface the matrix fails and the aggregates are separated and

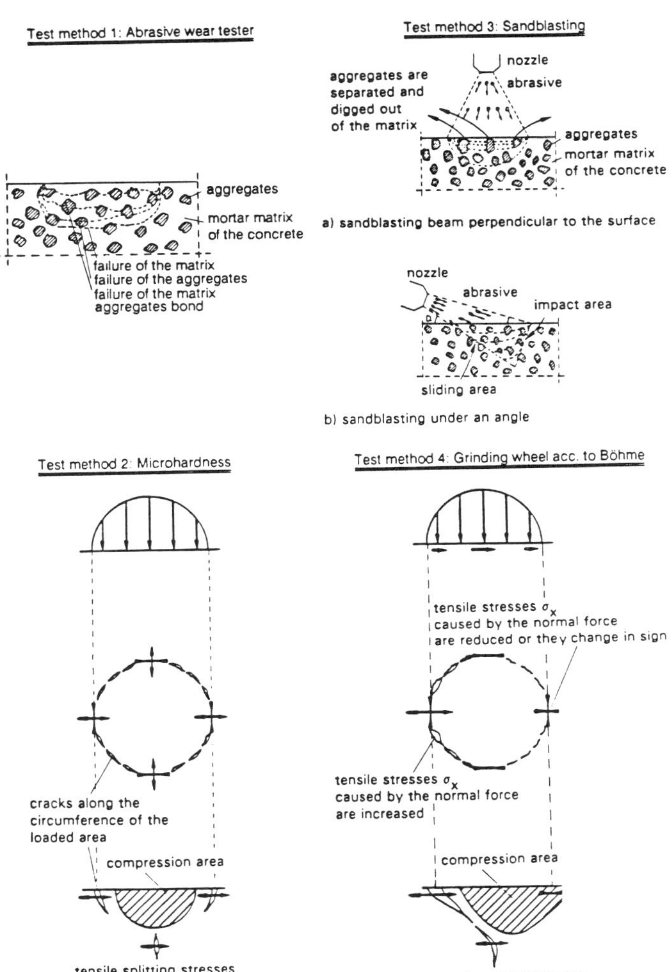

Fig. 14. Failure mechanisms of different exposure conditions

digged out of the matrix. The abrasion resistance is governed primarily by the strength of the matrix as well as by the bond properties between matrix and aggregates (Fig. 14; 3. Sandblasting). As a consequence the parameters influencing the matrix properties such as w/c and curing are particularly pronounced.

For the case of uniform grinding (test method 4) the stress state generated depends on the coefficient of friction between grinding material and concrete surface. In addition, also the aggregate particles are directly exposed. Therefore, also the abrasion resistance of the aggregate particles themselves plays a major role in the total behaviour of the concrete.

6 Summary

Depending on the particular type of exposure various stress states may be generated in concrete surfaces resulting in different failure mechanisms. When evaluating these exposure conditions one should distinguish between exposures leading primarily to severe stresses in the matrix, in the aggregates or in the interface. Different test methods place different emphasis on these various conditions.

In the experimental investigation on abrasion resistance of concrete it was found that an increase of the duration of curing, reduction of the water/cement-ratio, the use of high-strength cements or of additions which substantially increase matrix strength together with the choice of hard aggregates led to the highest abrasion resistance. The extent to which abrasion was reduced by such measures, however, depends on the type of test method. Therefore, the ranking of various types of concrete mixes in which various approaches to improve their abrasion resistance have been employed depends on the type of test method.

Therefore, when choosing a test method for a particular application particular attention has to be paid to the actual conditions to which the concrete in the structure is exposed.

7 References

[1] Bathe, K.-J. and Ramaswamy, S. (1979) On Three-Dimensional Nonlinear Analysis of Concrete Structures. **Nuclear Engineering and Design, Vol. 52, pp. 385-405**

[2] Föppl, L. (1936) Der Spannungszustand und die Anstrengung des Werkstoffes bei der Berührung zweier Körper.**Forschung auf dem Gebiet des Ing.-wesen, H. 5, pp. 209-221**

[3] Hamilton, G.M. and Goodman, L.E. (1966) The Stress Field Created by a Circular Sliding Contact. **J. of Applied Mech. pp 371-376**

[4] Hertz, H. (1882) Über die Berührung fester elastischer Körper und über die Härte. **Aus den Verhandlungen des Vereins zur Beförderung des Gewerbeflusses, Berlin**

[5] Hilsdorf, H.K. and Kunterding, R. (1988) Beanspruchung der Ober flächen von Stahlbetonsilos durch Schüttgüter. **Tagungsband: Silos-Forschung und Praxis, SFB 219, pp. 371-386**

[6] Kunterding, R. Beanspruchung der Oberflächen von Stahlbetonsilos durch Schüttgüter. **Teil I: Arbeitsund Ergebnisbericht 1985-1986 Teil II: Arbeitsund Ergebnisbericht 1987-1989 SFB 219, Universität Karlsruhe, Institut für Massivbau und Baustofftechnologie**

[7] Kunterding, R. Beanspruchung der Oberflächen von Stahlbetonsilos durch Schüttgüter. **Diss. Univ. Karlsruhe, vorauss. Juni 1990**

[8] Smith, J.O. and Liu, C.K. (1958) Stresses Due to Tangential and Normal Loads on an Elastic Solid with Application to some Contact Stress Problems. **J. of Applied Mech., pp. 157-165**

[9] State of the art report on high strength concrete (1984) ACI **Commitee 363 ACI 363 P-84 ACI-Journal pp. 362-411**

[10] High strengh concrete / State of the art (1989) **FIB-CEB Working Group on HSC**

56 MONITORING OF SURFACE MOISTURE BY MINIATURE MOISTURE SENSORS

P. NORBERG
Materials and Structures Division, The National Swedish Institute for Building Research, Gävle, Sweden

Abstract
A new electrolytic cell with resistance grids of Au has been developed and tried out according to the NILU WETCORR method in a climatic chamber. A commercial dew sensor, Murata HOS103, was also included in the evaluation programme. Several specimens of each sensor have been tested under varying temperature and relative humidity while mounted on a substrate of coil coated sheet metal. Although quite different in character, both sensors have shown acceptable properties in terms of individual variability, sensitivity, reproducibility and long-term stability. Regression analysis has provided an empirical relationship between on one side the current through the Au/Au-sensor, and on the other side the ambient relative humidity and the surface temperature. The equation works well for temperatures from 0 to 30 deg C and RH-values in the range of 50 to 100%. The dew sensor, HOS103, was found to react drastically to moisture loads exceeding 90 or 95% RH. Surface moisture measurements using the two types of sensors evaluated, can give valuable information about the microclimate on and near surfaces of buildings. Such knowledge is important when describing the interaction between the microclimate and the degradation of the materials.
Keywords: Surface Moisture, Moisture Sensors, Measuring Technique, Relative Humidity, Temperature, Time-of-Wetness, Reproducibility.

1 Introduction

Surface moisture plays a critical role in the deterioration of building materials, whether they are metallic or non-metallic. The deposition of moisture on surfaces of buildings not only depends on atmospheric conditions, such as precipitation, relative humidity, air temperature and prevailing wind, but also on factors such as building design, orientation, surface temperature and, not least, the material itself. The influences of all these factors make predictions of the microclimate on and near surfaces of buildings extremely uncertain from ambient meteorological conditions solely. Consequently, direct measurement of surface moisture using a miniature sensor would probably result in the most adequate data, enabling meaningful correlations with the material degradation.

In the past 30 years, electrochemical cells have been used for studies of atmospheric corrosion of metals and alloys. Many of these studies have involved measurement of the so-called time-of-wetness (TOW), a quantity that shows under which atmospheric conditions electrochemical reactions of some magnitude can occur on the surface of the sensor. A review of the different approaches and their results have been made by Mansfeld (1981).

The electrochemical cells or sensors have often consisted of galvanic couples, e g Cu/steel, Cu/Au and Zn/Au, which give rise to an electric potential difference in the presence of an electrolyte, such as surface moisture. The magnitude of either the potential difference, Sereda et al (1982), or the galvanic current, Kucera and Mattsson (1974), will to some extent reflect the corrosion rate of the anodic metal and also permit TOW to be estimated. TOW is in this context defined as the time during which the potential difference or the galvanic current exceeds a certain threshold value.

Another approach by Kucera and Mattsson (1974) involved electrolytic cells with only one metal, e g Cu or Zn, to which an external constant voltage was applied. The resulting current in this type of cell again is a vague measure of the corrosion rate but TOW can be defined similarly as above by selecting an appropriate current threshold. The development of this method has continued in Scandinavia during the past 15 years, to a large extent within the frames of joint Nordic research programmes, e g Haagenrud et al (1982). Further efforts made by the Norwegian Institute for Air Research (NILU) led to the so-called NILU WETCORR method, involving an automatic six-channel current integrator and the use of miniature Cu/Cu cells, Haagenrud et al (1984).

More recent collaboration between NILU and the National Swedish Institute for Building Research (SIB) has aimed at extending the NILU WETCORR concept to measurements of surface moisture and TOW on building materials and structures in general, Haagenrud et al (1985) and Svennerstedt (1987). This has called for some modification of the TOW-concept but also of the moisture sensor. Since the interest in this context is focused on the surface moisture measurement as such, and not necessarily with respect to atmospheric corrosion, a more inert sensor metal than Cu should be used. To this end, SIB/NILU have developed a new sensor which is equipped with resistance grids of Au.

The present paper aims at characterizing the function of the new Au/Au moisture sensor. In addition, a commercial miniaturized dew sensor from Murata has been considered. Both types of sensors were tested in parallel in a climatic chamber in which the temperature and relative humidity were varied over broad ranges. The sensors have been evaluated for their individual variability, sensitivity, reproducibility, long-term stability and dependence on the orientation relative to the gravitational force. Finally, some possible applications for this measuring technique will be exemplified and discussed.

2 Experimental

2.1 Moisture sensors

The new Au/Au-sensor is made by means of conventional thick-film technology using Au-paste which is screen-printed onto a sintered alumina backing. The Au-paste is a mixture of mainly Au-powder, a low-melting glass and some metal oxides. The printed circuit on the backing is hardened by heat treatment successively up to about 850 C, which causes the Au-paste to homogenize and stick to the backing.

The sensor measures 22 x 31 mm in total and has a thickness of 0.63 mm. The active sensor grid is 14 x 19 mm and consists of 74 alternating anodes and cathodes having a width of 127 μm, which is also the width of the spacing. In order to facilitate for the electric connections to be applied, a Pd/Pt/Au-paste is used for printing the soldering substrate.

Since the Au-paste involves glass as a binding agent, the sensor surface will to a large extent consist of glass after the hardening process. In this state the sensor is not particularly sensitive to moisture. In order to improve the sensitivity, the superficial glass layer has to be removed selectively, thereby uncovering the conducting metal. This was successfully made by etching the entire sensor in 40% HF (Hydrofluoric acid) for 5 min at room temperature. The etching treatment can probably be shortened to 1 or 2 min since almost all the glass had been removed after 5 min, leading to a pronounced reduction in mechanical stability of the Au grid. As an alternative, mechanical polishing with a glass-fibre pencil might work well, but this has yet to be tested.

Soldering of the electric connections was made at slightly reduced temperature using a soldering pencil, since excessive heat may damage the substrate. The soldered joints were then covered in two steps by polyurethane and nitrile rubber coating for protection and stabilization of the connection.

The other type of sensor that has been considered is a dew sensor, type HOS103, available from Murata, Japan. HOS103 only measures 6 x 7 x 0.7 mm and seems also to be built up on alumina backing. The sensing surface consists of some kind of black polymer coating applied on a six-finger metal grid. According to a brochure from Murata (1988) the sensor is particularly sensitive to high moisture levels but features reversed characteristics in that the resistance increases with increasing moisture loads. When condensation occurs on the sensor surface the resistance increases to more than 200 kohm. Under dry conditions, however, the resistance is typically 4 kohm.

2.2 Measuring principle and instrumentation

The instrument that is used together with the moisture sensors works according to the NILU WETCORR concept, Jonsson (1989). It is equipped with 16 channels devoted either to surface moisture, temperature or relative humidity. Each moisture sensor is driven by a constant voltage of 100 mV, with polarity reversal every 30 seconds to avoid net polarization. The resulting current, which has a lowest resolution of 0.1 nA, is measured every second and the average over one minute, or longer if preferred, is stored on tape (DEC-tape TU58). There is room for approximately 65 000 values on one cassette, which means that when all channels are connected and values are stored every minute, the cassette has to be replaced within approximately 68 hours.

The data are then transferred to a mini computer (VAX 6310) for data file editing, evaluation and plotting. At this stage it is possible to choose an appropriate current limit in order to define TOW and also to calculate TOW for a given measuring period.

2.3 Evaluation of the moisture sensors in a climatic chamber

The response of the Au/Au- and HOS103-sensors were studied under controlled conditions in a climatic chamber of make Brabender, type KSW 61/10H. Six moisture sensors of each type were mounted on two pieces of coil coated sheet metal in aluminium, using Scotch double stick adhesive. The sensor set-up was equal for both sheets, i e three sensors of each type per sheet. The sensors were oriented with the connections downwards, as this would probably represent the most common mounting in practice. The two sheets were in turn placed side by side along one of the side walls of the climatic chamber. The temperature was measured by copper-constantan thermocouples, both in the air and on the surface of one of the sheet metals. In the latter case the thermocouple was stuck onto the surface with epoxy glue. The relative humidity and temperature of the air were also measured employing a Rotronic MP-100 probe.

All 16 channels of the instrument were thereby occupied. Throughout the following test programme, minute averages of all channels were stored on tape.

The climatic chamber was set to alter every 4 hours between two combinations of temperature and relative humidity. After typically 3 days, two new combinations were chosen. The temperature and relative humidity were varied between 0 and 30 deg C, and approximately 60 and 100%, respectively. In total, 20 combinations were employed in this part of the programme. Close to zero temperatures and below, the climatic chamber could not be regulated properly, since it depends on a conventional psychrometer with wet and dry thermometers.

In most measuring situations, as was indicated earlier, the sensors would probably be mounted on a wall with the connections downwards. Sometimes, however, other orientations may be necessary, e g on roofs. To that end, a few tests were made also with the sensors mounted horizontally, both on the floor of the climatic chamber and on the ceiling. A third alternative was to place the sensors on the wall but turned 90 degrees so that the sensor fingers were vertical. These tests were made by changing the orientation of only one of the coil coated sheets, leaving the other in the original orientation as a reference. In this way, the possible influence of the gravitation on the sensor current should hopefully be revealed by a change in mean current ratio between the two sets of sensors.

All average values of the current were calculated for each channel from one of the 4 hour periods during which the temperature and relative humidity were kept constant. More specifically, the average was formed by the first 30 minutes of the last hour in that particular period. Mean and standard deviation were calculated for the six moisture sensors of each type.

In addition to the current averages that are stored on tape, the instrument permits on-line registration of the consecutive second values, including sign, of one channel at a time. This feature was utilized to study the response of the sensors to the polarity reversal during a two minute period for different moisture loads.

3 Results

3.1 Influence of polarity reversal

Fig. 1 shows the current-time behaviour during two minutes for one of the Au/Au-sensors when exposed to different moisture loads at 20 deg C. As expected, the current increases with increasing relative humidity. It should be noted, however, that on reversal of the polarity, a current transient occurs which becomes more pronounced as the moisture load increases. Bearing in mind that Fig. 1 shows the logarithm of the current, the contribution from the transient will dominate the minute average already at moderate moisture loads.

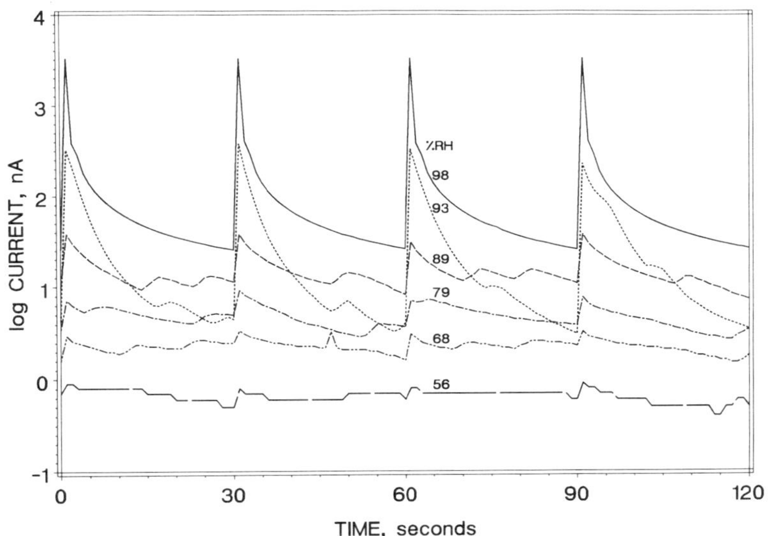

Fig.1 Current-time curves during 2 minutes for Au/Au-sensor exposed to different moisture loads at 20 deg C. Polarity reversal every 30 seconds.

Under the same conditions, on the other hand, the HOS103-sensor seems almost unaffected by the reversing polarity, see Fig. 2. It is interesting to note that the current hardly changes until the relative humidity is well above 90%, emphasizing the dew sensitivity of this sensor.

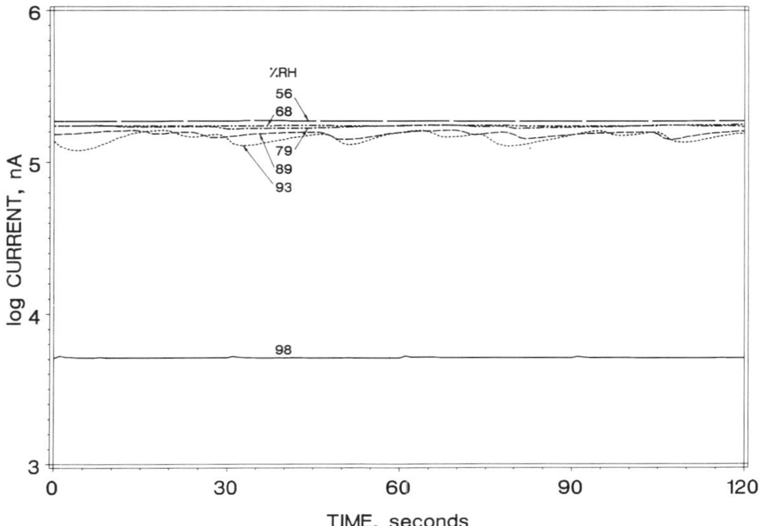

Fig.2. Current-time curves during 2 minutes for Murata HOS103-sensor exposed to different moisture loads at 20 deg C. Polarity reversal every 30 seconds.

3.2 Influence of relative humidity and temperature

Fig. 3 shows the mean and standard deviation of the current for the Au/Au-sensors as a function of the relative humidity at temperatures between 0 and 30 deg C. On the whole the current streches from about 250 nA at the saturation point down to less than 1 nA at 60% RH. The effect of an increased temperature is relatively moderate and results only in a minor increase in current.

Visual extrapolation of the curves in Fig. 3 indicates that the detection limit (0.1 nA) for surface moisture corresponds to a relative humidity of less than 50% even at 0 deg C. In order to better clarify the empirical dependence of the current on temperature and relative humidity, multiple regression analysis was performed using software from SAS Institute Inc. Of the various attempts made in order to find an appropriate function that would fit the data, the following gave the best agreement:

$$\log i_w = 3.439 + 3.409 * 10^{-4} * (RH)^2 - 1388/(T_s + 273.16) \quad (1)$$

where i_w, RH, and T_s are, respectively, the cell current in nA, the relative humidity in % and the surface temperature of the coil coated sheet metal in deg C. In the regression analysis the current for each individual Au/Au-sensor was used, in total 120 current values. From the analysis of variance it may be concluded that Equation (1) fits very well to the observed data, as indicated by the coefficient of determination (R^2 = 0.9571).

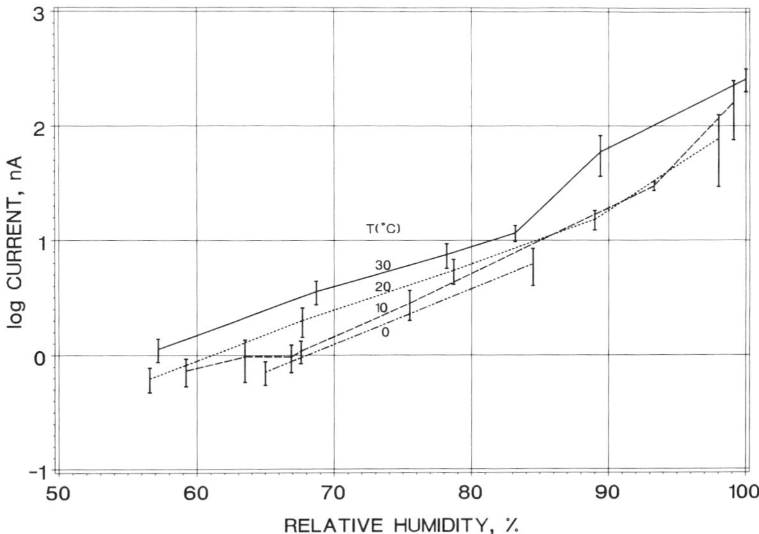

Fig.3. Relationship between current and relative humidity for Au/Au-sensors at different temperatures. Mean and standard deviation are shown.

The agreement between predicted and observed values is also shown in Fig. 4, where the 95% confidence limits are plotted for the individual predicted values.

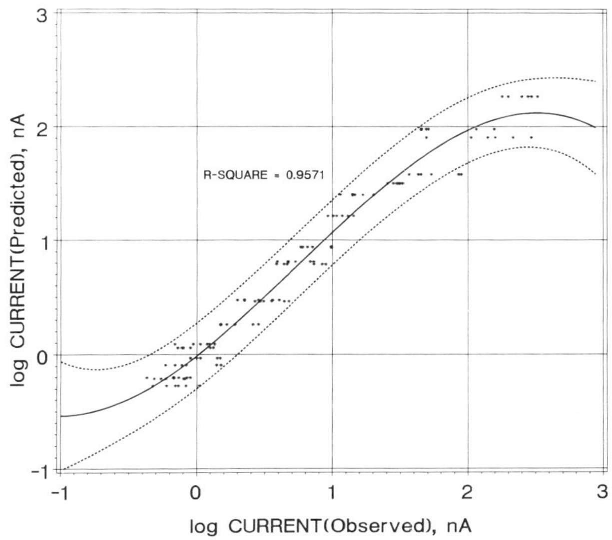

Fig.4. Comparison of observed and predicted current values for the Au/Au-sensors according to Equation 1.

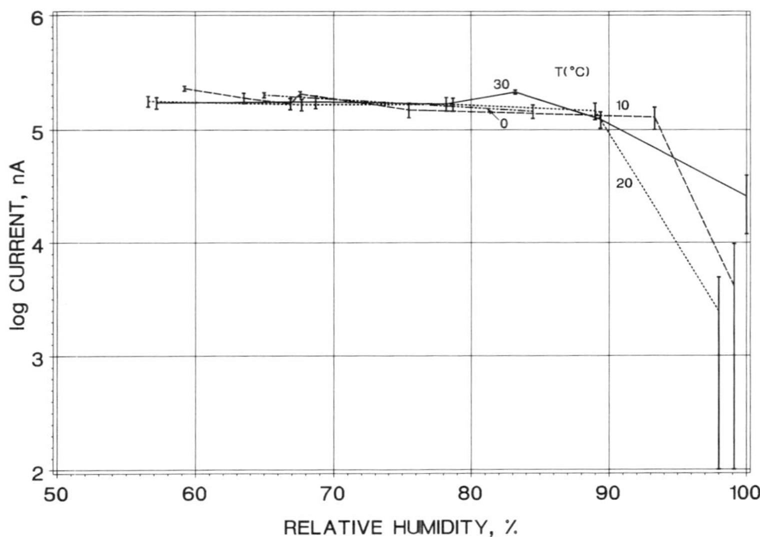

Fig.5. Relationship between current and relative humidity for Murata HOS103-sensors at different temperatures. Mean and standard deviation are shown.

Fig. 5 corresponds to Fig. 3 but concerns the HOS103-sensors under the same conditions. Below approximately 90% RH, the current is virtually independent of both temperature and relative humidity and stays at a level of about 1.8-2.0 * 10^5 nA. Above this RH-limit, however, the current drastically drops 1-2 decades when approaching the saturation point of the vapour concentration. Unfortunately, the number of data points is insufficient in this transition range, making it difficult to exactly state at what relative humidity the current starts to drop. Nor is it possible to tell what influence the temperature might have on the current under these conditions.

3.3 Influence of orientation

The possible effect of sensor orientation was studied at two combinations of temperature and relative humidity, namely, 10 deg C/ 80% RH and 30 deg C/ 90% RH. The relative difference in average current between the sensors that were moved about and the ones kept as references, was calculated for the Au/Au- and HOS103-sensors, respectively. For the sake of comparison, the mean and standard deviation of this quantity was also calculated for all the tests described in the previous section, all involving wall mounted sensors with horizontal sensor fingers. The results for the various orientations are summarized in Table 1.

Table 1. Relative influence of sensor orientation on the current

Orientation	T_s(deg C)	RH (%)	Rel. diff. (%) Au/Au	HOS103
Wall (hor)	29.66	83.2	-17.8	+1.9
Wall (ver)	29.67	78.8	-14.3	+3.0
Ceiling	29.64	79.1	-17.9	+2.2
Floor	29.54	80.8	-11.8	+1.4
Wall (hor)	9.97	94.5	+5.9	-1.7
Wall (ver)	9.99	90.7	+4.3	-0.8
Ceiling	9.97	91.1	+5.4	-1.0
Floor	10.04	91.3	-1.0	-0.5
Wall (horizontal, all tests)		Mean	-20.8	-10.6
		Std dev	22.4	26.2

The mean value shown above for all tests involving wall mounted sensors with horizontal fingers, should be close to zero if all the sensors were identical and if they were exposed to the same conditions in the climatic chamber. Since the difference is rather large, -20.8 and -10.6%, respectively, for the two types of sensors, and with an even greater spread, it seems reasonable to assume that this is mainly the result of different conditions within the chamber. In this case the reference sensors were always closer to the back wall while the ones that were moved about were closer to the door at the front. This implies that the small differences observed when changing the orientation of the sensors are well within the spread of the inherent variation in conditions within the chamber. Consequently, from a practical point of view the effect of varying orientations seems negligible, at least in comparison with other uncertainties involved.

3.4 Long-term stability

As far as the Au/Au-sensors are concerned, no signs of changes in moisture sensitivity have been noted in the course of this evaluation programme. Immediately after the activation with hydrofluoric acid the sensors seemed to be extremely sensitive but this effect held only for one or two days. The inert material in the Au/Au-sensors should provide for an intrinsic stability, and changes in the current should be due only to external factors, such as deposition of pollutants, pH-changes and mechanical damage. Since the sensors have only been exposed to rather mild conditions so far ageing effects should not be expected.

The HOS103-sensors are probably more sensitive to the environment since the manufacturer state that even ordinary tap water might damage the function. In the present study, however, the dew sensors have been working well. This implies that, at least for short-term studies, the HOS103-sensor might be useful enough.

4 Discussion

4.1 Function of sensors

The use of electrolytic cells in studies of atmospheric corrosion phenomena has mostly involved DC voltage without reversing the polarity. Haagenrud et al (1984), studied the influence of alternating DC voltage on Cu/Cu-cells and found that on reversal a 5-6 fold increase in integrated current resulted. The consequences of these findings were never discussed any further, despite the fact that the NILU WETCORR method from that time made use of polarity reversal as standard. In the present study these effects have been studied in more detail, since the instrument employed allows continuous monitoring of the cell current.

From electrochemistry it is well known that the current through an electrochemical cell depends, among other things, on the nature of the applied voltage, Metals Handbook (1987), according to the following expression:

$$i = C_{dl} \frac{dE}{dt} \tag{2}$$

where i is the resulting current, C_{dl} is the interfacial capacitance associated with the electrochemical double layer and dE/dt is the time rate of change in applied voltage. The results shown in Fig. 1 strongly suggest that the measured current mainly derives from the charging of the double layer capacitance, associated with the Au/electrolyte interface. As pointed out earlier, these transients account for the major part of the minute average even at rather low relative humidities. Immediately after each reversal, the current rapidly decreases to a much lower level, which to a large extent may be attributed to the DC resistance of the electrolyte. By shortening the reversing interval from 30 seconds to 5 or 6 seconds, the sensitivity would increase considerably, especially at high moisture loads.

4.2 TOW-criteria

Atmospheric corrosion of e g steel and zinc becomes significant when the relative humidity exceeds 80-85% RH. This has been the basis of the recent definition of TOW, namely, as the time during which the air temperature is above 0 deg C and, at the same time, the relative humidity exceeds 80% RH, ISO/DP 9223 (1986). Although the relative humidity is the most important factor determining TOW, the correlation is generally rather poor as far as buildings and building components are concerned, Svennerstedt (1989). Consequently, the need for in-situ measurements with miniature moisture sensors seems essential for this research area.

Using the meteorological definition above of TOW in combination with Equation 1 in Section 3.2, the corresponding current criterion for the new Au/Au-sensors can be approximated to 3.5 nA. This threshold current may be used as a reference or when no better criterion is known. In the general case, however, the current limit should be unique for each type of material depending on e g degree of porosity, ability for capillary suction and prevailing degradation mechanism. This work, using the present technique, remains to be done.

4.3 Applications

As already indicated, surface moisture measurements using miniature sensors can provide essential data for the description and prediction of in-service behaviour of building materials. Although measurements on the envelope of buildings, including windows and doors, would represent the most common application for this technique, numerous other interesting objects exist of which a few are exemplified here.

Recently, a pilot-study of the surface moisture variations experienced in different parts of high-humidity compartments of dwellings was accomplished, Norberg and Sjöström (1990). Data from such studies can be of help when choosing materials and designs for e g bathrooms.

Surface condensation and mould growth can occur in many other locations of dwellings. In times of energy conservation, it is common to have thermal insulation laid on top of the ceiling in the roof space. This increases the risk of condensation in the roof. Moisture sensors can be used to monitor the conditions in such remote parts of buildings.

Lindberg (1988), has employed the NILU WETCORR method to study the moisture permeability of different types of paints. The surface of the sensors was painted and then exposed to outdoor environment. Further studies along this line should involve the influence of e g colour and ultra-violet radiation on permeability changes of the paint film.

5 Conclusions

The new Au/Au-sensor and the commercial dew sensor Murata HOS103 have been evaluated under climatic chamber conditions. Utilizing the NILU WETCORR method with alternating DC voltage +/- 100 mV every 30 seconds the following regression formula for the current through the Au/Au-sensors would be valid:

$$\log i_w = 3.439 + 3.409 * 10^{-4} * (RH)^2 - 1388/(T_s + 273.16)$$

The dew sensor HOS103 reacts only to relative humidities exceeding 90-95% RH and is virtually independent of temperature.

The function of the Au/Au-sensor is based on the current needed to charge the double layer capacitance associated with the Au/electrolyte interface.

In the course of the evaluation programme both sensors have shown acceptable properties with respect to individual variability, sensitivity, reproducibility, and long-term stability. The sensors are not sensitive to orientation with reference to the gravitational force.

The present technique can be used to enlarge the knowledge about the prevailing moisture conditions on surfaces of buildings. Such data will most likely play an important part in future studies of the durability of building materials.

6 References

Haagenrud, S. Kucera, V. and Gullman, J. (1982) Atmospheric corrosion testing with electrolytic cells in Norway and Sweden. **Int. symp. on Atmospheric Corrosion**, Hollywood, Florida, Oct 1980 (ed W.H. Ailor) N.Y. Wiley, pp. 669-693.

Haagenrud, S.E. Henriksen, J.F. Danielsen, T. and Rode, A. (1984) An electrochemical technique for measurement of time of wetness. **Proc. 3rd Int. Conf. on Durability of Building Materials and Components**, Espoo 12-15 Aug 1984, pp. 384-401.

Haagenrud, S. Henriksen, J. and Svennerstedt, B. (1985) Våttidsmålinger på treplater - Prøvestudie med NILU-WETCORR-metoden (Time-of-wetness measurements on wood - Pilot study with the NILU WETCORR method). **NILU OR 17/85**, (In Norwegian).

ISO/DP 9223 (1986) Corrosion of metals and alloys. **Classification of Corrosivity Categories of Atmospheres.**

Jonsson, P. (1989) Våttidsutrustning - teknisk beskrivning (TOW-equipment - technical description). The National Swedish Institute for Building Research, **Research Report TN:9**, (In Swedish).

Kucera, V. and Mattsson, E. (1974) Electrochemical technique for determination of the instantaneous rate of atmospheric corrosion, in **Corrosion in Natural Environments, ASTM STP 558**, American Society for Testing and Materials, pp. 239-260.

Lindberg, B. (1988) Mätning av våttid på målade byggnadsmaterial (Measurement of TOW on painted surfaces of building materials). Scandinavian Paint and Printing Ink Research Institute, **NIF-Report T 2-88 M**, (In Swedish).

Mansfeld, F. (1981) Evaluation of electrochemical techniques for monitoring of atmospheric corrosion phenomena, in **Electrochemical Corrosion Testing, ASTM STP 727** (eds F. Mansfeld and U. Bertocci), American Society for Testing and Materials, pp. 215-237.

Metals Handbook (1987) Ninth Edition, **Volume 13, Corrosion**, ASM International, p 213.

Murata (1988) Sensor. **Catalogue No SG01E-5**, Murata MFG Co, Ltd, Kyoto, Japan.

Norberg, P. and Sjöström, Ch. (1990) Time-of-wetness measurements in high-humidity compartments of dwellings. Paper submitted to the **Int. CIB W67 Symp. on Energy, Moisture, Climate in Buildings**, Rotterdam, 3-6 Sept 1990.

Sereda, P. Croll, P.J. and Slade, H.F. (1982) Measurement of the time-of-wetness by moisture sensors and their calibration, in **Atmospheric Corrosion of Metals, ASTM STP 767** (eds S.W. Dean,Jr. and E.C. Rhea), American Society for Testing and Materials, pp. 267-285.

Svennerstedt, B. (1987) Time of wetness measurements in the Nordic countries. **Proc. 4th Int. Conf. on Durability of Building Materials and Components**, Singapore, 4-8 Nov 1987, pp. 864-869.

Svennerstedt, B. (1989) Ytfukt på fasadmaterial (Surface moisture on façade materials). The National Swedish Institute for Building Research, **Research Report TN:16**, (In Swedish).

57 PROPOSED METHODS OF TEST FOR CHEMICAL RESISTANCE OF CONCRETE AND CEMENT PASTE IN AGGRESSIVE SOLUTIONS

K. SHIRAYAMA
Kogakuin University, Tokyo, Japan
A. YODA
Ashikaga Institute of Technology, Ashikaga, Japan

Abstract
This paper summerized the draft of the standard testing methods for chemical resistance of concrete in aggressive solutions and that of cement paste, proposed by the Chemical Resistance Working Group of the Research Committee on the Safety of Structural Materials, organized in the Japan Testing Center for Construction Materials, as well as the related researches performed by the members of the Working Group from 1978 to 1982, on the many factors influencing the chemical resistance of concrete, mortar and cement paste. It also introduces an example of the recent experimental studies, making use of this proposed methods.
Keywords: Chemical Resistance, Test Method, Concrete, Cement Mortar, Cement Paste, Aggressive Solution.

1 Introduction

There have been many studies on chemical resistance of concrete in the past, but, in many cases, it is difficult to directly compare these test results with one another, because they had been obtained under different test conditions. Aiming at improving such circumstance by means of standardizing the methods of test for chemical resistance of concrete, the Chemical Resistance Working Group (C.R.W.G.) in the Research Committee on the Safety of Structural Materials, organized in the Japan Testing Center for Construction Materials, has proposed a draft of standard testing methods for chemical resistance of concrete in the aggressive solutions and that of cement paste in 1983, after the systematic research performed from 1978 to 1982, on many factors influencing the chemical resistance of concrete, mortar and cement paste.

The C.R.W.G. was composed of 15 members as follows; K. Shirayama (Univ. of Tukuba), Y, Kasai (Nihon Univ.), Y. Yoda (Ashikaga Inst. of Tech.), H. Ikenaga (Chiba Inst. of Tech.), H. Einaga (Univ. of Tukuba), K. Suzuki (Tokyo Metropolitan Univ.), H. Nishi (Onoda Cement Co.), N. Kashino (Building Res. Inst.), T. Maruiti (Shimizu Corporation), M. Makita (Public Works Res. Inst.), N. Otuki (Port & Harbour Tech. Res. Inst.), A. Suzuki (Conc. Pole & Pile Ass.), M. Yamamoto (Agcy. of Ind. Sci. & Tech.), H. Takahashi (Ready Mixed Conc. Ass.) and S. Kumahara (Japan Testing Center for Constr. Mat.).

The C.R.W.G. began its work by investigating the requirements for the standard methods of test for chemical resistance of concrete, and arranged them as follows:

(a) The test methods should be usable for the purpose of making a general comparison of chemical resistance between concretes made with different materials, aiming at clarifying influence of materials on the chemical resistance.

(b) The test methods should also be usable for the purpose of making a comparison of chemical resistance between concretes for specific applications in practice.
(c) The results obtained by the test methods should be reproducable.
(d) The test period of time should not be too long.
(e) The cost of the test should not be expensive, and no need for special apparatus and techniques to make the test.
(f) The test methods should be effective to analyse mechanism and rate of deterioration in the chemical attack.

The C.R.W.G. tried to propose the test methods for chemical resistance of concrete and cement paste in aggressive solutions, which satisfy these requirements as far as possible.
The proposed test methods are going to be introduced in the following paragraphs with the related experimental studies performed by the members of the C.R.W.G.

2 Properties of the test specimens to be measured

Having investigated the deterioration mechanism of concrete by chemical solution as shown in Fig. 1, the C.R.W.G. proposed the type of properties of the test specimens to be measured as shown in Table 1. Fig. 1 was sent to the RILEM Commission 32-RCA (Resistance of Concrete to Chemical Attack), and appeared in the report of the Committee by M. Regourd (1984), though it was fairly modified and improved. The properties enclosed by heavy lines in Fig. 1, represent those which can be measured or observed.

3 Test specimens specified in the draft

3.1 Shape, size and number of test specimens

In the draft of standard test methods, the test specimens for strength of concrete and cement paste are specified as shown in Table 2. The test specimens of mortar are not specified in the draft, but those recommended by the C.R.W.G., are also shown in Table 2. The test specimens for length change are same as those for flexural strength in shape and size, so length change can be measured by use of the flexural strength specimens to which gauge plugs or glass plates are attached. Dynamic modulus of elasticity shall be also measured by use of the test specimens

Table 1. Properties to be measured in the chemical resistance test

Specimens	properties to be tested in principle	Properties to be tested if necessary
Concrete	Visual inspection, weight change, length change, dynamic modulus of elasticity, flexual and compressive strength, depth of carbonation	pH at dif. depth, depth of deterioration, porosity
Paste	Visual inspection, weight change, compressive strength	Porosity

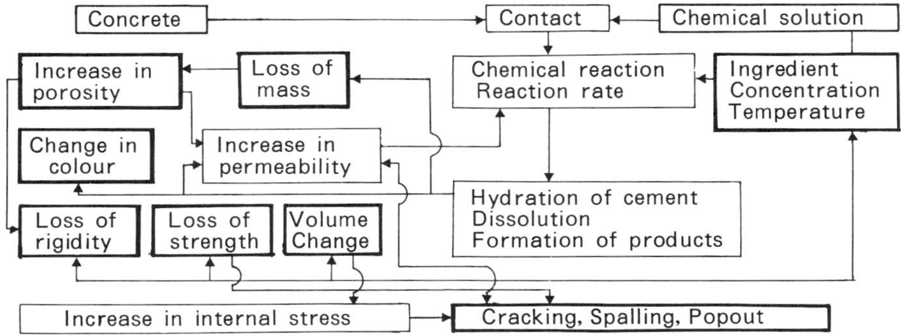

Fig. 1. Deterioration mechanism of concrete by chemical solution

Table 2. Test specimens

Test prop.	Item	Concrete	(Mortar)	Cement paste
Compressive strength	Shape	Cylinder, h.:d.=2:1	Cylinder, h.:d.=2:1	Cube
	Size	100 $\phi\times$ 200mm*	50 $\phi\times$ 100mm	20×20×20mm
	Number	More than 3	More than 3	More than 6
Flexural strength	Shape	Prism, W.:L =W:≧ (3W + 80mm)	Prism	
	Size	100×100×400mm*	40 × 40 × 160mm	
	Number	More than 3	More than 3	

*In case of the concrete with coarse aggregate less than 30mm in max. size

for strength. Visual inspection is made and properties other than strength, length change and elastic modulus of elasticity are measured by use of the test specimens for strength and/or specimens for length change if they are made separately.

3.2 Preparation of test specimens

The methods of preparation of concrete specimens are dependent upon the purpose of test described in 1 Introduction (a) and (b).

In case of test for the purpose of making a general comparison of chemical resistance between concretes as described in 1(a), concrete specimens shall be made in accordance with JIS A 1132 (Method of Making and Curing Concrete Specimens).

The specimens of cement paste shall be made in accordance with the following conditions:

- Mix proportion — For 6 specimens, 500g of cement and 250g of water shall be used.
- Specimen moulds — A gang mould for 6 specimens with a sub-mould put on it, shall be used. The sub-mould is used in order to prevent excess cement paste from flowing out of the mould during placing of the cement paste. The sub-mould shall be 20mm in depth.

Table 3. Curing of test specimens

Protective coating*		Demoulding 20 ~ 48 hr.**	7	21	24	26 28
			Age of specimen in days			
Not applied		Water at 20 ± 2°C	Sealed, air at 20 ± 2°C	Unsealed, air at 20 ± 2°C, R.H. 60 ± 5%		Water at 20 ± 2°C
Applied		Water at 20 ± 2°C	Sealed, air at 20 ± 2°C	Unsealed, air at 20 ± 2°C, R.H. 60 ± 5%	Coated, water at 20 ± 2°C	

*See 3.3, **The demoulding time is different according to the type of specimen.

- Mixing — The cement paste shall be mixed by use of the mixer specified in JIS R 5201 (Physical Testing Method of Cement) for 2 minutes.
- Placing, demoulding — The cement paste shall be placed into the the moulds up to 15mm over the top face of the moulds. The excess paste shall be scraped off even with the face of the mould after curing in a moist cabinet for more than 5 hours. The moulds shall be removed at least 20 hours after placing.

The test specimens of concrete and cement paste, made as described above, shall be cured as shown in Table 3.

In tests for the purpose of comparing the chemical resistance between concrete mixes for specific applications in practice as described in 1(b), the specimens shall be made and cured under the conditions simulating the anticipated work as closely as possible.

As for mortar specimens, it is recommended to make them in accordance with JIS R 5201 using mixproportion determined as described in 4.1(b) and to cure them as shown in Table 3, though the test methods of mortar are not specified in the draft.

3.3 Protective coating of test specimens

If deterioration or deformation of the specimens is estimated to be excessive, the faces of strength test specimens to which the load is to be applied, and the faces around gauge plugs or glass plates of the length change specimens, shall be protected employing the proper inert materials. As the proper coating materials, the type 1 tar-epoxy resin paint specified in JIS K 5664, or the solid paraffin having the melting point between 37.8°C and 43.3°C, specified in JIS K 8754, melted in a container immersed in water at about 70°C, are recommended.

If the chemical attack is extremely severe, the appropriate caps can be used with these coating materials for the strength test specimens.

When coating the specimens, the faces which are not to be coated, shall be covered with adhesive tape before coating. This tape is taken off after coating is finished.

The coated concrete specimen for flexual strength test is schematically shown in Fig. 2.

Table 4. Test results of the cement used as the common material

Type of cement	σ_{28} (MPa)	Chemical composition % SiO_2	Al_2O_3	Fe_2O_3	CaO	MgO	SO_3
Ordinary P. c.	40.2	22.0	5.4	3.1	65.1	1.5	2.0
Sul. resist. P. c.	33.3	22.4	3.9	4.6	64.8	1.6	1.9
Blast fur. slag c.	36.3	28.6	11.9	2.3	48.5	4.6	2.0

4 Experimental investigation on the test specimens

4.1 The common testing conditions

The members of C.R.W.G. systematically performed many experimental studies on the chemical resistance of concrete, mortar and cement paste under the common testing conditions as follows:

(a) Materials The properties of the cements used as common materials in the tests are shown in Table 4. As for concrete aggregates, river sand, having the fineness modulus of 2.8, and river gravel, having the maximum size of 25mm were used. For mortar, the Toyoura standard sand was used.

(b) Mix proportions The mix proportions of concretes were determined to be 60% in water cement ratio, and 12 ~ 15cm (1cm=10mm) in slump. The mix proportions of mortars were determined to be 55% in water cement ratio, 170 ~ 190mm in flow.

(c) Preparation of test specimens The test specimens were made in accordance with the methods specified in the draft, as shown in 3.2, and then cured in water at 20 ± 2°C up to the age of 7 days, followed by curing in air at 20°C and relative humidity of 60% up to the age of 28 days.

(d) Immersion in the chemical solutions After the 28 days' curing, test specimens were completely immersed in the several types of aqueous solution (5% sulphuric acid, 5% hydrochloric acid, 10% soduim sulphate, and so on) at about 20°C.

Here follows the results of the experimental studies on test specimens, performed under these common conditions. As the notation please refer to 8.

4.2 Studies on shape and size of test specimens

(a) Influence on loss of mass It is expected that the loss of mass expressed by percentage of the initial mass of specimen ($\Delta W\%$) will be estimated by the

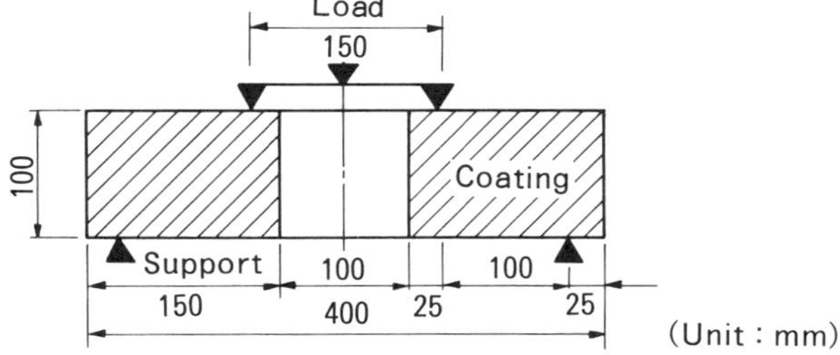

Fig. 2. Coated concrete specimen for flexural strength test under loading

Table 5. Properties of solutions in the Kusatu hot spring

Fountain	Temp.	pH	main dissolved constituent (mg/l)				
			CO_2	SO_4	Cl	H_2SiO_3	Al^{3+}
Bandaikou	83°C	1.0	1718.0	1463.0	647.2	263.0	24.0
Shirahata	63°C	1.1	1963.0	1439.0	95.7	393.6	191.6

following equation, if the specimens were uniformly deteriorated and scaled off from the surface.

$$\Delta W = \{\Delta d \cdot \rho \cdot S)/(V \cdot \rho)\} \times 100 = \{\Delta d(S/V)\} \times 100 \qquad (\%) \qquad (1)$$

The value of S/V is 0.45 for 100 × 100 × 400mm prism, and 0.50 for 100 ϕ× 200mm cylinder, so the ratio of the former to the latter will be 0.9. Yoda's test results on the concrete test specimens immersed in several solutions for 28 days or 91days, are shown in Fig. 3. It indicates that the test results are not much different from this ratio, though, to confirm this, it is required to make further experiment using specimens of wider range in S/V. 'Bandaikou' and 'Shirahata' in Fig. 3, are the names of fountains in the Kusatu hot spring, and the properties of solutions from them are shown in Table 5.

(b) Influence on strength In case of cylinder, loss of compressive strength ($\Delta R\%$) expressed by the percentage of the strength of concrete without degration may be estimated by the following equation.

$$\Delta R = [\{\pi D^2 - \pi(D - 2 \cdot \Delta d)^2\}/\pi D^2] \times 100 \qquad (\%)$$

$$= [1 - \{1 - (2 \cdot \Delta d/D)\}^2] \times 100 \qquad (\%) \qquad (2)$$

were D : diamter of the specimen (mm)
Δd : depth of deteriorated portion from the surface of the specimen (mm)

Assuming that Δd is same, independent from the diameter of the specimen, relation between ΔR and $\Delta R'$ for the specimen of D and D' in diameter respectively,

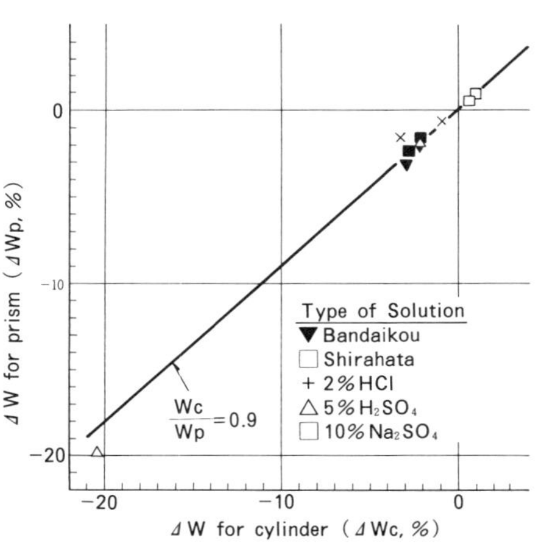

Fig. 3. Loss in mass and type of specimen

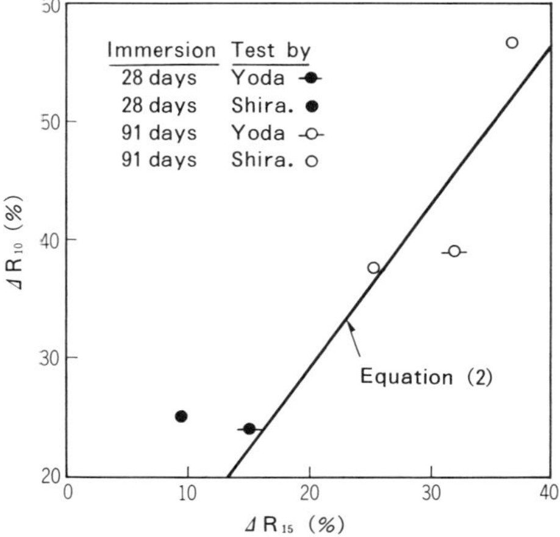

Fig. 4 Loss in strength and size of specimen

Table 6. Test results of the concrete test specimens after immersion in the Kusatu hot spring for 91 days

Curing	Strength at 28 days (MPa)		Ratio of properties after immersion					
			based on the initial value (Z_s)*			based on the water cured specimens (Z_w)*		
			Strength		Young's	Strength		Young's
	Com.	Flex.	Com.	Flex.	modulus	Com.	Flex.	modulus
W	27.9	3.6	0.90	0.94	0.92	0.74	0.61	0.79
WS	23.2	3.5	0.93	0.93	0.90	0.72	0.62	0.79
WA80	29.2	3.8	0.85	0.92	0.86	0.77	0.61	0.77
WA60	30.2	4.1	0.84	0.91	0.83	0.76	0.65	0.74
S	30.7	4.3	0.85	0.90	0.85	0.79	0.74	0.78

*As for Z_s, Z_w, see 8.(b)

is obtained by means of substituting D and D' for D in the equation (2) at an equal Δd.

Yoda and Shirayama compared the loss of strength in concrete cylinders having diameter of 150mm and those of 100mm, immersed in 2% hydrochloric acid during 28 or 91 days, on the basis of the strength of concrete cylinders cured in water at 20°C, and the test results are shown in Fig. 4. A curve in Fig. 4, represents the relation between ΔR for cylinder of 100mm in diameter (ΔR_{10}) and that of 150mm in diameter (ΔR_{15}), calculated by use of equation (2). It can be seen the test results roughly agree with this curve. Shirayama also found that Δd calculated from ΔR by use of equation (2) was about 1/1.5 of the depth of carbonation, and about 1/1.7 of the penetration depth of chlorine ions.

Influence of size of test specimens on the loss of flexural strength was not tested this time, but it might be larger than that on the compressive strength.

4.3 Study on number of test specimens

Kumahara tested the furexural strengths of the uncoated concrete test prisms immersed separately one by one in 5 ℓ of 5% sulphuric acid for 28 days. The specimens were made for three different water-cement ratios and two different curing conditions prior to immersion. Numbers of one set of specimens were three, five and six. The tests results showed that the range of coefficients of variation were 1 ~ 12% for three specimens, 2 ~ 9% for five specimens and 2 ~ 8% for six specimens. In case of five specimens, the range of coefficients of variation was reduced to less than 3%, by negrecting the maximum value and the minimum value. So, it may be recommended that the strength test specimens consist of set of a minimum of five test specimens, considering that the variation of compressive strength is generally less than that of flexural strength.

4.4 Studies on curing prior to immersion

A part of test results by Yoda on the influence of curing conditions of the test specimens prior to immersion into the Kusatu hot spring (See Table 5.) on the properties of concrete are shown in Table 6. The test specimens were made of the Portland cement concrete of the same mix proportion, and at the age of 28 days immersed into water at 20°C as well as the Shirahata fountain for 91 days. The

Table 7. Ratios Z_w of the compressive strengths of cement pastes after immersion in the test mediums for 4 weeks

Solution :	5% sodium sulphate						Sulphuric acid		
Cement :	O.P.		S.P.		B		O.P.	S.P.	B
W/C (%) :	30	50	30	50	30	50	50	50	50
Curing W	0.81	1.00	1.00	1.07	0.98	0.96	0.86	0.83	0.80
Curing D	0.83	0.95	1.02	1.08	0.94	0.75	0.68	0.82	0.59

methods of curing prior to immersion are shcwn below with their symbols.

W : In water at 20°C.

WS : In water at 20°C up to the age of 7 days, then sealed and in air at 20°C.

WA80 : In water at 20°C up to the age of 7 days, then in air at 20 °C and R.H. 80%.

WA60 : In water at 20°C up to the age of 7 days, then in air at 20°C and R.H. 60%.

S : Sealed and in air at 20°C.

From Table 6, the results may be summerized as follows:

(a) The range of differences of Z_s and Z_w (see 8.) caused by the curing conditions are roughly less than 0.1 as far as this experiment.

(b) Curing in water prior to immersion is not always the most effective method to make the strength ratio Z_w maximum.

(c) The ratios based on the strengths of the specimens immersed in water Z_w show the chemical resistance of concrete more clearly than those based on the initial strengths before immersion, Z_s.

The influence of curing prior to immersion on the compressive strengths of cement pastes was tested by Nishi. The test specimes were prism of 20 × 20 × 30mm in size, made of one of the three types of cement, ordinary Portland cement (O.P.), blast furnace slag cement, containing 30 ~ 60% of slag (B) and sulphur resisting Portland cement (S.P.). As for test solutions, 5% sodium sulphate and sulphuric acid maintained at pH 1.0 were used. Water was also used for test medium.

Types of curing prior to immersion used in this test are shown below with their symbols.

W : In water for 28 days

D : In water for 4 days, then sealed and in air for 4 days

The ratios of compressive strengths Z_w were obtained based on the strengths of the specimens immersed in water.

A part of the test results are shown in Table 7. From this results, it can be seen that the insufficient curing prior to immersion such as '*D*' in this test, might considerably decrease the chemical resistance or values of the strength ratio of cement paste, especially in the case of using the blast furnace slag cement.

4.5 Studies on protective coating of test specimens

The performance of the tar-epoxy coating over the concrete specimens applied according to the method specified in the draft were found satisfactory after the test by Yoda. In this test, the specimens were exposed to such aggressive solution as 'Bandaikou' fountain, 1.0 in pH and 83°C in temperature, for three months.

The paraffin coating was found also effective at moderate temperature up to the loss of exposed faces of specimens mounted to nearly 20mm in depth.

5 Methods of immersion for specimens in the test solutions specified in the draft

In case of test for the purpose of making a comparison of chemical resistance between the concretes for specific applications in practice as described in 1(b), the specimens shall be immersed in the solution simulating the anticipated service conditions as closely as possible.

In case of test for the purpose of making a general comparison of chemical resistance between the concretes made with different materials, as described in 1(a), the test specimens shall be immersed in the solutions under the following conditions.

(a) Test solutions
As for typical test solutions, the following aqueous solutions at 20 ± 2°C are mentioned; 2 wt.% hydrochloric acid and 5 wt.% sulphuric acid as typical acid solutions and 10 wt.% sodium sulphate and magnesium sulphate as typical saline solutions. In addition water shall be used as the base for comparison.

(b) Setting of the test specimens
The test specimens shall be set so as to leave the spaces of more than 30mm between specimens one another, as well as between specimens and bottom of the container, and to be sunk completely in the solution.

(c) Control of the test solutions
The test solutions shall be controlled so as to keep them in pH at their initial values. They shall be completely renewed at least once every two weeks for the first one month from start of immersion. The containers of the solutions shall be tightly covered by the inert coatings.

6 Experimental investigation on the methods of immersion

6.1 Study on change of test solution; effect of volume

Kumahara immersed 100 × 100 × 400mm concrete prisms in 5% sulphuric acid solution of different volume, namely, 3.3, 5 and 7 ℓ per prism and measured the change in pH of the solution. The results showed that the changes in pH for 7 ℓ and 5 ℓ were about 0.3 and there was not much difference between these two, but for 3.3 ℓ, the increase in pH was greater than 1.0 after 4 weeks immersion. So, it is recommended to use at least same volume of solution as the total volume of specimens immersed in it.

6.2 Study on change of test solution; difference by measuring point

Nishi immersed 100 × 100 × 400mm concrete prisms in pairs in 5% sulphuric acid solution and 5% solution of 35% hydrochloric acid, and measured the changes in concentration at two different points as shown in Fig. 5. The results are also shown in Fig. 5, and it is noticed that the differences of concentrations in the test solution between positions of the measuring points relative to the specimens are considerable.

6.3 Studies on influence of position of specimens on their deterioration

Kasai prepared mortar specimens according to the common testing conditions stated in 4.1 but cured in water up to the age of 14 days, then immersed horizontally in 2% hydrochloric solution for 8 weeks and 5% sulphuric acid solution so as to keep the under surfaces of specimens at a depth of 50mm or 200mm. The change of weight, cross sectional area and flexural strength after immersion were measured and their ratios to the initial values Z_s were determined. The results showed remarkable influence of depth of immersion on these ratios. Particularly the ratios of flexural strength for specimens immersed at a depth of 200mm were nearly half of those at a depth of 50mm. So it is recommended to regularly change the position of specimens up and down.

Effects of partial immersion and repeated immersion of specimens in the test solutions were also tested by Ikenaga and Kumahara and it was found that these methods of immersion tended to accelerate the deterioration compaired with complete successive immersion.

7 Methods of measurement for properties of specimens specified in the draft and related studies

It is specified in the draft that length change, dynamic modulus of elasticity and strength shall be measured according to the corresponding JIS and the other properties not specified in JIS shall be measured according to the methods specified in the draft, as outlined below.

(a) Depth of carbonation shall be measured at the cross section of specimen employing 1% phenolphthalein solution.

(b) pH at the different depth shall be measured at the surfaces of the specimen of different depth exposed by abrasive machining, employing digital pH meter with suitable flat surface glass electrode. A typical example of pH profile for mortar specimens thus obtained by Einaga is shown in Fig. 6. It can be seen that the slight deterioration of mortar was observed up to the depth of 8.0–8.5 in pH.

(c) Depth of deterioration shall be determined from the difference of cross sectional area of the sound part between before and after immersion.

(d) Porosity shall be measured by use of mercury porosimeter of 100 MPa in maximum pressure, and pore volume distribution against pore radius shall be determined. Methods of sampling and drying of specimens for the porosity test were specified in detail based on the study by Suzuki.

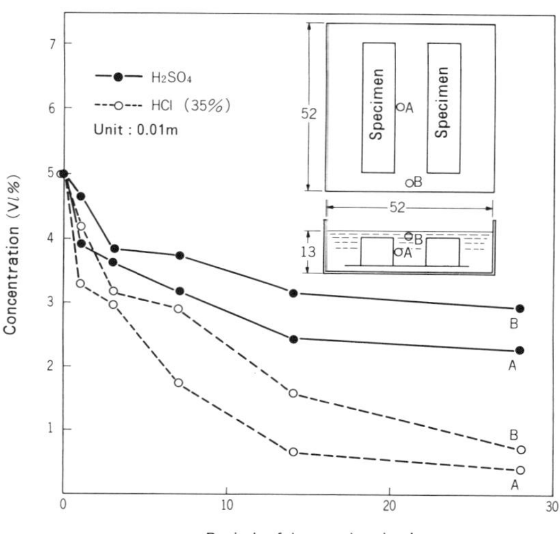

Fig. 5. Concentration of solution at different points

8 Calculation of test results specified in the draft

As described in 4.4 (c) for example, change of properties due to chemical attack of the specimens immersed in solutions are shown more clearly by comparing with that of specimens immersed in water. Therefore the methods for calculating test results specified in the draft are prepared taking this into consideration, as outlined below.

(a) Notation

X_{soi}, X_{woi}: Test result for each specimen just before immersion in test solution and in water respectively

X_{so}, X_{wo} : Average of X_{soi} and X_{woi} respectively

X_{si}, X_{wi} : Test result for each specimen just after immersion in test solution and in water respectively

X_s, X_w : Average of X_{si} and X_{wi} respectively

L_{si}, L_{wi} : Change for each specimen during immersion in test solution and in water respectively:

$$L_{si} = X_{si} - X_{soi} \qquad L_{wi} = X_{wi} - X_{woi}$$

L_s, L_w : Average of L_{si} and L_{wi} respectively

Y_{si}, Y_{wi} : Percentage of change for each specimen:

$$Y_{si} = 100(X_{si} - X_{soi})/X_{soi} \qquad Y_{wi} = 100(X_{wi} - X_{woi})/X_{woi}$$

Y_s, Y_w : Average of Y_{si} and Y_{wi} respectively

(b) As for compressive strength and flexural strength, calculate the following two ratios: $Z_s = X_s/X_{so} \qquad Z_w = X_s/X_w$.

(c) As for weight and dynamic modulus of elasticity, calculate Y_s, Y_w and percentage of relative change U: $U = Y_s - Y_w$

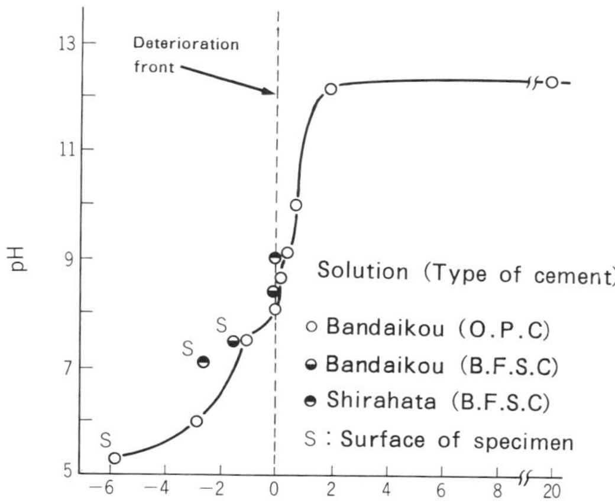

Fig. 6. A typical example of pH profile

Table 8. Z_w for compressive strength of superplasticized high strength concrete specimens after immersion for one year

C : B: S*	Slump (cm)	Air (%)	Water (kg/m^3)	X_{wo} (MPa)	X_w (MPa)	Z_w^{**} 2% HCl	5% H_2SO_4	10% Na_2SO_4
100: 0: 0	4	0.8	126	74.4	92.8	0.60	0.65	0.87
70:20:10	22	1.0	126	99.1	120.3	0.65	0.70	0.93
85: 0:15	22	0.9	140	110.6	116.4	0.62	0.69	0.90
80: 0:20	22	1.2	140	105.9	108.9	0.63	0.69	0.89

*C: Ordinary Portland cement, B: Fine slag powder, **See 8(a)(b)
S: Silica fume Water/(C + B + S) = 22 wt.%

(d) As for length change, calculate relative change L: $L = L_s - L_w$.

(e) As for properties other than mentioned above, suitable values for evaluating chemical resistance shall be selected among X_{so}, X_{wo}, X_s, X_w, Y_s, Y_w, L_s, L_w, Z_s, Z_w, U and L, and their calculated values shall be reported.

9 An example of the recent experimental studies

Yoda prepared 100 $\phi \times$ 200mm cylinder specimens with river sand and river gravel superplasticized high strength concretes, cured in water up to an age of 7 days, then kept in air at R.H. 80%. The specimens were immersed in the test solutions at an age of 28 days, and tested according to the methods specified in the draft. A part of the tests results are shown in Table 8.

10 Conclusion

As described above, various factors influencing on the test results of chemical resistance of concrete such as shape, size, curing conditions of the test specimens, methods of immersion for specimens in the test solutions were fairly clarified by our systematic research. We expect the draft of standard methods of test for chemical resistance of concrete in aggressive solutions, proposed based on the fruits of this research, including properties to be measured, methods of preparation and coating for specimens, calculation of test results and so on, will be put to effective use for evaluation of chemical resistance of concrete

11 References

JMC Working Group on Chemical Resistance Test (1984), Research on the method for chemical resistance of concrete in aggresive solutions. Cement Concrete, 443, 31–39, 444, 29–38.

Regourd, M. (1984) 32-RCA: Resistance of concrete to chemical attack. Materials and Structures, 14–80, 130–137.

58 APPLICATION OF FTIR SPECTROSCOPY AND XPS TO POLYMER/MINERAL JOINT SYSTEMS

S. WAGENER, H. HÖCKER
Lehrstuhl für Textilchemie und Makromolekulare Chemie der RWTH-Aachen, Aachen, Germany

Abstract
Model polymer/mineral joint systems were used to demonstrate the feasibility of Fourier transform infrared (FTIR) spectroscopy and X-ray photoelectron spectroscopy (XPS). The results indicate that the used polymers can be definitely identified on quartz and kaolinit surfaces and the polymer content is quantificable by both techniques. It is shown that the use of time-dependent FTIR-spectroscopy is most suitable to investigate the curing behaviour of polymers under varying conditions. Furthermore, the application of FTIR-spectroscopy to interface studies is illustrated.
Keywords: FTIR-spectroscopy, X-ray Photoelectron Spectroscopy, Polyurethane, Quartz, Kaolinit, Surface analysis, Interface analysis

1 Introduction

A major concern in the development of advanced composite materials technology is the problem of composite reliability. Polymer/mineral joints find practical applications in various areas as in protective coatings, structural composites or impregnations, and in order to ensure long-life durability of a polymer/mineral joint various technological tests are carried out.
The effectiveness of a modification is commonly judged according to practical and service-life related parameters. In this way symptoms for macroscopically observable failures are diagnosed and give hints to the quality of a modification. For a specific optimization it is furthermore necessary to interprete the macroscopic behaviour in order

to understand the reasons for failures. Therefore, the aim must be a characterization of the physico-chemical behaviour of the joint system before and after e.g. environmental stresses to provide information on the molecular changes within the composite material. This demand implies a molecular approach to the problem and requires sensitive analytical techniques.

In this paper we discuss the characteristics of Fourier transform infrared spectroscopy (FTIR-spectroscopy) and X-ray photoelectron spectroscopy (XPS) and their application to studies of polymer/mineral joint systems.

2 Spectroscopic techniques

A large number of analytical techniques may be considered appropriate to studies of composite materials but some restrictions must be imposed. Taking into account that studies of polymer/mineral joints may imply consideration of different problems, eg. adhesion phenomena or degradation processes upon ageing. The following demands for a suitable analytical technique are required:

- surface sensitivity (information depth limited to several monolayers)
- universal application according to sample composition, geometry and topography
- detailed chemical information on a molecular level
- quantification
- no sample damage during data acquisition (free of artifacts)
- reproducibility.

Up to now there is no technique fulfilling all the requirements. With regard to the above mentioned aspects and to the stage of commercially available techniques a combination of Fourier transform infrared spectroscopy (FTIR-Spectroscopy) and X-ray photoelectron spectroscopy (XPS) seem to be best suited to investigate surface and interface phenomena.

2.1 FTIR-spectroscopy

The powerful combination of specific compound identification and functional group characterisation led to infrared spectroscopy becoming the dominant tool for investigating the molecular structure of chemical compounds.

The physical basis for FTIR spectroscopy is the excitation of molecular vibrations in the infra-red region of the electromagnetic spectrum. A discrete amount of energy is required to excite a certain vibration so that each functional group absorbs infra-red light of a characteristic wavelength. The position of an absorbtion band depends on the mass of the atoms involved, the binding strength and surrounding chemical groups. Measurements are made in the central infra-red region covering the range of wavelengths between 2.5 and 5.0 corresponding to wavenumbers between 4000 and 200 cm^{-1}. The depth of information obtained is dependant on the sampling techniques employed. Transmission techniques are generally used to obtain bulk analysis whereas reflection techniques are employed for surface analysis. Depending on the particular sampling conditions the information depth ranges between 0,3 and 10 µm.

2.1 X-ray photoelectron spectroscopy

X-ray photoelectron spectroscopy is one of the most frequently used techniques for the analysis of the uppermost atom layers of solids (information depth amounts to about 50 atom layers corresponding to about 10 nm). XPS involves the measurement of kinetic energies of electrons ejected by interaction of a molecule with x-rays. Basically, the kinetic energy of a photoelectron is given by the difference between the energy of the irradiated X-ray photon and the binding energy of the photoelectron in the electron shell. The kinetic energy is characteristic of each element so that the determination of the elemental composition of the surface layers (except hydrogen and helium) can be detected. In contrast to energy dispersive X-ray analysis (EDX) the advantage of this technique is that according to chemical shifts information about the binding state of an element is provided. The binding state of an element depends on its chemical surrounding, and therefore different characteristic groups can be determined quantitatively.

3 Experimental

3.1 Aims

In combination the main features of FTIR-spectroscopy and XPS fulfil the demands of their application to surface and interface science of polymer/building material composites. Although their efficiency has been established already in many different research areas dealing e.g. with adhesion problems the problem orientated capability is strongly dependant on the sample properties. The initial studies described here were done therefore with the purpose of developing sampling preparation techniques and demonstrating feasibility to polymer/mineral joint systems.

Investigations of failure surfaces, interfaces (polymer/ mineral) and polymer/mineral joint systems presuppose primarily a definite identification of the organic/polymer component on the mineral surface and secondly the possibility of quantifying this content. Precisely defined model joint systems are best suited to prove these basic requirements. The first systems chosen for these studies were composites of cold curing polyurethane resins and different mineral powders.

3.2 Materials and sample preparation

3.2.1 Minerals

Quartz with a particlesize < 200 µm was purchased from Quarzwerk Dörentrop (FRG) and Kaolinit from Fa. Erbslöh & Co (Geisenheimer Kaolinwerke - FRG). Basalt obtained from Fa. Vama (Basaltwerke, Alsdorf Griesberg, FRG) was ground to a particlesize < 200 µm. The elemental composition of the minerals was determined by means of X-ray fluorescence analysis and is summarized in table 1.

Table 1. Oxide composition of quartz, kaolinit and basalt by weight% determined by X-ray fluorescence analysis

Oxide	Minerals		
	Quartz (weight%)	Kaolinit (weight%)	Basalt (weight%)
SiO_2	100,0	58,6	46,2
Al_2O_3		30,8	17,4
Fe_2O_3		1,3	12,4
CaO		0,1	9,4
MgO		0,6	3,3
TiO_2		0,8	3,1
K_2O		7,4	1,8
P_2O_5		0,1	1,6
MnO			0,2

The minerals were dried for 2 days at a temperature of 60°C and were used without further purification.

3.2.2 Adhesives

The polyurethanes studied in this work were supplied by Bayer AG, FRG. They are isocyanate functionalized prepolymers on the basis of isophorondiisocyanate designated as SZL 3023 and hexamethylendiisocyanate adducts designated as SZL 3024 and LS 2550. For some investigations the catalyst dibutyltin dilaurate (p.A.) was employed to shorten the curing time of the polyurethanes. Butylacetate(p.A.) was used as solvent.

3.2.3 Sample preparation

Polyurethane films were obtained by keeping a solution prepared of 5 g prepolymer 0,05g catalyst and 11,7 g butylacetate in a teflon dish for 7 days at standard climate (23°C/50% humidity). The polymer/mineral joints were prepared by adding 10 ml of the solution with varying prepolymer concentration, from 0,2% to 11% by weight to 3 g of quartz and kaolinit and storing these samples for 10 days at standard climate (23°C/50% humidity). For the investigation of the curing behaviour pure SZL 3024, SZL 3024 with

0,1% catalyst by weight, and 1/1 mixtures by weight of SZL 3024 with different minerals were applied to potassium bromide pellets and kept in a drying cupboard at 30°C. The corresponding FTIR measurements were carried out over a period of 48 days.

3.3 Infrared analysis

The infrared analysis were carried out in a Nicolet 60SX Fourier Transform Infrared Spectrometer. The curing behaviour was investigated in transmission. IR absorption spectra of the polyurethane films were obtained employing the attenuated total reflection (ATR) technique using a KRS-5 crystal and the analysis of the minerals were carried out by means of the diffuse reflection technique. All spectra were taken by 250 scans with a resolution of 4 cm^{-1} using a TGS-detector.

3.4 XPS analysis

The XPS measurements were carried out in a X-Probe™ spectrometer, model 206 manufactured by Surface Science Instruments (Mountain View, CA, USA). Al $K_{\alpha 1,2}$ radiation (1486.6 eV) was used to obtain XPS spectra at a power of 175 W for survey spectra and 70W for elemental spectra. All measured binding energies were charge corrected to the C 1s photolectron line at a binding energy of 285.0 eV attributed to the aliphatic carbon (C-C, C-H). Under the used operation conditions the FWHM is 1.55 eV for survey spectra and 0.9 eV for elemental spectra determined for the Au $4f^{7}/_{2}$ photoelectron line.

4 Results and discussion

4.1 Qualitative and quantitative characterization of polyurethanes on mineral powders

FTIR- and XPS spectra of the analysed polymer/mineral joints exhibit the spectroscopic features of both partners: the polymer and the mineral. In order to separate the spectroscopical information into the polymer and mineral parts it is necessary to characterise the parent compounds as well. Their data form the basis of all other results since precise band assignment, proper data manipulation (difference spectroscopy) and chemical changes due to the creation of a joint system or ageing processes can only be achieved by comparing the spectra of the joints with standards. Consequently, the spectra of the parent compounds are used as references.

4.1.1 FTIR-spectroscopy

Fig. 1 shows the FTIR absorption spectra of kaolinit (1), SZL 3024 (2) and the difference spectrum (3) obtained by subtracting kaolinit from the joint SZL 3024/kaolinit. The difference spectrum (3) representing the polymer on the

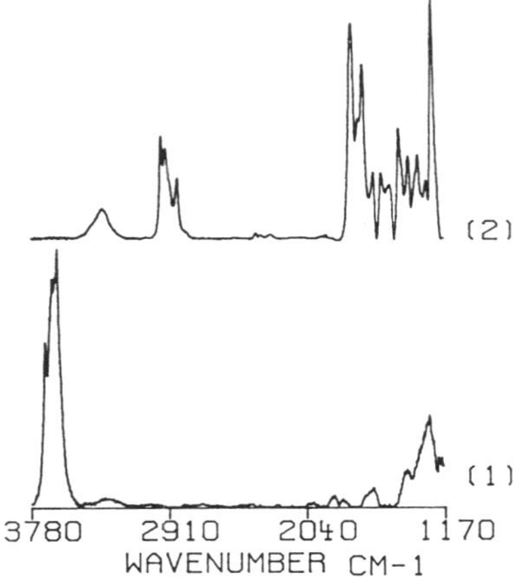

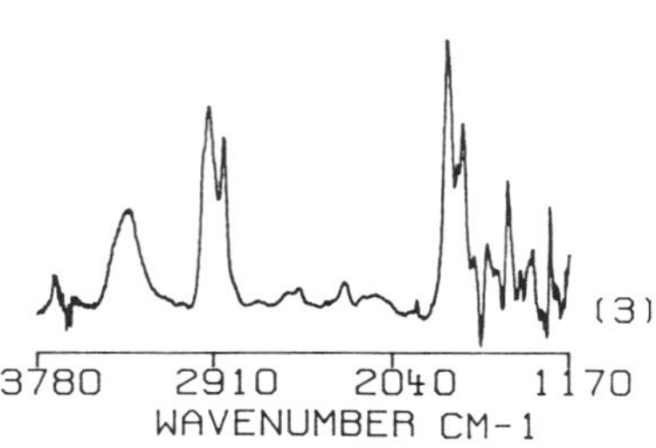

Fig. 1. FTIR absorption spectra of kaolinit (1), SZL 3024 (2) and SZL 3024 on kaolinit (3)

clay mineral surface exhibits the same absorption pattern as the pure polymer system. This is a clear evidence for its presence on the kaolinit surface.

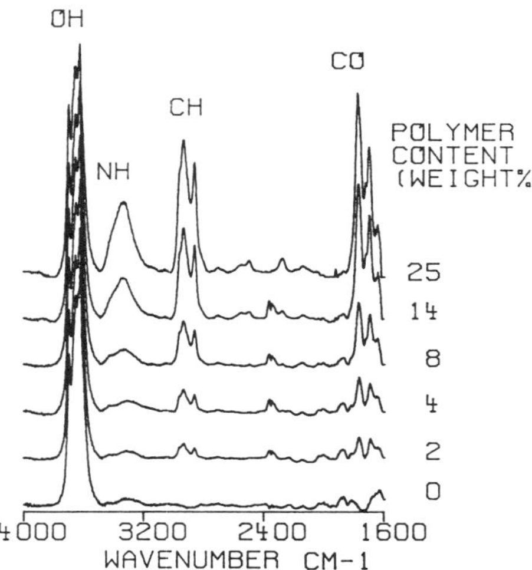

Fig 2. Absorption spectra of SZL 3024/kaolinit joint systems with varying polymer content

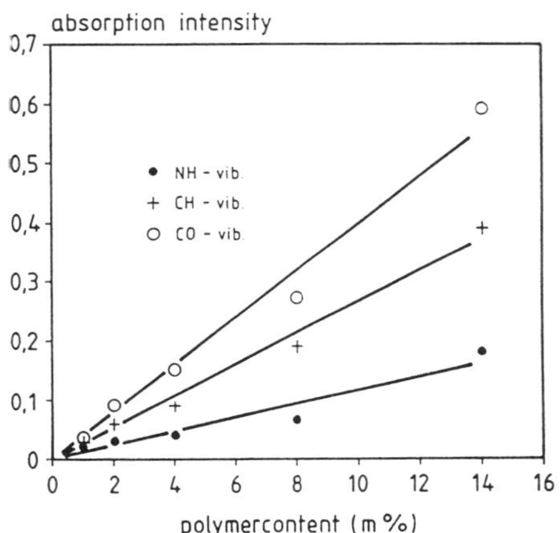

Fig 3. Absorption intensity of the NH-, CH- and CO-vibration in dependence of the SZL 3024 content in a kaolinit joint

To prove the possibility of quantification the obtained FTIR-spectra of SZL 3024/kaolinit joints with varying SZL 3024 content were normalized to the intensity of a mineral specific absorption band (internal standard), the stretching vibrations of hydroxyl groups at 3643 cm^{-1}. These standardized spectra are illustrated in Fig. 2. In the region presented the polyurethane shows three characteristic absoption bands; first, the NH stretching vibration at 3330 cm^{-1}, second the CH stretching vibration at 2925 cm^{-1} and last the CO deformation vibration at 1762 cm^{-1}. Their absorption intensities decrease upon diminishing polymer content. A quantification is derived by correlating band intensities and polymer content. This correlation is presented in Fig. 3. For this system - SZL 3024/Kaolinit - a linear correlation is observed and the detection limit for the polyurethane is reached at a polymer content of 2% by weight. According to different absorption coefficients the detection limits vary for the three functional groups and as the results indicate the absorption band best suited for quantification is the CO stretching vibration.

4.1.2 XPS

The elemental composition of the surface layers of untreated quartz is given by silicon (Si), oxygen (O) and carbon (C) due to atmospheric contamination. As shown in Fig. 4a the O/Si ratio is not exactly stoichiometric. A polymer modification of a mineral surface will change its composition as further elements in particular N and C are introduced. The polyurethane treatment of quartz powder should therefore result in an increase in the C and N and a decrease of the O and Si amount. These changes are observed for the treated quartz as shown in Fig. 4b which confirms the presence of the polymer at the mineral surface.

Although the polymer treatment can be detected without problems, the observed changes in the elemental composition are not specific for a polyurethane system since many polymers consists of N and C. The more characteristic informa-

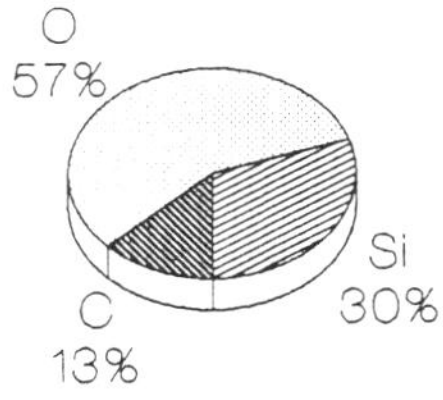

Fig. 4a. Elemental composition of quartz (atom%) determined by XPS

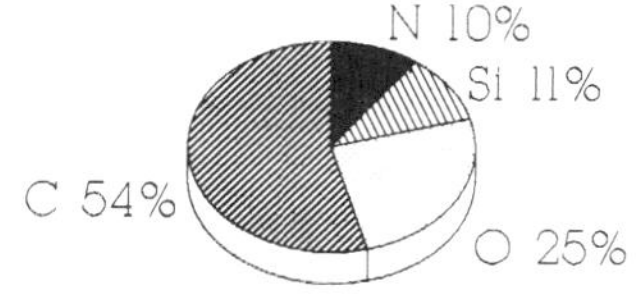

Fig. 4b. Elemental composition of a LS 2550/quartz joint determined by XPS

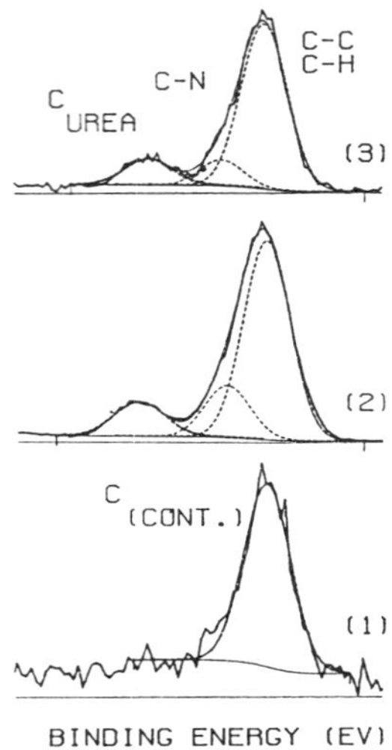

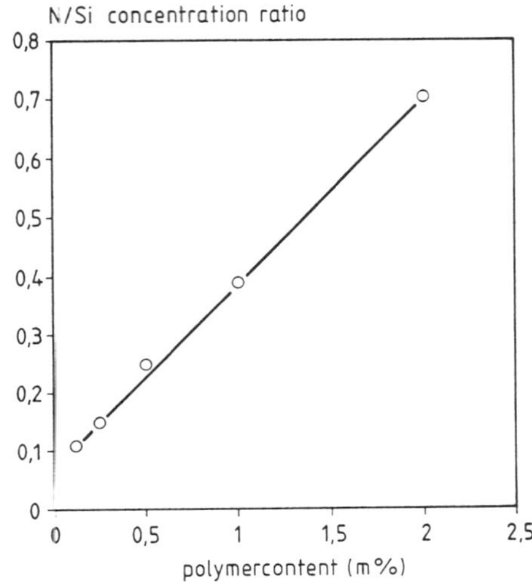

Fig. 5. C 1s element spectra of quartz (1), LS 2550 (2) and LS 2550/quartz joint (3) obtained by XPS

Fig. 6. N/Si concentration ratio of LS 2550/quarz systems in dependence of the LS 2550 content

tion is derived from the elemental spectra resolving the different binding states of the investigated elements. Fig. 5 shows the C 1s spectra for quartz, LS 2550 and LS 2550/quartz joint. Three different binding states are characteristic of this polymer: first, C-C, C-H at 285 eV; second, C-N at 286,2 eV and third, C_{urea} at 289,4 eV. These binding states are also detected for the joint system but not for the mineral and this gives clear evidence for the polyurethane treatment.

For this joint system a quantification of the polymer content by XPS can be derived from the ratio of the N to Si concentration as N is a polymer specific and Si a mineral specific element. The results obtained for LS 2550/quartz joints with varying polymer content are illustrated by Fig. 6. In the range from 0,2 to 2% polymer content a linear correlation between N/Si ratio and polymer content is observed.

Identification and quantification of the polymer on a mineral surface can be provided by both techniques. The type of information and the detection limit are specific for each technique. It therefore depends on the particular sample and problem to decide which technique is required.

4.2 Curing

In many applications as for crack repair, coatings and impregnations for building materials there is a need for cold curing polymer systems. They are applied in a more or less fluid condition and develop their properties, e.g. mechanical strength, only after curing. However, the curing process is influenced by various parameters as the application conditions or the surface pretreatment of the building material and due to the curing conditions polymer reactivity and properties may change.

In the work presented it was of primary interest to investigate the catalytic effect (positive or negative) of mineral powders on the curing behaviour of SZL 3023.
SZL 3023 is an aliphatic isocyanate functionalized prepolymer. Curing occurs under the influence of humidity and moisture; the addition of H_2O to isocyanates leads to a urea bridged polyurethane. As illustrated in Fig. 7 the curing process can be observed by means of time depending FTIR-spectroscopy due to the decrease of the NCO stretching vibration at 2270 cm^{-1} and the formation of the urea group indicated by the deformation vibration at 1640 cm^{-1} (right shoulder occurring beside the absorption at 1700 cm^{-1}).
Fig. 8 shows the curing of SZL 3023 in the course of time with and without a catalyst, quartz and kaolinit. The curing process is accelerated in the presence of the mineral powder although not to the extent of the used commercial polyurethane catalyst. The positive catalytic effect of kaolinit is more developed than that of quartz. This fact can be attributed to the higher Bronstedt and Lewis

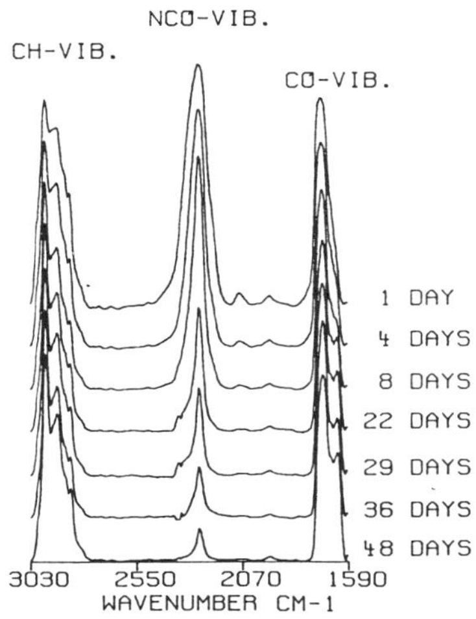

Fig. 7. Time depending absorption spectra of SZL 3024

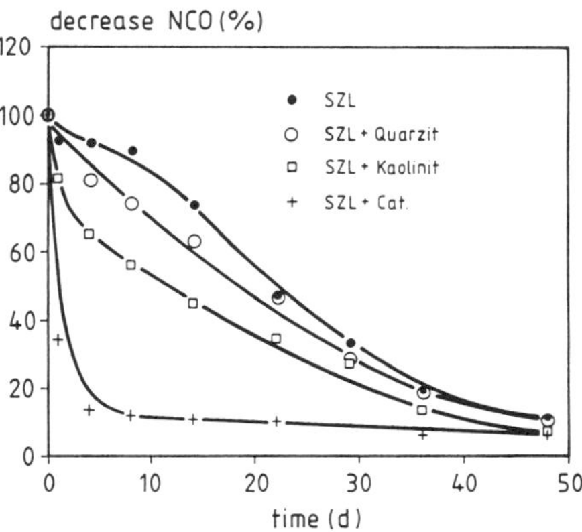

Fig. 8. Curing time determined from the decrease of the NCO-absorption band in dependence of the curing conditions

acidity of the kaolinit surface since it is known that acids promote this particular polyaddition reaction. However, the influence of various curing parameters may be observed in the same manner by means of time-dependent FTIR-spectroscopy.

4.3 Interactions between organic compounds and minerals
The material design and the control of the interface between polymers and minerals are known to be important to achieve durability of the adhesive bond to various environmental attacks since failure is most likely to occur at the interface. Knowledge of the nature of the bonding in the interface is therefore one essential requirement in order to improve interface properties.

One possibility to achieve an analytical access to an interface is to apply only several monolayers of the adhesive to a mineral surface as this is the region where interface reactions occur. These interactions between the adhesive layers and the mineral surface can be investigated by FTIR-spectroscopy since the position of an absorption band depends among other parameters on the chemical surrounding of this particular group. If changes arise in the vibrational spectra of the adhesive coumpound on the mineral surface compared to that of the reference they are assigned to the interface interactions. Fig. 9 presents the absorption spectra of diethylentriamin (deta) - a typical compound used in formulations of reactive epoxy resins -

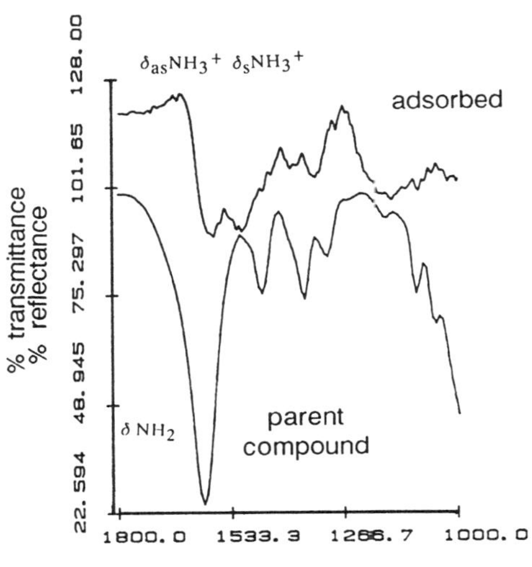

Fig. 9. Absorption spectra of diethylentriamin and diethylentriamin adsorbed to basalt

and deta adsorbed on basalt. The characteristic functional group is the amine; the corresponding absorption band for the NH-deformation vibration occurs at 1598 cm^{-1}. Adsorbed at a basalt surface changes due to the amine vibration are observed. The absorption band at 1598 cm^{-1} disappears and two new bands turn up. The absorption band at 1581 cm^{-1} can be assigned to the asymmetric and that at 1497 cm^{-1} to the symmetric deformation vibration of a protonated amine group. Based on these shifts for the amine group a ionic type of bonding due to the formation of an acid-base complex between the amine the basalt surface is suggested.

5 Conclusion

A molecular approach is required to understand the reinforcement and degradation mechanisms in practical application areas of polymer/mineral joints such as protective coatings and impregnations of building materials. The results obtained by employing FTIR-spectroscopy and XPS to the characterization of polyurethane/mineral joint systems have demonstrated that these techniques are most suitable to provide molecular information due to the chemical features of a polymer within a joint. Failure analysis implies the detection of molecular changes at the surface or interface of a composite material. The aim of this work was therefore to establish the possibilty of a definite identification and quantification of an organic or polymer compound on a mineral surface. It is shown that a polyurethane treatment of a quartz or kaolinit surface is qualitatively and quantitatively detected by both spectroscopic techniques. According to different information depths which are covered by FTIR-spectroscopy and XPS the detection limits differ. In general it will depend on the particular sample and problem which of these techniques is required.

Optimization of a polymer/mineral joint system is basically achieved by a modification of the polymer system. The properties of a polymer are rather well known but applied to a mineral properties may change due to the interaction with the mineral surface. In this study the curing behaviour of polyurethane in the presence of quartz and kaolinit was investigated by means of time-dependent FTIR-spectroscopy. The results establish a positive catalytic influence of the minerals. This effect is more developed for kaolinit and can be attributed to the higher Bronstedt and Lewis acidity of the clay mineral surface. The influence of other curing parameters such as condition of the mineral surface (contamination, moisture) can be investigated in the same manner.

The interface is in fact the least understood part of a polymer/mineral joint but it is also the region where often failure occurs. Surface analysis methods applied to interface studies are able to provide invaluable information concerning the manner in which the organic (monomer or polymer) compound interacts with the mineral surface. As demonstrated in this study FTIR-spectrosocopy is a powerful tool to investigate the interactions in an interface. Due to the shifts of the NH-deformation vibration, the spectrum of a polyamine adsorbed to basalt suggests an ionic type of bonding in the interface region.

The results discussed in this paper represent the very first steps we took but they reveal the broad analytical feasibility offered by FTIR-spectroscopy and XPS. It is of minor importance which particular aspect for a polymer/mineral joint system is examined. Studies concerning e.g. adhesion phenomena, ageing problems or product development require the same analytical approach: the observation of molecular structures and their changes under varying conditions. It is shown that this information can be provided by the means of FTIR-spectroscopy and XPS. Although these techniques permit this possibility, intensive research efforts will be essential in the future to correlate the results obtained from these studies with the macroscopic properties of a polymer/mineral joint system.

6 References

Brewis, D.M. and Briggs, D. (1985) **Industrial adhesion problems.** John Wiley & Sons, New York

Briggs, D. and Seah, M.P. (1988) **Practical Surface Analysis by Auger and X-ray Photoelectron Spectroscopy.** John Wiley & Sons, New York

Clark, D.T. (1978) The investigation of polymer surfaces by means of ESCA. **Polymer Surfaces.** John Wiley & Sons, New York

Conley, R.F. and Lloyd, M.K. (1971) Adsorption studies on kaolinites - II - adsorption of amines. **Clays and Clay Minerals**, 19, 273-282

Dieterich, D. (1987) Poyl(urethane). **Houben-Weyl**, 20, 1561-1757

EP 0 259 644 (1988) Bayer AG

Griffiths, de H. (1986) Fourier Transform Infrared Spectrometry. **Chemical Analysis**, 83

Nguyen, T. (1985) Application of fourier transform infrared spectroscopy in surface and interface studies. **Prog. Org. Coat.**, 13, 1-34

Theng, B.K.G. (1982) Clay activated organic reactions. **Development in Sedimentology**, 35, 197-237

Acknowledgement

This work was sponsored by the BMFT (Federal Minister of Research and Technology).

59 ASSESSING CARBONATION IN CONCRETE STRUCTURES

L.J. PARROTT
British Cement Association, Slough, UK

Abstract
This paper examines various measurements that lead to a clearer picture of carbonation in a structural element than that provided by testing solely with a phenolphthalein indicator solution. The measurements include periodic determinations of relative humidity and air permeability in a cavity drilled into the cover concrete. These are coupled with thermal analysis, pH measurements and water sorption tests on the drill powder. Examples of results from measurements on the external face of a concrete column are presented. The concrete was very permeable and seasonal wetting was limited; these data were consistent with the deep level of carbonation observed. Tests on the drill powder demonstrated the presence of a transition zone from full to zero carbonation. Such results help understanding of observed depths of carbonation and provide guidance for subsequent maintenance and repair.
Keywords: Carbonation, Permeability, Relative Humidity, Alkalinity, Thermal Analysis, Water Sorption, In-situ Measurements.

1 Introduction

Atmospheric carbon dioxide diffuses into concrete through empty pores in the binder matrix and reacts with the hydrated cement. The reaction lowers the alkalinity of the pore fluid : this can destroy the passive oxide layer on any adjacent steel reinforcement and leaves the steel susceptible to corrosion, Kashino (1984), Saeki et al (1984), Fukushi (1985) and Parrott (1990). Corrosion, being an electro-chemical process, is also dependent upon the conductivity of the concrete surrounding the steel and this, in turn, increases with concrete moisture content. However at moisture contents close to saturation the corrosion rate diminishes due to an insufficient supply of oxygen, Tuutti (1980), Gonzalez (1982).

Normally the rate at which the carbonation reaction front penetrates the concrete is slow, an alkaline environment is maintained around the reinforcement and no corrosion is observed. However, with inadequate cover, porous cover or where a particularly long service life is required, the carbonation front may penetrate to the steel and result in corrosion with associated cracking and spalling; e.g. see Parrott (1987). Repair, maintenance or replacement following this type of damage is expensive and improved methods of assessing carbonation at an earlier stage can reduce overall costs.

This paper examines various measurements that lead to a clearer picture of carbonation in structural concrete than that provided by testing solely with a phenolphthalein indicator solution. The measurements include periodic determinations of relative humidity and air permeability in a cavity drilled into the cover concrete. These are coupled with pH measurements, aggregate content determination, thermal analysis and water sorption tests on the drill powder. Test results are discussed in relation to the likely progress of subsequent carbonation and corrosion.

2 Experimental

2.1 Preparation of cavities

The measurements in this paper relate to two locations on a south facing column of a building in South-East England, Figure 1. The depth and position of the reinforcement was determined using a covermeter and three cavities, 20mm diameter by 35mm deep were drilled at each location taking care to avoid the reinforcement.

Fig.1. Column with four plugs

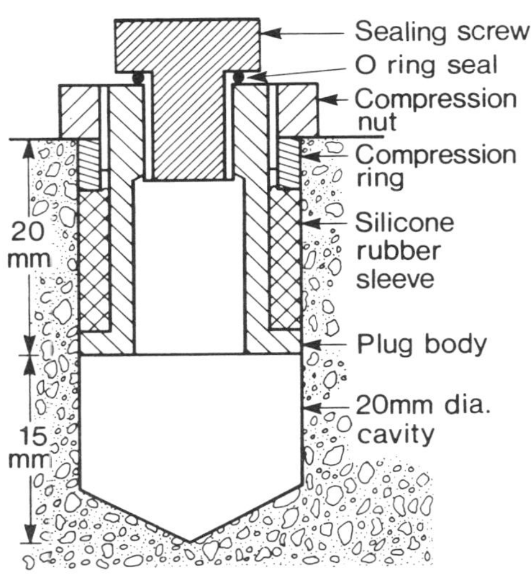

Fig.2. Plug and cavity in concrete

The cavities were drilled in about 2.5mm stages so that a small sample of the drill powder could be deposited on a paper tissue, lightly sprayed with a phenolphthalein indicator solution and any colour change noted. The bulk of the drill powder was collected using a collection chute and seven small sample jars. Each jar eventually contained the combined drill powder from three cavities for each 5mm depth range. These samples were used for aggregate content, pH, water sorption and thermal decomposition measurements. The cavities were carefully cleaned after each stage of drilling to remove all dust and avoid contamination of the next sample. When a depth of 35mm was reached the cavity walls were lightly sprayed with phenolphthalein indicator solution to obtain a check on the carbonation depth. Finally the cavities were fitted with a stainless steel plug that had an expanding silicone rubber sleeve to provide a gas-tight seal, Figure 2. Where the depth of carbonation exceeded 35mm a fourth cavity was drilled, as shown in Figure 1, to obtain samples of non-carbonated concrete.

2.2 Relative humidity and air permeability

Relative humidity was measured in each cavity at about one month intervals using a small capacitive probe. The use of the probe has been described in an earlier publication, Parrott (1988). Each probe reading is effectively calibrated during use by comparing it with the reading over an appropriate saturated salt solution. The standard deviation of readings was typically less than 2% relative humidity. The reductions of temperature at night lower the saturation vapour pressure of water. Over a period of 12 hours the moisture loss from the concrete would be small, Parrott (1988), and the relative humidity in a cavity would rise : for example reductions of temperature from 15 to 10 or 7.5°C could increase the relative humidity from 60 to 83 or 99%. Variability of relative humidity readings from this source were minimized by taking readings within two hours of mid-day. Also the probe and salt solutions were stored adjacent to the concrete for at least one hour prior to measurement so that temperature equilibrium was achieved.

Air permeability was determined by pressurizing the cavity and observing the drop in pressure with time. Details of the method have been described in earlier publications, Chen and Parrott (1989) and Parrott and Chen (1990). Coefficients of variation around 2% can be achieved for periodic testing of a single cavity. This figure rises to 4-16% for different cavities in sound concrete and becomes greater still if the concrete is cracked, Parrott and Chen (1990). A foam of liquid soap solution was applied to the stainless steel plug and the local concrete so that the passage of permeating air could be directly observed. If leakage around the seal between the plug and the concrete was detected the plug was removed and refitted with a suitable sealing medium. Normally patches of fine bubbles were observed on the concrete surface within 40mm of the centre-line of the cavity.

Table 1. Indicator solutions for pH measurement, Banyai (1972)

Indicator solution	pH (range)	Colour change
Nitrazine yellow	6.6 (6.4-6.8)	Yellow > blue
Phenol red	7.3 (6.4-8.2)	Yellow > red
Diphenol purple	7.8 (7.0-8.6)	Yellow > violet
Cresol red	7.9 (7.0-8.8)	Yellow > violet/red
α-naphtholphthalein	8.0 (7.3-8.7)	Yellow > blue
m-cresol purple	8.2 (7.4-9.0)	Yellow > violet
Phenolphthalein	9.0 (8.2-9.8)	Colourless > magenta
Thymolphthalein	9.9 (9.3-10.5)	Colourless > blue*
Brilliant orange	11.3 (10.5-12.0)	Yellow > red*
Tropaeolin 0	11.9 (11.1-12.7)	Yellow > red*
Titan yellow	12.5 (12.0-13.0)	Yellow > red

*Not readily visible on powdered concrete.

2.3 Drill powder tests

Phenolphthalein is the usual indicator for carbonation measurement but it was possible to test each drill powder samples with a range of indicators so that an approximate pH value could be assigned. As with the phenolphthalein solution the indicators shown in Table 1 were prepared using industrial methylated spirit and distilled water so that they contained 15% by weight of water. About 10mg of drill powder was deposited on paper tissue, lightly sprayed with indicator solution and any colour change during the first minute after spraying was noted.

The aggregate content of each drill powder sample was determined using 2.5g samples and the method described in British Standard 1881 : Part 124 (Section 5.9.3). Previous tests with local siliceous aggregates suggested that the insoluble residue represents 95% of the aggregate weight. Aggregate contents were used to express thermal analysis results in g/g ignited cement and sorption results in g/g dry cement paste.

Thermogravimetric analysis was conducted on 20mg samples using a computer controlled thermobalance, Killoh (1984). The heating rate was 5°C/minute and the atmosphere around the sample was dry nitrogen. Weight losses in various temperature ranges representing bound water from the cement gel, calcium hydroxide decomposition and calcium carbonate decomposition were determined according to Parrott and Killoh (1989).

Water desorption isotherms were determined by conditioning 2.5g samples over saturated salt solutions in vacuum desiccators at 20°C. The method and interpretation of results has been reported by Hagymassy et al (1972). The conditioning period was four days for each relative humidity and the final, dry condition was obtained by storing over silica gel for 2 days and then drying in an oven at 105°C for one day.

3 Results and discussion

3.1 Alkalinity and cover

At the top of the column the depth of neutralization indicated by phenolphthalein at the time the cavities were drilled was 22mm. This is shown in Figure 3 to be consistent with the pH results obtained subsequently using other indicators. The pH tests required visual judgement but similar results were obtained by different operators. Further use of these tests should permit a better assessment of their utility. The cover at the top of the column was 30mm or more so there was not yet any great risk of reinforcement corrosion and no local cracking was observed.

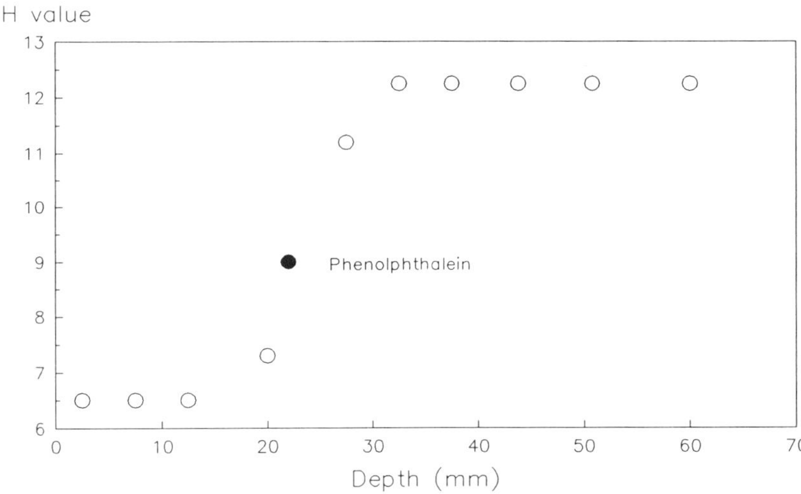

Fig.3. pH measured on drill powder from top of column.

The depth of phenolphthalein neutralization at the base of the column determined at the time the cavities were drilled was 50mm. Since the cover was only 31mm there was an obvious risk of corrosion and indeed cracks in the concrete parallel to the reinforcement were visible. Although the rate of steel corrosion increases rapidly with a reduction of pH below 10, Saeki et al (1984), the relatively dry condition of the concrete had a moderating effect. This will be considered further in Section 3.3.

3.2 Drill powder analyses

Aggregate contents plotted against depth from the exposed surface, Figure 4, show the reduced values near the surface that were expected from the work of Kreijger (1984) on particle packing. The average for aggregate content value suggests that the weight ratio of aggregate to cement was 4.4 ± 0.4, a figure typical of structural concrete at the time of construction.

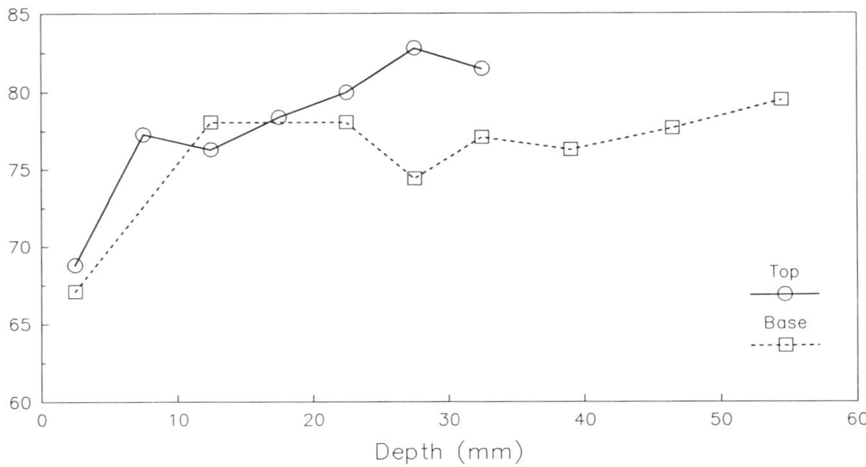

Fig.4. Aggregate content of drill powder from top and base of column.

Thermogravimetric results are shown in Figure 5. The calcium hydroxide losses increase where the carbonate losses decrease as would be expected from the carbonation reaction. The depths of phenolphthalein neutralization, shown as vertical lines, correspond to an intermediate level of carbonation. The gel water losses show changes with depth that parallel those for calcium hydroxide. The thermogravimetric results show regions of partial carbonation rather than a steep carbonation front; similar data were reported in a previous investigation by Parrott and Killoh (1989). The variations for the top of the column are consistent with the pH results shown in Figure 3.

Water desorption results for the top of the column, Figure 6, exhibit regular variations with increasing depth from the exposed surface, although the curves are virtually parallel above 33% relative humidity. The results in Figure 7 suggest that the reduced sorption capacity with increasing depth was closely related to the extent of carbonation. Pihlajavaara (1968) also reported a reduced sorption capacity with carbonation that is presumably due to a reduced surface area of the cement gel, Hilsdorf et al (1984). Poor curing and consequent gradients of cement hydration could also have contributed to the reduced sorption capacity near the exposed surface.

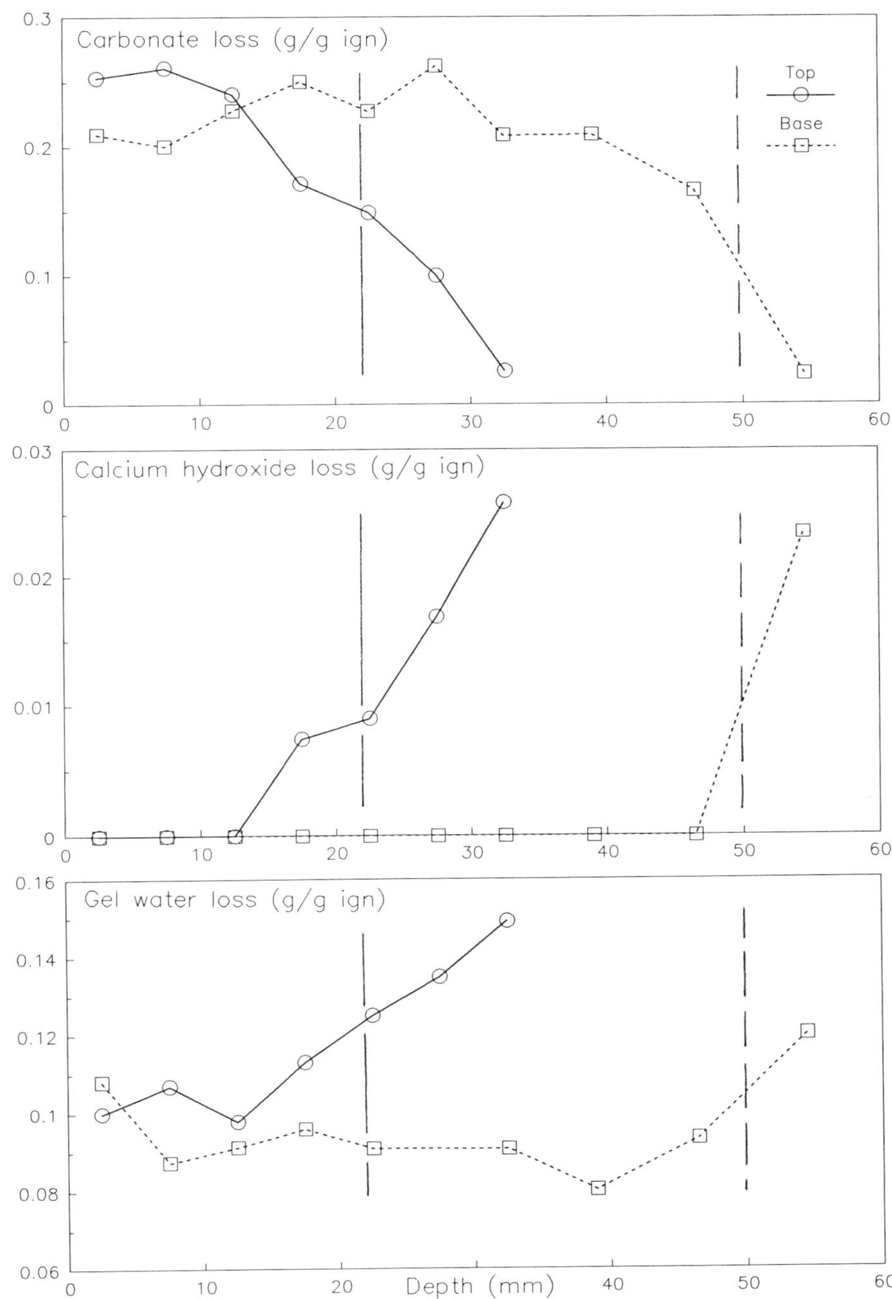

Fig.5. Carbonate, calcium hydroxide and gel losses versus depth from surface. Vertical lines show phenolphthalein neutralization.

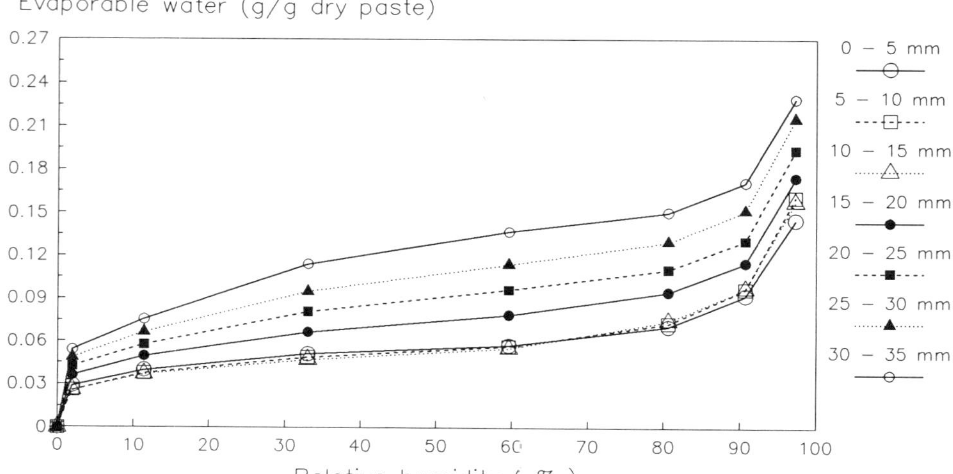

Fig.6. Water desorption of drill powders from different depths at top of column.

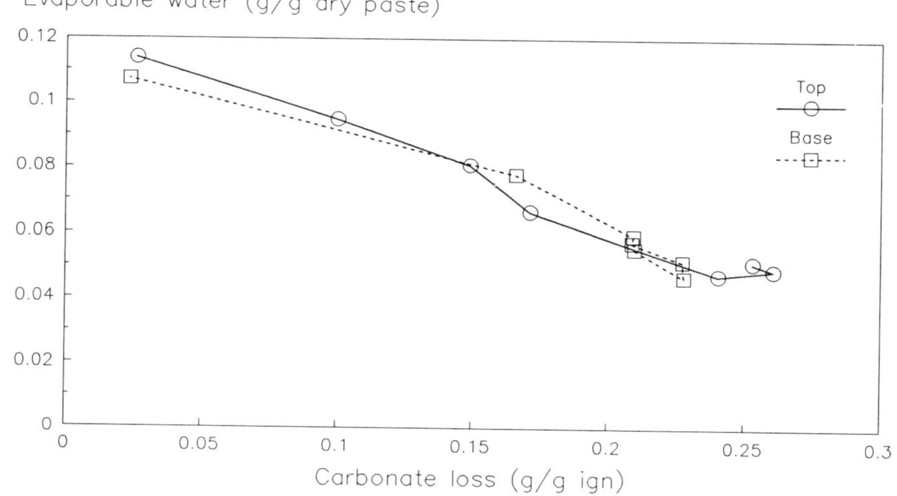

Fig.7. Evaporable water held at 33% RH versus carbonate loss for each depth at top and base of column.

3.3 Relative humidity and air permeability

The relative humidity results exhibited only small differences between cavities at the top and at the base of the column, the standard deviation of readings being about 1.5% relative humidity. During the first year of observation the relative humidity at the base of the column was slightly higher on average than that at the top of the column, Figure 8, possibly because the top was protected from direct rainfall by an overhanging roof slab. The relative

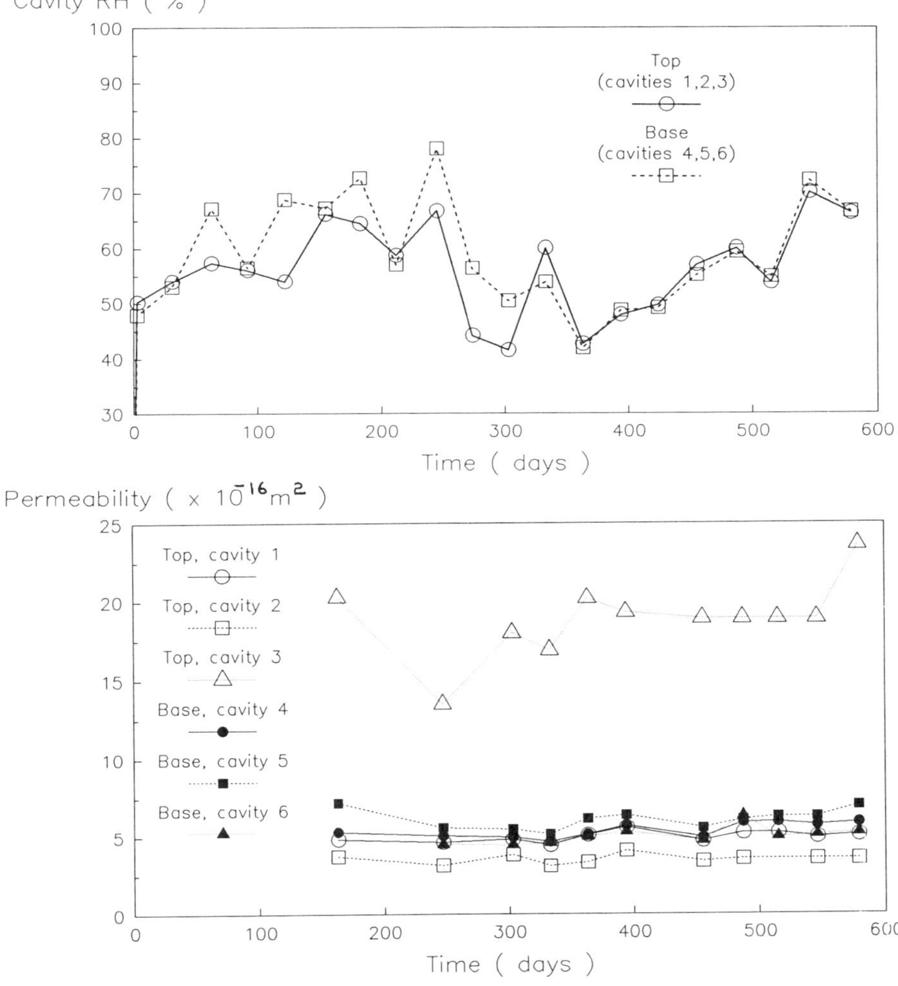

Fig.8. Relative humidities and permeabilities in cavities at top (1,2 & 3) and base (4, 5 & 6) of column

humidity values, generally between 40 and 70%, were low relative to measured ambient values probably because of the sheltered, south facing aspect and the warmth of the heated building.

Zero time in Figure 8 corresponded to mid-May and the small cyclic fluctuations of relative humidity corresponded approximately to annual changes of climatic conditions. The relative humidities were close to the optimum for the highest rate of carbonation, Parrott (1987). However they were also low enough to ensure that corrosion of reinforcement proceeded at a slow rate, Tuutti (1980), in spite of the depth of carbonation exceeding the cover at the base of the column.

The consistently dry condition of the concrete would be associated with very little blocking of pores by capillary condensation, so it is not surprising that the air permeability results in Figure 8 exhibit only small changes with time. The permeability values are typical of very low strength concrete with a cube strength at the end of the curing period of 10 to 12 MPa, Parrott and Chen (1990). On this basis the estimated depth of carbonation for the 39 year old column is 52mm, Parrott (1990), a figure that is very close to the measured value of 50mm at the base of the column. The consistently high permeability values for cavity 3 were associated with coarse bubbles of soap solution on the concrete surface and fine cracks in the concrete around the cavity. Also the 28mm depth of phenolphthalein neutralization for cavity 3 was higher than the 19mm depths for cavities 1 and 2.

The concrete at the base of the column was more deeply carbonated than that at the top. This could not be attributed to a drier condition of the concrete according to the relative humidity results in Figure 8. The aggregate content, water desorption and thermogravimetric analyses did not indicate any major differences between the concretes at the top and base of the column. The permeability results in Figure 8 indicate that the fully compacted concrete at the base of the column may be slightly more permeable than that at the top and this would help to explain differences of carbonation depth.

The present condition of the base of the column with its deep carbonation, low alkalinity and permeable concrete indicates that if the concrete should become wet there would be a high risk of more intensive reinforcement corrosion. Indeed extensive cracking has occurred at the base of an adjacent column where poor drainage of rainwater from the roof has caused wetting of the concrete. The lack of cracking elsewhere suggests that the relative humidity results in Figure 8 may be representative of most columns. Further measurements on other columns seem desirable. It would also seem advisable to ensure that water drainage is properly controlled and to institute a regular visual inspection of each column. There seems to be a good case for repairing cracked concrete, after any necessary treatment of local reinforcement and then coating the columns to minimize ingress of water and carbon dioxide.

4 Conclusions

The assessment of carbonation in a reinforced concrete column was greatly aided by conducting measurements additional to the usual ones for depth of phenolphthalein neutralization and cover. Thermogravimetric and water desorption measurements on drill powder indicated regions of partial carbonation. Measurements of pH on the drill powder also indicated regions of partial carbonation and low alkalinity. The cavities formed by drilling, 20mm diameter and 35mm deep, were sealed to permit periodic measurements of relative humidity and air permeability of the cover concrete. The relative humidity results indicated that the concrete was fairly dry. This was conducive to deep carbonation but moderated the rate of reinforcement corrosion in carbonated concrete. The concrete was found to be permeable and this helped to explain the deep levels of carbonation.

5 References

Banyai, E. (1972) Acid-base indicators. Chapter 3 of **Indicators** (ed E. Bishop), Pergamon Press, pp. 65-176.

Chen Zhang Hong and Parrott, L.J. (1989) Air permeability of cover concrete and the effect of curing. **British Cement Association Report.** Number C/5, October, 25 pages.

Fukushi, I. (1985) Carbonation and durability of reinforced concrete buildings. **Cement and Concrete.** (Japan), July, No. 461, 8-16.

Gonzalez, J. Vazquez, A. and Andrade, C. (1982) Les effets des cycles d'humidite sur la corrosion des armatures galvanisees. **Materials and Structures.** Vol.15, No. 88, 271-278.

Hagymassy, J. Odler, I. Yudenfreund, M. Skalny, J. and Brunauer, S. (1972) Pore Structure analysis by water vapour adsorption. **Journal of Colloid and Interface Science.** Vol.38, No. 1, 20-34.

Hilsdorf, H. Kropp, J. and Günter, M. (1984) Carbonation, pore structure and durability. **Proc. RILEM Seminar,** Hannover, 182-196.

Kashino, N. (1984) Investigation into the limit of initial corrosion occurrence in existing reinforced concrete structures. **Proc. Conf. Durability of Building Materials and components.**, Espoo, Finland, Vol.3, 176-186.

Killoh D. (1984) Computer controlled thermobalance. **Analytical Proceedings.** Vol. 21, March, 100-102.

Kreijger, P. (1984) The skin of concrete : composition and properties. **Materials and Structures.** Vol.17, No. 100, 275-283.

Parrott, L.J. (1987) A review of carbonation in reinforced concrete. **British Cement Association Report** C/1-0987.

Parrott, L.J. (1988) Moisture profiles in drying concrete. **Advances in Cement Research.** Vol.1, No.3, 164-170.

Parrott, L.J. (1990) Carbonation, corrosion and standardization. **Proc. Conf. The Protection of Concrete.** September, Dundee, Scotland. To be published.

Parrott, L.J. and Chen Zhang Hong (1990) Some aspects influencing air permeation measurements in cover concrete. **To be published.** 15 pages.

Parrott, L.J. and Killoh, D. (1989) Carbonation in a 36 year old, in situ concrete. **Cement and Concrete Research.** Vol.19, No. 4, 649-656.

Pihlajavaara, S. (1968) Some results of the effect of carbonation on the porosity and pore size distribution of cement paste. **Materials and Structures.** Vol.1, No. 6, 521-526.

Saeki, N. Takada, N and Fujita, Y. (1984) Influence of carbonation and sea water on corrosion of steel in concrete. **Trans. Jap. Concrete Institute.**, Vol.6, 155-162.

Tuutti, K. (1980) Service life of structures with regard to corrosion of embedded steel. **Performance of concrete in marine environment.** ACI Publication SP-65, 223-236.

Acknowledgements

The data presented in this paper are part of a programme on the effect of moisture upon carbonation, jointly supported by The Building Research Establishment and The British Cement Association. The assistance with the experimental work of Mr J Martin, Mr D Killoh and Mr P Pearson is gratefully acknowledged.

60 APPLICATIONS OF OPTICAL FIBER SENSOR SYSTEMS FOR MONITORING PRESTRESSED CONCRETE STRUCTURES

R. WOLFF, H.-J. MIESSELER
Strabag Bau-AG, Cologne, Germany

Abstract
Innovative use of glass fiber composites as an alternative to conventional steel reinforcement has made it possible for optical fiber sensors to be integrated in the prestressing tendons during production or directly into the concrete, thereby enabling constant monitoring of the structure to be carried out on a long-term basis.
This paper deals with the principles, development, initial applications, and future prospects of this type of sensor monitoring for concrete structures which thus become 'intelligent'.
Keywords: Glass Fiber Composites, Optical Fiber Sensors, Prestressing Tendons made of Glass Fiber Bars

1 Introduction

The demands to be met by modern man-made structures today are becoming increasingly diversified. In addition to the actual use, architectural viewpoints play an ever greater role. In the field of bridge construction these facts are nowadays demonstrated by structures which are becoming increasingly audacious. This is particularly evident in the greater span widths of bridges.
Spectacular cases of deterioration in the past have shown the necessity ty for loadbearing structures to be controlled on a permanent basis in order to guarantee their durability over a long period of time. In prestressed structures, cracks in the concrete and changes of the stress state in the tendons must be observed. Bridges, for example, can be monitored in this way by using optical fiber sensor systems which are incorporated in the high-performance fiber composite prestressing element or directly into the concrete.

2 Sensor technology

Owing to ever increasing requirements and demands on engineering building structures, the question of durability of the materials used and life expectancy of the structure is becoming of ever greater interest. Apart from actual new construction works, investment for the preservation of building structures accounts for a not inconsiderable portion of the funds available (regular investigation of bridges in the DIN

1076 class is therefore stipulated).

2.1 The principle of monitoring with optical fiber sensors

In telecommunications engineering, the optical fiber sensor is first and foremost a signal transmitter. Prerequisite for its high importance is the excellent light permeability achieved. In dependence with the mechanical stresses of the optical fiber sensor, light attenuation which is undesirable in telecommunications engineering is used as a sensor effect in the monitoring of building structures. In contrast to telecommunications engineering, efforts made in the development of the optical fiber sensors are directed to attaining the highest possible measuring signal as a result of mechanical changes in the optical fiber sensor.

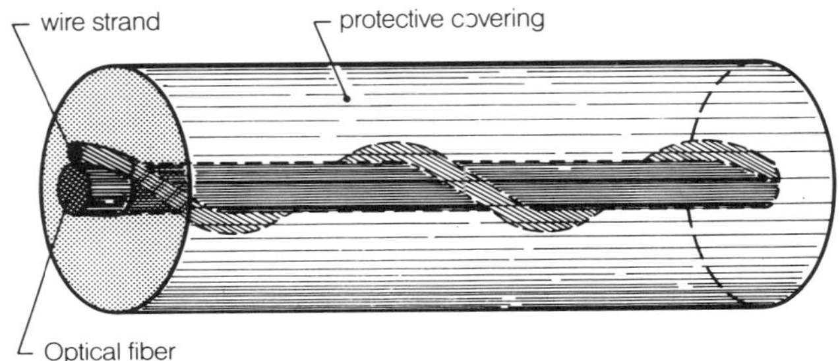

Fig. 1. Functional principle of the optical fiber sensor

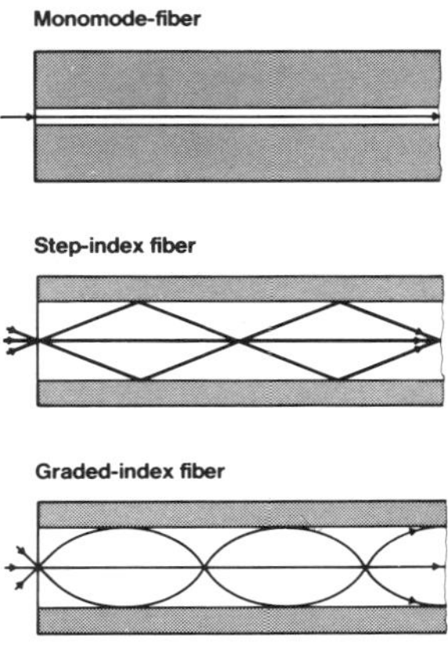

Fig. 2. Different types of optical fiber sensors

The gradient optical fiber sensor, with a decreasing index of refraction from the core radius, is usually employed for such sensor applications. The inner core is surrounded by a light-reflective outer sheath. It is also permeable to light but with a lower index of refraction. When a light ray is conducted through an optical fiber sensor, strew losses occur in areas of micro-bending. This resultant light loss is measured as a change of attenuation and expressed in decibels.

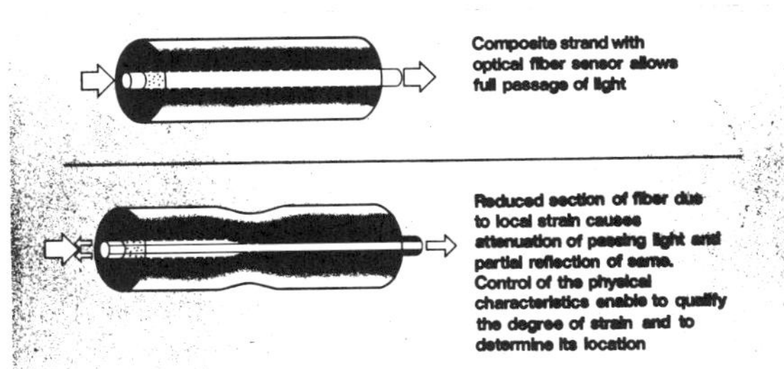

Fig. 3. Strain control with optical fiber sensors

As the optical fiber sensor is provided with a thin wire coil (Fig. 1), use is made of the knowledge that micro-bending can also be produced as a result of radial compression. When longitudinal tension occurs, the wire coil exerts radial pressure on the optical fiber sensor from a specific lay, thus causing micro-bending. This in turn results in corresponding changes of attenuation, thereby turning the optical fiber sensor into an optical fiber extension sensor.

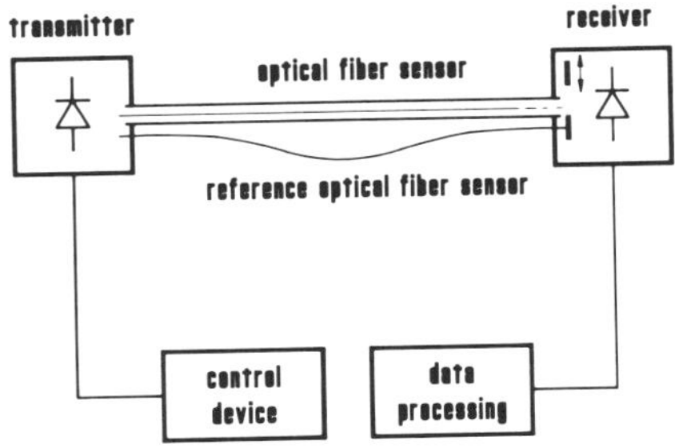

Fig. 4. Diagram of temperature compensated attenuation measuring process

The optical fiber extension sensor shows changes of attenuation as a function of the applied strain. As the stresses in concrete structures do not occur evenly and localised irregularities, such as for example cracks, can also develop, local defects should also be detectable in addition to the integral attenuation measuring process. To this end, the increased attenuation signal occurring at the point of interference can be recorded with the back-scatter measuring technique known from telecommunication engineering. The type of defect and also localised changes of stress are detectable by superimposing the attenuation curves of the loaded and unloaded sensor.

Clarification of the measurements:

- the control unit
  produces electrical signals for transmitter modulation

- the transmitter
  converts the control unit's electrical signals into light signals, bunching these in the extension sensor and reference fiber

- the receiver
  receives the light signals of the extension sensor and reference fiber, converting them into electrical signals

- the measuring electronics
  process the receiver's electrical signals, recording them in an appropriate manner

3 Investigations on Structural Elements for the Testing of Optical Fiber Sensors

There are two basically different possibilities for the monitoring of building structures using optical fiber sensors. Firstly for the monitoring of concrete structures, for example in the case of crack occurrence and its further development, and secondly to monitor the glass fiber composite tendons. To monitor cracks in concrete building structures; in new building structures sensors specially manufactured for this purpose are embedded directly into the concrete at locations on the bearing structure which the statics have shown to be particulary liable to stress. This enables statements to be made at a very early stage regarding any rehabilitation measures which might required, thereby reducing the cost for such works considerably. In existing structures, particularly structures which are already defective, these optical fiber sensors are post applied to the structural elements to be monitored, so that the cracks which have already occurred can be further observed, and repairs already carried out monitored in respect to the bearing capacity of the element. In glass fiber composite prestressing tendons, the optical fiber sensors are integrated directly into the fiber composite in order to show that these tendons are intact.

To prove that these optical fiber sensors fulfil the demands placed on them, comprehensive investigations were first carried out in the laboratory to determine basic applicability. The next stages were tests carried out at the University of Ghent with small prestressed beams having spans of 2.0 m and an overall height of 60 cm (Figs. 5 and 6). All sensor applications developed to date, that is sensors integrated in prestressing tendons, embedded directly in the concrete, and applied subsequently, were investigated in the tests. The beams were prestressed with two single-bar glass fiber prestressing tendons each, with and without bond.

Fig. 5. Cross-section of a 2 m beam

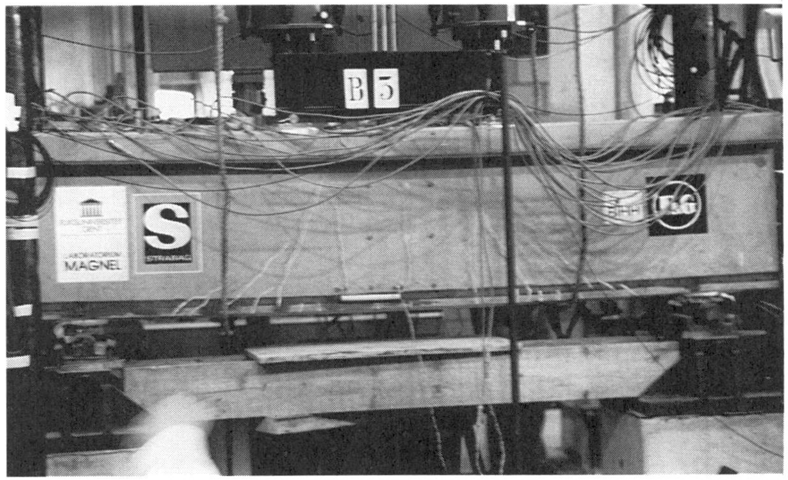

Fig. 6. View of 2 m beam

The final practical applicability test, also executed at Ghent University, was the investigation on a prestressed beam with a length of 20.0 m and a overall height of 1.0 m. The suitability of these sensors for the required applications and chosen methods of application could hereby be proven. This beam had limited prestressing comprising a 19-bar glass fiber prestressing tendon with a working load of 660 kN. In each case a total of 3 sensors were integrated into the prestressing tendon, 8 embedded in the concrete, and one subsequently integrated in grooves milled in the concrete.

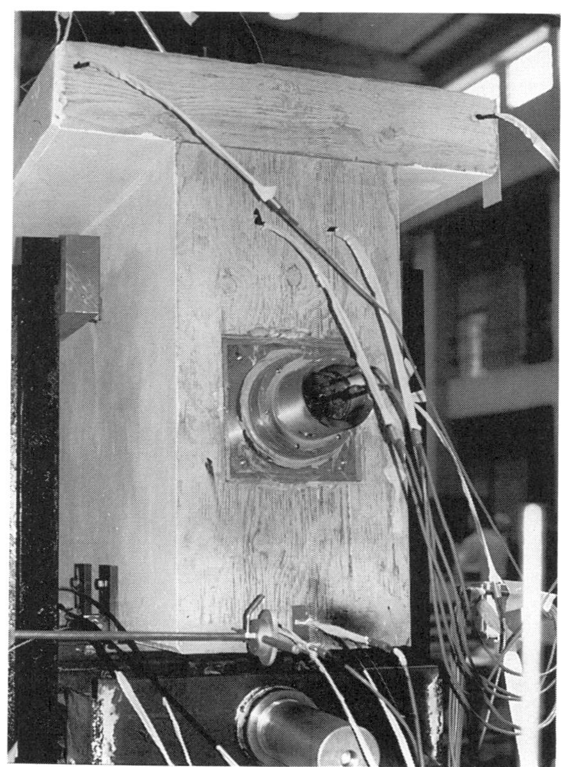

Fig. 7. Cross-section of 20 m beam

Both the transition from non-cracked to cracked state and exceeding of the concrete reinforcement's yield point could be determined with the aid of optical fiber sensors (Fig. 9). Similarly, these changes could be detected by optical fiber sensors in the prestressing tendon, up to failure of the beam (Fig. 10). The conventional measuring procedures, intended to control the results of the sensor measurements, showed concurring results. The application maturity of this state-of-the-art monitoring method could therefore be impressively substantiated.

Fig. 8. View of 20 m beam

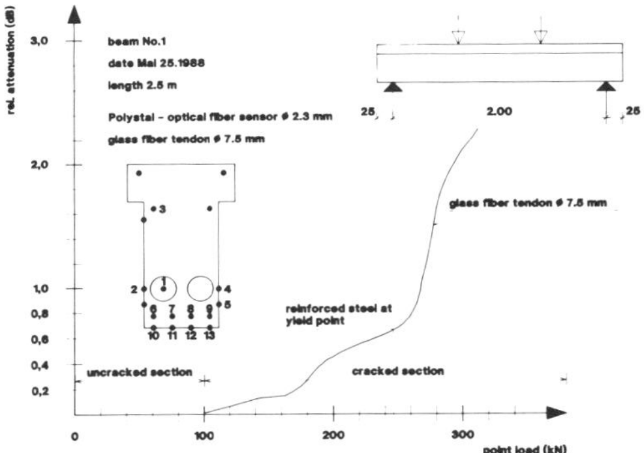

Fig. 9. Results of beam testing with optical fiber sensors in the tendon

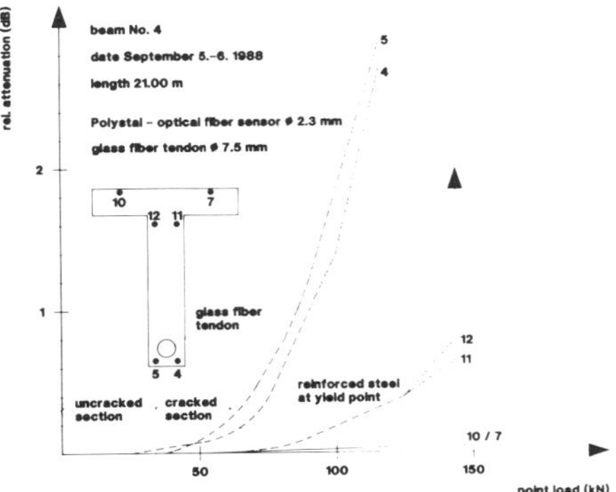

Fig. 10. Result of beam testing with optical fiber sensors in the concrete

4 Applications

4.1 Ulenbergstrasse Bridge

In the Ulenbergstasse Bridge, the world premier for a high-load road traffic bridge (bridge classification 60/30) constructed with glass fiber prestressing tendons, some prestressing tendons were provided with optical fiber sensors, in addition to those embedded directly in the concrete. This latest type of monitoring for building structures has been carried out here for the first time since the opening of the bridge for traffic in July 1986.

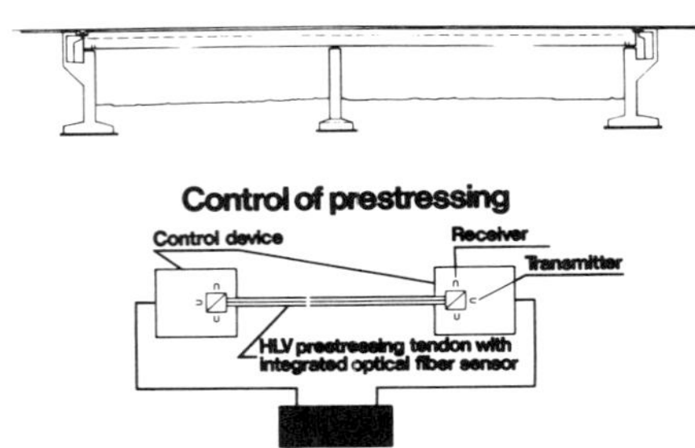

Fig. 11. Measurement principle

As an example, Fig. 12 shows the results for a period of one week in August 1987.
This gives a comparison of the temperatures constantly measured within the bridge interior by means of the sensors' attenuation, as well as elongation measured at the beginning and end of the bridge.
Measurements carried out over a period of almost three years indicate no alteration of the bridge structure (Figs. 11 and 12), thereby convincingly con-firming a completely normal behaviour for the structure.

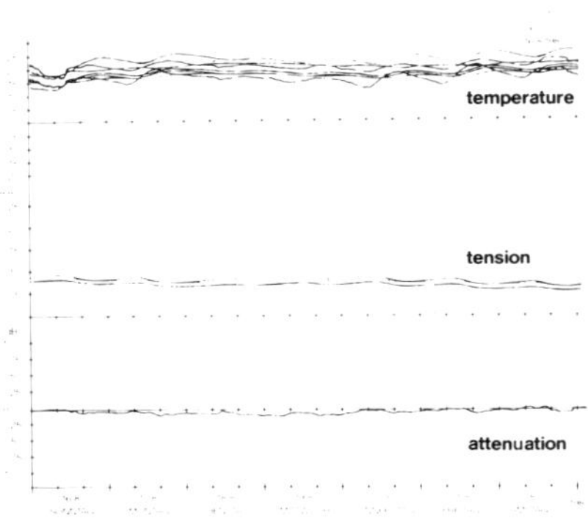

Fig. 12. Measurement results at the Ulenbergstrasse Bridge

4.2 Marienfelde Bridge, Berlin

Construction of the pedestrian bridge at Berlin-Marienfelde, a two span T-beam concrete bridge, is the first example of a fully monitored bridge structure. Designed as having external prestressing, the bridge's prestressing is achieved by seven 19-bar glass fiber composite prestressing tendons. In addition to all the prestressing tendons with integrated optical fiber sensors, a series of these sensors were embedded into the concrete or applied subsequently.

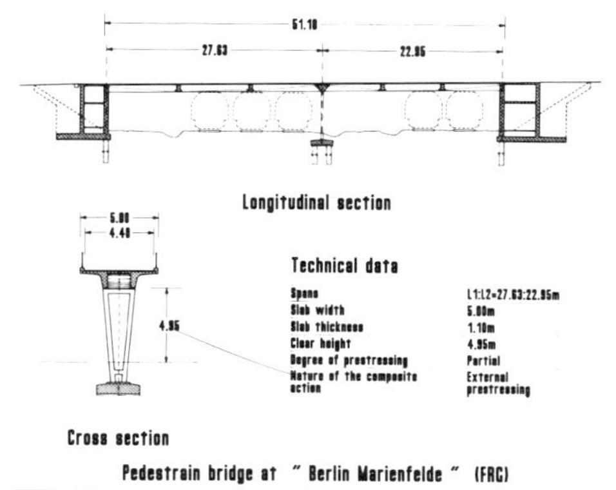

Fig. 13. Technical data, Marienfelde Bridge, Berlin

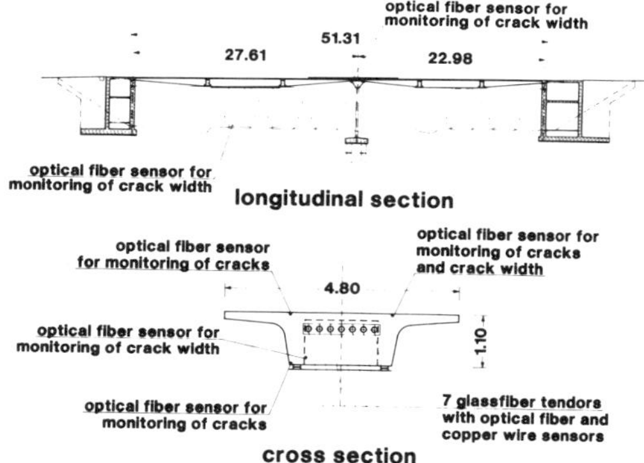

Fig. 14. Measurement Layout for the Marienfelde Bridge, Berlin

A load investigation was carried out on the bridge in November 1989. During this test a total load of 250 concrete slabs (weight per slab 1 to) were laid on the bridge in five layers, this load corresponding to twice the traffic load. Placing of the slabs led to an alteration of the strain state of the tendons as well as of the complete superstructure. This was in turn very impressively recorded with the aid of the optical fiber sensors.

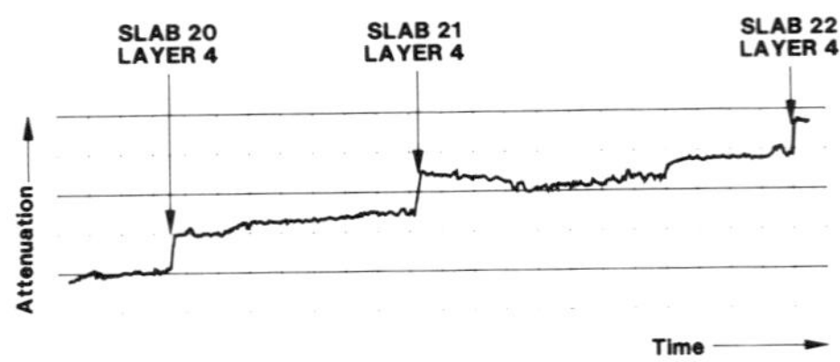

Fig. 15. Reaction of the optical fiber sensors on removal of the concrete slabs from the centre of the long span

With sensors of a particularly high sensitivity, assertions could be made regarding the dynamic behaviour of the bridge.

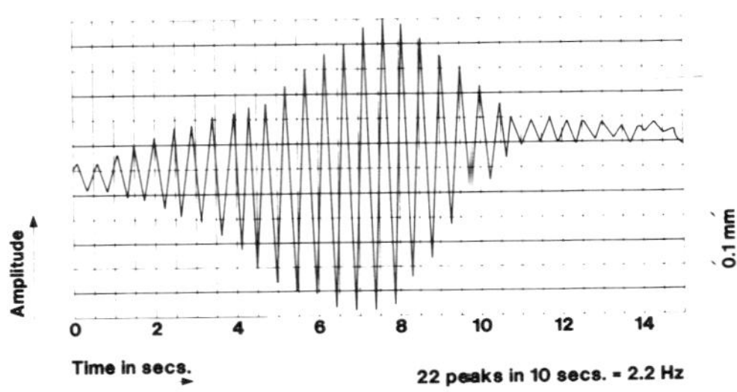

Fig. 16. Characteristic vibration of the bridge caused by specific stimulation

4.3 Schiessbergstrasse Bridge

This three span, solid concrete slab bridge (bridge classification 60/30), with two spans of 16.30 m and one of 20.40 m, is designed with limited prestressing comprising 27 glass fiber prestressing tendons (working load 600 kN) and post-bonding. Three glass fiber bars per tendon are provided with optical fiber sensors and there are to be four additional such sensors integrated directly into the concrete on the upper and four on the lower side of the slab.

Construction works will commence in March 1990.

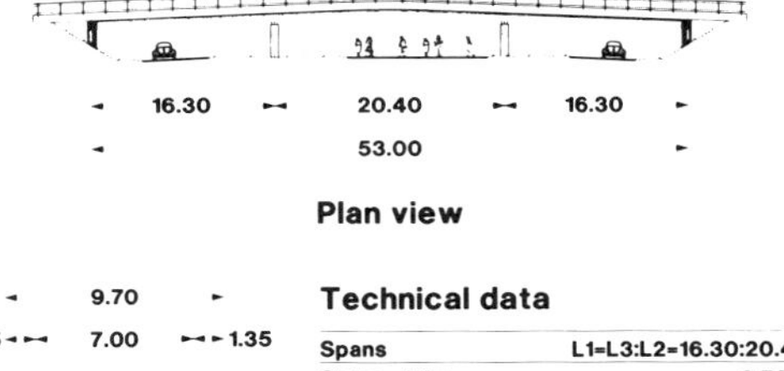

9.70

1.35 7.00 1.35

Cross section

Technical data

Spans	L1=L3:L2=16.30:20.40
Slabs widht	9.70m
Slabs thickness	1.12m
Clear height	3.00m
Load class (DIN 1072)	60/30
Degree of prestressing	Limited
Nature of the composites action	post-tensioning with subsequent bond

Fig. 17. Schiessbergstrasse Bridge, longitudinal section

5 Future Prospects

The current constantly increasing requirements and demands on heavy building structures make the question of the utilised materials' durability and the useful life of such structures in general to be of ever greater and wider interest. Compared with essentially new construction, investment for the preservation of building structures claims a not inconsiderable proportion of the funds available. The possibility of integrating sensors, on the one hand directly into the bar materials during production, and on the other the embedding of such sensors in the concrete, opens completely new horizons for the permanent monitoring of building structures.

6 References

Rehm, G. and Franke, L. (1979) Synthetic resin bonded glass fiber bars as reinforcement in concrete construction. German Committee for Reinforced Concrete, issue 304, 3-18.

Weiser, M. and Preis, L. (1982) Use of synthetic resin bonded glass fiber bars as reinforcement in construction engineering. Bauwirtschaft, issue 24-43, 90-94.

Waaser, E. and Wollf, R. (1986) A new material for prestressed concrete, HLV-Heavy-Duty Composite Bars comprising glass fibers. 'beton' 36, issue 7, 245-250.

Levacher, F. K. and Miesseler, H.-J. (1988) Monitoring tensile forces with integrated optical fiber sensors. Paper at the IABSE Congress, Helsinki, 313-318.

Franz, A. and Miesseler, H.-J. (1989) Berlin-Marienfelde Research - External prestressing and monitoring of the structure with integrated sensors. Paper at the German Concrete Congress, Hamburg.

Miesseler, H.-J. and Lessing, R. (1989) Monitoring of load bearing structures by means of optical fiber sensors. Paper at the IABSE Congress, Lisbon, 853-858.

61 PROGRESSIVE CONCRETE AND MORTAR DETERIORATION AS MEASURED BY COMPUTER-CONTROLLED MULTIPLE SONIC PULSE METHOD

P.P. HUDEC, W. WANG
Geology Department, University of Windsor, Windsor, Ontario, Canada

Abstract
Various sonic methods have long been used to monitor the deterioration of concrete. Sonic methods are based on the impedance of sound velocity in the specimen by the cracks formed during the deterioration process.
As part of an on-going study of alkali reactivity, sonic pulse velocity measuring equipment was developed which is unique and suitable for small specimens. The equipment is computer controlled - the timing of pulse velocity, and the accumulation, processing, and storing of data is fully automated. The sample of mortar is placed under constant load, and a 30 microsecond sonic pulse is generated and timed. One hundred pulses are measured; the velocity data is statistically cleaned, and the resultant mean velocity is stored on floppy disk media. Spreadsheet template is then used for further processing and comparison of data. Three specimens exposed to the same conditions test are used for comparison and control, and the results are averaged.
The sonic results have been used to monitor the progress of alkali reactivity, freeze-thaw deterioration, and uniaxial compressive strength changes; good correlations were found to exists between these parameters and the sonic measurements. Changes in sonic velocity rather than absolute velocity were used. The velocity decreases as the sample is subjected to either freeze-thaw or alkali reactivity. The aggregate type used in the mortar has a major influence both on absolute and on changes in the velocity. Although the correlations were significant, the sonic method is not as reliable as direct testing of the specimen; salt and mineral crystallization in the pores is thought to affect the velocity.
Keywords: Sonic pulse velocity, Alkali reactivity, Freeze thaw durability, Statistical treatment, Computer spreadsheet, Strength, Mortar, Aggregate

1 Introduction

While conducting research into aggregate, mortar and concrete durability under freeze thaw and alkali reactivity conditions, it was thought that some form of monitoring of the specimen condition as it goes through the various tests may be usefull. The available sonic equipment was not suitable for the small size samples used in the research, and new equipment had to be designed. The availability of an older generation of microcomputers that could be dedicated to instrument control and monitoring prompted the design of equipment that could, with the help of the computer, accurately measure the velocity of the sonic pulses in the specimen. By comparing the velocities at different stages of the experiment, it was hoped that the progress of the deterioration could be monitored.

2 Instrument and software design and description

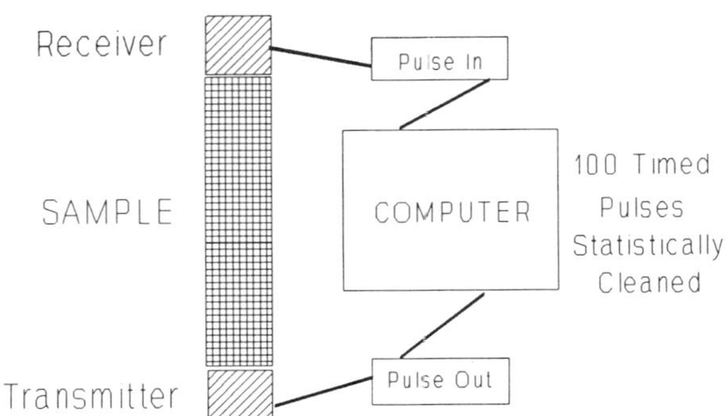

Figure 1 Schematic diagram of the Multiple Sonic Pulse apparatus.

The pulse velocity equipment (Figure 1) consists of two transducers: one serves as a transmitter of the pulse, the other as a receiver. Both transducers are controlled by a Radio Shack Model 4 microcomputer via a 'black box' interface. The computer clock performs the timing functions: initiates the pulse, and times its passage through the specimen. Software was written to both control the pulse frequency and the number of pulses, to retrieve the measured times, and to perform statistical cleaning and averaging of the results.

One hundred pulses are sent and measured. During acquisition, the data is temporarily stored in a RAM array. The one hundred time intervals are averaged, and those exceeding one standard deviation either side of the mean are discarded. New mean is calculated, and the data is stored on a floppy disk for later transfer to a spreadsheet on an IBM compatible for further analysis. Three cores from the same sample are used in analysis.

3 Sample Preparation and Description

The specimens used were 25mm (1 inch) diameter mortar cores containing various type of aggregates being tested for their alkali reactivity and the effect of various chemical treatments on the reactivity. The alkali reactive aggregates were chert from Putnam area, greywacke from Sudbury area, and carbonate from Ottawa area, all in Ontario.

Mortars containing different aggregate and different additives were cast into a 12x10x8 cm block, and quick-cured in water at 80°C. The mortar mix proportions and aggregate gradations were according to ASTM C-227 specifications. Three cores of 25mm diameter and approximately 65mm long were cut from each block. The ends were squared, ground, and 'dimpled' (for alkali expansion measurements).

4 Sonic Pulse Velocity equipment evaluation and calibration

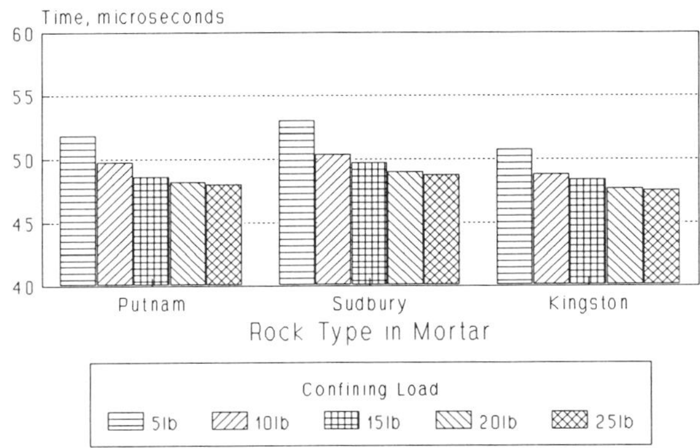

Figure 2 Effect of load and aggregate type on time interval.

The samples are held in the instrument by linearly applied compressive force. The effect of the compressive force and the repeatability of the measurements were determined by measuring the identical sample 30 separate times, each involving re-mounting of the sample in the instrument. These results, and the effect of the aggregate type are given in Figure 2. The results show that the loading force has a significant influence on the time of passage of the sonic pulse through the specimen. As the load increases, the time decreases. The relationship is exponential - small changes in light loading produce the largest differences, and as the load is increased, the differences grow smaller. As a consequence of this testing, a load of 15lbs, carefully applied and measured by a torque wrench, was used throughout the experiment.

The aggregate type also had a major influence on the sonic pulse velocity. Since the mortar preparation was similar and parallel (in terms of additives), the only variable was the aggregate type. This suggests that sonic methods are most useful for relative comparisons rather than absolute measurement of, for instance, strength or durability parameters. Thus, absolute velocities, although somewhat indicative, are probably not as useful as **relative change in velocity**, expressed as percent of the initial or starting velocity. Consequently, all results reported are as percent change in velocity.

5 The testing sequence

The usual sequence of testing involved:

1. Determination of initial velocity (in some cases)
2. Accelerated alkali reactivity testing
   (measurement of velocity in middle and end of test)
3. Freeze-thaw testing
   (measurement of velocity at the beginning and end of test, and in some at every cycle of the 5-cycle test)
4. Uniaxial compressive strength testing

6 Test results and discussion

6.1 The effect of alkali reactivity on sonic pulse velocity

An accelerated alkali reactivity (AR) testing in 80°C 1N NaOH solution was done on all the samples. Sonic velocity was determined before, during, and after the test. The results are shown in Figure 3. Two aggregate types are shown: Putnam chert and Sudbury greywacke. Chert is known as a fast reacting aggregate, whereas the greywacke reacts more slowly. This is shown both by the AR expansion rate (um/day), and by the differences in the change of sonic velocity. The progress of the alkali reactivity is shown by the 10 day and 22 day sonic measurements. The results are an average of 11 samples of chert and 9 samples of greywacke mortars.

AR Velocities

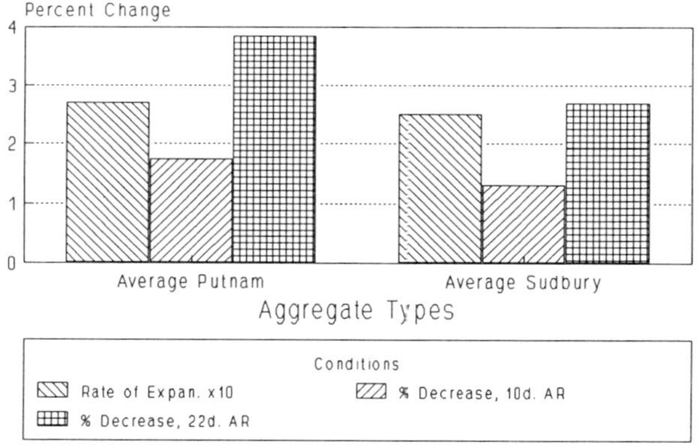

Figure 3 Effect of alkali reactivity on sonic pulse velocity.

As expected, the velocity decreases as alkali reaction proceeds. Sonic velocity measurements confirm that internal cracking due to AR slows the sonic pulse.

The cracking is the results of expansion of silica gel formed during the AR p-rocess.

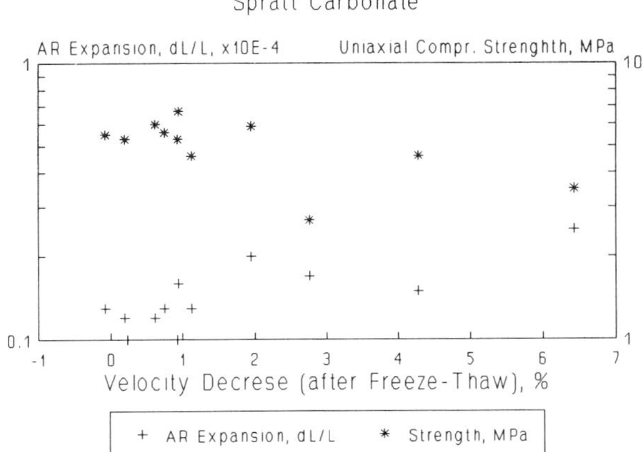

Figure 4 Expansion increases and stren-gth decreases as velocity decreases.

Figure 4 shows the relationship between the velocity decrease and both the AR expansion and the terminal uniaxial strength of the Spratt carbonate mortar. The strength of the mortar decreases with the velocity decrease; at the same time, the expansion measured during AR is shown increasing. This indicates that the parameters measured have the expected trends.

6.2 The results of Freeze-Thaw cycles on sonic pulse velocity

Freezing and thawing tests were run on the mortar cores that were first exposed to AR. While this may give different results compared to freeze-thaw tests run on fresh cores, relative comparisons can be drawn. In any case, concrete and mortar exposed in nature undergoes both alkali reactivity and freezing and thawing. AR probably aids in increased freeze-thaw deterioration. Sonic measurements serve to monitor the degree of deterioration.

Freezing and thawing tests were done by saturating the cores in a 3% NaCl solution, and then freezing the cores in air in a closed container on a sponge saturated with the chloride solution. This was found to be a severe, and reproducible test for aggregates, mortars, and concrete in prior experiments. The freeze-thaw loss was determined by weighing the amount of spalled and scaled material from the cores, and calculating this as a percent of the original weight. The sonic pulse velocity was also measured at the beginning, during, and at the end of the cycles. Velocity was also measured on the dry and saturated cores both before and after the freeze-thaw experiment. It was noted that the velocity was always substantially higher in the wet state than in the dry state. Sound travels well through water; in addition, water in the mortar pores probably sets up thixotropically, adding to the rigidity of the system.

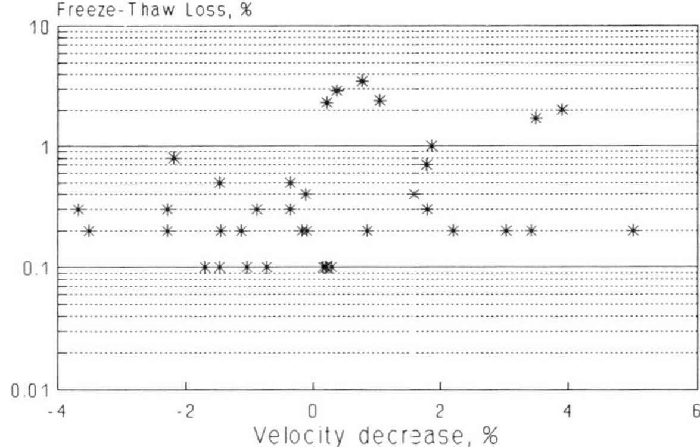

Figure 5 Freeze-thaw loss increases as the velocity decreases.

Figure 5 illustrates the results obtained when the percent of spalled and scaled material was compared to the decrease in the sonic pulse velocity. The relationship is not a strong one, but a general trend can be observed that as the velocity decreases, the freeze-thaw loss increases. The freeze-thaw loss in this case is principally a surface phenomenon, not affecting significantly the body of the mortar. The pulse travels through the body, and thus is not affected.

It was also observed that when the spalled amount and the remaining core mass were added together, the resulting mass was usually greater than the original mass of the core. Obviously, salt was crystallizing within the pores of the mortar, which would also provide bridging and rigidity to increase the sonic velocity.

7 Velocity and change in velocity relationship

The absolute pulse velocity is affected by the aggregate type. However, if this is removed as a variable, the absolute velocity has an influence on subsequent velocity decrease. Figures 6 and 7 illustrate this point. In both the carbonate and greywacke aggregate mortars, the initial velocity determines the percent decrease in the velocity as the sample goes through its various tests. This can be explained by the relationship of the velocity to compressive strength - the higher the strength, the higher the velocity. Higher strength mortars resist the degradation during AR and freeze-thaw better than lower strength mortars. In this instance, the variability of strength is due to various admixtures.

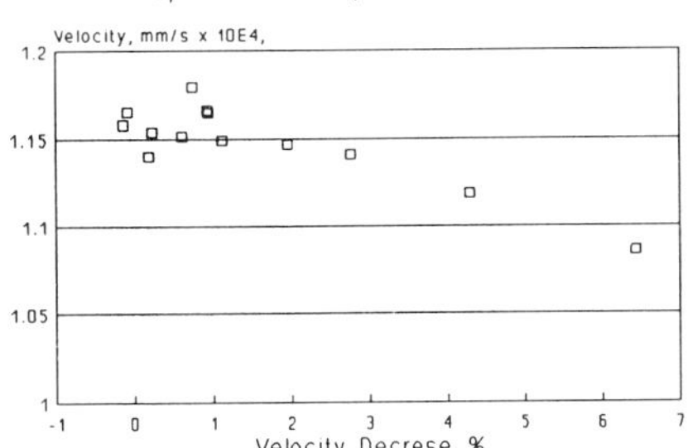

Figure 6 Effect of initial velocity on velocity change

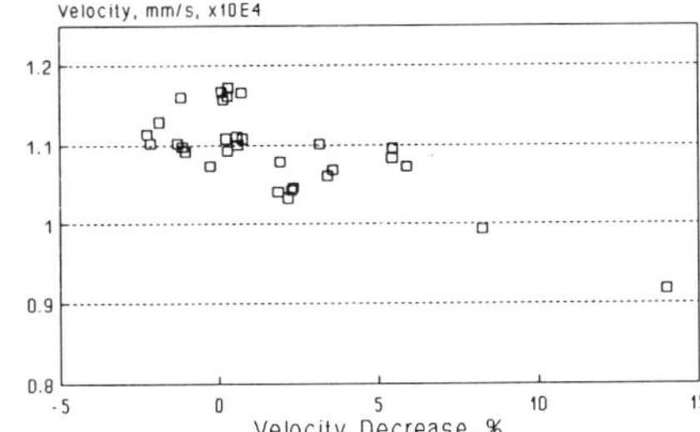

Figure 7 Effect of initial velocity on velocity change.

8 Summary and conclusions

The summary of the properties measured is given in Figure 8. The properties are expressed as arithmetic means. The figure shows that the mortars with the

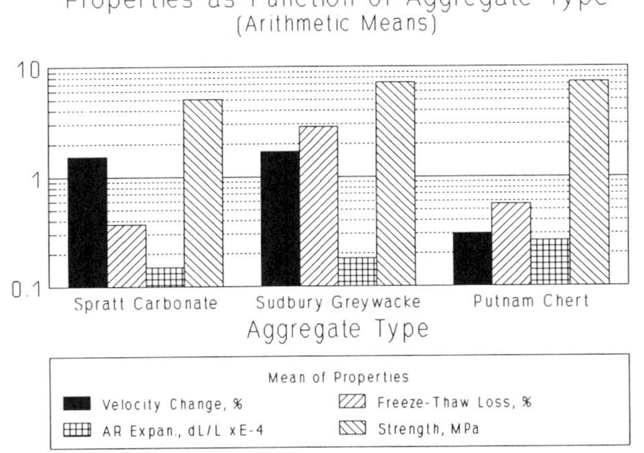

Figure 8 Summary of Measured properties by aggregate type

silica-rich aggregate behave similarly, whereas the carbonate containing mortars have somewhat different relationships among the means of different properties. In particular, the relationship of the velocity change to the freeze-thaw loss is reversed for the carbonate mortar. This emphasizes the observation that the test results tend to be aggregate specific, and some caution must be taken in using the results as means of determining the quality and durability of the aggregate.

The sonic pulse velocity measurements were shown to be sensitive to the aggregate type present in the mortar. Although only three aggregate types were used, it is probable that this applies to others. In addition to aggregate type, admixtures, proportions, method of preparation and curing all have an influence on the internal structure of concrete and mortar, and will influence the velocity of sound waves. Sonic methods are probably best suited for relative comparisons rather than as means of evaluation.

The sonic pulse method described in the report is well suited to following the progress of alkali reaction, and can supplement expansion measurements. Other sonic methods will probably work equally well. The procedure developed here is quick, and the equipment relatively inexpensive. Any computer, with proper electronic interface, can be used. The total cost of equipment other than the computer is estimated at less than $250.00. This does not include the development or assembly time.

9 Acknowledgement

The work described in this paper was supported by a research grant from the National Research and Engineering Council of Canada (NSERC).

62 SURFACE MOISTURE ON FACADE MATERIALS

B. SVENNERSTEDT
Swedish University of Agricultural Sciences, Lund, Sweden

Abstract
Familiarity with the prevailing moisture conditions on material surfaces is of considerable significance in estimating the degradation process of materials on facades of buildings. A thin film of moisture can arise as a result of rain, condensation or high relative humidity.

In order to characterize the prevailing conditions of surface moisture on a material surface, the dimensions, thickness of moisture film and moisture time can be used. The thickness of moisture film shows the level of surface moisture or the amount of moisture on the material surface. The moisture time shows the duration of moisture on the surface.

Measurements of surface moisture can be carried out with the so called current measurement technique. The measurement technique is based on the fact that the current generated in an electrochemical cell loaded with moisture is registered.

This paper will discuss both theoretical and experimental aspects concerning surface moisture on facade materials. In the first part of the paper relationships between the characteristic dimensions of surface moisture on facade materials and the influential factors of the surrounding atmosphere will be presented.

In the second part results of surface moisture measurements on facade materials will be presented and discussed. The measurements have been carried out in two Swedish climatic regions, on facades facing north. The facade materials of coated sheet metal and weather-resistant chip board have been studied.
Keywords: Surface Moisture, Facade Materials, Moisture Film, Moisture Time, Surface Moisture Measurements.

1 Introduction

In order to estimate the process of decomposition of building material, it is of great importance to be familiar with the inner and outer stresses of the material. The outside climatic stresses play a very significant role in the decomposition process of material used on a building´s outermost surfaces such as the facade. The factors which decide whether material decomposition proceeds slowly, or is accelerated, are present in what takes place in the microclimate surrounding various buildings. By microclimate is meant in general those meteorological variations on the material surface. The most important factors determining microclimate are surface temperature, ultraviolet radiation, surface moisture and extent of pollution on the material surface.

It was within corrosion research that researchers first arrived at an insight into the importance of surface moisture level and duration. Vernon discovered as early as the 1930's that corrosion speed of metallic material increased considerably when humidity exceeded a certain critical point.

In the 1950's and later, several researchers concurrently developed different techniques of measuring the actual time it took for Vernon's "critical state of moisture" to be exceeded. A new concept was introduced in order to quantify moisture duration on a material surface. The concept was defined initially for metal surfaces as that time during which there is enough moisture on the surface for the speed of corrosion to become significant.

2 Theoretical description of surface moisture conditions on facade materials

2.1 Variables describing surface moisture conditions

In the following, by surface moisture condition, is meant the moisture state on the material surface. The moisture state can be characterized by two variables, which describe the level and duration of surface moisture.

The level of surface moisture refers to the quantity of moisture on a material surface. Usually, a moisture film is formed when the surface is loaded with moisture. The thickness of the moisture film can be used as the variable, which describes the level of surface moisture.

The thickness of the moisture film varies according to the moisture load. When there is a relative humidity lower than 100 %, there is an adsorbed moisture film and the moisture film is formed only by thickness of some water molecules. When relative humidity is equal to 100 % or there is rain or condensation, the moisture film is thick enough to be seen by the naked eye.

The duration of surface moisture concerns the time when there is a moisture film of a certain thickness on the surface. To quantify the duration, the concept moisture time can be used as the second variable describing the surface moisture conditions. The definition of moisture time can be expressed as follows: "Moisture time is that time during which the moisture state exceeds a threshold value.

2.2 Connection between moisture on a facade surface and moisture in surroundings of the building

As mentioned above, the moisture state on a facade surface can be characterized by the thickness of the moisture film. By studying the moisture balance of a facade surface, a theoretical connection between the thickness of the moisture film and the affecting factors can be established.

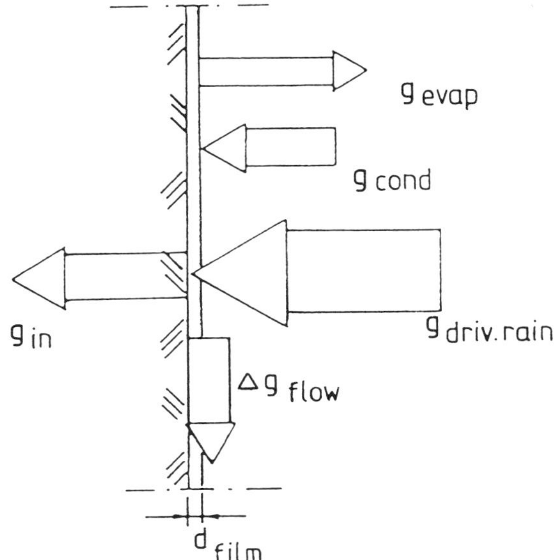

Figure 1. Moisture balance of facade surface.

Observing the moisture balance in Figure 1 it can be seen that the facade surface receives the moisture quantity $g_{driving\ rain} + g_{cond}$. The driving rain against facades varies considerably from place to place. Through measurements, the quantity and intensity of driving rain can be ascertained. The quantity of condensation can be calculated if the surface temperature, the air vapour concentration and the moisture coefficient of the surface are known.

From the facade surface the moisture quantity $g_{evap} + g_{in} + g_{flow}$ is emitted. The quantity of the evaporated moisture g_{evap} can be calculated with the same expression as for condensation calculations. For porous facade material some moisture is sucked into the material, g_{in}. For compacted facade material this moisture quantity can be neglected. When there is an obvious moisture film on the surface some moisture can also flow off the surface and this quantity can be expressed in g_{flow}.

The difference between the received and emitted moisture is that moisture which remains on the surface. On that base the following expression can be set up:

$$\frac{d(d_{film})}{dt} * \sigma = (g_{driving\ rain} + g_{cond}) - (g_{evap} + g_{in} + g_{flow}) \qquad (1)$$

where d_{film} = the thickness of the moisture film
σ = the density of water.

This is the base equation from which the thickness of the moisture film can be calculated in most cases. The equation can be used both for an adsorbed moisture film and an obvious moisture film.

The concept of moisture time can be used to quantify the duration of surface moisture. From the equation above it is obvious that the moisture film is mainly affected by driving rain and condensation.

The total moisture time, $t_{moisture}$, consists of the sum of moisture times for all periods of moisture loading. For every period of driving rain or condensation, times of film increasing and film decreasing are taken into consideration.

3 Surface moisture measurements on facade materials

3.1 Measurement method

The method for measuring surface moisture is based on an electrochemical measurement technique. The measurement equipment **consists** of an electrochemical cell, a zero resistance ampmeter and a voltage source with which the impressed voltage can be varied. A general arrangement of the device for measuring surface moisture is shown in Figure 2.

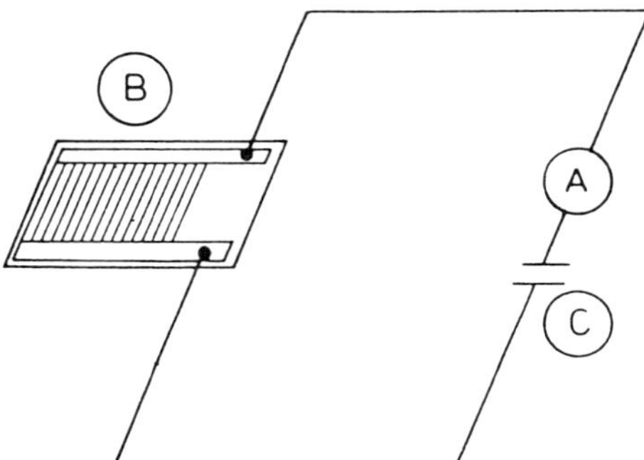

Figure 2. General arrangement of the device for measuring surface moisture. (A) is an ampmeter, (B) is an electrochemical cell and (C) is a voltage source.

The current flowing in the cells when they are exposed to moisture can continuously be measured and registered by an instrument developed by the National Swedish Institute for Building Research (SIB). With the SIB-instrument the cell current can be recorded during long time periods.

Using microprocessor technique, the instrument is made automatic with regard to data storage and data processing. The time, when the cell current exceeds a fixed current threshold value, is defined as the measured moisture time (Figure 3).

The electrochemical cell is built up with a thin material film on a ceramic backing. The most common material for the cell is copper. The dimensions of the cell are 20 x 30 mm and the material film is about 25 μm thick (Figure 4).

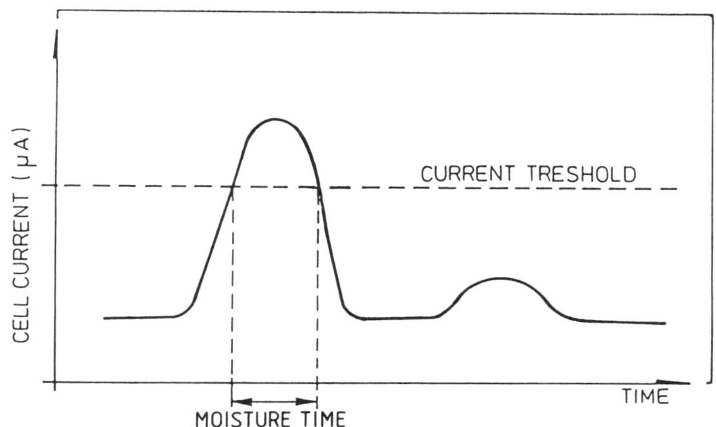

Figure 3. Definition of measured moisture time.

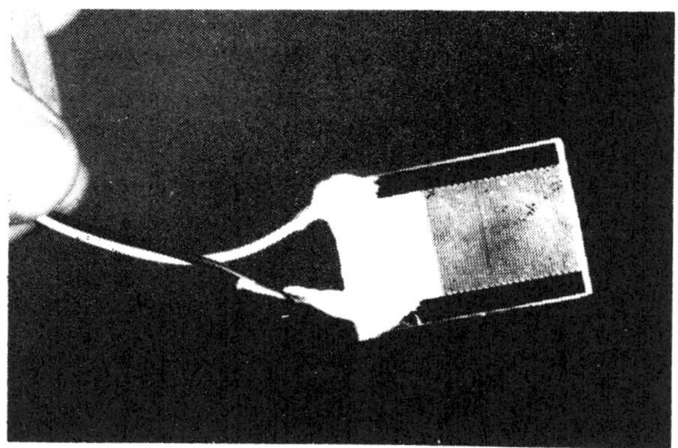

Figure 4. Copper-copper cell.

3.2 Examples of Swedish measurement results of surface moisture on facade materials

Measurements in the middle of Sweden

During 1984 field measurements were performed in Gävle on metallic and non-metallic facade materials. The metallic test material was coil-coated steel and the non-metallic material was chip board. Specimens were installed vertically on a wall in the direction of north at the exposure station of SIB. During two months in 1984 measurements of the current of copper-copper cells under atmospheric exposure were performed. The moisture time (MT) for the test period has been calculated with a current threshold of 0.1 μA and the results are given in Table 1.

Table 1. Surface moisture data for facade materials facing north in the middle of Sweden, 1984

	Materials Chip board			Coil coated steel		
Measuring period	"Wet" period (h)	"Dry" period (h)	MT (%)	"Wet" period (h)	"Dry" period (h)	MT (%)
August 1984 (7.8 - 14.8, 17.8 - 31.8)	31	526	6	25	527	5
September 1984 (1.9 - 30.9)	212	508	29	200	520	28

The test period between August and September 1984 was a very moist period in Gävle. In August the total precipitation was 51 mm and in September it was extremely wet with a total precipitation of 214 mm. The average relative humidity varied between 50 % RH and 70 % RH in August and 50 % RH and 95 % RH in September. The average air temperature varied between 10°C and 22°C for the whole test period.

Measurements in the south of Sweden

In the south of Sweden measurements have been carried out on a north facade of a test house during the winter period 1987/88. The test house was a new farm building situated between Lund and Malmö. Specimens of coil coated steel and chip board have been fastened vertically on the north facade. Surface moisture measurements have been performed by using copper-copper cells.

The measurement period has been between the end of November 1987 and the end of February 1988. Surface moisture has been calculated with a current threshold of 0.1 μA. In Table 2 results of measurements during the month of January 1988 for the north facade are shown.

The winter period 1987/88 was a mild and wet period in the south of Sweden. During the month of January 1988 the precipitation in Lund was 104 mm. The average air temperature in Lund was 3.2^{o}C with a highest value of 9.2^{o}C and a lowest value of -1.3^{o}C.

Table 2. Surface moisture data for facade materials facing north in the south of Sweden, 1988

Measuring period	Materials Chip board "Wet" period (h)	"Dry" period (h)	MT (%)	Coil coated steel "Wet" period (h)	"Dry" period (h)	MT (%)
January 1988 (1.1 - 31.1)	609	135	82	422	322	57

4 Conclusions

As pointed out in the beginning of this paper the surface moisture conditions for facade materials are of great importance when describing the micro climate. The surface moisture conditions can be characterized by two variables. Firstly the quantity of moisture on a material surface can be described by the thickness of the moisture film. Secondly the duration of surface moisture can be described by the moisture time variable . Theoretically there can be established expressions which give the connection between moisture on a facade surface and moisture in the surrounding of the building.

Surface moisture data can be measured. The measurement method is based on an electrochemical measurement technique, which uses electrochemical cells. When the cells are exposed to moisture they give off a current response, which can be continuously measured and registered.

Swedish measurements on both metallic and porous facade materials have shown a great difference in surface moisture data. For metallic material (coil coated steel) the moisture time data vary between 5 and 57 % of the exposure period. In the middle of Sweden (Gävle) during late summer 1984 moisture time was only 5 %. In the south of Sweden (Alnarp) moisture time was only 57 % during January 1988.

Moisture time data for porous materials (particle board) show a greater variation than the same data for metallic materials. During late summer 1984 moisture time in the middle of Sweden (Gävle) was 6 %. In the south of Sweden (Alnarp) moisture time was 82 % during January 1988.

The main types of moisture load from the outdoor climate are precipitation and surface condensation. These weather factors give a high cell response. When discussing precipitation, driving rain is the most important weather factor. A very high relative humidity does affect the cell but not to the same degree as the other types of moisture load.

5 Reference

Svennerstedt, B . (1989). Surface moisture on facade materials. The National Swedish Institute for Building Research. TN:16. Gävle. Sweden.

63 MODELS FOR THE EVALUATION OF THE SERVICE LIFE OF BUILDING COMPONENTS

A. LUCCHINI
Dipartimento di Ingegneria dei Sistemi Edilizi e Territoriali, Politecnico di Milano, Milan, Italy

Abstract:
The problem of the evaluation of the service life of building components is difficult to solve.
This problem depends on many variables which are difficult to evaluate and to control.
This is due to the following reasons:
- when a building component is made of different materials (which is often the case) the interaction between each material and the environment is influenced by the presence of the other materials;
- for each material of the component the detection of the agents which are significant for its deterioration and of the caracteristic properties which are sensitive to those agents as well as the assessment of the deterioration mechanisms, depend on the component's complexity and functioning;
- the assessment of the deterioration mechanisms requires consideration of all possible interactions (synergisms) that the materials provided with the assigned functions may involve.
The paper aims at making the above-mentioned problem less complex by introducing a simple model which allows identification of the significant variables and a model for the process of service life prediction in building components.
Key words: Design for Durability, Service Life Prediction of Building Components.

1 Introduction

The paper is to be understood as a contribution to the theoretical evaluation of the performance over time and service life of building components. The paper reports on the results of a study whose aim is to work out a methodology for the design of the quality over time of the building.

In the following:

- a model for the recognition of "duration characteristic properties" in a material with an

assigned function;
- a model for the process of service life prediction of building components.

2 Model for the recognition of "duration characteristic properties" in a material with an assigned function

Let us consider the case of a material (x) with a specific function (z), exposed in an environment (y).
We shall see that:

- the material (x) is provided with n(x) characteristic properties, among which $n'(x) \leq n(x)$ are sensitive to the environment;
- the environment (y) contains m(y) agents, among which $m'(y) \leq m(y)$ are deterioration factors for (x);
- the fulfilment of the function (z) by the material (x), involves exclusively that part of characteristic properties: $n''(x) \leq n(x)$ which is functional to it. Such properties can therefore be defined as the material's (x) characteristic properties, functional to the function (z), or simply, the **functional characteristic properties** of the material (x). Among these all those belonging to the part sensitive to environmental action, that is $n^*(x) = n'(x) \cap n''(x)$, are the **duration characteristic properties** (see Fig.1).

The functional characteristic properties and especially the duration characteristic properties as defined above, are extremely important. Indeed they allow the detection of a limited number, $c = f(n^*(x), n'(y))$, of the possible combinations agents/characteristic properties, made up by **deterioration factors/duration characteristic properties.** The analysis of the deterioration phenomena affecting the material relies only on such combinations. Therefore the amount of combinations deterioration factors/duration characteristic properties defines the real complexity of the problem of service life prediction for a material exposed to an environment with an assigned function.

3 Model for the process of service life prediction in building components

The model sets the phases of the process for service life prediction in building components. It envisages (see also Figs. 2, 2.1.1, 2.1.2, 2.2 and 2.3):

a **definition phase** for the preparation of:

- a list of agents (the whole group) with their intensity range, as an expression of the triggering environment;
- a list of the technological requirements and relevant levels of expression connotating the typological category to which the component belongs, significant for the service life prediction;
- a list of characteristic properties of the materials making up the component, mentioning characteristic parameters and relevant values at "0 time";

a **preparatory phase for the analytical evaluation** for the preparation of:

- a list of the performance specifications for the given triggering environment and for the considered connotating requirements;
- a list of the basic functions (the expression in terms of function of the connotating requirements significant for service life prediction);
- the lists of the analytical functions into which each basic function has to be structured;
- the functional structure of the component expressed in terms of functional model;
- the list of the parameters important to the control of the single performances;
- the list of the characteristic properties functional to each basic function;
- the correlation analytical functions/functional characteristic properties of the component's materials, expressed in terms of objectual model;

whose results are:

- a list of duration characteristic properties;
- a list of deterioration factors;
- the correlations between duration characteristic properties and deterioration factors.

It must be noted that a correct setting up of the preparatory phase calls for the evaluation of the potential changes in the deterioration factors and/or in the duration characteristic properties due to the use of more materials for the fulfilment of a basic function and/or the simultaneous fulfilment of more functions.

Should the changes be still unknown, experimental investigation may be needed.

By carrying out the preparatory phase as mentioned above, data are made explicit, which, altough being potentially available, are not taken into consideration and used by designers and therefore remain latent in drawings and plan descriptions;

a **verification phase** for the choice of the procedure (analytical or experimental) to be followed in order to evaluate the fatigue of the component and its fatigue resistance;

a **process of analytical evaluation** envisaging the following steps:

- for each formerly recognized combination duration characteristic property/deterioration factor, the evaluation of: deterioration factor intensity, degree of exposition to the deterioration factor, sensitivity of the duration characteristic property to the deterioration factor, deterioration mechanism, expected effect and intensity of the effect;
- recognition of the critical function and critical material;
- service life prediction.

a **process of experimental evaluation** which has to be organized in steps established by the Joint Committee CIB W80/RILEM 71 PSL methodology, later RILEM Reccomendation, with the following amendments:

- application starting from the preparatory phase, since it is believed that the definition and preparation phases of the herein proposed analytical evaluation provide enough information also for the experimental evaluation;
- insertion, between the 'pre-testing' and the 'testing' stage of a check phase for the feasibility of the experimental evaluation;
- performance after the conclusion of the experimental evaluation of an analytical evaluation, followed by a report and discussion on the results.

4 References

A.A.V.V. (1989) **Comportamento nel tempo dei sistemi tecnologici edilizi: Metodologia per la stima della propensione all'affidabilità di soluzioni tecniche a repertorio e verifica dell'operabilità della metodologia condotta su un repertorio tipo di soluzioni tecniche**, MPI 40%, Report n° 2

CIB W 80-RILEM 100 TSL Joint Committee, (1990) **Feedback from Practice of Durability Data - Inspection of Buildings**, National Swedish Institute for Building Research, Gävle, Sweden

Lucchini, A. (1989) **Metodologia per la valutazione dell'affidabilita' di elementi tecnici edilizi fuori sistema e problematica della durabilita**, Doctoral Thesis, Milan and Turin Polytechnics

Lucchini, A. (1990) **Organizzazione di un approccio sistematico interdisciplinare e modelli del processo per la previsione della service life.**

Consiglio Nazionale delle Ricerche, Comitato di Ingegneria e Architettura, Roma
Lucchini, A. (1989) An approach to Building Pathology, in **Seminar on Building Failures"**, The National Swedish Institute for Building Research, Gävle, Sweden
Maggi, P. N.- Croce, S.- Gottfried, A.- Morra, L. and Lucchini, A. (1989) Evaluation of Building Components Long-Term Quality, in **Quality for Building Users Throughout the World**, XIth CIB Congress, Paris
Master, L. W. (1987) Service Life Prediction, A State-of-the art-report. CIB Journal, Sept/Oct 1987, 292-296
RILEM, (1989) **Systematic Methodology for Service Life Prediction of Building Material and Components**, RILEM Reccomendation

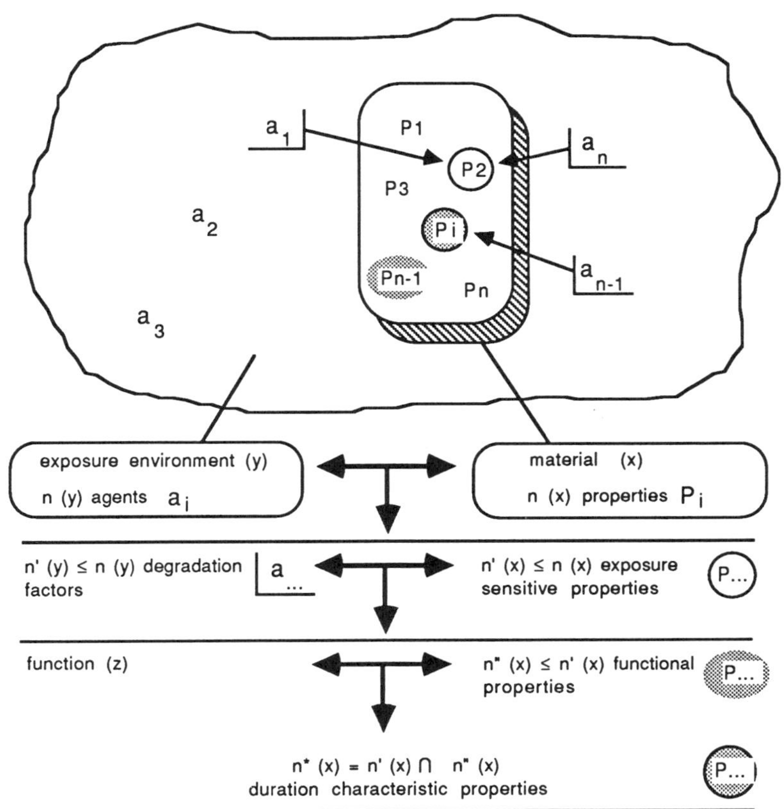

Fig.1. Model for the recognition of duration characteristic properties in a material with an assigned function

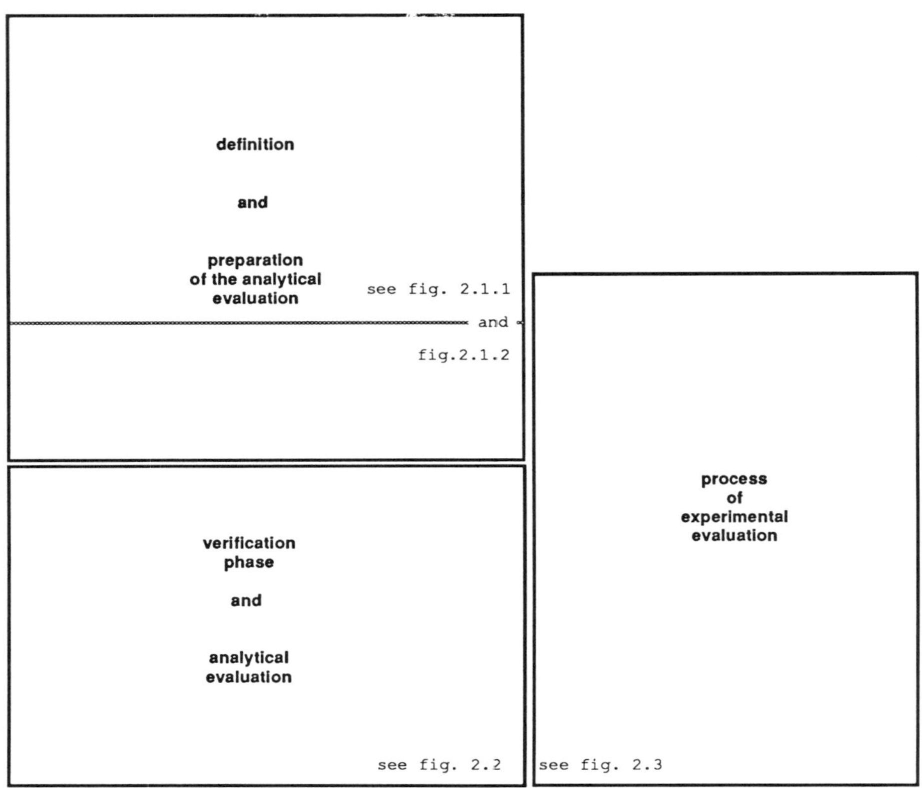

Model for the process of service life prediction of building components
definition
and
preparation
of the analytical
evaluation
see fig. 2.1.1
and
fig.2.1.2
verification
phase
and
analytical
evaluation
see fig. 2.2
process
of
experimental
evaluation
see fig. 2.3

Fig. 2

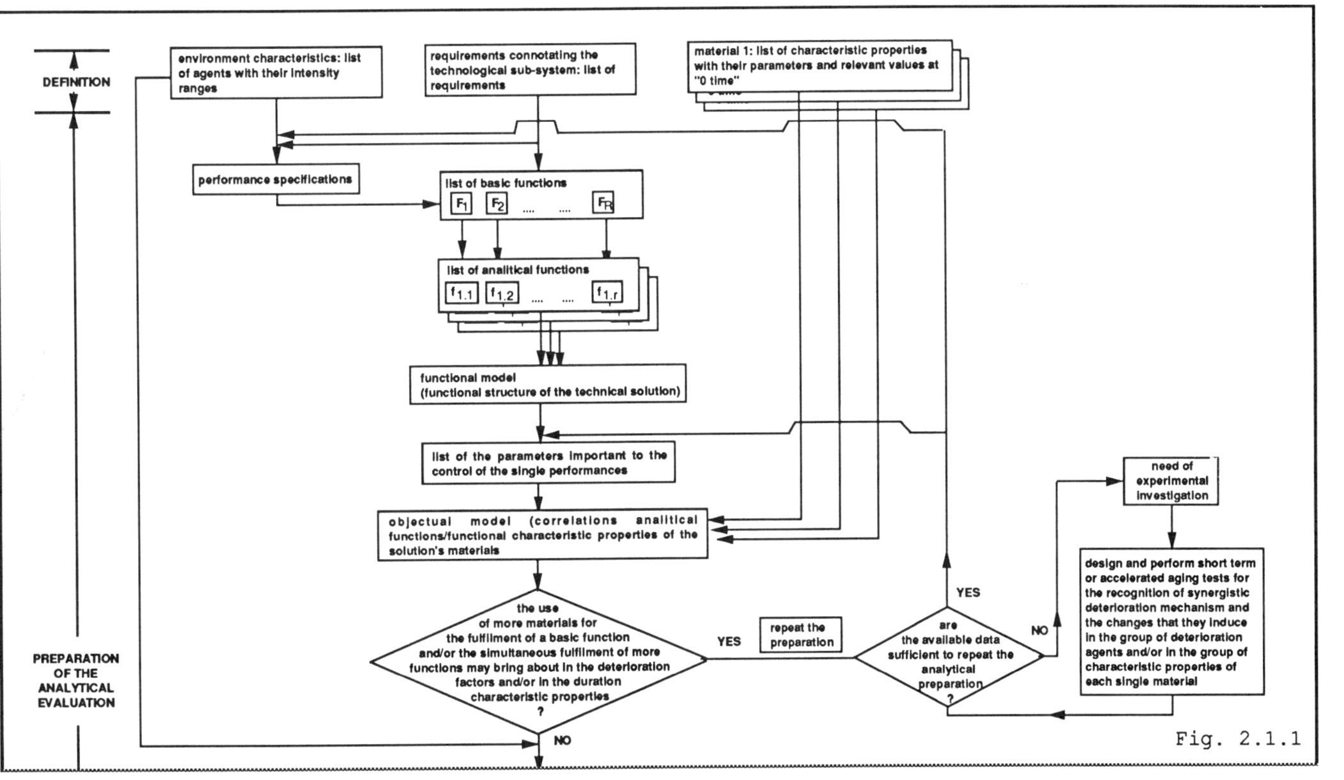
DEFINITION
PREPARATION OF THE ANALYTICAL EVALUATION
environment characteristics: list of agents with their intensity ranges
requirements connotating the technological sub-system: list of requirements
material 1: list of characteristic properties with their parameters and relevant values at "0 time"
performance specifications
list of basic functions
F1 F2 .... FR
list of analitical functions
f1.1 f1.2 .... f1.r
functional model (functional structure of the technical solution)
list of the parameters important to the control of the single performances
objectual model (correlations analitical functions/functional characteristic properties of the solution's materials
the use of more materials for the fulfilment of a basic function and/or the simultaneous fulfilment of more functions may bring about in the deterioration factors and/or in the duration characteristic properties ?
YES
NO
repeat the preparation
are the available data sufficient to repeat the analytical preparation ?
YES
NO
need of experimental investigation
design and perform short term or accelerated aging tests for the recognition of synergistic deterioration mechanism and the changes that they induce in the group of deterioration agents and/or in the group of characteristic properties of each single material
Fig. 2.1.1

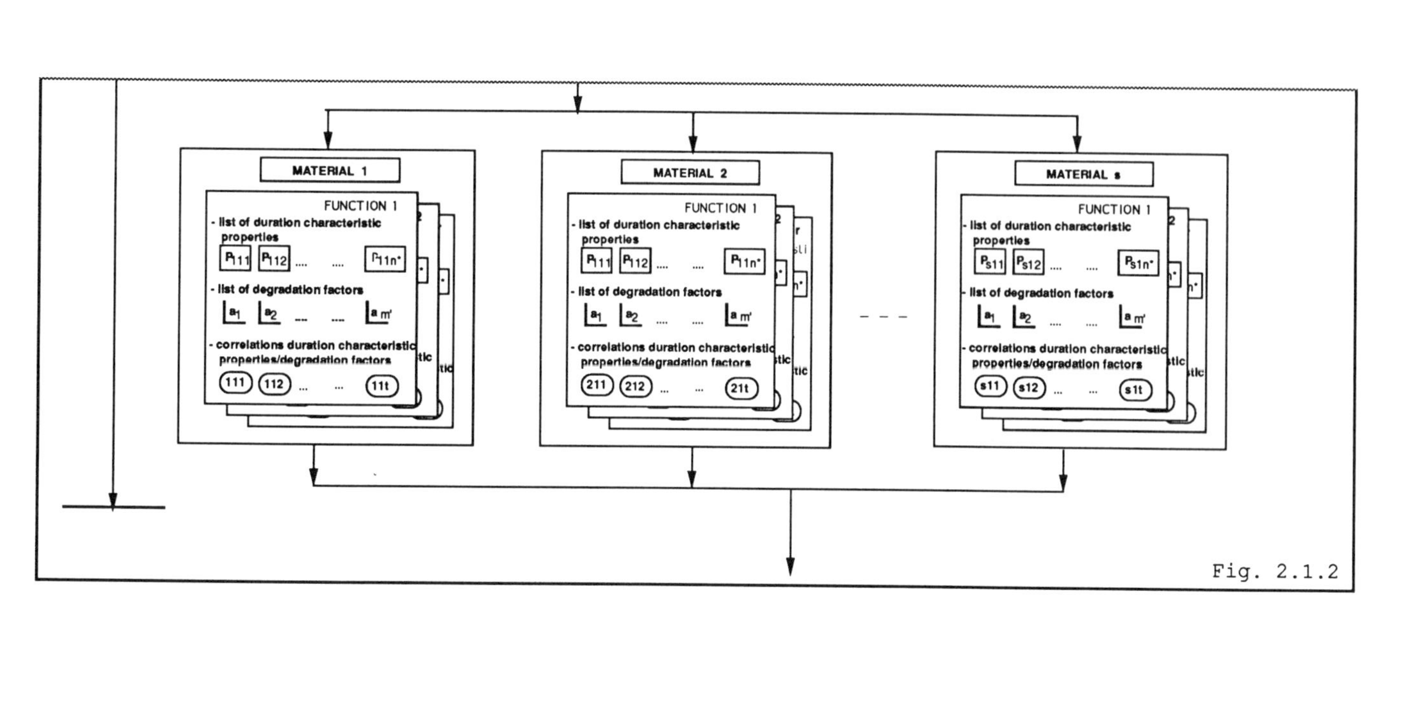
MATERIAL 1
FUNCTION 1
- list of duration characteristic properties
P111 P112 .... P11n*
- list of degradation factors
a1 a2 .... a m'
- correlations duration characteristic properties/degradation factors
111 112 ... ... 11t
MATERIAL 2
FUNCTION 1
- list of duration characteristic properties
P111 P112 .... P11n*
- list of degradation factors
a1 a2 .... a m'
- correlations duration characteristic properties/degradation factors
211 212 ... ... 21t
MATERIAL s
FUNCTION 1
- list of duration characteristic properties
Ps11 Ps12 .... Ps1n*
- list of degradation factors
a1 a2 .... a m'
- correlations duration characteristic properties/degradation factors
s11 s12 ... ... s1t

Fig. 2.1.2

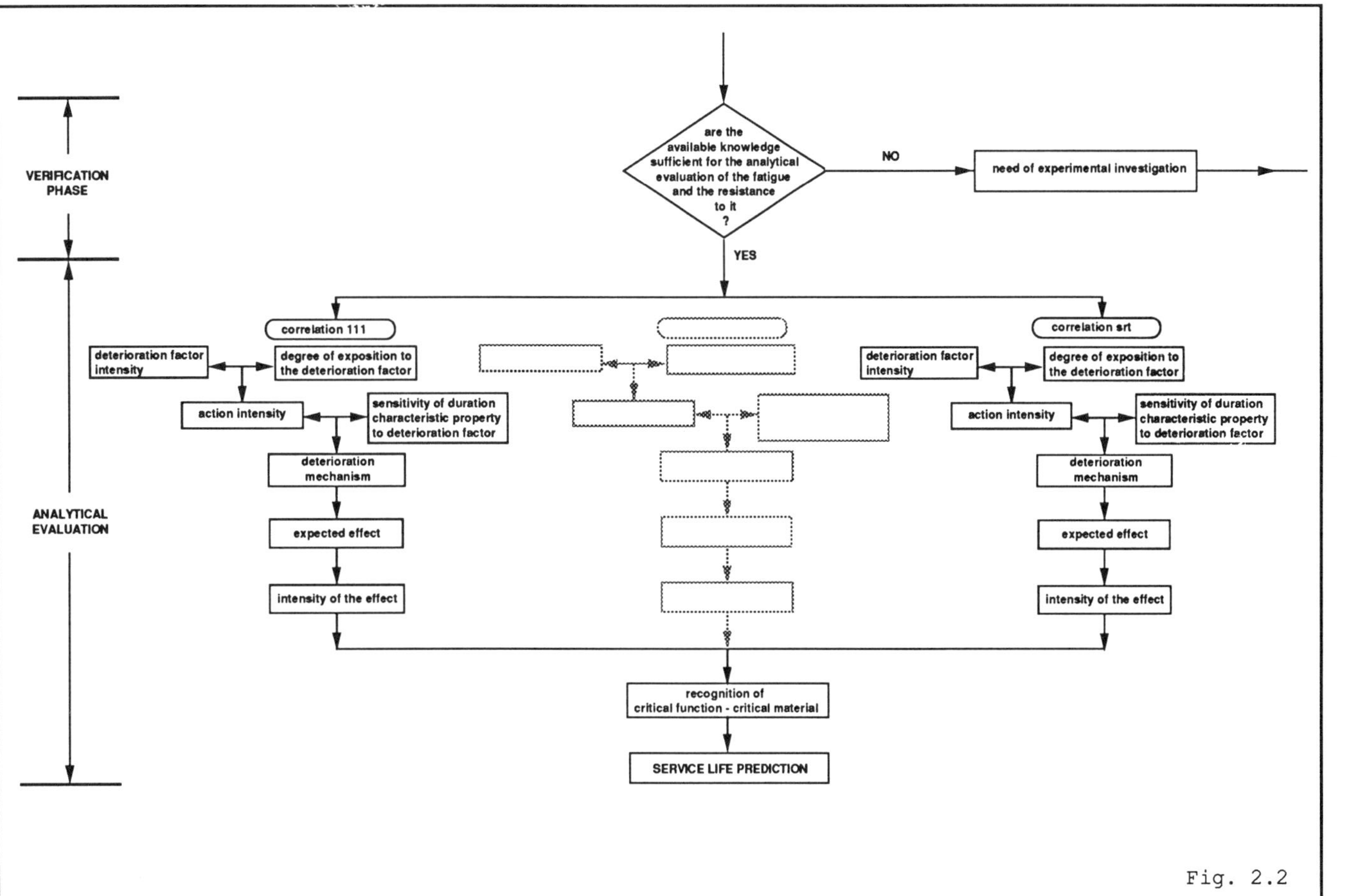

VERIFICATION PHASE
ANALYTICAL EVALUATION
are the available knowledge sufficient for the analytical evaluation of the fatigue and the resistance to it ?
NO
need of experimental investigation
YES
correlation 111
deterioration factor intensity
degree of exposition to the deterioration factor
action intensity
sensitivity of duration characteristic property to deterioration factor
deterioration mechanism
expected effect
intensity of the effect
correlation srt
deterioration factor intensity
degree of exposition to the deterioration factor
action intensity
sensitivity of duration characteristic property to deterioration factor
deterioration mechanism
expected effect
intensity of the effect
recognition of critical function - critical material
SERVICE LIFE PREDICTION

Fig. 2.2

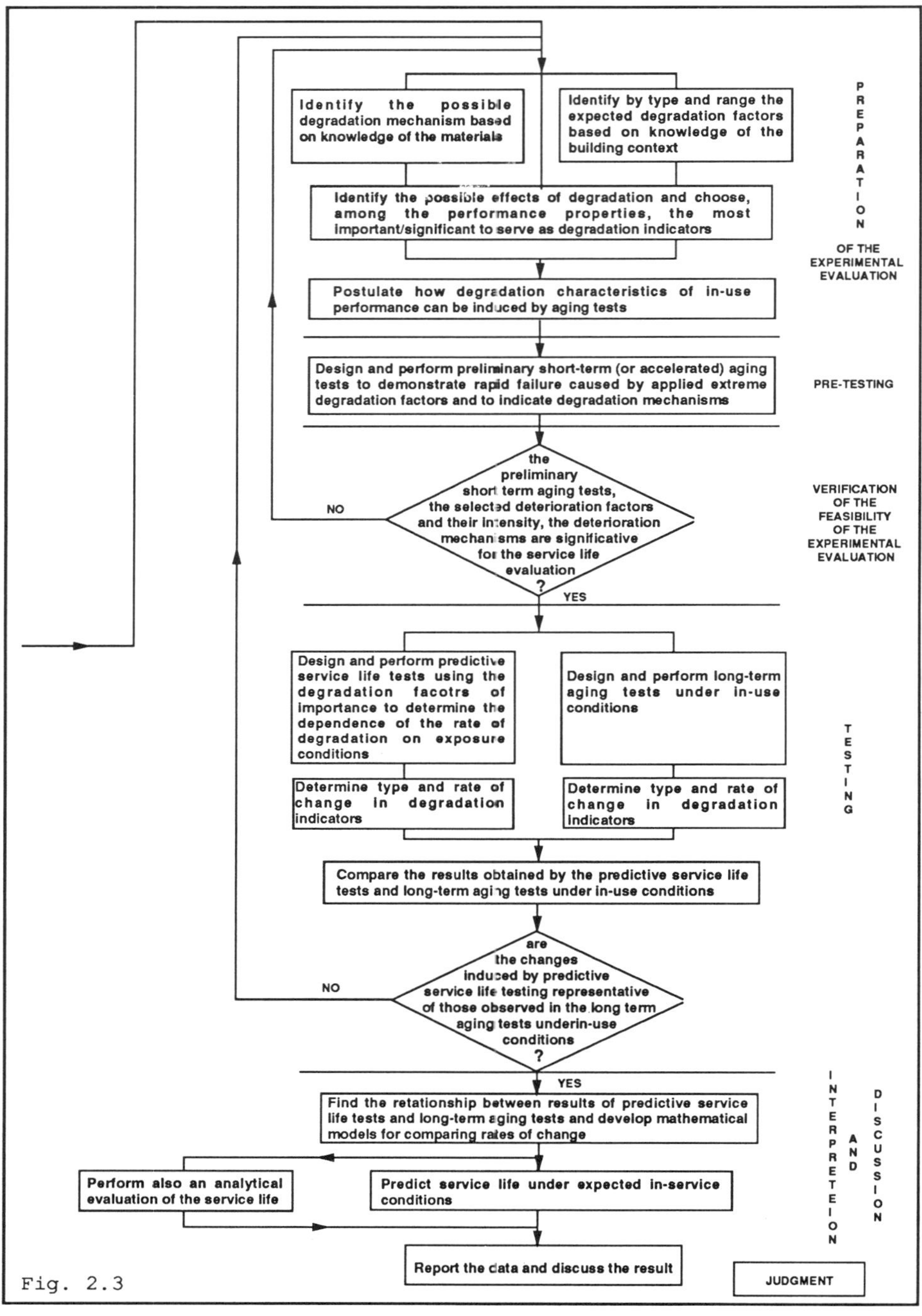
Identify the possible degradation mechanism based on knowledge of the materials
Identify by type and range the expected degradation factors based on knowledge of the building context
PREPARATION OF THE EXPERIMENTAL EVALUATION
Identify the possible effects of degradation and choose, among the performance properties, the most important/significant to serve as degradation indicators
Postulate how degradation characteristics of in-use performance can be induced by aging tests
Design and perform preliminary short-term (or accelerated) aging tests to demonstrate rapid failure caused by applied extreme degradation factors and to indicate degradation mechanisms
PRE-TESTING
the preliminary short term aging tests, the selected deterioration factors and their intensity, the deterioration mechanisms are significative for the service life evaluation ?
NO
YES
VERIFICATION OF THE FEASIBILITY OF THE EXPERIMENTAL EVALUATION
Design and perform predictive service life tests using the degradation facotrs of importance to determine the dependence of the rate of degradation on exposure conditions
Design and perform long-term aging tests under in-use conditions
Determine type and rate of change in degradation indicators
Determine type and rate of change in degradation indicators
TESTING
Compare the results obtained by the predictive service life tests and long-term aging tests under in-use conditions
are the changes induced by predictive service life testing representative of those observed in the long term aging tests underin-use conditions ?
NO
YES
Find the retationship between results of predictive service life tests and long-term aging tests and develop mathematical models for comparing rates of change
INTERPRETEION AND DISCUSSION
Perform also an analytical evaluation of the service life
Predict service life under expected in-service conditions
Report the data and discuss the result
JUDGMENT

Fig. 2.3

SESSION F

COMPOSITE ACTION AND COATINGS

64 LONG-TERM DURABILITY OF GRC COMPOSITES

K. KOMLOS, M. VANIS, B. BABAL, J. KOZANKOVA
Academy of Sciences, Technical University, Bratislava, Czechoslovakia

Abstract
This paper deals with the study of long term resistance of glass fibres in the Alkaline medium of Portland cement. AR glass fibres: REZAL, ESAP, and SVUS 16, developed in the United Kingdom, were incorporated in to the Portland cement matrix. REZEL and ESAP fibres are of a low zirconium dioxide content, SVUS 16 and Cem-FIL 1 fibres are of a high zirconium dioxide content. The long term ageing process was studied. One half of the specimens was stored in water, and the rest were dry cured. The modulus of rupture was determined at the age of 28, 90, 180, 360, and 720 days. The composite texture was studied by the SEM method. The investigated glass fibre types have shown a high alkali resistance, mostly when dry curing was applied. Within the testing period of two years, glass fibres of low, as well as of high zirconium dioxide content, show similar behaviour in the cement matrix.
Keywords: Glass Fibres, Alkaline Environment, Alkali Resistance, Fibre Corrosion, Interfacial Bond, Cement Composites, Flexural Strength, Durability.

1 Introduction

Much attention has been paid to the research and development of glass fibre reinforced cements and concretes (GRC) in the last two decades. These composite materials came into existence as a practical construction material due to their many advantageous properties, e.g. increased impact and flexural strength, higher durability under different loadings, as well as at higher temperatures etc. The grc material was successfully introduced in different fields of the building industry, e.g. as cladding panels (wall units, window units, spandrels, mullions, and column covers), further as permanent shuttering, ventilation ducts, noise barriers, garden and street furniture, such as planters, litter bins and street signs, and as junction boxes for electric cables, Hannant (1978).

Through the extension of fibre reinforced cement composites (FRC) to the field of building materials plays a particular role of its own in the already wide and well established area of application for composite materials, it should be borne in mind that glass fibre reinforced concrete is a relatively new construction material, and that as yet not much is known about its behaviour in comparison with the knowledge we have of other materials.

The function of the fibres in the relatively brittle cement matrix is to delay and control the tensile cracking of the material so that an unstable uncontrolled tensile crack growth is transformed into a slow and controlled crack growth. It is this unique characteristic of fibre reinforcement that gives the composite properties of post-cracking tensile resistance, increased tensile strain capability and enhanced energy absorption.

In Czechoslovakia much attention has been paid to the problem of alkali resistance of glass fibres and the result of these studies has been the development of two types of alkali resistant glass fibres. Glass fibres REZAL and ESAP are of low zirconium dioxide content, SVUX 16 belongs to the high zirconium dioxide content group.

2 Properties of glass fibres in cement composites

All commercial glass fibre compositions are based on silicate glasses; these are, almost inevitably, subject to some interaction with a strongly alkaline environment such as that which exists in Portland cement. At the most fundamental and micro level it has been important on fibre strength and the relationship between fibre strength changes and composite behaviour.

Direct knowledge of the strength of the fibre reinforcement is an important factor in understanding the behaviour of any composite material, although the strength of the composite may be in fact be influenced by a number of factors, including the strength of the matrix and the matrix fibre interfacial bond, in addition to the fibre strength.

The corrosion of glass fibre is governed by the pH value, but it is also influenced by the temperature, humidity of the environment, and last but not least by the glass composition.

In order to reduce the effect of alkali attack on the glass fibre, two methods were chosen. For the first suitable glass compositions were developed. For the second procedures leading to a decrease in alkalinity of the cement matrix were studied. Together with these effective coating materials were developed, in order to protect the glass fibre surface against alkali attack.

In recent years several alkali resistant (AR) glass composition were developed (e.g. in Great Britain, Germany, USA, and Japan). The most well known AR glass fibre technology is the Cem-FIL, developed by Pilkington Brothers and BRE in Great Britain.

In general, the opinion prevails that the presence of ZrO_2 in AR glass fibres is important. Its optimum content is studied, and its advantageous combination with other oxides is investigated in several countries.

The first attempt to reduce the alkalinity of the matrix was made by the application of an aluminous cement. For the reinforcement of this cement E-glass fibres were applied (brothers Birjukovic in the USSR, and Elkalite procedure in Great Britain). Recently, different additives were used in order to reduce the alkalinity of Portland cements. So different pozzolans having a high specific surface, such as bentonite, Kaolin, silica-fume, etc. were applied. The French producer L'Avenir from Lyon applies metakaolin as an additive.

Much attention has been paid to the development of AR glass fibres in Czechoslovakia too. These efforts resulted in the development of high zirconium (above 10% ZrO_2), and low zirconium (up to 5% ZrO_2) glass fibres. The composition of these fibres is given in Table 1.

Table 1. Glass fibre chemical composition

Fibre type	Components (weight %)							
	SiO_2	ZrO_2	BaO	Na_2O	Al_2O_3	CaO	Fe_2O_3	TiO_2
REZAL	59.29	4.97	9.94	10.4	5.1	10.3	-	-
ESAP	65.70	4.95	6.80	10.5	4.6	7.3	0.15	-
SVUS 16	58.30	11.40	-	14.3	-	8.6	-	7.4
Cem-FIL 1	62.00	16.70	-	14.8	0.8	5.6	-	0.1

SVUS 16 glass fibres were developed at the State Glass Research Institute in Hradec Kralove, REZAL glass fibres at the Department of Silicate Technology of the Slovak Technical University in Bratislava, and ESAP glass fibres at the Glass Research Institute in Trencin. Recently a new type of glass fibres named VVUS 17-F with the coating C-54 was developed.

3 Durability evaluation of GRC composites

The durability of GRC products depends in general on the durability of the glass fibres reinforcing these materials. In the case when Portland cement is used as matrix, the durability manifests the alkali resistance of glass fibres.

In recent years, several alkali resistance tests were introduced. The best known procedure, in which the specimens are cured under hydrothermal conditions, enables prediction of the retention of the glass fibre tensile strength after embedding the fibre in the cement matrix. This accelerated procedure, is the SIC (Strand-In-Cement) method, Proctor et al. (1982).

Along with the accelerated tests of the GRC composites at higher temperatures, Komlos et al. (1989), where according to the Arrhenius principle, an increase of the reaction rate in the fibre-matrix interface contact zone starts, long-term durability tests were carried out. These long term durability tests are more important for building practice because they show the behaviour of GRC products in real conditions.

4 Experimental programme

4.1 Materials and mixes

A portland cement type 400 was used to manufacture the cement paste. As fibre reinforcement REZAL, ESAP, SVUS 16 and Cem-FIL 1 alkali resistant glass fibres were applied. Their chemical composition is given in Tab.1. The fibre length was 36mm. The water:cement ratio of the cement paste was 0.370. The following weight fractions of glass fibres were incorporated into the cement paste: 0.4; 1.0; 1.6; 2.2; 3.3; and 4.4%. The batches were mixed in a laboratory pan mixer, and the fibres were incorporated into the cement paste in such a way as to ensure a uniform and random distribution in the cement matrix.

4.2 Casting, curing and testing

The mixes were compacted in steel moulds on a vibrating table (50 Hz; 0.35mm). Three specimens 40 x 40 x 160 mm could be manufactured simultaneously in these moulds. Together with the fibre reinforced specimens, plain specimens were cast. For each investigated parameter six specimens were manufactured. The values reported are the mean strength values determined on these specimens.

After casting the specimens were cured in moulds for 48 hours in a moist room (20°C, 90% R.H.), and after demoulding half of the specimens were stored in a dry environment (20°C, 60% R.H.), and the other half were stored in water (20°C). The specimens were tested at the age of 28, 90, 180, 360, and 720 days. The specimens were tested in flexural strength (three point loading being used). A 25 kN testing machine with a constant loading rate of 6 mm/min. was applied.

4.3 Results and discussion

For the determination of the durability of GRC composites a simple method was introduced. This method is based on the estimation of the retention of the strength ratio, i.e., the ratio of the flexural strength of fibre reinforced specimens to the flexural strength of the plain cement. matrix.

The results of the investigations carried out are summarized in Figures 1 to 4. In these figures the relationship between the flexural strength ratio and the age of the tested specimen is plotted. The investigations carried out have shown that the specimens stored in water exhibit considerably lower strength than dry cured specimens. The strength of composites of low glass fibre content, stored for 360 days in water, can decrease to the strength of the cement matrix. This phenomenon is in good agreement with results published abroad, as well as with our results obtained earlier. The results obtained show - more pronounced in the case of dry curing - that with the increase of fibre weight fraction, we obtain an increase in composite strength. The drop in strength around the composite age of 90 days, in the case of dry curing (see Fig. 1 and 2), can be affected by non-uniform shrinkage or by partial debonding in the fibre-matrix interface, initiated also by the shrinkage process.

After the strength tests were carried out, samples were taken investigated under the Scanning Electron Microscope (SEM). According t say that the fibre shape remained unchanged, even in the case of water curing, and that the fibre surface is covered with smaller and larger particles of hydration products (see Fig 5 to 8).

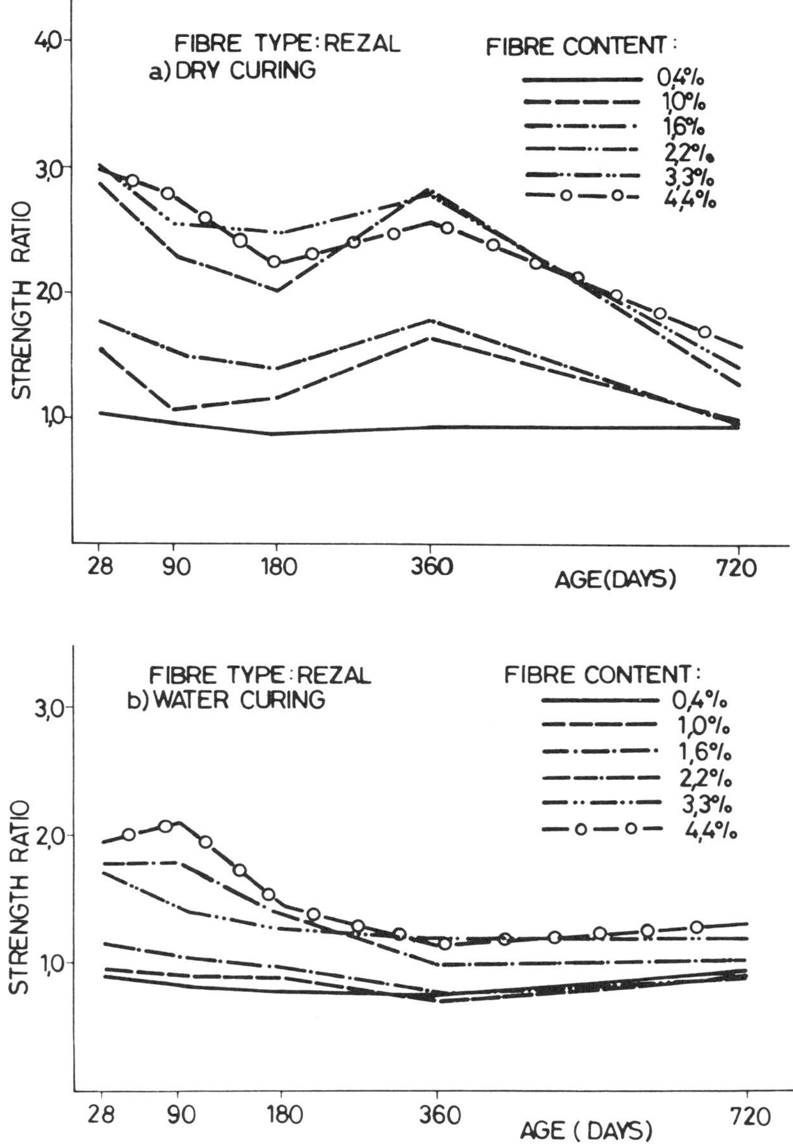

Fig.1. Strength ratio versus composite age relationship fibre type: REZAL, a/dry curing, b/water curing

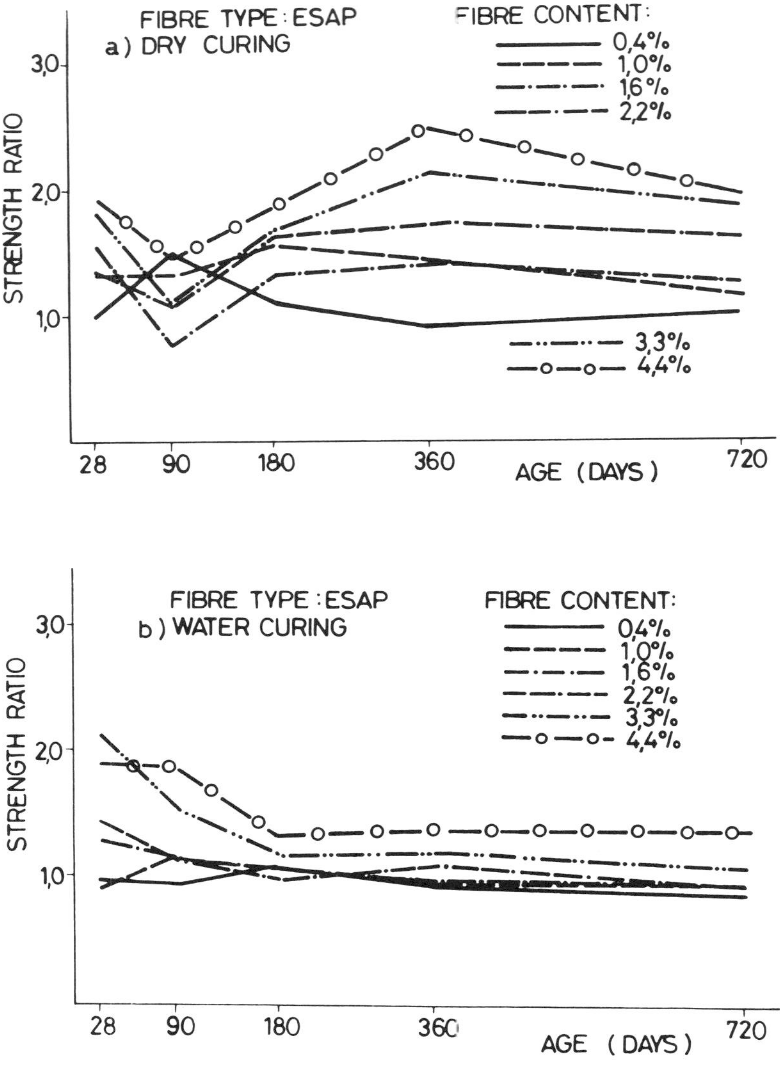

Fig.2. Strength ratio versus composite age relationship fibre type: ESAP, a/dry curing, b/water curing

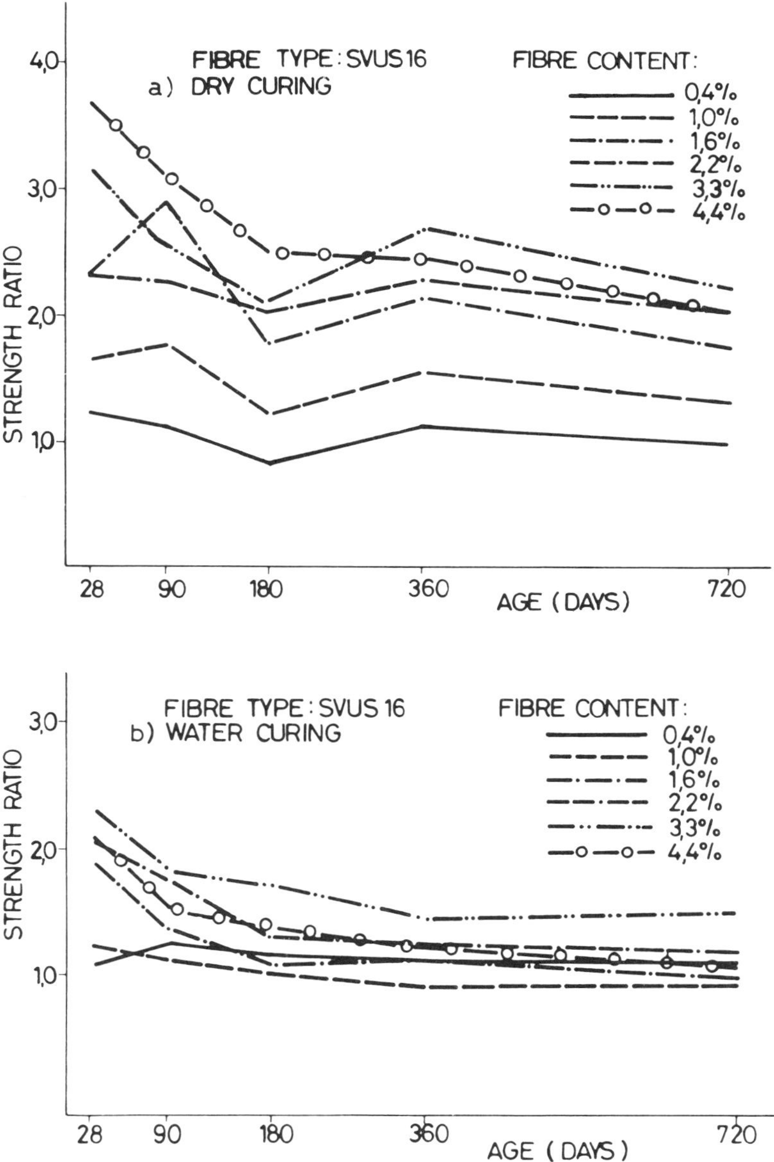

Fig.3. Strength ratio versus composite age relationship fibre type: SVUS 16, a/dry curing, b/water curing

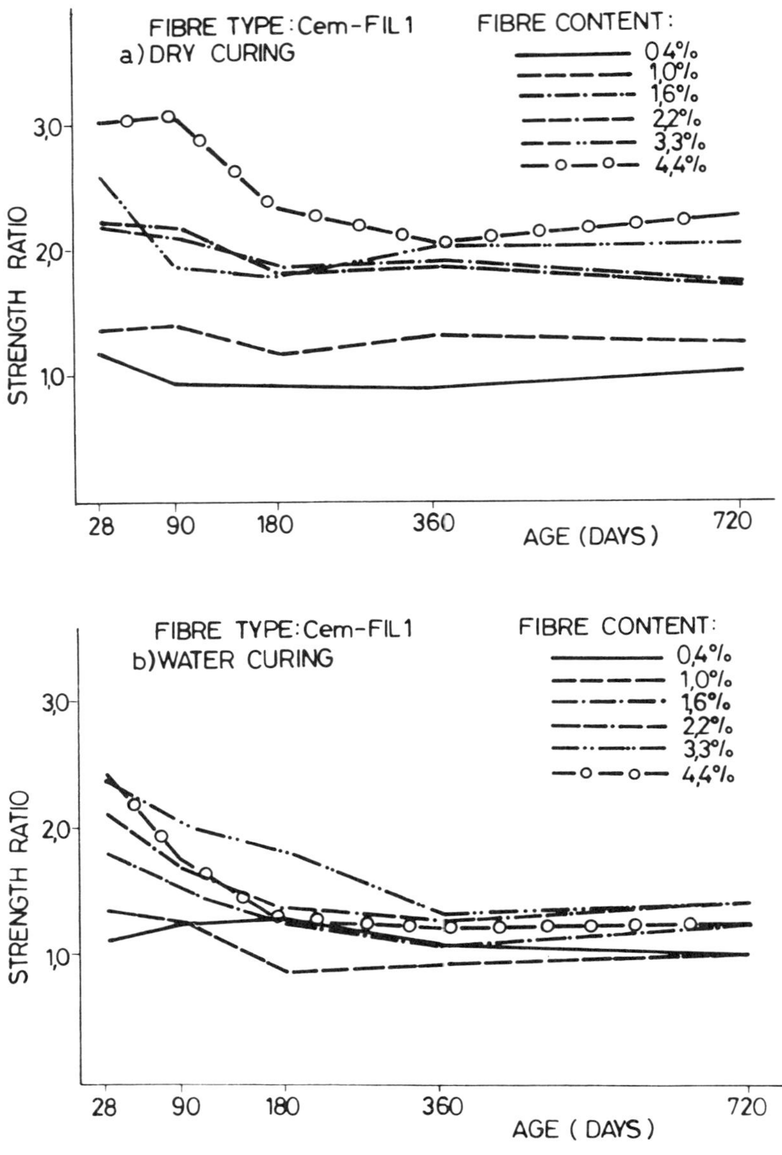

Fig.4. Strength ratio versus composite age relationship fibre type: Cem-FIL 1, a/dry curing, b/water curing

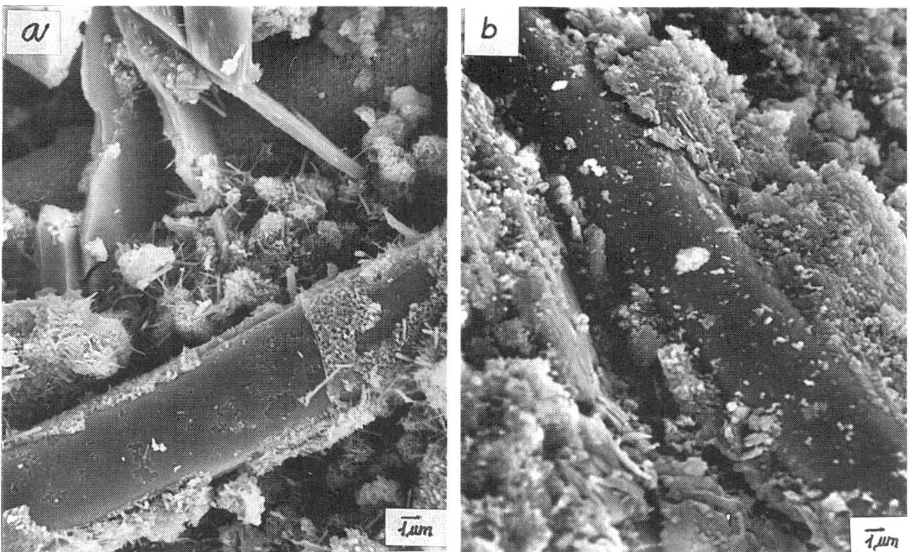

Fig.5. SEM of REZAL fibres in the cement matrix /720 days, fibre content: 4.4%/ a/dry curing, b/water curing

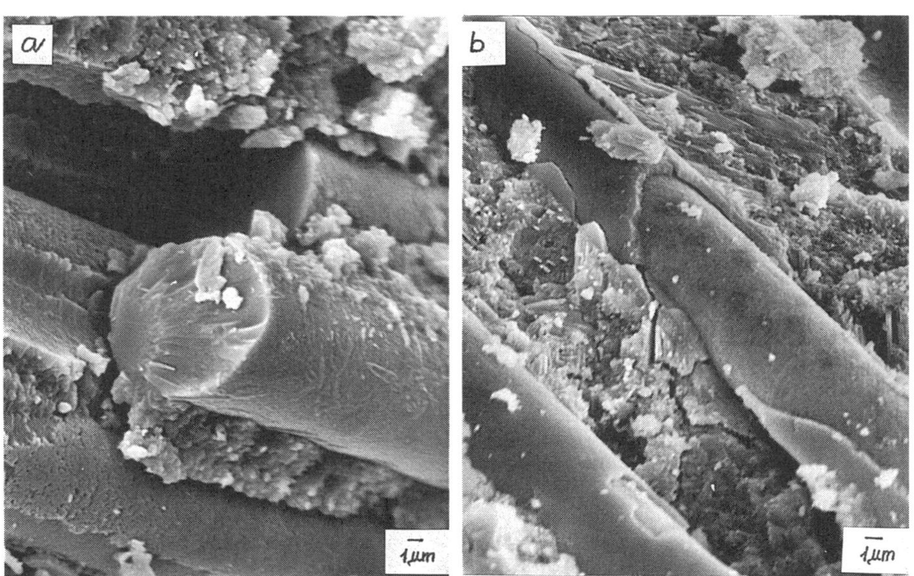

Fig.6. SEM of ESAP fibres in the cement matrix /720 days, fibre content: 4.4%/ a/dry curing, b/water curing

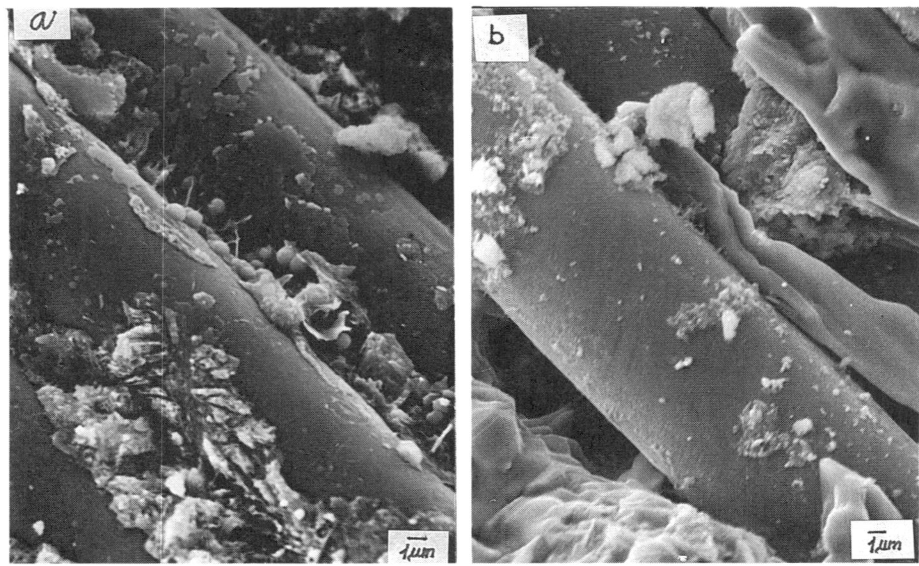

Fig.7. SEM of SVUS 16 fibres in the cement matrix /720 days fibre content: 4.4%/ a/dry curing, b/water curing

Fig.8. SEM of Cem-FIL 1 fibres in the cement matrix /720 days fibre content: 4.4%/ a/dray curing, b/water curing

5 Conclusions

The applied test procedure has proved to be a valuable and flexible experimental tool enabling comparisons to be made between glass fibre strength retention and glass fibre type.

Water storage has proved to be a more aggressive medium than dry curing. This fact was determined by other authors previously too.

The investigations have shown that even low zirconium glass fibres show a relatively high alkali resistance. This fact enables their application in different structural areas.

6 References

Hannant, D.J. (1978) Fibre cements and Fibre Concretes. Wiley Interscience Publication, Chichester, U.K.

Komlos, K. Babal, B. Vanis, M. and Kozankova, J. (1989). Untersuchungen der Bestandigkeit von Glasfaserbetonen. Tiefbau, 31, 150-158.

Proctor, B.A. Oakley, D.R. and Litherland, K.L. (1982) Developments in the assessment and performance of GRC over 10 years. Composites, 13, 173-179.

65 COMPARATIVE RESISTANCE TO SOFT WATER ATTACK OF MORTARS AND FIBRE-CEMENT PRODUCTS MADE WITH OPC AND BLENDS OF OPC AND MILLED QUARTZ, PFA AND GGBS

J.E. KRÜGER, S. VISSER
Division of Building Technology, CSIR, Pretoria, Republic of South Africa

Abstract
Mortar prisms and fibre-cement (FC) pipes were made with ordinary portland cement (OPC), and with blends of OPC and ground granulated blastfurnace slag (GGBS), OPC and fly ash (FA), and OPC and milled quartz (MQ). The prisms were either cured under ambient conditions or autoclaved in saturated steam at elevated temperatures.

The prisms and sections (rings) of the pipes were immersed in aggressive soft water containing 500 ppm CO_2. Half the number of prisms and rings was brushed every two weeks to remove loose, disintegrated material and weighed. The other half was not brushed, but was also weighed every two weeks.

Using the loss in mass as a measure of the degree of soft water attack, the results showed that autoclaved mortars and FC pipes containing FA or GGBS had significantly greater resistance to soft water than either of their unautoclaved equivalents. They also had greater resistance than both autoclaved and unautoclaved mortars and FC pipes containing either OPC alone or OPC and MQ. Their improved resistance to soft water attack is attributed to the formation of well-crystallised aluminian tobermorite and hydrogarnet in the autoclave. Scanning electron microscope and X-ray diffraction data are presented in the paper as evidence.

A description is given of an experimental pipeline (incorporating the various FC pipes) which has been installed in a water reticulation, which will be monitored to establish whether the performance of the pipes in the laboratory will be matched in the field.
Keywords: Soft Water, Ambient Curing, Autoclaving, Mortar, Fibre-cement, Portland Cement, Milled Quartz, Fly Ash, Ground Granulated Blastfurnace Slag

1 Introduction

Very pure or soft water corrodes cement-bonded products like concrete, mortar and fibre-cement products (see Figure 1). The main cause of the aggressiveness of natural waters is the presence of dissolved carbon dioxide and humic acid.

In addition, researchers at the National Building Research Institute found that the leaching process in fibre-cement pipes also frees sulphate ions that will penetrate the pipe wall, react there

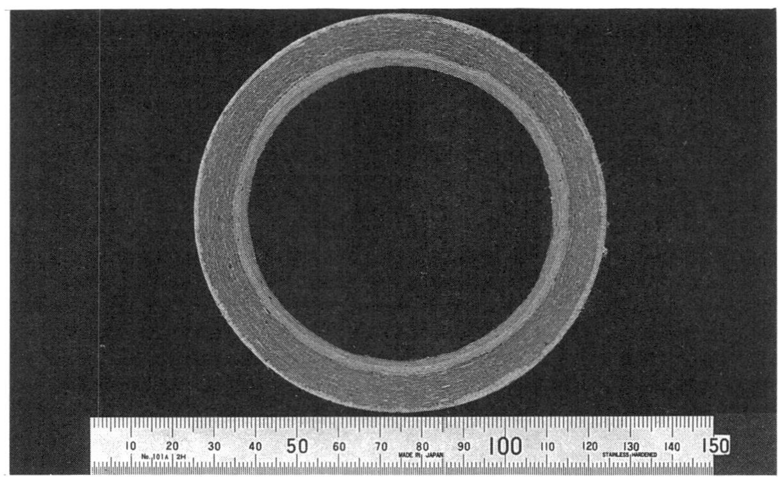

Fig.1. A section of a fibre-cement pipe after 20 years in a water reticulation

with $Ca(OH)_2$ and alumina to form ettringite between layers; the ettringite will then place the outer circumference under hoop stress. The stress may lead to longitudinal cracking in small diameter FC pipes (South African Fibre-Cement Association 1986) (Figure 2).

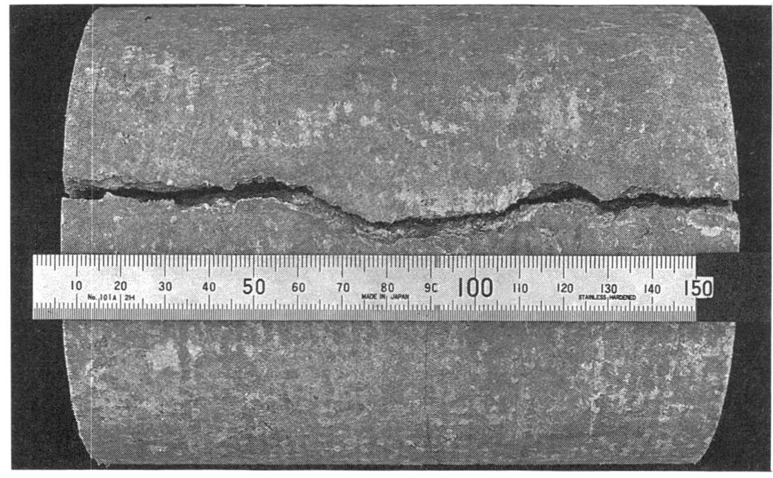

Fig.2. A longitudinal crack which developed in a 100 mm (inside diameter) fibre-cement pipe after 22 years in a water reticulation

The deterioration of cement-bonded products in contact with soft water is a well-known, world-wide problem and there are numerous publications on the topic. Most of the publications deal with the mechanism of attack, the assessment of the aggressiveness of water towards such products and the formulation of indices for measuring the degree of aggressiveness, such as the aggressiveness index of the American Water Works Association (AWWA) in the AWWA Standard C400 - 77, the Langelier Index (LI) of calcium carbonate saturation, the DIN aggressiveness index and others (International Standards Organisation 1982; Basson 1989).

One way of protecting cement-bonded products against soft water attack is to treat the water with lime or limestone to increase its temporary hardness to above 40-50 mg/l $CaCO_3$ (Van Aardt 1986). There appears to be little one can do to improve significantly the soft water resistance of structures themselves other than to impregnate them with aqueous solutions of magnesium or zinc fluosilicate and other chemicals, or by applying an appropriate coating (Van Aardt 1986); such methods are temporary, at best. Some researchers have suggested the partial replacement of portland cement with GGBS (Van Aardt 1986; Efes 1980; Gutt and Harrison 1977) and pozzolans like FA (Lea 1970, Van Aardt 1986 and Gutt and Harrison 1977) and the use of limestone aggregate (Hime et al. 1986).

This paper describes the investigaton of the soft water resistance of mortar and fibre-cement pipes in which GGBS or FA replaced part of the OPC.

2 Experimental work

2.1 Experiment with mortar

Mortar prisms, each measuring 25 mm x 25 mm x 285 mm were made with 3 parts of a quartz sand, 1 part of one of the cementitious materials mentioned in Table 1, and 0,46 of one part of water, by mass. The flow of the mortars, determined in accordance with ASTM C 109 - 68 (American Society for Testing and Materials 1987) varied between 90% and 110%.

The chemical composition of the OPC, GGBS and FA is presented in Table 2. The milled quartz (MQ) was 98% pure quartz. The GGBS, FA and MQ had a specific surface (Blaine) of 3 800, 5 550 and 4 150 cm^2/g, respectively. The fineness of the FA (retention on a 45 μm sieve) was 10%, by mass.

Table 1. Cementitious materials used in mortar prisms

Blends (by mass)
OPC alone
50% OPC plus 50% GGBS
70% OPC plus 30% FA
60% OPC plus 40% FA
60% OPC plus 40% MQ

Table 2. Chemical composition of OPC, GGBS and FA

Component	OPC	GGBS	FA
SiO_2	23,5	34,0	48,5
CaO	63,8	31,2	5,23
Al_2O_3	4,67	13,19	38,00
Fe_2O_3	2,80	0,43	2,97
MgO	1,88	17,26	1,08
SO_3	1,74	0,01	0,37
S	-	0,24	-
Na_2O	0,05	0,24	1,01
K_2O	0,48	0,76	0,50
Mn_2O_3	-	1,42	-
L.O.I*	1,14	2,48	2,46

* Loss on ignition

The prisms were stored in a fog room at 23 °C for 24 hours and then demoulded. After demoulding, half the number of prisms made with each cementitious material was cured under water at 23 °C for 27 days (unautoclaved) and the other half autoclaved at a steam pressure of 1 MPa (184 °C) for 7 hours and then stored in water at 23 °C for 26 days. The autoclaved and unautoclaved prisms were separated into two identical groups with respect to cementitious material. Every two weeks, the prisms from one group were brushed with a stiff nylon brush in an attempt to simulate the action of running water carrying grit, and weighed. The other group of prisms was not brushed (simulating standing water), but it was weighed. Each prism was placed in a separate container in soft water (distilled water containing 500 mg/l free carbon dioxide) to prevent mutual interference. The free carbon dioxide content of the water dropped to about 100 mg/l during 24 hours. The water was changed daily, except over weekends.

2.2 Experiments with fibre-cement pipes

Fibre-cement pipes with an internal diameter of 150 mm, a length of 5 m and a wall thickness of 15 mm, were commercially manufactured from each of the cementitious materials mentioned in Table 1. Half the number of pipes was cured under water at ambient temperature for 28 days. The other half was autoclaved at a steam pressure of 0,8 MPa (175 °C) for 10 hours. Manufacturing, curing and autoclaving were done at the factory. Four rings, 60 mm long, were cut from the pipes made with each of the cementitious materials and immersed in soft water as mentioned in Section 2.1. Two of the rings were brushed fortnightly; all four were weighed.

Two-metre lengths of the various pipes have, for future monitoring purposes, been installed in an 83 m long section of a water reticulation which conveys soft water with a Langelier saturation index of -1,7 at a rate of about 900 kl/day. Each of the cementitious materials mentioned in Table 1 is represented by eight

2 m lengths of pipe, four unautoclaved and four autoclaved. One length of a bitumen-dipped pipe made with 50% OPC plus 50% GGBS, unautoclaved, has also been installed in the section.

3 Results

The loss in mass of the mortar prisms stored in the soft water for 35 weeks and brushed fortnightly are given in Table 3 and presented graphically in Figure 3.

Table 3. Loss in mass (g) of brushed mortar prisms in soft water

Time (weeks) in soft water	Cementitious material in mortar prisms									
	OPC alone		50% OPC + 50% GGBS		70% OPC + 30% FA		60% OPC + 40% FA		60% OPC + 40% MQ	
	A	B	A	B	A	B	A	B	A	B
3	0	0	2	1	5	3	4	3	2	2
7	1	0	6	3	10	5	9	6	5	5
11	3	4	12	9	17	7	15	8	9	10
15	6	7	16	15	23	10	21	11	15	17
19	12	13	24	22	30	12	28	12	24	24
23	19	21	33	30	37	15	34	16	35	35
27	24	27	41	37	43	18	40	19	45	41
31	29	33	48	44	50	20	34	20	54	49
35	34	39	56	50	58	22	50	24	65	57

A Cured under water at 23 °C for 28 days
B Autoclaved at 1 MPa steam pressure for 10 hours and cured under water at 23 °C for 27 days

The loss in mass of the fibre-cement pipe rings stored in soft water for 35 weeks and brushed fortnightly is given in Table 4 and presented graphically in Figure 4.

The unbrushed mortar prisms and fibre-cement pipe rings showed similar losses, except that the rate of loss was, of course, much lower than that in the case of brushed prisms and rings.

Table 5 gives the thickness of pipe wall (as a percentage of the original thickness) that remained of the various brushed pipe rings after 35 weeks of exposure to soft water. Figure 5 shows the difference between the wall thickness of a pipe ring made with OPC cured in water at 23 °C for 28 days and that of an autoclaved pipe ring made with 70% OPC plus 30% FA.

Autoclaved mixtures of 70% OPC plus 30% FA, and 60% OPC plus 40% MQ, were both subjected to X-ray diffraction analysis and scanning electron microscopy to determine the nature of the hydration products that form, and to explain the reason for the improved resistance of autoclaved products in which part of the OPC had been replaced with FA. The results are presented in Figures 6 to 8.

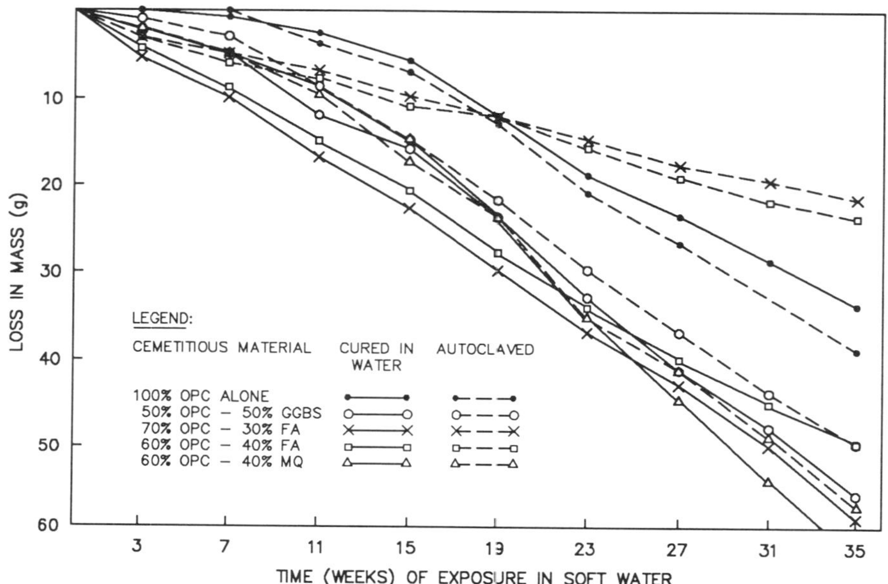

Fig.3. Loss in mass of mortar prisms stored in soft water and brushed fortnightly

Table 4. Loss in mass (g) of brushed fibre-cement pipe rings in soft water

Time (weeks) in soft water	Cementitious material in pipes									
	OPC alone		50% OPC + 50% GGBS		70% OPC + 30% FA		60% OPC + 40% FA		60% OPC + 40% MQ	
	A	B	A	B	A	B	A	B	A	B
1	2	1	3	1	1	1	1	1	0	1
5	17	20	16	10	17	9	17	8	23	16
9	40	47	37	25	45	25	46	21	55	39
13	69	73	62	46	76	43	80	34	99	68
17	99	100	88	68	110	64	108	48	137	102
21	129	127	113	90	143	80	144	61	183	137
25	167	157	136	107	184	100	180	75	232	169
29	200	185	160	130	220	117	214	86	276	209
35	256	230	186	152	285	139	267	104	346	270

A Pipes cured under water for 28 days
B Pipes autoclaved at 0,8 MPa for 10 hours

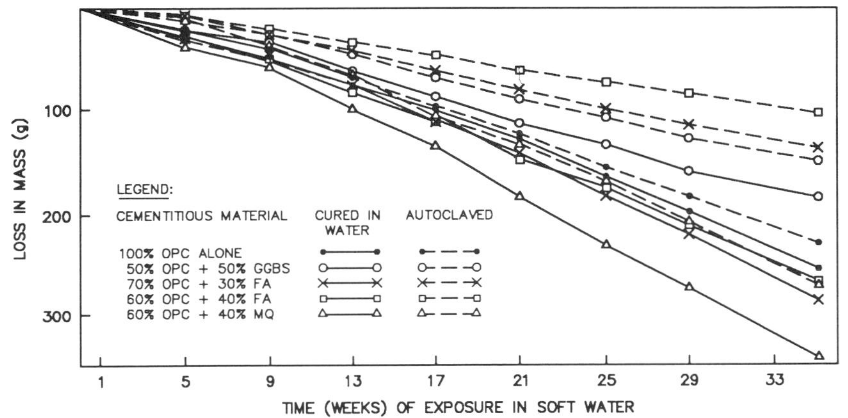

Fig. 4. Loss in mass of fibre-cement pipe rings stored in soft water and brushed fortnightly

Table 5. Comparison of wall thicknesses of pipes after 35 weeks in soft water and brushed fortnightly

Cementitious material	Percentage of wall thickness remaining after 35 weeks	
	Cured under water	Autoclaved
OPC alone	80	79
50% OPC plus 50% GGBS	85	88
70% OPC plus 30% FA	76	88
60% OPC plus 40% FA	79	92
60% OPC plus 40% MQ	71	77

4 Discussion

Any carbon dioxide in water that is not required to stabilise calcium bicarbonate that may be present in the water, is aggressive. The aggressive carbon dioxide will commence by leaching free lime from cement-bonded products and subsequently the cement hydrates and calcium carbonate as well.

The results presented in Table 3 and Figure 3 for the mortar prisms show that the binders can, after 35 weeks of exposure, roughly be divided into three groups with regard to their resistance to soft water attack. Autoclaved mortars made with OPC plus FA had good resistance to soft water; autoclaved and unautoclaved mortars made with OPC had medium resistance to soft water; unautoclaved mortar made with OPC plus FA as well as both autoclaved and unautoclaved mortars made with either OPC plus GGBS, or OPC plus MQ,

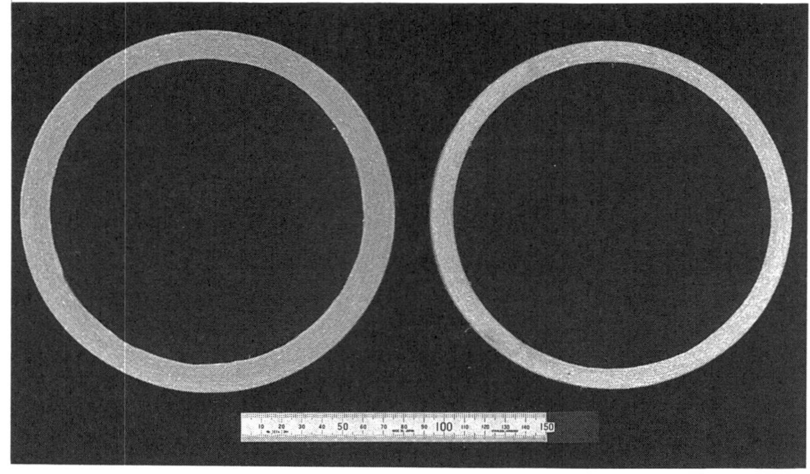

(a) (b)

Fig. 5. Difference between the wall thickness of (a) an autoclaved ring made with 70% OPC plus 30% FA and (b) a water-cured pipe ring made with OPC

had poor resistance. In fact the partial replacement of OPC with FA, GGBS or MQ reduced the resistance, except in the case of the autoclaved mortars containing FA. The latter mortars showed the least deterioration after 35 weeks of exposure to soft water, although their deterioration up to 19 weeks was greater than that of the mortars made with OPC.

The results presented in Tables 4 and 5 and in Figure 4 show that in the case of the FC pipe specimens, the best resistance was obtained with autoclaved pipes made with 60% OPC plus 40% FA, followed by autoclaved pipes made with 70% OPC plus 30% FA, and 50% OPC plus 50% GGBS. The resistance of the autoclaved and unautoclaved pipes made with OPC and the unautoclaved pipes made with OPC plus FA was poor. The pipes made with 60% OPC plus 40% MQ, autoclaved and unautoclaved, were the least resistant.

It is not clear why a relatively good performance was obtained with the GGBS in the case of the unautoclaved FC pipes, considering its poor performance in the mortar specimens. It is probably attributable to impermeability of the pipe walls; in other experiments the authors found that, other factors being equal, unautoclaved FC pipes made with OPC plus GGBS were denser than those made with OPC or with OPC plus FA.

There was a significant difference between the resistance of the autoclaved and the unautoclaved mortar and FC pipes made with OPC plus FA. The improvement in resistance effected by autoclaving is attributed to the formation of well-crystallised aluminian tobermorite and hydrogarnet during autoclaving, the aluminium being produced by the FA. In the unautoclaved products gel-like hydration

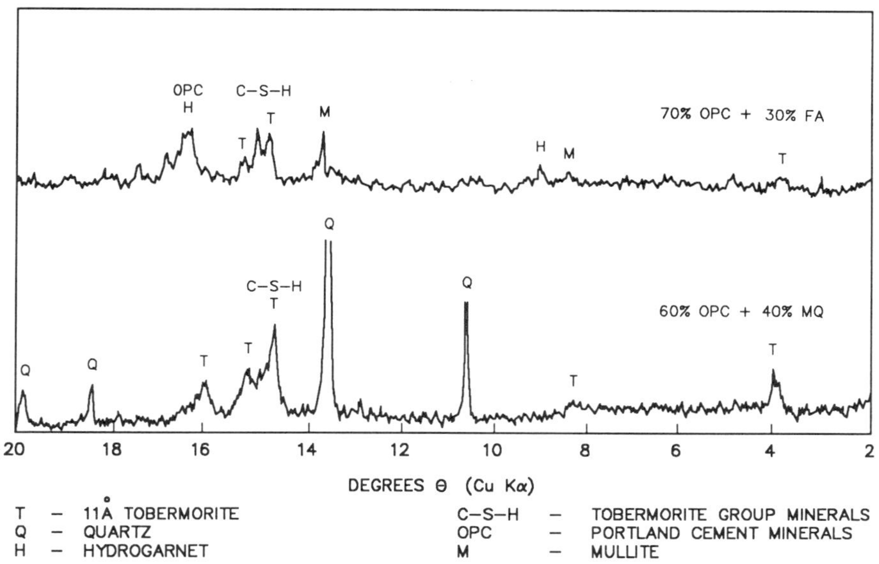

Fig.6. X-ray diffractograms for autoclaved mixtures of 70% OPC plus 30% FA and of 60% OPC plus 40% MQ

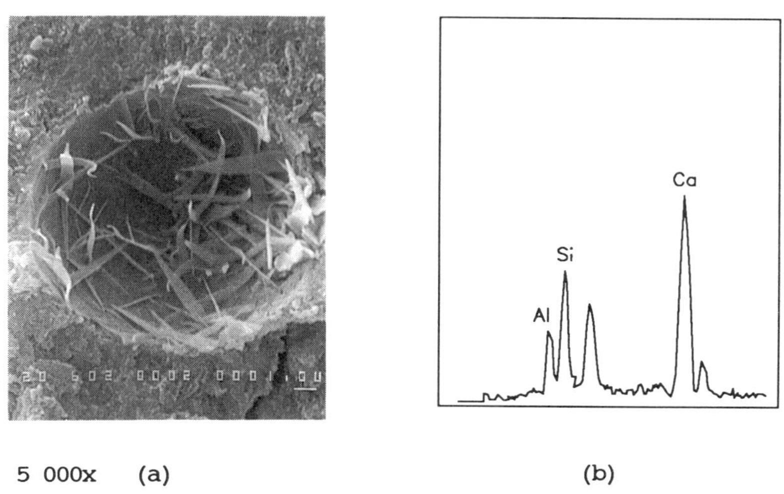

Fig.7. (a) Electron micrograph and (b) an EDXA pattern of tobermorite crystals in an autoclaved mixture of 70% OPC plus 30% FA

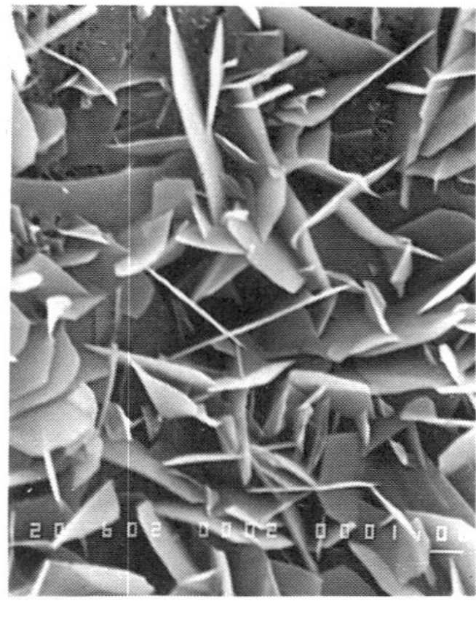

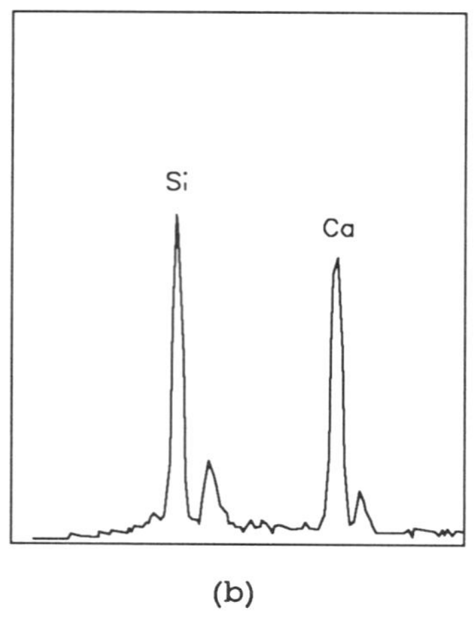

5 000x (a) (b)

Fig.8. (a) Electron micrograph and (b) an EDXA pattern (b) of tobermorite crystals in an autoclaved mixture of 60% OPC plus 40% MQ

products, which are vulnerable to soft water, are formed. The hydrogarnet that forms during autoclaving is unlikely to have any binding effect, but is nevertheless regarded as a stable mineral that would not readily be attacked by soft water. The presence of aluminian tobermorite and hydrogarnet in autoclaved blends of OPC plus FA has been confirmed by X-ray diffraction and electron microscopy. If one compares the X-ray diffractogram for a mixture of 70% OPC plus 30% FA, autoclaved at 1 MPa for 8 hours, with that for a mixture of 60% OPC plus 40% MQ, both presented in Figure 6, it is seen that while hydrogarnet is present in the former mixture, it is absent from the latter.

The electron micrographs shown in Figures 7(a) and 8(a) show that the autoclaved mixtures of 70% OPC plus 30% FA, and 60% OPC plus 40% MQ, both contain tobermorite. However, the EDXA patterns presented in Figures 7(b) and 8(b) show that while the tobermorite crystals in the former mix contain aluminium in their structure, aluminium is absent from the tobermorite crystals in the latter mixture.

The improved soft water resistance of the autoclaved FC pipes containing GGBS is probably also attributable to the formation of aluminian tobermorite and hydrogarnet in the autoclave (Krüger 1976). The formation of magnesian serpentine due to hydration of the slag might also have played a role (Visser et al. 1975).

5 Conclusions

(a) The soft water resistance of portland cement-bonded products can, in general, be significantly improved by replacing 30 to 40% of the cement with fly ash and then autoclaving them. The improvement is attributed to the formation of relatively stable aluminian tobermorite and hydrogarnet in the autoclave.

(b) The soft water resistance of both autoclaved and unautoclaved FC pipes could be improved by replacing about 50% of the cement with ground granulated blastfurnace slag. In the case of the unautoclaved pipes it is possibly due to improved impermeability of the pipe walls, and in the case of the autoclaved pipes to the formation of aluminian tobermorite and hydrogarnet, but the formation of magnesian serpentine in the autoclave is also a possibility.

Acknowledgements

(1) A tribute is due to the late J H P van Aardt who, while at the then National Building Research Institute, of the CSIR, initiated the work reported in this paper and did much to explain the increased resistance of autoclaved mortar and fibre-cement made with FA or GGBS to soft water.

(2) Our thanks go to:
The Manager of the Building Materials Technology Programme and the Director of the Division of Building Technology of the CSIR, for their support and encouragement.

Everite Limited, for making and supplying the FC pipes, Rocla (Pty) Ltd for supplying portal culvert boxes, the Pretoria Municipality for supplying a flow meter and valves, and the Tzaneen Municipality for making the site for the experimental pipeline available.

References

American Society for Testing and Materials (1987). Standard test method for compressive strength of hydraulic cement mortars. ASTM C 109 - 87. **1988 Annual Book of ASTM Standards**, Vol 04.01, pp. 58-61.

Basson, J.J. (1989). **Deterioration of Concrete in Aggressive Waters - Measuring Aggressiveness and Taking Counter Measures.** Portland Cement Institute, Midrand, South Africa, 22 pp.

Efes, Y. Influence of blast-furnace slag on the durability of cement mortar by carbonic acid attack - problems connected with tests on corroded specimens, in **Durability of Building Materials and Components.** ASTM STP 691, (Eds. P.J. Sereda and G.G. Litvan), American Society of Testing and Materials, pp. 364-376.

Gutt, W.H. and Harrison, W.H. (1977). Chemical resistance of concrete, Building Research Establishment. **Concrete** 11(5), pp. 35-37.

Hime, W.G., Erlin, B. and McOrmond, R.R. (1986). Concrete deterioration through leaching with soil-purified waters. Cement Concrete and Aggregates, **CCAGADP**, Vol. 8, No 1, pp. 50-51.
International Organisation of Standardization (1983). First draft proposal for a technical report. Reaction of asbestos-cement pipes to aggressive waters ISO/TC77, 11 pp.
Krüger, J.E. (1976). Contributions to the knowledge of the characteristics of vitreous blast-furnace slag with a high magnesia content. DSc Thesis, University of Pretoria, 1976, pp. 88-92.
Lea, F.M. (1970). **The Chemistry of Cement and Concrete**, Third Edition, Chapter 12: Action of acid and sulphate waters on portland cement. Edward Arnold (Publishers) Ltd., pp. 338-345.
Oberholster, R.E., van Aardt, J.H.P. and Brandt, M.P. (1983). Durability of cementitious systems, in **Structure and Performance of Cements** (ed. P. Barnes), Applied Science Publishers, pp. 394-415.
South African Fibre-Cement Association (1986). **Fibre-cement pipes.** SAFCMA Braamfontein, 2 pp.
Van Aardt, J.H.P. (1986). Concrete and other hydraulic-cement products in aggressive environments, in **Fulton's Concrete Technology.** Portland Cement Institute, Midrand, South Africa, pp. 410-413.
Visser, S., Krüger, J.E., van Aardt, J.H.P. and Brandt, M.P. (1976). XRD, DTA and EM data for autoclaved glasses, minerals and mechanical mixtures corresponding in composition to some of the minerals encountered in portland cement and granulated blast-furnace slag. CSIR Research Report Bou 33, pp 65-73.

66 LONG-TERM DURABILITY OF GRC

A.J. MAJUMDAR, B. SINGH
Building Research Establishment, Watford, UK

Abstract
Long-term strength properties of grc made from ordinary Portland cement and Cem-Fil AR glass fibres have been determined after 17 and 20 years. Similar results have been obtained for grc made from Cem-Fil 2 after 9 years of natural weathering on the BRE site at Garston.

The results indicate that after the same number of years of exposure the strength of grc in large components may be considerably higher than the strength of the material determined from small coupons. For example, the bending strength, ultimate tensile strength, and impact strength of grc in 1m by 1m vertical panels are 20 MPa, 7 MPa, and 5 kJ/m^2 respectively after 17 years of weathering whereas the corresponding values from coupons are 15 MPa, 5 MPa and 3 kJ/m^2.

It is also clear that when Cem-Fil 2 fibres are used in place of Cem-Fil, the pseudo-ductility of the grc composites is retained for a far longer period than with composites made from Cem-Fil; for instance, the average ultimate failure strain of Cem-Fil 2/grc was found to be 4700 microstrain after 9 years of weathering whereas the corresponding value for Cem-Fil/grc is only 150-200 microstrain after 10 years. Consequently the energy absorbing capacity of Cem-Fil 2/grc is expected to remain significant for a reasonable length of time. Microstructural differences between Cem-Fil 2/grc and Cem-Fil/grc on weathering help to explain the differences between these two types of composites.
Keywords: Glass fibre reinforced cement, Alkali-resistant glass fibres, Long-term properties, Microstructure.

1 Introduction

The first commercial alkali-resistant (AR) glass fibre for cement reinforcement, trademarked Cem-FIL, was launched by Pilkington Brothers PLC in 1971 following pioneering work by Majumdar and Ryder (1968) a few years earlier. Using the spray-dewatered method several glass reinforced cement (grc) boards containing nominally 5 wt% of 34-38 mm long Cem-FIL fibres were fabricated at the Building Research Establishment at that time using neat ordinary Portland cement (OPC) as the matrix. After moist curing in the Laboratory for the first seven days the boards were cut into specimen coupons, 150 x 50 mm and approximately 10 mm thick, and distributed to three different environments - dry air (40% rh) at 20°C, under water at 20°C and natural weathering on the BRE site at Garston. The specimen coupons were tested for their mechanical properties after pre-determined intervals and the results up to 10 years have been published (Building Research Establishment 1976, 1979). A limited number of specimens were kept for a longer time study and these have been tested recently, 20 years after they were made. The results are given in this report.

Fig. 1. Weathering of grc sheets at Garston

In order to assess the long-term weatherability of grc panels small timber frames were erected on the BRE site at Garston and 1m x 1m grc sheets were attached to these structures as shown in the photograph (Fig 1), simulating a vertical wall and a shallow roofing element. From time to time, up to a period of 17 years, panels were removed from the timber frames, sawn into 150 x 50 mm specimens and tested for strength and impact resistance. The glass fibre used in these grc composites was also Cem-FIL AR fibre, and the cement OPC.

In 1979, Pilkingtons introduced their second generation AR glass fibre Cem-FIL 2 which is considered to be more durable in the cement environment than its predecessor Cem-FIL. GRC boards were made at BRE by the spray-dewatering method using a nominal 5 wt% of 32 mm long fibre as the reinforcement and 70% OPC 30% sand as the matrix. Specimen coupons from these boards have been weathered on the BRE site at Garston for a period of 9 years. Results from these specimens are also included in this paper.

2 Experimental

The bending and tensile properties of grc composites were determined by standard BRE procedures described previously (Singh et at 1978) using an Instron testing machine. The impact strength of the materials was measured using the Izod pendulum method.

The microstructure of the various types of grc kept in different environments was examined by the Cambridge 'Stereoscan' Mark IIa scanning electron microscope (SEM). Specimens were cut from broken mechanical test specimens and the fractured surfaces coated with a layer of either gold or carbon.

3 Results

The mechanical properties of Cem-FIL/OPC composites following exposure of up to 20 years in three different environments are given in Table 1. In Table 2, the properties of grc panels made from Cem-FIL/OPC and weathered on the BRE site as wall and roof elements up to 17 years are listed. Results from small coupons kept near the panels are also included. The properties of Cem-FIL 2/grc made using 70% OPC 30% sand as the matrix and weathered on the BRE site up to 9 years are given in Table 3, comparative results on grc made from Cem-FIL are also listed. The tensile stress-strain diagrams of some selected composites after long-term exposure to different environments are shown in Fig 2.

The microstructure of various grc composites exposed to different environments for various lengths of time is illustrated in Fig 3.

TABLE 1. PROPERTIES OF SPRAY-DEWATERED OPC/GRC AT VARIOUS AGES (5 wt% Cem-FIL AR GLASS FIBRE)

Property	Total range for air and water storage conditions at 28 days	5 years			17 years	20 years	
		Air*	Water+	Weathering	Weathering	Air*	Water+
(a) Bending							
MOR (MPa)	35-50	30-35	21-25	21-23	13-18	30-32	13-15
LOP (MPa)	14-17	10-20	16-19	15-18	11-14	10-14	13-14
(b) Tensile							
UTS (MPa)	14-17	13-15	9-12	7-8	4-7	10-14	-
BOP (MPa)	9-10	7-8	7-9	7-8		~ 5	
Young's modulus (GPa)	20-25	20-25	28-34	25-32	25-36	24	-
(c) Impact strength (Izod) (kJ/m²)	17-31	18-21	4-6	4-7	2-5	17-22	~ 2

* At 40 per cent relative humidity and 20°C, + At 18-20°C. MOR = modulus of rupture
LOP = limit of proportionality UTS = ultimate tensile strength BOP = bend over point

4 Discussions

It has been pointed out in several previous publications, for instance, Building Research Establishment IP 36/79, that Cem-FIL/OPC composites are very strong and tough when manufactured and these good properties are retained over the long term if the composites are kept in relatively dry air. However, if the composites are exposed to wet environments these properties are substantially reduced and grc becomes a brittle material. The results in Table 1 confirm the earlier findings. When kept dry, Cem-FIL/OPC composites have remained strong and tough over a period of 20 years. In many indoor applications, therefore, this type of grc will be a suitable material.

When exposed outdoors in the form of 1m x 1m panels, grc made from OPC and Cem-FIL AR fibres has retained a higher proportion of its initial strength and toughness after 17 years than the small coupons treated similarly. The MOR values of both wall and roof panels are considerably higher than the corresponding LOP values and the UTS of 7 MPa (Table 2) is probably significantly higher than that of the matrix obtaining in these samples.

TABLE 2. PROPERTIES OF Cem-FIL/OPC COMPOSITE - PANELS AND COUPONS WEATHERED AT GARSTON

Environment	Age	Vertical (wall) panel					Horizontal (roof) panel					Coupon				
		MOR (MPa)	LOP (MPa)	IS (kJ/m²)	UTS (MPa)	E (GPa)	MOR (MPa)	LOP (MPa)	IS (kJ/m²)	UTS (MPa)	E (GPa)	MOR (MPa)	LOP (MPa)	IS (kJ/m²)	UTS (MPa)	E (GPa)
Damp cure in laboratory	28 days	36	-	20	-	-	36	-	20	-	-	36	-	20	-	-
Natural weathering at BRE, Garston	1 year	28	-	10	10	-	29	-	10	10	-	28	-	11	-	-
	5 years	26	18	8	8	36	21	17	8	6	35	21	18	6	7	35
	9 years	22	17	7	-	-	21	18	5	-	-	19	16	4	7	31
	17 years	20	14	5	7	31	19	16	4	7	35	15	13	3	5	31

When Cem-FIL 2 is used in place of Cem-FIL the weathering behaviour of grc is markedly improved as is clear from the comparative results given in Table 3. Although the tensile failure strain of Cem-FIL 2/grc after 9 years of weathering is less than half of the 28 day value, a value of ~ 0.5% is still very useful for a construction material. The tensile stress-strain curves shown in Fig 2 illustrate the points made above for the three different cases.

TABLE 3. PROPERTIES OF GRC MADE FROM Cem-FIL AND Cem-FIL 2

Environment	Age	Cem-FIL 2/OPC grc (70% OPC 30% sand matrix						Cem-FIL/OPC grc (70% OPC 30% matrix)					
		MOR (MPa)	LOP (MPa)	IS (kJ/m²)	UTS (MPa)	E (GPa)	Failure strain (microstrain)	MOR (MPa)	LOP (MPa)	IS (kJ/m²)	UTS (MPa)	E (GPa)	Failure strain (microstrain)
Damp cure in laboratory	28 days	38	12	21	15	30	11000	36	18	19	14	38	8250
Natural weathering at BRE Garston	2 years	36	11	20	13	39	8000	28	18	8	10	37	430
	4 years	33	15	14	13	36	7310	22	15	5 years 5	8	41	240
	9 years	30	14	13	11	44	4710	18	16	10 years 6	6	39	160

The microstructures of grc of various ages kept in different environments and shown in Fig 3, A - L are instructive in helping to

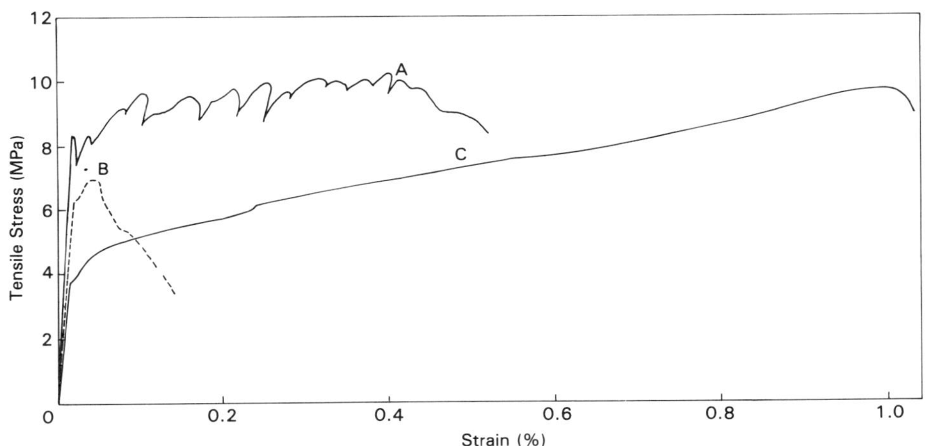

Fig. 2. Typical tensile stress-strain curves for GRC made from (A) Cem-FIL2, 70% OPC 30% sand after 9 years in natural weather, (B) Cem-FIL, OPC (vertical panel) after 17 years in natural weather and (C) Cem-FIL, OPC, stored in air 40% rh 20%C for 20 years

explain the changes in the properties of the material with time.

In Fig 3B the fracture face of a 20 year old grc sample kept in relatively dry air is shown. The photo is very similar to Fig 3A of young grc described by Stucke and Majumdar (1976). Even after 20 years the fibres appear to be undamaged and the possibility of substantial pull-out remains. Under these conditions the strength and toughness of grc will remain largely unaltered (See Table 1 and Fig 2). In Figs 3C and 3D the microstructure of grc under water storage shows a plain rather than fibrous texture typical of a brittle material. Fracture surfaces shown in Figs 3E and 3F refer to grc weathered for 5 and 17 years respectively. The microstructure is closer to that of water stored rather than air-stored grc.

The microstructure of the grc material taken from the 1m x 1m vertical (wall) panel after 17 years of weathering is shown in Fig 3G. The surface appears to be a little more fibrous than that of the corresponding coupon (Fig 3F) and this may explain the better properties obtaining in the panel. In Fig 3H, the microstructure seen on the freshly fractured surface of grc made using Cem-FIL 2 and weathered for 9 years is illustrated. The microstructure is similar to that of air-stored Cem-FIL/grc (Figs 3A and 3B) in the sense that there is little sign of fibre breakage. However Cem-FIL 2 fibre strand has remained integral (Fig 3H) whereas Cem-FIL fibre strands are usually separated into filaments. This difference is considered to be due to the special surface treatment provided for Cem-FIL 2 fibres. The microstructure shown in Fig 3H and also in Figs 3A and 3B is that of strong and tough grc.

The microstructure of grc that has become progressively weaker and embrittled with age is shown in Fig 3 I, J and K. Photographs I and J correspond to Cem-FIL/grc kept at 90% rh and under water, both at 20°C for 20 years and K refers to the same material exposed to weather for 17 years. Evidence of fibre breakage at or near the crack face is clear from these photographs. In Fig 3K, it is also seen that the glass fibres are surrounded by a deposit of cement hydration products, most probably $Ca(OH)_2$. These photographs are very similar to that of the five year water-stored Cem-FIL/OPC composite studied by Stucke and Majumdar(5) and reproduced in Fig 3L. Fibre fracture is seen clearly in this photo also and this type of microstructure is consistent with the poor performance of Cem-FIL/OPC composites in wet conditions. In this respect, the microstructure of Cem-FIL 2/grc is radically different and for this reason it is expected that such composites may remain strong and tough over a long period of time in wet environments.

5 Conclusions

After 20 years, grc made from Cem-FIL and OPC retains most of its initial strength and toughness if the composites are kept in relatively dry air. Under wet conditions, however the initial properties of grc show very substantial reductions in the long term and the material becomes brittle. There is some indication that the rate of deterioration of properties on weathering may be slower in components made of grc than in specimen coupons.

When Cem-FIL is replaced by Cem-FIL 2 as the reinforcement, the rate of deterioration of grc properties on weathering is much slower. The 9 year results on Cem-FIL 2/grc give grounds for cautious optimism.

The properties of grc kept in different environments over long periods can be related to its microstructure.

6 Acknowledgement

The authors are indebted to Dr S Duerden for some of the SEM photos. This paper is part of the research programme of the Building Research Establishment and it is published by permission of the Director.

7 References

Building Research Establishment (1976), A study of the properties of CemFIL/OPC composites. BRE Current Paper CP 38/76.

Building Research Establishment (1979) Properties of grc: 10 year results. BRE Information Paper IP 36/79.

Majumdar, A.J. and Ryder, J.F. (1968) Glass fibre reinforcement of cement products. Glass Technol., 9, 78-84.

Singh, B. Walton, P.L. and Stucke, M.S. (1978) Test methods used to measure the mechanical properties of fibre-cement composites at the Building Research Establishment in **Testing and test methods of fibre cement composites** (ed R.N. Swamy) Construction Press, Lancaster 377-387.

Stucke, M.S. and Majumdar, A.J. (1976) MIcrostructure of glass fibre reinforced cement composites. J. Mater. Sci., 11, 1019-1030.

Fig. 3. A-F Microstructure of grc at various ages. All micrographs refer to grc made from Cem-FIL AR glass fibres and OPC

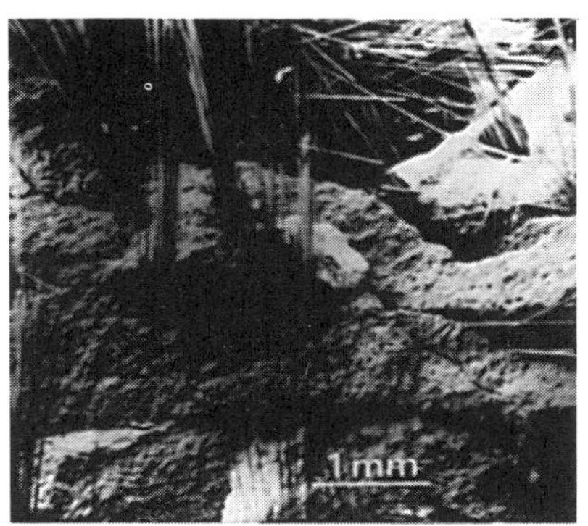

A Fracture surface of grc kept for 90 days in dry air (40% rh) at 20°C

B Fracture surface of grc kept for 20 years in dry air (40% rh) at 20°C

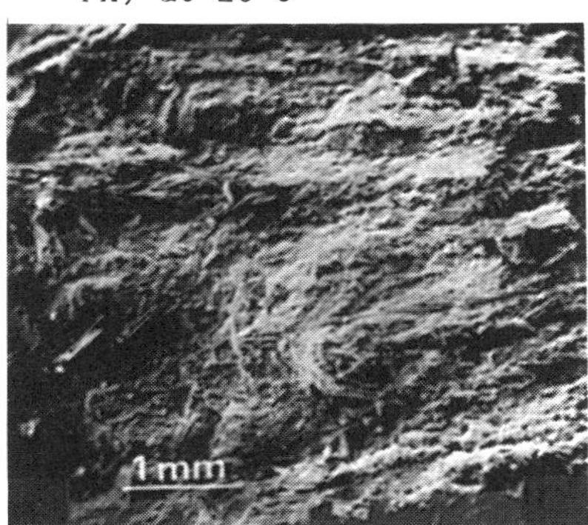

C Fracture surface of grc after 5 years in water at 20°C

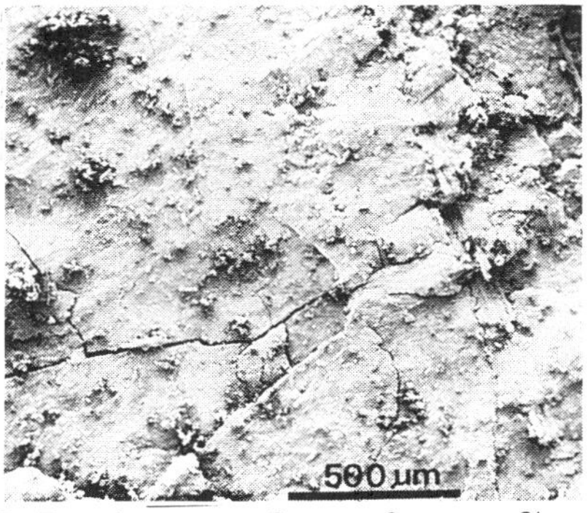

D Fracture surface of grc after 20 years in water at 20°C

E Fracture surface of grc naturally weathered for 5 years

F Fracture surface of grc naturally weathered for 17 years

Fig. 3. G-L Microstructure of grc at various ages. All micrographs except H refer to grc made from Cem-FIL AR glass fibres and OPC. H refers to grc made from 70% OPC 30% sand and CemFIl 2 AR glass fibre

G Fracture surface of grc naturally weathered for 17 years (vertical panel)

H Fracture surface of Cem-FIL 2/ grc naturally weathered for 9 years

I Microstructure of grc after 20 years in air (90% rh) not 20°C

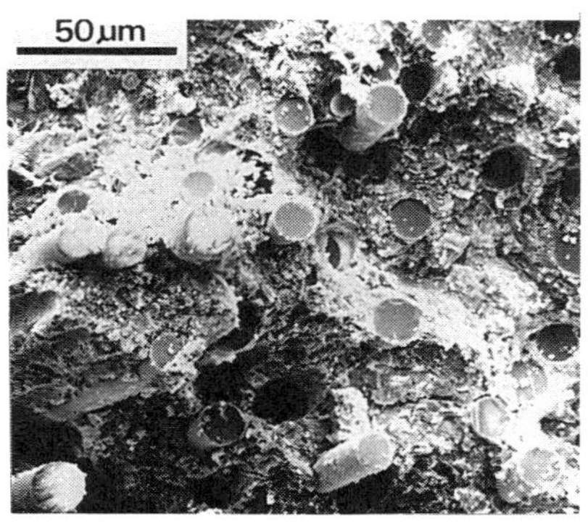

J Microstructure of grc after 17 years natural weathering

K Microstructure of grc after 20 years in water at 20°C

L Microstructure of grc after 5 years in water at 20°C

67 PROBABILISTIC APPROACH TO SERVICE LIFE PREDICTION OF FERROCEMENT ROOF SLABS

S.T. QUEK, S.L. LEE, P. PARAMASIVAM
Department of Civil Engineering, National University of Singapore, Singapore

Abstract
In view of the uncertainties which exist in the service life prediction of ferrocement roofing slabs due to the inherent variability in the material properties, fabrication process and loading, a simple probabilistic solution is proposed. In this approach, the service life of a typical slab with respect to a specified minimum first-crack strength is formulated as a function of the probability of the actual strength being less than the specified minimum. The loading considered is that due to differential temperature across the thickness of the slab. In order to obtain data for analysis, a preliminary experimental program was carried out in which the mean deterioration in strength with life and the uncertainty in determining the actual strength of a typical slab were estimated. The variation of the expected load acting on the slab during its life, characterized by a probability density function, is incorporated using Miner's cumulative damage rule. Within the load range considered, the results showed that ferrocement possesses good fatigue properties and is suitable for use as a secondary roofing material.
Keywords: Ferrocement, Fatigue, Probability, Uncertainty, Service life, First-crack, Residual strength.

1 Introduction

In Singapore, ferrocement slabs are being used as secondary roofing elements to provide thermal insulation as well as to minimize the deterioration of the primary roof arising from the effects of environmental loading such as changes in temperature and wetting and drying of the exposed surfaces. The durability of these secondary roofing slabs is of practical interest, particularly with respect to the service life cycle in relation to the actual load range that the slabs are experiencing.

The fatigue strength of ferrocement slabs subjected to cyclic bending stresses have been studied by various researchers (Balaguru et al 1979, Karasudhi et al 1977, McKinnon and Simpson 1975, Paramasivam et al 1981, Picard and Lachance 1974) in which the test specimens were cycled to failure to obtain stress-life (S-n) curves for different types and percentages of reinforcement. The inherent disadvantages of this procedure is the lengthy time required for each specimen to fail

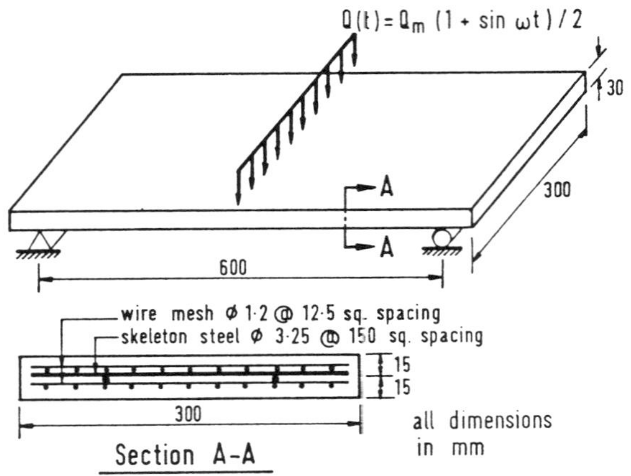

Fig.1. Typical ferrocement test specimen

especially when the cyclic stress level is low, which in the case of roofing slabs is the level of practical interest. In such studies, the recommended design strength and service life of the components should realistically account for any uncertainties which may exist. The uncertainties may arise from the inherent variability of the materials used and in the fabrication process, and the random nature of the environmental loadings and should be properly quantified (Ang and Munse, 1975).

In this study, a probabilistic treatment to provide a consistent and rational basis for specifying the strength and the corresponding service life of ferrocement roofing elements in relation to the differential temperature caused by the actual environment is presented. To obtain the relevant statistics, a preliminary experimental program was carried out in which the residual first-crack and ultimate strength of each test specimen were evaluated after it had been subjected to a predetermined number of cycles at a selected stress range. It is therefore unnecessary to cycle each specimen to failure. The temperature loadings used were estimated by monitoring the temperature difference across the thickness of a typical slab on the roof. Using Miner's cumulative damage rule, the probability of exceedance of the specified minimum strength with respect to the service life is computed. For a prescribed probability level, the service life corresponding to the design strength specified may also be estimated.

2 Reliability considerations

For the purpose of this study, a typical ferrocement roofing slab subjected to sinusoidal loading at mid-span as shown in Fig. 1 is considered. The primary parameter of interest is the residual first-crack strength of the specimen denoted by R_c(kN), after it has been subjected to some n cycles of loading having a minimum and a maximum amplitude of 0 and Q(kN) respectively, although the results for the corresponding residual ultimate strength, R_u(kN), are also presented. The parameters R_c, R_u and Q will be normalized by R_{co}, R_{uo} and R_{co} as r_c, r_u and q, respectively, where R_{co} and R_{uo} are the first-crack and ultimate strength of the specimen if it has not been cyclically loaded. The normalized residual strength, r (which will be used henceforth to denote either the first-crack or the ultimate strength), is not constant for fixed of values for q and n. This is due to the variability in the materials, fabrication and loading processes and r should thus be treated as a random variable. Let $F_r(r|q,n)$ denote the probability distribution function of r conditioned on given values of q and n. The probability that a slab will have a residual strength below some allowable value, r_a, after it has been cyclically stressed from 0 to q for n number of cycles is thus given by

$$p_{f|q,n} = P(r<r_a|q,n) = F_r(r_a|q,n) \tag{1}$$

In the case where $F_r(r|q,n)$ can be approximated by the normal distribution with a mean of $\mu_{r|q,n}$ and standard deviation, $\sigma_{r|q,n}$, the allowable strength is related to the probability of the slab having a residual strength below the specified allowable by

$$r_a = \mu_{r|q,n} + \sigma_{r|q,n} * \Phi^{-1}(p_{f|q,n}) \tag{2}$$

where Φ^{-1} is the inverse cumulative normal distribution function which is tabulated in standard probability books.

The magnitude of the cyclic stresses acting on the slab during its intended life is generally a randomly varying quantity. For the case of secondary roofing elements, the cyclic stresses arise from bending as a result of differential temperature across the thickness of the slab which varies with the weather. The variation in the actual differential temperature can be monitored over a period of time and a probability density function for the normalized stress, denoted as, $f_q(q)$, may be fitted. This can then be convoluted with Eq. 1 to obtain an overall probability of failure as a function of n only. Assuming that the ferrocement is approximately linear elastic within the working stress range and the modulus of elasticity does not deteriorate significantly with n, Miner's linear rule for cumulative fatigue damage may be used. The amount of fatigue damage may be assumed as proportional to the number of cycles that the specimen has been previously stressed, the proportional constant being dependent on the stress level. Hence, Eq. 1 may be modified to account for stress variations throughout the life of the element as

$$\int_{\text{all } q} F_r(r_a|q,n).f_q(q).dq = p_{f|n} \tag{3}$$

$$\sum_i F_r(r_a|q_i,n).P_q(q_i) = p_{f|n} \tag{4}$$

where Eq. 3 is for the case in which q is modelled as a continuous random variable and Eq. 4 is used when the variation of q is divided into discrete ranges.

Note that in this approach, given any two of the three variables, r_a, n and $p_{f|n}$, the third may be determined. In order to compute $p_{f|n}$, it is necessary to obtain

(a) the mean deterioration of residual strength as a function of n for different values of q,

(b) the statistics of r, in terms of a probability density function, and

(c) the probability density function of q.

3 Experimental program

All the tests were conducted using the Instron cyclic machine on rectangular slabs 700 mm by 300 mm with a thickness of 30 mm. The reinforcements consisted of one layer of skeletal steel in the form of welded mesh sandwiched between two layers of fine galvanized welded wire mesh as illustrated in Fig. 1. The nominal values of the dimensions and properties are tabulated in Table 1. The specimens were cast and taken out of the mould after 24 hours and subsequently moist-cured at about 28 ± 2 °C and 80 ± 10 % relative humidity for 28 days, after which they were air-dried before testing.

In this program, three properties were monitored, namely, the residual first-crack strength, R_c, ultimate strength, R_u, and modulus of elasticity within the working range, E. The cyclic test for each slab

Table 1. Ferrocement Data

Items		Nominal values
Wire Mesh:	Diameter	1.2 mm
	Grid size	12.5 x 12.5 mm
	Ultimate Strength	330 MPa
	Modulus of Elasticity	180 000 MPa
	Volume fraction	0.0063
Skeletal Steel:	Diameter	3.25 mm
	Grid size	150 x 150 mm
	Ultimate Strength	660 MPa
	Modulus of Elasticity	200 000 MPa
	Volume fraction	0.0018
Mortar:	Cube Strength (28 days)	45 MPa
	Modulus of Elasticity	28 000 MPa
	Tensile Strength	4.5 MPa
	Cement:sand:water (wt.)	1:2:0.47

specimen was performed in the following sequence. First, the specimen was loaded statically up to 1 kN, which is below the virgin nominal first-crack strength, to estimate the modulus of elasticity at n=0, E_o. Next, the specimen was subjected to n cycles of constant amplitude sinusoidal loading, ranging from 0 to Q kN, at a rate of 5 Hertz. This rate was chosen since previous fatigue studies (ACI Committee 215, 1974) have shown that for concrete, the fatigue strength is not significantly affected by the loading rate if cycled within the range of 1 to 15 Hz provided the maximum stress level is less than about 75 percent of the static strength. Three values of Q were used, namely, 0.75, 1.05 and 1.30 kN. These values are below the first-crack strength, which is of critical concern for secondary roofing slabs or in applications where permeability is one of the major considerations. The values of n selected were 3,000, 10,000, 30,000 and 100,000 cycles. For roofing slabs, one cycle would correspond to a time period of one day since the temperature is maximum during the day time and minimum during the night. The selected values of n would therefore correspond to a life of approximately 8, 27, 82 and 273 years, respectively. At the end of the cyclic loading, the specimen was immediately tested for each of the three properties, R_c, R_u and E, by loading the specimen to failure under static load test condition.

For each n and Q, three specimens were tested from which an average value of each of the three parameters under study, R_c, R_u and E was obtained. In order to capture the variation of the results due to the inherent variability of the materials used, fabrication process, and loadings, many specimens have to be tested. Due to time and cost constraints, two sets of nine specimens each were tested, at Q=1.05 kN and n=30,000 and 100,000 cycles, respectively. For proper control, the elements were cast in four batches of 18 specimens each, one-third of which were used as control specimens to determine the actual first-crack and ultimate strength of the uncycled slabs. The control specimens were tested at the beginning and at the end of each of the four series of experiments corresponding to the four batches. The distribution of the specimens used is summarised in Table 2.

4 Experimental results

To ensure consistency in the interpretation of the results, the values

Table 2. Distribution of Specimens Used

No. of cycles	Batch A	Batch B	Batch C	Batch D
0 (beginning)	3	3	3	3
3 000	3	3	3	0
10 000	3	3	0	3
30 000	3	3	9	0
100 000	3	3	0	9
0 (end)	3	3	3	3
Total No.	18	18	18	18

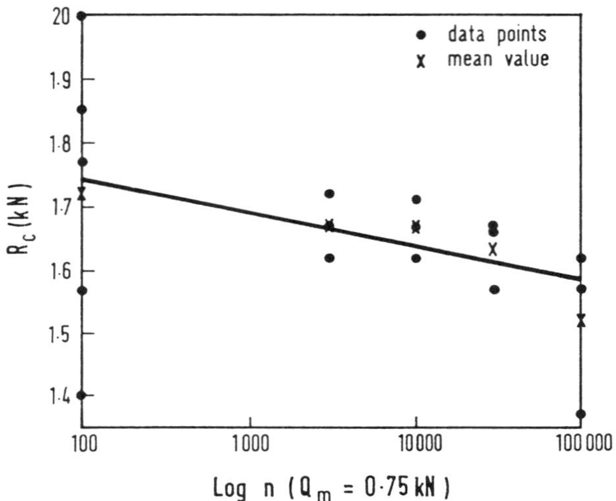

Fig.2. R_c(kN) vs log n for Q=0.75 kN

of the properties monitored were analyzed separately for each of the 4 batches. For example, in a batch of 18 specimens, 12 were tested at Q=0.75 kN for four different values of n using 3 specimens each; in addition, static strength tests using 3 uncycled specimens each, at the beginning and at the end of the experiment for this batch, were carried out. The individual values as well as the mean for each n, were then plotted on a graph as shown in Fig. 2, from which a best-fit line was drawn. A linear relationship between the strength and log n was adopted because of simplicity and the data do not distinctly indicate otherwise. The criterion used for deciding the line of fit was that the line should be as close to the mean values as possible, yet within the range of all the values obtained. This allows some amount

Table 3. Relationships between $\bar{r}$ and n

	q	linear-log	log-log
first-crack:	0.43	$\bar{r}_c = 1.07 - 0.035 \log n$	$\bar{r}_c = 1.07/n^{0.016}$
	0.60	$\bar{r}_c = 1.10 - 0.050 \log n$	$\bar{r}_c = 1.13/n^{0.025}$
	0.85	$\bar{r}_c = 1.15 - 0.075 \log n$	$\bar{r}_c = 1.19/n^{0.037}$
ultimate:	0.43	$\bar{r}_u = 1.02 - 0.010 \log n$	$\bar{r}_u = 1.03/n^{0.0056}$
	0.60	$\bar{r}_u = 1.03 - 0.015 \log n$	$\bar{r}_u = 1.04/n^{0.0072}$
	0.85	$\bar{r}_u = 1.04 - 0.020 \log n$	$\bar{r}_u = 1.05/n^{0.0095}$

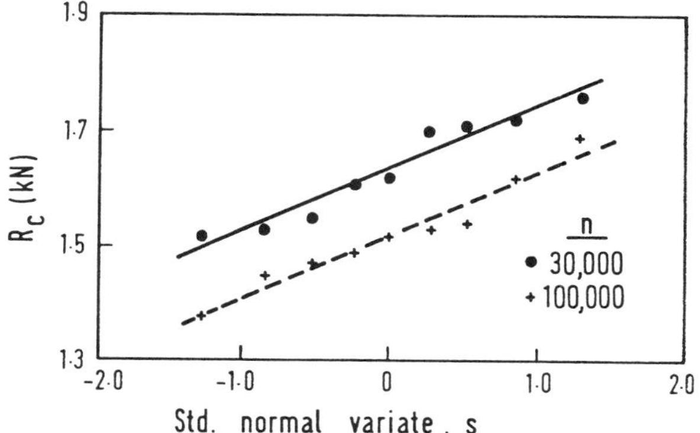

Fig.3. R_c(kN) vs std. normal variate, s

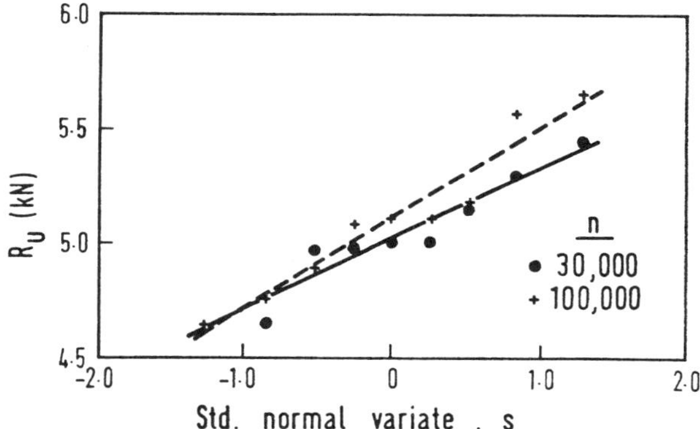

Fig.4. R_u(kN) vs std. normal variate, s

of judgment which is not possible if the standard regression analysis was performed. The overall mean first-crack and ultimate strength, R_{co} and R_{uo} respectively, of the uncycled slabs for this batch were estimated and used for normalizing the results. This process was similarly carried out for the other batches and the mean deterioration in the normalized first-crack and ultimate strength were obtained. A similar study showed that the results fitted as well using a log-log relationship and the two sets of relationships, summarized in Table 3, agree to within 1 percent. In view of the life span considered, the results indicate the good fatigue resistance of the class of specimen under study. The deterioration in the first-crack strength is more significant than that of the ultimate strength. There is no necessity to extrapolate the curve to higher values of n as it is unlikely that the slabs in practice will be used longer than 99 years.

The inherent variability in R_c and R_u were estimated from statisti-

Table 4. Comparision between normal and lognormal distribution fit

Data Set	Statistical Test			
	Reg. Coef. (Prob. Plot)		Kolmogorov-Smirnov Test	
	Normal	Lognormal	Normal	Lognormal
30,000(cr.)	0.973	0.971	0.207	0.210
100,000(cr.)	0.976	0.979	0.108	0.108
30,000(ult.)	0.968	0.967	0.204	0.213
100,000(ult.)	0.966	0.970	0.155	0.166

Note: Critical value of K-S test at 5% level of significance is 0.432

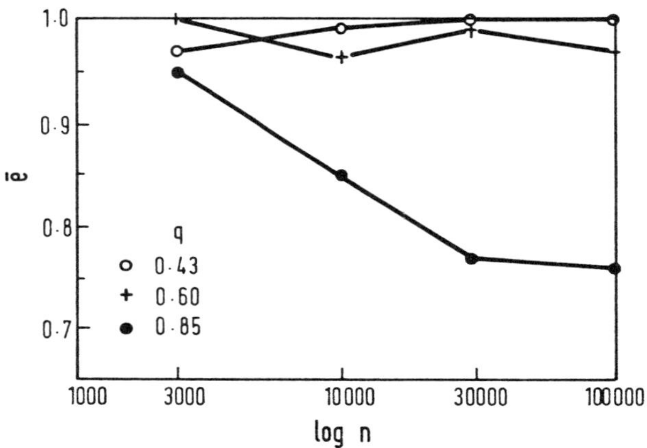

Fig.5. $\bar{e}$ vs log n

cal fitting of the results from specimens cycled at Q=1.05 kN for n = 30,000 and 100,000 cycles, as shown in Figs. 3 and 4. The four sets of data plotted showed that the normal distribution may be adopted for R_c and R_u. A comparative study was made of the fit using the normal distribution with the lognormal distribution and presented in Table 4. Due to the fact that only 9 data points were used for each set of data, the statistical results, in terms of the regression coefficients and the Kolmogorov-Smirnov values, do not show distinctly which distribution is better.

A plot of the mean of e(=E/E_o) versus log n is given in Fig. 5 showing that there is no significant deterioration in the mean value of e with n up to about 60% of the load corresponding to the virgin first-crack strength. One can thus assume linearity and the super-

Table 5. Load range, q

q	rel. freq.
< 0.45	0.86
0.45-0.65	0.12
0.65-0.90	0.02

position principle may be applicable, justifying the use of Miner's rule to estimate the cumulative damage when random load range is considered, provided the proportion of load above the 60% level is insignificant. At the 85% level, minor deterioration was observed.

5 Uncertainty analysis

In order to carry out the reliability computations discussed above, the major sources of uncertainty must be identified and quantified. The main sources of uncertainty in fatigue analysis are contributed by the actual environmental loadings, q, the predicted residual strength, r, and the measure of service life, n.

The uncertainty in q is explicitly accounted for through the use of a probability density function, $f_q(q)$, as a weighting function in Eq. 4. This function may be obtained by monitoring the temperature difference across the thickness of a typical slab on the roof. For a differential temperature of ΔT across the thickness of the slab, the stress acting on the extreme fibres is given by $E\alpha\Delta T/2$, where E is the modulus of elasticity and α the coefficient of thermal expansion of the specimen. By fixing thermal couples on the ferrocement roofing slab of a typical building, the data showed that a maximum temperature difference of $10^{o}C$ across the thickness during a 24 hours cycle is not unusual. During the hot season, a maximum difference of $13^{o}C$ or more may be obtained. Differential temperatures of $10^{o}C$ and $13^{o}C$ correspond to q values of approximately 0.4 and 0.52 respectively. The relative frequency of the stress range, q, in Table 5 is adopted in relation to the three values of q used in the experiment.

By comparison with q and r, the uncertainty in n can be assumed negligible for secondary roofing slabs. This is because a cycle of significant temperature variation corresponds to a time period of one day in the equatorial region. For other regions, a reasonably good mean value of n can be estimated using meteorological data over a number of years which can be obtained from the weather bureau.

The uncertainty in r comprises of:

(a) its inherent variability. For a given n and q, the inherent variability of r is approximated by its coefficient of variation (cov), δ_r, which is the standard deviation normalized by the mean of r. Using the results of Figs. 3 and 4, an average value of 0.07 for δ_r over the range of n considered is obtained. Theoretically, one should use as many samples as possible to minimize statistical uncertainty in the estimation of the coefficient of variation.

(b) uncertainty in $\bar{r}$. In this study, the log-log relationships, $\bar{r} = c\, n^{\alpha}$, given in Table 3 are used, which are approximate due to the fact that limited samples are used and the experimental conditions are imperfect. Since $\bar{r}$ for each n and q are averaged over limited samples, the variability of $\bar{r}$ is not negligible. Based on random sampling theory, the cov of $\bar{r}$ is given by $\delta_r/\sqrt{m}$, where δ_r is the cov due to the inherent variability of r and m is the number of samples used to estimate that particular mean value. The cov of $\bar{r}$ is quite constant for all the data obtained, having an average value of 0.04. A second source of uncertainty arises from the fact that although the relationships in Table 3 are for fixed values of q, it is impossible experimentally to ensure that q is constant for each batch of tests for two reasons. First, although the load is pre-set in the cyclic test machine, there is a slight variation in the magnitude over n cycles in the experimentation due to purely mechanical reasons. This was monitored in the course of the experiment and estimated to be small. Its contribution to the uncertainty in $\bar{r}$ due to q is within a cov of 0.02. The second factor is due to the fact that q is normalized by the first-crack load which varies between specimens even within the same batch of specimens cast. The amount of uncertainty can be estimated using the values of the first-crack load obtained from the controlled specimens and found to have a cov of less than 0.06.

The total uncertainty in r, is estimated by taking the root-sum-square of the individual cov, and an approximate value of 0.10 is obtained.

6 Prediction of service life

Based on the results of Tables 3 and 5, the total uncertainty in r obtained in the previous section and assuming r to be normally distributed, the probability of failure with respect to different allowable values of r, r_a, for some specified n can be computed using Eq. 4. The results are plotted in Figs. 6 and 7.

Fig. 6 shows the increase in the probability of failure with time for different allowable first-crack strength ratio. From the results, the following can be deduced:

(a) for a probability level of 0.001, if the design first-crack strength ratio is 0.63, the replacement period is 25 years; and
(b) for a given design first-crack strength ratio, say 0.6, the decrease in the level of confidence (measured in terms of probability value) from T = 1 to T = 25 years is about an order of magnitude.

Fig. 7 is only of academic interest since the results obtained are based on the assumption that the specimens have not cracked during its lifetime. Realistically, once the specimen has cracked, the deterioration in the ultimate strength would be more rapid than depicted in Fig. 7 since other environment factors such as carbonation and corrosion will become significant, which is not accounted for in this study.

As a matter of interest, a comparison is made in Figs. 6 and 7 between the cases when the normal and lognormal distributions are

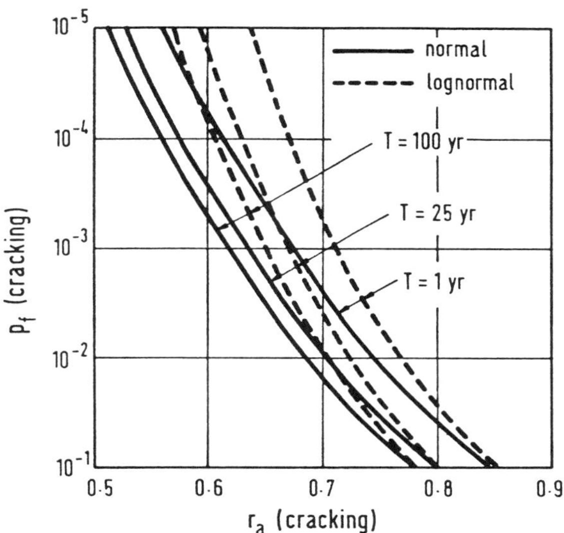

Fig.6. p_f(first-crack) vs r_a(first-crack)

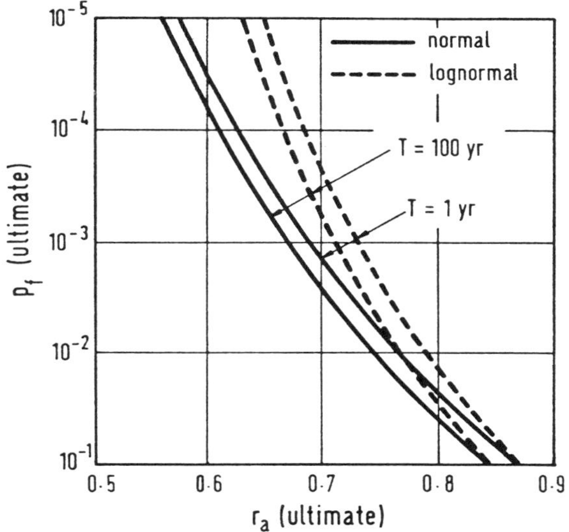

Fig.7. p_f(ultimate) vs r_a(ultimate)

assumed for r. It can be seen that the normal distribution gives conservative results. However, unless sufficient data are obtained or substantiated by physical reasons, it is not possible to ascertain which is a truly better model.

7 Concluding remarks

A simple probabilistic approach to the study of the fatigue strength of ferrocement roofing slabs, including the results from a preliminary experimental program are presented. The results in this paper show that ferrocement has good fatigue properties within the stress range considered and is therefore an attractive secondary roofing material. Through quantification of the uncertainty which exist in modelling and prediction, the service life of the specimen designed with a predetermined first-crack strength can be estimated once a target probability of exceedance of the specified allowable strength is prescribed.

8 Acknowledgements

This study is part of an on-going program under the National University of Singapore research grant RP880623.

9 References

ACI Committee 215 (1974) Considerations for Design of Concrete Structures Subjected to Fatigue Loading. **ACI J.**, Vol. 71, No. 3, 97-120.

Ang, A.H-S. and Munse, W.H. (1975) Practical Reliability Basis for Structural Fatigue. **Preprint 2492, ASCE Nat. Struc. Eng. Conf.**

Balaguru, P.N. Naaman, A.E. and Shah, S.P. (1979) Fatigue Behaviour and Design of Ferrocement Beams. **J. Struc. Div., ASCE**, Vol. 105, ST7, 1333-1346.

Karasudhi, P. Mathew, A.G. and Nimityongskul, P. (1977) Fatigue of Ferrocement in Flexure. **J. Ferrocement**, Vol. 7, No. 2, 80-95.

McKinnon, E.A. and Simpson, M.G. (1975) Fatigue of Ferrocement. **J. Test. & Eval.**, Vol. 3, No. 5, 359-363.

Paramasivam, P. Das Gupta, N.C. and Lee, S.L. (1981) Fatigue Behaviour of Ferrocement Slabs. **J. Ferrocement**, Vol. 11, No. 1, 1-10.

Picard, A. and Lachance, L. (1974) Preliminary Fatigue Tests on Ferrocement Plates. **Cement & Conc. Res.**, Vol. 4, 967-968.

68 ADHESION OF SILICONE SEALANTS

W. GUTOWSKI, A. GERRA, L. RUSSELL
CSIRO Division of Building, Construction and Engineering, Highett, Victoria, Australia

Abstract

It is shown in this paper, that the strength of adhesion between the sealant and substrate depends upon surface properties of these materials expressed in terms of specific components of their surface energies and the acid-base interactions.

The predictive model regarding the relationship between the bond strength and the energy of acid-base interaction developed may be applicable in both dry and humid/wet environments.

Also, a new procedure for assessing sealant adhesion is described.
Keywords: Silicone Sealants, Adhesion, Acid-base Interactions, Shear Strength, Tensile Strength, Peel Strength, Fracture Energy.

1 Introduction

The success of silicone sealants in modern technologies has its basis in their capacity to form strong chemical bonds with the surface of the typical substrates used in curtain walls, e.g. aluminium, glass and granite. Resultant adhesive forces exerted across the interface exceed the cohesive forces between the sealant molecules and, thus, perfect initial adhesion is assured in the system. The emergence of various finishes applied to the surface of structural members, (e.g. a variety of polymeric coatings on aluminium, metallic and ceramic coatings on glass), however, results in significant variations of surface properties of the substrate which can lead to undesirable reduction of sealant adhesion.

2 Energy of interaction and force of interaction

Any interfacial interactions can be conveniently analysed in terms of the energy (U) or enthalphy (ΔH) of interaction, since the interaction force and hence the strength of the adhesive bond is defined as the first derivative of the interaction energy with relation to the separation distance 'r' between the atoms or molecules, i.e.

$$F = -\frac{dU}{dr} \tag{1}$$

In adhesion science, the (negative) energy of interaction between materials 1 and 2 in immediate contact is known as the thermodynamic work of adhesion, W_A, which can be estimated using the following fundamental Dupré equation:

$$W_A = \gamma_1 + \gamma_2 - \gamma_{12} \quad (2)$$

where γ_1 and γ_2 are the surface energies of materials 1 and 2 in contact, and γ_{12} is the interfacial energy.

An alternative expression for the estimation of work of adhesion was developed by Good and Girifalco (1987), i.e.

$$W_A = 2\,\Phi(\gamma_1\,\gamma_2)^{1/2} \quad (3)$$

where Φ is the interaction parameter given by

$$\Phi = (d_1 d_2)^{1/2} + (p_1 p_2)^{1/2} \quad (4)$$

In Eq. (4), d and p are the dispersive and non-dispersive fractions of total surface energy of material 1 or 2, i.e.

$$d = \gamma^d/\gamma \quad \text{and} \quad (5)$$
$$p = \gamma^p/\gamma \quad (6)$$

According to Fowkes (1978), the total work of adhesion W_A comprises terms associated with the dispersive, i.e. W_A^d, and acid-base interactions, i.e. W_A^{ab}, components

$$W_A = W_A^d + W_A^{ab} \quad (7)$$

The acid-base component of the thermodynamic work of adhesion can be calculated using the following expression:

$$W_A^{ab} = W_A - 2(\gamma_1^d/\gamma_2^d)^{1/2} \quad (8)$$

Interactions related to hydrogen bonds are a sub-set of acid-base interactions.

3 Experimental

3.1 Materials

Glass, aluminium and steel, as well as a range of engineering plastics (such as polymethyl methacrylate (PMMA), acrylonitrile-butadiene-styrene terpolymer (ABS), nylon 6-6, high impact polystyrene (PS), vinyl copolymer, acetal, polypropylene – natural (PP) and filled (PP Filled), polyethylene – low density (LDPE), high density filled (HDPE), and ultra-high molecular weight (UHMW–PE)), were selected to cover a broad range of variability of surface properties of substrates expressed in terms of their surface energies γ_1, polar (γ_1^p) and dispersive (γ_1^d) components. Three silicone sealants, denoted A, B and C, were chosen for this work. Sealants A and C are qualified as structural, whilst B is a general purpose sealant.

3.2 Surface preparation

All substrates were thoroughly cleaned by washing with ethyl alcohol. After overnight drying at room temperature they were wiped three times with ethyl alcohol (ABS, PS, vinyl copolymer, PMMA) or with MEK (all other substrates) and allowed to dry for one hour prior to bonding.

3.3 Cure of specimens

All bonded assemblies were allowed to cure for four weeks prior to testing under the following conditions: two weeks at 50% RH, 23°C; one week at 98% RH, 38°C; and one week at 50% RH, 23°C.

3.4 Test methods

Peel strength: This was determined using the specimens prepared in accordance with the ASTM C-794 standard modified slightly with regard to dimensions of substrates which were 25 mm wide and 100 mm long. Two specimens were tested for each experimental point at the strain rate equal to 50 mm/min.

Shear strength: This was determined by single lap-shear specimens, as illustrated in Fig. 1(a). Three specimens were used for each experimental point. The strain rate during testing was 10 mm/min.

Tensile strength: This was determined using specimens illustrated in Fig. 1(b). Three specimens were used for each experimental point. The strain rate during testing was 10 mm/min.

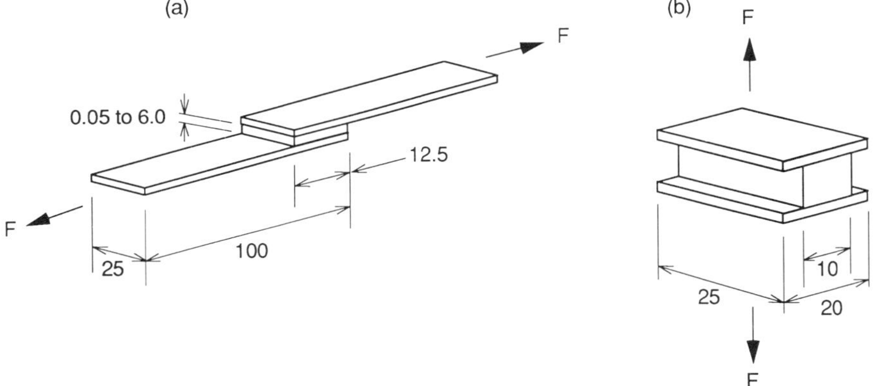

Fig. 1. Geometry of specimens used throughout experiments: (a) single lap-shear specimen; (b) tensile specimen.

Surface energy of substrates and sealants: This was determined from the wettability studies using the following Eqs (Kinloch 1987, Kaelble and Uy 1970):

$$0.5\ \gamma_2^{(1)}\ (1 + \cos\theta^{(1)}) = \left(\gamma_1^{d}\gamma_2^{d(1)}\right)^{1/2} + \left(\gamma_1^{p}\gamma_2^{p(1)}\right)^{1/2} \tag{9}$$

$$0.5\ \gamma_2^{(2)}\ (1 + \cos\theta^{(2)}) = \left(\gamma_1^{d}\gamma_2^{d(2)}\right)^{1/2} + \left(\gamma_1^{p}\gamma_2^{p(2)}\right)^{1/2} \tag{10}$$

which are solved for unknown parameters γ_1^d and γ_1^p; the dispersive and non-dispersive components of the total surface energy of the substrate

$$\gamma_1 = \gamma_1^d + \gamma_1^p \tag{11}$$

The parameters θ_1 and θ_2 in Eqs (9) and (10) are the equilibrium contact angles exhibited by the test liquids 1 and 2 deposited on the substrate's surface. The test liquids of known surface properties, i.e., γ_2, γ_2^d and γ_2^p, used during the experiments were water, formamide and glycerol. For more particulars regarding the determination of a solid's surface energy, see Kinloch (1987), Wu (1982) and Kaelble and Uy (1970).

4 Results and discussion

4.1 Sealant's surface energy with reference to acid-base interactions

It is shown in the literature (Klosowski 1988) that the surface energy of a cured silicone sealant (poly-dimethylsiloxane) determined at the sealant/air interface is 21.7 mJ/m^2 and its polarity (at this particular surface) is 0.05. Similar results are obtained by these authors, as shown in Table 1 below, for all sealants investigated.

It is suggested in this paper, that the above properties of the 'cured sealant/air' interface, and particularly γ_2^p (non-dispersive component of sealant's surface energy), are not relevant from the viewpoint of adhesive bond formation between the sealant and the substrate. The process of the formation of adhesive bond between the silicone sealant and substrate takes considerable time whilst the sealant is still in its uncured state. During this process, the mobility of molecules and flexibility of chains allows them to obtain preferential orientation with regard to the substrate's reactive sites taking part in acid-base interactions. It is assumed in this work that the polarity of the uncured sealant is the major factor contributing to the process of strength development.

It is apparent from the data presented in Table 1 that all uncured sealants investigated exhibit high polarity, i.e., p = 0.41, as compared with that exhibited by the same sealants cured in contact with air. This is indicative of the sealant's capacity to form bonds attributed to the acid-base interactions with appropriate substrates. High polarity of the uncured sealant is shown to be in sharp contrast with the essentially non-polar character of the cured sealant at the surface exposed to the air.

Table 1. Surface properties of silicone sealants (Gutowski 1990)

Interface	γ_2 (mJ/m^2)	γ_2^p (mJ/m^2)	Polarity 'p'
Sealant A			
uncured sealant/N_2	14.21	5.80	0.41
cured sealant/air	18.09	0.40	0.02
cured sealant/LDPE[a]	18.39	2.13	0.11
cured sealant/PS[b]	18.10	7.25	0.41
cured sealant/ABS[b]	21.12	6.92	0.33
Sealant C			
uncured sealant/N_2	18.10	7.42	0.41
cured sealant/air	17.75	0.50	0.03
cured sealant/LDPE[a]	24.50	7.55	0.31
cured sealant/Acetal[a]	23.85	7.63	0.32
cured sealant/PS[b]	20.30	8.12	0.40

[a] Sealant peeled off from the substrate.
[b] Substrate dissolved in MEK.

4.2 Relationship between the bond strength and surface properties of bond components

Surface energies of all substrates used in experiments, and energy of interaction (W_A and W_A^{ab}) between these substrates and sealant C are given in Table 2.

Table 2. Surface properties of substrates used in bonding with sealants A and C and relevant thermodynamic work of adhesion W_A and energy of acid-base interactions W_A^{ab} (values of W_A and W_A^{ab} relevant to bonds with sealant C) (Gutowski 1990)

Material	γ_1 (mJ/m²)	γ_1^p (mJ/m²)	γ_1^d (mJ/m²)	W_A (mJ/m²)	W_A^{ab} (mJ/m²)
Glass	59.0	33.6	25.4	64.51	31.57
Aluminium	38.5	15.4	23.1	52.80	21.38
Nylon 6-6	39.44	16.11	23.33	53.44	21.86
ABS	39.97	12.68	27.29	53.56	19.21
Vinyl copolymer	43.36	13.99	29.37	55.78	20.36
PS	41.00	12.44	28.56	54.34	19.41
PMMA	37.84	11.15	26.69	52.27	18.25
HDPE	37.26	9.58	27.68	51.17	16.79
UHMW–PE	28.47	4.33	24.14	43.41	11.29
Acetal	29.68	4.12	26.56	43.43	11.06
PP	24.36	3.82	20.54	40.32	10.70
PP filled	24.16	3.73	20.43	40.14	10.59
LDPE	26.34	2.65	23.69	40.66	8.85

W_A and W_A^{ab} calculated according to Eq. (8) using the following sealant properties: $\gamma_2 = 18.09$ mJ/m²; $\gamma_2^p = 7.42$ mJ/m²; $\gamma_2^d = 10.68$ mJ/m² (all sealant properties refer to its uncured state).

Figure 2 illustrates the relationship between the shear and peel strength and the energy of acid-base interactions (W_A^{ab}) for the range of substrates bonded with the silicone sealant A. The pattern of this relationship is identical for both testing modes. It is noticeable that the bond strength increases monotonically with the increase of the energy of acid-base interactions.

Interesting observations are made with regard to the failure mode of the systems investigated. The lowest strength and 100% adhesive failure (delamination at the sealant/substrate interface) occur with the substrates exhibiting the lowest polar component of their total surface energy, e.g. LDPE with $\gamma_1^p = 2.65$ mJ/m² (see Table 2). As the value of γ_1^p for subsequent materials increases, the resultant bond strength also increases monotonically for such plastics as PP, Acetal and UHMW PE. An interesting transition point has been observed in bonds made with HDPE which exhibited 100% adhesive failure at 0.2 mm sealant bead thickness, and the mixed failure mode (i.e. 35% adhesive/65% cohesive failure) in the case of 6.0 mm thick sealant bead (see Fig. 2).

The same failure mode occurs in shear and peel tests for all systems investigated, with the exception, again, of HDPE which gives 100% cohesive failure in peel tests. This

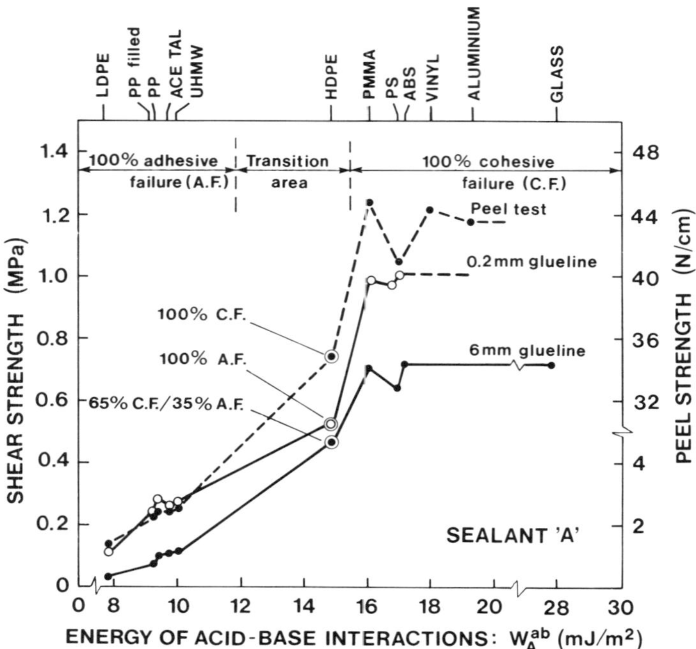

Fig. 2. The relationship between the strength (shear and peel) and energy of acid-base interactions, W_A^{ab}, for the bonds made with the sealant A and variety of organic and inorganic substrates (Gutowski 1990).

phenomenom will require further investigations, since it may indicate that the peel test is less severe under certain conditions than either the tensile or shear test. The observed transition between the adhesive and cohesive modes of failure probably occurs at the point where the energy of acid-base interactions, W_A^{ab}, equals about 14 to 15 mJ/m^2. All other systems with the energy of acid-base interactions greater than approximately 18 mJ/m^2 exhibit 100% cohesive failure within the sealant, whether the substrate is of an organic nature (plastics such as PS, PMMA, ABS, Nylon 6-6) or an inorganic nature (aluminium, glass), within the scope of the experiment.

The relationship between the strength of bonds made with sealant C and relevant energies of acid-base interactions is shown in Fig. 3. The pattern of this relationship is similar to that obtained with sealant A, as illustrated earlier in Fig. 2. All substrates exhibiting low energy of acid-base interactions with the sealant show 100% adhesive failure. The transition in this case occurred at W_A^{ab} equal to about 19 mJ/m^2 for bonds with ABS compared with sealant A where the transition occurred for HDPE at W_A^{ab} = 16.79 mJ/m^2. All bonds made for the 'ABS/sealant C' system failed 100% cohesively at any thickness of the sealant bead within the range 0.05 to 3.0 mm. When the bead thickness approached 6 mm, however, the failure mode changed to the mixed one exhibiting the average 20% adhesive/ 80% cohesive mode. Peel test specimens for this system broke 100% cohesively. It can be summarised for this sealant that, again, there is a monotonic strength increase with the increase of the energy of acid-base interactions across the sealant/substrate interface. All systems exhibiting W_A^{ab} up to about 19 mJ/m^2 show 100% adhesive failure with the transition (mixed: cohesive/adhesive mode) observed for the system that exhibits W_A^{ab} = 19.4 mJ/m^2, and above 20 mJ/m^2 100% cohesive mode occurs.

4.3 New peel test for assessing sealant adhesion

The ASTM C-794 Standard entitled 'Standard Test Method for Adhesion-In-Peel of Elastomeric Joint Sealants' is widely used for the assessment of the adhesion of sealants. In this test, a peel adhesion specimen is prepared (see Fig. 4) in which a reinforcement mesh is

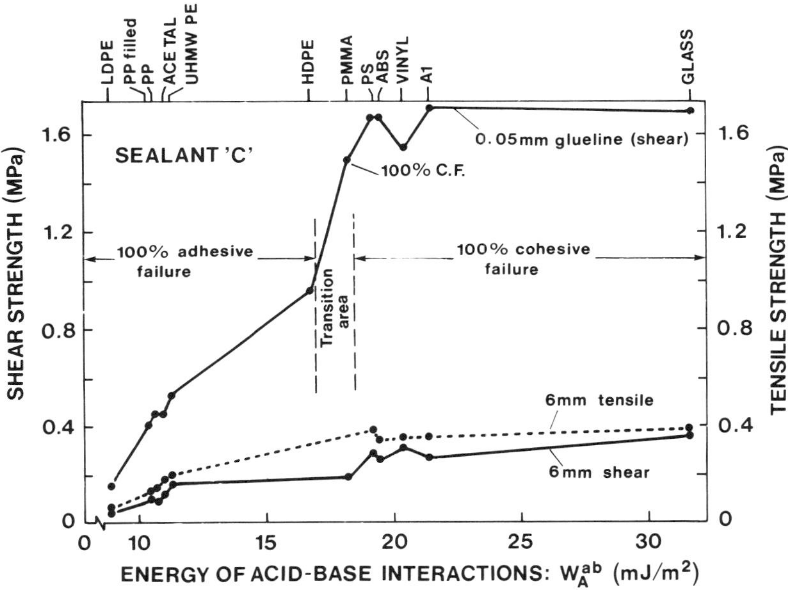

Fig. 3. The relationship between the strength (shear and tensile) and energy of acid-base interactions, W_A^{ab}, for the bonds made with the sealant C and variety of organic and inorganic substrates.

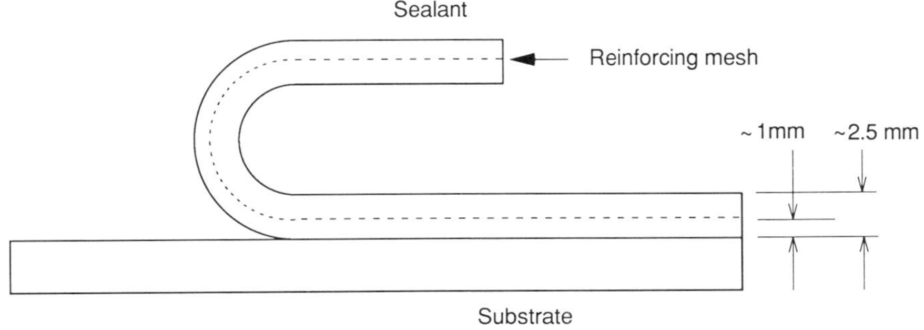

Fig. 4 . Peel adhesion specimen prepared in accordance with ASTM C-794 Standard.

embedded in sealant at a distance of about 1 to 1.5 mm from the substrate. After an appropriate cure time the specimen is peeled at 180° at the crosshead speed equal to 5 cm/min.

It has been established after extensive trials that the thickness and peel rate parameters of the ASTM test are not always acceptable for detecting potential adhesion problems in the joints between sealants and substrates exhibiting poor adhesion.

Consider the data for high density polyethylene (HDPE) for a range of thicknesses as seen in Fig. 2, and the data in Table 3 which show the influence of the peel rate on the strength of a bond and more importantly on the failure mode.

Table 3. Influence of the peel rate on the failure mode and strength of the bond between sealant A and HDPE

Rate (cm/min)	Peel strength (N/cm)	Failure mode
0.05	9.2	100% adhesive
0.5	11.4	100% adhesive
2.5	19.6	100% cohesive
5.0	21.1	100% cohesive
25.0	23.4	100% cohesive
50.0	22.8	100% cohesive

It is apparent that a very slow rate, i.e. 0.05 to 0.5 cm/min, should be used instead of the existing 5 cm/min in order to obtain meaningful results regarding the sealant/substrate adhesion since, at any higher rates, the failure mode changes from 100% adhesive to 100% cohesive. However, such a slow rate is frequently unacceptable for practical reasons.

Further studies on the improved peel test showed that a very thin layer of sealant between the mesh and substrate (as obtained by smearing sealant thinly before embedding mesh into it), provides the possibility of an objective assessment of adhesion, independent of the peel rate within range 0.002 to 100 cm/min.

4.4 Influence of water on bond strength

Tensile specimens, as illustrated in Fig. 1, were prepared using sealant C and the following substrates: LDPE, PP, Acetal, UHMWPE, PS, ABS, Vinyl, Nylon 6-6, and Glass. After one month cure, 50% of the specimens were immersed in H_2O for two weeks, whilst the remainder were left to be tested in dry conditions.

Figure 5 illustrates the resultant relationship between the strength of 'wet' and 'dry' bonds and W_A^{ab}, the energy of acid-base interactions calculated using Eq. (8). It is apparent from this Figure that the strength of 'wet' bonds is slightly less (except for PS) than those tested in dry conditions, and that the general relationship 'strength *v.* W_A^{ab}' is similar whether the bonds are maintained in a dry or wet environment prior to testing. The influence of water on the strength of the bonds made with a variety of sealants and substrates is a subject of further research.

5 Conclusions

(a) It is shown in this work that the strength of adhesion between silicone sealants and a range of organic and inorganic substrates investigated can be attributed to the acid-base interactions between the substrate and sealant.

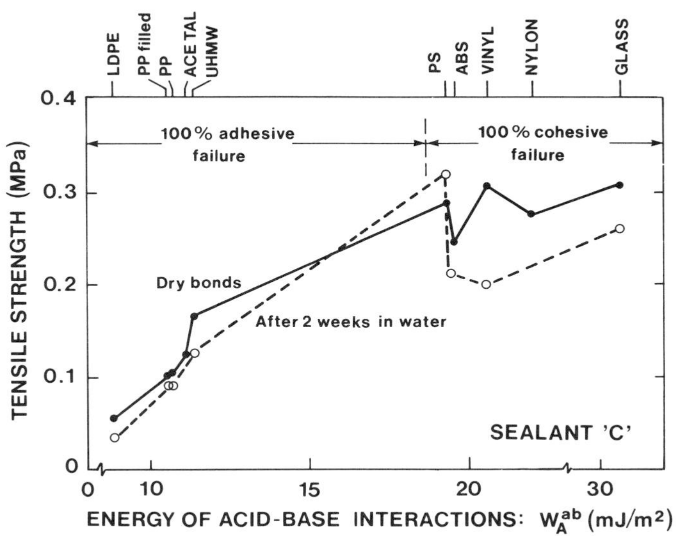

Fig. 5. Influence of water immersion on the strength *v.* W_A^{ab} characteristic for a range of substrates bonded with sealant C.

(b) The strength of bonds (whether tested in shear, tensile or peel) increases monotonically with the increase of the energy of acid-base interactions, W_A^{ab}, whose value can be quantified by Eq. (8).

(c) In practical terms, the value of the energy of acid-base interactions is directly related to the value of the polar component of total surface energy of the substrate (γ_1^p) and the sealant in its uncured state (γ_2^p). This indicates that, in order to increase the strength of adhesion of a given sealant to the substrate of interest, it is necessary to increase the polarity of the substrate. The data show, that the 100% cohesive failure within the sealant can be obtained if the value of γ_1^p is greater than about 12–13 mJ/m².

(d) Within the scope of the experiments there are three areas relateing to failure mode in the relationship 'strength *v.* W_A^{ab}' for a range of sealants investigated:

(i) 100% adhesive failure for $W_A^{ab} = 0$ to 12 mJ/m²,
(ii) mixed mode (adhesive/cohesive failure) for $W_A^{ab} = 12$ to 19 mJ/m²,
(iii) 100% cohesive failure within sealant for $W_A^{ab} \geq 19$ mJ/m².

(e) Sealant molecules at the sealant/substrate interface exhibit significant orientation effects due to the acid-base interactions between nondispersive groups present at the substrate's surface and those of the sealant during formation of adhesive bond. The quantity of these groups exposed at the sealant/substrate interface is dependent upon the polarity of the substrate surface.

(f) Peel specimens for adhesion assessment should be prepared so that the distance between the substitute and reinforcing mesh (screen or airplane cloth) is about 0.05 to 0.1 min. Such a specimen can be tested at any rate between 0.002 and 100 cm/min to assess sealant adhesion to the substrate.

6 Acknowledgments

The authors wish to acknowledge with appreciation the help of Mrs Magda Morehouse in determining surface energies of materials investigated. Also, sponsorship by the 'Structural Glazing Consortium' is gratefully acknowledged.

7 References

Fowkes, F.M. (1978) Acid-base interactions in polymer adhesion, in **Physico-Chemical Aspects of Polymer Surfaces**, 2 (ed. K.L. Mittal), Plenum Press, New York and London.

Good, R.J. and Girifalco, L.A. (1987) **Journal of Physical Chemistry**, 64, 561.

Gutowski, W.S. (1990) Adhesive properties of silicone sealants. **Symposium on Building Sealants: Material Properties and Performance, ASTM STP 1069**. (ed. Thomas F. O'Connor), American Society for Testing and Materials, Philadelphia.

Kaelble, D.H. and Uy, K.C. (1970) **Journal of Adhesion**, 2, 51.

Kinloch, A.J. (1987) **Adhesion and Adhesives**, Chapman and Hall, London, New York.

Klosowski, J.M. (1988) Durability of silicone sealants, in **Adhesives, Sealants and Coatings for Space and Harsh Environments** (ed. L.H. Lee), Plenum Press, New York and London

Wu, S. (1982) **Polymer Interface and Adhesion**, Marcell Dekker, New York, Basel.

69 DURABILITY OF STEEL FIBRE REINFORCED CONCRETE WHEN HELD IN SEWAGE WATER

M.A. SANJUAN, A. MORAGUES, B. BACLE, C. ANDRADE
Insitute of Construction Sciences 'Eduardo Torroja', Madrid, Spain

Abstract
The most important properties of steel fibre concretes are toughness, impact, abrasion and post-crack resistance. These energy absorbing characteristics can be particularly effective in resisting the aggressive enviroment surrounding concrete structures.

In this study we try to establish how the durability of concrete can be modified by addition of normal steel, stainless steel or galvanized steel fibres when subjected to different aggressive media. The aggressive solution in the present research was selected for its likeness to common practical cases: sewage water containing ammonia, sulphides and chlorides.

The results obtained of durability of composite material after 400 days are described.

Calcium nitrite, used as a possible corrosion inhibitor, does not appear to defend the steel against the aggressive solution's attack in the conditions of the test, moreover, the attack appears to be greater when nitrites are present in the specimens.

Normal steel fibres do not resist the aggressive solution, however, the behaviour of stainless and galvanized steel seems to be better.
Keywords: Steel Fibre Reinforced Concrete, Corrosion, Sewage water, Chlorides Difusion.

1 Introduction

For over 50 years SFRC has been a major fibre composite material to the world's concrete industry. Steel fibres have been used with the intention of increasing resistance to cracking and impact, and in some cases the use of fibres distributed in the concrete mix acts as a substitute for the traditional corrugated steel reinforcement. The mechanical behaviour has been rather satisfactory, particularly where critical points of a structure require high resistance to impact, cavitation, shearing, and wear and tear, Zollo (1985).

The metallic fibres designed for addition to concrete are normally elongated elements of varied cuts whose shape favours a greater adhesion to the mortar or concrete mix. They can be of carbon steel, stainless steel, and may also have an anti-corrosive heat galvanization treatment.

One of the concerns about this type of fibre concrete is possible corrosion of the fibres and the possible deterioration in contact with sewage waters. The corrosion appears mainly at the concrete surface developing unaesthetic spots. The corrosion behavior of steel fibres in SRFC has not been as methodically studied as those of reinforcing and prestressing steel, and the literature contains ccontradictory results, Mangat and Gurusamy (1987), Byfors and Skarendahl (1987). In general it may be observed that high dosages or cement of stainless steel fibers were employed in those studies where the fibers did not corrode in contact with aggressive environments.

In the present paper, some preliminary results of the behaviour of SFRC in contact with sewage water, are presented.

2 Experimental Method.

Cylindrical pipe-shaped specimens were made measuring 150 x 300 mm with a wall thickness of 50 mm, whose composition is shown in table 1.

Two water/cement ratios, 0.45 and 0.55, were employed. In some cases an inhibitor was employed (3% of calcium nitrite by weight of cement) while other sets of specimens contained a plastifier added in the proportion of 1% of weight of cement.

Carbon steel, stainless steel and galvanized steel fibres were added having an aspect ratio of 30/.50, 30/.40 and 40/.60, respectively.

Table 1. Steel Fibre Reinforced Concrete dosage

Material	Dosage (Kg/m^3)
Ordinary Portland Cement	350
Sand (0 - 6 mm)	820
Gravel (6 - 12 mm)	1110
Water (without plastifier)	190
Water (with plastifier)	154
Steel Fibres	35
Slump	5 cm.

A total of forty specimens was fabricated, which means 10 series of four identical specimens. In table 2, the series are numbered.

Table 2. Series of specimens

W/C	no fibres	normal fibres	normal fibres with[$Ca(NO_2)_2$]	Stainless Steel Fibres	Galvanized Steel Fibres
0.55	1	2	3	4	5
0.44 (with plastifier)	6	7	8	9	10

The specimens were demoulded 24 hours after mixing and cured under water for 28 days. Two specimens were subsequently placed in a container of an aggressive solution simulating sewage water. This aggressive medium was a solution of Sodium Sulphide (2.9 g/l) and Ammonium Chloride (29 g/l). The other half of the specimens were held in potable water as a blank test. The layout of the specimens in both media is given in figure 1.

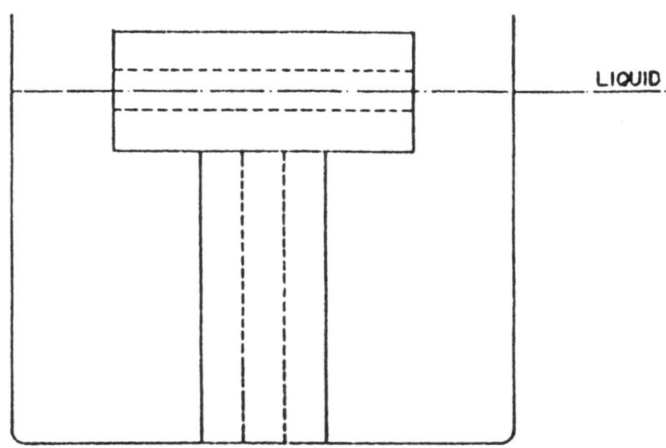

Fig.1. Arrangement of specimens in the test tank.

The vertical specimen was totally submerged and the horizontal one was partially immersed.

The study of the diverse specimen's behaviour in their respective storage media is carried out over time through the visual inspection of the modifications occurring in the specimens and by analysis of the concentration of the ions in the aggressive solution. Two ages were studied: 60 and 400 days of immersion.

3 Results

3.1 Storage in tap water

The concrete of all series of specimens remained unaltered after 400 days of immersion in tap water.

Regarding the state of fibres only the carbon steel fibre specimens, both without and with calcium nitrite, showed light rust spots on their surface. The galvanized and stainless steel fibres did not show oxidation signs.

3.2 Storage in agrressive solution

The concrete of the specimens partially submerged in the aggressive solution began to crack at the borders after two months of immersion. After 400 days, the concrete shows generalized cracking and spalling of the cover (Fig.2).

Regarding the state of the fibres, specimens with carbon steel fibres and W/C ratio of 0.55 both without and with nitrite (Fig.3), are all attacked and exhibited greater fibre oxidation in the partially submerged specimens than in the fully submerged ones.

While the oxide stain extends around the fibre on the surface of the concrete in the submerged speccimens, the rust grows as an eruption in the non-immersed specimens (Fig.4).

In the first case the oxide colour was clearer than in the second one.

Fig.2. Plain concrete held in sewage water

Fig.3. Specimens held in the aggressive solution partially submerged: 1. without fibres; 2. SFRC; 3. SRFC with calcium nitrite.

Fig.4. SFRC from left to right: partially submerged (in potable and sewage water) and totally submerged (in potable and sewage water).

The use of a plastifier with therefore a lower Water/Cement ratio (W/C = 0.44), improves, in all cases studied, the behaviour. After two months, all the specimens remain unaltered. Nevertheless, after 400 days, the fibres show oxidation symptoms although the concrete remains in good condition (Fig.6). Therefore, the use of a plastifier improves the durability of the concrete reinforced with carbon steel fibres.

Fig.5. Stainless steel fibres reinforced concrete with plastifier.

Fig.6. SFRC with plastifier and calcium nitrite.

4 Discussion

Although one year is still a short time to predict longterm behaviour, however, different performances can be remarked at this time.

The concrete itself when partially submerged in the aggressive solution exhibited severe surface spalling. This damage was most severe in the portions of the specimen exposed to the air. This result suggests that the mechanism of damage is a salt recrystallization, thus continuous transport of water with dissolved salts is occuring through the concrete. (Fig.7), and then these salts are precipitated in the region of evaporation. The expansive forces due to salt crystallization induce cracking and spalling.

Cracking evolution may be the main initial cause for a later acceleration of this deleterious process. As a consequence, the diffusion rate of substances dissolved in water will be increased.

Figure 8 shows the same mechanism of dissolution transport but without evaporation in the fully immersed specimens case.

Hydrogen sulphide was formed in the aggressive solution (formed at pH = 6)(Table 3)

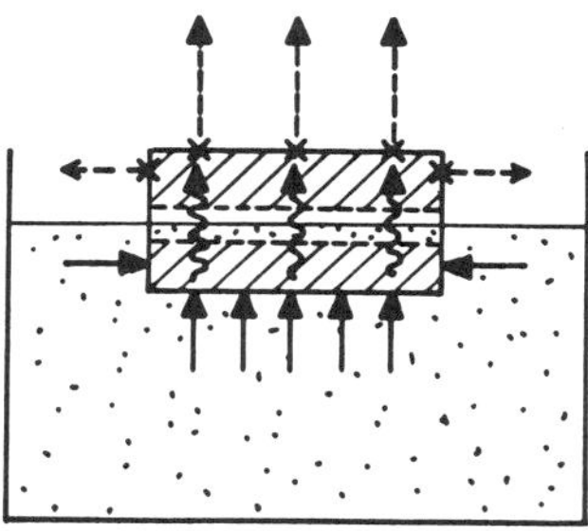

——▶ Water transport by hydraulic pressure and capillary suction

〰▶ Transport of water and dissolved agents

---▶ Evaporation of water

XXX Crystallization of dissolved agents

Fig.7. Water transport mechanism.

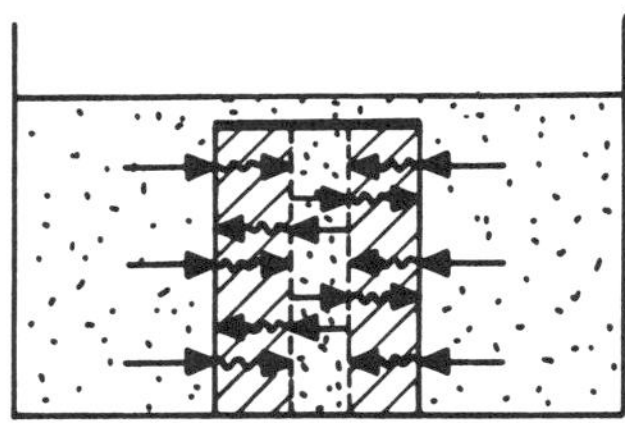

——▶ Water transport by hydraulic pressure and capillary suction

〰▶ Transport of water and dissolved agents

Fig.8. Water transport mechanism.

Table.3. Sulphur species as a function of pH

pH	% H_2S	% HS^-
5.0	99.0	1.0
5.5	96.9	3.1
6.0	90.7	9.3
6.5	75.6	24.4
7.0	49.5	50.5

Depending on chemical equilibrium it may escape from the solution, however, sulphide and bisulphide corrosion of the fibres located on the surface can occur.

In the case of carbon steel FRC specimens held in potable water, staining always appears over a period of time.

As expected, the corrosion was greater in the fibres of the specimens held in sewage water. At pH>5 black FeS was formed, Burriel (1983). The corrosion reactions are:

$$Fe^{2+} + H_2S \text{ --------> } FeS + 2H^+ \quad (1)$$
$$Fe^{o} + 2SH^- \text{ --------> } FeS + H_2 + S^{2-} \quad (2)$$

Acid attack on the concrete enables an acceleration of the process through depassivation, and chloride ions may penetrate via the pores and cracks to the interior of the concrete.

If the type of oxide produced is capable of spreading through the adjacent pores without causing excessive pressures, fissures are not created. When corrosion products are formed rapidly and diffusion does not take place, fissures may result in the fibres which have surface contact, thus giving rise to a spalling or splitting of the cover. (Bentur and Diamond, 1985). Calcium nitrite, used as possible corrosion inhibitor, does not appear to defend the steel against the attack of aggressive solution in the conditions of the test, moreover, the attack appears to be greater when nitrite ions are present in the specimens.

Galvanized and stainless steel fibres held in potable water exhibited good performance. In sewage water the galvanized fibres showed a greater oxidation. The behaviour of stainless steel fibres in sewage water appears to be the best.

The use of plastifier, which leads to a lower water /cement ratio, improved the resistance of the concrete very much. Low water/cement ratio is a parameter influencing the durability of the concrete in an aggressive enviromment, because this ratio influences the permeability of concrete decisively.

As expected the corrosion was more intense in partially immersed conditions than when the specimens were completely submerged, because of the lack of oxygen in the latter case.

Finally, figure 9 shows the intensity of damage in the specimens partially immersed in sewage water.

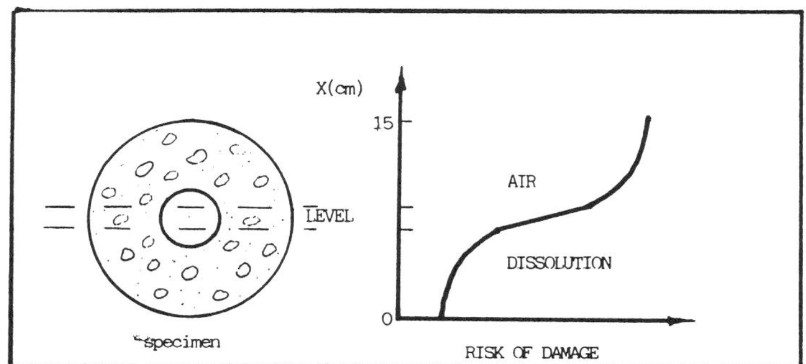

Fig.9. Risk of Damage.

5 Conclusions

The results presented in this study are based on about fifteen months exposure in potable and sewage water. Based on this information the following points concerning the durability of SFRC held in sewage water can be made:

1. Corrosion of carbon steel fibres is very fast. Corroded fibres cannot avoid the developing of the cracks. Moreover, their oxide can favour crack propagation.

2. Galvanized and stainless steel fibres act on the cracks decreasing their propagation.

3. The use of plasticiser greatly increases the durability of the specimens in the aggressive solution.

4. Calcium nitrite, added as a possible corrosion inhibitor, creates internal stresses by crystallization near the surface inducing cracking and spalling of the concrete.

6 Acknowledgement

The authors would like to thank the collaboration of DRAMIX in providing the fibres for this study.

7 References

Bentur, A. and Diamond, S. (1985). Cracking processes in steel fibre reinforced cement paste. Cement and Concrete Research, 15, 331-342.

Burriel, F. Lucena, F. Arribas, S. and Hernandez, F.(1983). Química Analítica Cualitativa. Editorial Paraninfo, Madrid, Spain.

Byfors, K. and Skarendal, A. (1987). Service life prediction of steel fibre concrete water tanks. Conf.on Durability of Building Materials and Components. Singapore.

Mangat, P.S. and Gurusamy, K. (1987). Long-term properties of steel fibre reinforced marine concrete. Materials and Structures, 20, 273-282.

Zollo, R.F. (1985). An overview of progress in applications of SRFC. Sweden Joint Seminar (NSF -STU). Swedish Cement and Concrete Research Institute Publications, Stockholm, Sweden.

70 SOME STUDIES ON THE USE OF CONCRETE COATINGS TO PROTECT REBARS FROM CORROSION

C. ALONSO, A.M. GARCIA, C. ANDRADE
Institute of Construction Sciences 'Eduardo Toroja', Madrid, Spain

Abstract

The corrosion of steel reinforced concrete structures is a costly problem which can be mitigated using a variety of protection methods. One of them is the use of impermeable concrete coatings. Concrete coatings as methods for the long term protection of rebar and concrete itself, should be considered in terms of their barrier properties ie, preventing access of corrosive enviroments to the rebars.

In the present paper barrier properties against carbonation, chloride penetration and water permeability are studied for several types of coatings. Coatings were applied on mortar specimens and then submitted to accelerated tests.

The advantages of the methodologies used are their simplicity and the ability to obtain comparative data in a relatively short time.

Results also indicate that epoxy coatings have higher resistance against aggresive penetration although in drying periods they retain humidity for longer which could allow higher corrosion rates.

Keywords: Concrete Coatings, Rebar Corrosion, CO_2 and Chlorides Permeability.

1 Introduction

It is well known that concrete protects rebars from corrosion, but in the presence of some types of aggressive species they may corrode.

The corrosion of rebars can seriously affect the service life of structures which are designed to fulfil a series of requirements for long periods of time, either structural or aesthetic (Weber, 1983) (Andrade, 1989).

The aggressive agents that more frequently initiate corrosion of rebars are: atmospheric CO_2 that neutralizes the concrete and destroys the passive layer in the whole surface of the rebar, and the chlorides ions that locally depassivate the steel.

The damage caused by the corrosion of rebars assumes a high economic cost (Sitter 1986). Due to this fact several methods of protection are rapidly developing, focused on stopping or at least to diminishing the corrosion problem.

Among all protection methods employed, concrete coatings is one of most widely used (Lindberg, 1987) (Pfeifer, 1984). From 1970s concrete coatings have been used not only with a decorative function but also to slow down the diffusion of aggressive species from outside.

The relevant increase in the use of concrete coatings has produced the appearance in the market of a great number of them. However there are no standardized tests to evaluate their protective capacity and make easier an appropriate selection for each need (Browne, 1987) (Baba, 1987).

In this paper some improvements of a methodology previously described are presented (Andrade, 1988). The tests allow classification of coatings taking into consideration their resistance to CO_2 and chlorides penetration. The problem associated with the use of coatings are also evaluated.

2 Experimental

2.1 Materials

Mortar specimens of 20x55x80 mm. and 50x50x50 mm. size were made with Ordinary Portland cement (c/s = 1/3 and w/c = 0.5). The curing was made for 28 days at 100% R.H. and 20 ± 2ºC. Later they were dried for 15 days at 50% R.H. After that each specimen was covered with two coats of paints. The coatings tested were:

1 Blank (free of coatings)
2 Colloidal Silica based (Col SB)
3 Epoxy Resin based (ERB)
4 Cement based (CB)
5 Acrylic Resin based (ARB)
6 Ethylen Polymer Resin based (ETPRB)
7 Methyl Metacrylate Resin based (MMRB)

The prismatic specimens also have embedded two steel bars and a graphite rod for corrosion control.

2.2 Procedure

The cubic specimens were exposed to accelerated carbonation by bubbling 100% CO_2 through a chamber with 60% R.H.

A set of prismatic specimens were exposed to a semi-accelerated carbonation using 1% CO_2 for 43 days, going on with 5% CO_2 for 163 days and finally 100% CO_2.

The evolution of the carbonation process was controlled determining the changes in weight of the specimens and also measuring the corrosion potential and corrosion current evolution on the rebars.

Other set of prismatic specimens were submitted to wetting and drying cycles (20 hours immersion and 4 hours drying). The aggressive media used were NaCl saturated solution (brine) for 75 days followed by 0.2M NaCl for 170 days.

The potential measurements were made using a saturated calomel electrode. The corrosion current was determinated by means of the polarization resistance method (Stern, 1957).

3 Results

3.1 Carbonation

The results obtained from measuring the changes in weight of the specimens during the accelerated and semi-accelerated carbonation are presented in figures 1 and 2 respectively.

From figure 1 it can be deduced that specimens coated with CB and ARB have a high increase in weight, with exponential trend during the first 3 days, followed to a slower increase. They behave slightly better than without coating. The weight of the specimens with ERB and ETPRB changes much more slowly. MMRB is in an intermediate position.

The classification obtained from these results is:

ERB > ETPRB > MMRB > ARB $\simeq$ CB > blank

More protective to CO_2 ---------> Less protective to CO_2

This was also confirmed with phenolphthalein after 8 days carbonation and measuring the corrosion potential and corrosion current on the rebars.

In figure 2 are the changes in weight during the semi-accelerated carbonation. A decrease in weight of the specimens is detected with 1% CO_2. The carbonation process is too slow compared with the drying effect due to the low humidity in the chamber.

With 5% CO_2, speciemens ERB and ETPRB still lost weight, which implies they are more resistante to CO_2 penetration, whereas the rest increase. When phenolphthalein was used here it confirmed that ERB and ETPRB were partially carbonated and the rest were completely carbonated.

With 100% CO_2 ERB and ETPRB suffer a high increase in weight. When they reach a constant value the final of the carbonation is considered.

This result will be compared with those of natural

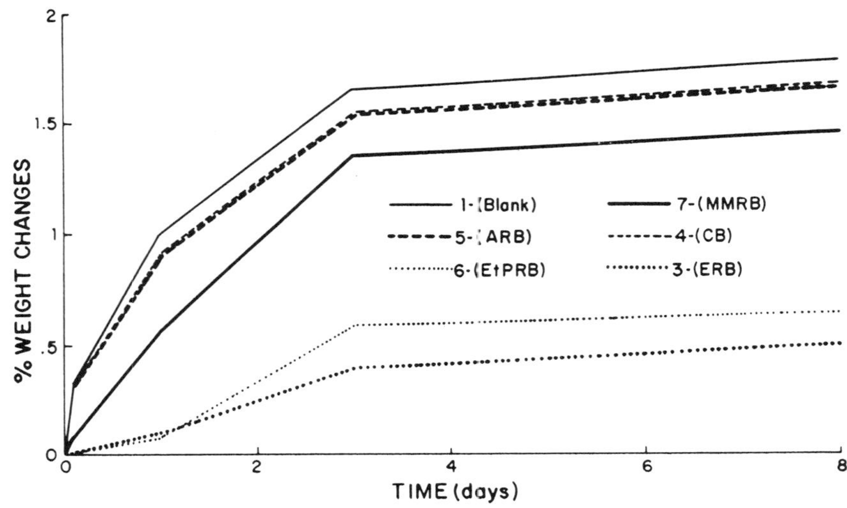

Figure 1.-Weight changes of the coated specimens during accelerated carbonation.

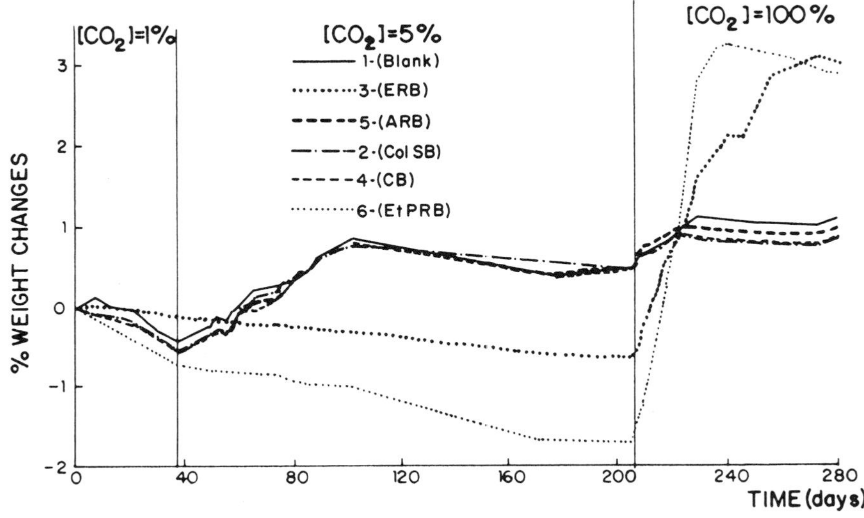

Figure 2.-Weight changes of coated specimens during semi-accelerated carbonation.

exposure after 6 months it was obtained:

type of coating -	blank	CB	ARB	MMRB	ETPRB	ERB
carbonation depth (mm)	2-3	2-3	1.5-2	$\simeq$1	1	0

3.2 Chlorides penetration

In figures 3 and 4 are the results of corrosion current and corrosion potential measured during the test.

When the specimens were in the wetting and drying cycles in brine it was observed that after 75 days the corrosion current (figure 3) does not reach values higher than $2.10^{-3}\mu A/mm^2$ being the rebars in the specimens ERB and ETPRB which less corrode.

The shaded zone between 10^{-3} - $2.10^{-3}\mu A/mm^2$ separates dangerous corrosion from a non corroding current.

The situation offered more information when the specimens were immersion in 0.2M NaCl. The specimen without coating and those with Col. SB, CB, ARB reach corrosion current values between 2.10^{-3} - $5.10^{-3}\mu a/mm^2$. The ETPRB 2.10^{-3} - $7.10^{-4}\mu A/mm^2$ and the lower ones for ERB.

The corrosion potential (figure 4) gives in brine cycles values around - 200 mV for the specimens ERB and ETPRB, whereas the rest of the specimens become more negative than - 350 mV after 30 days.

When immersion in 0.2M NaCl those with the coatings Col SB, CB, ARB and blank are in the region of - 600 mV, ETPRB are only under -350 after 50 days. The ERB is always around -150 mV.

4 Discussion

The use of concrete coatings to protect rebars from corrosion implies selection among a variety of products. But sometimes the only information known is that given by the manufacturer, basically composition properties. But the coatings need to be tested in a realistic way related to their use: CO_2 resistance, chlorides and water penetration.

The methodology here used to achieve these objetives is simple, realistic and non - destructive.

4.1 Carbonation resistance

The control of the changes in weight during accelerated and semi-accelerated carbonation is a reliable method to determine the protective capacity of the coatings against CO_2 penetration.

The first one gives good and fast information about

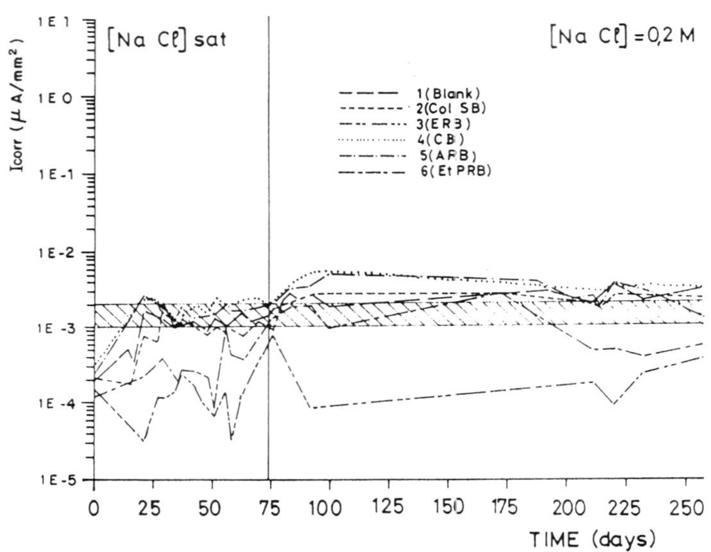

Figure 3.-Icorr evolution measured on rebar embedded in coated mortar specimens. Submetted to drying/immersion cycles. (NaCl solution).

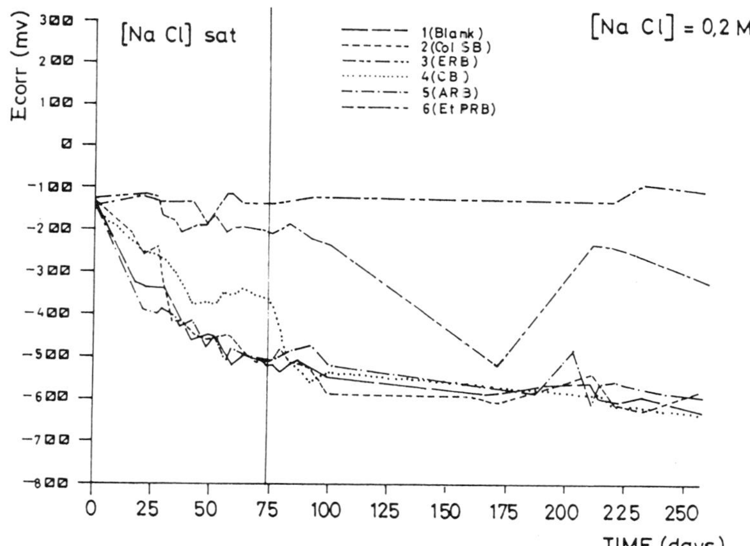

Figure 4.-Ecorr evolution of rebars. Drying/inmersion cycles. (NaCl solution).

coatings and also classify them. The semi-accelerated takes longer and is less selective, but simultaneously informs about water permeability. Here, the slowness of the process allows the specimens to reach an inner equilibrium of humidity with that of outside and therefore the changes in weight are a result between the evaporated water and the penetration effect of CO_2.

The use of electrochemical techniques analysis how rebars react against the presence of CO_2 and humidity content (Garcia 1990).

The use of phenolphthalein as indicator of the advance of carbonation has allowed us to prove that it is also uniform in the accelerated test like natural exposure. However it is a destructive method.

4.2 Chloride diffusion

The resistance of coatings to chloride diffusion may be determined with a test similar to those used here, measuring the changes of corrosion potential and corrosion current on the rebars.

The low corrosion current measured after 75 days in brine cycles could be related with the final high levels of chlorides at the surface of the rebar. (Goñi 1989) proved that for $[Cl^-] > 1M$ appears a decreasing attack (lower corrosion current than that theoretically possible) because the high ionic strenght in the media diminishes the ionic movility.

It seems that the aggressive solutions should be less concentrated in NaCl than used but provably higher than 0.2. It would be more convenient to use a concentration that allows high mobility of Cl^- ion and the depassivation of the rebar in a reasonably short time.

A coating is impermeable to chlorides diffusion if it is to water penetration. This implies that a highly impermeable coating is only recommended when the concrete is dry. But if the concrete needs to breath the coating should be permeable to water vapour.

4.3 Classification of the coatings

Concerning the evaluation of the coatings the main requirement should accomplish, if they are going to be used for concrete in order to avoid rebar corrosion, is that they must work as a barrier against CO_2 an chlorides.

The barrier property mainly depends on the coating pore size. Therefore it is basic to test the painting on concrete or mortar because the application form and the concrete surface finishing are crucial in the final resistance.

Finally to say that the classification obtained here does not means that all coatings in the market with similar base will behave alike.

5 Conclusions

- The accelerated carbonation is a reliable and short test which allows to know the CO_2 resistance of concrete coatings.
- The semi-accelerated carbonation partially classifies coatings. However it gives simultaneous information of water permeability.
- The chlorides diffusion resistance of concrete coatings may be tested through cycles of drying/immersion in NaCl solutions, using rebars as sensors of the penetration of the aggressive.
- The classifications of the coatings tested are:

CO_2 permeability: blank $\simeq$ Col SB>CB $\simeq$ARB>MMRB>ETPRB>ERB
Cl^- permeability: blank $\simeq$ Col SB $\simeq$ CB $\simeq$ ARB>ETPRB>ERB
Water permeability: blank $\simeq$ Col SB $\simeq$ CB $\simeq$ ARB>ETPRB>ERB

6 References

Andrade, C. Alonso, C. Bacle, B. and Rodriguez, J. (1988) Accelerated testing methodology for evaluating carbonation and chloride resistance of concrete coatings. FIP Symposium 61 - 67.

Andrade, C. Alonso, C. Gonzalez, J.A. and Rodriguez,J (1989) - Remaining service life of corroding structures.Symposium. Durabiity of structures (IABSE), 359 - 364.

Baba, A. and Senbu, O. (1987) - A predictive procedure for carbonation depth of concrete with various types of surface layers, 4th Int. Conf. on Durab. of Build. Mater and Comp. 679 -685.

Browne, R.D. and Robery, P.C. (1987) - Practical experience in the testing of surface coatings for reinforced concrete. 4th Int. Conf. on Durab. of Build. Mater and Comp, 325 - 333.

Garcia, M. Andrade, C. and Alonso, C. (1990) - Evaluation of resistance of concrete coatings against carbonation and water proof penetration Conf. The protection of concrete. Dundee, Sept.

Goñi, S. Moragues, A. and Andrade, C. (1989) - Influence of the conductivity and the ionic strength of synthetic solutions which simulate the aquouse phase of concrete in the corrosion process. vol. 39, nº 215, 19 - 28. Materiales de Construcción.

Lindberg, B. (1987) - Protection of concrete against aggressive atmospheric deterioration by use of surface treatment (painting). 4th. Int. Conf. on Durab. of Build Mat. and Comp. 309 - 316.

Pfeifer, D.W. and Perenchio, W.F. (1984) - Cost - effective protection of rebars against chlorides, sealers or overlays. Concrete Construction, May. 503 - 507.
Sitter, W. (1986) - Interdependence between technical service life, prediction of service life of concrete structures. Oct. Bolonia.
Stern, M. and Geary, A.L. (1957) - Theoretical Analysis of the shape of polarization curves. J. of Electr. Soc. vol. 104 (1) 56.
Weber, H. (1983) - Methods for calculating the progress of carbonation and the associated life expectancy for reinforced concrete components. Bentonwerk + Fertigteil + Technik vol 8, 508 - 514.

Acknowledgments

Authors are grateful to Spanish CICYT for the finantial support of this research.

71 A STUDY ON DURABILITY OF THICK ANTI-CORROSIVE COATINGS

N. MURAKAMI, S. SHIRAISHI, T. HIRANO
Takenaka Technical Research Laboratory, Japan

Abstract
Thick anticorrosive coating finish materials have been changed with respect to performance variations. These new types consist of phthalic acid, chlorinated rubber and urethane coatings. Currently, fluorinated coatings have become applicable for certain situations.

This paper describes the results of our experimental study with attention to the functions of different coatings. The weather resistance and anticorrosive properties of different coatings were evaluated according to the results of an accelerated test, a natural weathering test and a weathering test in actual use.

In conclusion, it was found that urethanes and fluorinated coatings are superior to the other coatings and that undercoat controlling grades and deteriorated degrees of conventional coatings have been a great effect on the functions by new coatings.
Keywords: Anticorrosive coating, coating materials, weathering test, salt spray, chalking, discolouration, durability, gloss, adhesion.

1 Introduction

Anticorrosive coating finish, which is achieved mainly by coating a synthetic resin based paint on an oily rush-resisting paint such as minimum, has been used as an efficient anticorrosive means for structures.

Thick anticorrosive coatings were developed to meet the requirement for extended recoating cycle associated with the trend of structures built on an increased scale. Historically, these materials originated from phthalic acid coatings developed in the nineteen-fifties, followed by chlorinated rubber coatings and then urethane coatings. Presently, normal temperature drying fluorinated coatings are commercially available as highly durable paints, Japan Steel Construction Society, (1988).

The two major requirements of thick anticorrosive coatings are good appearance and corrosion resistance; the former corresponds to high weather resistance and the latter to high durability.

In this paper, the authors present the results of assessment of various thick anticorrosive coatings as to retention of good appearance and corrosion resistance in both accelerated and natural weathering tests and rank them by type and by brand.

2 Experimentation

2.1 Accelerated weathering tests

Four types of coating materials were used for the experiments: phthalic acid, chlorinated rubber, urethane and fluorinated coatings. Evaluation was conducted for two painting colours: white (Munsell Colour Scale N9.5/0) and red (Munsell Colour Scale 5R4/13). Four paint brands of respective major paint manufacturers were selected for the study. Note that SS41 (rolled steel for ordinary structures) after blast treatment was used as the support.

Accelerated weathering treatment was conducted using a Sunshine Weatherometer. Treatment time was 300-1500 hours for phthalic acid and chlorinated rubber coatings and 300-5000 hours for urethane and fluorinated coatings.

Coating assessment as to the parameters listed below was made in accordance with the "Guideline of the Society of Steel Construction of Japan: Manual for the investigation of Paint Films on Steel Bridge" (1982).

The parameters were; Rust, Blistering, Scaling, Staining, Chalking, Discolouration/fading, Gloss, Adhesion properties.

Discolouration/fading and gloss were determined respectively using a colourimeter and a gloss meter Chalking was measured using a chalking tester and rated in four grades. Adhesion properties were assessed in four grades by cross-cut method and the rectangular arrangement method. (See Fig. 1)

[Chalking scores]

Score (RN)	Description
3	Almost no changes from initial state.
2	Slight whitening.
1	Considerable whitening but initial color tone remains.
0	Significant whitening and initial color tone is unestimatable.

[Scoring by the cross-cut method]

Score (RN)	3	2	1	0
Scaling state				Still severer scaling

Fig. 1. Evaluation criteria (Chalking, Adhesive strength).

2.2 Salt spray test

Supports were coated with the same materials and method as those used in the accelerated weathering tests. Then, samples were prepared by making a cut about 50 mm long with a depth such that it extended from the coating surface to the base surface to monitor the occurrence and progress and expansion of rust.

Salt spray treatment was conducted in accordance with JISK 5400. The assessed parameters were rust, blistering, and scaling. The same assessment methods as in the accelerated weathering tests were used.

2.3 Natural weathering tests

Samples were prepared with the same materials and method as those in the accelerated weathering tests.

Samples were exposed to open air at a seashore in the southern part of Osaka Prefecture, Japan, and observed for paint film deterioration. The items, method and parameters were the same as those in the accelerated weathering tests.

2.4 Weathering tests in actual use

A lofty chimney was used for the weathering tests in actual use. The subjects were four coating materials of four manufacturers. Testing was conducted for a paint colour of white alone.

The parameters and determinations were made in the same manner as in the accelerated weathering tests.

3 Results and Discussion

3.1 Accelerated weathering resistance

Visual inspection revealed no rust, blistering, cracking, scaling or other abnormalities in 5000 hours of weathering. Nor was there any difference between the paint colours red and white. Paint film chalking was rated in the four grades shown in Table 6. Durability differed among the coating types, which ranked as to the severity of chalking in the descending order of phthalic acid coating, chlorinated rubber coating, urethane coating and fluorinated coating. As for paint colour difference, white colour tended to be more susceptible to chalking than red colour. (See Table 1)

Table 1. Results of accelerated weathering tests (chalking).

Type of coating	Weathering time (Hr) 300	900	1500	2100	3000	5000
Phthalic acid coating	3	3	1	-	-	-
	3	3	2	-	-	-
Chlorinated rubber coating	3	3	3	3	0	-
	3	3	3	2	2	-
Urethane coating	3	3	3	3	2	2
	3	3	3	3	2	1
Fluorine coating	3	3	3	3	3	3
	3	3	3	3	3	3

Discolouration/fading depended largely on paint colour. Specifically, red colour showed greater colour differences in comparison with white colour. Durability was compared on the basis of ΔE values of 3 to 6, which represent clear colour differences in the NBS unit system, discolouration/fading tended to increase with weathering time in the case of white colour. However, there were almost no differences among the coating types. In the case of red colour, there were significant differences among the coating types, with noticeable discolouration/fading noted in chlorinated rubber coating. Durability by coating type is as follows: Fluorinated coatings > phthalic acid coatings ≃ chlorinated rubber coatings ≃ urethane coatings.

Initial gloss value was 80 to 90 irrespective of the type of coating material, with a tendency to decrease noticeably with weathering time irrespective of colour. In terms of gloss retention, the tested materials ranked in the descending order of fluorinated coatings > urethane coatings > chlorinated rubber coatings > phthalic acid coatings. The fluorinated coatings had a gloss retention of

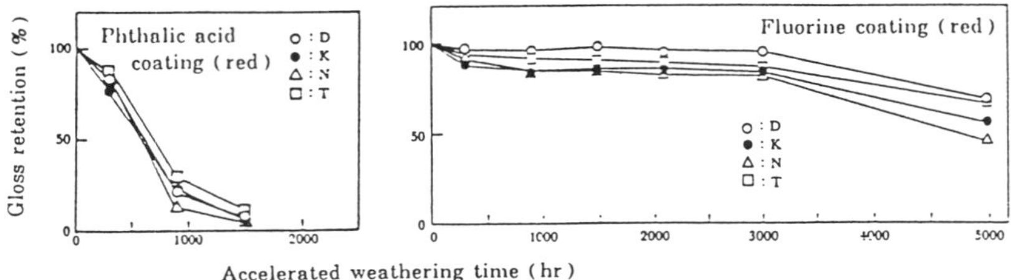

Figure 2. Fluctuations of gloss in accelerated weathering treatment.

about 80% even after 3000 hours of weathering, which decreased to about 50% after 5000 hours of the same treatment. This suggests that fluorinated coatings possesses much higher durability in comparison with the other types of coats. (See Fig. 2)

Adhesion was rated by the cross-cut method in four grades. There were no differences between the paint colours in coating materials. As for differences among the coating materials, there were no significant decrease in adhesion after 3000 hours of weathering treatment except for phthalic acid coatings. The phthalic acid coatings showed a very noticeable scaling state (grade 0) after 900 to 1500 hours of treatment.

3.2 Salt spray

No changes occurred in the uncut portion in all but phthalic acid coating, with a very good paint film condition obtained. On the other hand, blistering occurred in brand T in phthalic acid coatings after 500 hours of treatment. Blistering occurred in the other three brands after 1000 hours of treatment. (See Figure 3)

Almost the same deterioration behaviour was noted among the coating types and brands in the cut portion. In the initial state following 100 hours of treatment, no changes occurred in brand D or N, both of which are organic zinc based coatings, while minor blistering was present around the cut portion in brands K and T, both of which are inorganic zinc based coatings. When treatment time exceeded 200 hours, there were no differences between the organic and inorganic zinc based coatings, ie., blistering occurred in both

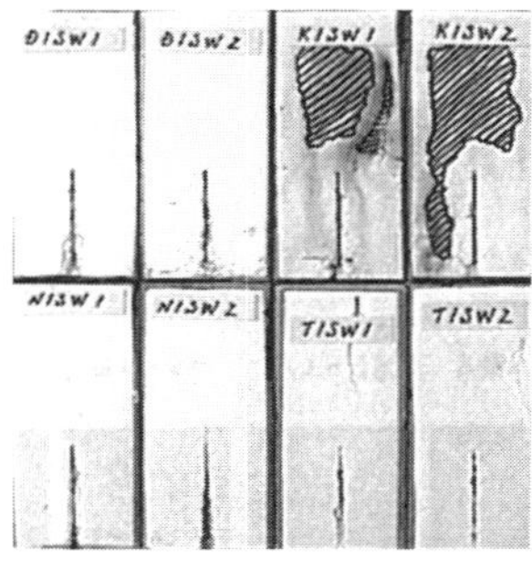

Figure 3. Phthalic acid coating after 2000 hours salt spray treatment.

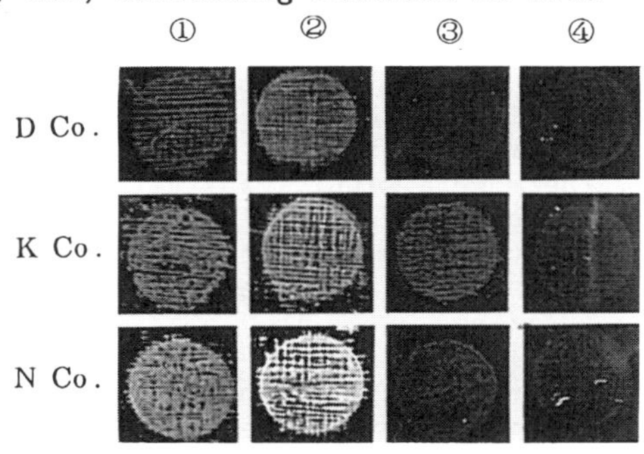

Figure 4. Chalking in deteriorated coats 1) Phthalic acid; 2) Chlorinated rubber; 3) Urethane 4) Fluorinated coating.

cases. However, the organic zinc based coatings tended to have slightly greater blisters around the cut portion in comparison with the inorganic zinc base coatings. Also noted was a tendency of the organic zinc based coatings to produce larger amounts of rust liquid from the cut portion. These phenomena were noted irrespective of the type and brand of coating material. This suggested that inorganic zinc based coatings, in comparison with organic zinc based coatings, show greater amounts of zinc elution due to contact of their base component zinc with water, and have stronger anticorrosive effects due to the difference in ionization tendency between iron and zinc.

3.3 Natural weathering

No samples showed rust, blistering, cracking, scaling nor other abnormalities during the natural weathering tests.

The progress of chalking varied widely depending upon the type of coating and paint colour. In the case of white colour, noticeable chalking occurred in phthalic acid coatings after a year weathering and in chlorinated rubber coatings after two years weathering. Chalking, though slight, appeared in urethane coatings as well after two years weathering. No abnormalities occurred in fluorinated coatings, when comparing white and red colours, the chalking rate in white colour was estimated to be about two times that in red colour. This can be associated with the pigment concentration of the overcoating paint.

Discolouration/fading was largely affected by the type of coating and paint colour. Specifically, much higher colour difference values were obtained in red colour than with white colour. The four tested types ranked in the descending order of phthalic acid coatings > chlorinated rubber coatings, urethane coatings > fluroinated coatings. Generally, there was no significant differences among the paint colours or brands. Comparisons of coating stainability on the basis of gloss values before and after washing revealed that the four types are almost equivalent to each other.

Comparisons of changes in colour difference and gloss retention between natural and accelerated weathering tests revealed almost the same behaviour, suggesting that a year of natural weathering is equivalent to several hundred to 1000 hours of accelerated weathering. (See Fig. 5)

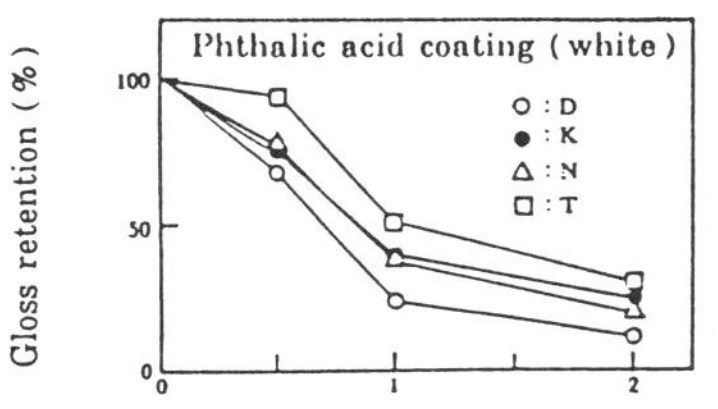

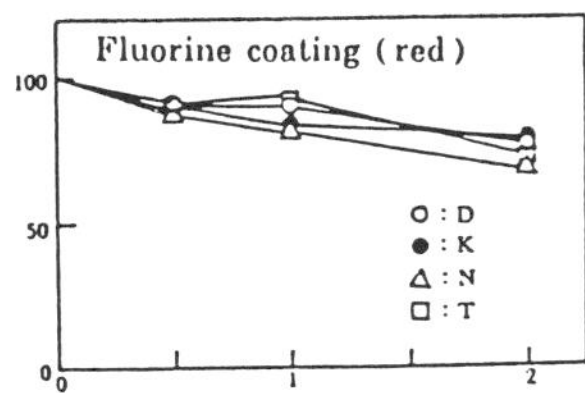

Outdoor weathering time (year)

Figure 5. Fluctuations of gloss after natural weathering treatment.

3.4 Weathering in actual use

No samples showed rust, blistering, cracking, scaling nor other abnormalities during the weathering tests in actual use. The

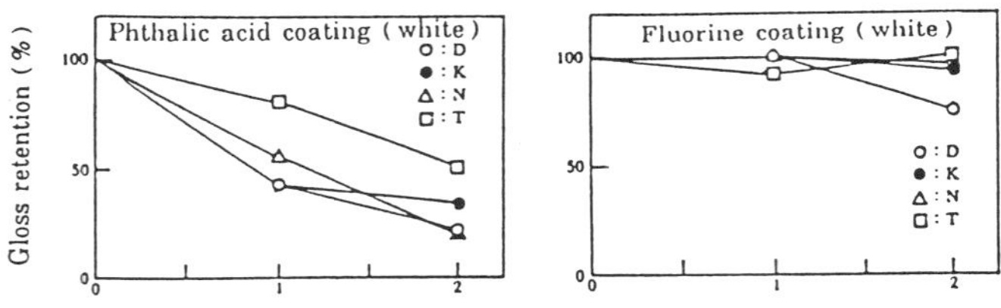

Figure 6. Fluctuations in colour difference after weathering in service.

progress of chalking was similar to that in natural weathering tests, but the degree was higher. Specifically, severer chalking occurred in phthalic acid and chlorinated rubber coatings than in natural weathering tests. Chalking occurred in urethane coatings as well. Slight staining occurred on the surface of fluorinated coatings, but it was within the normal range. (See Figure 4)

Visual inspection revealed that staining occurred to almost the same degree in all but chlorinated rubber coatings. The chlorinated rubber coating involved significant differences among the brands, with noticeable staining noted in brands D and T. This staining, accompanied by yellowing due to pain film deterioration, was attributed to chemical changes due to solution of paint components such as plasticizers.

Fluctuations in colour difference among the coating types were similar to those found in the accelerated weathering and natural weathering tests. Specifically, noticeable changes occurred in phthalic acid and chlorinated rubber coatings, while minor changes occurred in urethane and fluroinated coatings. Data comparisons between the weathering in actual use and natural weathering tests suggested that deterioration was accelerated by weathering at high temperatures (70 to 80°C).

Talking about gloss retention, the results obtained were similar to those obtained in the accelerated and natural weathering tests. It should be noted, however, that it is difficult to make comparisons under the same conditions due to differences in substrate smoothness, and so general behaviour alone is discussed here. (See Figure 6)

Judging from the findings shown above, it can be said that coating durability as rated by appearance is about one to two years for phthalic acid coatings, four to five years for urethane coatings and over several years for fluorinated coatings. Careful consideration should be given to the aspects of staining and yellowing before using chlorinated rubber coatings. From the viewpoint of corrosion resistance, the obtained data strongly suggest that the use of high grade thick anticorrosive coatings such as urethane coatings and fluorinated coatings contribute to extension of service life, although it remains impossible to predict service life.

4 Conclusions

Assessment of thick anticorrosive coatings as to durability in accelerated weathering tests, salt spray tests, natural weathering tests and weathering tests in actual use revealed the superiority of urethane coatings and fluorinated coatings over phthalic acid coatings and chlorinated rubber coatings. Evaluation of durability in terms of appearance revealed that their service life is at most two to three years. Phthalic acid coatings were found inferior in corrosion resistance as well, with relatively shorter service life.

Aside from differences among the material types, paint colour has significant effects on weather resistance. Dispersion among the brands is also wide. This suggests that in thick anticorrosive coating finish, much consideration must be given to coating material type, paint colour and brand for appropriate choice.

5 References

Japan Steel Construction Society, (1988) **Practice of anticorrosive coatings.** Sankaido Co. Ltd, Japan.

Japan Steel Construction Society, (1982) **Guidelines of the Society of Steel Construction -Manual for the Investigation of paint Films on Steel Bridge-.**

72 DURABILITY OF NON-ASBESTOS FIBRE REINFORCED CEMENT

J.M. WEST
Building Research Establishment, Watford, UK

Abstract
Asbestos cement replacement roofing products made from cementitious materials and reinforced with fibres such as polyvinyl alcohol, cellulose and glass, have been tested in the form of flat coupons to determine the materials properties after 5 years of weathering. Two products were also tested after 4 or 5 years either in the tropics or in water at 80°C.

In samples cut from corrugated roofing sheets, both tensile and bending strength and stress at the limit of proportionality are often considerably lower in wet specimens, but not the Izod impact strength. The tensile strength perpendicular to the plane of the board for some three year old samples including in this case slates and flat sheets, was found to be higher in the denser materials and tended to increase with weathering. It appears that the three early types of corrugated sheeting are showing reasonable durability in terms of maintaining their strength up to five years. The bending strength is rather low in one case, and the limit of proportionality can initially be rather poor but in all cases this improves later.
Keywords: Asbestos cement alternatives, Polyvinyl alcohol fibre, Glass fibre, Cellulose fibre, Durability, Bending strength, Tensile strength, Limit of proportionality.

1 Introduction

In the last decade asbestos containing products have increasingly become a health issue, causing the asbestos cement (AC) industry to consider alternatives. During this time it has marketed a variety of building products with a similar appearance but reinforced with alternative fibres which are considered to be much less of a health hazard than asbestos, as described by West and Majumdar (1989). These alternatives have mostly been made from a premixed slurry as with AC. However a variety of alternative building components were developed a decade earlier at BRE before the health issue was so prominent. Various hydraulic binders and a spray suction process were used to make glass fibre reinforced cement (GRC) and gypsum (GRG), the latter being particularly good for lasting impact strength and fire protection in drier environments. The more recent products have also used artificial fibres such as glass or fibres of textile polymers or of natural origin such as cellulose to reinforce cement and autoclaved hydrated calcium silicate products respectively, as described by Hodgson (1987) and Studinka (1989).

It is a known fact within the roofing industry that failures have occurred in fibre cement products far in excess of those found in its predecessor AC, and when this first became apparent BRE was asked to keep a watching brief in this field. As a result several of these products were studied at BRE from 1983 onwards. This paper is mainly concerned with corrugated sheet based on these materials ie. a cement matrix which is reinforced with fibres (FRC). Corrugated sheet was and still probably is the main AC product in buildings being applied largely as roofing. The replacement by other products, particularly those just mentioned and by metal based alternatives has sharply reduced the use of AC products in the UK in recent years.

2 Source and description of the products tested

Where possible material was purchased locally and this applies to slate and flat sheet. However all but one of the following corrugated sheets came from the UK manufacturers because of supply difficulties. The cellulose reinforced autoclaved sheets with a hydrated calcium silicate matrix had a 6 inch profile. The first of these sheets had a smaller pattern on the lower face and has been coded C-Sm in Table 1 while the second was coded C-Lg and seems to have slightly poorer properties. This may affect comparisons. Studies of the two textile reinforced sheets indicate that they contain fibres of polyvinyl alcohol (PVAL) and some cellulose. The one coded P6 had a 6 inch profile but the second P12 had a 12 inch profile, and a different brand name. The GRC sheet had a 12 inch profile, was a prototype and has been given the code letter G.

3 Procedure

3.1 Weathering and other treatments

The corrugated sheeting at BRE was supported on frames set at 45° facing south in an open position with ties to stop displacement by the wind. After 5 years sections of sheeting were taken for testing. The two small sheets weathered at Freetown, Sierra Leone were similarly mounted but subjected to tropical heat with very humid conditions alternating with seasonal dry dusty desert winds. Sheet C-Sm (cellulose) was weathered for 5 years (code SL5y) and sheet G (glass) for 4 years (code SL4y). For comparison data have been included from younger material, which was cut into test specimens or coupons before exposure. These coupons and artificial slates were weathered flat on wire mesh. Data from a few coupons remaining in other conditions after earlier tests have also been included.

3.2 Preparation for the tests

To obtain coupons from the corrugated sheeting, strips were cut wet by running a diamond saw along the ridges and furrows. These were trimmed to give 50mm wide flat strips and were then cut across at

150mm intervals. The coupons were numbered, then randomised before being allotted to tests in groups of 6 if possible. Before being tested coupons were conditioned for several days at 65% relative humidity (RH) and 20°C before dry testing, or were placed in water at 20°C for 1 day before wet testing. A few coupons from P12 were cut across the corrugations. This is denoted by a T in Table 1.

3.3 Test methods

The tests were made by methods developed at BRE for characterising the properties of GRC as described by Singh et al. (1978). These tests were designed to measure materials properties of large numbers of coupons rather than the performance of components. For example, they are used to measure the dry four point bending strength or modulus of rupture (MOR) of coupons for a span of 135mm, as shown in Table 1. To accommodate a wide range of different FRC roofing materials, performance specification is favoured in standards under preparation. This in turn favours the use of bending moment per metre width (moment/m) so this has also been calculated from these recent tests. This will only be directly comparable to values obtained by testing flat material as the values obtained on corrugated sheeting, by standard tests, depend on the profile. Other data derived from the bending tests are the limit of proportionality (LOPb), and the modulus of elasticity in bending (Eb). The 8 ends of 4 coupons from each bending test were broken in an Izod test by using a 12.1J pendulum, to obtain impact strengths (IS). If sufficient coupons were available some were given a tensile test, at a rate of 1000 microstrain per minute controlled with the aid of a 50mm gauge length extensometer. Peak tensile stress (TS), LOPt and Young's modulus (Et), were calculated. Most standards specify soaking test material in water. If material was available wet tests were done on a group of coupons immersed for 1 day at 20°C. Tensile strength perpendicular to the plane of the sheets (PTS) was also measured by bonding pairs of aluminium blocks on to 50mm x 50mm squares of material using epoxy adhesive. Freely articulated grips were used to pull on these blocks.

4 Results and discussion

Physical properties obtained from testing the corrugated sheets are given in Table 1. These are expressed as means of several test results, the number of tests being given below the bending and tensile strengths along with the coefficient of variation. The t-distribution was used to test for the significance of the differences between mean values, such as MOR values before and after weathering. The mean values were taken to be significantly different if the probability that the samples were drawn from the same population was less than 0.01. The material will be identified by either its reinforcing fibre or the board code explained in the description above. Rubber jaws were used in the most recent tests and this may well have enabled a slightly higher peak stress to be achieved.

The MOR has fallen during weathering for most products except P6 (PVAL). Both PVAL products were only weathered in the UK and started with a low MOR which appeared to have risen up to 2 years followed

Table 1. Properties of profiled fibre cement in relation to weathering

Storage	Property	Units	Cellulose			Textile (PVAL)			Glass	
			C-Sm	C-Sm	C-Lg	P6	P12	P12	G	G
	MOR	**(MPa)**	-	**30.9**	-	**21.2**	**15.2**	-	**23.7**	-
7 days	n, c.var.	(%)	-	6,11	-	6,4	6,4	-	6,10	-
minimum	LOPb	(MPa)	-	21.2	-	9.6	8.8	-	9.8	-
at	Eb	(GPa)	-	12.4	-	12.9	12.6	-	9.9	-
65% RH	**Peak TS**	**(MPa)**	-	**19.0**	-	**8.1**	**6.6**	-	**9.8**	-
and	n, c.var.	(%)	-	6,8	-	6,8	6,7	-	6,5	-
20°C	**LOPt**	**(MPa)**	-	**10.3**	-	**3.8**	**4.5**	-	**4.0**	-
	Et	(GPa)	-	14.4	-	13.9	15.6	-	12.6	-
	Izod IS	(kJ/m²)	-	1.5c	-	3.1c	3.2c	-	5.4c	-
2 years	**MOR**	**(MPa)**	-	**30.3**	-	**23.4**	**17.1**	**17.3**	**20.7**	-
UK	n, c.var.	(%)	-	3,3	-	6,2	6,6	6,11	6,8	-
weather	LOPb	(MPa)	-	27.9	-	18.4	16.8	17.1	15.4	-
tested	**Peak TS**	**(MPa)**	-	**18.6**	-	**7.1**	**9.0**	-	**9.6r**	-
after	n, c.var.	(%)	-	3,13	-	6,20	4,8	-	6,5	-
storage	**LOPt**	**(MPa)**	-	**15.9**	-	**3.3**	**8.8**	-	**6.4**	-
at	Et	(GPa)	-	16.7	-	20.1	19.2	-	13.4	-
65% RH	Izod IS	(kJ/m²)	-	2.0c	-	3.8c	3.5c	3.6	3.6c	-
			SL5y							SL4y
5 Years	Thickness	(mm)	5.8	6.2	5.9	6.5	T7.5	7.2	7.7	6.6
of	Moment/m	(N)	148	185	158	165	T135	145	190	137
weather	**MOR**	**(MPa)**	**26.1**	**29.3**	**27.2**	**23.5**	**14.4**	**16.7**	**19.3**	**18.7**
then	n, c.var.	(%)	6,9	6,5	6,10	6,6	4,20	6,7	6,6	6,3
tested	LOPb	(MPa)	23.6	28.7	26.1	22.0	14.3	16.7	17.1	18.6
after	Eb	(GPa)	8.8	12.9	11.8	16.8	12.1	15.3	11.8	13.7
storage	**Peak TS**	**(MPa)**	-	**20.0r**	**15.6r**	**9.3r**	-	**7.9r**	**10.2r**	**9.9r**
at	n, c.var.	(%)	-	4,7	4,8	6,3	T	5,13	6,3	4,14
65% RH	**LOPt**	**(MPa)**	-	**17.2**	**13.3**	**8.5**	-	**6.7**	**8.8**	**9.8**
	Et	(GPa)	-	16.2	15.7	20.1	-	19.7	14.5	15.6
	Izod IS	(kJ/m²)	1.8	2.7	2.3	3.1	5.1	3.6	2.9	1.7
7 days	**MOR**	**(MPa)**	-	**23.3**	-	**21.8**	**13.8**	-	**22.1**	-
in	LOPb	(MPa)	-	12.3	-	10.5	6.3	-	7.8	-
water	Eb	(GPa)	-	9.5	-	12.8	10.6	-	10.4	-
at	**Peak TS**	**(MPa)**	-	**10.7**	-	**7.5**	**5.0**	-	**8.7**	-
20°C	**LOPt**	**(MPa)**	-	**5.2**	-	**4.5**	**2.8**	-	**3.4**	-
	Et	(GPa)	-	13.0c	-	15.8c	12.4c	-	11.3c	-
			C-Sm/80°C	**C-Sm**	**C-Lg**	**P6**	**P6/80°C**	**P12**	**G**	**G/20°C**
5 years	Thickness	(mm)	6.2c	6.2	5.9	6.4	6.3c	7.2	7.7	6.7c
at BRE	Moment/m	(N)	202	158	127	142	93	107	170	107
either	**MOR**	**(MPa)**	**31.6**	**24.7**	**21.9**	**20.5**	**13.9**	**12.3**	**17.4**	**14.5**
in UK	n, c.var.	(%)	6,7	6,17	6,7	6,3	6,9	6,5	6,8	6,8
weather	LOPb	(MPa)	24.8	19.0	14.8	15.0	12.2	12.0	12.9	11.4
or in	Eb	(GPa)	10.8	10.4	9.2	15.4	12.6	13.9	10.4	12.0
water	**Peak TS**	**(MPa)**	**16.9r**	**11.3r**	**11.4r**	-	**6.6r**	**5.8r**	**7.5r**	**6.0r**
	n, c.var.	(%)	6,10	4,5	4,5	-	6,6	6,16	5,5	5,11
all	**LOPt**	**(MPa)**	**7.3**	**8.7**	**8.1**	-	**5.7**	**5.7**	**6.0**	**5.8**
tested	Et	(GPa)	13.2	12.9	12.9	-	15.1	17.1	14.0	18.4
wet	Izod IS	(kJ/m²)	2.1c	1.9	2.2	3.2	3.1c	4.3	3.3	5.1c

Key: C-Sm=small pattern, C-Lg=large pattern, SL=Sierra Leone, 4y=4 years, 5y=5 years, P6=6"profile, P12=12"profile, c=given treatment eg. weathering as coupons (otherwise sheet cut into coupons after weathering), r=rubber faced jaws, T= Coupons cut transversely.

by relatively little change up to 5 years. After 5 years at 80°C in water there was a significant loss of MOR (at the 0.001 significance level). The weathered cellulose products show a slight fall in MOR in the UK and a moderate fall (only significant at the 0.05 level) in the tropics but even then are still the strongest products overall including when wet and show very good wet strength after 5 years at 80°C in water. The GRC corrugated sheet shows a sharper decline in strength in the first 2 years weathering, as is usual with GRC, but this decline is only significant at the 0.05 level. During the full weathering period it shows a significant fall in MOR in the UK and in the tropics, but this was still not the weakest product in bending and tensile strengths. After 5 years at 20°C in water there was a significant loss of MOR and some loss of tensile strength. Litherland et al. (1981), describe a hot water test to predict such losses specifically for glass, but this is not appropriate for the other fibres mentioned. Because of the initial strength loss in GRC, it must be formulated with a higher initial strength than is likely to be needed with other fibres. Tensile strength for these FRC materials is at a lower value of calculated stress than MOR.

Values of stress at the limit of proportionality (LOP) in bending have risen significantly in every material up to 2 years and somewhat after this, except for PVAL (P12) in the UK and cellulose in the tropics. The fall in the latter is only significant at the 0.05 level and this value is still higher than for the other products. Tensile LOPt shows a similar rise except for a low value of P6 at 2 years. All the materials tested wet showed an increase which was significant at the 0.001 level. The high values for the cellulose products when dry show the material in a good light as LOPt is considered to be the stress at which the matrix starts cracking and is therefore one of the most important properties of FRC. However when wet this value is only moderate. This observation and the very low initial LOPt values for the PVAL and glass products (which are even lower when wet) may be one reason for failures to occur in these materials. On the other hand one possible contribution to failure, ie. stress caused by drying shrinkage, will tend to be lower when the material is uniformly damp. If these products can survive installation the increase in LOP could become useful. The stress at LOP is likely to be more closely related to the matrix formulation than to the fibre type.

Values of Young's modulus show a similar but less marked tendency to increase with weathering except again for cellulose in the tropics. Modulus in bending (Eb) tends to be an underestimate because of compliance in the testing equipment and shifting of the neutral axis, the latter also tending to increase calculated stress at the MOR. The 5 year results for modulus in bending may be on the low side as they have not been corrected for compliance.

Changes in the properties, such as strength loss, could come from a change in either the matrix, fibre or both. Work at BRE has shown PVAL fibre to be quite resistant to hot wet conditions so that the weakening of P6 after 5 years at 80°C in water is interesting, and also contrasts with the good strength of the cellulose product.

Impact strength and strain to failure, which tends to be closely correlated with it, are low with these products.

When considering these results it should be borne in mind that the formulation of individual products is likely to have changed in the last five years and that failures can result partly from restraint at fixings or overlaps which were absent on the BRE exposure site.

Table 2 has perpendicular tensile strength (PTS) values which are higher in high density products and tend to increase on exposure. This might be expected as PTS is largely a measure of the strength of the matrix but superficial weakness is found with cellulose reinforcement, perhaps due to damage of the cellulose by sunlight.

Table 2. Perpendicular tensile strength comparisons of dry FRC (MPa)

Test	Treatment	Cell flat	GRC flat	Cell corr	PVAL corr	PVAL corr	AC slate	PVAL slate	GRC slate
PTS	Air	1.7	3.0	3.6	3.3	3.9	5.9	3.6	6.4
PTS 3yrs	Weather	s2.1	-	s1.3	2.8	3.1	>6.6	4.2	5.6
Density(g/cc)	Air	1.4	1.8	1.6	1.8	1.8	2.1	2.0	2.0

s= superficial failure on the exposed face, >= adhesive failure

5 Conclusions

These 5 year results show little change in strength after UK weathering except for the GRC which loses some strength as expected but is still not the weakest product. With 4 or 5 years of the harsher tropical exposure, small but in one case significant, falls in dry strength have occurred. The cellulose product still remains strongest. The important property of the tensile LOP, which is rather low initially in the products reinforced with PVAL and glass, has increased for all the materials. The corrugated sheets show low impact strength (in comparison to material produced by the spray process). Perpendicular tensile strength measurements have shown that with cellulose the surface layer is weakened during exposure.

Most wet specimens gave considerably lower test values than when dry, particularly for the cellulose type. These materials are therefore likely to be more vulnerable to cracking when damp.

6 References

Hodgson A.A. (1987) **Alternatives to Asbestos and Asbestos products** Anjalena Publications Ltd. Crowthorne UK.

Litherland K.L. Oakley D.R. and Proctor B.A. (1981) The use of accelerated ageing procedures to predict the long term strength of GRC composites. **Cement and Concrete Res.**, 11, 455

Singh B. Walton P.L. and Stucke M.S. (1978) Testing and test methods of fibre cement composites. **Proc. RILEM Symp. Sheffield**, 377-387. Construction Press Ltd. Lancaster

Studinka J.B. (1989) Asbestos substitution in the fibre cement industry. **The International J. of Cement Composites and Lightweight Concrete**, 11, 73-78.

West, J.M. and Majumdar, A.J. (1989) Alternatives to asbestos cement products in buildings. **CIB XIth International Congress, Paris.** June 19-23, Theme II, Volume II, 497-504.

73 TEN-YEAR DURABILITY TEST RESULTS ON EXTERNAL WALL MASONRY COATING SYSTEMS

T. NIREKI
Building Research Institute, Ministry of Construction, Tsukuba, Japan
K. OMATA
Japan Masonry Coating Manufacturers' Association, Tokyo, Japan
N. HIRAMA
Asia Industry Co. Ltd, Kuki, Japan
T. INOUE
Kowa Chemical Industry Co. Ltd, Tokyo, Japan
H. KABEYA
Misawa Homes Institute of Technology, Tokyo, Japan

Abstract
This paper deals with the durability characteristics of masonry coating systems under ten-year outdoor exposure tests and also under various laboratory tests, especially with a view towards the protective performance of the coating layers - adhesive strength, hardness, tensile strength, elongation and the effectiveness for the prevention of carbonation of the wall components.
Keywords: Outdoor exposure test, Masonry coating system, Protective performance, Carbonation, Ductility

1 Introduction

In Japan, there has been a wide-spread application of masonry coating systems, using a mixture of inorganic aggregates and various types of resins (epoxy, polyurethane etc.) with admixtures, having an overall thickness of some 0.3 to 15mm and applied by spray or roller to various types of external wall surfaces such as concrete, autoclaved aerated concrete, cement mortars, concrete blocks, inorganic sheets, etc.

The total production of materials for masonry coating has reached some 510,000 tons current at 1988, including those for interior use. (Some 60% is assumed to have been used for maintenance work.) The quality of these materials has been standardized in several Japanese Industrial Standards (JIS) and also their application methods have been specified in Japanese Architectural Standards Specifications (JASS).

The durability characteristics of the systems are clearly dependent upon the components of the mixture, particularly the type of resin, pigment, fine aggregates and their mixing ratio. Given this situation, the Building Research Institute (BRI) started a research programme in 1970 on the durability performance of masonry coating systems and a joint research project with the Japan Masonry Coating Materials Manufacturers' Association, Nireki (1983), aimed at obtaining more practical data for designers or practitioners in view of the service life of coating systems.

2 Performance Tests

2.1 Objective Coating Systems

Types of coating system on the market at present can be divided into the following four categories;

Type 1 : Mixture of mainly inorganic materials, white Portland cement, dolomite plaster, sodium silicate, silicate sand, perlite, etc., plus pigment. Thickness 0.3 - 15mm

Type 2 : Mixture of acrylic emulsion, polyvinyl acetate emulsion with silicate sand, perlite, fine white marble, plus pigment. Thickness 0.3 - 15mm

Type 3 : Mixture of acrylic, polyvinyl, epoxy emulsion or solvent type epoxy, polyurethane, with silicate sand, limestone sand. Thickness 0.3 - 8.0mm

Type 4 : Mixture of acrylic or chloroprene rubber, emulsion or solvent type of polyurethane with calcium carbonate powder, silicate sand, etc. Thickness 1.5 - 8.0mm

Among these, Type 1 is inherited "Cement Lithin Finishing" or "Stucco". Type 4 is a rather newly developed system which has relatively high waterproofing and also flexibility to follow the movements of substrates.

Each type of system was applied to each of asbestos cement sheeting, mortar, concrete blocks and autoclaved aerated concrete. Outdoor test specimens were from 150 x 300mm to 450 x 1800mm, those for laboratory tests being mainly 60 x 150mm, Nireki (1986).

2.2 Outdoor Exposure Tests

Testing was carried out under the following conditions;

(1) ordinary exposure methods - 30 and 90 degrees to the horizontal facing south, vertical facing north, and horizontal
(2) EMMAQUA, as specified in ASTM E 838-81
(3) under artificial dew condensation
(4) alternate elongation and shrinkage - for coating layers
(5) application of pre-determined strain - for coating layers

2.3 Laboratory Tests

Testing was carried out under the following conditions;

(1) sunshine-type carbon arcs
(2) fluorescent lamps - combination of two different lamps in cabinets
(3) fluorescent lamps - as above but at normal temperature
(4) under elevated temperature - surface temperature is the same as in (1)
(5) cyclic loading at low frequency after the treatments (1) to (4)

3 Results and Discussion

From the vast amount of data obtained from the tests, the following are the condensed results which would be useful for a rational selection of coating system.

3.1 Surface Appearance

According to visual inspection over the ten year test period, no terminal deterioration such as cracking, blistering or peeling has yet been observed in any type of system.

Some factors and their levels were wilfully adapted for investigation or to ensure that they were preferential factors. On the basis of these factors, the following conclusions could be drawn. Characteristics of surface colour vary dependent upon the type of pigment in the system, even with inorganic pigment. The tendency for colour change can be evaluated by applying the sunshine-type acceleration test within 300 hours. Soiling by dust or mould varies according to the surface texture, moulding occurring on the north-facing surface more so than on that facing south. Chalking advances within five years. The exposure angle in the outdoor tests was most severe at 0 degrees to the horizontal. A more ductile top coat ensures a higher level of masonry coating system, especially for Type 4.

3.2 Protective Performance Against Carbonation

Performance against carbonation of cementitious substrates has become an important requirement for a masonry coating system, especially for concrete, cement mortar or aerated lightweight concrete products.

Type 1, a mixture of cement and aggregates, carbonated as in Figure 1 and a protective role is not to be expected in the thinner coating layers such as 11 or 12. On the other hand, a thick system can act as an effective coating layer for substrata even after ten years. In systems 13 to 17, partial carbonation shows on the surface of the mortar even though the whole coating layer is not fully carbonated. This is because the carbonation progresses through the local thinner layer as the result to obtain rough texture of surface appearance.

The same characteristics were observed in Type 2 systems.

Type 3 is a mixture of resin and powdered additives with top coating, the protective performance of which seems to be superior to Types 1 and 2. However, the results show a different characteristic among this group as in Figure 1. Taking 7 and 22/29 as examples, the former has an uneven, crater-like texture which produces a partial thin layer, whilst in the latter examples, the undercoating is evidently effective. It can also be said that the performance of the top coating (usually a solvent paint-type coating) plays an important role in the prevention of carbonation and in the service life of the total coating system.

3.3 Ductility

Concrete wall elements or cement mortar often show movement which sometimes proceeds cracking. Hence, ductility has become one of the important requirements for surface coating systems acting as a barrier against water or vapour in the external environment. In this context, systems of Types 1 or 2 can follow a limited amount of movement, but Type 3 and especially Type 4 have been developed for following the movement of substrata.

Considering ductility over time, a typical model of the relationship between load and elongation in Type 4 systems can be seen in Figure 2. A summary of the behaviour in outdoor exposure tests shows that although the load itself increases, the capacity for elongation decreases over time.

Figure 3 shows one of the actual results obtained from the outdoor exposure tests on the relationship between elongation and the thickness of the Type 4 system. It is clear that ductility is dependent upon the thickness of the layer. The same tendency has been obtained in the laboratory tests, showing that the ambient temperature is a prominent factor for minimizing elongation. As to the practical selection of Type 4, the required amount of ductility should be taken into account, the range being from 0.1mm (said to be the minimum for watertightness) to over 0.5mm (usual cracks observed in the field).

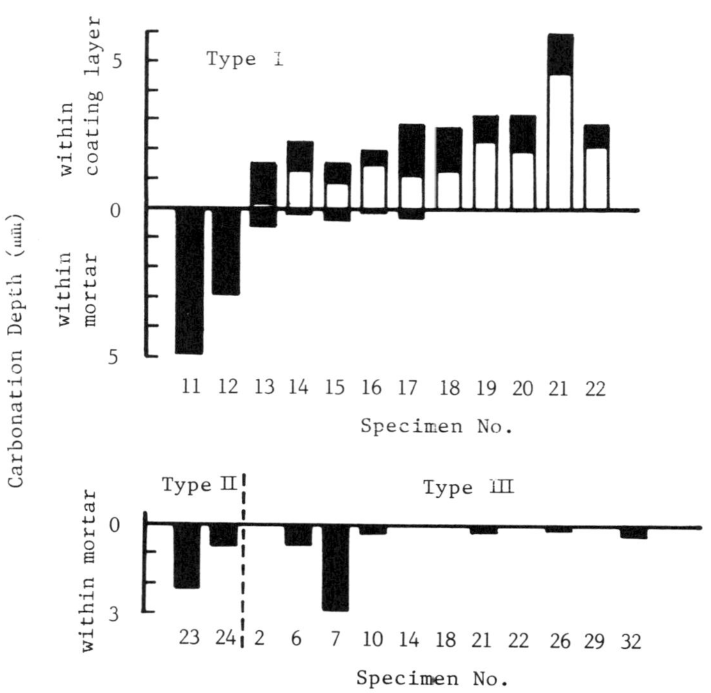

Fig. 1 Carbonation depth after 10 years outdoor exposure

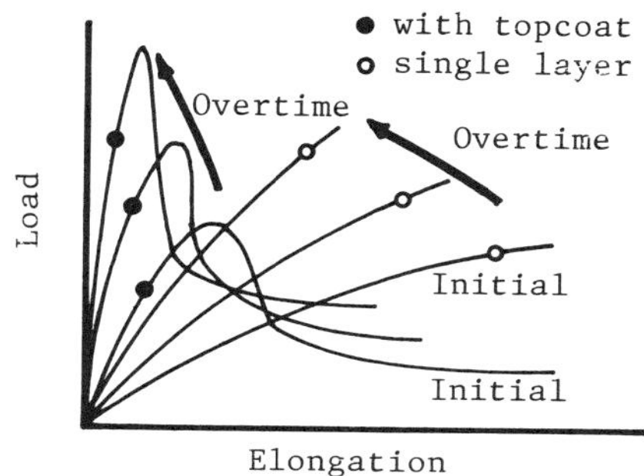

Fig.2 Typical behaviour of load -elongation overtime

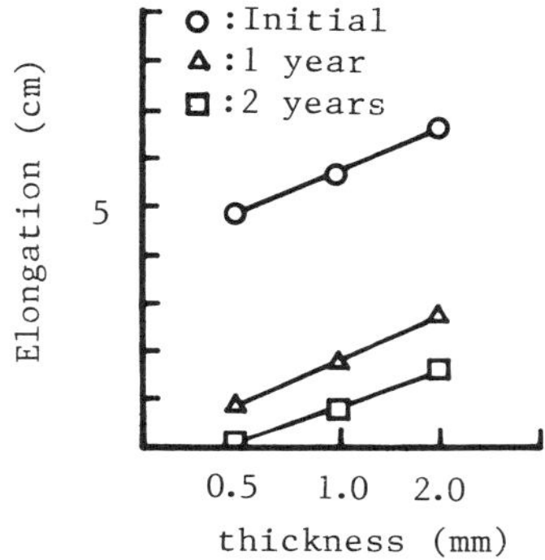

Fig. 3 **Relation between elongation** & thickness of coating layer

4 References

Nireki, T. et al., (1983) Performance evaluation on masonry coating systems for establishing the prediction of service life time, in Proceedings of CIB Congress 86 (Sweden), vol 2, pp 183-193.

Nireki, T. et al., (1986) Performance over time of exterior coating systems, in Proceedings of CIB Congress 86 (Washington), vol 6, pp 2220.

74 DURABILITY OF AERATED LIGHTWEIGHT CONCRETE PANELS WITH SURFACE COATING SYSTEMS

T. NIREKI
Building Research Institute, Ministry of Construction, Tsukuba, Japan
H. KABEYA
Misawa Homes Institute of Technology, Tokyo, Japan
N. HIRAMA
Asia Industry Co. Ltd, Kuki-City, Japan
T. INOUE
Kowa Chemical Industry Co. Ltd, Tokyo, Japan

Abstract
This paper shows the experimental results of 14 years of outdoor exposure tests on full-sized aerated lightweight concrete (ALC) panel walls with masonry coating systems. In the outdoor exposure test, visual observation was carried out on surface appearance (discolouration, cracking etc.) and X-ray analysis was carried out on measurement of water content, CO_2 content and the crystal component in ALC. As well as outdoor exposure tests, some laboratory tests were done under accelerated conditions to help the analysis of the results of the outdoor tests. On the basis of the results obtained it was concluded that;

(a) Recoating of surface finishing is required to maintain satisfactory use of the panel and all the sealants should be replaced.
(b) The increase of calcite ($CaCO_3$) clearly shows the advancing of carbonation. Carbonation leads to the cracking of ALC due to carbonation shrinkage.
(c) Water content in the panels is related to the rate of carbonation.

Keywords: Aerated lightweight concrete, Durability, Outdoor exposure test, Protective coating systems, Carbonation, Cracking, Tobermorite, Calucite.

1 INTRODUCTION

Production of aerated lightweight concrete (ALC) panels in Japan had been started on the basis of technology translation from European countries and they have come to be widely applied as various building components, the total amount of production reaching some 300 million square metres at present, a figure which will increase in coming years. Research into the durability of ALC products themselves had been carried out [1][2] and it had been recognized that their protective coating systems played an important role. Since 1974, the Building Research Institute(BRI) has been conducting outdoor exposure tests aimed at evaluating the overall durability characteristics of ALC panel wall systems with various protective systems as well as the sealing materials for panel joints.

The focal points of this paper are;

(a) Performance over time of the protective coating systems and sealing materials
(b) Deterioration of ALC panels, freezing and thawing or likely cracks
(c) Degree of carbonation of ALC and its on shrinkage cracking
(d) Effect of water on the rate of carbonation

2 Outdoor exposure test

2.1 Specimens and test conditions

Four walls, each with four full-sized ALC panels (10cm in depth) in vertical and horizontal application, facing north and south, were exposed at the BRI outdoor exposure site (36°07'N, 140°04'E, 29m from sea level) from 1974. (See Fig. 1)

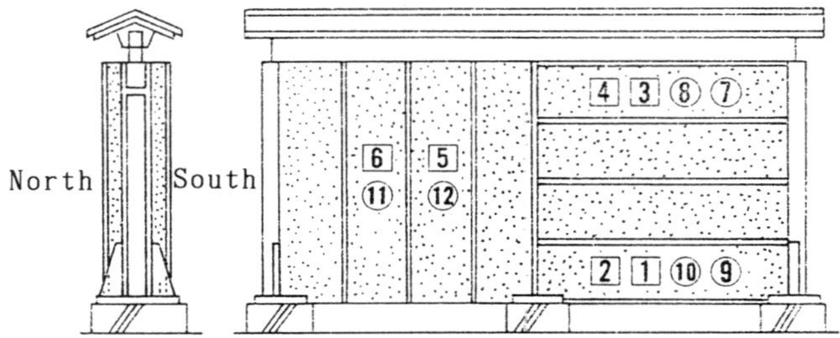

Fig. 1 Outline of outdoor exposure test for ALC panel

2.2 Evaluation items

a) Visual observation - deterioration characteristics of ALC panel, surface coating layer (acrylic masonary coating), sealant(acrylic)
b) Change of crystal components - investigation by X-ray diffraction analysis
c) Carbonation - measurement of the amcunt of CaO by differential thermal analysis
d) Water content - amount of water and its distribution

3 Results and discussion

3.1 Visual observation

The acrylic masonary coating system, a layer 1.0 to 2.0mm thick of fine aggregate (white marble) and acrylic resin with admixtures widely applied for the coating of ALC panels, shows fine cracks within the layer and exposure of the fine aggregates due to the splashing of the rain, but remains in fairly satisfactory condition.

On the other hand, the coating around the jointed part of the panel and on the sealant shows rather excessive deterioration, as shown in photograph 1.

As for the ALC panels themselves, a fair amount of damage due to freeze-thaw action can be observed, as shown in photograph 2. No damage due to the corrosion of the reinforcing bars is yet observed. However, fine cracks can be seen around the corner of the panel onto the front surface for which a finer diagnosis is required to clarify their causes. Shrinkage and stiffening of the sealant is evident which leads to the deterioration of the surface coating system and of the ALC panels.

Photo 1 Deterioratio of protective coating

Photo 2 Cracks at the edge of ALC panel

It has been pointed out that ALC and its inner reinforcements would deteriorate in time with alternating wet and dry conditions. When the required performance of the surface coating system or sealant decreases, the carbonation of ALC is progressive.

Apart from the results of the outdoor exposure tests, the information obtained from the field conditions survey, conducted on a seven year old ALC building, showed panel-cracking, dislodgement of the panel edge (this due mainly to freeze-thaw action as observed in the outdoor exposure test) and the deterioration of protective surface coating and also of sealing materials.

The cracks can be classified into two probable types, one follows the line of the reinforcing bars (see photograph 2) and the other seems to bear no relation to the bars. The main reason for the former case is clearly identified to be the corrosion of reinforcing bars treated initially with a resinous type of anticorrosive coating system or a cement-latex type system. The reason for the other type of cracking seemed to be alternate expansion and shrinking under conditions of water absorption and drying, or due to external force. Initially, the ALC itself is liable to absorb water and the amount of drying shrinkage is $2\text{-}4 \times 10^{-4}$m. On the other hand, the amount of strain needed to generate the cracks is some 5×10^{-4}m. Therefore, cracking due to drying shrinkage would hard to be yielded.

3.2 X-ray diffraction analysis

The stability of ALC relies on the ratio of CaO to SiO_2 and this condition of not remaining CaO (free lime) is a requirement in the production process of ALC. Figure 2(abcve) shows the results of X-ray analysis on new ALC and Figure 2(below) shows the results of the same analysis on ALC which has been exposed for 14 years.

In Figure 2 Tobermorite and α-quartz (crystallized silicate) can be observed. This means that ALC shoulc be quite a stable material under an ordinary building ambient atmosphere. In contrast, the existence of $CaCO_3$ can be clearly observed in Figure 2(below), which is evidence that ALC underwent carbonation in an atmosphere containing CO_2. The same pattern emerged from samples taken from the above mentioned field survey.

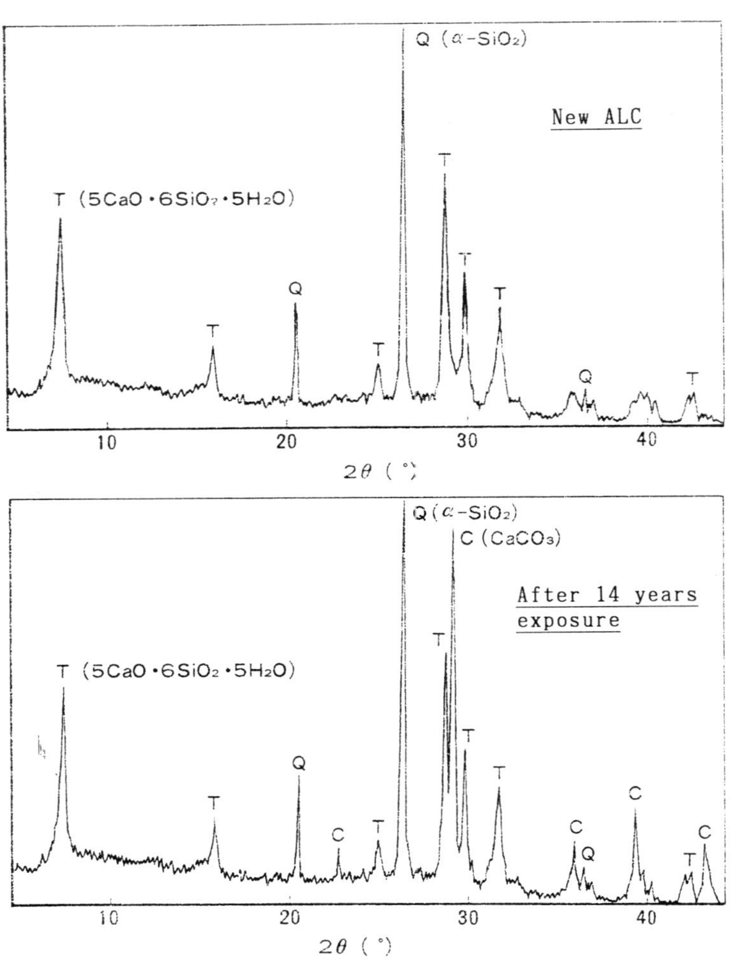

Fig. 2 X-ray diffraction pattern of ALC

3.2.1 Laboratory test for carbonation of ALC

Laboratory tests were carried out to investigate the shrinkage and carbonation of ALC on the basis of data obtained from the outdoor exposure test and the field survey. Actual test conditions were CO_2 content 5%, 20°C, 80% RH, in due consideration of the fact that the only component related to the carbonation of ALC is CaO, the degree of carbonation being obtained by measuring the amount of $CaCO_3$ in the aged ALC, an already established testing condition for the carbonation of ordinary concrete[2]. The relation between the amount of shrinkage and carbonation is illustrated in Figure 3 and it could be concluded that ALC is susceptible to cracking when the degree of carbonation reaches 40-50%.

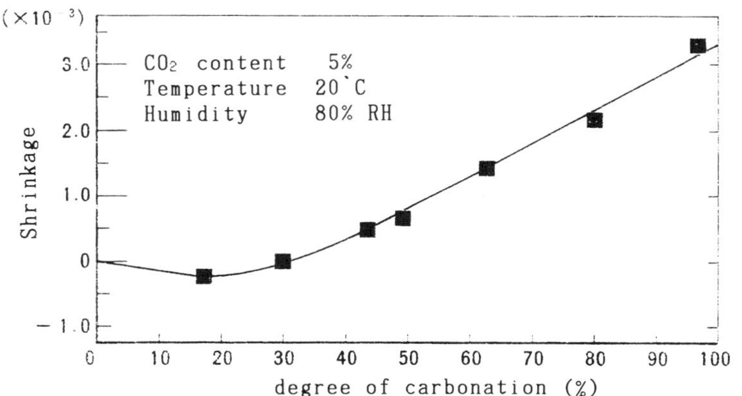

Fig. 3 Relationship between shrinkage and carbonation of ALC

3.2.2 Carbonation of exposed ALC panel specimen

The CO_2 content in the exposed ALC panel is shown in Figure 4 as the result of differential thermal analysis of a core specimen (30mm dia) extracted from the panel and sliced every 10mm through the depth of the panel.

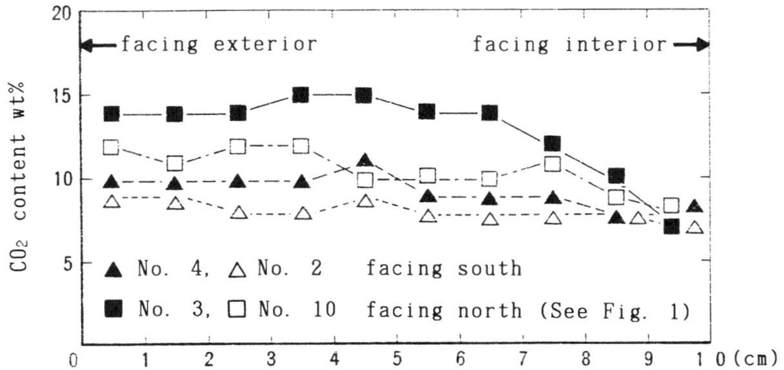

Fig. 4 CO_2 content in the exposed ALC panel

The distribution of carbonation can be summarized as follows;

a)Some water content is preferential, as mentioned in 2 above.
b)Carbonation begins at the external surface and a different rate cannot be distinguished on the internal surfaces of both panels.
c)Carbonation is more developed in north-facing panels than in those facing south.
d)Water content in ALC has a close connection with the progress of cabonation - 5.5% in north-facing panels, 3.0% in south-facing panels.
e)The upper part of a panel shows a higher degree of carbonation compared with the lower part.

4 Conclusion

On the basis of the experimental analysis, the following conclusions can be drawn.

1)ALC exhibits shrinkage and brittleness over time due to the reactions of components on ALC under the existence of appropriate amounts of CO_2 and water.
2)Advancement of carbonation produces carbonation shrinkage which leads to cracks in the ALC.
3)Cracks due to carbonation shrinkage result in early deterioration of the protective coating layer, which in turn means a failure of the first line of defence against extrinsic agents. As for the effectiveness of the protective coating systems, the results obtained from an accelerated test in the laboratory show performance differences among types of coating systems (see Figure 5). Test conditions were 20°C, 60% RH, CO_2 content 10% (wilfully doubled for acceleration). Cross-linked polyurethene is clearly superior to the acrylic emulsion type in view of protecting against carbonation. However, complete surface sealing often leads to blistering of the coating layer due to vapour pressure from inside of the panel. Thus, proper vapour permeability is a key to the selection of the coating system.

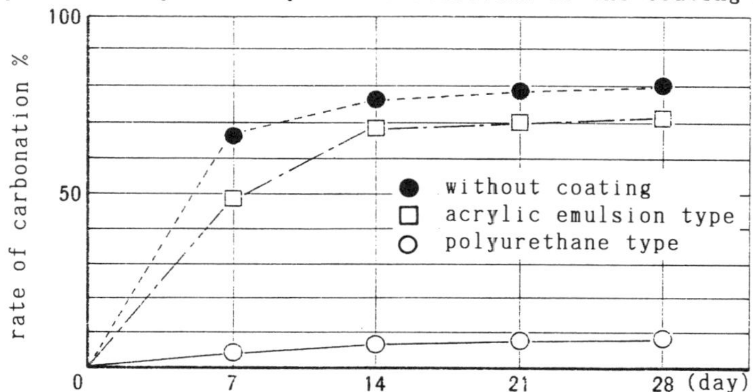

Fig. 5 Effect of coating system against carbonation of ALC

4) Water penetration through the ALC or cracks always yields a fatal deterioration, especially the corrosion of reinforcing bars and freeze-thaw action in colder regions. Also, consideration should be given to the recent phenomenon of acid rain - a pH of around 4.0 not being extraordinary in Japan - which would act as a prominent corrosive agent of reinforcing bars in ALC.
5) The service life time of the protective coating system adopted here could be said to be 13 years providing that periodic local repair to the corners or around the joints is ensured. That of the sealant is at most 8 years, especially in view of the "preventive maintenance". After the 14 year exposure test, the surface had been recoated by eight different types of protective coating system and various types of sealant had been applied after the removal of earlier ones in order to evaluate the effectiveness of repair, giving data for the maintenance planning of ALC wall panels.

References

Nakano, S. and Kabeya, H. (1983) Durability performance of surface coating for the newly developed aerated autoclaved concrete components, in Proceedings of 9th CIB Congress (Sweden) vol 4, pp. 425-434.

Abe, M. (1989) Experimental study on the prediction of carbonation rate of concrete, in Proceedings of AIJ Annual Meeting, Japan pp. 613-614.

75 CONTRIBUTION TO THE DESIGN FOR DURABILITY OF THE BUILDING TECHNOLOGICAL SYSTEM: A METHODOLOGY FOR THE EVALUATION OF THE RELIABILITY OF THE FUNCTIONAL MODELS

P.N. MAGGI, S. CROCE, M. REJNA, P. BOLTRI, A. LUCCHINI
Dipartimento di Ingegneria dei Sistemi Edilizi e Territoriali, Politecnico di Milano, Italy
L. MORRA, A. GOTTFRIED, B. DANIOTTI
Dipartimento di Ingegneria dei Sistemi Edilizi e Territoriali, Politecnico di Torino, Italy

Abstract:
The reliability of the over time performance of a building component depends first of all on the reliability of its functional model.
The paper presents a methodology for the theoretical evaluation of the reliability of functional models.
This methodology is useful for:
- the recognition of the presence in a functional model of incompatibilities between functions which can generate disturbance factors;
- the theoretical evaluation of a reliability index for each functional model (evaluation according to scoring criteria). This can be done also with functional models for the design of new building components when reliability evaluation cannot be supported by experimental data;
- the comparison of the reliability index of different functional models and consequently for the optimization of design choices.
Key words: Design of Durability, Reliability of the Functional Model of Building Components.

1 Introduction

The methodology aims at the prediction of service life and reliability of building components already designed and manufactured by the industry.

Such prediction is based on analytical evaluation of the components according to scoring criteria leading to the expression of propensity indexes.

This paper focuses on the methodology for the reliability propensity evaluation, which can provide an evaluation parameter, in substitution of the proper reliability value referred to the mean time before failure (service life).

Such evaluation, expressed in relative terms as to the single catalogue to be investigated, is homogeneous with the criteria underlying the evaluations of perfomances at zero time resulting from research carried out by the same teams in the last few years (*)

2 Description of the methodology contents

The working method proposal envisages the analysis of the functioning models of the building components belonging to the catalogue to be investigated.

This calls for the reconstruction of the mentioned models starting from the solutions and according to the sequence shown in figure 1.

As to the stages shown in figure 1, it has to be noted that:

- connotating requirements are requirements of technological behaviour according to which performances at zero time have been evaluated(**);
- each connotating requirement is met by a basic function;
- each basic function is met by a given number of analytical functions(see fig.2);
- the reconstruction of the elementary functional models, that is, of the fulfilment model of each basic function, involves the assessment of the control parameters for each analytical function and of the corresponding significance ranges;
- the reconstruction of the above mentioned models starts with the objectual model of each component, by trying to detect the location of the analytical functions on the model's layers, then recognizing the structure of the corresponding elementary functional model;
- the evaluation of the reliability propensity of each elementary functional model is based on the application of judgement criteria.

  The following aspects have been considered of fundamental importance to the definition of such criteria:

  - the simplicity of the model in terms of functional location and of qualitatively diversified analytical functions;
  - the wearying of the model in terms of functional load;
  - the redundance of the model in terms of simultaneous presence in the model of various orders of analytical functions, each able to ensure the fulfilment of the basic function.

(*) refer to the paper "Evaluation Method of the Quality of Building Components" presented in this conference

(**) P.N. Maggi, S.Croce et al., (1988) Evaluation of the Performance Propensity of out of system Building Technical Components in **Performance Requirements in Building.** Luxembourg

Five judgement criteria have been worked out, by which the reliability propensity of each elementary functional model can be evaluated according to a dimensionless relative score system (see fig.3)

- the reconstruction of the overall functional models is achieved by combining the elementary functional models referring directly to the objectual model of each considered building component;
- the evaluation of the reliability propensity of the overall functional model is based on the application of judgement criteria:
  - the simplicity of the model expressed in terms of:
    - . number of functional locations
    - . number of analytical functions
    - . number of singular attribution of functions to single locations
  - the wearying of the model expressed in terms of:
    - . functional load
    - . variability of the functional load
    - . critical load
    - . functional load balance
  - the distribution of functions expressed in terms of:
    - . serial connotations of analytical functions
    - . parallel connotations of analytical functions

Fig.3 shows an example of overall functional model.

3 References

A.A.V.V. (1983) **Correlabilità tra prove di laboratorio e comportamento in servizio per il controllo prestazionale nel processo edilizio: sperimentazione su pavimentazioni sottili.** Quaderno n.6 Dipartimento di Ingegneria dei Sistemi Edilizi e Territoriali. Politecnico di Torino .

A.A.V.V. (1987) **Comportamento nel tempo dei sistemi tecnologici edilizi.** MPI 40%.Report n.1. Milan.

A.A.V.V.(1989) **Comportamento nel tempo dei sistemi tecnologici edilizi: Metodologia per la stima della propensione all'affidabilità di soluzioni tecniche a repertorio e verifica dell'operabilità della metodologia condotta su un repertorio tipo di soluzioni tecniche.**MPI 40%.Report n.2. Milan.

A.A.V.V. (1989) Evaluation of building components long-term quality in **Quality for building users throughout the world.** Paris.

Boltri P. (1987) **Metodologia per la progettazione della qualità tecnologica del sistema edilizio.** Doctoral Thesis. Milan and Turin Polytechnics

Lucchini A.(1989) **Metodologia per la valutazione dell'affidabilità di elementi tecnici edilizi fuori sistema e problematica della durabilità.** Doctoral Thesis. Milan and Turin Polytechnics

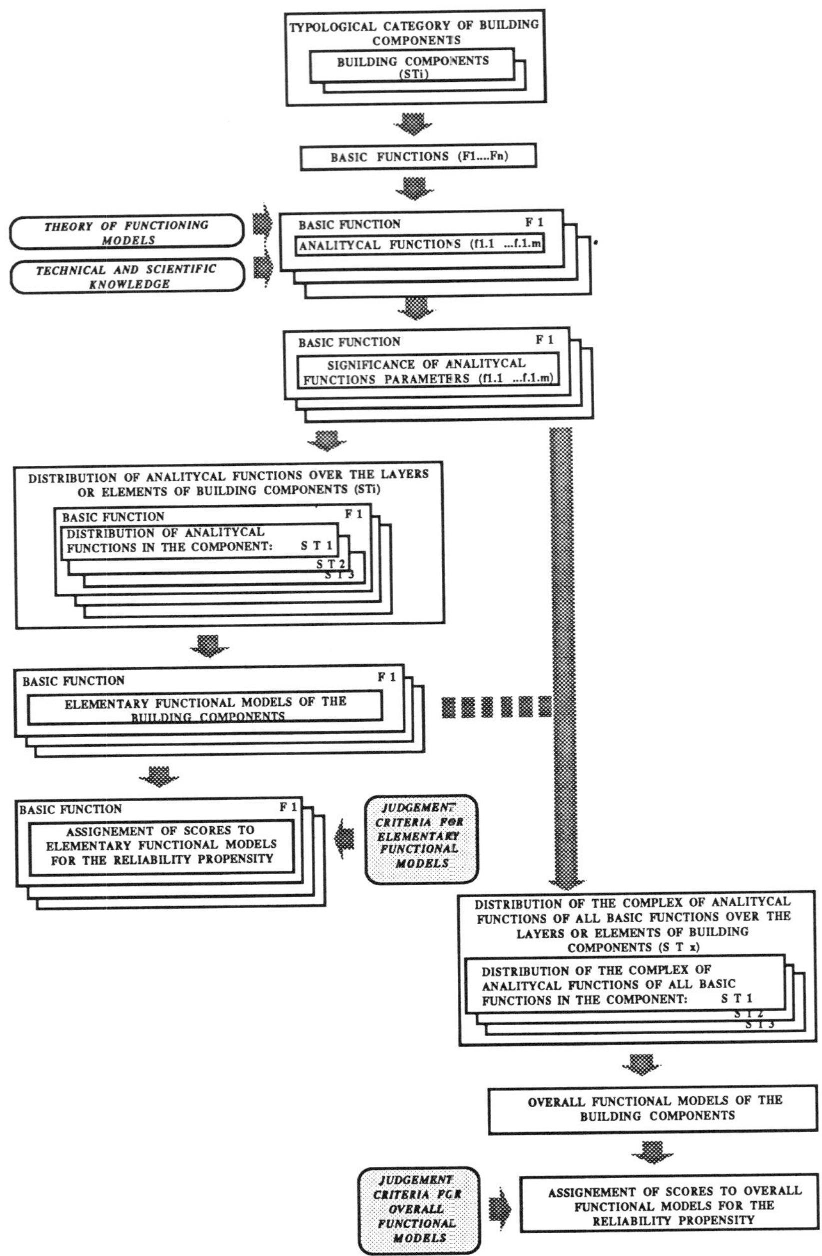

Fig.1. Procedure for the evaluation of the reliability propensity

		NON-BEARING EXTERNAL WALLS	PV

CORRELATIONS BETWEEN BASIC AND ANALYTICAL FUNCTIONS FOR THE TECHNICAL ELEMENT SUB-CLASS	B

	F1 TO CONTROL INTERSTITIAL CONDENSATION	F2 TO CONTROL SUPERFICIAL CONDENSATION	F3 TO CONTROL THE DYNAMICS OF TEMPERATURES IN THE HOT SEASON	F4 TO CONTROL THE DYNAMICS OF TEMPERATURES IN THE COLD SEASON	F5 TO INSULATE FROM AIRBORNE NOISE	F6 TO INSULATE THERMALLY	F7 TO RESIST TO HANGING LOADS	F8 TO CONTROL WATER INFILTRATION
fa to resist to radiative heat flows		●	●	●		●		
fb to resist to convective heat flows		●	●	●		●		
fc to resist to conductive heat flows	●	●	●	●		●		
fd to provide heat bild up			●	●				
fe to foster convective heat flows			●					
ff to foster water re-evaporation	●	●						●
fg to resist to water vapour permeation	●							
fh to provide water vapour permeation	●							
fi to provide constant resistance to water vapour permeation	●							
fl to provide constant resistance to heat transmission	●							
fm to hinder by mass the transmission of sound waves					●			
fn to interrupt the transmission of sound waves in solids					●			
fo to provide visco-elastic dampening in solid mean					●			
fp to provide air-tightness					●			
fq to provide acoustic absorption					●			
fr to provide compressive strengh							●	
fs to provide bending strengh							●	
ft to provide resistance to water flow								●
fu to provide opposition to capillary water permeation								●

Fig.2. Correlation basic functions (Fi) / analytical functions (fi), for the typological category of building components "external walls"

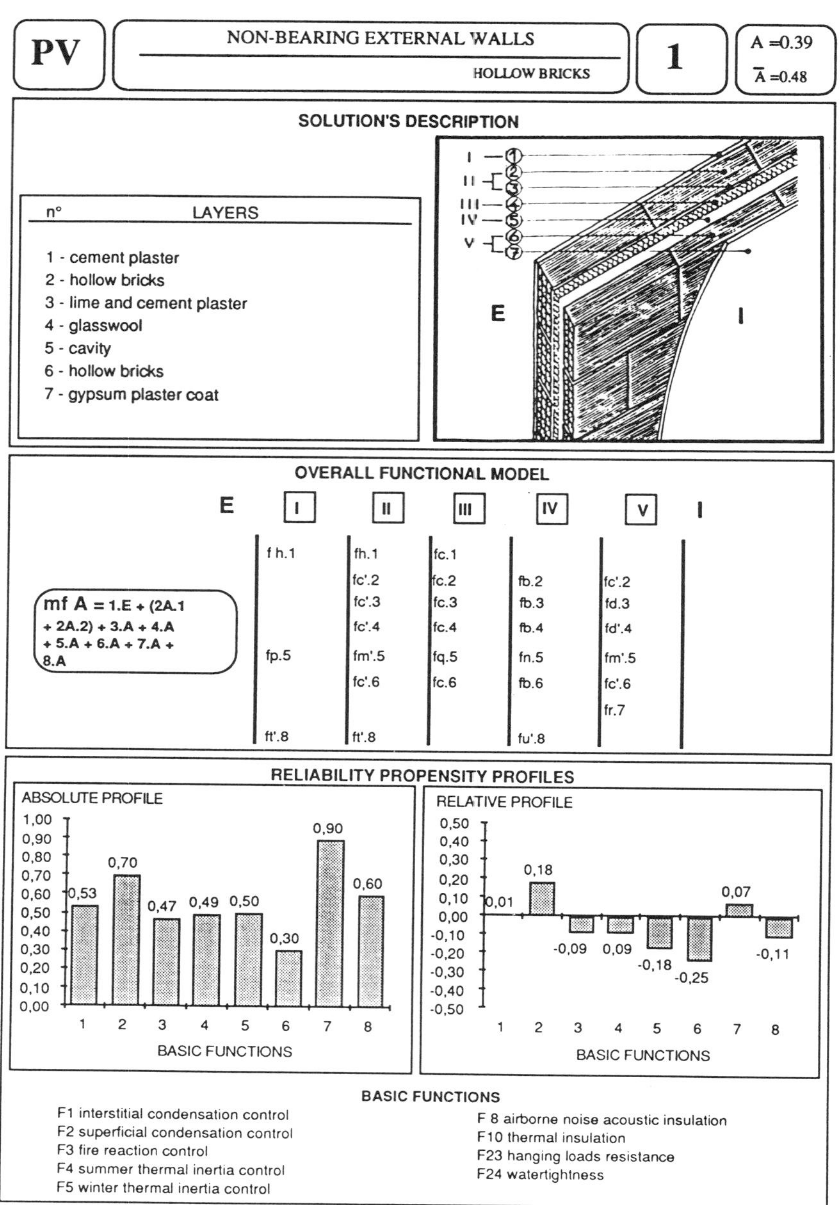

Fig.3. Card of the reliability propensity evaluation of each elementary functional model (see the profiles) and of the overall functional model (see indexes A and $\bar{A}$) of a building component belonging to the external walls typological category

INDEX

This index has been compiled from the keywords assigned to the individual papers by the Concrete Information Service of the British Cement Association. The assistance of the Association in this is gratefully acknowledged. The numbers refer to the opening page of the papers.